Environmental
Nanotechnology

Environmental Nanotechnology

Applications and Impacts of Nanomaterials

Mark R. Wiesner, Ph.D., P.E.   Editor

Jean-Yves Bottero, Ph.D.   Editor

Second Edition

New York   Chicago   San Francisco
Athens   London   Madrid
Mexico City   Milan   New Delhi
Singapore   Sydney   Toronto

Library of Congress Control Number: 2016951109

McGraw-Hill Education books are available at special quantity discounts to use as premiums and sales promotions or for use in corporate training programs. To contact a representative, please visit the Contact Us page at www.mhprofessional.com.

**Environmental Nanotechnology:
Applications and Impacts of Nanomaterials,
Second Edition**

1 2 3 4 5 6   QVS   20 19 18 17 16

ISBN 978-0-07-182844-4
MHID 0-07-182844-3

The pages within this book were printed on acid-free paper.

Sponsoring Editor
  Lauren Poplawski

Editorial Supervisor
  Stephen M. Smith

Production Supervisor
  Lynn M. Messina

Acquisitions Coordinator
  Lauren Rogers

Project Manager
  Apoorva Goel,
  Cenveo® Publisher Services

Copy Editors
  Priyanka Sinha and Surendra Shivam,
  Cenveo Publisher Services

Proofreader
  Manish Tiwari, Cenveo Publisher Services

Indexer
  Robert Swanson

Art Director, Cover
  Jeff Weeks

Composition
  Cenveo Publisher Services

Contents

v

About the Contributors

Mark R. Wiesner (Chaps. 1, 4, 6, 9, 12, 13) holds the James B. Duke Chair in Civil and Environmental Engineering at Duke University, where he has appointments in the Pratt School of Engineering and the Nicholas School of the Environment. He serves as Chair of the Department of Civil and Environmental Engineering and Director of the National Science Foundation's Center for the Environmental Implications of NanoTechnology (CEINT). His work has focused on applications of emerging nanomaterials to membrane science and water treatment and an examination of the fate, transport, and effects of nanomaterials in the environment. From Omaha, Nebraska, Dr. Wiesner holds a B.A. in mathematics and biology from Coe College, an M.S. in civil and environmental engineering from the University of Iowa, and a Ph.D. in environmental engineering from the Johns Hopkins University, and he did postdoctoral work in chemical engineering at the Ecole Nationale Supérieure des Industries Chimiques (ENSIC). He is a musician (electric and double bass), a 2004 de Fermat Laureate, and the 2011 recipient of the Clarke Water Prize, and he was elected to the U.S. National Academy of Engineering in 2015.

Jean-Yves Bottero (Chaps. 1, 6, 12) is CNRS Research Director Emeritus at CEREGE (Centre Européen de Recherche et d'Enseignement des Géosciences de l'Environnement, Aix-Marseille Université, CNRS, IRD, Collège de France) and former director of Labex SERENADE [Laboratory of Excellence for Safe(r) Ecodesign Research and Education Applied to Nanomaterial Development], which gathers 15 French partners, 13 academic partners, and 2 industry partners active in nanoproducts and end of life. He also holds an appointment as adjunct professor at Duke University. His research addresses physicochemical phenomena of surfaces and particles. Dr. Bottero's early work addressed the structure of materials used in water treatment at the nanometric scale, and most notably demonstrated for the first time the existence of the Al_{13} species that controls the chemistry of the now widely used "polyaluminum" coagulants. He has worked extensively on topics ranging from particle aggregation and membrane filtration to solid waste disposal and reuse. More recently, Dr. Bottero has been a senior spokesman in Europe on advancing the agenda for research on possible environmental and health impacts of nanomaterials, serving as president of the Ecotechnology Committee of l'Agence Nationale de La Recherche, a member of the expert committee "Assessment of the Risks Associated with Nanomaterials" at the French Agency for Food, Environmental and Occupational Health & Safety (ANSES), and an expert for the OECD regarding the effects of nanomaterials on the effluent treatment in the wastewater treatment plant and agriculture application.

Mélanie Auffan (Chaps. 7, 12) is a CNRS Research Scientist at the CEREGE (Centre Européen de Recherches et d'Enseignement des Géosciences de l'Environnement, Aix-Marseille Université, CNRS, IRD, Collège de France) in Aix en Provence. She holds a doctoral degree from the University of Paul Cézanne in Aix-Marseille and did postdoctoral work at Duke University. She is member of the steering committees of iCEINT (International Consortium for the Environmental Implications of Nano Technology) as well as the Center for the Environmental Implications of NanoTechnology (CEINT). Her research addresses the physicochemical properties and surface reactivity of nanoparticles in contact with living organisms. Since 2012, she is part of the Labex SERENADE.

Claire Auplat (Chap. 14) is a Professor in innovation and strategy at Novancia Business School in France and a Director of Research at Paris-Dauphine PSL University. She holds a Ph.D. in innovation management from Imperial College Business School and a Ph.D. in policy from Paris Sorbonne University. Her research explores the links between increasing environmental awareness, innovation, institutional processes, and entrepreneurial dynamics. She has worked on institutional change and the emergence of new industries and practices in the field of nanotechnologies for nearly 15 years, first at Rice University (in the United States), then at Sciences Po's Chair of sustainable development (France) and at the Imperial College business school (UK), and now at the SERENADE LABEX (France) where she launched the NEIS research chair on Nanotechnology Eco-design Innovation and Strategy. She is a member of the French National Research Agency (ANR) steering committee for the Major Societal Challenge program and she is a leading expert in France on the managerial implications of the Safer by Design (SBD) approach for nanotechnology.

Andrew R. Barron (Chaps. 2, 9) is the Ser Cymru Chair at Swansea University's College of Engineering, where his research involves the application of nanotechnology to fundamental problems in energy research. He also is the Charles W. Duncan, Jr.–Welch Chair of Chemistry and Professor of Materials Science at Rice University. He spent 8 years on the faculty at Harvard University. He received his Ph.D. from the Imperial College of Science and Technology, University of London, and served as a postdoctoral research associate at the University of Texas. Dr. Barron currently sits on the editorial boards of three chemistry and materials science journals, is a fellow of the Royal Society of Chemistry, and is the 1997 recipient of the Humboldt Senior Service Award. Research in the Barron Group at Rice University focuses on the chemistry and materials science of carbon nanomaterials. Research initiatives encompass problems across applications in energy research and at the border between nanoscale science and biology.

Daniel Borschneck (Chap. 3) is CNRS research engineer in CEREGE (Centre Européen de Recherches et d'Enseignement des Géosciences de l'Environnement, Aix-Marseille Université, CNRS, IRD, Collège de France). He obtained his Ph.D. in physics applied to archaeology (condensed matter physics) at the University of Bordeaux in 1998. He is in charge of an analytical platform combining x-ray microanalysis and 3D imaging. His research concerns the multiscale physicochemical characterization of heterogeneous materials with x-ray-based techniques such as diffraction, microfluorescence, micro and nanotomography. In the field of nanotechnology, his research focuses on the location and the identification of nanoparticles or aggregate of nanoparticles in complex media such as plants, roots, soft tissues, and cells using 3D x-ray imaging at the micro and nanoscale.

Jonathan Brant (Chaps. 3, 6) is an Associate Professor in the Department of Civil and Architectural Engineering at the University of Wyoming, and a licensed professional environmental engineer in Wyoming. He obtained a Ph.D. in environmental engineering at the University of Nevada, Reno, and did postdoctoral work at Duke University. Dr. Brant's research focuses on developing novel physicochemical separation processes for water and wastewater applications. Improvements in these technologies are based on his work with engineered nanomaterials and our improved understanding of, and ability to manipulate, materials at the nanoscale. Dr. Brant has authored over 30 peer-reviewed articles on these subjects and is the Director of the University of Wyoming's Center of Excellence in Produced Water Management, whose mission is to develop sustainable technologies and strategies for oil and gas produced waters.

Bernard Cathala (Chap. 2) is currently a Senior Researcher at the Biopolymer, Interaction and Assemblies (BIA) laboratory of the French National Institute for Agricultural Research (INRA) in Nantes. He heads the team "nanostructured assemblies" and is deputy director of BIA lab. He received his Ph.D. in 1995 in organic chemistry and physical chemistry from the University of Toulouse. His research focuses on of biopolymers organization at nanoscale to elaborate new bio-based materials. More precisely, his current research interests are the chemistry and physical chemistry of nanocelluloses (both cellulose nanocrystals and nanofibrillated cellulose) dedicated to the fabrication of smart coatings and dispersed systems stabilized by nanocelluloses. He has directed more than 10 academic and industrial funded research projects and he has filed 5 patent applications.

Corinne Chaneac (Chap. 2) has been a professor in the Department of Chemistry at Pierre and Marie Curie University Paris since 2009. She works in the Laboratoire Chimie de la Matière Condensée of Paris (CMCP) as senior scientist. Her research focuses on sol-gel chemistry for the synthesis of nanoparticles and nanomaterials with tunable properties. She is director of the Center of Competence in Nanoscience for the Ile-de-France region (C'Nano), and she is a member of the International Consortium for the Environmental Implications of Nanotechnology (iCEINT).

Perrine Chaurand (Chap. 3) holds an undergraduate degree in Chemical Engineering and Doctoral degree in Geosciences of Environment. Since 2007, she has worked as a research engineer at the CEREGE (Centre Européen de Recherche et d'Enseignement des Géosciences de l'Environnement, Aix-Marseille Université, CNRS, IRD, Collège de France) where she has been a member of GDRi iCEINT, Labex SERENADE and Equipex NANO-ID collaborations. Her expertise is in physicochemical characterization of heterogeneous and divided solids using a multiscale approaches and determination of speciation of trace elements. She is in charge of the x-ray imaging platform at the CEREGE which offers the capability to perform 2D chemical and mineralogical microanalysis and mapping (micro-XRF and micro-XRD) and 3D imaging at micro- and nano-scales (micro and nano x-ray computed-tomography).

Benjamin P. Colman (Chap. 12) is an Assistant Professor of Aquatic Ecosystem Ecology in the Department of Ecosystem Conservation Sciences at the University of Montana. Dr. Colman is ecosystem ecologist who is broadly interested in carbon and nutrient cycling in ecosystems. During the course of his career he has studied the drivers of biogeochemical cycles in ecosystems from the arctic to the tropics and along the continuum from upland ecosystems to freshwater and marine aquatic ecosystems. Much of Dr. Colman's work focuses on the changes in biogeochemistry that arise from human-driven chemical perturbations, including elevated nitrogen deposition from fossil fuel combustion, saltwater intrusion into historically freshwater wetlands as a result of rising sea-levels, and manufactured nanomaterials. Dr. Colman earned a B.A. in Chemistry from Carleton College, and a PhD in Ecology, Evolution, and Marine Biology from the University of California, Santa Barbara. He was a Postdoctoral Associate and Research Scientist in the Biology Department at Duke University. While at Duke, Dr. Colman worked on perfecting the art of mesocosm experiments as a member of the Center for the Environmental Implications of NanoTechnology.

Richard T. Di Giulio (Chap. 5) is Professor of Environmental Toxicology in the Nicholas School of the Environment at Duke University. He also serves at Duke as Director of the Integrated Toxicology and Environmental Health Program, Director of the Superfund Research Center, and Co-Principal Investigator for the Center for the Environmental Implications of NanoTechnology. Dr. Di Giulio has published extensively on subjects including biochemical and molecular mechanisms of adaptation and toxicity, biomarkers for chemical exposure and toxicity, and effects of chemical mixtures and multiple stressors. His current work focuses on mechanisms by which polycyclic aromatic hydrocarbons (PAHs) and nanomaterials perturb embryonic development in fish models (zebrafish and killifish), the evolutionary consequences of hydrocarbon pollution on fish populations, and the ecological and human health impacts of coal extraction and fly ash disposal. Additionally, he has organized symposia and workshops and written on the broader subject of interconnections between human health and ecological integrity. Dr. Di Giulio serves as an advisor for the Scientific Advisory Board of the U.S. EPA and is associate editor for *Environmental Health Perspectives*. Dr. Di Giulio received a B.A. in comparative literature from the University of Texas at Austin, an M.S. in wildlife biology from Louisiana State University, and a Ph.D. in environmental toxicology from Virginia Polytechnic Institute and State University. He is an active member of the Society of Environmental Toxicology and Chemistry (SETAC) and the Society of Toxicology (SOT).

Nicholas Geitner (Chap. 3) is currently a Postdoctoral Research Associate for the Center for the Environmental Implications of NanoTechnology in Duke University's Department of Civil and Environmental Engineering. After completing his Ph.D. in physics at Clemson University with a focus in nano-biophysics, his work has focused on the fundamental interactions between nanomaterials and environmental constituents. This includes how such fundamental interactions affect nanomaterial attachment to environmental surfaces and the implications for transport and trophic transfer as well as the detection of nanomaterials in complex environmental media using hyperspectral imaging. Theoretical work includes modeling of nanomaterial transport and trophic transfer and molecular dynamics simulations of nanoparticle interactions with environmental molecules.

Laure Giamberini (Chap. 5) is a professor of aquatic ecotoxicology at the University of Lorraine, France (LIEC, UMR CNRS 7360). Her research addresses (1) development and validation of cellular and individual biomarkers in different freshwater invertebrate species, (2) metal (ionic and nanoparticles) metabolism in invertebrates, and (3) environmental parasitology. She is a research coordinator in Labex R21 concerning environmental dissemination and impacts of strategic metals. Currently she is lab leader for two national and one international ANR programs, P2N Mesonnet (with GDRi iCEINT, Duke University) and Cesa Nanosalt in the frame of nanoecotoxicology. She is also involved in a French-Canadian ANR project IPOC, concerning the interaction between pollution and climate change. She is the author of 47 articles in international and national peer-reviewed journals, proceedings, and book chapters. She has presented some 85 communications at international and national conferences. She has been responsible for license "Life Sciences" for 10 years. She is a regular expert for Belgium FNRS and French ANR, CNRS, and ANSES (French Agency for Food, Environmental and Occupational Health & Safety).

Christine Ogilvie Hendren (Chap. 13) is a research scientist and Executive Director of the Center for the Environmental Implications of NanoTechnology (CEINT). Dr. Hendren synthesizes CEINT research across the various disciplines and entity members of the center, with a focus on facilitating efficient interdisciplinary communication in a continual drive toward environmentally relevant approaches to assessing environmental risks. She leads center efforts to develop the CEINT NanoInformatics Knowledge Commons (NIKC) to collect, integrate and interrogate nanomaterial information to advance the nanomaterial environment, health and safety (nanoEHS) field. Current NIKC efforts focus on investigating key parameters to forecast potential environmental risk profiles of engineered nanomaterials, and on developing targeted approaches to collect and integrate complex nanomaterial datasets within the center and with a global network of data partners. Dr. Hendren co-chairs the Nanotechnology Cancer Informatics Program's Nanotechnology Working Group, and serves as the U.S. co-chair for the NNCO US-EU Risk Assessment Community of Research. Dr. Hendren's research interests include combining risk assessment of new technologies with decision support; linking and leveraging emerging data to support decisions under conditions of great uncertainty; establishing a new career path for interdisciplinary scientists with expertise in communicating across disciplinary and institutional boundaries to effectively address wicked problems; and integrating vital aspects of the humanities including epistemology, history, and storytelling with the teaching and practice of environmental risk assessment.

Michael Hoffmann (Chap. 4) is the Theodore Y. Wu Professor of Environmental Science at the California Institute of Technology. He received a B.A. in chemistry from Northwestern University and a Ph.D. in chemical kinetics from Brown University. In 1973, he was awarded an NIH postdoctoral training fellowship in environmental engineering science at Caltech. From 1975 to 1980, he was a member of the faculty at the University of Minnesota and since 1980 has been a member of the faculty at Caltech (Engineering and Applied Science). Dr. Hoffmann has published more than 220 peer-reviewed professional papers and is the holder of seven patents in the subject areas of applied chemical kinetics, aquatic chemistry, atmospheric chemistry, environmental chemistry, catalytic oxidation, heterogeneous photochemistry, sonochemistry, electrochemistry, and pulsed-plasma chemistry. He is a member of the National Academy of Engineers. In 1991, Dr. Hoffmann received the Alexander von Humboldt Prize for his research and teaching in environmental chemistry. In 1995, he was presented with the E. Gordon Young Award by the Chemical Society of Canada in recognition of his work in the field of environmental chemistry. He has also served as a Distinguished Lecturer at the Hebrew University (Jerusalem), the University of São Paulo (Brazil), and the University of Buenos Aires. In 2001 Dr. Hoffmann was presented with the American Chemical Society Award for Creative Advances in Environmental Science and Technology, and in October 2003 he received the Jack E. McKee Medal for groundwater protection from the Water Environment Federation.

Ernest (Matt) Hotze (Chap. 4) is currently the Administrative Director of the NSF-sponsored Nanosystems Engineering Research Center for Nanotechnology-Enabled Water Treatment (NEWT) headquartered at Rice University. He previously served as managing editor with the American Chemical Society Publications group in Washington, D.C. During his time with ACS Dr. Hotze worked with the journal editors, business teams and authors of *Environmental Science & Technology*, *Environmental Science & Technology Letters*, *Journal of Agricultural and Food Chemistry* and *ACS Sustainable Chemistry & Engineering*. Hotze holds a Ph.D. from Duke University and an M.S. from Rice University.

Jean Pierre Jolivet (Chap. 2) is professor emeritus at Pierre and Marie Curie University Paris. He was group leader of the inorganic nanomaterials research group in the Laboratoire Chimie de la Matière Condensée of Paris (CMCP). His research activities have focused on the chemistry of oxides and more specifically on the preparation in aqueous solution of nanoparticles with controlled properties and the surface chemistry of these nano objects.

Jérôme Labille (Chap. 6) is a researcher with French National Research Center (CNRS) at the Geosciences and Environment lab (CEREGE). He obtained a Ph.D. in physical chemistry of geosciences in the French Institute INPL, where his research focused on the characterization of the interactions between minerals and bacterial polysaccharides from soil, and their colloidal fate. He did subsequent postdoctoral work at the Analytical Center for Biophysicochemistry of the Environment (CABE) in Geneva, where he studied the transport and bioavailability of nanometric elements in organic gels. He has been working at CEREGE for 3 years on the environmental fate of manufactured nanoparticles, considering and characterizing the numerous conditions that control their bioavailability and toxicity, such as surface reactivity, colloidal dispersion, and interaction with organics or pollutants.

Sophie Lanone (Chap. 11) obtained her Ph.D. in 1998, and, after a 2-year postdoctoral position at Yale University School of Medicine (Pulmonary Department, directed by Pr. Jack Elias), she joined Inserm as a permanent researcher. She is now an Inserm research director at the Mondor Institute of Biomedical

Research (IMRB/Inserm U955, Créteil, France). Sophie Lanone is a member of the NanoScience Competence Center (C'Nano) *Nanoscience & Society Task Force* since 2009.

For the past 10 years, her research has been mainly dedicated to study the respiratory effects of environmental contaminants at the molecular, cellular, and organ levels, particularly those of nanomaterials such as carbon nanotubes and metallic nanoparticles. She identified the presence of welding-related metal nanoparticles (Fe, Mn, Cr) in the fibrotic alterations of lung tissue from welders, and demonstrated that such nanoparticles induced an inflammatory response that could be linked to the fibrotic response in these patients.

Clément Levard (Chaps. 3, 7) is a CNRS Research Scientist and physical chemist at CEREGE (Centre Européen de Recherches et d'Enseignement des Géosciences de l'Environnement, Aix-Marseille Université, CNRS, IRD, Collège de France). He is interested in the physicochemical properties of nano-structured materials in an environmental context. In particular, he is interested in both the behavior of manufactured and natural nano-structured materials in natural systems and in the properties of these materials in order to develop eco-designed nano-products. To develop these aspects, he specialized in synchrotron-based techniques using x-rays including total diffusion with pair distribution function analysis, standing waves and absorption spectroscopy. He is a member of synchrotron SOLEIL Peer Review Committee "Ancient Materials, Environment and Earth". He is also an expert for OECD regarding the stability of nano-structured materials in environmental waters. He has published 35 peer-reviewed articles in international journals in the field.

Gregory V. Lowry (Chaps. 7, 8) is a professor of environmental engineering at Carnegie Mellon University in Pittsburgh, Pennsylvania where he holds the Walter J. Blenko, Sr. Chair of Civil & Environmental Engineering. He serves as Deputy Director of the Center for the Environmental Implications of NanoTechnology (CEINT) and associate editor of the journal *Environmental Science: Nano*. His research interests are broadly defined as transport and reaction in porous media, with a focus on the fundamental physical/geochemical processes affecting the fate of inorganic and synthetic organic contaminants and engineered nanomaterials in the environment. He is an experimentalist and works on a variety of application-oriented research projects developing novel environmental technologies for restoring contaminated sediments and groundwater, including reactive nanoparticles for efficient in situ remediation of entrapped NAPL and innovative sediment caps for in situ treatment and management of PCB-contaminated sediments.

Armand Masion (Chap. 3) is an environmental spectroscopist. He is currently a research director with the French national research center at CEREGE (Centre Européen de Recherche et d'Enseignement des Géosciences de l'Environnement, Aix-Marseille Université, CNRS, IRD, Collège de France), which he joined in late 1995 as a research scientist after his Ph.D. in geosciences (1993) and a postdoctoral research position at the Savannah River Ecology Lab in Paul Bertsch's group (1993–1995). His early career focused on the colloid chemistry of hydrolysable metal ions. Since the early 2000s, he has been working on the synthesis and the environmental fate of nanoparticles and nanomaterials.

Manuel D. Montaño (Chap. 3) is a Postdoctoral Associate at Duke University investigating how nanomaterials influence the fate and behavior of molecular contaminants present in nano-enabled products. He completed his Ph.D. in 2014 at the Colorado School of Mines, where his doctoral work focused on the development of analytical techniques for the detection and characterization of engineered nanomaterials in the environment; specifically working to develop single particle ICP-MS (spICP-MS) as a viable option for the routine analysis of nanoparticles in complex biological and environmental matrices. His research interests include further developing analytical instrumentation such as spICP-MS and field flow fractionation for accurate and precise determination of engineered nanomaterials in the environment, quantifying the release of plasticizers and polymer additives from nano-enabled composites using liquid chromatography tandem mass spectrometry, developing methods for the extraction and quantification of carbon nanotubes in sediment using near infrared fluorescence spectroscopy, and developing accurate analytical techniques for the differentiation of engineered and naturally occurring nanoparticles in environmental samples.

Nancy Ann Monteiro-Riviere (Chap. 11) is a Regents Distinguished Research Scholar and University Distinguished Professor of Toxicology at Kansas State University, and Director of the new Nanotechnology Innovation Center of Kansas State (NICKS). She was a professor of investigative dermatology and toxicology at the Center for Chemical Toxicology Research and Pharmacokinetics, North Carolina State University

(NCSU), for 28 years. Dr. Monteiro-Riviere is also a professor in the Joint Department of Biomedical Engineering at UNC–Chapel Hill/NCSU and research adjunct professor of dermatology at UNC–Chapel Hill School of Medicine. She received an M.S. and Ph.D. from Purdue University. She completed postdoctoral training at the Chemical Industry Institute of Toxicology in Research Triangle Park, North Carolina. Dr. Monteiro-Riviere was president of the Dermal Toxicology and In Vitro Toxicology Specialty Sections of the National Society of Toxicology and currently serves as chairperson of the Board of Publications. She is a Fellow of the Academy of Toxicological Sciences and of the American College of Toxicology. She serves on several toxicology editorial boards and national panels, including many in nanotoxicology. Dr. Monteiro-Riviere has published 300 manuscripts and book chapters. Her current research focuses on the cutaneous absorption and toxicity of engineered nanomaterials (quantum dots, fullerenes, and carbon nanotubes) and their effects on human health.

Catherine Mouneyrac (Chap. 5) is a professor of marine ecotoxicology at the Université Catholique de l'Ouest (France). She teaches graduate and undergraduate courses in animal physiology, aquatic ecology, and ecotoxicology. She gained her Ph.D. from the University of Lyon I (France) in the physiology of fish nutrition, and then a D.Sc. in aquatic ecotoxicology from the University of Nantes (France). She is the dean of the Faculty of Sciences and the head of the emerging contaminants research team of the MMS (Sea, Molecules, Health) unit at L'UNAM. Dr. Mouneyrac's general field research concerns the response of aquatic organisms to natural and chemical stress. She is actively involved in research to elucidate the mechanisms involved in potential toxicity by nanomaterials and endocrine disruptors toward estuarine and marine organisms. At the interface of fundamental and applied research, she aims to fulfill the gap between ecological (bio-indicators) and ecotoxicological (biomarkers) approaches, the final objective being to help environmental diagnosis. More precisely, she studies biomarker tools allowing extrapolating from suborganismal (biochemical biomarkers, energy reserves, reproduction processes) and organismal (biological indices, behavioral biomarkers) responses to effects occurring at higher levels of biological organization (population). Dr. Mouneyrac has participated in the conception and realization of numerous national, European, and international programs. She has published more than 75 peer-reviewed papers and book chapters in the field of aquatic ecotoxicology. She is a referee for numerous international journals on environmental pollution. She is member of different research networks in environmental pollution assessment. Dr. Mouneyrac is part of the expert committee on the assessment of the risks related to physical agents, new technologies, and development areas and the working group "Assessment of the Risks Associated with Nanomaterials" at the French Agency for Food, Environmental and Occupational Health & Safety (ANSES). She has been selected as a senior decision maker to follow the national study course of the Institut des Hautes Études pour la Science et la Technologie, whose supervision is the responsibility of the ministries of higher education and research and education in France.

Jérôme Rose (Chaps. 3, 9) serves as CNRS Senior Scientist (DR1) at the CEREGE Laboratory (Aix-Marseille University, CNRS, IRD joint lab). He is heading the INTERFAST team of the CEREGE since 2006 with appointments as adjunct faculty at Rice and Duke Universities (USA). Dr. Rose was the 2006 recipient of the bronze medal for excellence of research (CNRS-France). He is the CNRS-INSU national representative for the large geo-physical-chemical scale facilities and characterization platforms since 2013. Dr. Rose is the director of the SERENADE French ANR funded consortium (LABEX) dealing with safe(r) by design nanotechnologies since 2014. He headed up one of the first French National multidisciplinary programs addressing the potential health effect of nanoparticles, entitled "Surface Properties of Manufactured Nanoparticles: Mutagenetic Effects, Mechanisms of Bio-degradation and Accumulation" (2003–2005). More recently his group has determined the molecular and thermodynamic mechanisms responsible of the enhanced reactivity of iron nanoparticles smaller than 10 nm. The team also co-discovered double wall Ge-imogolite nanotubes. The team unraveled the environmental exposure to nanoparticles from various commercial products by taking into account the various life stages of commercialized products. He is employing intensively synchrotron-based techniques to study mechanisms at a molecular level. He published more than 120 papers in peer reviewed journals.

Catherine Santaella (Chap. 5) has been a research scientist at the CNRS (National Center for Scientific Research) in France since 1991. Currently, she has a position at the Laboratory of Microbial Ecology of the Rhizosphere and Extreme Environment at the Biosciences and Biotechnologies Institute of Aix-Marseille (BIAM, UMR 7265). Her research addresses plant-bacteria interactions in fluctuating environments and ecotoxicology. Her work focuses on molecular, cellular and functional responses in the rhizosphere induced by environmental stresses and response patterns to nanomaterials and metals at the level of the bacterial cells or communities in a soil-plant-bacteria system. She is a research coordinator (1) and partner (2) of national programs in the framework of the Labex SERENADE and Lab leader for 1 international ANR (National Research Agency) program, P2N Mesonnet (with GDRi iCEINT, Duke University) on nanoimpacts. She is the author of international scientific publications in international peer-reviewed journals (57) as well as proceeding (5), book chapters (6), and patents (8). She has presented 78 communications in international and national conferences. She is a regular expert for the OMNT (Observatory of Micro and Nano Technologies) and for French ANR, CNRS, and ANSES (French Agency for Food Environmental and Occupational Health and Safety).

Fabienne Schwab (Chaps. 5, 12) is a visiting scientist at France's Geoscience and Environment Lab CEREGE (Centre Européen de Recherches et d'Enseignement des Géosciences de l'Environnement, Aix-Marseille Université, CNRS, IRD, Collège de France) on a fellowship awarded by the Swiss NSF. She previously did post-doctoral work at Duke University and holds a doctoral degree from ETH Zurich, where she performed experimental research on sorption chemistry and phytoplankton (nano)ecotoxicology of organic micropollutants and engineered nanomaterials. Her research is devoted to advance analytical applications, and to characterize and quantify plant (nano)ecotoxicology and nanomaterial transport and fate in the environment.

Alain Thiéry (Chap. 12) obtained his doctorate Ph.D. (Docteur d'Etat ès sciences) from the University Aix-Marseille in 1987. He is full professor at the University of Aix-Marseille, head of the major research field "Biomarkers, Biotechnology, Ecotoxicology and Human Health" in the IMBE [Mediterranean Institute of marine and terrestrial Biodiversity and Ecology, Aix-Marseille University, CNRS, IRD]. He has worked for about 30 years on freshwater invertebrates, ecophysiology and ecology, and founded the team "Biomarkers & Bioindicators of the Environment" [BBE] in 2006. His research interest is in the impact of pollutions mainly nanoparticles on freshwater invertebrates and community assemblages, using various biocenotic techniques including mesocosm conception and ecotoxicological tests.

Antoine Thill (Chap. 3) is a scientist at the Laboratoire Interdisciplinaire sur l'Organisation Nanometrique et Supramoléculaire (LIONS) in the Department of Materials Science of the Commissariat de l'Energie Atomique in Saclay (Paris). His present field of interest comprises to the investigation of the synthesis, structure and dynamics of nanoparticles in complex systems using x-ray-based scattering, spectroscopic and microscopic techniques. Lately, he has been particularly involved in several research programs on imogolite-like materials during which he discovered the DW Germanium based imogolite nanotubes and made progress on the formation mechanism of imogolite-like nanoparticles. He is the co-editor of a book on nanotubular clay minerals.

Sally Tinkle (Chap. 14) joined the Science and Technology Policy Institute (STPI) in Washington, D.C., in April 2013 in order to bring her expertise in human health research, policy, and administration, especially as it relates to emerging technologies and environmental exposures, to the policy arena. Before joining STPI, she served as the Deputy Director of the National Nanotechnology Coordination Office of the National Science and Technology Council, where she led strategic planning and implementation of the National Nanotechnology Initiative, especially for environmental, health, and safety research. As a Senior Science Advisor in the Office of the Deputy Director, National Institute of Environmental Health Sciences, National Institutes of Health, Dr. Tinkle worked on health issues related to biofuels and the bioeconomy, application of global earth observations to human health monitoring, and environmentally induced pulmonary health conditions, as well as nanotechnology. She received her Ph.D. from the Department of Physiology at the University of Colorado School of Medicine and was a postdoctoral fellow at the National Jewish Health Center's Department of Occupational and Environmental Health Science, Denver, Colorado.

Olga V. Tsyusko (Chap. 5) is an Assistant Research Professor in the Department of Plant and Soil Sciences at the University of Kentucky. She received her B.S. in biology from Uzhgorod National University in Ukraine and then moved to the United States, where she earned her Ph.D. in toxicology/ecology at the University of Georgia. Her postdoctoral training in Molecular Biology was completed at the Savannah River Ecology Laboratory in SC. The focus of her current research is on environmental toxicogenomics and toxicogenetics, examining effects and toxicity pathways induced by various environmental stressors, including engineered nanomaterials and ionizing radiation, in soil invertebrates and plants. She is a member of the Center for Environmental Implications of NanoTechnology (CEINT).

Jason M. Unrine (Chap. 5) is an Assistant Professor in the Department of Plant and Soil Sciences at the University of Kentucky with a secondary appointment in the Department of Toxicology and Cancer Biology. Prior to this he served as a research scientist at the University of Georgia Savannah River Ecology Laboratory, where he also undertook Ph.D. and postdoctoral in toxicology and environmental analytical chemistry, respectively. He earned his B.S. in biology from Antioch College. His research focuses on understanding the fate, transport, bioavailability and adverse ecological and human health effects of trace-elements and metal-based manufactured nanomaterials. He is a member of the steering committee and leader of the cellular and organismal impacts theme of the Center for Environmental Implications of NanoTechnology (CEINT), Associate Editor at *Environmental Chemistry* and is an executive board member for the International Society of Trace Element Biogeochemistry (ISTEB).

Vladimir Vidal (Chap. 3) has Ph.D. in physics and materials science (surfaces and interfaces), and has long studied structures and properties of nanostructured coatings for x-ray optics. After working as beamline scientist and instrumental developer on KMC-2 in BESSY II synchrotron, he joined the "Conservatoire des Arts et Métiers" to work on physic metrology at the "Institut National de Métrologie" (LNE/CNAM). At the CEREGE (Centre Européen de Recherches et d'Enseignement des Géosciences de l'Environnement, Aix-Marseille Université, CNRS, IRD, Collège de France) he applies x-ray characterization techniques to study nanomaterials and their impact the environment from nano- to macro-metric scales.

Peter Vikesland (Chap. 10) is a Professor of Civil and Environmental Engineering at Virginia Tech. He serves as the director of the Virginia Tech Institute for Critical Technology and Applied Science (ICTAS) Center for Sustainable Nanotechnology (VTSuN) and its affiliated Interdisciplinary Graduate Education Program (IGEP). Dr. Vikesland's work has focused on the development of nanomaterial-enabled sensing technologies and on the evaluation of the environmental implications of carbonaceous and metallic nanomaterials. He obtained his B.A. in chemistry from Grinnell College and his M.S. and Ph.D. in civil and environmental engineering from the University of Iowa. Following a postdoctoral fellowship at The Johns Hopkins University, Dr. Vikesland joined the Virginia Tech faculty in 2001.

Haoran Wei (Chap. 10) has been a Ph.D. student in the Department of Civil and Environmental Engineering at Virginia Tech since 2013. He is a graduate research assistant in Dr. Peter Vikesland's research group. Haoran's research interest is to develop advanced analytical approaches for environmental pollutant detection. He is presently working on a nano-sized pH probe that can provide high spatial resolution in a confined environment. Haoran obtained his bachelor's degree in environmental engineering from Shandong University in 2010. After that, he went to Tsinghua University and received his master's degree in environmental engineering in 2013. There, he synthesized nano-sized and nano-porous adsorbents for emerging contaminant removal.

Environmental Nanotechnology

Overview

PART 1

Overview

CHAPTER 1
Nanotechnology and the Environment

Nanotechnology and the Environment

Mark R. Wiesner

Department of Civil & Environmental Engineering, Duke University, Durham, North Carolina, USA

Jean-Yves Bottero

CEREGE UMR 6635 CNRS–Europole de l'Arbois, Aix-en-Provence Cedex, France

Advances in information technologies, materials science, biotechnology, energy engineering, and many other disciplines—including environmental engineering—are converging at the quantum and molecular scales. This molecular terrain is common ground for interdisciplinary research and education that will be an essential component of science and engineering in the future. Much like the digital computer and its impact on science and technology in the twentieth century, the tools that serve as portals to the molecular realm will serve as both instruments of discovery and rallying points for social interaction between researchers from many disciplines. In this setting, environmental engineers and scientists will take on new roles in collaborating with materials scientists, molecular biologists, chemists, and others to address the challenges of meeting society's needs for energy and materials in an environmentally responsible fashion.

Nanotechnology is defined as a branch of engineering that deals with creating objects smaller than 100 nm in dimension. Behind this definition is a vision of building objects atom by atom, molecule by molecule [1] by self-assembly or molecular assemblers [2]. Activities spawned by a "nanomotivated" interdisciplinarity will affect the social, economic, and environmental dimensions of our world, often in ways that are entirely unanticipated. We focus here on the potential impacts of nanomaterials on human health and environment. Many of these impacts will be beneficial. In addition to a myriad of developments in medical science, there is considerable effort underway to explore uses of nanomaterials in applications such as membrane separations, catalysis, adsorption, and analysis with the goal of better protecting environmental quality.

However, along with these innovations and the growth of a supporting nanomaterials industry, there is also the need to consider impacts of nanomaterials on environment and human health. Past technological accomplishments such as the development of nuclear power, genetically modified organisms, information technology, and synthetic organic chemistry have generated public cynicism as some of the consequences of these technologies, often environmental, become apparent. Even potable water disinfection, the single most important technological advance with regard to prolonging human life expectancy, has been found to produce carcinogenic by-products. Some groups have called on industry and governments to employ the precautionary principle while conducting more research in toxicology and transport behaviors [3, 4]. The precautionary principle, often associated with the western European approach to regulation, might be summarized as "no data, no market." In contrast, the risk-based approach that has come to typify regulatory development in the United States, might be reduced to the philosophy of "no data, no regulation." Both approaches require reliable data.

In this book, we consider the topic of nanomaterials through the lens of environmental engineering. A key premise of our approach is that the nanomaterials industry is an emerging case study on the design of an industry as an environmentally beneficial system throughout the life cycle of materials production, use, disposal, and reuse. One element of this socioindustrial design process is an expansion of the training and practice of environmental engineering to include concepts of energy and materials production and use into environmental engineering education and research.

Origin and Organization of This Book

The origin of this book was a dialogue between the Wiesner and Bottero groups that began during Wiesner's sabbatical leave at the Centre Européen de Recherche et d'Enseignement des Géosciences de l'Environnement (CEREGE) in Aix-en-Provence in 1998. Our motivation in 1998 was a perceived need to introduce a materials science and life cycle dimension to environmental engineering curricula, and this need resonated with a growing interest in nanotechnology. Although efforts were being made at that time to import newly developed nanomaterials to environmental applications, apart from futuristic scenarios of "grey-goo"–producing nanobots [5], the environmental impacts of nanomaterials produced in the near term had not yet been addressed.

With financial support from the Office of Science and Technology of the French Consulate, we organized the first-ever public forum addressing the environmental implications of nanotechnologies in December 2001. This event brought together nanochemistry and environmental researchers from various laboratories in France and the United States to speculate on health and environmental impacts of nanomaterials. The first edition of this book, published in 2007, compiled information in environmental applications and implications of nanotechnology based on well-established work in the areas of materials development, photocatalysis, and colloid science as well as newer, and more speculative efforts evaluating the possible impacts of nanomaterials of human health and the environment. In 2008, a new center headquartered at Duke University and led by Mark Wiesner and Greg Lowry, the Center for the Environmental Implications of NanoTechnology (CEINT) was funded by the National Science Foundation (NSF) and Environmental Protection Agency (EPA). CEINT included the CEREGE (Center for Economic Research and Graduate Education) and other French laboratories as international partners through a GDRI-iCEINT (Groupe De Recherche International: international CEINT) and the CEINT effort was mirrored in 2011 by funding from the French government to create a center headquartered at the CERGE, SERENADE (Safe(r) and Ecodesign Research and Education applied to Nanomaterial Development), led by Jean-Yves Bottero and Jerome Rose. Now, 17 years after initiating work on the topic of environmental nanotechnology, and a decade after the publication of the first edition of this book, the body of literature has grown to the extent that a revision to the initial book is required. This revision represents the collaborative work of CEINT and SERENADE researchers.

Properties and Principles of Nanomaterials

Nanoscale particles have long been present in environmental systems and play a significant role in the function of many natural processes. Examples of naturally occurring nanoscale particles include those resulting from mineral weathering (e.g., iron oxides and silicates) and emissions from combustion processes (i.e., carbon soot). Advances in the area of nanoscience have provided means of characterizing and manipulating naturally occurring nanomaterials as well as manufacturing engineered nanoparticles (e.g., metal oxides, carbon nanotubes, and buckminsterfullerene). It is now possible to control the chemical and physical properties of nanoparticles and to tailor them for specific applications. In Chapter 2, methods for fabricating nanomaterials and some of the unique properties of these materials are presented. Methods for structural and chemical characterization are presented in Chapter 3.

Nanoparticles are nearly all surface. As an approximation, a 4-nm-diameter solid particle has more than 50 percent of its atoms at surface. One gram of single-walled carbon nanotubes (SWNTs) has over $10~m^2$ of available surface area. This results in nanoparticles being highly surface reactive and implies that their behavior will, to a great degree, be mediated by interfacial chemical interactions. These interactions should therefore be governed by the characteristics of these surfaces.

Because atoms at interfaces behave differently, nanomaterials are likely to have unique properties compared to larger bulk materials of the same general composition. Also, as the size of particles of a given material approaches the nanoscale, material properties such as electrical conductivity, color, strength, and reactivity may change. These changes are in turn related to underlying effects of size that include quantum confinement in semiconductor particles, surface plasmon resonance in some metal particles, and superparamagnetism in magnetic materials. Greater reactivity, and the ability of some nanoparticles to act as electron shuttles or, in other cases, as photocatalysts, holds particular interest in environmental applications. The photocatalytic properties of mineral and fullerene nanomaterials are presented in Chapter 4. The ability of some nanomaterials to photocatalytically produce reactive oxygen species (ROS) that may be used to oxidize contaminants or inactivate microorganisms may also present a risk of toxicity to organisms.

For nanomaterials to present a risk, there must be both a hazard, such as toxicity, and potential for exposure to these materials. The interfacial chemical properties of nanoparticles in aqueous media affect particle aggregation and deposition processes that in turn affect exposure. Nanoparticle stability and transport are important in determining whether or not these materials can be removed by water-treatment technologies, or whether nanoparticles have a high potential for mobility in the environment. The chemistry of nanoparticles is examined from a colloid science perspective, considering factors that may be different at the nanoscale in determining the nanoparticle stability and transport in Chapter 6. Nanomaterials may also be transformed in environmental and in physiological systems, altering their exposure and hazard. Transformations are discussed in Chapter 7.

Environmental Applications of Nanomaterials

The products of nanochemistry are being been used to create new generations of technologies for curing environmental maladies and protecting public health. Water pollution control, groundwater remediation, potable water treatment, and air quality control are being advanced through nanomaterial-based membrane technologies, adsorbents, and catalysts. The use of nanomaterials as photocatalysts is discussed in Chapter 4. Groundwater remediation, water treatment, and sensor development are among the applications in which nanomaterials are finding their way into environmental engineering practice. The remediation of contaminated groundwater is a costly problem that has been approached in environmental engineering using both pump-and-treat and *in situ* technologies. In most cases, pumping contaminated groundwater to the surface to remove contaminants and reinjecting the treated water has proven to be both, cost-prohibitive and incapable of meeting cleanup goals. As a result, *in situ* treatments such as biodegradation have been explored extensively. Physical-chemical approaches to *in situ* treatment have included the use of zerovalent iron and catalysts to promote redox reactions that degrade contaminants. Nanomaterials have been developed to promote such reactions at high rates; however, successful application of this technology will require a high degree of control of nanoparticle mobility, reactivity, and ideally, specificity for the contaminant of interest. Background on groundwater remediation and the development of nanomaterials for *in situ* treatment is presented in Chapter 8.

Membrane technologies are playing an increasingly important role as unit operations for environmental quality control, resource recovery, pollution prevention, energy production, and monitoring. In water treatment they can be used for a wide spectrum of applications, ranging from particle removal to organic removal and desalination. Nanoscale control of membrane architecture and the use of nanomaterials to modify membrane properties has been the focus of considerable research over the last 20 years. In Chapter 9, principles of membrane processes are summarized and examples of the use of nanomaterials to create new membrane systems are described. Chapter 10 describes several cases where nanomaterials have been used to enable new sensors.

Potential Impacts of Nanomaterials on Organisms and Ecosystems

Possible risks associated with nanomaterial exposure may arise during nanomaterial fabrication, handling of nanomaterials in subsequent processing to create derivative products, product usage, and as the result of postusage or waste disposal practices. The quantities of nanomaterials produced per year are large and increasing rapidly. However, the quantities of engineered nanomaterials produced are orders of magnitude smaller than those produced by natural and incidental processes such as combustion. Naturally occurring nanoparticles are ubiquitous, distributed throughout the atmosphere, oceans, soil systems, terrestrial water systems (groundwater and surface water), and in and/or on most living organisms [6]. Sometimes there is great similarity between engineered, natural, and incidental nanomaterials. For example, although first discovered in the laboratory, buckyballs were ultimately found to come from natural and incidental sources as well as geologic deposits [7, 8], candle soot, and meteorites [9].

The properties of nanomaterials that may create concern in terms of environmental impact (such as nanoparticle uptake by cells) are often precisely the properties desired for beneficial uses such as in medical applications. While it may be desirable from the standpoint of medical therapeutics to have nanoparticles readily taken up by cells, this same trait may also have negative implications in an environmental context. Early work evaluating nanomaterial toxicity was sometimes performed with ill-conceived methodologies that included the application of unrealistically high concentrations of nanomaterials or exposure pathways, or the use of solvents that produced toxic by-products [10]. Exposure via inhalation has been an important concern from the beginning, particularly in the context of nanomaterials fabrication. Early experiments conducted by washing the lungs of laboratory animals with solutions containing the carbon nanotubes formed nanotube aggregates within the lungs and subsequent suffocation [11]. Methodologies for handling nanomaterials have developed greatly over the last decade along with a well-established community of nanotoxicologists. There is now an extensive body of literature that includes entire journals dedicated to the topic. This literature includes developmental results in addition to simple evaluation of mortality, often using model organisms in embryonic and larval stages that allow for the examination of specific tissue uptake and impacts on the reproductive, renal, and digestive systems, development of the cardiovascular system, and developmental changes in the nervous systems [12–14]. Work on the toxicity of nanomaterials, focused on possible human exposure, is reviewed in Chapter 11 while ecotoxicity is the topic of Chapter 5. The use of mesocosms and microcosms in studying ecosystem impacts of nanomaterials is described in Chapter 12.

The matter of determining whether or not a substance is "dangerous" involves not only determining the material's hazard, but also its exposure or probability of coming in contact with an organism. The approach informally adopted by researchers to evaluate nanomaterial risk can be summarized as varying its properties and observing the resultant changes in its exposure or hazard. The goal of this approach is implicitly to relate risk directly to nanomaterial properties. However, with the exception of the trivial case of predicting that nanomaterials made from toxic or highly reactive materials should be toxic, this approach has yielded few actionable guidelines for predicting nanomaterial risk. Chapter 13 describes an alternative approach of evaluating nanomaterials using functional assays, and a framework for combining information on exposure and hazard to analyze the risks of nanomaterials. An analysis of risk implies that the resulting information will be used for informed action, whether it be on the part of a consumer deciding which products to buy or a government deciding whether or not regulatory intervention is needed. The development of policies and regulations through a systematic process that engages a range of stakeholders is domain of risk governance. The final chapter of this book introduces concepts around the topic of risk governance as it applies to nanotechnology.

Nanotechnology as a Test Case

Nanotechnology is only one in a continuous stream of emerging technologies. On the horizon are developments in metamaterials, synthetic biology, artificial photosynthesis, and biofuels, to name just a few emerging technologies, which will require an examination of their sustainability.

Elements shared by emerging technologies in the generic sense include high degrees of uncertainty and an evolving profile of perceived risk and benefits. At the heart of the problem is the question: "How do you forecast risk to human health and the environment for an industry that doesn't exist yet?" A process for forecasting the risk of an emerging technology should include the ability to generate forecasts and associated levels of uncertainty that provide both long-term guidance and relevance for the issues of immediate concern that typically announce the arrival of a new technology. The appropriate level of investment in such proactive research around an emerging technology is perhaps impossible to determine. Recent investments in research of this nature in the nanotechnology sector have been estimated to be no more than 5 percent of the total investment under the U.S. National Nanotechnology Initiative (NNI). Reported public investments by E.U. countries appear to be of a similar magnitude. Well-established components of the private sector that may be users of nanomaterials have also shown interest in investing in such research. However, the credibility of private-sector–financed research in this area has been questioned by environmental nongovernmental organizations (NGOs) [15] and the public may be reluctant to validate data produced by such efforts. The European Union is moving toward regulation of nanomaterials under its Registration, Evaluation, Authorisation and Restriction of Chemical substances legislation (REACH) which places greater responsibility on industry to identify at an early stage the intrinsic properties of chemical substances that may present risk and to provide safety information on new substances entering European markets. The European Chemicals Agency is formulating technical instructions to help companies include nanomaterials in their registration dossiers under REACH and other compliance obligations for substances they make or import. Japan is moving in a similar direction under its Chemical Substances Control Law that, as amended, requires companies to report hazard and use information for chemicals manufactured in or imported into Japan. While the U.S. position may differ from that of other developed nations, constituencies in all cases appear to be calling for more reliable data.

These data are a public good that is not readily provided by the private sector. Governments must take the lead in producing this public good in the form of impartial data on the environmental and health implications of emerging technologies. Such data are needed to assure the public of the credibility of the data and to provide information to the private sector, particularly start-ups; to assure the flow of investment capital and troubleshoot potential problems with proposed products. The development of new technologies and a precautionary, proactive examination of the possible impacts of these technologies are not activities in opposition. An evaluation of the possible impacts of emerging technologies on human health and the environment is essential to the adoption and advancement of the technologies themselves. It is an insurance policy that encourages commercialization of new technologies by protecting companies and investors against damages, liability, and loss of investment and above all, an investment in the well-being of future generations.

Acknowledgments

Work to produce this book and portions of the research results presented in this book were supported by grants from the U.S. National Science Foundation (NSF), the U.S. Environmental Protection Agency (EPA), France's Centre National de la Recherche Scientifique (CNRS) and the French Agence Nationale de Recherche (ANR). The NSF and EPA supported much of this work under NSF Cooperative Agreement EF-0830093 and DBI-1266252, Center for the Environmental Implications of NanoTechnology (CEINT). Any opinions, findings, conclusions, or recommendations expressed in this material are those of the author(s) and do not necessarily reflect the views of the NSF or the EPA. This work has not been subjected to EPA review and no official endorsement should be inferred. The CNRS provided support through GDRi-iCEINT International Consortium for the Environmental Implications of NanoTechnology in collaboration with Duke University. Support from the ANR was provided through grants ANR-3-CESA-0014/NANOSALT, ANR-10-NANO-0006/MESONNET, and the Labex SERENADE (ANR-11-LABX-0064) funded by the "Investissements d'Avenir" French Government program of the French National Research Agency (ANR) and through the A*MIDEX project (ANR-11-IDEX-0001-02). Part of this work was also supported by the Swiss National Science Foundation (PBEZP3-140058 and P300P3-158517).

References

1. Feynman, R., "There's Plenty of Room at the Bottom" (lecture, American Physical Society meeting at Caltech, Pasadena, CA, December 29, 1959).
2. Drexler, K.E., "Molecular Engineering: An Approach to the Development of General Capabilities for Molecular Manipulation," *Proceedings of the National Academy of Sciences USA,* 78: 5275–5278, 1981.
3. Commission, E., *Nanotechnologies: A Preliminary Risk Analysis on the Basis of a Workshop March 2004* in *Nanotechnologies: A Preliminary Risk Analysis,* R.A.U.P.H.a.R.A. Directorate, Editor. 2004, European Commsion Community Health and Consumer Protection Directorate General of the European Commission: Brussels. 11–29.
4. Commission, E., *Opinion on the Appropriateness of Existing Methodologies to Asses the Potential Risks Associated with Engineered and Adventitious Products of Nanotechnologies.* 2005, Scientific Committee on Emerging and Newly Identified Health Risks. 41–58.
5. Drexler, K.E., "Engines of Destruction," in *Engines of Creation: The Coming Era of Nanotechnology,* Anchor Books: New York, NY. 1986.
6. Hochella, M.F.J., et al., "Nanominerals, Mineral Nanoparticles, and Earth Systems." *Science* 319 (2008): 1631–1635.
7. Heymann, D., L.P.F. Chibante, and R.E. Smalley, "Determination of C-60 and C-70 Fullerenes in Geologic Materials by High-Performance Liquid-Chromatography." *Journal of Chromatography A* 689(1) (1995): 157–163.
8. Heymann, D., et al., "Fullerenes in the Cretaceous-Tertiary Boundary-Layer." *Science* 265(5172) (1994): 645–647.
9. Becker, L., et al., "Fullerenes in Meteorites and the Nature of Planetary Atmospheres," in *Natural Fullerenes and Related Structures of Elemental Carbon,* F.J.M. Rietmeijer, Editor. Springer, Netherlands. 2006, 95–121.
10. Sayes, C.M., et al., "The Differential Cytotoxicity of Water-Soluble Fullerenes." *Nano Letters* 4(10) (2004): 1881–1887.
11. Warheit, D.B., et al., "Comparative Pulmonary Toxicity Assessment of Single-Wall Carbon Nanotubes in Rats." *Toxicological Sciences* 77 (2004): 117–125.
12. Wassenberg, D.M. and R.T. Di Giulio, "Synergistic Embryotoxicity of Polycyclic Aromatic Hydrocarbon Aryl Hydrocarbon Receptor Agonists with Cytochrome P4501A Inhibitors in *Fundulus heteroclitus.*" *Environmental Health Perspectives* 112 (2004): 1658–1664.
13. Oxendine, S.L., et al., "Adapting the Medaka Embryo Assay to a High-Throughput Approach for Developmental Toxicity Testing." *Neurotoxicology* 27 (2006): 840–845.
14. Roy, N.M. and E. Linney, "Zebrafish as a Model for Studying Adult Effects of Challenges to the Embryonic Nervous System," in *Source Book of Models for Biomedical Research,* P.M. Conn, Editor. Humana Press, Inc., Totowa, NJ. 2007.
15. Powell, K., "Green Groups Baulk at Joining Nanotechnology Talks: International Council on Nanotechnology Accused of Industry Bias." *Nature* 432(5) (4 November 2004). Published online (news@nature.com) 3 November 2004.

Principles and Methods

CHAPTER 2

Nanomaterials Fabrication

Jean Pierre Jolivet

Chemistry Laboratory of Condensed Matter in Paris, University P. Curie, UMR7574 (UPMC/CNRS), Paris, France

Bernard Cathala

Interactions Assemblies Biopolymers Research Unit, National Institute of Agricultural Research, Nantes, France

Andrew R. Barron

Energy Safety Research Institute, College of Engineering, Swansea University, Wales, United Kingdom

Corinne Chaneac

Chemistry Laboratory of Condensed Matter in Paris, University P. Curie, UMR7574 (UPMC/CNRS), Paris, France

The ability to fabricate nanomaterials or nanoparticles with specific size, shape, and crystalline structure has inspired the application of nanochemistry to numerous fields, including catalysis, optics, electronics, and medicine. The synthesis of well-defined nanoparticles has resulted in several prominent milestones in the progress of nanoscience, including the discovery of fullerenes [1], graphene [2], and carbon nanotubes [3, 4], the synthesis of well-defined quantum dots [5–7], and the shape control of semiconductor nanocrystals [8]. However, despite an expansion in the methods of nanoparticles synthesis, it is still difficult to generalize underlying physical or chemical principles behind existing synthesis strategies to any arbitrary nanomaterial. A general, mechanistic understanding of nanoparticle formation that might guide the development of new materials remains lacking [9]. Though the synthesis of nanoparticles with control over size, shape, and size distribution has been a major part of colloid chemistry for decades, it remains an intensely studied topic as is evident by a substantial body of literature. In this chapter, we provide an overview of the main methods that have proved to be successful for the fabrication of several classes of nanomaterials: specifically, oxides, chalcogenides, metals, fullerenes, graphene, and organic nanoparticles like cellulose.

Specificity and Requirements in the Fabrication Methods of Nanoparticles

Ultradispersed systems, such as dispersions of nanoparticles, are intrinsically thermodynamically metastable, in large part due to the very high interfacial areas. Nanoparticle surface area represents a positive contribution to the free enthalpy of the system. If the activation energies are too high, spontaneous evolution of a nanoparticle dispersion can occur causing an increase in nanoparticle size or the formation of nanostructured domains and leading to the decrease of the surface area. Consequently, it follows that:

- An ultradispersed system with a high surface energy can be only kinetically stabilized.
- Ultrafine powders cannot be synthesized by methods involving energies that exceed a threshold, but rather through methods of "soft chemistry" that maintain the forming particles in a metastable state.
- Additives and/or synthesis conditions that reduce the surface energy are needed to form nanoparticles stabilized against sintering, recrystallization, and aggregation.

Synthesis methods for nanoparticles are typically grouped into two categories:

1. The first involves division of a massive solid into smaller portions. This "top-down" approach may involve milling or attrition (mecanosynthesis), chemical methods for breaking specific bonds (e.g., hydrogen bonds) that hold together larger repeating elements of the bulk solid, and volatilization of a solid by laser ablation, solar furnace, or some other method, followed by condensation of the volatilized components.

2. The second category of nanoparticle fabrication methods involves condensation of atoms or molecular entities in a gas phase or in solution. This is the "bottom-up" approach in which the chemistry of metal complexes in solution holds an important place. This approach is far more popular in the synthesis of nanoparticles, and many methods have been developed to obtain oxides, chalcogenides, and metals.

The liquid-phase colloidal synthetic approach is an especially powerful tool for convenient and reproducible shape-controlled synthesis of nanocrystals—not only because this method allows for the resulting nanocrystals to be precisely tuned in terms of their size, shape, crystalline structure, and composition on the nanometer scale, but also because it allows them to be dispersed in either an aqueous or a nonaqueous medium. Moreover, these nanoparticles can be modified in liquid suspension by treatment with various chemical species for application and use in a diverse range of technical or biological systems.

Oxides

The most widespread route to fabrication of metal oxide nanoparticles involves the bottom-up approach by the precipitation in aqueous solution from metal salts. Organometallic species can also be used in hydrolytic or nonhydrolytic pathways, but due to their cost and the difficulty in manipulating these compounds, they are used less frequently and primarily for high-tech applications. An alternative top-down approach has been demonstrated for aluminum and iron oxide nanoparticles; however, it is possible that this methodology could be extended to other oxides.

From Molecular Species to Nanoparticles

Variations on this approach include the hydroxylation of metal cations in aqueous solutions, the use of metal alkoxides, nonhydrolytic routes such as those employing metal halides.

Hydroxylation of Metal Cations in Aqueous Solution and Condensation: Inorganic Polymerization

The metal cations issued for the dissolution of salts in aqueous solution form true coordination complexes in which water molecules form the coordination sphere. The chemistry of such complexes, and especially their acid behavior, provides a framework for understanding how the solid (oxide) forms via inorganic polycondensation [10, 11]. The binding of water molecules to a cation involves an orbital interaction allowing an electron transfer from a water molecule to a cation following Lewis's acid-base concept of the coordination bond. Such a transfer drives the electronic density of water molecules toward the cation and weakens the O-H bond of the coordinated water molecules. They are consequently stronger Brønsted acids than the water molecules in the solvent itself, and they tend to be deprotoned spontaneously according to the hydrolysis equilibrium:

$$[M(H_2O)_n]^{z+} + h\ H_2O \Leftrightarrow [M(OH)_h(H_2O)_{n-h}]^{(z-h)+} + h\ H_3O^+ \tag{2.1}$$

or by neutralization with a base:

$$[M(H_2O)_n]^{z+} + h\ HO^- \Leftrightarrow [M(OH)_h(H_2O)_{n-h}]^{(z-h)+} + h\ H_2O \tag{2.2}$$

In these equilibria, h is the hydroxylation ratio of the cation.

The acidity of such cations strongly depends on the strength of the M-O bond due to the electron transfer from oxygen toward the metal. The acidity can be related to the polarizing character of the cation which is given by the ratio of formal charge (oxidation state) to its size. The equilibrium constant of the first step of hydroxylation (h = 1) for many cations where d stands for the M-O distance can be empirically expressed by [10]:

$$\log K \approx -20 + 11(z/d) = -\Delta G° \cdot RT \tag{2.3}$$

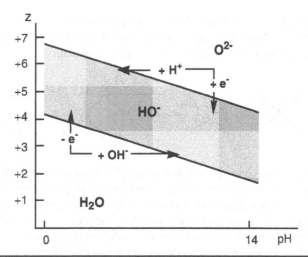

FIGURE 2.1 Composition of the coordination sphere of a cation as function of its formal charge, z, and the pH of the medium [13]. Possible initiation methods of condensation reactions are depicted.

The hydrolysis rate of cations in aqueous solution also depends strongly on the pH of the medium because the equilibrium involves the transfer of protons. From the acidity constants of medium-sized cations [10, 12], a charge-pH diagram was established [13] in which three domains are plotted (Figure 2.1). The lower domain corresponds to the existence of aqua-cations $[M(H_2O)_n]^{z+}$, the upper to oxo-anions $[MO_n]^{(2n-z)-}$, and the intermediate domain corresponds to hydroxylated complexes containing at least one hydroxo ligand. These domains are separated by two lines corresponding to $h = 1$ and $h = 2n - 1$, respectively. Therefore, condensation can be initiated by varying the pH of the solution by addition of a base on an acid (element M with $z \leq 4$) or by addition of an acid on a base (element M with $z > 4$). Condensation can also be initiated via redox reactions with elements having several stable oxidation states (Figure 2.1).

Hydrolysis is, strictly speaking, a neutralization reaction carried out by the water molecule in Equation (2.1). For this reaction, it has been shown [10]:

$$\Delta H^{\circ} = (75.2 - 9.6\ z)\ \text{kJ mol}^{-1},$$

$$\Delta S^{\circ} = (-148.4 + 73.1\ z)\ \text{J mol}^{-1},\ \text{and}$$

$$\Delta G^{\circ}_{298} = (119.5 - 31.35\ z)\ \text{kJ mol}^{-1} \tag{2.4}$$

The reaction is spontaneous ($\Delta G^{\circ} < 0$) for elements of charge equal to or greater than 4. Therefore, at room temperature, tetravalent elements do not exist as purely aquo complexes, even in strongly acidic medium. For elements with $z \leq 4$, ΔG° becomes negative only if the temperature is higher than 298 K. Therefore, it is necessary to heat the solution in order to carry out hydrolysis of the cation (forced hydrolysis or thermohydrolysis). The monomeric, electrostatically charged, hydroxylated species $[M(OH)_h(H_2O)_{n-h}]^{(z-h)+}$ or $[MO_{n-h}(OH)_h]^{(2n-z-h)-}$ are generally observed in solution only at very low concentrations. More often, they condense and form soluble polynuclear species, polycations ($h < z$), and polyanions ($h > z$) respectively, in which the cations are bounded by hydroxo or oxo bridges [11]. These entities are generally of molecular size, although giant polyanions of molybdenum containing up to 368 Mo ions have been recently synthesized [14]. The condensation of neutral complexes with $h = z$ is, in general, not limited and continues to the formation of a solid. Hydroxylated complexes condense via two basic mechanisms of nucleophilic substitution, depending on the nature of the coordination sphere of the cations. The cation must also have an electrophilic character high enough to be subjected to the nucleophilic attack. Condensation of aquohydroxo complexes proceeds by elimination of water and formation of hydroxo bridges (olation) [see Equation (2.5)]:

$$\text{H}_2\text{O—M—OH} + \text{—M—OH}_2 \longrightarrow \text{H}_2\text{O—M—OH—M—OH}_2 + \text{H}_2\text{O} \tag{2.5}$$

For oxohydroxo complexes, there is no water molecule in the coordination sphere of the complexes and therefore no leaving group. Condensation has to proceed in this case via a two-step associative mechanism leading to the formation of oxo bridges (oxolation) [see Equation (2.6)]:

$$\tag{2.6}$$

Usually, during condensation, the nucleophilic character of hydroxo ligands cancels in polycations, and the electrophilic character of the cation cancels in polyanions. Condensation of electrically neutral ions ($h = z$) continues always indefinitely until there is precipitation of a solid (hydroxide, oxyhydroxide, or more or less hydrated oxide) or of a basic salt in the presence of complexing ligands. Elimination of water from noncharged complexes never leads to a sufficient change in the average electronegativity to cancel the reactivity of functional groups. In theory, an hydroxide $M(OH)_z$ is formed via endless condensation of aquohydroxo complexes. However, the hydroxide may not be stable. Its spontaneous dehydration, more or less rapid and extensive, generates an oxyhydroxide $MO_x(OH)_{z-2x}$ or a hydrated oxide $MO_{z/2}(H_2O)_x$. The reaction is associated with structural changes in order to preserve the coordination of the cation. Usually, elements with a +2 charge precipitate as hydroxides, and those with a +3 charge as oxyhydroxides (the final stage of evolution is the oxide). Those of higher charge form oxides of various level of hydration [15]. This sequence is a clear illustration of the increasing polarization of the hydroxo ligands by the cation, which is associated with the covalent nature of the M-O bond.

Particle morphology may vary depending on synthesis conditions. Moreover, aging in aqueous solution may bring about significant dimensional, morphological, and structural changes. In order to understand how small particles form and what role the experimental parameters play on their characteristics and on evolution, it is useful to review the kinetic aspects of condensation reactions.

The precipitation of a solid involves four kinetic steps [16–19]:

1. Formation of the zero-charge precursor $[M(OH)_z(H_2O)_{n-z}]^0$, which is able to condense and form a solid phase. Hydroxylation of the cation is a very fast acid/base reaction, but the rate of formation of the zero-charge precursor in solution can largely vary depending on whether the reaction starting from cationic complexes for example, takes place through addition of a base, thermohydrolysis, or slow thermal decomposition of a base such as urea.

2. Creation of nuclei, through condensation (olation or oxolation) of zero-charge precursors. The condensation rate is a function of precursor concentration, and as long as it is small at the onset of cation hydroxylation, the rate is almost zero (zone I, Figure 2.2a). Beyond a critical concentration C_{min}, the condensation rate increases abruptly and polynuclear entities, the nuclei, are formed in an "explosive" manner throughout the solution (zone II, Figure 2.2c). Indeed, nucleation is an abrupt kinetic phenomenon because, since its order is high compared to the precursor concentration, it is either extremely fast or nonexistent within a narrow concentration range (Figure 2.2b,c). If the rate of generation of the precursor is significantly smaller than the condensation rate, nucleation sharply reduces the precursor concentration and the condensation rate decreases equally rapidly. When the precursor condensation is again close to C_{min}, formation of new nuclei is no longer possible.

3. Growth of the nuclei through addition of matter, until the primary particle stage is reached. This step follows the same chemical mechanisms as nucleation: olation or oxolation. Nuclei grow until the precursor concentration reaches the solution saturation (in other words, the solubility limit) of the solid phase (zone III, Figure 2.2b,c). Growth, having kinetics of first or second order, is a somewhat faster process. The number, and therefore the size, of the primary particles that form from a given quantity of matter is linked to the relative nucleation and growth rates (Figure 2.2a). In order to obtain particles of homogeneous size, it is necessary that the nucleation and growth steps be separated to ensure that a single nucleation stage takes place, and that growth, via accumulation of all remaining matter, be controlled. This implies that the nucleation rate should be much greater than the rate at which the precursor is generated. Under these conditions, nucleation is very brief and clearly decoupled from the growth phase. If the nucleation rate is not high enough compared to the rate of generation of the precursor, precursor concentration remains higher than C_{min}

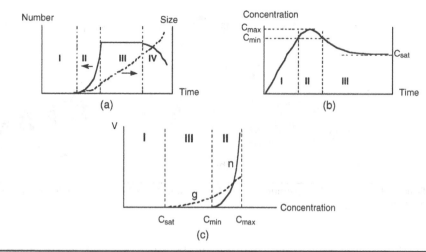

FIGURE 2.2   Change (a) in the number and sizes of particles formed in solution, and (b) in concentration of the soluble precursor of the solid phase during precipitation [19]. Condensation rate, which is zero for $C < C_{min}$, becomes infinite for $C \geq C_{max}$. C_{sat} is the solubility of the solid phase. (c) Nucleation (n) and growth (g) rates as a function of precursor concentration in solution.

throughout the reaction, and nucleation and growth are simultaneous. The growth of the first nuclei is much larger than that of the younger ones, which leads to a large particle size distribution.

4. Nucleation and growth steps form particles under kinetic control following a reaction path of minimum activation energy under conditions imposed to the system (acidity, concentration, temperature), but the products are not necessarily thermodynamically stable. Aging of the suspensions, which may take place over a very large timescale (hours, days, or months), allows the system to tend toward or reach stability, and it is often associated with modifications of some physical or chemical characteristics of the particles. "Ostwald ripening" leads to an increase in the average particle size and possible aggregation (zone IV, Figure 2.2a). Aging may also trigger a change in morphology and crystalline structure or even cause crystallization of amorphous particles. In fact, aging is one of the most important phenomena that must be considered, because it determines the characteristics of the particles after precipitation.

Control of Particle Size, Crystalline Structure, and Morphology

There are different techniques to obtain a solid. Inhomogeneities in the hydrolysis products often present during such a missing procedure may result in random condensation and the formation of an amorphous solid with an ill-defined chemical composition. Such a result is exemplified by the case of ferric ions. They precipitate quasi-instantaneously at pH ≥ 3 into a poorly defined, highly hydrated phase, called two-line ferrihydrite [20]. (This phase takes its name from its X-ray diffraction pattern, which exhibits only two broad bands.) In similar conditions, Al^{3+} ions form a transparent amorphous gel [21]. At pH ≥ 2, Ti^{4+} ions form an amorphous oxyhydroxide with a composition near to $TiO_{0.3}(OH)_{3.4}$ [21]. These solids are formed of very small-sized particles, around 2 to 3 nm in diameter, and are strongly metastable. They evolve spontaneously in suspension more or less quickly to form crystalline nanoparticles, with possibly an increase in particle size, releasing simultaneously the lattice energy (and decreasing the surface energy) to decrease the free enthalpy of the system. The acidity of the suspension during evolution is the most important parameter to control crystalline structure and the size of the final particles. Two distinct mechanisms are involved in the transformation. When the suspensions are aged at a pH where the solid is partially soluble, the concentration in solution may be enough to feed nuclei of a more stable crystalline phase. A transfer of matter occurs via the solution from the soluble amorphous phase toward a less soluble crystalline phase during a slow dissolution-crystallization process allowing formation of well-crystallized particles. Such a process is involved in the formation of goethite, α-Fe(O)(OH), during aging of ferrihydrite in suspension at pH < 5 or pH > 10. Rod-like particles of mean dimensions $150 \times 25 \times 15$ nm are obtained (Figure 2.3). These particles behave as nematic lyotropic liquid crystals exhibiting very interesting magnetic properties [23]. In similar ranges of acidity, the aluminate gel is transformed into platelets of hydroxide $Al(OH)_3$, gibbsite at pH < 5 and bayerite at pH > 8 [21]. In a rather acidic medium (pH < 1), amorphous

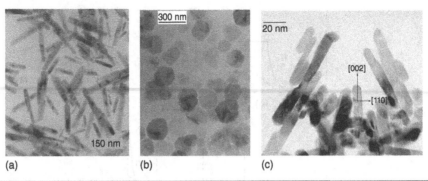

Figure 2.3 Particles of (a) goethite α-Fe(O)(OH), (b) gibbsite Al(OH)$_3$, and (c) rutile TiO$_2$ synthesized in aqueous medium.

titanium oxyhydroxide are transformed into elongated TiO$_2$ rutile nanoparticles. In these examples, the final size of particles depends on the acidity of the medium: the particle size increases when the acidity is strong.

If the suspensions are aged at an acidity where the solubility of the solid is very low or at a minimum, the concentration of soluble species in equilibrium with the solid phase does not allow an efficient transport of matter, and crystallization of the early amorphous material will occur more easily by a transformation *in situ*, in the solid state. Nanoparticles of hematite, α-Fe$_2$O$_3$, are so obtained from ferrihydrite at $6 \leq$ pH ≤ 8 [24]. Very small nanoparticles of boehmite, γ-Al(O)(OH), (around 300 m$^2 \cdot$ g^{-1}) are similarly obtained by aging of aluminate gels at the same pH range (6–8) [21]. Although boehmite is not the most thermodynamically stable phase at room temperature, it is probably kinetically stabilized because the system is constrained to evolve without heating and transforms on the lowest activation energy path. Between pH 2 and pH 7, where the solubility of titania is very low, the amorphous solid is transformed into TiO$_2$ anatase nanoparticles [22]. Over this acidity range, the particle size of anatase depends on the pH of precipitation and aging. Precipitation by addition of a base at room temperature may also lead to stable crystalline nanoparticles without involving any transformation by the above mechanisms. For instance, magnetite Fe$_3$O$_4$ is easily obtained by coprecipitating aqueous Fe^{3+} and Fe^{2+} ions with x = 0.66 [23]. Iron ions are distributed into the octahedral (Oh) and tetrahedral (Td) sites of the face-centered cubic (fcc) stacking of oxygen according to ([Fe^{3+}]$_{Td}$[Fe^{3+}Fe^{2+}]$_{Oh}$O$_4$). Magnetite is characterized by a fast electron hopping between the iron cations on the octahedral sublattice. Crystallization of spinel is quasi-immediate at room temperature and electron transfer between Fe^{2+} and Fe^{3+} ions plays a fundamental role in the process [26, 27]. In effect, maghemite, γ-Fe$_2$O$_3$, [Fe^{3+}]$_{Td}$[Fe$^{3+}_{5/3}$V$_{1/3}$]$_{Oh}$O$_4$ (where V stands for a cationic vacancy) does not form directly in solution by precipitation of ferric ions, but a small proportion of Fe^{2+} ($\leq$10 mol %) induces the crystallization of all the iron into spinel. Electron mobility brings about local structural rearrangements and drives spinel ordering. Besides this topotactic process, crystallization of spinel can also proceed by dissolution crystallization, resulting in two families of nonstoichiometric spinel particles ([Fe^{III}]$_{Td}$[Fe$^{III}_{1+2z/3}$Fe$^{II}_{1-z}$V$_{z/3}$]$_{Oh}$O$_4$) with very different mean size [26]. The relative importance of these two pathways depends on the Fe^{2+} level in the system and the end products of the coprecipitation are single phase only for $0.60 \leq z \leq 0.66$. Different interfacial ionic and/or electron transfers that depend on the pH of the suspension can be involved in the transformation. In basic media, the oxidation of magnetite proceeds by oxygen reduction at the surface of the particles (electron transfer only) and coordination of oxide ions, while in acidic medium and anaerobic conditions [28], surface Fe^{2+} ions are desorbed as hexa-aquo complexes in solution (electron and ion transfer) according to:

$$[Fe^{3+}]_{Td}[Fe^{2.5+}_2]_{Oh}O_4 + 2H^+ \rightarrow 0.75\,[Fe^{3+}]_{Td}[Fe^{3+}_{5/3}V_{1/3}]_{Oh}O_4 + Fe^{2+}_{aq} + H_2O \qquad (2.7)$$

The comparison with the cases where M^{2+} is different from Fe^{2+} emphasizes the role of electron mobility between Fe^{2+} and Fe^{3+} ions in the crystallization process. With other divalent cations, intervalence transfers are negligible and a spinel ferrite forms only by dissolution-crystallization [25]. With z = 0.66, corresponding to stoichiometric magnetite, the mean particle size is controlled in the range of 2 to 12 nm by the conditions of the medium, pH and ionic strength (I), imposed by a salt (8.5 $\leq$ pH $\leq$ 12 and 0.5 $\leq$ I $\leq$ 3 mol $\cdot$ L^{-1}) (Figure 2.4) [29]. Such an influence of acidity on the particle size is relevant to thermodynamics rather than kinetics (nucleation and growth processes). Acidity and ionic strength act on the electrostatic surface

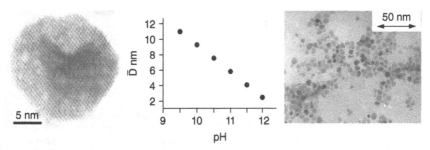

Figure 2.4 Electron micrographs of magnetite particles synthesized by precipitation in water and particle size variation against pH of precipitation [28].

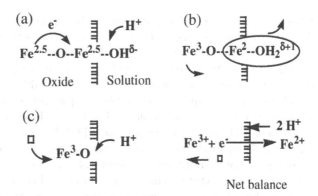

Figure 2.5 The transformation of magnetite in maghemite is due to the solubilisation of lattice FeII [28].

charge (Figure 2.5). This leads to a change in the chemical composition of the interface, inducing a decrease of the interfacial tension, γ, as stated by Gibbs's law, $d\gamma = -\Gamma_i d\mu_i$, where Γ_i is the density of adsorbed species i with chemical potential μ_i. Finally, the surface contribution, $dG = \gamma dA$ (A is the surface area of the system), to the free enthalpy of the formation of particles is lowered, allowing the increase in the system surface area [30].

A very interesting technique for obtaining oxide nanoparticles is the thermolysis (or forced hydrolysis) of acidic solutions. Heating of a solution to approximately 50 to 100°C enables, particularly with trivalent and tetravalent elements (Al, Fe, Cr, Ti, Zr, etc.), a homogeneous hydrolysis in conditions close to thermodynamic equilibrium [31]. Under such conditions, the slow speed of formation of the hydrolyzed precursors allows decoupling of the nucleation and growth steps, from a kinetic standpoint. As a result, narrow particle-size distributions can be obtained.

Thermolysis at 90 to 100°C of acidic ferric solutions (pH $\leq$ 3) forms hematite [11, 32]. In these conditions acidity facilitates oxolation leading to oxide. The acidity and the nature of the anions are, however, crucial for the control of the size of particles. At low concentration of chloride (C < 10^{-3} mol · L^{-1}), six-line ferrihydrite forms initially [32, 33]. It transforms into hematite during thermolysis, but the particle size depends strongly on the acidity of the medium [20]. At high concentration of chloride, akaganeite, β–Fe(O)(OH), is first formed [34]. This metastable phase is slowly transformed into hematite during thermolysis and large (μm-sized) polycrystalline particles with various morphologies are obtained depending on the nature of anions in the medium [35–37]. Thermolysis at 95°C of aluminum nitrate solutions for one week produces exclusively boehmite, γ-Al(O)(OH). The change in the acidity over a large range allows modification of the shape of the nanoparticles. At pH = 4 to 5, heating produces fibers or rods around 100 nm in length. The fibers are formed by aggregation of very small platelets 3 nm thick and 6 nm wide, exhibiting (100) lateral faces and (010) basal planes. The particles synthesized at pH = 6.5 are pseudohexagonal platelets 10 to 15 nm wide and 4 to 5 nm thick with (100) and (101) lateral faces, while those synthesized at pH = 11.5 are diamond-shaped, 10 = 25 nm wide (Figure 2.6). Such a change in the nature of lateral faces of particles results from the change in surface energy induced by the variation in the electrostatic surface charge density as a function of the pH [28]. This is an important feature of boehmite particles, because they are the precursor of γ-alumina, γ-Al$_2$O$_3$, largely used as a catalyst [21].

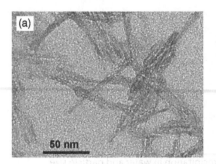
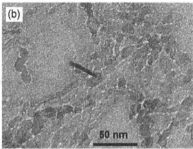
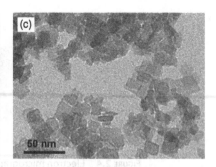

FIGURE 2.6 TEM micrographs of boehmite nanoparticles synthesized at (a) pH of 4.5, (b) pH of 6.5, and (c) pH of 12.

Thermolysis of strongly acidic $TiCl_4$ solutions enables a very efficient structural and morphological control of titanium oxide nanoparticles. After heating for one day at 90°C, $TiCl_4$ in concentrated perchloric acid solutions ($HClO_4$, 1–5 M) forms mixtures containing various proportions of the different TiO_2 polymorphs (anatase, brookite, and rutile). After heating for one week, the metastable phases, anatase and brookite, disappear through transformation into rutile with very different shapes depending on the acidity. It is thus possible to adjust the aspect ratio of rutile nanoparticles from around 5:1 to 15:1. When $TiCl_4$ is thermolyzed in concentrated hydrochloric acid (HCl, 1–5 M), brookite nanoplatelets are stabilized when the stoichiometries of Cl/Ti and H^+/Ti are optimized [38]. Brookite is currently obtained in the form of large particles [39]. Nanoparticles of brookite apparently are never obtained except as by-products of various reactions [40]. Quasi-quantitative synthesis of brookite nanoparticles seems to result from a specific precursor, $Ti(OH)_2Cl_2(H_2O)_2$, containing chloride as ligands in the early complexes formed in solution. As long as chloride ions are present in suspension, brookite nanoparticles remain stable, while if chloride ions are replaced by perchloric anions, brookite transformation into rutile is complete after several hours at 90°C [38].

These examples underscore the versatility of oxide nanoparticle chemistry in an aqueous medium. A strict control of the reaction bath is consequently critical to obtain well-defined nanoparticles. It is, however, interesting to distinguish two sorts of physico-chemical conditions in these syntheses. In moderately acidic or basic media, the sign and the density of the electrostatic surface charge of particles varies as a function of pH due to proton adsorption-desorption equilibria. A semiquantitative model [29] works well to account for this size effect for anatase and magnetite and to explain the change in shape of boehmite nanoparticles. In strongly acidic media used for thermolysis, the solubility of the solid is high because the surface is totally protonated and the ionic atmosphere near the surface of particles during formation is very likely high and constant, so that the surface energy is always low. Depending on their structure, some faces can be energetically favored, but dissolution-crystallization processes very likely play a role in the growth of particles. Other parameters such as thermolysis temperature, concentration, and presence of specific ligands have to be taken into account in the describing nanoparticle formation.

Hydrolysis of Metallo-Organic Compounds

Metallo-organic compounds, and especially metal alkoxides [41, 42], are largely involved in so-called sol-gel chemistry of oxide nanomaterials [43]. Metal alkoxides are also precursors of hybrid organic-inorganic materials, because such compounds can be used to introduce an organic part inside the mineral component [44–46]. Sol-gel chemistry mainly involves hydrolysis and condensation reactions of alkoxides $M(OR)_z$ in solution in an alcohol ROH, schematically represented as:

$$M(OR)_z + z\ H_2O \rightarrow M(OH)_z + z\ ROH \rightarrow MO_{z/2} + z/2\ H_2O \tag{2.8}$$

These two reactions, hydroxylation and condensation, proceed by nucleophilic substitution of alkoxy or hydroxy ligands by hydroxylated species according to:

$$M(OR)_z + x\ HOX \leftrightarrow [M(OR)_{z-x}(OX)_x] + x\ ROH \tag{2.9}$$

If X represents an organic or inorganic ligand, the reaction is a complexation. In organic medium, both hydrolysis and condensation follows an associative mechanism in forming intermediates species in transition states in which the coordination number of the metal atom is increased. That explains why the

Alkoxide	Gel Time (h) at RT			
	pH 7	pH 1 (HCl, HNO₃)	pH 9 (NH₃)	DMAP
Si(OMe)₄	44			
Si(OEt)₄	242	10	10	5 min
Si(OBu)₄	550			

TABLE 2.1 Gelation Time of Silicon Alkoxides as a Function of Alkoxy Groups Si(OR)₄ at Several Values of pH in Water and in 4-(Dimethylamino)Pyridine (DMAP)

reactivity of metal alkoxides toward hydrolysis and condensation is governed by three main parameters: the electrophilic character of the metal (its polarizing power), the steric effect of the alkoxy ligands, and the molecular structure of the metal alkoxide. Generally, the reactivity of alkoxides toward substitutions increases when the electronegativity of metal is low and its size is high. That lowers the covalence of the M-O bond and enhances the reaction rates. Silicon alkoxides are weakly reactive in the presence of water (χ_{Si} = 1.74) while titanium alkoxides (χ_{Ti} = 1.32) are very sensitive to moisture. Alkoxides of very electronegative elements such as O=P(OEt)₃ (χ_{P} = 2.11) are quite inert and do not react with water in normal conditions. The reactivity of metal alkoxides is also very sensitive to the steric hindrance of the alkoxy groups. It strongly decreases when the size of the OR group increases. For instance, the rate constant, k, for hydrolysis of Si(OR)₄ at 20°C decreases from 5.1×10^{-2} L·mol⁻¹s⁻¹ for Si(OMe)₄ to 0.8×10^{-2} L·mol⁻¹s⁻¹ for Si(OBu)₄ and the gelation time is increased by a factor of 10 (Table 2.1).

The acidity of the medium also influences the rate of hydrolysis and condensation reaction to a great extent as well as the morphology of the products. In an excess of water and in acidic medium (pH ≤ 4), the silicon alkoxides form transparent polymeric gels while in basic medium (pH ≥ 8), the condensation is also accelerated relatively to the reaction in neutral medium (Table 2.1) and leads to perfectly spherical and monodispersed particles of hydrated silica, as exemplified by Stöber's method [47]. These variations reflect the acid or basic catalysis of the involved reactions. It is possible to explain the overall structure of the silica polymer by considering at an early stage of condensation, a branched chain of silicic tetrahedra containing several types of groups:

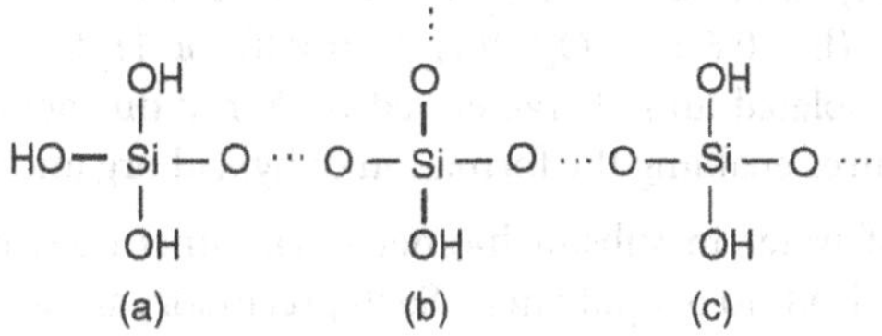

such a chain being formed by the oxolation reaction:

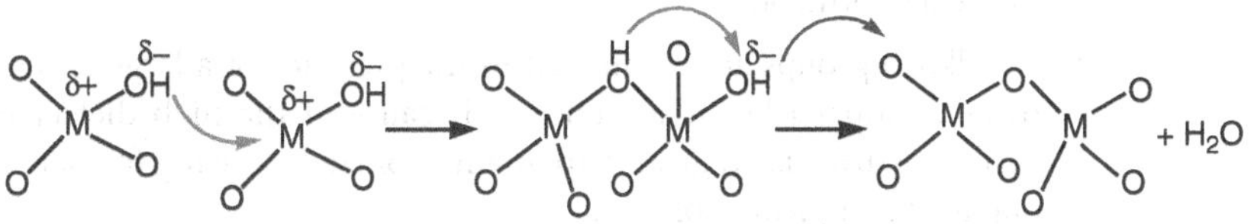

One may consider three different reactive groups: terminal or monocoordinated (A), branched or tricoordinated (B), and middle or dicoordinated (C). Using the partial charges model [11], it is possible to estimate the relative partial charges on the sites A, B, and C (Table 2.2):

Site	δ(Si)	δ(OH)
A	+0.50	−0.06
B	+0.58	+0.06
C	+0.54	0

TABLE 2.2 Partial Charge Calculated of the Various Sites into a Chain of Silica Tetrahedral

In an alkaline medium, catalysis involves the first step of the condensation mechanism—that is, nucleophilic attack by the anionic forms (or OH$^-$). It must take place preferentially on sites with the highest partial charge—in the middle of the chain (sites B and C), leading to cross-linked polymers forming dense particles, in agreement with experimental observation. In an acidic medium, catalysis impacts the second step of condensation. Elimination of the proton from the alcohol bridge in the transition state is eased by the protonation of an OH ligand, which favors formation of the leaving group (aquo ligand). The OH groups concerned are those located at the ends of chains, which bear the highest negative partial charge, or even those of the Si(OH)$_4$ monomer. As a result, poorly cross-linked and poorly condensed chains are formed. Therefore, the morphology of the particles is heavily dependent upon the conditions of acidity in which condensation takes place. The catalysis of silica condensation may also be affected by nucleophilic activation using additives such as dimethylaminopyridine (DMAP, see Table 2.1). The reactivity of metal alkoxides is also deeply influenced by their molecular structure and complexity that depends on the steric hindrance of the alkoxo ligands OR, especially for the transition element alkoxides. Due to the fact that the oxidation state, z, is generally smaller than the coordination number of the metal, it inhibits coordination of the metal in the monomeric M(OR)$_z$ species. For instance, this occurs in the case of titanium alkoxide Ti(O^{i}Pr)$_4$, which is a monomer in isopropanol. The coordination of titanium is only four and the reaction with water leads to instantaneous precipitation of heterogeneous and amorphous titania particles. With ethoxy ligands, titanium forms oligomeric species [Ti(OEt)$_4$]$_n$ (n = 3 in benzene, n = 2 in EtOH) in which the titanium coordination is higher (n = 5 in the trimer, n = 6 in the dimer) because of the formation of a solvate [Ti(OEt)$_4$]$_2 \cdot$ (EtOH)$_2$. Monodispersed spherical particles have been synthesized by controlled hydrolysis of a diluted solution of Ti(OEt)$_4$ in EtOH [48]. The monodispersity clearly results from slower hydrolysis and condensation reactions with less reactive precursors allowing decoupling of the nucleation and growth steps. It is however possible to control the reactivity of low coordinated titanium in the presence of specific ligands. For instance, hydrolysis at 60°C of titanium butoxide Ti(OBu)$_4$ in the presence of acetylacetone forms monodispersed 1- to 5-nm TiO$_2$ anatase nanoparticles [49]. A very elegant design of the shape of anatase nanospheres and nanorods is obtained by controlling the rate of hydrolysis of Ti(O^{i}Pr)$_4$ at 80°C in the presence of oleic acid. In a general way, the rate of reactions and the nature of condensed species obtained depend also on the hydrolysis ratio defined as h = H$_2$O/M.

- Molecular clusters are formed with very low hydrolysis ratios (h < 1). The condensation reactions are relatively limited. Hydrolysis of [Ti(OEt)$_4$]$_2 \cdot$ 2(EtOH) forms soluble species such as Ti$_7$O$_4$(OEt)$_{20}$(h = 0,6); Ti$_{10}$O$_8$(OEt)$_{24}$(h = 0,8); or Ti$_{16}$O$_{16}$(OEt)$_{32}$(h = 1). A variety of such clusters have been isolated and characterized by X-ray diffraction. They can assemble themselves into nanostructures enabling the formation of hybrid organic-inorganic materials [50].

- Addition of water in substoichiometric amounts does not allow the substitution of all alkoxo ligands that leads to oxopolymers. Such precursors are well designed for obtaining coatings or thin films. The residual OR groups can react with surface hydroxyl groups of the substrate forming covalent bonds. The films are strongly adhesive and the organic residues can be then eliminated by thermal treatment.

- All alkoxo groups are eliminated in the presence of a large excess of water (h >> 10), leading to oxide nanoparticles in suspension. Because of the high dielectric constant of the medium, the surface hydroxylated groups are mainly ionized allowing formation of sols or gels similar to those obtained in aqueous solution.

Nonhydrolytic Routes to Oxide Nanoparticles

Nonhydrolytic sol-gel chemistry has proved to be a promising route to metal oxides [51]. In nonaqueous media in the absence of surfactant, one possibility is the use of metal halides and alcohols [52]. This approach is based on the general reaction scheme:

$$\equiv\text{M-X} + \text{ROH} \rightarrow \equiv\text{M-OH} + \text{RX}$$

$$\equiv\text{M-OH} + \equiv\text{M-X} \rightarrow \equiv\text{M-O-M}\equiv + \text{HX} \tag{2.12}$$

The recent isolation of a series of amine substituted alcohol complexes [53] has allowed for an estimation of the change in the acidity of alcohols upon coordination to a metal. Complexation of a protic Lewis base (e.g., ROH, R$_2$NH) results in the increase in Brönsted acidity discerned by a decrease in pK$_a$ of about 7 for the α-proton. This activation of the coordinated ligand by increasing the formal positive charge

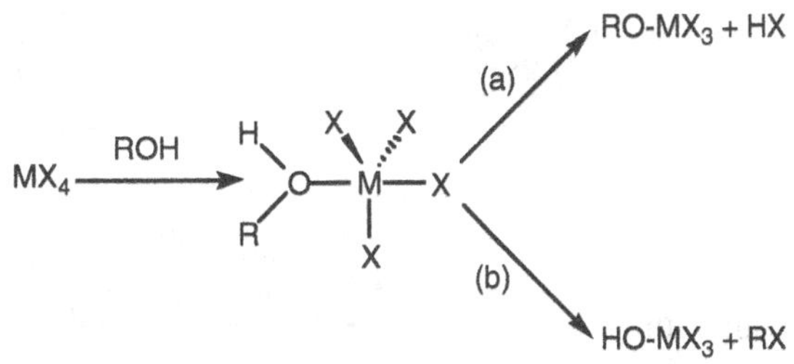

Scheme 2.1 Possible reactions for the alcoholysis of MX_4.

on the α-substituent is analogous to the activation of organic carbonyls toward alkylation and/or reduction by aluminum alkyls [54–56].

While reaction of primary and secondary alcohols with tetrachlorosilane is the usual method for preparing tetraalkoxysilanes [41] the same reaction with tertiary alcohols and benzylic alcohols form silica and the corresponding alkyl halide, RCl. The two modes of reaction involve initially the coordination of a lone pair of electrons of an alcoholic oxygen atom to the silicon center, followed by the cleavage of either the hydroxyl or alkoxyl group (Scheme 2.1). Electron-donor substituent groups in the alkyl radical direct the process to hydroxylation [pathway (b) with the liberation of RCl] by favoring the nucleophilic attack of chloride on the carbon group, due to its increased cationic character.

Hydroxylated species so formed react with unsolvolyzed compound according to:

$$\equiv M\text{-}OH + X\text{-}M\equiv \rightarrow \equiv M\text{-}O\text{-}M\equiv + HX \tag{2.13}$$

Benzyl alcohol seems to be well designed for synthesis of various oxide nanoparticles. Nanoparticles of anatase with size varying from 4 to 8 nm are obtained at temperatures from 40 to 150°C with different concentrations of $TiCl_4$ [57]. In similar conditions, $VOCl_3$ forms nanorods (approximately 200×35 nm) of vanadium oxide and WCl_6 forms platelets (approximately 30–100 nm, thickness 5–10 nm) of tungsten oxide [58].

Alkyl halide elimination can also occur between metal chloride and metal alkoxide following the reaction:

$$\equiv M\text{-}Cl + RO\text{-}M\equiv \rightarrow \equiv M\text{-}O\text{-}M\equiv + RCl \tag{2.14}$$

Such a reaction between TiX_4 and $Ti(OR)_4$ in heptadecane in the presence or trioctylphosphine oxide (TOPO) at 300°C produces spherical nanoparticles of anatase, around 10 nm in diameter [59]. Here, TOPO acts as a nonselectively adsorbed surfactant, which slows down the rate of reaction, allows the control of particle size, and avoids the formation of other TiO_2 polymorphs. In the presence of the mixed surfactant system [60], TOPO and lauric acid (LA), with increasing ratios LA/TOPO, a spectacular control of the shape of anatase nanorods is obtained (Figure 2.7). The specifically strong adsorption of LA onto (001) faces slows down the growth along (001) directions, thereby inducing growth along (101) directions that results in the formation of rods.

Another nonhydrolytic synthesis of oxide nanoparticles involves thermal decomposition of metal organic complexes in solution in the presence of surfactant. One of the most studied approaches involves the thermolytic decomposition of an inorganic complex at high temperatures. Two approaches include: the decomposition of $Fe(acac)_3$ or $FeCl_3$ and $M(acac)_3$ salts [61–63], and the decomposition of $Fe(CO)_5$ and $M(acac)_2$ salts [64, 65]. For simple oxides (e.g., Fe_3O_4), the precursor [e.g., $Fe(acac)_3$] is added to a suitable solvent heated to a temperature that allows for the rapid decomposition of the precursor. The choice of temperature and the temperature control (i.e., variation of the temperature during the reaction) are important in defining the resulting nanoparticle size and size distribution. By this method, highly uniform nanoparticles can be obtained.

In addition to simple metal oxides (M_xO_y) a range of mixed metal oxides can also be prepared. For example, nanospheres and nanocubes of cobalt ferrite can be obtained from cobalt and iron acetylacetonates, $Co(acac)_2$ and $Fe(acac)_3$ in solution in phenyl ether and hexadecanediol in the presence of oleic acid and oleylamine [66]. Heating at 260°C forms $CoFe_2O_4$ spherical nanocrystals with a diameter of 5 nm. These nanocrystals serve as seeds for a new growth as the second step of the synthesis, giving perfect nanocubes

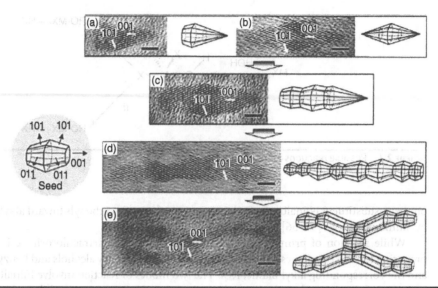

Figure 2.7   HRTEM analyses of TiO$_2$ anatase nanoparticles and simulated three-dimensional shape (3D) of (a) a bullet, (b) a diamond, (c) a short rod, (d) a long rod, and (e) a branched rod. The long axes of the nanocrystals are parallel to the c-axis of the anatase structure, while the nanocrystals are faceted with (101) faces along the short axes. Hexagon shapes [the (010) projection of a truncated octagonal bipyramid] truncated with two (001) and four (101) faces are observed either at the one end or at the center of the nanocrystals. The branched shape is a result of the growth along (101) directions starting from the hexagon shape. Scale bar is 3 nm [60].

from 8 to 12 nm, depending on the conditions. Nanocubes in the 8-nm range can also be used as seeds to obtain spheres. Variations in the morphology of numerous oxide nanocrystals, including nanocrystals of Fe, Co, Mn ferrites, Co$_3$O$_4$, Cr$_2$O$_3$, MnO, NiO, ZnO, and others, have been obtained by pyrolysis of metal carboxylates in the presence of different fatty acids (oleic, myristic) [67–69].

Metal carbonyl complexes are also interesting precursors to synthesize uniform metal oxide nanoparticles. Thermal decomposition at 100°C of iron pentacarbonyl, Fe(CO)$_5$, in octyl ether in the presence of oleic acid forms iron nanoparticles which are then transformed to monodisperse spherical γ-Fe$_2$O$_3$ nanoparticles by trimethylamine oxide acting as a mild oxidant (Figure 2.8) [70]. Thermal decomposition of Fe(CO)$_5$ in solution containing dodecylamine as a capping ligand and under aerobic conditions forms also γ-Fe$_2$O$_3$ nanoparticles with diamond, sphere, and triangle shapes with similar 12-nm size [71]. Uniform-sized MnO nanospheres and nanorods are obtained by heating at 300°C the mixture of Mn$_2$(CO)$_{10}$ with oleylamine in trioctylphosphine (TOP) [72]. The size of nanospheres can be varied from 5 to 40 nm depending on the duration of heating, using phosphines both as solvent and stabilizing agent (Figure 2.8). With triphenylphosphine, 10-nm MnO particles can be obtained. If the surfactant complex is rapidly injected into a solution of TOP at 330°C, nanorods 8 × 140 nm of MnO are produced. In fact, these rods are polycrystalline. They are formed by aggregation of spheres with oriented attachment and having a core shell structure with a thin Mn$_3$O$_4$ shell. Heating of W(CO)$_6$ at 270°C for 2 hours in trimethylamine oxide in the presence of oleylamine forms uniform nanorods of tungsten oxide with an X-ray diffraction pattern matching the W$_{18}$O$_{49}$ reflections [73]. The lengths of the nanorods are controlled by the temperature and the amount of oleylamine.

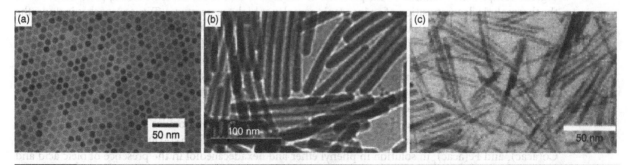

Figure 2.8   TEM image of (a) a two-dimensional hexagonal assembly of 11 nm γ-Fe$_2$O$_3$ nanocrystallites [71]; (b) 8- × 140-nm-sized MnO nanorods [72]; and (c) 75 ± 20 nm tungsten oxide nanorods [73].

From Minerals to Materials

As discussed earlier, precursor sol-gels are traditionally prepared via the hydrolysis of metal compounds. However, in the case of aluminum oxide nanoparticles, the relative rate of the hydrolysis and condensation reactions often makes particle size control difficult. The aluminum-based sol-gels formed during the hydrolysis of aluminum compounds belong to a general class of compounds: alumoxanes. The term *alumoxane* is often given to aluminum oxide macromolecules formed by the hydrolysis of aluminum compounds or salts, AlX_3 where $X = R$, OR, $OSiR_3$, or O_2CR; however, it may also be used for any species containing an oxo (O^{2-}) bridge binding (at least) two aluminum atoms, that is, Al-O-Al [74]. The structure of alumoxanes was proposed to consist of linear or cyclic chains (Figure 2.9) analogous to that of polysiloxanes [75]. In fact, alumoxanes exist as 3D cage structures [76–78]. For example, siloxy-alumoxanes, $[Al(O)(OH)_x(OSiR_3)_{1-x}]_n$, consist of an aluminum-oxygen nanoparticle ore structure (Figure 2.9c) analogous to that found in the mineral boehmite, $[Al(O)(OH)]_n$, with a siloxide substituted periphery [79–81]. Based on the knowledge of the boehmite-like nanoparticle core structure of hydrolytically stable alumoxanes, it was proposed that alumoxanes could be prepared directly from the mineral boehmite. Such a top-down approach represented a departure from the traditional synthetic methodologies.

The boehmite-like core structure (Figure 2.9c) could be prepared directly from the mineral boehmite. The type of capping ligand used in such a process must be able to abstract and stabilize a small fragment of the solid-state material. In the siloxy-alumoxanes, it was demonstrated that the "organic" unit itself contains aluminum, as shown in Figure 2.10a. Thus, in order to prepare the siloxy-alumoxane the "ligand" $[Al(OH)_2(OSiR_3)_2]^-$, would be required as a bridging group. However, the carboxylate anion, $[RCO_2]^-$, is an isoelectronic and structural analog of the organic periphery found in our siloxy-alumoxanes (Figure 2.10).

Initial syntheses were carried out using the acid as the solvent or xylene [82, 83], however, subsequent research demonstrated the use of water as a solvent and acetic acid as the most convenient capping agent [84]. A solventless synthesis has also been developed [85]. Thus, the synthesis of alumoxane nanoparticles may be summarized as involving the reaction between dirt (boehmite), vinegar (acetic acid), and water. The function of the acid is twofold. First, to cleave the mineral lattice and "carve out" nanoscale fragment, and second, to provide a chemical cap to the fragment (Figure 2.11).

The carboxylate-alumoxane nanoparticles prepared from the reaction of boehmite and carboxylic acids are stable in air and water. The soluble carboxylate-alumoxanes can be dip-, spin-, and spray-coated onto various substrates. The physical properties of the alumoxanes are highly dependent on the identity of the substituents. The size of the alumoxane nanoparticles is dependent on the substituents, the reaction

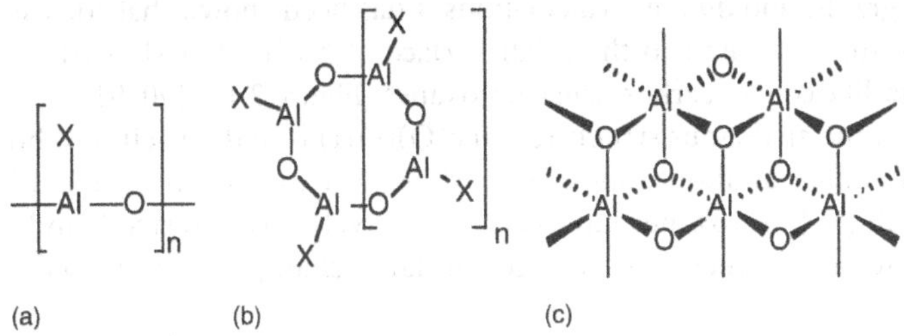

FIGURE 2.9 Structural models proposed (a and b) and observed (c) for aluminum oxide nanoparticles formed from the hydrolysis of aluminum compounds.

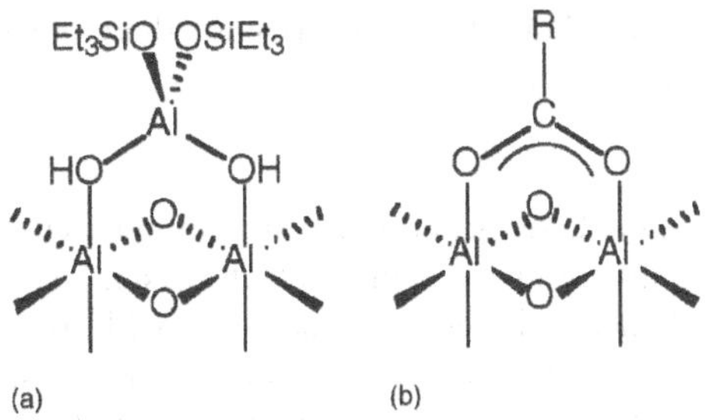

FIGURE 2.10 Structural relationship of the capping ligand for (a) siloxy and (b) carboxylate alumoxane nanoparticles.

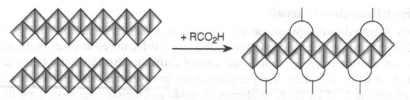

FIGURE 2.11   Pictorial representation of the reaction of boehmite with carboxylic acids.

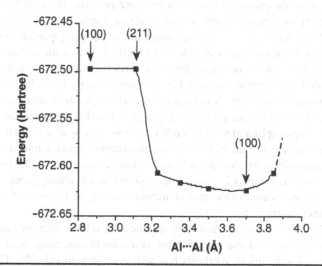

FIGURE 2.12   Total energy calculation of a carboxylic acid interacting with an Al_2 unit as a function of the Al ··· Al distance. The Al ··· Al distances present on the (100) and (211) crystallographic planes of boehmite are marked.

conditions (concentration, temperature, time, etc.), and the pH of the reaction solution [86]. Unlike other forms of oxide nanoparticle, the alumoxanes are not monodispersed but have a range of particle sizes. Also, unlike other metal oxide nanoparticles, the core of the alumoxane can undergo a low temperature reaction that allows for the incorporation of other metals (e.g., Ti, La, Mo, V, Ca). This occurs by reaction of metal acetylacetenoates $[M(acac)_n]$ with the carboxylate alumoxane [87–89]. Using a combination of X-ray crystallography and *ab initio* calculations it has been shown that the carboxylate ligand is therefore near perfectly suited to bind to the (100) surface of boehmite (Al ··· Al = 3.70 Å), and hence stabilize the boehmite-like core in carboxylate alumoxanes (Figure 2.12) [90, 91].

Given the analogous structure of Fe(O)(OH) (lepidocrocite) to boehmite, it is not surprising that the iron analog of alumoxane nanoparticles (i.e., ferroxanes) is readily prepared. First prepared by Rose et al. [92], ferroxanes have been extensively characterized, and have shown identical structural features to alumoxanes and undergo similar exchange reactions [93].

Semiconductor Nanoparticles (Quantum Dots and Quantum Rods)

The synthesis of semiconductors as nanoscale particles yields materials with properties of absorbance and fluorescence that differ considerably from those of the larger, bulk-scale material. Highly specific bands of absorbance or fluorescence arise from the quantum confinement of electrons that are excited in these materials when exposed to light. These materials are therefore of great interest in applications ranging from medical imaging to tagging and sensing.

Solution Processes

The most studied nonoxide semiconductors are cadmium chalcogenides (CdE; with E = sulfide, selenide, and telluride). CdE nanocrystals were probably the first material used to demonstrate quantum-size effects corresponding to a change in the electronic structure with size—that is, the increase of the band gap energy with the decrease in size of particles [5–7, 94, 95]. These semiconductor nanocrystals are commonly synthesized by thermal decomposition of an organometallic precursor dissolved in an anhydrous solvent containing a source of chalcogenide and a stabilizing material (polymer or capping ligand). Stabilizing molecules bound to the surface of particles control their growth and prevent particle aggregation. Although cadmium chalcogenides are the most studied semiconducting nanoparticles, the methodology for the formation of semiconducting nanoparticles was first demonstrated independently for InP and GaAs [96, 97]. In both cases it was demonstrated that the reaction of the metal halide with the trimethylsilyl-derived phosphine or arsine resulted in the formation of the appropriate pnictide and Me_3SiCl:

$$InCl_3 + P(SiMe_3)_3 \rightarrow InP + 3Me_3SiCl \qquad (2.15)$$

Although these initial studies were performed as solid-state reactions, carrying them out in high boiling solutions led to the formation of the appropriate nanoparticle materials. The most widely used development of the Barron/Wells synthetic method was by Bawendi and colleagues [8] in which an alkyl derivative was used in place of the halide. Dimethylcadmium $Cd(CH_3)_2$ is used as a cadmium source and bis(trimethylsilyl) sulfide, $(Me_3Si)_2S$, trioctylphosphine selenide, or telluride (TOPSe, TOPTe) serve as sources of selenide in trioctylphosphine oxide (TOPO) used both as solvent and capping molecule. The mixture is heated at 230 to 260°C. It is best to prepare samples over a period of a few hours of steady growth by modulating the temperature in response to changes in the size distribution as estimated from the absorption spectra of aliquots removed at regular intervals. Temperature is lowered in response to a spreading of size distribution and increased when growth appears to stop. Using this method, a series of sizes of CdSe nanocrystals ranging from 1.5 to 11.5 nm in diameter can be obtained (Figure 2.13). These particles, capped with TOP/TOPO molecules, are nonaggregated and easily dispersible in organic solvents forming optically clear dispersions.

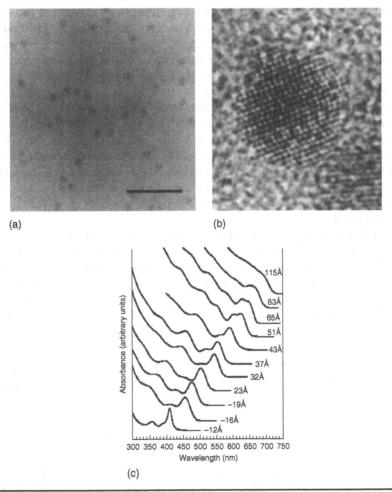

FIGURE 2.13   TEM image of (a) well-dispersed CdSe nanoparticles of around 3.5 nm in mean diameter (bar: 20 nm); (b) CdSe crystallite 8 nm in size showing stacking faults in the (002) direction (bar: 5 nm); (c) room temperature optical absorption spectra of CdSe nanocrystallites dispersed in hexane and ranging in size from 1.2 to 11.5 nm [8].

When similar synthesis are performed in the presence of surfactant, strongly anisotropic nanoparticles are obtained. From $Cd(CH_3)_2$ and TOPSe in hot mixture of TOPO and hexylphosphonic acid (HPA) rod-shaped CdSe nanoparticles can be obtained [98]. The role of adsorption of HPA is clearly demonstrated by the change in shape of nanoparticles with the increase in HPA concentration. An increase in the concentration of Cd precursor into the reactor also induces a higher aspect ratio of nanoparticles exhibiting quasi-perfect wurtzite crystalline structure [99]. A further slow addition of monomer sustains growth of (001) faces without additional nucleation giving rods exceeding 100 nm in length with aspect ratio over 30:1. The role of HPA seems to be to increase the growth rate of the (001) faces relative to all other faces. Further support for this argument comes from the formation of multipods, especially tetrapods. These are remarkable single-crystal particles consisting of a tetrahedral zinc blende core with four wurtzite arms (Figure 2.14). They are formed when a CdSe or CdTe nanocrystal nucleates in the zinc blende structure instead the wurtzite structure. The wurtzite arms grow out of the four (111) equivalent faces of the tetrahedral zinc blende core. A key parameter for achieving tetrapod growth is the energy difference between the two structural types, which determines the temperature range in which one structure is preferred during nucleation and the other during growth. For CdS and CdSe, the energy difference is very small and is difficult to isolate controllably the growth of one phase at a time. With CdTe, the energy difference is large enough that, even at the elevated temperatures preferred for wurtzite growth, nucleation can occur selectively in the zinc blende structure, the wurtzite growth being favored by using higher temperatures in the presence of surfactant.

Because $Cd(CH_3)_2$ is extremely toxic, pyrophoric, and explosive at elevated temperature, other Cd sources have been used. CdO appears to be an interesting precursor [100, 101]. The reddish CdO powder dissolves in TOPO and HPA or TDPA (tetradecylphosphonic acid) at about 300°C giving a colorless homogeneous solution. By introducing selenium or tellurium dissolved in TOP, nanocrystals grow at 250°C to the desired size. Nanorods of CdSe or CdTe can also be produced by using a greater initial concentration of cadmium as compared to reactions for nanoparticles. The evolution of the particle shape results from a diffusion-controlled growth mechanism. The unique structural feature of (001) facets of the wurtzite structure and the high chemical potential on both unique facets makes the growth reaction rate along the c-axis much faster than that along any other axis. The limited amount of monomers maintained by diffusion is mainly consumed by the quick growth of these unique facets. As a result, the diffusion flux goes to the c-axis exclusively, which is the long axis of the quantum rods [102]. This approach further enables large-scale production of Cd chalcogenide quantum dots and quantum rods with controllability of their size, monodispersity, and aspect ratio. This approach has been successfully applied for synthesis of numerous other metal chalcogenides, including ZnS, ZnSe, and $Zn_{1-x}Cd_xS$ [103]. CdS nanorods have also been obtained from $Cd(S_2CNCH_2CH_3)_2$, an air-stable compound, thermally decomposed in hexadecylamine HDA at around 300°C [104]. Various shapes of CdS nanocrystals are obtained in changing the growth temperature. Rods are formed at elevated temperatures (300°C) and armed rods (bipods, tripods) are obtained as the growth temperature is decreased to 180°C. Around 120°C, tetrapods of four-armed rods are dominant.

A similar procedure, using $Mn(S_2CNCH_2CH_3)_2$, enables formation of MnS nanocrystals with various shapes including cubes, spheres, monowires, and branched wires (bi-, tri-, and tetrapods) [103]. Nanorods of diluted magnetic semiconductors, $Cd_{1-x}Mn_xS$, have also been obtained by this procedure [105]. Various shapes of PbS nanocrystals have similarly been produced from $Pb(S_2CNCH_2CH_3)_2$ [104]. NiS nanocrystals, elongated along the (110) direction, were prepared by solventless thermal decomposition of a mixture of nickel alkylthiolate and octadecanoate [106]. Similarly, Cu_2S nanorods or nanodisks are obtained by

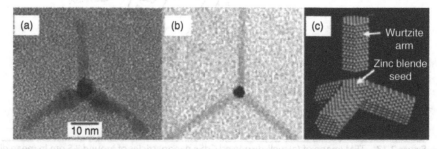

Figure 2.14   Tetrapod-shaped nanocrystals of (a) CdSe and (b) MnS formed by epitaxial growth of four wurtzite arms on a tetrahedral zinc blende seed results in tetrapod-shaped nanocrystals (c) [98, 99].

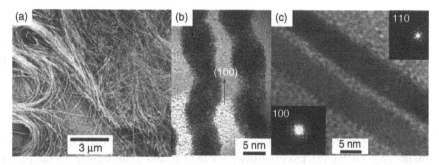

FIGURE 2.15 High-resolution (a) SEM and (b) TEM images of PbSe nanowires grown in solution in the presence of oleic acid. (c) high-resolution TEM images of PbSe nanowires formed in the presence of oleic acid and n-tetradecylphosphonic acid. Selected area electron diffraction from single nanowires imaged along the (100) and (110) zone axes (insets to c). (d, e) TEM image of PbSe nanorings (bar 5 nm) [108].

solventless thermal decomposition of a copper alkylthiolate precursor [107]. Finally, a very interesting design of nano-objects with advanced shapes results from oriented attachment of nanoparticles. PbSe nanowires of 3.5 to 18 nm in diameter and 10 to 30 mm in length are obtained from the reaction between lead oleate with TOPSe at 250°C in solution in diphenylether in the presence of TDPA [108]. In the presence of hot (250°C) hexadecylamine in diphenyl ether, lead oleate and TOPSe forms PbSe nanorings resulting very likely from a similar oriented attachment of nanoparticles (Figure 2.15).

Nanoparticles from the Vapor Phase
The chemical vapor deposition (CVD) of semiconductors from molecular precursors has been extensively studied. One class of precursors is the so-called single-source precursors, all required elements are in the same molecule. The use of single-source precursors allows for the structure of films grown by CVD to be controlled by the structure of the precursor molecule employed [109–111]. Such a process requires the precursor structure to remain intact during deposition [112]. However, vapor phase molecular cleavage can alter film morphology as well as influence phase formation. It has been observed that a major consequence of precursor decomposition in the vapor phase is cluster formation, leading to a rough surface morphology [113]. While particulate growth during CVD is often an undesirable component of the film deposition process, it is possible to prepare highly uniform well-defined nanoparticles of InSe and GaSe from the vapor phase thermolysis of precursor molecules under CVD conditions.

CVD growth using the cubane precursors, [(^{t}Bu)GaSe]$_4$ (Figure 2.16a) and [(EtMe$_2$C)InSe]$_4$ (Figure 2.16b) carried out on single-crystal KBr substrates in a hot-walled CVD system allows for the formation of spherical nanoparticles [114].

The InSe grown at 290°C from [(EtMe$_2$C)InSe]$_4$ consist of spheres with a mean diameter of 88 nm. The electron diffraction pattern exhibits well-defined rings, consistent with apolycrystalline hexagonal InSe. As with InSe, those grown from [(^{t}Bu)GaSe]$_4$ at 335°C consist of pseudospherical nanoparticles, however, unlike the InSe films, these appear as "strings of pearls" which retain their connectivity and remain intact after being floated from the growth substrate. From analysis of the micrograph, the mean GaSe particle diameter is 42 nm. A similar "strings of pearls" morphology has been observed for the solution decomposition of Co$_2$(CO)$_8$ in the presence of soluble organic polymers [117]. The small sizes and pseudospherical geometry of the InSe and GaSe particles suggests a vapor phase component is operative during the deposition of these metal selenide films. The fragmentation of the precursors in the vapor phase

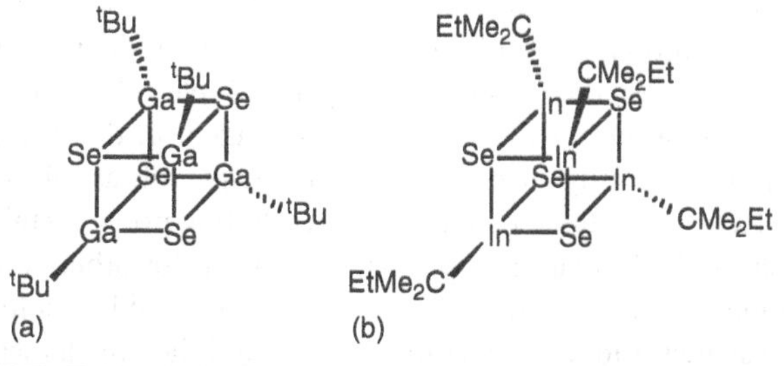

FIGURE 2.16 The structure of cubane single-source precursors used to synthesis GaSe and InSe nanoparticles [115, 116].

may combine to form particles that will remain in the vapor phase until their size is such as to "precipitate" out of the gas stream. The deposition of particles from the vapor phase due to gravitational or thermophoretic forces have been studied for SiC [118]. Factors that may control the particle size include: precursor concentration, vacuum versus atmospheric growth, growth temperature, and thermal stability of the precursor.

Metallics, Bimetallics, and Alloys

The most currently synthesized metallic nanoparticles include two categories of metals: noble or precious metals (Ag, Au, Pd, Pt), less extensively Ru and Cu, and ferromagnetic metals (Co, Fe, Ni). Silver, gold, and copper are essentially used for their color, yellow to red, due to their plasmon resonance located within the visible domain of the electromagnetic spectrum. Palladium, platinum, and ruthenium are largely used for heterogeneous catalysis. Cobalt, iron, and nickel are interesting as magnetic nanomaterials for various applications such as information storage in recording devices, ferrofluids, and microwave composite materials.

The two main routes for the synthesis of metallic nanoparticles are the reduction of metallic salts in solution, involving a large variety of salts and reducing agents, and the decomposition of zerovalent metal compounds. Whatever the reaction involved, the formation of monosized nanoparticles is achieved by a combination of a low concentration of solute and a protective layer (polymer, surfactant, or functional groups) adsorbed or grafted onto the surfaces. Low concentrations are needed to control the nucleation/growth steps and polymeric layers reduce diffusion causing diffusion to be the rate-limiting step of nuclei growth, resulting in uniformly sized nanoparticles. The protective layer also limits or avoids irreversible aggregation of nanoparticles.

Reduction Mechanism

The basic reaction involved in the production of metallic nanoparticles is the reduction of metal cations in solution:

$$M^{z+} + Red \rightarrow M^0 + Ox \tag{2.16}$$

The driving force of the reaction is the difference, $\Delta E°$, between the standard redox potentials of the two redox couples implicated, $E°(M^{z+}/M^0)$ and $E°(Ox/Red)$. The value of $\Delta E°$ determines the composition of the system through the equilibrium constant K given by:

$$\ln K = nF \, \Delta E°/RT \tag{2.17}$$

The reaction is thermodynamically possible if $\Delta E°$ is positive, but practically, its value must be at least 0.3 to 0.4, otherwise the reaction proceeds too slowly to be useful. Thus, highly electropositive metals (Ag, Au, Pt, Pd, Ru, Rh) with standard potentials $E°(M^n+/M^0) > 0.7$ V react with mild reducing agents while more electronegative metals (Co, Fe, Ni) with $E°(M^n+/M^0) < -0.2$ V need strongly reducing agents and have to be manipulated with care because the metallic nanoparticles are very sensitive to oxidation. Some widely used reactions are listed in Table 2.3. Complexation of cations in solution plays an important role on their reducibility and, as the stability of complexes increases, reduction is more difficult (Table 2.4).

A very large majority of redox systems in aqueous media is also strongly dependent on pH because the nature of the metal complex in solution is highly dependent on the acidity of the medium. Consequently, the control of metal complex chemistry in solution by complexation and/or by adjusting pH allows control of the reactivity of the species in a given metal-reducing agent system. It is also important to note that reduction can be performed in aqueous or nonaqueous media. However, synthesis in a nonaqueous medium presents serious difficulties because the metallic salts are generally very weakly soluble. Organometallic compounds can be used with suitable reducing agents, which are soluble in a specific medium. In some cases, the solvent can play a multiple role—for instance, primary alcohols (methanol, ethanol), which act as both solvent and reducing agent in the synthesis of Au, Pd, Pt nanoparticles, and polyols (e.g., glycerol, diethyleneglycol) can play a triple role as solvent, reducing agent, and stabilizer in the preparation of a lot of metallic nanoparticles, including noble metals and transition metals. Whatever the reaction involved, the formation of particles occurs, as for other systems (oxides, chalchogenides), in two steps: nucleation and growth. Since the metal atoms are highly insoluble, as soon as they are generated they aggregate by a stepwise addition, forming clusters and then nuclei when they reach a critical size. This step is favored by a high supersaturation of metal atoms, and consequently by a high rate of reduction or

Metal Species	E° (V)	Reducing Agent	Conditions	Rate
Au^{3+}, Au^+, Pt^{4+}, Pt^{2+}, Pd^{2+}	≥ 0.7	organic acids, alcohols, polyols	≥70°C	Slow
Ag^+, Rh^{3+}, Hg^{2+}, Ir^{3+}		aldehydes, sugars	<50°C	Moderate
		hydrazine, H_2SO_3	Ambient	Fast
		$NaBH_4$, boranes,	Ambient	Very fast
Cu^{2+}, Re^{3+}, Ru^{3+}	0.7 and ≥0	polyols	>120°C	Slow
		aldehydes, sugars	70–100°C	Slow
		hydrazine, hydrogen	<70°C	Moderate
		$NaBH_4$	Ambient	Fast
Cd^{2+}, Co^{2+}, Ni^{2+}, Fe^{2+}	<0 and ≥−0.5	polyols	>180°C	Slow
In^{3+}, Sn^{2+}, Mo^{3+}, W^{6+}		hydrazine	70–100°C	Slow
		$NaBH_4$, boranes	Ambient	Fast
		hydrated e^-, radicals	Ambient	Very fast
Cr^{3+}, Mn^{2+}, Ta^{5+}, V^{2+}	<0.6	$NaBH_4$, boranes	T, P > ambient	Slow
		hydrated e^-, radicals	Ambient	Fast

TABLE 2.3   Guidelines for the Choice of Reducing Agents and Reaction Conditions in the Precipitation of Metal Particles [119]

decomposition of precursor species in solution. The growth of nuclei proceeds at first by stepwise addition of new metal atoms formed in solution, leading to primary particles. Particle growth may continue by further addition of metal atoms controlled by diffusion toward the surface, by incorporation onto the surface, or by coalescence of primary particles and formation of larger secondary ones. In order to avoid coalescence and limit the growth and also to form stable dispersions of nanoparticles, protective agents are introduced into the reaction medium, allowing, as already indicated before, formation of a surface layer limiting the diffusion and growth, and also the nanoparticle aggregation.

Nanoparticles of Noble or Precious Metals

Gold and silver nanoparticles have been synthesized and studied extensively for a long time. Colloidal gold, as described in 1857 by M. Faraday [120], is probably the first monodispersed system ever reported in the literature (apart from carbon black used for Egyptian ink in the time of the Pharaohs!). Reduction of chloroauric acid, $HAuCl_4$, in aqueous solution by citrate at 100°C forms spherical nanoparticles. Their mean diameter increases from around 10 to 20 nm when the concentration of chloroauric acid decreases [121]. Here, citrate acts as a reducing agent and also as a stabilizer. Organic ligands, with a group such as phosphine, having a strong affinity for metal are often used to stabilize metallic nanoparticles. For gold, thiol derivatives are strong stabilizers and the reduction of Au(III) ions by citrate or borohydride in the presence of a thiol ligand gives uniform Au nanoparticles, the Au/thiol ratio controlling the mean size of the nanoparticles [122]. Silver nanoparticles are similarly obtained by reduction of silver nitrate by ferrous citrate in an aqueous medium, their stabilization in solution resulting from silver citrate adsorption [123]. Very uniform silver nanoparticles have also been obtained by reduction in nonaqueous media [124]. Aqueous $AgNO_3$ was vigorously mixed with chloroform containing tetra-n-octylammonium bromide, $[(C_8H_{17})_4N]Br$, acting as a catalyst for phase transfer. 1-Nonanethiol was first added to the gray organic phase collected followed by an aqueous solution of sodium borohydride ($NaBH_4$). A stable dispersion of nearly spherical 1-nonanethiol–capped silver nanoparticles in chloroform was obtained.

Couple	log K	E° (V)
$Ag^+ + e \rightarrow Ag^0$	–	+0.80
$Ag(NH_3)_2^+ + e \rightarrow Ag^0 + 2\ NH_3$	−7.2	+0.38
$Ag(S_2O_3)_2^{3-} + e \rightarrow Ag^0 + 2\ S_2O_3^-$	−13.4	+0.01
$Ag(CN)_3^{2-} + e \rightarrow Ag^0 + 3\ CN_3^-$	−22.2	−0.51

TABLE 2.4   Influence of Silver Ion Complexation on the Redox Potential

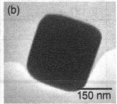

FIGURE 2.17   (a) SEM images of several partially developed gold tetrahedra. (b) SEM image of gold nanocubes [127].

Gold and silver nanoparticles are also obtained by various methods such as γ-radiolytic reduction [125] or photochemical reduction [126], always in the presence of various protective agents. Gold and silver nanoparticles obtained in this fashion are generally spherical. However, the shape of nanoparticles can be varied and controlled using different methods. Gold truncated tetrahedra, octahedra, icosahedra, and cubes have been obtained by the polyol process (Figure 2.17) [127]. Chloroauric acid and polyvinylpyrrolidone (PVP) are introduced in boiling ethylene glycol. Ethylene glycol serves as both solvent and reducing agent, PVP stabilizes the particles and also, in conjunction with the concentration of the gold precursor, controls their shape. A low concentration of silver ions in the medium orients the process toward the formation of gold nanocubes.

Another interesting seeding growth method has been used to produce nanorods [128]. The basic principle for the shape-controlled synthesis involves two steps: first, the preparation of spherical gold nanoparticles of around 3.5 nm in diameter (seeds), by reduction at RT of $HAuCl_4$ by $NaBH_4$ in the presence of citrate [citrate serves only as a capping or protective agent because it cannot, at room temperature, reduce Au(III)]; second, growth of the seeds in rod-like micellar environments acting as templates [129]. The growth solution contains $HAuCl_4$, CTAB (cetyltrimethylammonium bromide), and ascorbic acid to which the seed solution is added. Ascorbic acid is a mild reducing agent, which cannot reduce the gold salt in the presence of micelles without the presence of seeds. Gold nanorods are so obtained with various aspect ratios (Figure 2.18). The aspect ratio is controlled by varying the ratio of metal salt to seed, if some Ag^+ ions are present [130]. Silver ions are not reduced under these experimental conditions, and their role in controlling the shape of gold nanorods is not yet completely understood. Replacement of citrate by CTAB in the seed formation step and the use of a binary surfactant mixture improve the procedure in terms of selectivity in particle shape and amount of gold nanorods [131]. A similar procedure with silver nitrate allows the formation of silver nanorods and nanowires [132]. More generally, templating syntheses involving surfactants have been used to control the morphology of various metallic materials [129].

Silver nanowires were also obtained by an intriguing sono-assisted self-reduction template process [131]. A solution of a suitable precursor, the adduct Ag(hexafluoro-acetylacetone)-tetraglyme (tetraglyme = 2,5,8,11,14-pentaoxadecane) in ethanol/water 1:1 is loaded into the pores of an anodic aluminum oxide membrane consisting of ordered hole arrays of 200 nm in diameter and 60 μm in thickness, which is used as a noninteracting template (Figure 2.19). The Ag self-reduction process is activated by sonication for 2 hours at 45°C. This process does not require any reducing agent and directly forms crystalline silver nanowires. They are recovered by dissolving the templating membrane in NaOH solution at RT. The nanowire dimensions are determined by the pore dimensions and thickness of the membrane.

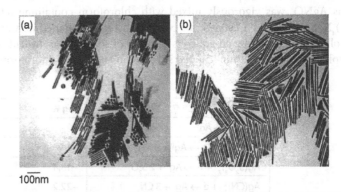

FIGURE 2.18   (a) Shape-separated 13 aspect ratio gold nanorods and (b) shape-separated 18 aspect ratio gold nanorods [128].

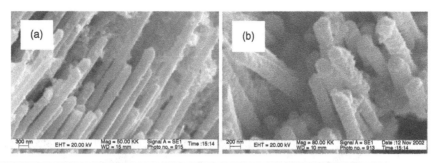

FIGURE 2.19   SEM images of free-standing Ag nanowires at low (a) and high (b) magnification [133].

Nanoparticles of Ferromagnetic 3D Transition Metals (Co, Fe)

The strategy used for the synthesis of 3D transition metals nanoparticles is analogous to that involved for precious metals, but with harder conditions because the redox potential of 3D transition metal elements is much lower. The strategy involves either injecting a strong reducing agent into a hot nonaqueous solution of a metal precursor containing surfactants, or injecting a thermally unstable zerovalent metal precursor into a hot solution containing stabilizers. Adjusting the temperature and the metal precursor–to–surfactant ratio controls the nanoparticle size. Higher temperatures and larger metal precursor–to–surfactant ratios produce bigger nanoparticles. Surfactants not only stabilize the nanoparticles in dispersion but also prevent or limit their oxidation [134]. The following illustrative examples show the variety of methods developed.

Reduction of cobalt salts by polyalcohols (polyols) at temperatures between 100 and 300°C in the presence of stabilizers produce Co nanoparticles with diameters of 2 to 20 nm [134, 135]. In a typical synthesis, cobalt acetate and oleic acid are heated at 200°C in diphenylether in the presence of trioctylphosphine (TOP). Reduction is started by addition of 1,2-dodecanediol solubilized in hot and dry diphenylether and heating at 250°C for 15 to 20 minutes. With a Co-to–oleic acid molar ratio of 1 and Co-to-TOP molar ratio of 2, 6 to 8 nm crystalline hexagonal closed pack structure (hcp) Co particles are obtained. Increasing the concentration of oleic acid and TOP by a factor of 2 yields smaller, 3 to 6 nm nanoparticles [134].

Reduction by superhydride LiBEt$_3$H of anhydrous cobalt chloride at 200°C in solution in dioctylether in the presence of oleic acid and alkylphosphine yields Co nanoparticles with a complex ε crystalline structure (β-Mn). The average particle size is coarsely controlled by the type of phosphine. Bulky P(C$_8$H$_{17}$)$_3$ limits the growth and produces 2 to 6 nm particles, while less bulky P(C$_4$H$_9$)$_3$ lead to larger (7–11 nm) particles [136]. The combination of oleic acid and trialkylphosphine produce a tight ligand shell, which allows the particles to grow steadily while protecting them from aggregation and oxidation. In effect, trialkylphosphine reversibly coordinates the metal surface, slowing but not stopping the growth. Oleic acid used alone is an excellent stabilizing agent, but it binds so tightly to the surface during synthesis that it impedes particle growth.

Reduction by dihydrogen of the organometallic compound Co(η^3-C$_8$H$_{13}$)(η^4-C$_8$H$_{12}$) in tetra hydro furane (THF) in the presence of PVP forms very small Co nanoparticles with face centered cubic (fcc) crystalline structure, around 1 and 1.5 nm depending on the temperature, 0 and 20°C, respectively [137]. The decomposition at 150°C under H$_2$ of the same organometallic precursor, but in anisole and in the presence of oleic acid and oleylamine (1:1 Co/oleic acid and Co/amine molar ratios) produces initially (3 hours of reaction) spherical 3-nm Co nanoparticles. After 48 hours, nanorods 9 × 40 nm are obtained [138]. The role of surfactants is very important in controlling the shape of particles. Increasing the concentration by a factor of 2 for oleic acid produces very long (micron range) nanowires of 4 nm in diameter, whereas the nature of the amine changes drastically the dimensions of nanorods having an aspect ratio varying between 1.7 and 22. The combination of lauric acid and hexadecylamine produces monodisperse 5 × 85 nm nanorods forming spontaneously 2D crystalline superlattices [139].

Cobalt nanoparticles have also been obtained from micellar media. Reverse micelles are water droplets dispersed in oil and stabilized by a surfactant monolayer, typically sodium bis-(2-ethylhexyl)sulfosuccinate [Na(AOT)]. The size of the micelles depends on the amount of solubilized water and varies from 0.5 to 1.8 nm. In the liquid, collisions between droplets induce exchange between water pools [140]. By mixing two micellar solutions, one containing Co(OAT)$_2$ and the other one NaBH$_4$, well-crystallized fcc Co nanoparticles are formed with a mean size around 6.4 nm. Extraction of nanoparticles with TOP forms a protective layer against aggregation and oxidation.

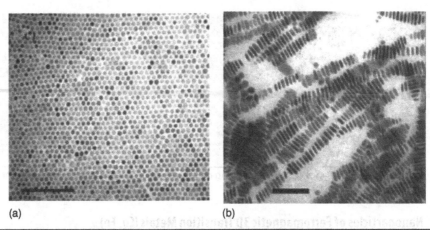

(a)   (b)

Figure 2.20 (a) Spherical Co nanocrystals synthesized in the presence of both oleic acid and TOPO and (b) Co nanodisks synthesized in the presence of amine. The disks are stacked in columns because of magnetic interactions. Bars are 100 nm [141].

Pyrolysis around 180°C of cobalt carbonyl $Co_2(CO)_8$ dissolved in anhydrous o-dichlorobenzene–containing surfactant (mixtures of oleic acid, lauric acid, and TOP) produces monodisperse ε-Co nanoparticles ranging in size from 3 to 17 nm. The particle size is tuned by the reaction temperature and the composition of surfactant [140]. In the presence of linear amines, hcp-Co nanodisks are formed, coexisting with ε-Co nanospheres (Figure 2.20) [139]. The length and diameter of the disks are controlled by variation of the reaction time following nucleation as well as by variation of the precursor-to-amine surfactant ratio. It has been observed that the length of the linear amine carbon chain controls the dimensions of Co disks (lower disks are obtained with shorter chains) whereas tri-substituted amines R_3N hinder the formation of disks. This suggests the R-NH_2 function is responsible for disk formation by selective adsorption, and steric interactions among neighboring adsorbed molecules may have an impact on the growth rate of the (001) faces.

It is difficult to synthesize iron nanoparticles by the polyol process because in the conditions of the process, Fe(0) results from disproportionation of Fe(II) whereas Co(II) and Ni(II) are quantitatively reduced [142]. Other methods to synthesize iron nanoparticles have been used. For instance, reduction of $FeCl_2$ in THF by $V(C_5H_5)_2$ in the presence of PVP forms PVP-stabilized α-Fe nanoparticles 18 nm in mean size. Sonolysis of $Fe(CO)_5$ in solution in anisol in the presence of poly(dimethylphenylene oxide) (PPO) yields nonagglomerated spherical Fe nanoparticles with a mean size of around 3 nm. Interestingly, the smaller particles (< 2.5 nm) have the bcc (body-centered cubic) structure of α-Fe and are superparamagnetic, whereas the larger ones ($\geq$ 2.5 nm) adopt the fcc structure of γ-Fe and are antiferromagnetic or paramagnetic. These γ-Fe nanoparticles could result from thermal gradients and very high local temperatures during sonication with rapid quenching avoiding recrystallization, or from the presence of interstitial carbon inside the particles that would stabilize the γ-Fe phase.

Very beautiful iron nanocubes were obtained from the decomposition of the organometallic compound $Fe[N(SiMe_3)_2]_2$ (Figure 2.21) [143]. Heating at 150°C under dihydrogen pressure for 48 hours the solution of complex in mesitylene in the presence of hexadecylamine and oleic acid forms a black precipitate containing monodisperse 7 nm bcc-Fe nanocubes. These nanocubes are included in bigger cubes forming extended 3D superlattices.

Thermal decomposition of $Fe(CO)_5$ in TOPO at 340°C under an Ar atmosphere produces spherical 2 nm Fe nanoparticles easily dispersible in pyridine [144]. These nanoparticles can be further transformed to nanorods. $Fe(CO)_5$ solubilized in POP is added to a hot suspension (320°C) in TOPO of spherical 2 nm Fe nanoparticles. This operation yields a black solid, which is washed with acetone to remove the surfactant and then dispersed in pyridine containing didodecyldimethylammonium bromide (DDAB). After refluxing for 12 hours, the supernatant contains 2×11 nm bcc-Fe nanorods (α-Fe). An increase in the concentration of DDAB increases the aspect ratio of rods. While diameter remains close to 2 nm, the length may be increased up to 22 nm. Such a transformation of nanospheres to nanorods seems to be caused by aggregation and by the strong binding of DDAB on the growing aggregates. After the aggregation of two particles, the third one will be bound on the top instead the central part of the aggregate where DDAB is strongly bounded. Then, further aggregation generates a unidimensional nanostructure.

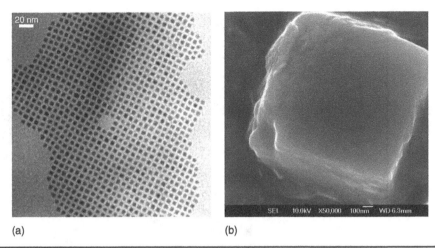

(a)   (b)

FIGURE 2.21   (a) TEM micrograph of 2D assembly of iron nanocubes and (b) SEM of 3D superlattice of nanocubes [143].

Bimetallics and Alloys

Bimetallic nanoparticles, composed of two different metals, are of interest in the improvement of catalytic properties [145] and in development of magnetic properties [146]. For instance, the ordered alloys Co-Pt and Fe-Pt are particularly interesting for magnetic recording because of their very high magnetocrystalline anisotropy, making these materials especially useful for practical applications such as magnetic memory devices as well as in biomedicine. The synthesis of bimetallics is generally made by coreduction of metal salts. Coreduction is the simplest preparative method and it is very similar to that used for monometallic nanoparticles. Successive reduction, carried out to prepare core-shell structured nanoparticles, is of little importance and will not be discussed here.

Au-Pt bimetallic nanoparticles have been obtained by citrate reduction of the mixture of tetrachloroauric acid, $HAuCl_4$, and hexachloroplatinic acid, H_2PtCl_6. The ultraviolet-visible (UV-vis) absorption spectra of the citrate stabilized sol of nanoparticles is not the simple sum of those of the two monometallic nanoparticles, indicating that the bimetallic particles have an alloy structure, as confirmed by X-ray diffraction and absorption spectroscopy [122]. A similar method produces citrate stabilized Pd-Pt nanoparticles. Polymer stabilized Pd-Pt nanoparticles have been prepared by simultaneous reduction of palladium chloride $PdCl_2$ and hexachloroplatinic acid by refluxing the alcohol/water (1:1 v/v) mixed solution in the presence of PVP at about 90°C for 1 hour [145]. Similarly, PVP-stabilized Pd-Rh nanoparticles were obtained by reduction in alcohol.

The polyol process [135] appears to be an efficient way to synthesize bi- or polymetallic nanoparticles of 3D metals, when heterogeneous nucleation controls the size of ferromagnetic nanoparticles. The addition of small amounts of a platinum or silver salt, which reduces at a lower temperature than 3D elements, forms nuclei for the growth of cobalt, nickel, or iron. Polymetallic spherical particles of the alloys $Co_xNi_{(100-x)}$ can be synthesized by precipitation from cobalt and nickel acetate dissolved in 1,2-propanediol with an optimized amount of sodium hydroxide. The number of nuclei depends on the relative amount of platinum, allowing the control of the particle size of $Co_xNi_{(100-x)}$ alloys over a very large range, from micrometric to nanometric [142]. The sodium hydroxide allows the precipitation of metal as hydroxides or alkoxides before the reduction. Their slow dissolution takes place at lower temperatures than the reduction, and this step controls the growth of metallic particles in solution. In controlling the basicity of the medium of the Pt or Ru seed-mediated polyol process, $Co_{80}Ni_{20}$ nanoparticles with surprising anisotropic shapes can be obtained [147, 148]. These particles are hcp crystallized when x > 30 (for x < 30, the particles very rich in nickel are fcc structured and always isotropic in shape). At low hydroxide concentrations (molar ratio $[HO^-]/[Co + Ni] < 2$), growth is favored along the c-axis, forming nanowires 8 nm in diameter and 100 to 500 nm in length with cone-shaped ends (Figure 2.22). Theses wires appear linked to a core, forming a sea-urchin–like shape (Figure 2.22b). With molar ratios $[HO^-]/[Co + Ni] > 2$, the growth occurs preferentially perpendicular to the c-axis resulting in rods that are 20 to 25 nm in diameter and 75 to 100 nm in length. The particular shape of the ends of nanowires seen in Figure 2.22b probably results from a lower growth rate at the end of the reaction, when the Co(II) and Ni(II) concentration fall (as the result of their reduction) and when the molar ratio $[HO^-]/[Co + Ni]$ is increased. If it is assumed that the shape of wires

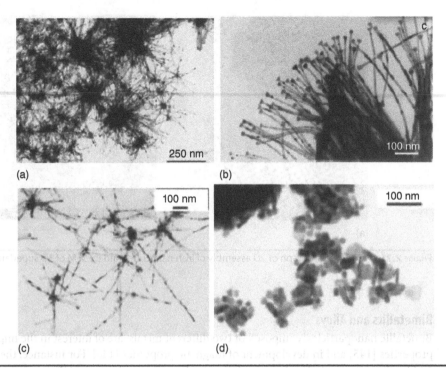

Figure 2.22   Particles $Co_{80}Ni_{20}$ obtained by polyol process with molar ratio $[HO^-]/[Co + Ni]$ equal to 1.25 (a, b), 1.8 (c), and 2.5 (d) [148].

is controlled by the growth rate, the head, formed at the end of the reaction, grows under conditions of low supersaturation inducing growth perpendicular to the c-axis. The polyol process has also been used to form Fe-Pt nanoparticles from the acetylacetonates $Fe(acac)_3$ and $Pt(acac)_2$ in ethylene glycol [148, 149].

A strong reducing agent such as hydrazine N_2H_4 has been used to reduce metal salts and to form Fe-Pt nanoparticles in water at low temperature. H_2PtCl_6 and $FeCl_2$ together with hydrazine and a surfactant such as sodium dodecyl sulfate (SDS) or CTAB, are mixed in water. Heating at 70°C allows reduction and forms fcc-structured Fe-Pt nanoparticles [150]. Reduction of $FeCl_2$ and $Pt(acac)_2$ mixtures in diphenyl ether by superhydride, $LiBEt_3H$, in the presence of oleic acid, oleylamine, and 1,2-hexadecanediol at 200°C has led to 4 nm Fe-Pt nanoparticles [61]. The initial molar ratio of the metal precursor allows control of the composition of the final particles more easily than in the polyol process. However, a major drawback of using borohydride is the contamination of the final product by boron.

Another process used to produce Fe-Pt nanoparticles is the thermal decomposition of $Fe(CO)_5$ and reduction of $Pt(acac)_2$ by 1,2-alkanediol [61]. The mixture is heated to reflux (297°C). Oleic acid and oleylamine are used for surface passivation and stabilization of particles. The composition of particles is controlled by the Fe-Pt ratio and fine-tuning of particle size between 2 and 5 nm is achieved by controlling the surfactant-to-metal ratio. Alternatively, to make larger Fe-Pt nanoparticles, a seed-mediated growth method has been used [61].

Carbon-Based Nanomaterials

Although nanomaterials had been known for many years prior to the discovery of C_{60}, the field of nanoscale science was really founded upon this seminal discovery and that of subsequent carbon nanomaterials. Part of the reason for this explosion in nanochemistry is that the carbon materials range from well-defined nano-sized molecules (i.e., C_{60}) to tubes with lengths in hundreds of microns range and now to graphene, a 2D carbon structure that has huge potential in the field of electronics and photonics/optoelectronics. Despite this range of scale, carbon nanomaterials have common reaction chemistry: that of the field of organic chemistry. This provides them with almost infinite functionality. A further advantage is that since every C_{60} molecule is like every other one, ignoring ^{12}C and ^{13}C isotope effects, C_{60} provides a unique monodispersed prototype nanostructure assembly with particle size of 0.7 nm. In no other nanomaterial system is there the ability to prepare true monodispersed material.

The previously unknown allotrope of carbon, C_{60}, was discovered in 1985 [1], and in 1996, Curl, Kroto, and Smalley were awarded the Nobel Prize in Chemistry for this discovery. The other allotropes of carbon are graphite (sp^2) and diamond (sp^3). C_{60}, commonly known as the "buckyball" or buckminsterfullerene, has a spherical shape comprising of highly pyramidalized sp^2 carbon atoms. The C_{60} variant is often compared to the typical white and black soccer ball, hence buckyball, however, this is also used for higher derivatives. Fullerenes are similar in structure to graphite, which is composed of a sheet of linked hexagonal rings, but they contain pentagonal (or sometimes heptagonal) rings that prevent the sheet from being planar. The unusual structure of C_{60} led to the introduction of a new class of molecules known as fullerenes, which now constitute the third allotrope of carbon. Fullerenes are commonly defined as "any of a class of closed hollow aromatic carbon compounds that are made up of 12 pentagonal and differing numbers of hexagonal faces." The number of carbon atoms in a fullerene range from C_{60} to C_{70}, C_{76}, and higher. Higher-order fullerenes include carbon nanotubes that can be described as fullerenes that have been stretched along a rotational axis to form a tube. Given the differences in the chemistry and size of fullerenes such as C_{60} and C_{70} as compared to nanotubes, these will be dealt with separately. However, it should be appreciated that they are all part of the fullerene allotrope of carbon. In addition there have also been reports of nanohorns [151] and nanofibers [152]. Nanohorns are single-walled carbon cones that can be filled with biological material or metal oxides.

Fullerenes and carbon nanotubes are not necessarily products of high-tech laboratories; they are commonly formed in such mundane places as ordinary flames, produced by burning hydrocarbons [153, 154], and they have been found in soot from both indoor and outdoor air [155]. However, these naturally occurring varieties can be highly irregular in size and quality because the environment in which they are produced is often highly uncontrolled. Thus, although they can be used in some applications, they can lack in the high degree of uniformity necessary to meet many needs of both research and industry.

Fullerenes

The vast majority of studies have involved the chemistry of C_{60} (IUPAC name is $(C_{60}\text{-}I_h)[5,6]$fullerene) and given its now commercial synthesis this is the most common fullerene studied. The spherical shape of C_{60} is constructed from 12 pentagons and 20 hexagons, has I_h symmetry, and resembles a soccer ball (Figure 2.23a).

The next stable homologue is C_{70} (Figure 2.23b), which has D_{5h} symmetry with a shape similar to a rugby ball or American football. This is followed by C_{74}, C_{76}, C_{78}, and so on, in which an additional six-membered rings are added. Mathematically and chemically, to make a stable fullerene, one has to follow two principles: Euler's theorem and the isolated pentagon rule (IPR). Euler's theorem states that for the closure of each spherical network, n (n ≥ 2) hexagons and 12 pentagons are required, while the IPR says no two pentagons may be connected directly with each other, as destabilization is caused by two adjacent pentagons. Fullerenes are composed of sp^2 carbons in a similar manner to graphite, but unlike graphite's extended solid structure, fullerenes are spherical and soluble in various common organic solvents. Due to their hydrophobic nature, fullerenes are most soluble in CS_2 ($C_{60} = 7.9$ mg·mL^{-1}) and toluene ($C_{60} = 2.8$ mg·mL^{-1}), an important requirement for chemical transformation. Although fullerenes have a conjugated system, their spherical aromaticity is distinctive from benzene due to the closed-shell structure. In contrast to benzene, which has all C-C bonds in equal length, fullerenes such as C_{60} have two distinct classes of bonds. As determined experimentally, the shorter bonds at the junctions of two hexagons ([6,6] bonds) and the longer bonds at the junctions of a hexagon and a pentagon ([5,6] bonds); see Figure 2.24. The alternation in bond length shows that the double bonds are located at the junction of

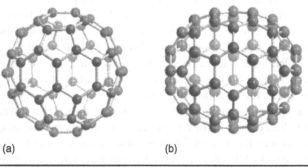

(a)   (b)

Figure 2.23 Molecular structures of (a) C_{60} and (b) C_{70}.

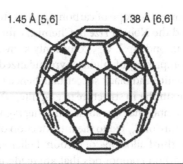

1.45 Å [5,6]   1.38 Å [6,6]

FIGURE **2.24**   Bond lengths of [5,6] and [6,6] bonds of C_{60}.

hexagons, while there is almost no double bond character in the pentagon rings in the lowest energy state. As expected from the I_h symmetry of C_{60}, there is by symmetry only one carbon type, which can be confirmed by ^{13}C NMR (Nuclear Magnetic Resonance) spectroscopy. In agreement with its D_{5h} symmetry, C_{70} has five ^{13}C signals in the ratio of 1:2:1:2:1.

Fullerenes represent a unique category of cage molecules with a wide range of sizes, shapes, and molecular weights. Most of the effort thus far has gone into the study of C_{60} fullerenes, which can now be prepared to a purity of parts per thousand. Because of the unique icosahedral symmetry of C_{60}, these molecules provide prototype systems for spectroscopy, optics, and other basic science investigations.

Synthesis of Fullerenes

The first observation of fullerenes was in molecular beam experiments, where discrete peaks were observed corresponding to molecules with the exact mass of 60 or 70 or more carbon atoms. In 1985, Harold Kroto (of the University of Sussex), James R. Heath, Sean O'Brien, Robert Curl, and Richard Smalley (from Rice University) published their observations along with the proposed structure for C_{60} [1]. Subsequent studies demonstrated that C_{60} was in fact ubiquitous in carbon combustion, and by 1991 it was relatively easy to produce grams of fullerene powder. Although the synthesis is relatively straightforward, fullerene purification remains a challenge to chemists and determines fullerene prices to a large extent.

The first method of production of fullerenes used laser vaporization of carbon in an inert atmosphere, but this produced microscopic amounts of fullerenes. In 1990, a new type of apparatus was developed by Krätschmer and Huffman in which carbon rods were vaporized in a helium atmosphere (approximately 100 torr) [156]. The subsequent black soot is collected and the fullerenes in the soot are then extracted by solvation in toluene. Pure C_{60} is obtained by liquid chromatography. The mixture is dissolved in toluene and pumped through a column of activated charcoal mixed with silica gel. The magenta C_{60} comes off first, followed by the red C_{70}.

Although many mechanisms have been described, only the "pentagon road" appears to explain high yields of C_{60}. In this proposed mechanism, the yield is high because clustering continues in a hot enough region to permit the growing clusters to anneal to the minimum energy path: one where the graphene sheet (a) is made up solely of pentagons and hexagons, (b) has as many pentagons as possible, while (c) avoiding structures where two pentagons are adjacent. If the pentagon rule structures really are the lowest energy forms for any open-carbon network, then one can readily imagine that high-yield synthesis of C_{60} may be possible. In principal, all one needs to do is adjust the conditions of the carbon cluster growth such that each open cluster has ample time to anneal into its favored pentagon rule structure before it grows further.

In the Krätschmer-Huffman experiment [156], carbon radicals are produced by the slow evaporation of the surface of a resistively heated carbon rod. After the Krätschmer-Huffman method was introduced, it was found at Rice University that a simple AC or DC arc would produce C_{60} and the other fullerenes in good yield as well, and this is now the method used commercially [157]. Even though the mechanism of a carbon arc differs from that of a resistively heated carbon rod (because it involves a plasma) the Helium gas pressure for optimum C_{60} formation is very similar. Thus, it is not so much the vaporization method that matters, but rather the conditions prevailing while the carbon vapor condenses. Adjusting the helium gas pressure, the rate of migration of the carbon vapor away from the hot graphite rod is controlled and thereby the carbon radical density in the region where clusters in the size range near C_{60} are formed.

A ratio between the mass of fullerenes and the total mass of carbon soot defines fullerene yield. The yields determined by UV-vis absorption are approximately 40 percent, 10 to 15 percent, and 15 percent in laser, electric arc, and solar processes, respectively. Productivity of a production process can be defined as a product between fullerene yield and flow rate of carbon soot. Interestingly, laser ablation technique has

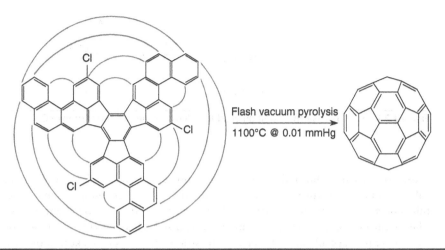

Scheme 2.2 Rational synthesis of C_{60} showing the proposed $C \cdots C$ connectivity.

both the highest yield and the low productivity and, therefore, a scale-up to a higher power is costly. Thus, commercial production of fullerenes is a challenging task. The world's first computer-controlled fullerene production plant is now operational at the MER Corporation, which pioneered the first commercial production of fullerene and fullerene products.

Despite the commercialization of C_{60}, interest in the "rational" synthesis has continued. Scott and coworkers have reported a 12-step synthesis of C_{60} [158]. Despite the low overall yield the important step is that C_{60} is the only fullerene produced. A molecular polycyclic aromatic precursor bearing chlorine substituents at key positions forms C_{60} when subjected to flash vacuum pyrolysis at 1100°C (Scheme 2.2).

Endohedral Fullerenes

Endohedral fullerenes are fullerenes that have incorporated in their inner sphere atoms, ions, or clusters. Endohedral fullerenes are generally divided into two groups: endohedral metallofullerenes and nonmetal-doped fullerenes. The first endohedral metallofullerenes were synthesized in 1985, called $La@C_{60}$. The @ sign in the name reflects the notion of a small molecule trapped inside a shell.

Doping fullerenes with electropositive metals takes place *in situ* during the fullerene synthesis in an arc reactor or via laser evaporation. A wide range of metals have been successfully encased inside a fullerene, including Sc, Y, La, Ce, Ba, Sr, K, U, Zr, and Hf. Unfortunately, the synthesis of endohedral metallofullerenes is unspecific because in addition to unfilled fullerenes, compounds with different cage sizes such as $La@C_{60}$ or $La@C_{82}$ are prepared. In addition, the synthesis of $Sc_3N@C_{80}$ in 1989 demonstrated that a molecular fragment could be encapsulated within a fullerene cage.

Endohedral metallofullerenes are characterized by the fact that electrons will transfer from the metal atom to the fullerene cage and that the metal atom takes a position off-center in the cage. The size of the charge transfer is not always simple to determine. In most cases it is between two and three charge units (e.g., $La_2@C_{80}$) or as large as six electrons (e.g., $Sc_3N@C_{80}$). These anionic fullerene cages are very stable molecules and do not have the reactivity associated with ordinary empty fullerenes. For example, the Prato reaction yields only the monoadduct and not multiadducts as with empty fullerenes (see Chemically Functionalized Fullerenes below). This lack of reactivity is utilized in a method to purify endohedral metallofullerenes from empty fullerenes [159].

Saunders and colleagues reported the existence of the endohedral $He@C_{60}$ and $Ne@C_{60}$ that form when C_{60} is exposed to a pressure of around 3 bars of the appropriate noble gases [160, 161]. Under these conditions it was possible to dope one out of every 650,000 C_{60} cages with a helium atom. Endohedral complexes with He, Ne, Ar, Kr, and Xe, as well as numerous adducts of the $He@C_{60}$ compound have also been proven [159] with operating pressures of 3000 bars and incorporation of up to 0.1 percent of the noble gases. While the isolation of single atoms of the noble gases is not unexpected, the isolation of $N@C_{60}$, $N@C_{70}$, and $P@C_{60}$ is very unusual. Unlike the metal derivatives, no charge transfer of the pnictide atom in the center to the carbon atoms of the cage takes place.

Chemically Functionalized Fullerenes

Although fullerene has a conjugated aromatic system, its reactivity is very different from planar aromatics, as all fullerene carbons are quaternary, containing no hydrogen, which renders characteristic substitution reactions of planar aromatics impossible. Therefore, only two types of primary chemical transformations

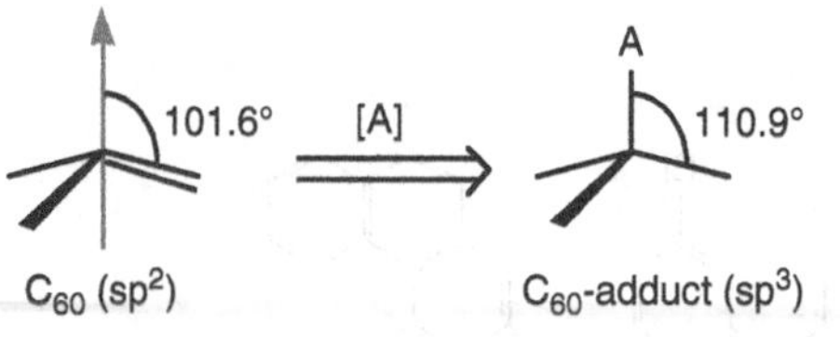

FIGURE 2.25 Strain release after addition of addend A to a pyramidalized carbon of C_{60}.

exist: redox reactions and addition reactions. Among these two, addition reactions have the largest synthetic value in fullerene chemistry as they can also function as a screening probe for the chemical properties of fullerene surfaces. Another remarkable feature of fullerene addition chemistry is their thermodynamics. The sp^2 carbon atoms in a fullerene are pyramidalized. This dramatic variation from planarity draws great strain energy, especially in C_{60} fullerene. The strain energy is circa 8 kcal·mol^{-1}, which is about 80 percent of its heat of formation. So the relief of strain energy to fullerene cage resulting from moderate number of addends bound to fullerene surface is the major driving force for addition reactions as reactions leading to sp^3 hybridized C atoms strongly relieve the local strain of pyramidalization, as shown in Figure 2.25.

Another important feature for fullerene carbons is its rehybridization of the sp^2 σ and the p π orbitals corresponding to the derivation from planarity. Calculations have shown that average hybridization at carbon in C_{60} is $sp^{2.278}$ and a fractional s character of 0.085, which results in the π orbitals extending further beyond the outer surface than into the interior of fullerene. This analysis implies that fullerenes, especially C_{60} are fairly electronegative molecules as well as having the low-lying π^* orbital with considerable s character. Indeed, the reactivity of fullerene can be considered as a fairly localized, but electron deficient, polyolefin void of substitution reaction. This trend in terms of electrophilicity allows for the ready chemical reduction and nucleophilic addition to fullerenes.

Theoretical calculations show that the LUMO (t_{1u} symmetry) and LUMO +1 (t_{1g} symmetry) of C_{60} molecular orbitals, exhibit a relatively low-lying energy and are triply degenerated. So C_{60} was predicted to be a fairly electronegative molecule that can be reduced up to hexanion. Cyclic voltammetry studies show that C_{60} can be reduced and oxidized reversibly up to six electrons with one-electron transfer processes. Indeed, reduction reactions were the first chemical transformation carried out to C_{60}. Fulleride anions can be generated by electrochemical method and then be used to synthesize covalent organofullerene derivatives by quenching the anions with electrophiles. Alkali metals can chemically reduce fullerene in solution and solid state. It is with the alkali metal–doped K_3C_6 that superconductivity was first found in fullerene materials. Alkaline earth metals can also intercalated with C_{60} by direct reaction of C_{60} with alkaline earth metal vapor to form M_xC_{60} (x = 3–6), which possess superconductivity as well. Besides, C_{60} can also be reduced by less ectropositive metals such as mercury to form C_{60}^- and C_{60}^{2-}. In addition, fulleride salts can also be synthesized with organic molecules to form fullerene-based charge transfer complexes. The well-known example of this type is [$TDAE^+$][C_{60}^-], which possesses remarkable electronic and magnetic behavior.

As stated above, geometric and electronic analysis predicted that fullerene behaves like an electro-poor conjugated polyolefin. Indeed, C_{60} and C_{70} undergo various nucleophilic reactions with carbon, nitrogen, phosphorous, and oxygen nucleophiles. C_{60} reacts readily with organolithium and Grignard compounds to form alkyl, phenyl, or alkanyl fullerenes. However, one of the most widely used nucleophilic additions to fullerene is the Bingel reaction (Scheme 2.3), where a carbon nucleophile was generated by

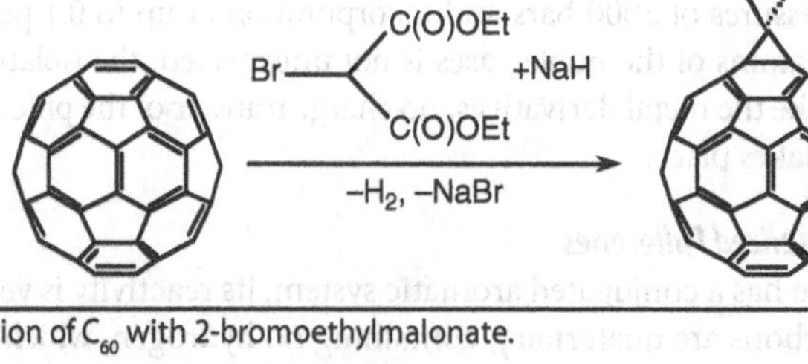

SCHEME 2.3 Bingel reaction of C_{60} with 2-bromoethylmalonate.

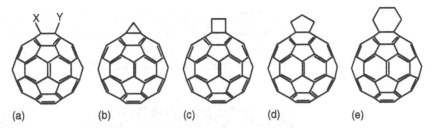

FIGURE 2.26 Geometrical shapes built onto a [6,6] ring junction: (a) open, (b) three-membered ring, (c) four-membered ring, (d) five-membered ring, and (e) six-membered ring.

deprotonation of α-halo malonate esters or ketones and added to form a clean cyclopropanation of C_{60} with a 30 ~ 60 percent yield. Later, it was found that the α-halo esters and ketones can be generated *in situ* with I_2 or CBr_4 and a weak base as 1,8-diazabicyclo[5.4.0]undec-7ene (DBU). This further simplified the reaction procedures. The Bingel reaction is considered one of the most versatile and efficient methods to functionalize C_{60}. Hundreds of fullerene derivatives have been made this way.

Cycloaddtion is another powerful tool to functionalize fullerenes. The great advantage of cycloaddition reaction is that the reaction generally occurs at the [6,6] bonds, which limits the possible isomers (Figure 2.26).

The dienophilic feature of the [6,6] double bonds of C_{60} enables the molecule to undergo various Diels-Alder reactions—that is, a [4+2] cycloaddition reaction. Another important feature of cycloaddition is that monoadducts can be generated in high yields and purified by flash chromatography on SiO_2. However, the best-studied cycloaddition reactions of fullerene are [3+2] additions with diazoderivatives (Scheme 2.5) and azomethine ylides (Prato reactions). In this reaction, azomethine ylides can be generated *in situ* from condensation of α-amino acids with aldehydes or ketones, which produce [1,3] dipoles to further react with C_{60} in good yields (40–60% percent) (Scheme 2.4). Hundreds of useful building blocks have been generated by those two methods. Interestingly, the Prato reactions have also been successfully applied to carbon nanotube and carbon onions to yield highly soluble carbon nanotube derivatives.

Successful reactions for producing fullerenes have also involved the use of Grignard reagents [162, 163], organolithiums [164], and other nucleophiles. Current research includes the derivatization of the fullerene cage for the addition of amino acids [165] or pharmaceuticals for biological applications [166].

The oxidation of fullerenes, such as C_{60}, has been of increasing interest with regard to applications in photoelectric devices, biological systems, and possible remediation of fullerenes [167]. It has also been shown that $C_{60}O$ will undergo a thermal polymerization [168, 169], in an analogous manner to that of organic epoxides. The oxidation of C_{60} to $C_{60}O_n$ (n = 1, 2) may be accomplished by a range of methods, including, photooxidation, ozonolysis, and epoxidation. With each of these methods, there is a limit to the isolable oxygenated product, $C_{60}O_n$ with n < 3. The only exception involves passing C_{60} through a corona discharge ionizer in the presence of oxygen, which allows for the detection of species formulated as $[C_{60}O_n]^-$ (n $\leq$ 30); however, the products were only observed in the MS [168]. Highly oxygenated fullerenes, $C_{60}O_n$ with 3 $\leq$ n $\leq$ 9, have been prepared by the Lewis base–enhanced catalytic oxidation of C_{60} with $ReMeO_3/H_2O_2$ [170, 171].

Carbon Nanotubes

Another key breakthrough in carbon nanochemistry came in 1993, when Iijima and Ichihashi reported the synthesis and observation of needle-like tubes made exclusively of carbon [3]. This material became known as carbon nanotubes. There are two types of nanotubes. The first that was discovered was multi-walled

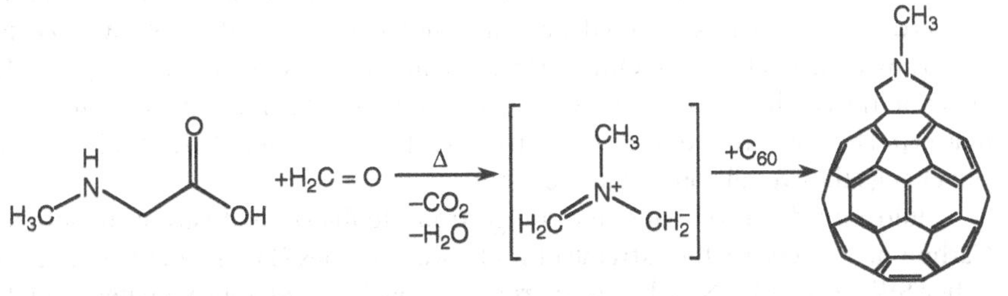

SCHEME 2.4 Prato reaction of C_{60} with N-methyglycine and paraformaldehyde.

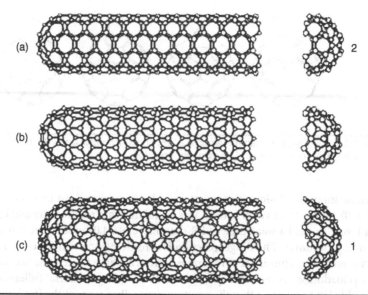

FIGURE 2.27 Structure of single-walled carbon nanotubes (SWNTs) with (a) armchair, (b) zigzag, and (c) chiral chirality.

nanotubes (MWNTs) resemble many pipes nested within each other. Shortly after MWNTs were discovered, single-walled nanotubes (SWNTs) were observed. Single-walled tubes resemble a single pipe that is potentially capped at each end. The properties of single- and multi-walled tubes are generally the same. Though SWNTs are believed to have superior mechanical strength and thermal and electrical conductivity; it is also more difficult to manufacture them.

Single-walled carbon nanotubes are by definition fullerene-based materials. Their structure consists of a graphene sheet rolled into a tube and capped by half a fullerene (Figure 2.27). The carbon atoms in a SWNT, like those in a fullerene, are sp^2 hybridized. They form a hexagonal ring network that resembles a graphene sheet. The structure of a nanotube is analogous to taking this graphene sheet and rolling it into a seamless cylinder. The critical differentiation parameters of individual carbon nanotubes are their diameter and chirality. Most of the presently used SWNTs have been synthesized by a pulsed laser vaporization method, pioneered by the Smalley group at Rice University. Their result represented a major breakthrough in the field.

The physical properties of SWNTs have made them an extremely attractive material for the manufacturing of nanodevices. SWNTs have been shown to be stronger than steel as estimates for the Young's modulus approaches 1 Tpa [172]. Their electrical conductance is comparable to copper with anticipate current densities of up to 10^{13} A·cm^{-2} and a resistivity as low as 0.34×10^{-4} Ω·cm at room temperatures. Finally, they have a high thermal conductivity (3000–6000 W·m·K^{-1}) [173–175].

The properties of a particular SWNT structure are based on its chirality. If a tube were unrolled into a graphene sheet, vectors (ma_2 and na_1) could be drawn starting from a carbon atom that intersects the tube axis (Figure 2.28). Then the armchair line is drawn. This line separates the hexagons into equal halves. Point B is a carbon atom that intersects the tube axis closest to the armchair line. The resultant vector of a_1 and a_2 is R and is termed the chiral vector. The wrapping angle (ϕ) is formed between R and the armchair line. If $\phi = 0°$, the tube is an armchair nanotube; if $\phi = 30°$, it is a zigzag tube. If ϕ is between 0° and 30°, the tube is called a chiral tube. The values of n and m determine the chirality or "twist" of the nanotube. The chirality in turn affects the conductance of the nanotube, its density, its lattice structure, and other properties. A SWNT is considered metallic if the value n − m is divisible by three. For example, an armchair tube is metallic in character. Otherwise, the nanotube is semiconducting [176]. Environment also has an effect on the conductance of a tube. Due to its highly delocalized π electrons, it is possible for a nanotube to accept electrons from or donate electrons to its environment [177, 178]. Molecules such as O_2 and NH_3 can change the overall conductance of a tube.

Multi-walled carbon nanotubes range from double-walled carbon nanotubes to carbon nanofibers. Carbon nanofibers are the extreme of multi-walled tubes (Figure 2.29). They are thicker and longer than either SWNTs or MWNTs, having a cross-sectional area of approximately 500 Å^2 and are between 10 to 100 mm in length. They have been used extensively in the construction of high strength composites [179].

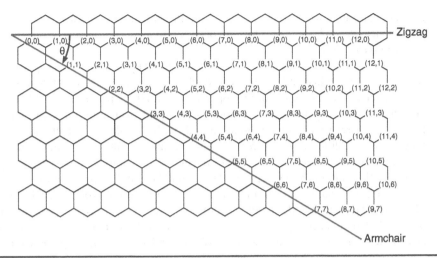

Figure 2.28 Chirality diagrams of a single-walled carbon nanotube.

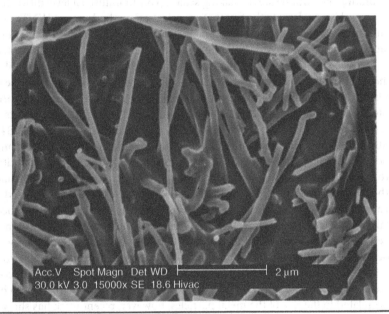

Figure 2.29 SEM image of vapor grown carbon nanofibers.

Synthesis of Single-Walled Carbon Nanotubes

A range of methodologies have been developed to produce nanotubes in sizeable quantities, including arc discharge, laser ablation, high-pressure carbon monoxide (HiPco), and vapor-liquid-solid (VLS) growth. It is worth noting that the latter is often referred to as chemical vapor deposition (CVD), however, this is not strictly correct. All these processes take place in vacuum or at low pressure with a process gases, although VLS growth can take place at atmospheric pressure. Large quantities of nanotubes can be synthesized by these methods; advances in catalysis and continuous growth processes are making SWNTs more commercially viable.

Nanotubes were first observed in 1991 in the carbon soot of graphite electrodes during an arc discharge that was intended to produce fullerenes. Because of the high temperatures caused by the discharge in this process, the carbon contained in the negative electrode sublimed. The fullerenes appear in the soot that is formed, while the CNTs are deposited on the opposing electrode. Tubes produced by this method were initially MWNTs. However, in 1993 Bethune et al. reported that with the addition of cobalt to the vaporized carbon, it was possible to grow SWNTs [180]. This plasma-based process is analogous to the more familiar electroplating process in a liquid medium. This method it produces a mixture of components and requires further purification to separate the CNTs from the soot and the residual catalytic metals. Producing CNTs in high yield depends on the uniformity of the plasma arc and the temperature of the deposit forming on the carbon electrode.

Higher yield and purity of SWNTs may be prepared by the use of a dual-pulsed laser. In 1995, Guo et al. reported that SWNTs could be grown through direct vaporization of a Co/Ni-doped graphite rod with a high-powered laser in a tube furnace operating at 1200°C [181, 182]. By this method, it was possible to grow SWNTs in a 50 percent yield without the formation of an amorphous carbon overcoating. Samples are prepared by laser vaporization of graphite rods with a catalyst of cobalt and nickel (50:50) at 1200°C in flowing argon, followed by heat treatment in a vacuum at 1000°C to remove the C_{60} and other fullerenes. The initial laser vaporization pulse is followed by a second pulse, to vaporize the target more uniformly and minimize the amount of soot deposits. The second laser pulse breaks up the larger particle ablated by the first pulse (that would result in soot formation) and feeds the products to the growing SWNT structure. The material produced by this method appears as a mat of "ropes," 10 to 20 nm in diameter and up to 100 μm or more in length. Each rope consists of a bundle of SWNTs, aligned along a common axis. By varying the growth temperature, the catalyst composition, and other process parameters, the average nanotube diameter and size distribution can be varied.

Although arc-discharge and laser vaporization are currently the principal methods for obtaining small quantities of high-quality SWNTs, both methods suffer from drawbacks. The first is that they involve evaporating the carbon source, making scale-up on an industrial level difficult and energetically expensive. The second issue relates to the fact that vaporization methods grow SWNTs in highly tangled forms, mixed with unwanted forms of carbon and/or metal species. The SWNTs thus produced are difficult to purify, manipulate, and assemble for building nanotube-device architectures for practical applications.

To overcome some of the difficulties of these high-energy processes, Smalley and coworkers developed a chemical catalysis method. In 1998 Smalley and coworkers reported the use of hydrocarbons as a carbon feedstock for single-walled tube growth [183]. Here molybdenum and iron/molybdenum catalysts were heated in a tube furnace to 850°C under 1.2 atm of ethylene. Previous reports utilizing a gas-phase growth reaction had produced multi- or single-walled tubes in very low yield [184, 185]. The use of CO as a feedstock led to the development by Smalley and coworkers of the high-pressure carbon monoxide (HiPco) procedure [186]. By this method, it was possible to produce gram quantities of SWNTs. The process involves injecting $Fe(CO)_5$ into a gas-phase reactor operating between 800 and 1200°C and 1 and 10 atm carbon monoxide. The HiPco method was better than previously reported gas-phase growth methods because it did not use a hydrocarbon as a feedstock. The key to the HiPco process is the formation of metal catalyst particles in the vapor phase, and it is thought that these are responsible for the SWNT growth, based upon an analogy with the growth by a catalyst on substrate. The growth of SWNTs from chemical processes was first reported in 1992, and can be likened to the VLS growth of SiC wiskers [187, 188]. During VLS growth a preformed catalyst particle (most commonly nickel, cobalt, iron, or a combination thereof) is placed on a substrate. The diameters of the nanotubes that are to be grown has been proposed to be related to the size of the metal particles; however, recent work has shown that this is not necessarily true [189]. VLS growth apparatuses are usually constructed of a tube furnace that is set up so gases can flow through the tube while it is being heated to high temperatures. As with other growth systems, multi-walled tubes were the first type of tubes grown [190]. In 1999, Dai and coworkers reported the large scale VLS growth of SWNTs using iron impregnated silicon nanoparticles and methane [191]. Methane was chosen as the feedstock because of its high thermal stability and its ability to retard the formation of amorphous carbon in the reactor. Since these reports, a wide range of precursors have been used for the catalyzed VLS growth of SWNTs. Recent approaches have involved the use of well-defined nanoparticles or molecular precursors [189, 192]. Many different transition metals have been employed, but iron, nickel, and cobalt remain the focus of most research. SWNTs grow at the sites of the metal catalyst; the carbon-containing gas is broken apart at the surface of the catalyst particle, and the carbon is transported to the edges of the particle, where it forms the SWNTs. The catalyst particles generally stay at the tip of the growing SWNT during the growth process, although in some cases they remain at the SWNT base, depending on the adhesion between the catalyst particle and the substrate. The length of the SWNTs grown in surface supported catalyst VLS systems appears to be dependent on the orientation of the growing tube with the surface. Within particular catalyst samples there are often two classes of tubes grown: short, straight SWNTs, and long, curved ones. It has been proposed that the straight SWNTs are a result of growth along the surface (Figure 2.30a) while the longer SWNTs are formed by growth out of the plane of the surface (Figure 2.30b). The growth rate of the former will be limited due to SWNT/surface interactions, while the later has unrestricted growth away from the surface [189]. Once the reaction run is complete

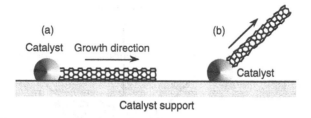

Figure 2.30   Schematic representation of supported catalyst SWNTs growth in which the SWNT grows parallel to the surface (a) or out from the surface (b).

(and the gas flow is removed) the SWNTs grown out of the surface will fall over. In the absence of additional factors, the rate of SWNT growth parallel to the surface is controlled by the frictional forces between the SWNT and the surface. By properly adjusting the surface concentration and aggregation of the catalyst particles, it is possible to synthesize vertically aligned carbon nanotubes—that is, as a carpet perpendicular to the substrate.

Of the various means for nanotube synthesis, the chemical processes show the most promise for industrial-scale deposition in terms of its price/unit ratio. There are additional advantages to the VLS growth of SWNTs. Unlike the above methods, VLS is capable of growing SWNTs directly on a desired substrate, whereas the SWNTs must be collected in the other growth techniques. The growth sites are controllable by careful deposition of the catalyst. Additionally, no other growth methods have been developed to produce vertically aligned SWNTs.

Chemical Functionalization of Carbon Nanotubes

The limitation on using carbon nanotubes in any practical applications has been their solubility; SWNTs have little to no solubility in most solvent due to the aggregation of the tubes. Aggregation is a result of the highly polarizable, smooth sides of the SWNTs forming bundles or ropes with a van der Waals binding energy of approximately 500 eV/μm of tube contact [193]. The van der Waals force between the tubes is so great that it takes tremendous energy to pry them apart. The insolubility of nanotubes makes it very difficult to make combinations of nanotubes with other materials, such as in composite applications. The functionalization of nanotubes—that is, the attachment of "chemical functional groups"—provides a strategy for overcoming those barriers. Functionalization can improve solubility and processibility, and will be able to link the unique properties of nanotubes to those of other materials. Through the chemical functional groups, nanotubes might take the interaction with other entities, such as solvents, polymers, nanoparticles, and other nanotubes. In functionalization of SWNTs, a distinction should be made between covalent and noncovalent functionalization. Covalent functionalization shows covalent linkage of functional groups onto the surface of nanotubes, either the sidewall or the cap of nanotubes. It is important to note that covalent functionalization methods have one problem in common: extensive covalent functionalization modifies SWNT properties by disrupting the continuous π–system of SWNTs.

Current methods for solubilizing nanotubes without covalent functionalization include highly aromatic solvents, super acids [194], DNA [195], polymers [196], or surfactants [197, 198]. These methods allow the sidewall of the nanotube to remain untouched, and conserve the tube's electronic structure. However, upon drying of the solution, bundles re-form. SWNTs may be made soluble in a range of organic solvents without sidewall functionalization via their reduction by Na/Hg amalgam in the presence of dibenzo-18-crown-6 [199]. The [Na(dibenzo-18-crown-6)]$_n$ [SWNT] complex shows solubility in CH$_2$Cl$_2$ and N,N-dimethylformamide (DMF) being comparable to surfactant dispersed SWNTs; however, measurable solubilities are also observed in hexane, toluene, and alcohols.

A noncovalent functionalization is mainly based on supramolecular interaction using various adsorption forces, such as van der Waals and p-stacking interactions. Covalent functionalization relies on the chemical reaction at either the sidewall or end of the SWNT. The high aspect ratio of nanotubes, sidewall functionalization is much more important than the functionalization of the cap. Direct covalent sidewall functionalization is associated with a change of hybridization from sp^2 to sp^3 and a simultaneous loss of conjugation [200]. Defect functionalization takes advantage of chemical transformations of defect sites already present. Defect sites can be the open ends and holes in the sidewalls, and pentagon and heptagon irregularities in the hexagon graphene framework. All these functionalizations are exohedral derivatizations. Taking the hollow structure of nanotubes into consideration, endohedral functionalization of SWNTs is possible—that is, filling the tubes with atoms or small molecules [200–202].

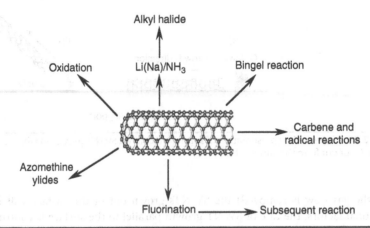

SCHEME 2.5 Schematic description of various covalent functionalization strategies for SWNTs.

Different application of nanotubes requires varied, specified modification to achieve processability and accessibility of nanotubes. Thus, the covalent functionalization can provide a higher degree of fine-tuning the chemistry and physics of SWNTs than noncovalent functionalization. Until now, a variety of methods have been used to achieve the functionalization of nanotubes (Scheme 2.5).

Functionalization of SWNTs using [1,3] dipolar addition of azomethine ylides, a method originally developed for modification of C_{60} [203]. Substituted pyrrolidine moieties were successfully introduced onto the surface of SWNTs. The functionalized SWNTs are soluble in most common organic solvents. The azomethine ylide functionalization method was also used for the purification of SWNTs. In 2001, under an electrochemical condition, a series of aryl diazonium salts were used to react with SWNTs to achieve functionalized SWNTs. Subsequently, SWNTs were functionalized by the diazonium ions *in situ* generated from the corresponding aniline [204, 205]. A solvent-free reaction appears to be the best chance for large-scale application of this method [206]. Here, SWNTs are reacted with a para-substituted aniline and isoamyl nitrate. This forms a diazonium salt *in situ* that reacts with the tube's sidewall. It is possible to control the amount of functionalization on the tube by varying reaction times and the amount of aniline used. It has been reported that this method leads to high functionalization (one group per every 10–25 carbon atoms or 8–12%).

Billups and coworkers have reported organic functionalization through the use of alkyl halides on tubes treated with lithium in liquid ammonia [207]. The reaction occurs through a radical pathway. In this reaction, functionalization occurs on every 17 carbons. Most success has been found when the tubes are dodecylated. These tubes are soluble in chloroform, DMF, and THF. Besides functionalization, there is the possibility of creating highly lithiated carbon materials. The lithium intercollates between the SWNTs to give a C:Li ratio of approximately 1 lithium atoms per 2.2 carbons.

The addition of oxygen moieties to SWNT sidewalls can be achieved by treatment with acid or wet air oxidation and ozonolysis [208]. The direct epoxidation of SWNTs may be accomplished by the reaction with either trifluorodimethyldioxirane, formed *in situ* from trifluoroacetone and Oxone (potassium peroxymonosulfate, $KHSO_5$) in $MeCN/H_2O$[4] or 3-chloroperoxybenzoic acid (m-CPBA)/CH_2Cl_2 [209], or using $ReMeO_3/H_2O_2$ catalysis (Scheme 2.6) [171]. Catalytic de-epoxidation allows for the quantitative analysis of sidewall epoxide and led to the surprising result that previously assumed "pure" SWNTs actually contain approximately 1 oxygen per 250 carbon atoms. Sidewall osmylation of SWNTs has been obtained by exposing the SWNTs to OsO_4 vapor under UV photoirradiation [210]. The covalent attachment of osmium oxide increased the electrical resistance of tubes by up to several orders of magnitude. Cleavage of OsO_4 resulted in the recovery of the original resistance.

In 1999, Margrave and coworkers reported the direct fluorination of a nanotube sidewall [211]. For this method, elemental fluorine was passed over the tubes at 150 to 325°C. The fluorination allows for the tubes to be soluble in alcohols after brief ultrasonication (1 mg · mL^{-1} in 2-propanol). It has been ascertained that in fluorination at the optimal temperature, C:F ratios of up to 2:1 can be achieved without disruption of the tubular structure. The fluorinated SWNTs (F-SWNTs) proved to be much more soluble than pristine SWNTs in alcohols, DMF, and other selected organic solvents. Investigation of the structure of F-SWNTs has been explored by density functional theory calculations and scanning tunneling microscopy

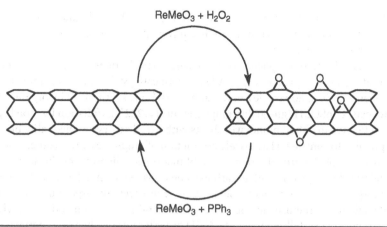

$ReMeO_3 + H_2O_2$

$ReMeO_3 + PPh_3$

SCHEME 2.6   Catalytic oxidation and de-epoxidation of SWNTs.

imaging [212, 213]. Scanning tunneling microscopy revealed that the fluorine formed bands of approximately 20 nm [213]. Calculations on the fluorinated (10,10) SWNTs with C_2F stoichiometry performed by using density functional theory revealed [1,2] addition is more energetically preferable than [1,4] addition. Recently, solid-state ^{13}C NMR has demonstrated the predominance of [1,2] addition and provided accurate quantification of the C:F ratio [214]. F-SWNTs make highly flexible synthons and subsequent elaboration has been performed with organo lithium, Grignard reagents, and amines [215–218].

Functionalized nanotubes can be characterized by a variety of techniques, such as atomic force microscopy, transmission electron microscopy, UV-vis spectroscopy, and Raman spectroscopy. Changes in the Raman spectrum of a nanotube sample can indicate if functionalization has occurred. Pristine tubes exhibit two distinct bands. They are the radial breathing mode (230 cm^{-1}) and the tangential mode (1590 cm^{-1}) [219]. When functionalized, a new band, called the disorder band, appears at approximately 1350 cm^{-1}. This band is attributed to sp^3-hybridized carbons in the tube. Unfortunately, while the presence of a significant D mode is consistent with sidewall functionalization and the relative intensity of D (disorder) mode versus the tangential G mode (1550–1600 cm^{-1}) is often used as a measure of the level of substitution. However, it has been shown that Raman spectroscopy is an unreliable method for determination of the extent of functionalization since the relative intensity of the D mode is also a function of the substituents distribution as well as concentration [217].

Graphene

Exfoliation of graphite into graphene, a carbon 2D monolayer, was demonstrated in 2004 by Geim and Novoselov that received the Nobel Prize in Physics in 2010 [2, 220]. Since then, graphene has caused excitement in the scientific community for its unique optical, electronic, and mechanical properties that make it a material of choice for many applications: electronics, fast and flexible optoelectronics, electrode-active material in the field of renewable energy (e.g., photovoltaic cells, fuel cells), or for composites. Graphene is hexagonal structure of a six-way bond between carbon atoms where carbon atoms are densely packed in a regular sp^2-bonded atomic-scale chicken-wire pattern (Figure 2.31). This crystal structure

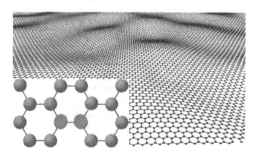

FIGURE 2.31   Graphene structure: a carbon monolayer obtained by graphite exfoliation.

holds a symmetry that is the key to electronic properties and quite exceptional, which is related to the fact that two carbon atoms occupy a strictly equivalent unit cell, offering a "new" degree of freedom compared to other 2D systems [221].

Today the term *graphene* covers a whole family of materials: graphene obtained by exfoliation of graphite; graphene produced by CVD (growth catalyst from the decomposition of a carbon derivative) or graphene oxide nanoparticles; and graphene ribbons synthesized by organic chemical processes. Different ways exist to obtain high-quality graphene sheets [222]. Exfoliation by scotch-graphite approach is actually very effective to obtain monolayer sheets with excellent crystal inequality but extremely low yield makes it impossible to consider a large-scale production of graphene by this route. Obtaining high-quality graphene is also possible by growth epitaxial (sublimation of silicon from the surface of silicon carbide) or CVD (catalytic decomposition of a carbonaceous gas on a metal surface). However, these surface-growth techniques do not allow us to obtain graphene sheets in large quantities and at low costs. Indeed, the epitaxial growth requires an enclosure under ultrahigh vacuum and extremely expensive starting material (SiC wafers of crystalline) while the CVD growth requires the use of a metal catalyst (nickel or copper) to be eliminated via a step of transferring to another surface. For larger quantities of graphene, the direct exfoliation techniques in a solvent and by insertion into the graphite before exfoliation step were developed more recently (Figure 2.32). Exfoliation of the graphite may be carried into organic and aqueous media (in the presence of surfactants) [223]. The exfoliation in an organic medium, like dimethylformamide or *N*-methyl-2-pyrrolidone, provides low-cost graphene sheets in solution and in concentrated solution (>1 mg·mL^{-1}). In aqueous media in the presence of surfactant molecules, higher concentrations (from about 5 mg·mL^{-1}) of sheets of a few hundred nanometers is obtained but mainly composed of multilayer (only 10–15% of monolayer). For obtaining the monolayer sheets, simply an additional sorting step can be performed. The nature of the surfactant used has a greater impact on the concentration of dispersions on the characteristics of exfoliated platelets. The methods of intercalation compounds between the planes of graphite followed by exfoliation step are numerous. Among these compounds, we can mention the alkali metals and chloride and iodine bromide. Exfoliation is then performed in an organic solvent (ethanol or NMP for alkali metals) or in a liquid medium in the presence of surfactants (chloride and iodine bromide). The development of these multistep processes for obtaining graphene sheets is in full progress.

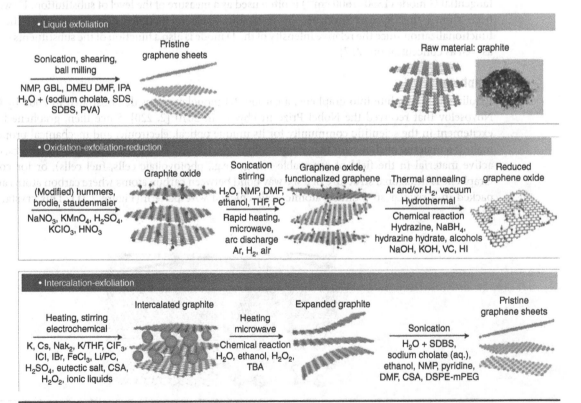

FIGURE 2.32 The different ways of graphene synthesis by exfoliation process [223].

Cellulose, an Organic Nanoparticle

The use of cellulose as a raw material has a very long history since it is involved in many applications and industrial products. Cellulose is found in wood, plant fibers, and many biological sources including algae, fungi, bacteria, and even in some marine animals (e.g., tunicates). Cellulose is used for energy production, building materials, clothing, paper, and numerous other applications. The major reason for its wide spectrum of uses is its abundance—it is the most available polymer on earth. The annual production of cellulose has been estimated at over 7.5×10^{10} tons. Cellulose has been used since time immemorial. Anselm Payen first isolated cellulose in 1838, and its physical and chemical aspects have been extensively studied since then. Cellulose is usually found as a fibrous, tough, and water-insoluble material that plays a crucial role in the structure of plant growth. It is the major component of the plant cell wall, and together with other polymers such as hemicellulose and lignin, forms complex composite materials responsible for the mechanical properties of plants. Regardless of its origin, cellulose is a high molecular weight homopolymer of β-1,4-linked anhydro-D-glucose. The condensation reaction between two glucopyranoses in their β forms leads to the creation of a C1-O-C4 link and the disaccharide, cellobiose. The creation of cellobiose induces the tilting of one of the glucose units for stereochemical reasons. Cellobiose is thus considered as the repeating unit of cellulose chains (Figure 2.33).

In nature, cellulose chains occur as compact semicrystalline fibrils due to the formation of intra- and interchain hydrogen bond networks and van der Walls interactions. The aggregation of cellulose chains is related to the biosynthesis process. Cellulose chains are synthesized at the surface of the plasma membrane by rosette-shaped enzymatic complexes comprising cellulose synthases. The polymerization and extrusion of cellulose chains is closely followed by the aggregation of the chains to form the crystalline structure. Typically, the microfibrils obtained have nanometric dimensions and, depending on the biological structure, the lateral dimensions of the crystallites usually vary from 2 to 50 nm. Some constraints may occur during the biosynthesis process, inducing the formation of less ordered zones. This results in the formation of amorphous zones distributed all along the elementary fibrils. At a higher hierarchical level, the microfibrils are assembled into fibrils (Figure 2.34).

Nanocelluloses are a family of products that correspond to the isolation of the elementary fibril elements, either by top-down or bottom-up approaches (Figure 2.34). Using the top-down strategy, two main categories of nanocelluloses can be obtained: cellulose nanocrystals (CNCs) and nanofibrillated cellulose (NFC) [224–228]. The first one corresponds to the isolation of the crystalline parts of the fibrils by selective hydrolysis of the amorphous region, whereas the latter arises from the mechanical delamination of fibers to isolate the elementary fibrils. The CNCs are stiff crystalline nanorods, whereas the NFCs are more flexible materials that have nanoscale diameters and a typical length of several micrometers. The third type of nanocellulose is obtained using the bottom-up approach due to the ability of some bacteria to biosynthesize cellulose as crystalline fibers, that is, bacterial cellulose (BC), from low-molecular-weight compounds (sugars, alcohols, acids, etc.) [226, 229, 230].

Cellulose Nanocrystals

CNC preparation consists of the controlled acid hydrolysis of elementary cellulose fibers. The first report of CNC isolation was published in 1952 by Rånby, who studied the hydrolysis of *Tunicine* cellulose by sulfuric acid [231]. Transmission electron microscopy images revealed the presence of needle-shaped particles, whereas electron diffraction demonstrated their crystalline structure. Since that time, CNCs have been extracted from various sources including wood, cotton linters, algae (e.g., *Valonia* and *Cladophora*), bacterial cellulose and various types of lignocellulosic sources [225, 226]. Our current understanding of the hydrolysis process is based on the presence of intrinsic disordered regions at regular intervals along the fiber. This assumption was actually confirmed by the neutron scattering studies of the cellulose fibers of

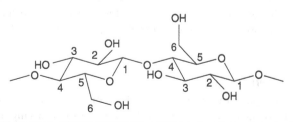

FIGURE 2.33 Chemical structure of cellobiose.

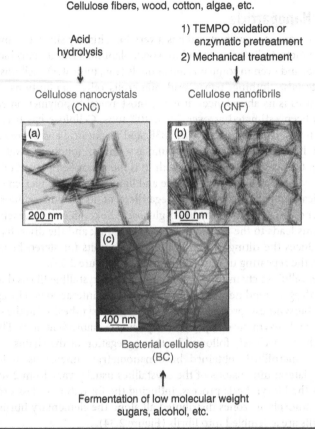

Cellulose fibers, wood, cotton, algae, etc.

Acid
hydrolysis

1) TEMPO oxidation or
enzymatic pretreatment
2) Mechanical treatment

Cellulose nanocrystals
(CNC)

Cellulose nanofibrils
(CNF)

Bacterial cellulose
(BC)

Fermentation of low molecular weight
sugars, alcohol, etc.

FIGURE 2.34 Main categories of nanocelluloses and their schematic fabrication process.

ramie in which labile hydrogen atoms were replaced by deuterium atoms, increasing the contrast between nonaccessible crystalline regions and accessible amorphous zones. The longitudinal periodicity determined by neutron scattering was found to be identical to the degree of polymerization of the CNCs obtained after hydrolysis, confirming the occurrence of regular disordered regions that were more sensitive to the acid hydrolysis [232]. Thus, the biosynthesis process has a clear influence on the final morphologies of the CNCs. The dimensions of CNCs are different depending on the biological sources and the conditions of hydrolysis. Both length and lateral dimensions are variable. For example, CNCs from wood are 3 to 5 nm in width and 100 to 200 nm in length, whereas higher dimensions are found for CNCs from the algae *Valonia* (i.e., 20 nm in width and more than 1–3 µm in length). The morphology of the CNCs is also dependent on their biological origin and fabrication process. For instance, CNCs obtained from bacterial cellulose present a ribbon-like shape, whereas *Valonia* presents a square morphology. As for CNCs from cotton, they are composed of associated parallel subunits, probably arising from the tight association of crystallites. In general, CNCs are obtained by hydrolysis with sulfuric acid, but hydrochloric acid is also used and, to a lesser extent, other mineral acids (e.g., phosphoric and hydrobromic acids). The use of sulfuric acid induces the formation of negative charges on the CNC surface due to the formation of sulfate esters, whereas in the case of hydrolysis with hydrochloric acid, the CNC surface remains neutral. Thus, in the first case, the CNC dispersion presents a good colloidal stability due to the electrostatic repulsions caused by the charged sulfate groups, whereas the neutral CNCs display a tendency toward aggregation and sedimentation. However, the presence of sulfate groups on the CNC surface reduces the thermostability of the CNCs.

Cellulose Nanofibrils

The general principle of the fabrication of NFCs consists in the fibrillation of cellulose fibers into nanoscale elements by applying intensive mechanical treatments to form a highly intricate web-like network of fibrils. NFC was first extracted from wood in the early 1980s when Herrick et al. and Turbak et al. succeeded in fibrillating wood fibers with a high-pressure homogenizer [233, 234]. Since then, various equipment and methodologies have been used to produce NFCs. Depending on the type of mechanical process used, the morphology of the NFCs obtained can vary. However, the major drawback of the mechanical process is the high consumption of energy to delaminate the nanofibrils. Thus, the effects of pretreatments prior to mechanical delamination have been evaluated and succeed in significantly decreasing energy consumption. Chemical treatments applied to produce pulp or cellulose from wood or lignocellulosic material can be considered as the first pretreatment of cellulose. Nevertheless, two other major strategies are used for the production of NFCs from purified pulp. The first one consists of a chemical modification of the fibers to introduce charges on the fibril surface, whereas the second one involves a controlled enzymatic action. Increasing the charge density of pulp fibers leads to charge repulsion, inducing a decrease in fiber friction and, consequently, less susceptibility to flocculation as well as a decrease in clogging during delamination. The introduction of charged groups can be achieved either by carboxymethylation or by oxidation reactions. Carboxymethyl groups can be introduced in their charged form to increase the swelling of the fibers and the electrostatic repulsion between the fibers in order to facilitate delamination. Another way to introduce charged carboxylated groups is the oxidation of primary hydroxyl groups of the glucose unit by the 2,2,6,6-tetramethylpiperidine-1-oxyl (TEMPO) radical. This reaction was found to be selective and efficient on insoluble fibers under aqueous and mild conditions, leading to a high level of fibrillation after mechanical treatment [235]. The second type of pretreament used to reduce energy consumption is the action of enzymes [236]. The complete hydrolysis of cellulose requires the action of several enzymatic systems. Therefore, the use of single purified endoglucanase induces modifications of the fibers rather than hydrolysis, resulting in a weakening of the cohesion of the fibers without severely degrading them. This results in the decrease of energy consumption during mechanical disintegration. Fine control of enzyme action through the quantity applied and the initial purification of the fibers is required to prepare NFCs with optimal properties. It should be emphasized that the quantity of residual hemicelluloses in the pulp is a crucial parameter for both the preparation of the NFCs and their properties since it can affect the interaction between fibers, as well as promote a higher dispersion of colloidal stability with respect to their content of charged groups. Subsequent to pretreatment, three main categories of technologies can be used to achieve the mechanical disruption of fibers. The homogenization procedure combines high pressure with mechanical shearing by pumping the raw cellulose materials into a valve that closes the adjustable gap. The rapid opening and closing of the valve leads to a decrease in pressure, increasing the velocity of the slurry and resulting in high shearing forces. The second type of technology is the microfluidizer. Fibrillation consists of injecting the cellulosic pulp under pressure through a Z-shaped chamber with a diameter that usually ranges from approximately 100 to 400 µm. The third major process is based on grinder devices that also allow the defibrillation of cellulose. The procedure consists in the injection of nonfibrillated cellulose onto two counter-rotating grinding stones. Some other techniques have also been also used to a somewhat lesser extent for the production of nanofibrils, including cryocrushing, refiners, and extruders. The final quality of NFCs can vary depending on the process chosen. Indeed, even if elementary fibrils are the main components of NFCs, the material may be inhomogeneous due to the presence of fibers, fiber fragments, fine particles, and the likes. The fraction of each component depends on the treatment applied to the fibers before homogenization and the forces of the mechanical treatment. In contrast to nanocrystalline cellulose (NCCs), the starting material seems to be less important for the final quality of NFCs. The more severe the delamination is, the more fibrillated the material will be. However, intense treatment reduces the length of the nanofibers and, consequently, the degree of network entanglement. A higher degree of fibrillation can be indicated by an increase in the transparencies of the NFC materials due to the generation of low-dimension fibrils with limited light scattering potential.

Bacterial Cellulose

In addition to plant cellulose, another source of cellulose has been extensively studied in the literature: bacterial cellulose (BC). Brown published the first report of bacterial cellulose at the end of the nineteenth century [237]. This discovery was made by studying the growth of a ferment used for the fabrication of vinegar, referred to as the "mother" of vinegar. When this ferment is grown on liquid culture media, the formation of a jelly-like mat occurs at the surface of the culture media. Elementary analysis of the pellicle led to the conclusion that it was actually cellulose produced by bacteria. Later studies demonstrated that various bacteria from different species were able to produce cellulose. BC occurs as ribbon-shaped nanofibrils with a width of 50 to 80 nm and a thickness of 3 to 8 nm, which can form an intricate 3D network. BC differs from the cellulose of higher plants since it has a higher crystallinity, up to 90 percent, whereas the crystallinity of plant cellulose is usually about 40 to 50 percent. BC is also found to be rich in I_α cellulose allomorphs, whereas most plant cellulose is I_β-rich. Another difference is that BC can be easily obtained in a highly pure form by simple mild alkaline extraction, since it is free of other contaminants found in plant cell walls such as hemicellulose or lignin. It is hypothesized that bacteria produce cellulose to protect themselves from UV rays and chemical environments, and to have easy access to oxygen since the BC mat helps the bacteria to float at the surface of the culture media. Many bacterial species have been shown to produce bacterial cellulose, and genetically modified bacteria have been produced to improve yield and quality. BC can be cultivated from various sources of carbon including low-molecular-weight sugars, alcohols, acids, and agro-industrial wastes. BC can be cultivated in static media and agitated media. Static cultivation yields smooth and regular mats, whereas in agitated culture media, filaments or spheres can be produced at higher yields.

Applications and Industrial Production of Nanocellulose

As mentioned before, cellulose nanoparticles have been known for a long time. However, over the two last decades, they have received an increasing amount of attention, demonstrated not only by the tremendous increase in academic reports, but also by the exponential increase of patents on nanocellulose [238]. This interest can be explained by both the general trend toward the use of renewable resources to replace petroleum-based materials, as well as the wide panel of applications for nanocelluloses. Moreover, many sources can be used to produce nanocelluloses worldwide. Their production does not compete with food production, and they offer remarkable functional properties such as low density, high mechanical performance, high specific surface, and good chemical reactivity. Nanocelluloses can therefore be seen as outstanding interaction platforms since they occur as stiff nanorods with chemically contrasted surfaces due to both the crystalline arrangement of cellulose chains and the different chemical functions existing on their surface. Depending on the types of nanocelluloses and the process used for their fabrication, they are able to interact with other components through hydrogen or van der Walls interactions via hydroxyl and CH groups, electrostatic interactions due to sulfate or carboxylic functions, as well as by covalent bonding as a result of the reactivity of hydroxyl groups. A nonexhaustive description of nanocellulose application fields is given below.

Due to their crystalline structure, nanocelluloses display very good mechanical properties and have therefore been used in nanocomposite materials such as reinforcing agents [224, 239]. However, in addition to mechanical reinforcement resulting from their nanoscale dimensions, nanocelluloses can be used for highly intricate networks that present barrier properties for use in packaging materials to improve oxygen and water permeability [240]. Nanocelluloses have also been used to develop dispersed systems such as high internal phase emulsions, foams, and aerogels. Indeed, NCCs, NCs, and BCs have all shown the ability to stabilize two-phase systems, either by direct adsorption at the hydrophobic/hydrophilic interface or by the formation of entangled networks that trap the dispersed phases [241–243]. Coatings based on nanocelluloses have also been described [244]. As a result of their nano-sized dimensions and their stiffness, optically active coatings have been developed. For example, antireflective coatings have been produced by controlling the nanopoporosity in order to adjust the refractive index of thin films to induce antireflective properties [245, 246]. When these films are deposited on reflective substrate, structural colors can be obtained, and the fabrication of films totally composed of biopolymers allows the detection of hydrolyzing enzyme activity as a result of the degradation of the films that decreases their thickness and results in the loss of the structural color [247, 248]. Optical properties can also be achieved using the self-assembling properties of NCCs, which can form chiral nematic phases that lead to iridescent films when dried. The optical properties can easily be tuned through the control of the repulsive interactions between the NCCs, either via a mechanical action such as sonication or the addition of a small amount of electrolytes [249]. Nanocellulose can also be used as a rheology modifier in various application fields, including the food industry. Indeed, due to its biological origin, BC has a significant potential as an ingredient in food. BC is already used in the food industry in Southeast Asia as a delicacy known as "Nata." Moreover, BC has been considered to be a GRAS (generally recognized as safe) ingredient in the United States since 1992. It has been successfully used as a multifunctional food ingredient that can act as a rheology modifier, a low-calorie ingredient, a water-retention agent, or an emulsifier for food emulsions [250]. Besides the use of nanocellulose itself, an emerging trend in research involving nanocelluloses is their combination with other nano-objects to create hybrid assemblies with optimized functionalities. For example, nanocelluloses have been associated with nanoclays, metallic nanoparticles, and carbon nanotubes [251–254].

Nanocelluloses are thus promising nanomaterial products, but their transition to everyday products has been limited up until now by the lack of industrial production. However, since 2012, a company called CelluForce has been developing the first industrial plan that would be capable of producing 1 ton per day of NCCs in Canada. In 2012, surface-modified NCC was added to the Domestic Substances List (DSL) in Canada. The DSL is an inventory of substances that may be imported and/or manufactured. An assessment of risk to the environment and to human health from this surface-modified NCC revealed no suspicion of toxicity. Many other pilot plants exist worldwide, especially in northern European countries (Sweden, Finland), Canada, the United States, Japan, and France as well. Numerous start-up projects are being developed together with investments from larger industrial companies. The availability of nanocelluloses, both NCCs and NFCs, will increase in the coming years, opening the way to industrial applications.

References

1. Kroto, H. W.; Heath, J. R.; O'Brien, S. C.; Curl, R. F.; Smalley, R. E. *Nature*. 1985, *318*, 162.
2. Novoselov, K. S.; Geim, A. K.; Morozov, S. V.; Jiang, D.; Zhang, Y.; Dubonos, S. V.; Grigorieva, I. V.; et al. *Science*. 2004, *306*, 666–669.
3. Iijima, S. *Nature*. 1991, *364*, 56.
4. Iijima, S.; Ichihashi, T. *Nature*. 1993, *363*, 603.
5. Berry, C. R. *Phys. Rev.* 1967, *161*, 848.
6. Rossetti, R.; Nakahara, S.; Brus, L. E. *J. Chem. Phys.* 1983, *79*, 1086.
7. Efros, A. A.; Efros, A. L. *Sov. Phys. Semicond*. 1982, *16*, 1209.
8. Murray, C. B.; Norris, D. J.; Bawendi, M. G. *J. Am. Chem. Soc.* 1993, *115*, 8706.
9. Jun, Y. W.; Choi, J. S.; Cheon, J. *Angew. Chem. Int. Ed.* 2006, *45*, 2.
10. Baes, C. F.; Mesmer, R. E. *The Hydrolysis of Cations*; J. Wiley and Sons: New York, 1976.
11. Jolivet, J. P. *Metal Oxide Chemistry and Synthesis. From Solution to Solid State*; Wiley: Chichester, 2000.
12. Sillen, L. G. *Stability Constants of Metal-Ion Complexes*; The Chemical Society: London, 1964.
13. Jorgensen, C. K. *Inorganic Complexes*; Academic Press: New York, 1963.
14. Müller, A.; Beckmann, E.; Bögge, H.; Schmidtmann, M.; Dress, A. *Angew. Chem. Int. Ed.* 2002, *41*, 1162.
15. Wells, A. F. *Structural Inorganic Chemistry*; 5th ed.; Clarendon Press: Oxford, 1991.
16. Nielsen, A. E. *Kinetics of Precipitation*; Pergamon Press: Oxford, 1964.
17. Sugimoto, T. *Adv. Colloid Interface Sci.* 1987, *28*, 65.
18. Haruta, M.; Delmon, B. *J. Chim. Phys.* 1986, *83*, 859.
19. LaMer, V. K.; Dinegar, R. H. *J. Amer. Chem. Soc.* 1950, *72*, 4847.
20. Jolivet, J. P.; Chanéac, C.; Tronc, E. *Chem. Comm.* 2004, 481.
21. Euzen, P.; Raybaud, P.; Krokidis, X.; Toulhoat, H.; LeLoaer, J. L.; Jolivet, J. P.; Froidefond, C. In *Alumina, in Handbook of Porous Materials*; Schüth, F., Sing, K. S. W., Weitkamp, J., Eds.; Wiley-VCH: Chichester, 2002, 1591–1677.
22. Pottier, J.; Cassaignon, S.; Chanéac, C.; Villain, F.; Tronc, E.; Jolivet, J.-P. *J. Mater. Chem.* 2003, *13*, 877–882.
23. Lemaire, B. J.; Davidson, P.; Ferré, J.; Jamet, J. P.; Panine, P.; Dozo, I.; Jolivet, J. P. *Phys. Rev. Let.* 2002, *88*, 125507.
24. Combes, J. M.; Manceau, A.; Calas, G. *Geochemica Cosmochem. Acta*. 1990, *54*, 1083.
25. Jolivet, J. P.; Tronc, E.; Chanéac, *Comptes Rendus de Chimie*. 2002, *5*, 659.
26. Jolivet, J. P.; Belleville, P.; Tronc, E.; Livage, J. *Clays Clay Miner.* 1992, *40*, 531.
27. Tronc, E.; Belleville, P.; Jolivet, J. P.; Livage, J. *Langmuir*. 1992, *8*, 313.
28. Jolivet, J. P.; Tronc, E. *J. Colloid Interface Sci.*, issue 49, 1988, *125*.
29. Vayssières, L.; Chanéac, C.; Tronc, E.; Jolivet, J. P. *J. Colloid Interface Sci.* 1998, *205*, 205.
30. Jolivet, J. P.; Froidefond, C.; Pottier, A.; Chanéac, C.; Cassaignon, S.; Tronc, E.; Euzin, P. *J. Mater. Chem.* 2004, *14*, 3281.
31. Matijevic, E. *Pure Appl. Chem.* 1980, *52*, 1179.
32. Cornell, R. M.; Schwertmann, U. *The Iron Oxides, Structure, Properties, Reactions, Occurrence and Uses*; VCH Weinheim Germany Publishers, 2003.
33. Schwertmann, U.; Friedl, J.; Stanjek, H. *J. Colloid Interface Sci.* 1999, *209*, 215.
34. Bottero, J. Y.; Manceau, A.; Villieras, F.; Tchoubar, D. *Langmuir*. 1994, *10*, 316.
35. Bailey, J. K.; Brinker, C. J.; Mecartney, M. L. *J. Colloid Interface Sci.* 1993, *157*, 1.
36. Sugimoto, T.; Muramatsu, A.; Sakata, K.; Shindo, D. *J. Colloid Interface Sci.* 1993, *158*, 420.
37. Matijevic, E.; Scheiner, P. *J. Colloid Interface Sci.* 1978, *63*, 509.
38. Pottier, A.; Chanéac, C.; Tronc, E.; Mazerolles, L.; Jolivet, J. P. *J. Mater. Chem.* 2001, *11*, 1116.
39. Keesmann, I. *Z. Anorg. Allg. Chem.* 1966, *346*, 31.
40. Arnal, P.; Corriu, J. P. R.; Leclercq, D.; Mutin, P. H.; Vioux, A. *Chem. Mater.* 1997, *9*, 694.
41. Bradley, D. C.; Mehrotra, R. C.; Gaur, D. P. *Metal Alkoxides*; Academic Press: London, 1978.
42. Turova, N. Y.; Turevskaya, E. P.; Kessler, V. G.; Yanoskaya, M. I. *The Chemistry of Metal Alkoxides*; Kluwer Academic Publishers: Boston, 2002.
43. Brinker, C. J.; Scherer, G. W. *Sol-Gel Science*; Academic Press: San Diego, 1990.
44. Sanchez, C.; Soler-Illia, G. J. A. A.; Ribot, F.; Lalot, T.; Mayer, C. R.; Cabuil, V. *Chem. Mater.* 2001, *13*, 3061.
45. Sanchez, C.; Ribot, F. *New J. Chem.* 1994, *18*, 1007–1047.
46. Sanchez, C.; Gomez-Romero, P. *Hybrid Functional Materials*; Wiley: Chicester, 2003.
47. Stöber, W.; Fink, A.; Bohn, E. *J. Colloid Interface Sci.* 1968, *26*, 62.
48. Barringer, E. A.; Bowen, H. K. *Langmuir*. 1985, *1*, 414.

49. Scolan, E.; Sanchez, C. *Chem. Mater.* 1998, *10*, 3217.
50. Rozes, L.; Steunou, N.; Fornasieri, G.; Sanchez, C. *Monatshefte Chemie.* 2006, *137*, 501.
51. Vioux, A. *Chem. Mater.* 1997, *9*, 2292.
52. Niederberger, M.; Pinn, N. *A Metal Oxide Nanoparticles in Organic Solvents: Synthesis, Formation, Assembly, and Application, Engineering Materials and Processes Series*; Springer Verlag: London, 2009.
53. McMahon, C. N.; Bott, S. G.; Barron, A. R. *J. Chem. Soc., Dalton Trans*, issue 18. 1997, 3129–3138.
54. Power, M. B.; Bott, S. G.; Clark, D. L.; Atwood, J. L.; Barron, A. R. *Organometallics.* 1990, *9*, 3086.
55. Power, M. B.; Bott, S. G.; Atwood, J. L.; Barron, A. R. *J. Am. Chem. Soc.* 1990, *112*, 3446.
56. Power, M. B.; Nash, J. R.; Healy, M. D.; Barron, A. R. *Organometallics.* 1992, *11*, 1830.
57. Niederberger, M.; Bartl, M. H.; Stucky, G. D. *J. Am. Chem. Soc.* 2002, *124*, 13642.
58. Niederberger, M.; Bartl, M. H.; Stucky, G. D. *Chem. Mater.* 2002, *14*, 4364.
59. Trentler, T. J.; Denler, T. E.; Bertone, J.; Agrawal, A.; Colvin, V. L. *J. Am. Chem. Soc.* 1999, *121*, 1613.
60. Jun, Y. W.; Casula, M. F.; Sim, J. H.; Kim, S. Y.; Cheon, J.; Alivisatos, P. *J. Am. Chem. Soc.* 2003, *125*, 15981.
61. Sun, S.; Anders, S.; Thomson, T.; Baglin, J.; Toney, M.; Hamann, H.; Murray, C.; et al. *J. Phys. Chem. B.* 2003, *107*, 5419–5425.
62. Zeng, H.; Rice, P.; Wang, S.; Sun, S. *J. Am. Chem. Soc.* 2004, *126*, 11458–11459.
63. Sun, S.; Zeng, H.; Robinson, D.; Raoux, S.; Rice, P.; Wang, S.; Li, G. *J. Am. Chem. Soc.* 2004, *126*, 273–279.
64. Chen, M.; Liu, J.; Sun, S. *J. Am. Chem. Soc.* 2004, *126*, 8394–8395.
65. Han, S.; Yu, T.; Park, J.; Koo, B.; Joo, J.; Hyeon, T.; Hong, S.; et al. *J. Phys. Chem. B.* 2004, *108*, 8091–8095.
66. Song, Q.; Zhang, Z. J. *J. Am. Chem. Soc.* 2004, *126*, 6164.
67. Yin, Y.; Gu, Y.; Kuskovsky, I. L.; Andelman, T.; Zhu, Y.; Neumark, G. F.; O'Brien, S. *J. Am. Chem. Soc.* 2004, *126*, 6206.
68. Jana, N. R.; Chen, Y.; Peng, X. *Chem. Mater.* 2004, *16*, 3931.
69. Sun, S.; Zeng, H.; Robinson, D. B.; Raoux, S.; Rice, P. M.; Wang, S. X.; Li, G. *J. Am. Chem. Soc.* 2004, *126*, 273.
70. Hyeon, T.; Lee, S. S.; Park, J.; Chung, Y.; Na, H. B. *J. Am. Chem. Soc.* 2001, *123*, 12798.
71. Cheon, J.; Kang, N. J.; Lee, S. M.; Lee, J. H.; Yoon, J. H.; Oh, S. J. *J. Am. Chem. Soc.* 2004, *126*, 1950.
72. Park, J.; Kang, E.; Bae, C. J.; Park, J. G.; Noh, H. J.; Kim, J. Y.; Park, J. H.; et al. *J. Phys. Chem. B.* 2004, 13594.
73. Lee, K.; Seo, W. S.; Park, J. T. *J. Am. Chem. Soc.* 2003, *125*, 3408.
74. Andrianov, K. A.; Zhadanov, A. A. *J. Polym. Sci.* 1958, *30*, 513.
75. Pasynkiewicz, S. *Polyhedron.* 1990, *9*, 429.
76. Barron, A. R. *Comments Inorg. Chem.* 1993, *14*, 123.
77. Harlan, C. J.; Mason, M. R.; Barron, A. R. *Organometallics.* 1994, *13*, 2957.
78. Landry, C. C.; Harlan, C. J.; Bott, S. G.; Barron, A. R. *Angew. Chem., Int. Ed. Engl.* 1995, *34*, 1201–1202.
79. Apblett, A. W.; Barron, A. R. *Ceramic Transactions.* 1991, *19*, 35.
80. Apblett, A. W.; Warren, A. C.; Barron, A. R. *Chem. Mater.* 1992, *4*, 167.
81. Landry, C. C.; Davis, J. A.; Apblett, A. W.; Barron, A. R. *J. Mater. Chem.* 1993, *3*, 597–602.
82. Landry, C. C.; Pappè, N.; Mason, M. R.; Apblett, A. W.; Tyler, A. N.; MacInnes, A. N.; Barron, A. R. *J. Mater. Chem.* 1995, *5*, 331–341.
83. Landry, C. C.; Pappè, N.; Mason, M. R.; Apblett, A. W.; Barron, A. R. In *Inorganic and Organometallic Polymers*; ACS Symposium Series: 1998; Vol. 572, 149.
84. Callender, R. L.; Harlan, C. J.; Shapiro, N. M.; Jones, C. D.; Callahan, D. L.; Wiesner, M. R.; Cook, R.; et al. *Chem. Mater.* 1997, *9*, 2418–2433.
85. Shahid, N.; Barron, A. R. *J. Mater. Chem.* 2004, *14*, 1235–1237.
86. Vogelson, C. T.; Barron, A. R. *J. Non-Cryst. Solids.* 2001, *290*, 216–223.
87. Callender, R. L.; Barron, A. R. *J. Am. Ceram. Soc.* 2000, *83*, 1777.
88. Harlan, C. J.; Kareiva, A.; MacQueen, D. B.; Cook, R.; Barron, A. R. *Adv. Mater.* 1997, *9*, 68.
89. Kareiva, A.; Harlan, C. J.; MacQueen, D. B.; Cook, R.; Barron, A. R. *Chem. Mater.* 1996, *8*, 2331–2340.
90. Koide, Y.; Barron, A. R. *Organometallic.* 1995, *14*, 4026–4029.
91. Bethley, C. E.; Aitken, C. L.; Koide, Y.; Harlan, C. J.; Bott, S. G.; Barron, A. R. *Organometallics.* 1997, *16*, 329–341.
92. Rose, J.; Cortalezzi-Fidalgo, M. M.; Moustier, S.; Magnetto, C.; Jones, C. D.; Barron, A. R.; Wiesner, M. R.; et al. *Chem. Mater.* 2002, *14*, 621–628.
93. Cortalezzi-Fidalgo, M. M.; Rose, J.; Wells, G. F.; Bottero, J. Y.; Barron, A. R.; Wiesner, M. R. *Mat. Res. Soc., Symp. Proc.*, issue 14, 2003, 800.
94. Rossetti, R.; Nakahara, S.; Brus, L. E. *J. Chem. Phys.* 1983, *79*.
95. Eychmüller, A. *J. Phys. Chem. B.* 2000, *104*, 6514.
96. Healy, M. D.; Laibinis, P. E.; Stupik, P. D.; Barron, A. R. *J. Chem. Soc., Chem. Commun.*, issue 79, 1989, 359.
97. Wells, R. L.; Pitt, C. G.; McPhail, A. T.; Purdy, A. P.; Shafieezad, S.; Hallock R. B. *Chem. Mater.* 1989, *1*, 4.
98. Manna, L.; Scher, E. C.; Alivisatos, A. P. *J. Am. Chem. Soc.* 2000, *122*, 12700.
99. Manna, L.; Milliron, D. J.; Meisel, A.; Scher, E. C.; Alivisatos, A. P. *Nature Mater.* 2003, *2*, 382.
100. Peng, Z. A.; Peng, X. *J. Am. Chem. Soc.* 2001, *123*, 183.
101. Peng, Z. A.; Peng, X. *J. Am. Chem. Soc.* 2002, *12*, 3343.
102. Peng, Z. A.; Peng, X. *J. Am. Chem. Soc.* 2001, *123*, 1389.
103. Zong, X.; Feng, Y.; Knoll, W.; Man, H. *J. Am. Chem. Soc.* 2003, *125*, 13559.
104. Jun, Y. W.; Lee, S. M.; Kang, N. J.; Cheon, J. *J. Am. Chem. Soc.* 2001, *123*, 5150.
105. Jun, Y. W.; Jung, Y. Y.; Cheon, J. *J. Am. Chem. Soc.* 2002, *12*, 615.
106. Lee, S. M.; Jun, Y. W.; Cho, S. N.; Cheon, J. *J. Am. Chem. Soc.* 2002, *124*, 11244.
107. Sigman, M. B., Jr.; Ghezelbash, A.; Hanrath, T.; Saunders, A. E.; Lee, F.; Korgel, B. A. *J. Am. Chem. Soc.* 2003, *125*, 16050.
108. Cho, K. S.; Talapin, D. V.; Gaschler, W.; Murray, C. B. *J. Am. Chem. Soc.* 2005, *12*, 7140.

109. MacInnes, A. N.; Power, M. B.; Barron, A. R. *Chem. Mater.* 1992, *4*, 11.
110. MacInnes, A. N.; Power, M. B.; Barron, A. R. *Chem. Mater.* 1993, *5*, 1344.
111. MacInnes, A. N.; Cleaver, W. M.; Barron, A. R.; Power, M. B.; Hepp, A. F. *Adv. Mater. Optics Electron.* 1992, *1*, 229
112. Cleaver, W. M.; Späth, M.; Hnyk, D.; McMurdo, G.; Power, M. B.; Stuke, M.; Rankin, D. W. H.; et al. *Organometallics.* 1995, *14*, 690.
113. Okuyama, K.; Huang, D. D.; Seinfeld, J. H.; Tani, N.; Matsui, I. *Jpn. J. Appl. Phys.* 1992, *31*, 1.
114. Stoll, S. L.; Gillan, E. G.; Barron, A. R. *Chem. Vapor Deposition.* 1996, *2*, 182.
115. Stoll, S. L.; Bott, S. G.; Barron, A. R. *J. Chem. Soc., Dalton Trans.* 1997, 1315.
116. Power, M. B.; Barron, A. R.; Hnyk, D.; Robertson, H. E.; Rankin, D. W. H. *Adv. Mater. Optics Electron.* 1995, *5*, 177.
117. Harle, O. L.; Thomas, J. R., *Dispersions of ferromagnetic cobalt particles. J. Appl. Polym. Sci.* 1966, *10* (12), 1915–27.
118. Allendoy, M. D.; Hurt, R. H.; Young, N.; Reagon, P.; Robbins, M. *J. Mater. Res.* 1993, *8*, 1651.
119. Goia, D. V.; Matijevic, E. *New J. Chem.* 1998, *22*, 1203.
120. Faraday, M. *Phil. Trans.* 1857, *147*, 145.
121. Turkevitch, J.; Stevenson, P. C.; Hillier, J. *Faraday Discuss. Chem. Soc.* 1951, *11*, 55.
122. Toshima, N.; Yonezawa, T. *New J. Chem.* 1998, 1179.
123. Jolivet, J. P.; Gzara, M.; Mazières, J.; Lefebvre, J. *J. Colloid Interface Sci.* 1985, *107*, 429.
124. He, S.; Yao, J.; Jiang, P.; Shi, D.; Zhang, H.; Xie, S.; Pang, S.; Gao, H. *Langmuir.* 2001, *17*, 1571.
125. Belloni, J.; Mostafavi, M.; Remita, H.; Marignier, J. L.; Delcourt, M. O. *New J. Chem.* 1998, 1239.
126. Henglein, A. *Chem. Mater.* 1998, *10*, 444.
127. Kim, F.; Connor, S.; Song, H.; Kuykendall, T.; Yang, P. *Angew. Chem. Int. Ed.* 2004, *43*, 3673.
128. Jana, N. R.; Gearheart, L.; Murphy, C. J. *J. Phys. Chem. B.* 2001, *105*, 4065.
129. Pileni, M. P. *Nature Mater.* 2003, *2*, 145.
130. Jana, N. R.; Gearheart, L.; Murphy, C. J. *Adv. Mater.* 2001, *13*, 1389.
131. Nikoobakht, B.; El-Sayed, M. A. *Chem. Mater.* 2003, *15*, 1957.
132. Jana, N. R.; Gearheart, L.; Murphy, C. J. *Chem. Commun.,* issue 105, 2001, 617.
133. Malandrino, G.; Finocchiaro, S. T.; Fragalà, I. L. *J. Mater. Chem.* 2004, *14*, 2726.
134. Murray, C. B.; Sun, S.; Doyle, H.; Betley, T. *Mater. Res. Bull.,* issue 26, 2001, 985.
135. Fievet, F.; Lagier, J. P.; Figlarz, M. *Mater. Res. Bull.* 1989, *14*, 29.
136. Sun, S.; Murray, C. B. *J. Appl. Phys.* 1999, *85*, 4325.
137. Osuna, J.; deCaro, D.; Amiens, C.; Chaudret, B.; Snoeck, E.; Respaud, M.; Broto, J. M.; et al. *J. Phys. Chem.* 1996, *100*, 14571.
138. Dumestre, F.; Chaudret, B.; Amiens, C.; Fromen, M. C.; Casanove, M. J.; Renaud, P.; Zurcher, P. *Angew. Chem. Int. Ed.* 2002, *41*, 4286.
139. Dumestre, F.; Chaudret, B.; Amiens, C.; Respaud, M.; Fejes, P.; Renaud, P.; Zurcher, P. *Angew. Chem. Int. Ed.* 2003, *42*, 5213.
140. Puntes, V. F.; Krishnan, K. M.; Alivisatos, P. *App. Phys. Lett.* 2001, *78*, 2187.
141. Puntes, V. F.; Zanchet, D.; Erdonmez, C. K.; Alivisatos, P. *J. Am. Chem. Soc.* 2002, *124*, 12874.
142. Toneguzzo, P.; Viau, G.; Acher, O.; Fievet-Vincent, F.; Fievet, F. *Adv. Mater.* 1998, *10*, 1032.
143. Dumestre, F.; Chaudret, B.; Amiens, C.; Renaud, M.; Fejes, P. *Science.* 2004, *303*, 821.
144. Park, S. J.; Kim, S.; Lee, S.; Khim, Z. G.; Char, K.; Hyeon, T. *J. Am. Chem. Soc.* 2000, *122*, 8581.
145. Toshima, N.; Yonezawa, T.; Kushihashi, K. *J. Chem. Soc. Faraday Trans.* 1993, *89*, 2537.
146. Sun, S. *Adv. Mater.* 2006, *18*, 393.
147. Chakroune, N.; Viau, G.; Ricolleau, C.; Fievet-Vincent, F.; Fievet, F. *J. Mater. Chem.* 2003, *13*, 312.
148. Ung, D.; Viau, G.; Ricolleau, C.; Warmont, F.; Gredin, P.; Fievet, F. *Adv. Mater.* 2005, *17*, 338.
149. Iwaki, T.; Kakihara, Y.; Toda, T.; Abdullah, M.; Okuyama, K. *J. Appl. Phys.* 2003, *94*, 6807.
150. Gibot, P.; Tronc, E.; Chanéac, C.; Jolivet, J. P.; Fiorani, D.; Testa, A. M. *J. Magn. Magn. Mater.* 2005, *555*, 290–291.
151. Harris, P. J. F.; Tsang, S. C.; Claridge, J. B.; Green, M. L. H. *J. Chem. Soc., Faraday Trans.* 1994, *90*, 2799.
152. Chambers, A.; Park, C.; Baker, R. T. K.; Rodrigues, N. M. *J. Phys. Chem. B.* 1998, *102*, 4253.
153. Liming, Y.; Saito, K.; Hu, W.; Chen, Z. *Chem. Phys. Lett.* 2001, *346*, 23–28.
154. Duan, H. M.; McKinnon, J. T. *J. Phys. Chem.* 1994, *98*, 12815–12818.
155. Murr, L. E.; Bang, J. J.; Esquivel, E. V.; Guerrero, P. A.; Lopez, D. A. *J. Nano. Res.* 2004, *6*, 241–251.
156. Krätschmer, W.; Lamb, L. D.; Fostiropoulos, K.; Huffman, D. R. *Nature.* 1990, *347*, 354–358.
157. Smalley, R. E. *Acc. Chem. Res.* 1992, *25*, 98–105.
158. Scott, L. T.; Boorum, M. M.; McMahon, B. J.; Hagen, S.; Mack, J.; Blank, J.; Wegner, H.; et al. *Science.* 2002, *295*, 1500.
159. Ge, Z.; Duchamp, J. C.; Cai, T.; Gibson, H. W.; Dorn, H. C. *J. Am. Chem. Soc.* 2005, *127*, 16292–16298.
160. Saunders, M.; Jiménez-Vázquez, H. A.; Cross, R. J.; Poreda, R. J. *Science.* 1993, *259*, 1428–1430.
161. Saunders, M.; Jimenez-Vazquez, H. A.; Cross, R. J.; Mroczkowski, S.; Gross, M. L.; Giblin, D. E.; Poreda, R. J. *J. Am. Chem. Soc.* 1994, *116*, 2193–2194.
162. Hirsch, A.; Soi, A.; Karfunkel, H. R. *Angew. Chem., Intl. Ed.* 1992, *31*, 766.
163. Hirsch, A.; Grösser, T.; Skiebe, A.; Soi, A. *Chem. Ber.* 1993, *126*, 1061.
164. Fagan, P. J.; Krusic, P. J.; Evans, D. H.; Lerke, S.; Johnston, E. *J. Am. Chem. Soc.* 1992, *114*, 9697.
165. Yang, J.; Barron, A. R. *Chem. Commun.* 2004, 2884–2885.
166. Zakharian, T.; Ashcroft, J.; Mirakyan, A.; Tsyboulski, D.; Benedict, N.; Weisman, B.; Wilson, L. J.; et al. *Electrochem. Soc. Proceedings.* 2004, *14*, 338.
167. Chikkannanavar, S. B.; Luzzi, D. E.; Paulson, S.; Johnson, A. T., Jr. *Nano Lett.* 2005, *5*, 151.
168. Smith, A. B. III; Tokuyama, H.; Strongin, R. M.; Furst, G. T.; Romanow, W. J.; Chait, B. T.; Mirza, U. A.; Haller, I. *J. Am. Chem. Soc.* 1995, *117*, 9359.
169. Britz, D. A.; Khlobystov, A. N.; Porfyrakis, K.; Ardavan, A.; Briggs, G. A. D. *J. Chem. Soc., Chem. Commun.* 2005, 37.

170. Tanaka, H.; Takeuchi, K.; Negishi, Y.; Tsukuda, T. *Chem. Phys. Lett.* 2004, *384*, 283.

171. Ogrin, D.; Barron, A. R. *J. Mol. Cat. A: Chem.* 2006, *244*, 267–270.

172. Treacy, M. M. J.; Ebbesen, T. W.; Gibson, J. M. *Nature.* 1996, *381*, 678.

173. Berber, S.; Kwon, Y. K.; Tomanek, D. *Phys. Rev. Lett.* 2000, *84*, 4613.

174. Thess, A.; Lee, R.; Nikolaev, P.; Dai, H. J.; Petit, P.; Robert, J.; Xu, C.; Lee, Y. H.; Kim, S. G.; Rinzler, A. G.; Colbert, D. T.; Scuseria, G. E.; Tomanek, D.; Fischer, J. E.; Smalley, R. E. *Science.* 1996, *273*, 483.

175. Girifalco, L. A.; Hodak, M.; Lee, R. S. *Phys. Rev. B.* 2000, *62*, 13104.

176. Bachilo, S. M.; Strano, M. S.; Kittrell, C.; Hauge, R. H.; Smalley, R. E.; Weisman, R. B. *Science.* 2002, *298*, 2361.

177. Kong, J.; Franklin, N. R.; Zhou, C. W.; Chapline, M. G.; Peng, S.; Cho, K. J.; Dai, H. J. *Science.* 2000, *28*, 622.

178. Collins, P. G.; Bradley, K.; Ishigami, M.; Zettl, A. *Science.* 2000, *287*, 1801.

179. Schadler, L. S.; Giannaris, S. C.; Ajayan, P. M. *Appl. Phys. Lett.* 1998, *73*, 26.

180. Bethune, D. S.; Klang, C. H.; deVries, M. S.; Gorman, G.; Savoy, R.; Vazquez, J.; Beyers, R. *Nature.* 1993, *363*, 605.

181. Guo, T.; Nikolaev, P.; Rinzler, A. G.; Tománek, D.; Colbert, D. T.; Smalley, R. E. *J. Phys. Chem.* 1995, *99*, 10694.

182. Guo, T.; Nikolaev, P.; Thess, A.; Colbert, D. T.; Smalley, R. E. *Chem. Phys. Lett.* 1995, *243*, 49.

183. Hafner, J. H.; Bronikowski, M. J.; Azamian, B. R.; Nikolaev, P.; Rinzler, A. G.; Colbert, D. T.; Smith, K. A.; et al. *Chem. Phys. Lett.* 1998, *296*, 195.

184. Peigney, A.; Laurent, C.; Dobigeon, F.; Rousset, A. *J. Mater. Res.* 1997, *12*, 613.

185. Dai, H.; Rinzler, A. G.; Nikolaev, P.; Thess, A.; Colbert, D. T.; Smalley, R. E. *Chem. Phys. Lett.* 1996, *260*, 471.

186. Nikolaev, P.; Bronikowski, M. J.; Bradley, R. K.; Rohmund, F.; Colbert, D. T.; Smith, K. A.; Smalley, R. E. *Chem. Phys. Lett.* 1999, *313*, 91.

187. Bootsma, G. A.; Gasson, H. J. *J. Cryst. Growth.* 1971, *10*, 223.

188. Westwater, J.; Gosain, D. P.; Tomiya, S.; Usui, S.; Ruda, H. *J. Vac. Sci. Technol. B.* 1997, *15*, 554.

189. Ogrin, D.; Colorado, R., Jr.; Maruyama, B.; Pender, M. J.; Smalley, R. E.; Barron, A. R. *Dalton Trans.*, issue 1, 2006, 229–233.

190. Fonseca, A.; Hernadi, K.; Nagy, J. B.; Bernaerts, D.; Lucas, A. A. *J. Mol. Catal. A: Chem.* 1996, *107*, 159.

191. Cassell, A. M.; Raymakers, J. A.; Kong, J.; Dai, H. *J. Phys. Chem. B.* 1999, *103*, 6484.

192. Anderson, R. E.; Colorado, R., Jr.; Crouse, C.; Ogrin, D.; Maruyama, B.; Pender, M. J.; Edwards, C. L.; et al. *Dalton Trans.*, issue 25, 2006, 3097–3107.

193. Thess, A.; Lee, R.; Nikolaev, P.; Dai, H. J.; Petit, P.; Robert, J.; Xu, C.; et al. *Science.* 1996, *273*, 483.

194. Davis, V. A.; Erickson, L. M.; Parra–Vasquez, A. N. G.; Ramesh, S.; Saini, R. K.; Kittrell, C.; Billups, W. E.; et al. *Macromolecules.* 2004, *37*, 154.

195. Zheng, M.; Jagota, A.; Strano, M. S.; Santos, A. P.; Barone, P.; Chou, S. G.; Diner, B. A.; et al. *Science.* 2003, *302*, 1545.

196. Tang, B. Z.; Xu, H. *Macromolecules.* 1999, *32*, 2569.

197. Moore, V. C.; Strano, M. S.; Haroz, E. H.; Hauge, R. H.; Smalley, R. E. *Nano Lett.* 2003, *3*, 1379.

198. Moore, V. C.; Strano, M. S.; Haroz, E. H.; Hauge, R. H.; Smalley, R. E. *Nano Lett.* 2003, *3*, 1379.

199. Anderson, R. E.; Barron, A. R. *J. Nanosci. Nanotechnol.* 2006 in press.

200. Hirsch, A. *Angew. Chem. Int. Ed.* 2002, *40*, 4002.

201. Bahr, J. L.; Tour, J. M. *J. Mater. Chem.* 2002, *12*, 1952.

202. Banerjee, S.; Hermraj-Benny, T.; Wong, S. S. *Adv. Mater.* 2005, *1*, 17.

203. Georgakila, V.; Kordatos, K.; Prato, M.; Guldi, D. M.; Holzinger, M.; Hirsch, A. *J. Am. Chem. Soc.* 2002, *124*, 760.

204. Bahr, J. L.; Yang, J.; Kosynkin, D. V.; Bronikowski, M. J.; Samlley, R. E.; Tour, J. M. *J. Am. Chem. Soc.* 2001, *123*, 6536.

205. Bahr, J. L.; Tour, J. M. *J. Mater. Chem.* 2001, *12*, 3823.

206. Dyke, C. A.; Tour, J. M. *J. Am. Chem. Soc.* 2003, *125*, 1156.

207. Liang, F.; Sadana, A. K.; Peera, A.; Chattopadhyay, J.; Gu, Z.; Hauge, R. H.; Billups, W. E. *Nano Lett.* 2004, *4*, 1257.

208. Mawhinney, D. B.; Naumenko, V.; Kuznetsova, A.; Yates J. T., Jr.; Liu, J.; Smalley, R. E. *J. Am. Chem. Soc.* 2000, *122*, 2383.

209. Ogrin, D.; Chattopadhyay, J.; Sadana, A. K.; Billups, E.; Barron, A. R. *J. Am. Chem. Soc.* 2006, *128*, 11322–11323.

210. Cui, J.; Burghard, M.; Kern, K. *Nano Lett.* 1993, *3*, 613.

211. Mickelson, E. T.; Huffman, C. B.; Rinzler, A. G.; Smalley, R. E.; Hauge, R. H.; Margrave, J. L. *Chem. Phys. Lett.* 1998, *296*, 188.

212. Kudin, K. N.; Bettinger, H. F.; Scuseria, G. E. *Phys. Rev. B.* 2001, *63*, 45413.

213. Kelly, K. F.; Chiang, I. W.; Mickelson, E. T.; Hauge, R. H.; Margrave, J. L.; Wang, X.; Scuseria, G. E.; et al. *Chem. Phys. Lett.* 1999, *313*, 445.

214. Alemany, L. B.; Zeng, L.; Zhang, L.; Edwards, C. L.; Barron, A. R. *Chem. Mater.* 2007, issue 19, 735–744.

215. Boul, P. J.; Liu, J.; Mickelson, E. T.; Huffman, C. B.; Ericson, L. M.; Chiang, I. W.; Smith, K. A.; et al. *Chem. Phys. Lett.* 1999, *310*, 367.

216. Saini, R. K.; Chiang, I. W.; Peng, H.; Smalley, R. E.; Billups, W. E.; Hauge, R. H.; Margrave, J. L. *J. Am. Chem. Soc.* 2003, *125*, 3617.

217. Zhang, L.; Zhang, J.; Schmandt, N.; Cratty, J.; Khabashesku, V. N.; Kelly, K. F.; A. R. Barron *Chem. Commun.*, issue 35, 2005, 5429–5430.

218. Zeng, L.; Zhang, L.; Barron A. R. *Nano Lett.* 2005, *5*, 2001.

219. Rao, A. M.; Richter, E.; Bandow, S.; Chase, B.; Eklund, P. C.; Williams, K. A.; Fang, S.; et al. *Science.* 1997, *275*, 187.

220. Novoselov K. S.; Geim A. K. *Nat. Mater.* 2007, *6*, 183.

221. Avouris P.; Dimitrakopoulos C.; *Mater. Today.* 2012, *15*, 86.

222. Park, S.; Ruoff, R. S. *Nat. Nanotechnol.* 2009, issue 4, 217–224.

223. Ren W.; Cheng H-M. *Nat. Nanotechnol.* 2014, *14*, 726.

224. Samir, M.; Alloin, F.; Dufresne, A. *Biomacromolecules.* 2005, *6* (2), 612–626.

225. Habibi, Y.; Lucia, L. A.; Rojas, O. J., *Chem. Rev.* 2010, *110* (6), 3479–3500.

226. Klemm, D.; Kramer, F.; Moritz, S.; Lindström, T.; Ankerfors, M.; Gray, D.; Dorris, A. *Angew. Chem. Int. Ed.* 2011, *50* (24), 5438–5466.

227. Tingaut, P.; Zimmermann, T.; Sebe, G. *J. Mater. Chem.* 2012, *22* (38), 20105–20111.
228. Lin, N.; Huang, J.; Dufresne, A. *Nanoscale.* 2012, *4* (11), 3274–3294.
229. Petersen, N.; Gatenholm, P. *Appl. Microbiol. Biotechnol.* 2011, *91* (5), 1277–1286.
230. Huang, Y.; Zhu, C. L.; Yang, J. Z.; Nie, Y.; Chen, C. T.; Sun, D. P., *Cellulose.* 2014, *21* (1), 1–30.
231. Rånby, B. G. *Arkiv for Kemi.* 1952, *4* (13), 241–248.
232. Nishiyama, Y.; Kim, U. J.; Kim, D. Y.; Katsumata, K. S.; May, R. P.; Langan, P. *Biomacromolecules.* 2003, *4* (4), 1013–1017.
233. Herrick, F.; Casebier, R.; Hamilton, J.; Sandberg, K. *J. App. Poly. Sci. Symp.* 1983, *37*, 797–813.
234. Turbak, A.; Snyder, F.; Sandberg, K. *J. App. Poly. Sci. Symp.* 1983, *37*, 815–827.
235. Saito, T.; Isogai, A. *Colloids Surf., A.* 2006, *289* (1–3), 219–225.
236. Pääkkö, M.; Ankerfors, M.; Kosonen, H.; Nykänen, A.; Ahola, S.; Österberg, M.; Ruokolainen, J.; et al. *Biomacromolecules.* 2007, *8* (6), 1934–1941.
237. Brown, A. *J. Che. Soc. Chem. Commun.* 1886, *49*, 432–439.
238. Charreau, H.; Foresti, M. L.; Vazquez, A. *Recent Pat. Nanotechnol.* 2013, *7* (1), 56–80.
239. Siro, I.; Plackett, D. *Cellulose.* 2010, *17* (3), 459–494.
240. Lavoine, N.; Desloges, I.; Dufresne, A.; Bras, J. *Carbohydr. Polym.* 2012, *90* (2), 735–764.
241. Capron, I.; Cathala, B. *Biomacromolecules.* 2013, *14* (2), 291–296.
242. Kalashnikova, I.; Bizot, H.; Cathala, B.; Capron, I., *Langmuir.* 2011, *27* (12), 7471–7479.
243. Cervin, N. T.; Andersson, L.; Ng, J. B. S.; Olin, P.; Bergstrom, L.; Wagberg, L. *Biomacromolecules.* 2013, *14* (2), 503–511.
244. Martin, C.; Nord, J. B. *Pulp Paper Res. J.* 2014, *29* (1), 19–30.
245. Podsiadlo, P.; Sui, L.; Elkasabi, Y.; Burgardt, P.; Lee, J.; Miryala, A.; Kusumaatmaja, W.; et al., *Langmuir.* 2007, *23* (15), 7901–7906.
246. Azzam, F.; Moreau, C.; Cousin, F.; Menelle, A.; Bizot, H.; Cathala, B. *Langmuir.* 2014, *30* (27), 8091–8100.
247. Cranston, E. D.; Gray, D. G. *Biomacromolecules.* 2006, *9* (7), 2522–2530.
248. Cerclier, C.; Lack-Guyomard, A.; Moreau, C.; Cousin, F.; Beury, N.; Bonnin, E.; Jean, B.; et al. *Adv. Mater.* 2011, *23* (33), 3791–3795.
249. Beck, S.; Bouchard, J.; Berry, R. *Biomacromolecules.* 2011, *12*, 167–172.
250. Shi, Z.; Zhang, Y.; Phillips, G. O.; Yang, G., *Food Hydrocol.* 2014, *35*, 539–545.
251. Olivier, C.; Moreau, C.; Bertoncini, P.; Bizot, H.; Chauvet, O.; Cathala, B. *Langmuir.* 2012, *28* (34), 12463–12471.
252. Galland, S.; Andersson, R. L.; Salajkova, M.; Strom, V.; Olsson, R. T.; Berglund, L. A. *J. Mater. Chem. C.* 2013, *1* (47), 7963–7972.
253. Hamedi, M. M.; Hajian, A.; Fall, A. B.; Hakansson, K.; Salajkova, M.; Lundell, F.; Wagberg, L.; et al. *ACS Nano.* 2014, *8* (3), 2467–2476.
254. Wu, C. N.; Saito, T.; Yang, Q. L.; Fukuzumi, H.; Isogai, A. *ACS Appl. Mater. Interfaces* 2014, *6* (15), 12707–12712.

Methods for Structural and Chemical Characterization of Nanomaterials

Jérôme Rose

CEREGE UMR 7330 CNRS–Aix-Marseille University, Marseille, France, and iCEINT CNRS–Duke University, Aix-en-Provence, France

Perrine Chaurand

CEREGE UMR 7330 CNRS–Aix-Marseille University, Marseille, France, and iCEINT CNRS–Duke University, Aix-en-Provence, France

Daniel Borschneck

CEREGE UMR 7330 CNRS–Aix-Marseille University, Marseille, France, and iCEINT CNRS–Duke University, Aix-en-Provence, France

Nicholas Geitner

iCEINT CNRS–Duke University, Aix-en-Provence, France, and Duke University, Durham, North Carolina, USA

Clément Levard

CEREGE UMR 7330 CNRS–Aix-Marseille University, Marseille, France, and iCEINT CNRS–Duke University, Aix-en-Provence, France

Armand Masion

CEREGE UMR 7330 CNRS–Aix-Marseille University, Marseille, France, and iCEINT CNRS–Duke University, Aix-en-Provence, France

Vladimir Vidal

CEREGE UMR 7330 CNRS–Aix-Marseille University, Marseille, France, and iCEINT CNRS–Duke University, Aix-en-Provence, France

Antoine Thill

CEA Saclay, Direction des Sciences de la Matière, Département de Recherche sur l'Etat Condensé les Atomes et les Molécules, LIONS, Gif-sur-Yvette, France

Jonathan Brant

Department of Civil & Architectural Engineering, University of Wyoming, Laramie, Wyoming, USA

Manuel D. Montaño

iCEINT CNRS–Duke University, Aix-en-Provence, France and Duke University, Durham, North Carolina, USA

Introduction

In this chapter, we survey several methods for analyzing nanomaterials in terms of structure, size, chemistry, concentration, and so on. Particle size influences an array of material properties and is therefore of concern. For example, the crystal properties of the material such as the lattice symmetry and cell parameters may change with size due to changes in surface free energy [1]. This is particularly true for particles where the number of atoms at the surface represents a significant fraction of the total number of atoms (e.g., <10–20 nm). Size may also affect the electronic properties of the materials due to the confinement of electrons, commonly discussed in terms of the quantum size effect, and the existence of discrete electronic states that give rise to properties such as the size-dependent fluorescence of CdS and CdSe nanoparticles or the electrical properties of carbon nanotubes.

The surface properties of nanomaterials play an important role in determining nanoparticle toxicity. The toxicity of CdS and CdSe nanoparticles, for instance, is completely controlled by their surface coating. The intracellular oxidation of bare CdSe nanoparticles results in the release of Cd^+ ions, the toxicity of which is well known. However, when CdSe nanoparticles are covered with organic molecules, the cytotoxicity is reduced (see Chapters 5 and 12). Adsorptive interactions involving nanomaterials, the effects of nanoparticle surface chemistry on particle stability and mobility (Chapter 6), and photocatalytic properties (Chapter 4) are also covered in greater detail later in this book.

Principles of Light-Material Interactions (Absorption, Scattering, Emission), Atomic Force Microscopy, Scanning Tunneling Microscopy

We present here a nonexhaustive list of characterization techniques based on the interaction between light and matter. Light (e.g., X-ray, UV-vis, infrared) can be viewed as both an electromagnetic wave (an oscillating electromagnetic field) and a particle called the photon. Within this wave-particle duality, the wave is characterized by its energy (E_0) and its wave vector ($\vec{k}_0$), where the wave vector has an amplitude inversely proportional to the wavelength (λ) and generally points in the direction that the wave propagates. The wavelength in turn can be expressed as a function of the energy. In a vacuum E (keV) = 1239,44/λ (nm). Like the light, electrons also have properties that resemble waves on one hand and particles on the other. The electron beam has properties similar to the photon beam with its own energy and wave vector. The interaction between light (a photon beam or an electron beam) and matter can occur in a variety of different scenarios, which can be summarized as follows (see also Figure 3.1):

- No scattering occurs and the light (or electron) is transmitted through the material with its initial characteristics ($\vec{k}_0$, E_0).

- Elastic scattering, that is, the wave vector is affected, but there is no change in energy. The incident beam is dispersed over a range of "angles" via diffraction processes. Modification of the wave vector occurs as a function of the spatial aspects of the matter (crystal structure, size, shape). This case gives rise to classic diffraction methods [X-ray diffraction (XRD), electron and neutron diffraction] and scattering techniques [small angle X-ray scattering (SAXS), light scattering, neutron scattering].

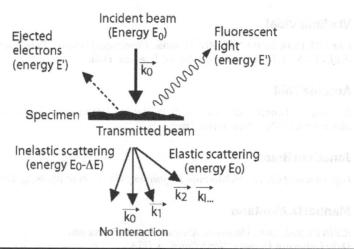

Figure 3.1   Examples of the various manners of interactions between light and electron beams with matter.

- Inelastic scattering, where the energy of light is affected by internal excitation processes at the molecular or atomic scale (e.g., electronic, nuclear, transitions) that shift the incident beam energy from an energy E_0 to an energy $E_0 \pm \Delta E$. Raman spectroscopy, Fourier transform infrared spectroscopy (FTIR), and Energy Electron Loss Spectroscopy (EELS) are techniques which measure this energy shift.

- Inelastic scattering with absorption is when, depending on the beam energy, a given material may partially absorb the incident beam. This absorption transfers the beam energy to the material that, in turn, reaches an excited state. The excited matter can then relax through several processes causing the emission of fluorescent photons (X-ray and light) or electrons (Auger and secondary electrons) having different energy. This absorption phenomena can be monitored in the following two ways:

    1. The transmitted beam (an attenuated part of the incident beam) can be collected to assess the absorption properties. For example, X-ray absorption spectroscopy (XAS) measures beam absorption at various energies. The absorption of the incident beam, depending on the nature and structure of the sample matter, can also be used in microscopy by analyzing the absorption contrast as is done in electronic, light, and X-ray microscopy [e.g., by X-ray computed tomography (CT)].

    2. Alternatively, emissions induced by relaxation of the absorbed energy can be monitored. Examples of characterization techniques utilizing that measure emissions during the relaxation step include X-ray fluorescence spectroscopy (XRF), luminescence spectroscopy, Auger spectroscopy, X-ray photoelectron spectroscopy (XPS), and nuclear magnetic resonance (NMR) imaging (in fact NMR measures the magnetic response of a material after a magnetic excitation).

Two other techniques that were developed in the 1980s that play a crucial role in characterizing nanosystems are based on a different type of operating principle: atomic force microscopy (AFM) and scanning tunneling microscopy (STM). These two techniques are not based on the interaction between a light source and matter. AFM and STM are advanced microscopy techniques developed to measure surface topography with angstrom-level resolution. In both cases the operating principle consists of scanning a probe in close proximity to the surface. The location between the probe and the surface is determined by the change in cantilever deflection and tunnel current in AFM and STM, respectively.

Structural Characterization

In the following sections we detail different techniques for characterizing structure in objects from the atomic scale to the dimension of nanoparticles. These techniques allow for determination of properties such as nanomaterial size, shape, and elemental structure.

Characterization of Atomic Structure

The characterization of the atomic arrangement of nanoparticles can be performed to determine both the long-range (i.e., the crystal parameters) and short-range order of a material's atomic structure. Of particular interest is the arrangement of atoms at the material's surface. The various characterization techniques that are available for these purposes may or may not be element specific. Therefore, care must be taken to select an appropriate characterization method. We first discuss the nonspecific techniques followed by the element-specific methods.

Nonspecific Techniques
X-Ray Diffraction

Operating Principles   X-ray scattering was first used to study the long-range order of the atomic arrangement in crystals in the early twentieth century (Wyckoff, 1963). The comparison of the diffraction pattern of unknown samples with the diffraction pattern of reference compounds from the database is necessary to identify the nature of minerals present in mixtures. Furthermore, information regarding the size of nanoparticles may also be derived from the XRD pattern, as particle size strongly affects the peak width. Particle size may be calculated based on the Sherrer formula, which states:

$$S = \lambda / \omega \cos(\theta) \qquad (3.1)$$

where S is the particle size, λ the wavelength of the beam, θ the diffraction angle, and ω is the width of the peak at half-maximum.

For nanoparticles, as for other types of materials, it is essential to determine both the crystallinity and polymorph type (i.e., two or more minerals having the same chemical composition, but different crystal structures). One of the best examples for illustrating this point is titanium dioxide (TiO_2). Due to the unique physical and chemical properties of TiO_2, it is widely used in industrial applications, particularly as a photocatalyst. The precise arrangement of the oxygen and titanium atoms may differ to create different crystal structures of "polymorphs" of TiO_2. Three different TiO_2 polymorphs exist—rutile, anatase, and brookite (Figure 3.2).

Limitations One of the main limitations of XRD when characterizing nanoparticles is that if the material loses its crystal structure (amorphization) and this process occurs without changing particle size, it will affect both the intensity and diffraction peak width in the same fashion.

Total Scattering XRD is used to probe the periodic structure of minerals (periodicity over distances above 100 Å), but a method has been recently "rediscovered" as a result of synchrotron light sources. This technique is based on the total scattering of particles and is analyzed by a relationship called the pair distribution function (PDF). This method essentially fills the gap between XAS (see later in the chapter) and XRD. The PDF technique has long been used for studying the nonperiodic structure of matter in noncrystalline materials (glasses) [2]. The local structure determined by the pair distribution function is the probability of finding an atom at a distance r from a reference z atom. These distances are probed by looking at scattered light as a function of angle, where larger angles probe smaller distances. The PDF transforms the signal obtained in the reciprocal space [wave vector space $Q = 4\pi \sin(\theta)/\lambda$ with θ the scattering angle and λ wavelength of the incident beam] to the real space (interatomic distance space). The spatial resolution is directly linked to the Q range scanned (>20 Å^{-1}). Therefore, to obtain a better interatomic size resolution it is necessary to measure data using high X-ray energy (>100 keV), that is to say, low wavelength (<0.12 Å). The high X-ray energy is a strong limitation of this technique, since it requires a synchrotron radiation source. The PDF analysis of ceria nanoparticles [3] illustrates the interest of such a tool.

Using neutron diffraction, the authors have shown that nanoscale ceria has Fenkel-type oxygen defects. The defects disappeared after a thermal treatment as shown in the post-thermal treatment PDF curve, which exhibits a higher intensity compared to the ceria before thermal treatment, for example, a higher number of interatomic distances (Figure 3.3).

Raman Spectroscopy

Operating Principles Raman spectra result from the scattering of electromagnetic radiation by the molecules in solid bulk materials. The energy of the incident light beam (usually in the visible region of the spectrum

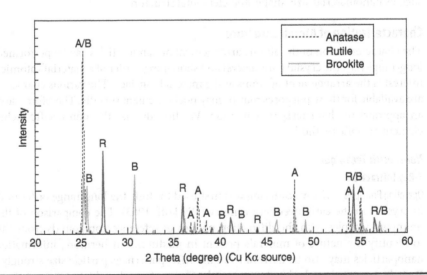

FIGURE 3.2 Theoretical X-ray diffraction patterns of the three polymorphs of TiO_2 particles (A: anatase, B: brookite, and R: rutile).

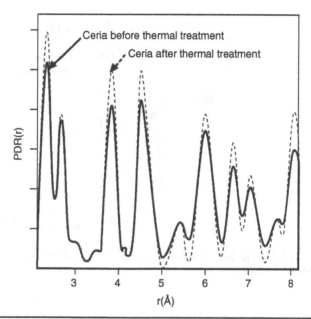

FIGURE 3.3   PDF of ceria before and after thermal treatment. (Adapted from [3].)

and sometimes in the ultraviolet zone) is slightly lowered or raised by inelastic interactions with the vibrational modes. It is a powerful tool for investigating the structural and morphological properties of solids at a local level [4]. In a simple approach, Raman spectra can fingerprint the nature or type of crystal phases.

Sample Preparation and Limitations   One very interesting point concerning Raman spectroscopy is that the sample can be solid or in solution. A limitation of Raman spectroscopy, however, is that the sensitivity of the technique is dependent upon the material being characterized.

Application in the Particular Case of Nanoparticles   Raman spectroscopy is frequently used in nanoparticle characterization and also interactions with biological media [5]. Like XRD, the position of Raman peaks can be considered as a fingerprint for different nanomaterials. In some cases, like that of TiO_2 for which the Raman peaks are particularly intense, it is possible to distinguish between the different polymorphs. For Anatase, six different peaks are recorded at 144 cm^{-1} (Eg), 197 cm^{-1} (Eg), 397 cm^{-1} ($B1g$), 518 cm^{-1} ($A1g$ and $B1g$, unresolved), and 640 cm^{-1} (Eg). On the other hand, for rutile, three different peaks are detected at 144 cm^{-1} ($B1g$), 448 cm^{-1} (Eg), and 613 cm^{-1} ($A1g$) (a fourth very weak band corresponding to the $B2g$ mode also exists at 827 cm^{-1}) [6]. In the particular case of nanoparticles, the signal is strongly affected by particle size as well as shape. This point will be detailed further later in this chapter.

Element-Specific Techniques: Local Structure and Oxidation State
X-Ray Absorption Spectroscopy
Operating Principles   X-ray absorption spectroscopy (XAS) is one of the most powerful techniques for probing the local atomic structure in a vast array of materials. XAS is a short-range order method that can be used regardless of the sample's physical state (crystalline, amorphous, in solution, or in a gas phase). Another important property of XAS is that it is an element-specific technique, in contrast to some spectroscopic methods such as Raman spectroscopy. The operating principle of this method requires that the incident X-ray beam energy be scanned from below to above the binding energy of the core -shell electrons of the target atom. By doing so, one observes an abrupt increase in the absorption coefficient corresponding to the characteristic absorption edge of the selected element (Figure 3.4). Depending on the electron that is excited, the absorption edges are named K for 1s, LI for 2s, LII, LIII for 2p, and so on.

The theory of XAS is described in detail elsewhere [7–9].Two characterization methods involving X-ray absorption are XANES and EXAFS, which focus on different portions of the absorption spectra. XANES yields information on chemical bonds, valence and symmetry, while EXAFS provides information

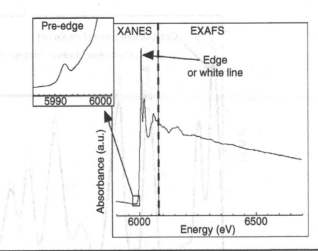

FIGURE 3.4   The Cr K-edge absorption spectrum of chromite ($CrFe_2O_4$) showing the XANES (X-ray absorption near edge structure) and EXAFS (extended X-ray absorption fine structures) parts.

on coordination number, chemical species, and distances. In the next section we will briefly introduce the theoretical basis of XAS and detail its use in characterizing the atomic arrangement of nanoparticles.

The energy position of the absorption edge and its shape reflect the excitation energy of the inner-shell electrons. The transition is always from core level to unoccupied states. The resulting excited photoelectron has generally enough kinetic energy to move through the material and this phenomenon can even occur in insulators. The presence of neighboring atoms around the central and excited atom leads to a modulation of the absorption coefficient due to interferences between outgoing and backscattered photoelectron waves. These modulations are present in the EXAFS zone. Oscillations can be extracted as a function of the photoelectron wave vector $k = \sqrt{2m_e(E - E_0)/\hbar^2}$, where m_e is the mass of electron, E is the energy, and E_0 is the binding energy of the photoelectron. The conventional EXAFS analysis based on single scattering was developed by Sayers et al. [10]:

$$\chi(k) = \frac{\mu(E) - \mu_0(E)}{\Delta\mu_0(E)} = -S_0^2 \sum_i \frac{N_i}{kR_i^2} |f_i(k)| e^{-2\sigma_i^2 k^2 e^{\frac{2R_i}{\lambda(k)}}} \sin[2kR_i + \phi_{ij}(k)] \tag{3.2}$$

where $\mu(E)$ is the measured absorption coefficient, $\mu_0(E)$ is a background function representing the absorption of an isolated atom, $\Delta\mu_0(E)$ is the jump in the absorption coefficient at the energy of the edge, S_0^2 is the amplitude reduction factor due to multielectronic effects, N_i is the coordination number, R_i is the interatomic distances between the central atom and the neighboring atom of type i, σ_i is a Debye-Waller factor describing the static and dynamic disorder in a Gaussian approximation, $|f_i(k)|$ is the amplitude of the backscattering wave from the neighbor of type i, $\lambda(k)$ is the free mean path of the photoelectron, that accounts for inelastic losses, and $\phi_{ij}(k)$ is the phase shift between the central ion j and its neighbors i. From Equation (3.2) it is possible to extract from EXAFS oscillations information such as the interatomic distances and the number and nature of surrounding atoms.

The pioneering work of Sayer et al. revolutionized the way EXAFS data are analyzed [10]. Because of the sinusoidal nature of EXAFS spectra, Sayer et al. used a Fourier transform to visualize the various electronic shells surrounding the central absorber. A pseudoradial distribution function (RDF) is obtained that provides the position of the different scatterers (Figure 3.5).

Sample Preparation   XAS has the ability to analyze samples under various physical forms (solid, liquid, and gas) with little preparation. In transmission mode, it is crucial to prepare pellets (for solid samples) with no pinholes and constant and appropriate thickness. It is also important that the size of the particles in the sample not be much larger than one absorption length (which is always the case for nanoparticles). For nanoparticles in suspension it is essential to prevent any settling during the measurement.

Limitations   The interpretation of EXAFS is limited in that it does not provide any information about bond angles between atoms. Moreover, it is not practically possible to distinguish between atoms that are in the same line in the periodic table ($Z \pm 2$) (see Figure 3.6).

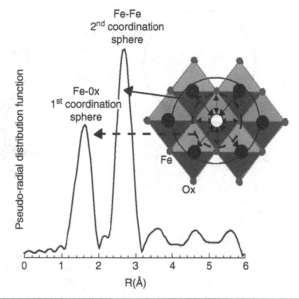

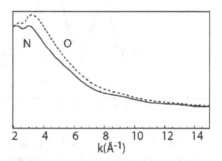

FIGURE 3.5 Radial distribution function of iron in lepidocrocite (γ-FeOOH).

FIGURE 3.6 Backscattering amplitude factor $|f_i(k)|$ for oxygen and nitrogen.

In XANES analysis, the theory is not as yet fully quantitative as in the case for EXAFS and requires different physical considerations. Here, X-rays from a synchrotron source may induce significant chemical changes within the sample (e.g., oxydo-reduction processes), which has obvious consequences for accurately characterizing the material of interest.

Application in the Particular Case of Nanoparticles   XAS can be helpful in determining the structural evolution of nanoparticles as a function of the nature of the ligands capping them. For example, Chemseddine et al. demonstrated that the presence of acetate or thiolate modifies the symmetry of CdS in the surface layers where CdS are octahedrally coordinated, whereas in the bulk on the particle the Cd is tetrahedraly coordinated [11]. As soon as the fraction of surface atoms becomes higher than 15 to 20 percent, XAS is sensitive enough to determine the modification of the surface site. For example, Auffan et al. were able to determine that iron atoms at the surface of nanomaghemites (γ-Fe$_2$O$_3$) coated with DMSA (di-mercaptosuccinic acid) were highly asymmetric due to ligand exchange between OH and SH groups from the DMSA (Figure 3.7). The resulting EXAFS spectra were then the combination of two Fe sites [12].

EXAFS can be a powerful tool for characterizing nanomaterials when used along with XRD. For very small particles (<3–4 nm) the XRD spectra are quite noisy, and accurate information concerning the minerals is difficult to extract. Using EXAFS can help in solving this limitation. The Debye-Waller factor determined by EXAFS modeling is related to the structural disorder of the particles. For example, Choi et al. observed an increase of the static disorder as TiO$_2$ particle size decreased [13]. Moreover, a volume contraction as particle size decreases has been highlighted by a decrease of the Ti-Ti interatomic distances. Using XAS it is also possible to identify the evolution of oxidation states for nanoparticles. For example,

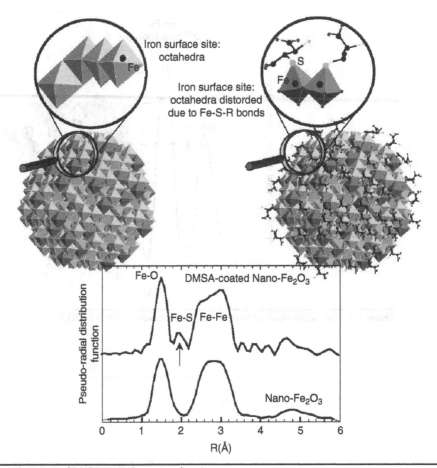

Figure 3.7   XAS spectra of DMSA-coated nanomaghemites.

Thill et al. (2006) were able to determine the surface reduction of CeO_2 nanoparticles when reacted with a suspension of *Escherichia coli*. In this study the nanoparticles had a diameter of around 8 nm with 30 percent of Ce atoms at the surface. It has been demonstrated that 30 to 40 percent of the cerium was reduced in the biological system. Indeed, the Ce^{4+} and Ce^{3+} XANES spectra are very different (Figure 3.8), allowing for their relatively easy differentiation. The XANES spectra can be decomposed using Gaussian individual components that can lead to a quantification of the $Ce^{3+}/(Ce^{4+} + Ce^{3+})$ ratio.

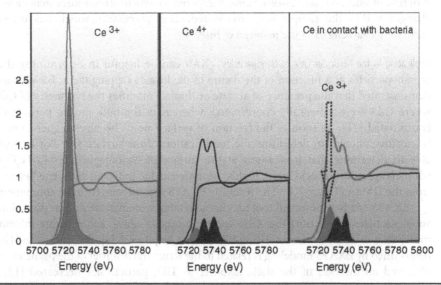

Figure 3.8   Results from XANES analysis at the Ce LIII-edge for Ce^{3+}, Ce^{4+} and CeO_2 in contact with bacteria. (Adapted from Thill et al., 2006.)

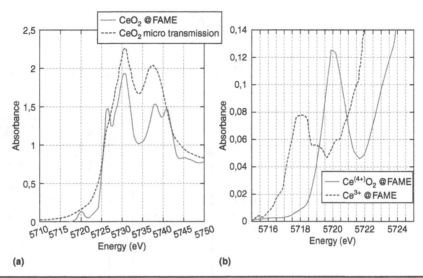

Figure 3.9   Ce LIII-edge XANES using conventional transmission mode and HERDF measurement on the Fame beamline at ESRF (Grenoble, France). (Courtesy of O. Proux.) Comparison between conventional transmission mode (a) and HERDF (b) where spectra of Ce^{3+} and Ce $^{4+}$ can clearly be seen shifting by at least 2 eV.

More recently high energy resolution fluorescence detection (HERFD) allows to further enhance XANES sensitivity (Figure 3.9). Using a detector with an energy resolution better than the core-hole lifetime broadening gives XANES spectra with sharper spectral features compared to conventional total fluorescence (TFY) spectra.

Mössbauer Spectroscopy   Mössbauer spectroscopy is a redox sensitive technique. Raj Kanel et al. have determined the kinetics of iron oxidation during arsenic (V) removal from groundwater using nanoscale zerovalent iron (ZVI) [118]. Mössbauer spectroscopy results confirmed that even before any contact between ZVI and As(V) solution, 81 percent of iron atoms are oxidized [maghemite + $Fe(OH)_2$] (Figure 3.10). The interaction with As(V) solution leads to an increase of the iron zerovalent state.

Operating Principles   The Mössbauer effect involves the interaction of γ radiation (i.e., the resonant absorption) with the nuclei of the atoms of a solid. Here, γ-rays are used to probe the nuclear energy levels related to the local electron configuration and the electric and magnetic fields of the solid. To date, Mössbauer spectroscopy has been mainly used to study Fe nanoparticles, but Au and Pt materials can also be studied by following nuclear transitions [14]. Like XAS, the Mössbauer spectroscopy is element specific. Mössbauer spectra consist of plotting the transmission of γ-rays as a function of their source velocity. A Mössbauer spectrometer consists of a vibrating mechanism that imparts a Doppler shift to the source energy and then to a source. In the absence of any magnetic field, the Mössbauer spectrum consists of one

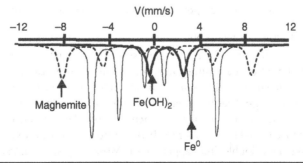

Figure 3.10   The Mössbauer individual component spectra of ZVI illustrating the maghemite, $Fe(OH)_2$, and Fe^0 constituents. (Adapted from [118].)

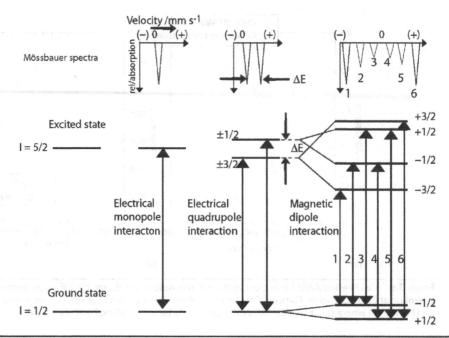

FIGURE 3.11 Resonant Mössbauer absorption in the presence and absence of a magnetic field.

or two absorption maxima between I1/2 and I3/2 nuclear levels (Figure 3.11). The difference between the ground and excited state levels is called the chemical or isomer shift, δ, which is described according to the following relationship:

$$\delta = \frac{4\pi}{5} Ze^2 R^2 \left(\frac{\delta R}{R}\right)\left[\left|\psi(0)\right|^2_{ABS} - \left|\psi(0)\right|^2_{SOURCE}\right] \qquad (3.3)$$

where Z is the nuclear charge, δR is the difference between the radii of the ground and excited states, R is the mean radius of the ground and excited states, and $\left|\psi(0)\right|^2_{ABS}$ and $\left|\psi(0)\right|^2_{SOURCE}$ are the electron density of the absorbant and source, respectively.

The different interactions that may occur between the nuclei and accompanying electrons and the information that may be derived from each can be summarized as follows:

- The shift due to monopole interactions provides information about the coordination number, valency, and spin state of the studied atom.

- The quadrupole shift provides information about the site distortion.

- The magnetic hyperfine field provides information about the valence and magnetic properties of the compound.

Sample Preparation In performing Mössbauer spectroscopy, samples must be prepared either as powders or as homogeneous thin-films (e.g., 5 mg/cm² of Fe). As with the case of powder-XRD it is important to avoid any textural effect of the powder. The samples do not require any vacuum. However, measurements are generally performed at low temperature (i.e., in the presence of liquid nitrogen).

Application in the Particular Case of Nanoparticles For nanoparticles some specificity exists when applying Mössbauer spectroscopy. When the grain size of fine particles is smaller than a critical grain size, Dc, they are composed of many single magnetic domains even though there is no external magnetic field. Therefore, the magnetic fields will not be stationed in a fixed direction like a bulk material, but rather, jump from one magnetization direction to another one giving them a greater magnetic susceptibility referred to as the superparamagnetic phenomenon. The Dc varies from one mineral to another. For instance, the Dc of α-Fe_2O_3 is 20 nm. If the grain size of iron oxide in composites is smaller than Dc, superparamagnetic doublet lines appear in the Mössbauer spectra. Using this property, Liu et al. combined XRD and Mössbauer spectroscopy on a Fe_2O_3-Al_2O_3 nanocomposite, and found that below 1373 K, the average grain size is below 20 nm (i.e., it exhibited a superparamagnetic behavior) [15, 16].

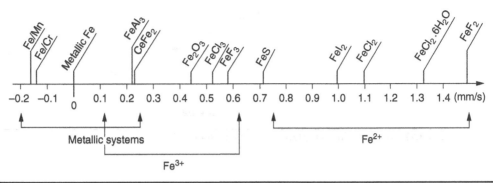

FIGURE 3.12 Chemical shifts of various iron species.

In the case of iron, Mössbauer spectroscopy is capable of differentiating redox states and also atomic sites. The chemical shifts of ^{57}Fe in various solid systems are summarized in Figure 3.12.

X-Ray Photoelectron Spectroscopy (XPS)

Operating Principle   XPS is based on the measurement of the energies of photoelectrons emitted as a result of the interaction between an incident X-ray beam with matter. As it measures the energies emitted from photoelectrons, XPS is classified as an electron spectroscopy technique. The main difference between electron and X-ray spectroscopy is the depth to which the surface is characterized. XPS is a surface or near-surface sensitive technique, as it is sensitive to a depth of between 1 to 5 nm. On the other hand, X-rays are used to characterize the bulk structure. However, for nanoparticles smaller than 20 nm, XPS can also be considered as a bulk sensitive technique.

Sample Preparation   XPS is classified as a surface sensitive technique, but only for relatively large surfaces. For analysis of nanoparticles with XPS, the best preparation method is to deposit them on a clean and flat surface. More than for TEM, XPS requires an ultrahigh vacuum chamber. It is therefore important to dry the system before analyzing it using XPS. When examining interactions between nanoparticles and living cells, sample preparation can be quite complicated as it is imperative to avoid any chemical modifications during drying (e.g., oxidation). Samples with large surface areas, or with volatile components, should ideally be dried and placed in a vacuum chamber prior to insertion into the XPS equipment.

Application to Nanoparticles   The ejected electrons correspond to core photoemission. The core-level peaks for a given atom can exhibit different binding energies due to symmetry or oxidation state effects. In particular, core-level photoemission can be very sensitive to changes in the oxidation state of an element. As an example, for TiO_2 optical irradiation can lead to the formation of charge carriers by optical absorption across the band gap. These charge carriers can directly participate in redox processes on the TiO_2 surface. XPS can distinguish the different Ti oxidation states—that is, the $Ti2p_{3/2}$ photoemission varies from 455.3 eV for TiO (Ti^{2+}) to 456.7 eV for Ti_2O_3 (Ti^{3+}), 457.6 eV for Ti_3O_5 ($2xTi^{3+}$, Ti^{4+}), and 458.8 eV for TiO_2 (Ti^{4+}) [116]. In the case of Mo metallic nanoparticles, XPS can probe for surface oxidation. Indeed, the two peaks of the three-dimensional (3D) photoemission vary from 231.6 and 228.3 eV to 235.8 and 232.7 eV for metallic and 6+ Mo oxidation state. The presence of the 235.8 and 232.7 eV peaks fingerprint the oxidation of Mo nanoparticles [117].

Chemistry, Structure, Transformation

Nuclear Magnetic Resonance

Operating Principles [17, 18]   Nuclear magnetic resonance (NMR) is one of the most versatile spectroscopy techniques for determining the speciation of a given element, the structure of molecules, diffusion coefficients, and so on. In this section, only the basic 1D experiment will be presented. When placed in a strong external magnetic field B_0, degenerate energy levels of a nucleus with a spin quantum number I are split into energy levels characterized by the magnetic quantum number m which has $2I + 1$ values, ranging from I to $-I(I, I - 1, I - 2 \dots -I)$. For example, ^{1}H has a spin $I = 1/2$; when placed in a magnetic field B_0, there are $2(1/2) + 1 = 2$ levels, where $m = 1/2$ and $m = -1/2$ (Figure 3.13).

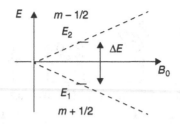

Figure 3.13   Nuclear spin energy level diagram for a spin =1/2.

For each level, the energy E is

$$E = -\gamma h B_0 m / 2\pi \tag{3.4}$$

where γ is the gyromagnetic ratio, characteristic of a nucleus, and h is Plank's constant.
Consequently,

$$\Delta E = -\gamma h B_0 (m_2 - m_1)/2\pi \tag{3.5}$$

From Equation (3.5), it appears that not every nucleus is observable by NMR; if $I = 0$, then $m = 0$ and thus $\Delta E = 0$. As a rule of thumb, when the mass number of a nucleus is odd, I is half integer (e.g., ^{1}H $I = 1/2$, ^{27}Al $I = 5/2$); when the mass number is even and the atomic number is odd, I is integer (e.g., ^{2}H $I = 1$); when both mass and atomic numbers are even, $I = 0$ (e.g., ^{12}C, ^{16}O).

For two adjacent energy levels, $m2 - m1 = -1$, and then,

$$\Delta E = \gamma h B_0 / 2\pi \tag{3.6}$$

The transitions between two energy levels a and b have the same probability. The difference in population of these levels is given by the Boltzmann distribution $\dfrac{N_a}{N_b} = \exp(-\Delta E / kT)$ where k is the Boltzmann's constant. An important consequence is that the signal depends on the number of transitions (i.e., the number of atoms), which makes NMR a quantitative method.

Since $\Delta E = h\nu$, the transition, or resonance, occurs when:

$$\nu_0 = \Delta E / h = \gamma B_0 / 2\pi \tag{3.7}$$

ν_0 is also called Larmor frequency.

For a given element, multiple resonances can be observed; indeed, the electrons in the molecular environment of a nucleus create a short-range magnetic field which adds to or opposes the applied external field B_0. Consequently, the magnetic field around the nucleus is no longer B_0, but B_1 which can written as

$$B_1 = B_0 (1 - \sigma) \tag{3.8}$$

where σ is the shielding constant. As a result, the resonance frequency is not ν_0 but ν_1. This difference in frequency is called chemical shift and is characteristic of a given molecular environment, thus providing speciation data. Usually, ν_1 is expressed in ppm (parts per minutes) with respect to reference compound so as to have a value independent of the experimental conditions. The chemical shift is

$$\delta = \frac{\nu_1 - \nu_{\text{ref}}}{\nu_{\text{ref}}} \times 10^6 \tag{3.9}$$

Modern Fourier transform spectrometers use a radiofrequency pulse to cause the resonance. Due to the Heisenberg uncertainty principle, the frequency width of the radiofrequency pulse (typically 1–10 µs) is wide enough to simultaneously excite nuclei in all local environments. In such experiments, the net magnetization M_0 which precesses around the z axis (i.e. B_0) flips into the xy plane with an angle θ depending on the duration of the radiofrequency pulse. The magnetization then precesses in the xy plane and returns

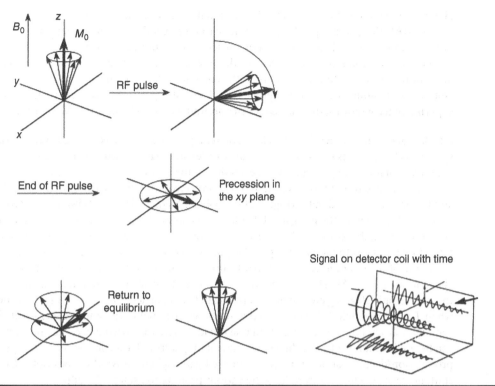

Figure 3.14   Schematic of the magnetization in a pulse NMR experiment.

to equilibrium. The signal recorded on the detector coil in the xy plane is a damped sine wave (Figure 3.14). This time domain free induction decay (FID) is converted into more readable frequency domain data by Fourier transform. The T_1 relaxation time (i.e., the spin-lattice relaxation time) is a measure of how fast the net magnetization recovers its value M_0 along B_0; the T_2 relaxation time is the time the transverse magnetization M_{xx} takes to return to its equilibrium value, that is, before the RF pulse. Generally, T_2 is shorter than T_1.

Line broadening, that is, the loss of resolution of NMR signals, depends on a number of parameters, especially in solid state experiments. However, some line-broadening causes can be avoided: the dipole-dipole coupling D which is proportional to $3\cos^2(\theta-1)$ can be circumvented by spinning the sample at $54.74°$ with respect to B_0, or the magic angle, for which $D=0$; this is known as MAS-NMR (magic angle spinning). Chemical shift anisotropy is partially lifted by spinning the sample at kilo-hertz rates, mimicking the molecular tumbling in liquids.

Sample Preparation   In first approximation, samples of nanoparticles/nanomaterials for NMR analysis should be supplied as dry homogeneous powders. However, samples in a "semidry" or gel-like state may qualify for a HR-MAS analysis. Also, some particles with a characteristic nucleus retaining all degrees of freedom (e.g., the central tetrahedral Al in the Keggin structure of Al_{13}) may be analyzed in the liquid; in that case, increasing line width or signal loss may monitor aggregation [19]. Samples are recovered without damage after analysis, which makes this technique attractive for the observation of "precious" samples.

Limitations and Remedies   The versatility and richness of NMR unfortunately comes with a rather poor sensitivity of the method. It depends on the gyromagnetic ratio γ and the abundance of a given isotope. Sensitivity is usually given relative to ^1H (= 1). Common nuclei such as ^{27}Al (100% natural abundance), ^{29}Si (4.7% abundance), and ^{13}C (1% abundance) have NMR sensitivities of 0.201, 3.7×10^{-4}, and 1.7×10^{-4}, respectively. Signal-to-noise ratio is improved by accumulating multiple scans. For a simple Bloch decay experiment, the repetition rate is controlled by the relaxation time T_1 which varies greatly with the observed nucleus and nature of the compound. For example, it is of the order of 0.1 s for routine proton NMR, and can reach 10 min for ^{29}Si in certain tailor-made nanoparticles [20]. The signal/noise ratio is also improved by transferring the magnetization from an abundant-spin (e.g., ^1H) to a dilute-spin system (e.g., ^{13}C or ^{29}Si); this is achieved in cross polarization (CP) experiments, the magnetization transfer occurring when the

Hartmann-Hahn condition is met [21]. For a half-spin system, this condition is $\gamma_1 B_1 = \gamma_2 B_2$. For ^{1}H and ^{13}C CP-MAS, the gain is typically around three to four. Another severe limitation of NMR is that it rapidly becomes inoperative when paramagnetic nuclei are present in the sample, as they broaden the resonance of neighboring atoms beyond detection, thereby causing substantial or complete signal loss. Finally, a limitation, which is technical in nature, is the current difficulty in analyzing low-frequency nuclei. Typical commercial multinuclear (tunable) probes are given with specifications down to the ν_0 of Ag under optimal experimental conditions, that is, frequencies around 4 to 5 MHz for a 100-MHz magnet.

Application in the Particular Case of Nanoparticles For nanoparticles, NMR is of high interest since this element-specific technique does not require long-range order and may be used to characterize complex nanoparticle/nanomaterial itself and particularly the evolution of the coating. In the highly publicized case of sunscreens, NMR was used to monitor the evolution of the protective AlOOH shell around the TiO$_2$ UV-absorbing core and to characterize the simulated aging of the Poly Di Methyl Silane (PDMS) organic layer which ensure the adequate dispersion of the inorganic UV filter in the lotion [22]. NMR is also used to qualify the successful synthesis of a number of materials, for example, ^{27}Al and ^{29}Si MAS-NMR give characteristic signals of imogolite formation and allow to monitor the possible presence of by-products and unreacted Al monomers [23–26].

Some of the limitations of the NMR technique can be used to the benefit of the user. As mentioned above in the case of Al$_{13}$ [19], the broadening of an initially sharp resonance can characterize the onset of the clusterization/aggregation of originally isolated nanoparticles. In the case of quadrupolar nuclei, that is, when $I > 1/2$, line width is very sensitive to the electric field gradient, a general rule being that loss of symmetry around the nucleus results in line broadening; as a consequence, the line width and its evolution can be used to monitor the binding environment of quadrupolar nuclei. For example, pH variations lead to protonation/deprotonation of hydroxyls in the ligand sphere of Al; this does not cause the resonance to shift, however, the line width reflects the various protonation states [27, 28].

When added intentionally and in (very) small quantities, the presence of paramagnetic elements can be beneficial. By inducing faster spin lattice relaxation, they enable shorter recycle delays between scans; in this case, it is an absolute requirement that the paramagnetic nuclei are distributed homogeneously throughout the sample. However, shorter T_1 times imply broader lines, therefore this technique should not be applied where spectral resolution is important. Another use of paramagnetics is the identification of their binding sites on a molecule/surface by monitoring the resonances of the host structure that are suppressed upon addition of paramagnetic nuclei [29]. This is also a way to identify certain redox numbers of metals: MnIII, CrIII, CuII, and so on are paramagnetic and will either suppress or not produce NMR signals.

Structure, Shape, Size, and Chemistry Characterization

SEM and TEM Spectroscopic techniques can provide detailed information about the structure and size of nanoparticles. The most common examples of electron microscopy techniques used for characterizing nanoparticles include scanning electron microscopy (SEM) and transmission electron microscopy (TEM). A potential drawback to these techniques is that in some cases particle shape can induce indirect modification of the spectroscopic signal and is thus a source of error in these types of measurements. Nevertheless, these microscopy techniques allow for direct visualization of nanoparticles, and thereby provide information about particle size, shape, and structure. With this in mind, both SEM and TEM imaging are highly versatile and powerful techniques for characterizing nanoparticles.

Operating Principle The general operating principles and the components that make up the respective instruments of TEM and SEM imaging are summarized in Figure 3.15.

The operational principle for a SEM is based on the scanning of finely focused beams of electrons onto a surface. When using an SEM, there are a number of different visualization techniques that can be used. During scanning, the incident electrons are completely backscattered, reemerging from the incident surface of the sample. Since the scattering angle is strongly dependent on the atomic number of the nucleus involved, the primary electrons arriving at a given detector position can be used to yield images containing both topological and compositional information. The backscattering mode is generally used on a polished section to minimize the effects of local topology and therefore obtain information on the composition of the sample. The high-energy incident electrons can also interact with loosely bound conduction band electrons in a sample. The amount of energy given to these secondary electrons as a result of these interactions is small, and so they have a very limited range in the sample (a few nanometers). Because of this, only secondary electrons that are emitted within a very short distance of the surface are able to escape

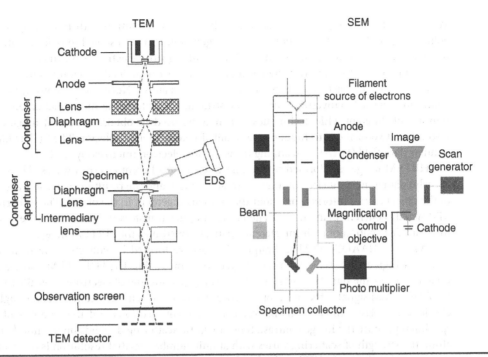

FIGURE 3.15 Schematics illustrating the operating principles of SEM and TEM microscopes.

from the sample. This means that the detection mode boasts high-resolution topographical images, making this the most widely used of the SEM modes. SEM can provide both morphological information at the submicron scale and elemental information at the micron scale. Recent developments in terms of electron source (field emission) have led to the development of high-resolution SEM. Using a secondary or backscattering electron image one can look at particles as small as 10 to 20 nm (Figure 3.16). Chemical information using EDX, however, is obtained at the micron scale and not for individual particles.

In contrast with SEM, TEM analyzes the transmitted or forward-scattered electron beam. Here the electron beam is passed through a series of lenses to determine the image resolution and obtain the magnified image (Figure 3.16b). The highest structural resolution possible (point resolution) is achieved upon use of high-voltage instruments (acceleration voltages > 0.5 MeV). Enhanced radiation damage, which may have stronger effects for nanostructured materials, must, however, be considered in these cases.

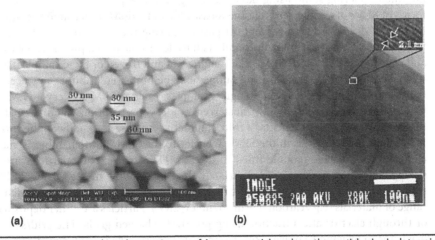

(a)   (b)

FIGURE 3.16 (a) An FESEM secondary electron image of Ag nanoparticles where the particle size is determined using the appropriate scale (courtesy of Volodimir Tarabara). (b) A TEM image of imogolite (single-walled aluminosilicate nanotube) (courtesy of Clément Levard).

With corrections it is possible to achieve subangstrom resolution with microscopes operating at lower voltages (typically, 200 keV), allowing the oxygen atoms to be resolved in oxides materials. On the other hand, as high resolution is achieved in TEM, resulting from electron wave interference among diffracted peaks and not only to the transmitted beam in the absence of deflection, a limitation to structural resolution can arise from nanoparticles with a very low number of atoms. Nevertheless, conventional TEM is the most common tool used to investigate the crystal structure of materials at the subnanometer scale. There exist a number of different TEM techniques that may be used to obtain structural images with atomic level resolution; two of these techniques are detailed below: high-resolution TEM (HRTEM) and high-angle annular dark field (HAADF) scanning transmission electron microscopy (STEM).

HRTEM images are formed by the interference of coherent electron waves. The object transmits the (nearly) planar incident electron wave, interacts with it, and the resulting electron wave ψ_e at the exit plane of the object carries information about the atom arrangement in the object. The ψ_e corresponds to a set of "diffracted" coherent plane waves. The electron optics transfers these waves to the image plane, and the intensity distribution of their interference pattern constitutes the HRTEM image.

With the HAADF-STEM techniques, images are formed by collecting electrons that have forward scatter at high angles, typically a few degrees or more, using HAADF-STEM. Unlike normal dark-field imaging, where the signal comes from elastic (Bragg) scattering of electrons typically to smaller angles, the HAADF-STEM signal is the result of inelastic electron scattering typically to larger angles. At high angles, elastic and inelastic interactions between the incident electrons and the columns of atoms within the specimen produce the image contrast. Since inelastic scattering depends on the number of electrons in an atom, the strength of scattering varies with atomic number. Spatial resolution is determined by the size of the focused incident electron probe. With electron beam sizes of less than 3 Å, imaging at atomic level resolution is possible. In a HAADF image, brighter spots represent the heavier atomic columns while the less intense spots indicate the lighter atomic columns.

Coupled to EDS (energy dispersive spectrometry) or to EELS (electron energy loss spectroscopy), TEM can provide information about the elemental distribution at a very low spatial scale (several nanometers). The energy resolution of EELS is 0.2 eV, while for EDX it is >140 eV at 6 keV. In addition to elemental information, EELS can be used to determine the electronic structure, bonding, and nearest neighbor distribution of the specimen atoms. Therefore, EELS can provide information on the speciation of elements. However, the very low intensity of the EELS signal at energies higher than 1500 to 2000 eV is a strong limitation to study the K-edge of elements with atomic number higher than the silicon. Therefore EELS is well adapted to study elements in a 100- to 1500 eV energy range corresponding to the K-edge of low-atomic-number elements like carbon and oxygen or the L-edge of transition metals like Fe or Cu.

Samples Preparation For conventional SEM measurements the sample must be dry and conductive. If the sample material is not conductive it must be coated with some material, usually carbon or gold, using a sputter coating device. In some instances it may also be advantageous to coat already conductive materials to improve image contrast. The conductive coatings are usually applied at a thickness of about 20 nm, which is too thin to interfere with dimensions of surface features. Prior to coating, the sample is mounted on a ring stand using nonconducting carbon tape. In many instances, as with most common forms of SEMs, it is necessary that the measurement be done under high vacuum. This requires that the sample be dry in order to prevent off-gassing during the measurement. However, recent advances in the design of SEM measurement chambers have led to the development of environmental SEM (ESEM) and cryogenic SEMs (cryo-SEM), for imaging wet and frozen or fixed samples, respectively. As opposed to conventional SEM, in ESEM the sample may be both wet and does not need to be conductive. It is therefore desirable for delicate biological samples. In cryo-SEM and cryo-TEM the sample is frozen or fixed using liquid nitrogen and transferred to a cryopreparation chamber that is held in vacuum. In the case of SEM measurements, a thin conductive coating is usually applied to allow high-resolution imaging or microanalysis in the SEM. Transfer to the SEM/TEM chamber is via an interlocked airlock and onto a cold stage module fitted to the SEM/TEM stage. Nanoparticles are particularly viable for study using TEM as they are appropriately thin and may be imaged using TEM-support grids. TEM-support grids are a fine mesh support that are commonly made of copper, which may be covered with a range of materials (e.g., carbon, Formvar, holey, SiO_2). Particles are either deposited through evaporation or through electrostatic attraction using positively charged grids. The grid/sample must be allowed to dry prior to imaging to prevent off-gassing once the sample is placed in the vacuum. When examining particle samples it is important to avoid aggregation during the drying step, which will inhibit analysis of individual particles.

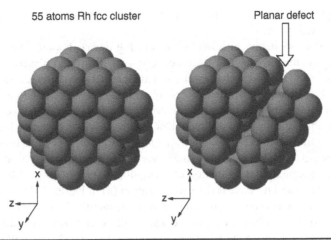

FIGURE 3.17   Images of a stable structure composed of 55 atoms and one with a planar defect that can be identified with HRTEM images. These two pictures show a hard-ball model of the structure. (Adapted from [30].)

Application in the Particular Case of Nanoparticles   HRTEM may be employed to provide extremely valuable information about the atomic structure of nanoparticles. For instance, Marın-Almazo et al. used HRTEM to determine the atomic arrangement of rhodium nanoparticles ($d = 1.8$ nm) [30]. The authors found these nanoparticles the (111) and (200) inter-planar distances corresponded to large minerals indicating that no distortion of the network existed. However, for the smaller clusters (<1.5 nm) some range stacking faults, dislocations, and twins were identified (as illustrated in Figure 3.17).

In another study by Yan et al. it was found that by coupling XRD analysis (Sherer equation) and TEM imaging, it was possible to determine the structure and size of ultrasmall gold nanoparticles ($d = 1$ nm) [31]. TEM can also be a powerful characterization technique for studying the dispersion and chemistry of nanoparticles in the environment. For instance, using TEM, Hiutsunomiya and Ewing found that airborne particulates ($d < 2$ μm) from coal-fired power plants contained 1 to 10 ppm of uranium using HAADF-STEM images and that the uranium was located in nanoparticles ($d < 10$ nm) as uraninite (UO_2) [32]. These nanoparticles were encapsulated in graphite, which may retard oxidation of the tetravalent uranium to the more mobile hexavalent form. Spatially resolved EELS has been used to study the morphology of carbon nanotubes [33]. EELS results demonstrated that even for tiny nanotubes the covalent nature of the chemical bonds is preserved, whereas near-field EELS pointed out the specific character of the surface valence electron excitation modes in nanotubes in relation with their curved anisotropy.

AFM/STM

Operating Principles   The invention of the atomic force microscope (AFM) in 1982 is considered one of the most important instrumental breakthroughs in the development of nanoscience. The AFM provides a means both to characterize the physical properties of materials at the atomic scale and to measure forces between surfaces with piconewton resolution. The operating principles of both the AFM and the STM may be described in terms of an optical lever acting as a sensitive spring. The optical lever operates by reflecting a laser beam off the end of a cantilever, typically made of silicon or silicon nitride, at the end of which is attached a tip or probe. Angular deflection of the tip causes a twofold larger angular deflection of the laser beam. The reflected laser beam strikes a position-sensitive photo-detector consisting of two side-by-side photodiodes. The difference between the two photodiode signals indicates the position of the laser spot on the detector and thus the angular deflection of the tip. Because the tip-to-detector distance generally measures thousands of times the length of the cantilever, the optical lever greatly magnifies the motions of the tip. Because of the approximately 2000-fold magnification in the measured deflection, the optical lever detection can theoretically obtain a noise level as low as 10^{-14} m/Hz$^{1/2}$.

The advantage of the AFM/STM over electron microscopes is that it is possible to measure in the z-axis in addition to both the x-axis and y-axis. In this way it is possible to get a 3D resolved image of a surface. Furthermore, using the AFM it is possible to measure interfacial forces between surfaces in both gaseous and liquid environments. For example, Brant et al. (2002 and 2004) used an AFM to characterize the surface morphology of water treatment membranes and to subsequently measure the interfacial forces between the membrane surface and various nanoparticles. This information may then be used to either optimize or prevent particle attachment to a given surface as in groundwater transport processes or engineered systems.

Although originally conceived as an imaging device, because the operating principle of the AFM is based on the measurement of force between a small tip and a surface with piconewton sensitivity, this method can also be used to characterize interactions between surfaces and nanomaterials. Force is measured by recording deflection of the free end of a cantilever as its fixed end approaches and is subsequently retracted from a sample surface. The interaction force occurring between the AFM probe and the sample surface is then calculated according to Hooke's law ($F = k\Delta z$) where F is the force; k is the cantilever spring constant; and Δz is the vertical deflection of the cantilever. A positive vertical deflection indicates repulsion while a negative one indicates attraction. Figure 3.18 represents a typical force curve generated by an AFM, where force is plotted as a function of separation distance (h). Initially the colloid probe is far from the sample surface and no force is detected [position (a) in Figure 3.18]. As the probe approaches the surface it encounters some type of force (repulsive or attractive) before contacting the surface [position (b) in Figure 3.18]. The probe is then pressed against the sample surface until a preset loading force is reached, and then the probe is retracted [position (c) in Figure 3.18]. The pull-off force is measured as the force required to separate the two surfaces (maximum negative deflection) [position (d) in Figure 3.18] and is used to approximate the strength of adhesion between the two surfaces.

A significant advantage of the AFM over other force measuring techniques, is its ability to operate in either air or water. AFM force measurements may be carried out using either a standard silicon nitride–tipped cantilever or a probe with attached material such as a colloid or nanoparticle.

By attaching nanomaterials such as a single-wall nanotube, information on interacting forces between the nanomaterial and an approaching surface may be measured directly. Alternatively, a colloid probe can be attached to the AFM tip and used to measure interactions with a lawn of deposited nanomaterials. Measured AFM force curves are then related to hypothetical interfacial forces arising from hydration, electrostatic repulsion, Lifshitz–van der Waals energies (Hamaker constant), and acid-base interactions. This is accomplished by modifying the solution chemistry to isolate the specific interactions or properties of interest. For example, the Lifshitz-van der Waals surface energy component of the nanoparticles can be determined by performing AFM force measurements in purely apolar solvents (e.g., cyclohexane). By measuring the interfacial interaction in a nonpolar solvent, both acid-base and electrostatic interactions are limited, thus isolating the van der Waals component. The Hamaker constant for the system (see Chapter III) is proportional to the AFM measured force and the square of separation distance. Therefore, the Hamaker constant may be calculated as the gradient of a plot of the square root of force versus separation distance for the sphere-plate geometry represented, for example, by a colloid probe approaching a nanomaterial lawn.

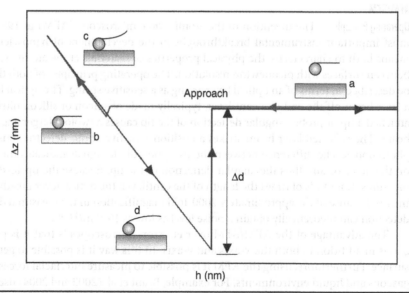

FIGURE 3.18   AFM force as a function of separation distance on approach [from (a) to (c)] and retraction [from (c) to (d)] from the surface.

Sample Preparation   For AFM-type measurement, no special sample preparation is required other than fixation to a sample stage, typically a small circular metal disk. Nonconducting carbon tape is commonly used for these purposes. One of the distinct advantages of AFM is that measurements may be carried out in both air and water. For analysis of powder materials they must be deposited onto a surface and fixated to prevent movement during imaging.

Limitations   Samples for AFM analysis are generally constrained to rather smooth samples with average roughness values of less than around several microns. This constraint results from the need to maintain contact between the scanning probe and the surface and limitations in the change in height that the probe can accommodate. Particles in size from around 1 μm to several 10s of microns may be attached to AFM cantilevers for force measurements. The minimum particle size continues to decrease with improving methods of particle attachment. For example, nanotubes are now being used as tethers for attaching particles to cantilevers serving to reduce the minimum particle size. The principle constraint here results from the use of epoxy to fix the particle to a tipless cantilever. Below a critical size, the epoxy covers the sample particle and thus alters the interaction chemistry.

Shape, Size, and Behavior in Liquid Phase
Scattering Experiments (Small Angle X-Ray Scattering, Light Scattering)

Scattering experiments may be used to characterize particle suspension *in situ*. Light or X-ray scattering experiments may be used to measure the size, shape, agglomeration state, and dynamic properties (diffusion coefficient) of nanoparticles in a liquid sample. These properties are averaged over the whole scattering volume, and thus over a large number of nanoparticles or nanoparticle aggregates. The statistical relevance and the nondestructive *in situ* nature of scattering experiments represent significant advantages of these techniques over the electron scattering techniques previously discussed. The operating principles of a typical scattering apparatus are illustrated in Figure 3.19. One monochromatic beam is incident on a sample. Typically, both the incident and scattered beam are shaped by optics adapted to the radiation characteristics (e.g., apertures, slits, and lens). Part of the incident beam crosses the sample unaffected and some is scattered. A detector measures at an angle θ the scattered intensity $I(\theta,t)$. The volume illuminated by the incident beam and analyzed by the detector is the scattering volume V.

From the general experimental setup shown in Figure 3.19, three different types of measurements can be performed: dynamic, static, and calibrated light scattering. Static scattering measures the value of $I(\theta,t)$ averaged over a period of time t, typically as a function of θ. Static scattering measurements can be used to obtain structural information such as size, shape, and agglomeration state. Dynamic scattering measures the instantaneous values of $I(t)$ over time at a fixed angle θ. The variability in scattering over time gives information on the diffusion coefficient due to Brownian motion of the scattering nanoparticles in suspension. Dynamic scattering is thus an indirect measurement of particle size, shape, and interactions. The third type of information given by scattering experiments is obtained by the calibration of the measured intensity to obtain the quantitative average number of scattered photons per unit solid angle [34]. This method yields, with almost no approximations or model assumptions, physical quantities such as the volume or specific surface averaged over the whole sample. Different radiation sources used for scattering experiments can yield complementary information on the sample. Light scattering is associated with

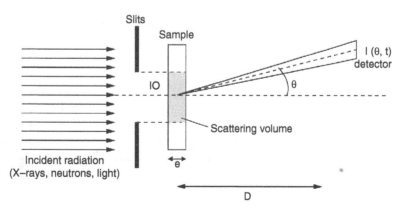

Figure 3.19 Illustration of the scattering principle upon which most scattering techniques and apparatuses are based.

variations in dielectric properties (or refractive index) [35, 36], X-rays are scattered by electrons [36, 37], and neutrons are scattered by nuclei [38].

Small Angle X-Ray Scattering

Operating Principle [39, 40]   SAXS is a technique that examines the structural features of objects between a few angstrom and ca. 200 nm in size, that is, it covers nearly three orders of magnitude. The incident X-ray beam is scattered by the electrons of the sample thus producing a scattering pattern from which parameters such as size, shape, specific surface area, and aggregate structure can be determined. Since the scattering centers are electrons, this technique is not element specific. The scattering vector $\vec{q}$ is the path difference between the incident beam $\vec{s_0}$ and the scattered beam $\vec{s}$ at a 2θ angle (Figure 3.20).

The modulus of the scattering vector is

$$q = \frac{4\pi \sin\theta}{\lambda} \tag{3.10}$$

where 2θ is the scattering angle and λ is the wavelength of the incident beam.

The scattered intensity is given by:

$$I(q) = 4\pi V \int_0^\infty r^2 \gamma(r) \frac{\sin qr}{qr} dr \tag{3.11}$$

where V is the volume and $\gamma(r)$ is the correlation function [41]. This function depends only on structural features of the sample, when $r \to \infty$, $\gamma(r) \to 0$.

As a matter of fact, the scattered signal depends on the electron contrast, that is, the difference in electron density between the sample itself and its solvent (e.g., air, water). This means that not every system is observable by SAXS; insufficient contrast will produce weak and unexploitable scattering

While SAXS gives access to a number of structural features, and in some cases even to the chemical speciation, extracting this information from scattering data usually requires more or less sophisticated modeling, which often implies a fairly precise knowledge of the nature of the sample as prerequisite. However, there are sample characteristics that can be derived from the scattering data without any a priori knowledge.

- *Size:* For the smallest angles, the intensity can be rewritten as:

$$I(q) = A \exp\left(-\frac{q^2 R_g^2}{3}\right) \tag{3.12}$$

  where A is a constant and R_g is the radius of gyration. This is known as Guinier's law. The radius of gyration R_g can be linked to particle dimensions of a given shape in the case of a homogeneous suspension. For example, in the case of a sphere $R_g^2 = 3r^2/5$ (r is the radius), for a cylinder $R_g^2 = r^2/2 + h^2/12$ (r is cylinder radius and h is length). For heterogeneous samples (multiple sizes and shapes, aggregates, etc.), Guinier's law yield the R_g of the largest particles/aggregates in the sample. Guinier's law is valid for $q^2 R_g^2 \ll 1$ in diluted samples with roughly globular shaped objects. In the case of anisotropic particles, the q range in which the law is applicable is even narrower.

- *Texture:* As indicated earlier, SAXS can provide a description of the structure of aggregated objects. In such a case, the structure is determined in terms of the fractal geometry introduced in the late 1970s (Mandelbrot). Indeed, the scattered intensity can be written as:

$$I(q) = \phi V_{\text{Part}} P(q) S(q) \tag{3.13}$$

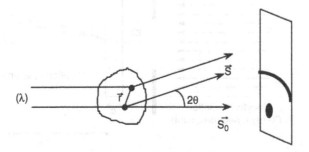

FIGURE 3.20   Schematic of the scattering by two point centers separated by r.

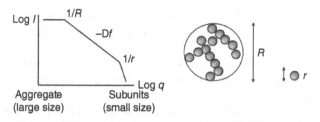

FIGURE 3.21 Determining the fractal dimension of an aggregate.

with ϕ the volume fraction, V the particle volume, $P(q)$ the particle form factor, and $S(q)$ the inter particle structure factor.

It has been shown that [42]:

$$S(q) \propto q^{-D_f} \tag{3.14}$$

where D_f is the fractal dimension. As a consequence, the fractal dimension can be derived by determining the slope of a simple log I versus log q plot (Figure 3.21) [43].

For an accurate determination of the fractal dimension D_f, the portion of the plot between $1/R$ and $1/r$ must be strictly linear and extend over at least one order of magnitude.

- *Surface area and interface structure:* The large q region of a scattering curve can be used as an alternate method to determine the specific surface. Porod's law states that

$$\lim_{q \to \infty} \frac{q^4 I(q)}{Q} = \frac{S}{V} \tag{3.15}$$

with S the surface, V the volume, and Q the Porod invariant which depends only on the electron contrast of the sample and is written as:

$$Q = \int q^2 I(q) dq \tag{3.16}$$

The applicability of Porod's law is limited to the systems having a sharp, well-defined interface with the surrounding medium. When the interface presents a nonsharp structure, SAXS allows to evaluate the thickness of this structure [41].

Sample Preparation In most cases, samples can be used "as is" for SAXS measurements. Very diluted samples may be difficult to analyze. Nevertheless, the sensitivity of modern synchrotron-based SAXS instrumentation is sufficient to monitor subsecond kinetics for systems in the 0.1 to 0.01 mmol concentration range. At the opposite, too concentrated samples will produce multiple scattering and consequently signal loss. This is usually not an issue, but in the cases where multiple scattering is suspected, a two or three times dilution can be used to verify that the scattering pattern is not modified by the concentration changes.

Limitations One of the main limitations of SAXS is that not only the structure of interest but the entire sample (including the sample holder) contributes to the scattering pattern. Consequently, it may become tricky to extract the desired information from the SAXS data. SAXS data treatment may be challenging. This is especially true for heterogeneous samples with little (or no) knowledge about their nature. For samples obtained under more controlled conditions, detailed structural information can be obtained by modeling.

Application in the Particular Case of Nanoparticles The size domain covered by SAXS corresponds to the size range in the ISO definition of nanoparticles. Thus, SAXS is relevant to every structural (or chemical) investigation regarding nanoparticles. This section deliberately will make no attempt to be comprehensive but rather highlight less intuitive aspects of the technique. One early nanomaterial produced at an industrial level is the Al_{13} tridecamer used as coagulant in water treatment. Sulfate precipitated Al_{13} has been investigated by XRD [44]. Suspensions obtained from Al chloride were examined by SAXS: the size of Al_{13} in the suspension is 12.4 Å in agreement with previous crystallographic data, and the fractal dimension of 1.8 of the aggregate is indicative of a cluster-cluster aggregation mechanism [45, 46]. The same approach has been made with Fe (III) hydrolysis showing that the aggregation of FeOOH nanoparticles followed fractal structures depending on the pH [47]. Although SAXS is not element specific, this technique can be used in certain cases to derive a (semi) quantitative speciation. Of course, this applies to model systems with some preexisting knowledge of the nature of the species. For example, the speciation of Al was determined from scattering data using a LCF approach where the individual Al species (monomer, dimer, trimer, and Al_{13}) were modeled as hard spheres with various spacings between them to simulate the presence of organics [47, 48]. This simple procedure was also used to confirm the speciation of Fe in Fe-PO_4 systems [49]. More rigorous modeling of course enhances the level of details to be extracted. *Ab initio* calculations revealed the roof tile shape of the precursors of Ge substituted imogolite nanotubes and ascertained the concentration-dependent single- versus double-wall structure of the tubes at the end of the growth process [26, 50]. An often overlooked application of SAXS is its imaging capabilities. Successful position resolved SAXS experiments have been reported since the mid-1990s [51]. More recent developments enabled high-resolution SAXS-based computed tomography which provides a level of structural details superior to absorption-based CT and covering a size range beyond the limits of electron microscopy.

Light Scattering   The same experiment as for SAXS can be performed by light scattering. However, for the visible light wavelength, the particles that can be successfully analyzed by static light scattering are of course much larger (>100 nm). Figure 3.22 shows the effect of adsorbed nanoparticles on *Synechocystis* bacteria. We see that once a specific coverage of the bacteria by CeO_2 nanoparticles is reached, the bacteria rapidly aggregate.

To observe nanoparticles in suspension, a different approach has to be performed. The Brownian motion of nanoparticles can be used to measure their radius by dynamic light scattering (DLS). When a monochromatic light source interacts with one nanoparticle whose size is small compared to the light wavelength, the beam is scattered in all directions. When two nanoparticles are scattering the same incident beam, the total scattering detected in one point of space also depends on the interference between the scattered light of the two nanoparticles and therefore of their respective distance. For an ensemble of nanoparticles, the scattered intensity observed on a screen shows an ensemble of bright and dark points, which are commonly referred to as speckles. If the nanoparticles are immobile in the solution, the speckles are fixed on the screen. However, as a result of Brownian motion, the intensity on the screen fluctuates due to variations of the interference part of the scattered intensity. These fluctuations are directly related to the diffusion coefficient of the nanoparticles.

The signal is analyzed by a correlator to obtain the intensity autocorrelation function $C(\tau) = \langle I(t)I(t+\tau)\rangle$. When $\tau = 0$, $C(\tau)$ is simply $\langle I(t)^2\rangle$, but when τ is sufficiently large, $I(t)$ and $I(t+\tau)$ are completely uncorrelated and $C(\tau) = \langle I(t)\rangle^2$. The characteristic time required for the autocorrelation function to go

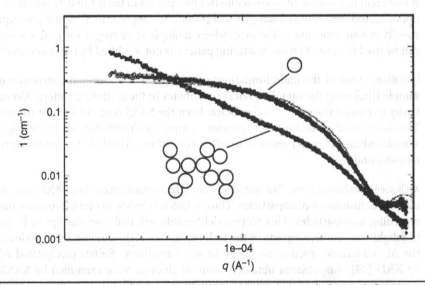

FIGURE 3.22   Small angle light scattering measurements of *Synechocystis* bacteria before and after contact with 50 ppm CeO_2 nanoparticles. The bacteria in contact with the nanoparticles form aggregates. (Courtesy of O. Zeyons.)

from $\langle I(t)^2 \rangle$ to $\langle I(t) \rangle^2$ is related to the diffusion coefficient of the nanoparticles. Therefore, for relatively dilute suspensions of noninteracting and monodisperse nanoparticles the following relationship is true:

$$C(\tau) \cong e^{-\Gamma \tau} \qquad (3.17)$$

where $\Gamma = Dq^2$ and D is the Stokes-Einstein diffusion coefficient $D = kT/6\pi\eta R$. For a polydisperse nanoparticle suspension, the autocorrelation function is then:

$$C(\tau) = \int P(D) e^{-Dq^2\tau} dD \qquad (3.18)$$

where $P(D)$ is the normalized intensity weighted diffusion coefficient distribution function. Several algorithms exist to extract $P(D)$ from the correlation function. So, in principle, information about the particle size distribution in a polydisperse suspension can be obtained. For concentrated and/or interacting nanoparticles, the diffusion coefficient is modified due to the hydrodynamic and thermodynamic (attraction and repulsion) interactions between the neighboring nanoparticles. The motion of the nanoparticles is then strongly coupled and the correlation function is no longer a simple exponential law. For typical particle size measurements, the suspension is diluted such that interparticle interactions are negligible. Therefore, DLS is not suited for concentrated samples or complex mixtures where the particles are likely to be interacting with one another.

Raman Spectra The Raman spectra for nanoparticles is modified, compared to that measured for larger particles, as a result of phonon confinement [52]. In the model developed to analyze the modification of the Raman peaks for nanocrystalline materials, the nanoparticles are considered as an intermediate case between a perfect infinite crystal and an amorphous material. The development of this model indicates that the Raman line of a perfect crystal is modified for nanoparticles by producing asymmetric broadening and peak shifts. For example, for TiO_2, the Raman peak at 142 cm^{-1} measured for large crystals shifts to 146 cm^{-1} for 8 nm particles and to 148 cm^{-1} for 5 nm particles. Simultaneously, the Full Width at Half Maximum (FWHM) increases from 10 to 18 cm^{-1} [53]. In separate experiments, Choi et al. observed a similar effect when examining TiO_2 anatase nanoparticles on the peak at 142 cm^{-1} [13].

XAS XAS can be used to characterize the size and shape of metallic nanoparticles. The parameter primarily reflecting the size and shape of metal particles is the average coordination number. This effect is dependent on the size and shape of the metal cluster. However, the estimation of particle geometry relies on an accurate description of the relation between particle size/shape and the average coordination number [54, 55]. In the case of face-centered cubic (fcc) metal, it has been shown that the coordination number for the first and second coordination spheres are sensitive to the size of the cluster up to 20 nm (Figure 3.23). The average coordination number of the third coordination sphere (N3) is more sensitive to the size of larger objects, but determining accurately the third coordination sphere requires a scan of EXAFS spectra at a high k value (up to 20 Å–1) [56], which is not always possible.

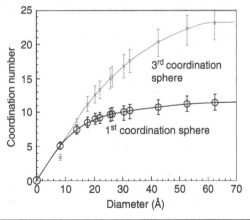

FIGURE 3.23 Average coordination of the first and third coordination spheres as a function of the size of the fcc particles [55].

Because metal oxides display a multitude of different crystal structures, no general correlation between EXAFS coordination numbers and average particle size/morphology has been published [57]. In fact, information can only be obtained for specific minerals [e.g., lepidocrocite (FeOOH)] [58]. XANES can in some cases be used to determine the nanoparticle size [59]. Indeed, the LIII white line for 5d metals (Pt...) is at the center of electronic charge transfer between metals that are present inside the cluster. Bazin showed that a strong correlation exists between the intensity of the white line and the size of the Pt cluster [61]. But the quantitative correlation between size and XANES shape remains difficult. EXAFS has been successfully applied to determine the size and structure of polycations during metal hydrolysis and the formation of gels or nanoparticles. Additional examples include iron [60–63], chromium [64, 65], gallium [66], titanium [67], and zirconium [68–72]. In the case of iron chloride, for instance, the very first steps of iron octahedra polymerization due to a pH increase (i.e., evolution of the hydrolysis ratio = [OH]/[Fe]) were determined. In fact, the Fe-Fe interatomic distances can be associated to iron octahedra linkages (face, edge, double corner, single corner sharing) as demonstrated by Manceau et al. (1988) [62]. Moreover, the number of neighbors at a given distance is associated to the length of iron octahedra polymers. Theoretical values for N for a number of different iron clusters are reported in Figure 3.24.

The use of XAS at the Fe K-edge can be used to characterize the different early polymerization steps of a metal salt (Figure 3.25). Using this tool it is possible to visualize the arrangement of iron clusters as they undergo hydrolysis.

3D Distribution of Nanomaterials within Complex Media (Bio and Nonbio Media)

X-Ray Computed Tomography   Detection and spatial distribution of nanoparticles, their aggregates and agglomerates (NOAA) in complex media such as natural samples or manufactured materials, biological media, or characterization of NOAA behavior in nanostructured material are still challenging. Microscopy techniques provide two-dimensional (2D) images of size and shape of nanoparticles at extremely high resolution can be coupled with chemical information, but are limited to regions very close to the sample surface. Besides, sample preparation for 2D observation (embedding, sectioning, and coating) may lead to artifacts. Internal images or reconstructed 3D images can be obtained but at the cost of destroying the sample (by using micromachining techniques, such as focused ion beams). Thus, a 3D nondestructive (noninvasive) and *in situ* investigation from which many 2D sections can be extracted is very interesting.

X-ray tomography (also frequently referred to as computed tomography, CT) is a radiographic imaging technique that provides 3D images on material's internal structure. The sample preparation is typically minimal, and for many materials this technique is nondestructive. The principle is based on the 3D computed reconstruction of a sample from 2D projections acquired at different angles around its axis of rotation [71].

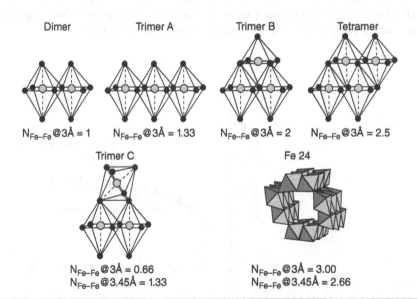

Dimer   Trimer A   Trimer B   Tetramer

$N_{Fe-Fe}@3Å = 1$ $N_{Fe-Fe}@3Å = 1.33$ $N_{Fe-Fe}@3Å = 2$ $N_{Fe-Fe}@3Å = 2.5$

Trimer C   Fe 24

$N_{Fe-Fe}@3Å = 0.66$ $N_{Fe-Fe}@3Å = 3.00$
$N_{Fe-Fe}@3.45Å = 1.33$ $N_{Fe-Fe}@3.45Å = 2.66$

FIGURE 3.24   Examples of 3D Fe clusters and their corresponding coordination numbers (N_{Fe-Fe} @3 Å = number of edge sharing, N_{Fe-Fe} @3.45Å = number of double corner sharing).

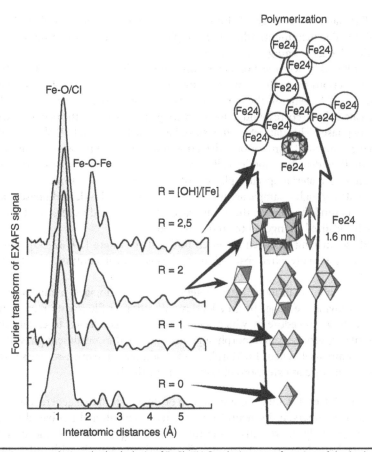

FIGURE 3.25 Nucleation steps during the hydrolysis of $FeCl_3.6H_2O$ solutions as a function of the hydrolysis ratio R.

Most early applications of CT (in the 1970s) were for medical imaging (visualization of the human body and brain), but the benefits of true 3D visualization of internal structure led to rapid adaptation of the technique in other fields. Applications in other wide variety of research domains such as materials sciences (wood technology; shape characterization of particles and pore space in powders, foams, and other porous materials; porosity and pore morphology in cement or mortar [73]; fracture and fatigue crack propagation of materials), paleontology and archeology [74], geosciences in general (soil science [75], marine science, structural geology [76]) and biology (animal morphology, structural study of bones and teeth [77], cellular and subcellular structure), as well as industrial applications (food engineering, electronic) followed shortly. Reviews have been published on specific CT applications in materials sciences [78–81], geosciences and geology [82–84], and biology [85, 86].

Operating Principle   The word *tomography* derives from the Greek "tomos," to slice or to cut, and "graph" an image or representation. The basis of X-ray tomography is X-ray radiography: an X-ray beam is sent on a sample and the transmitted beam is recorded on a 2D detector.

X-ray attenuation follows the Beer-Lambert law:

$$I = I_0 e^{-\mu x} \tag{3.19}$$

where I is the attenuated intensity after the X-rays have passed through an object of thickness x, I_0 is the incident radiation intensity, and μ is the linear attenuation coefficient (also known as LAC). The attenuation coefficient μ depends upon the composition and density of the sample, and also the incident beam energy.

$$I = I_0 exp\left[\sum_i (-\mu_i x_i)\right] \tag{3.20}$$

Linear attenuation coefficients [Equation (3.20)] for a large number of elements and composites and for a wide range of energies can be accessed via various online resources, for instance the National Institute of Standards database (http://www.nist.gov/pml/data/xraycoef/).

The classical way to get 3D information is to collect a large number of 2D radiographs (projections) at fixed angular increments while the sample is either rotated continuously or step scanned with a total rotational angle ranging from 180° to 360°.

A filtered back-projection algorithm can then be used to reconstruct the volume of the sample from this projections stack, that is, mathematically back-calculated the full 3D distribution of linear attenuation coefficients. In general, the darker voxels (isotropic volume element or 3D pixel) in the reconstructed grayscale volume correspond to low-density phases, whereas the brighter voxels are attributed to the high-density phases. This dataset is often stored as a stack of virtual 2D slices or 3D volume rendering for quick viewing. This most conventional detection mode is called the absorption mode.

Another detection mode, called phase contrast mode is based on the fact that the photon wave experiences material-specific phase shifts as it propagates through a sample; X-ray wave fronts that propagate through a sample are modulated in both amplitude and phase due to refraction and absorption effects. Behind the sample, the X-ray wave propagates and develops interferences that emphasize regions of high-localized changes in the refractive index.

Phase imaging has been reviewed in detail elsewhere [79, 87–90]. This contrast is efficient for edge detection especially when absorption only leads to weak contrast, for example, in soft tissues [87, 91] and fossils [92].

Extracting quantitative parameters from 3D images requires appropriate image processing (i.e., improving the image by grey-level modification or filtering to remove noise or subtract background), thresholding (procedure by which a grayscale range is affected to certain discrete material), and analysis (quantitative geometry and morphology measures of the constituent phases). The development of many 3D tools (commercial: Avizo FEI, Morpho+, VGStudiomax, and open-source software as ImajeJ, Pore3D, Imorph, etc.) has emerged as extensions of existing 2D methods.

Experimental Setup   CT imaging systems can be divided into two general classes based on their X-ray sources: lab-based scanners and synchrotron systems. Conventional lab-based CT system employed X-ray tube source, generally microfocus X-ray tube producing a divergent and polychromatic beam. CT requires energetic X-ray beam passing through dense and large sample, with sufficient counting rate on the detector. The beam energy is then optimized with the application of elevated accelerating voltage [usually from 40 to more than 100 kV (e.g., 180 kV for the GE Nanotom)]. During a CT scan, the sample rotates in between a stationary X-ray source and an X-ray–sensitive detector or a scintillator screen that converts the X-rays into visible light, which is then recorded by a 2D detector. The detector [usually charge-coupled device (CCD) camera] has typically 1024×1024 pixels. Scan duration varies from 30 min at low resolution to several days at very high resolution. In this setup, the conical X-ray beam makes geometrical magnification (M) possible: geometrical magnification is a function of the source-to-object and object-to-detector distances (r_{so} and r_{od}, respectively).

$$M = \frac{r_{so} + r_{od}}{r_{so}} \tag{3.21}$$

Total spatial resolution d_{total}, is function of detector pixel size r_D, X-ray spot size S, geometric magnification M, and the source-to-object and object-to-detector distances r_{so} and r_{od}, respectively.

$$d_{total} = \frac{\sqrt{r_D^2 + \left(S \frac{r_{od}}{r_{so}}\right)^2}}{M} \tag{3.22}$$

It should be noted that reconstructed voxel size is often confused with spatial resolution.

In order to maximize spatial resolution, the conventional micro-CT approach relies on maximizing geometric magnification. As a result, high-resolution analysis of larger sample is not possible with this system without taking a small subsample. Image resolution achievable with conventional lab-based system ranges from millimeter to micrometer scales. To overcome this problem, conventional systems coupled with optics to enhance resolution while preserving a comfortable working distance were developed. Synchrotron sources offer high-brilliance, collimated (almost parallel), and monochromatic beam. This allows the use of X-ray optical elements to achieve very high resolution for 3D nanoscale imaging (<100 nm; e.g., 30 nm at the Anatomix beamline, SOLEIL, France). X-ray optical elements used at synchrotron beamlines dedicated to CT [e.g., Fresnel zone plate (FZP), or Kirkpatrick-Baez optics (KB)] are detailed in the literature [80, 82, 93, 94]. Recent lab-based systems have benefited from synchrotron optical developments and are also now capable to offer nanometer resolution [95].

Sample Preparation   One of the main advantages of CT is the nondestructive nature of the technique. The sample is simply mounted on a high-resolution sample rotary stage and centered at the sample stage rotation axis. Sample is, for example, stuck on a needle or placed in a capillary. The first requirement concerns sample size. Indeed, X-rays have the ability to penetrate opaque objects but with some limits, mainly function of their X-ray energy, that is, accelerating voltage of the X-ray source. X-rays should pass through the sample with a sufficient counting rate on the detector. X-ray source with low energy is only dedicated to analyze small sample and/or with low density. It can be necessary (especially at high resolution) to take a small subsample. In addition to technical challenges to obtain very small sample, the representativeness of the sample may be limited by this operation. Representativeness of analyzed volume [called volume of interest (VOI)] is also a critical point. As the VOI should fit in the surface of the detector, a compromise must be found between the maximum VOI size and the spatial resolution. For example, FOV are in the millimeter range for CT at micrometer scale. The second main requirement is the high sample stability. Indeed tomography reconstruction requires a rigid sample, which does not move during the CT scan, particularly when going to very high spatial resolution. Sample drift or deformation should be avoided. Special care is needed for samples than can dry, that is, move during the scan. These samples should be subjected to a drying or resin-embedding step prior to analysis. Specifically, biological samples should be pretreated to fix all of the structural constituents (by chemical fixation and/or "transparent" resin embedding). Resin embedding rigidifies the sample structure at the nanometer scale but can introduce contrast loss due to resin X-ray attenuation [84]. This can be considered as destruction or modification of the original sample. Moreover, biological soft tissues exhibit low X-ray attenuation contrast. X-ray visualization of their microstructure requires staining by X-ray contrast enhancement agents containing high-atomic-number (high-Z) elements [86, 96, 97]. Main element probes that have been reported for visualizing biological microstructures by CT include osmium, gold, silver, iodine, platinum, mercury, tungsten, and lead.

Another "staining-free" method is critical point drying, which has been widely used for preparing electron microscopy samples. In this method, volatile component such as water contained in the sample are replaced with air voids, improving X-ray attenuation contrast of soft tissues. The critical-point-drying procedure reduces damaging dehydration effects by water or organic solvents but can nevertheless introduce some structural distortions [98].

Application in the Particular Case of Nanoparticles   As high spatial submicron resolution can now be reached with both lab-based systems and synchrotron beamlines, CT became a valuable technique for the 3D detection and localization of NOAA in manufactured materials (cement, paint, polymers) and in complex media (soils, plants, organisms, soft tissues). (See figure 3.26.) In the field of "safer-by-design" production or eco-conception, the accurate distribution of NOAA and their residues in these complex matrixes is of particular interest to evaluate their fate in the environment at each manufactured product life-cycle step, to understand the transfer mechanisms between the different environmental compartments, and to assess their potential toxicity. Mielke et al. studied the uptake and accumulation of TiO_2 nanoparticles (nano-TiO_2 with average

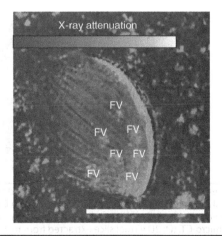

FIGURE 3.26   3D reconstructed image of a *T. thermophila* cell showing the internalization of nano-TiO_2 (dense voxels) in some food vacuoles (referred as FV). Scale bar is 25 µm. (Adapted from [99]. See also the movie http://www.bren.ucsb .edu/~holden/AEM_SI_2013_movie/UXRM029_UCSB_Protozoa_TiO2NP_VRT.mpg)

diameter of 37.5 nm) by *Tetrahymena thermophila* protozoan either by direct exposure of a TiO_2-rich growth medium or by ingestion of TiO_2-encrusted bacteria [99]. Using a Zeiss Xradia UltraXRM-L200 X-ray microscope in phase contrast mode, they performed *in situ* localization of nano-TiO_2 inside food vacuoles of *T. thermophila* cells with a 150 nm spatial resolution.

In a recent study, van den Brule et al. evaluated the respiratory biopersistence and the lung toxicity of high aspect ratio aluminogermanate tubular nanomaterials (called Ge-imogolites) with a double-wall (DW) structure and a mean length of 62 ± 19 nm, 60 days after intratracheal instillation in rats (1 mg Ge-imogolite per rat) [100]. Thanks to a multitechnique approach, combining 2D chemical and 3D imaging and local structural analysis, Ge-imogolites with intact local atomic structure were localized in dense fibrotic lung zones. Paraffin-embedded DW-treated lung cross sections were prepared and investigated at each step of the technical approach without any sample modification or destruction. Complementary information have been collected: Ge-enriched zones within treated lungs were located by micro-X-ray fluorescence (micro-XRF) and X-ray absorption spectroscopy (XAS) at Ge K-edge revealed that Ge detected in the treated lung exhibited an intact local atomic structure and Ge-rich lung areas were 3D scanned by micro-CT (using a Zeiss XRadia MicroXCT-400 X-ray microscope with a spatial resolution of 4.24 µm) to identify at high resolution the presence of Ge-imogolites (i.e., brilliant zones with high X-ray attenuation) in dense fibrotic alveolar zones (Figure 3.27).

CT performed at high spatial resolution is also a powerful tool to obtain the 3D morphology of a nanostructured material and to understand structural mechanisms controlling NOAA transfer. For example, Bossa et al. studied the weathering of cement incorporating TiO_2 nanoparticles (nano-TiO_2) for self-cleaning properties and the associated nano-TiO_2 released into the environment [74]. They performed leaching tests on cement to quantify the nano-TiO_2 particulate release and it's kinetic. Then they combined micro- and nano-CT analysis to evaluate the role of cement structural parameters (porosity and porous network morphology as pores connectivity, pores size distribution, etc.) on the nano-TiO_2 release and to identify the mechanisms controlling it. Thanks to the very high resolution achieved with the Zeiss Xradia UltraXRM-L200 system in absorption contrast mode, they were able to characterize in 3D the cement porous network with pores size down to 130 nm and highlighted a very thin altered layer at the cement surface with an increased porosity of about 70 percent. Image analyze of CT-reconstructed data with specifically developed algorithms allowed to determine the size of the throat between connected pores open to the surface. Results revealed that a throat size of 800 nm is the main structural parameter controlling nano-TiO_2 release in the very thin altered layer at cement surface.

Limitations in the Particular Case of Nanoparticle Localization

Spatial Resolution Detection and localization of NOAA in complex media and manufactured materials require very high spatial resolution. Indeed all features smaller than the voxel size cannot be easily distinguished on the reconstructed dataset. Even if the linear attenuation coefficient (μ) of nanoparticles exhibits a large increase compared to the μ of sample media, NOAA should reach a sufficient voxel occupancy value to give a sufficient contrast in the voxel value. It should be noted that the spatial resolution of CT cannot compete with the highest resolution of electronic microscopy.

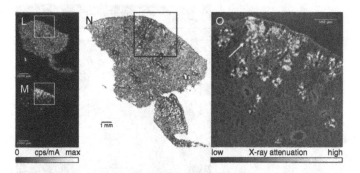

FIGURE 3.27 Paraffin-embedded lung cross section of a rat exposed to Ge-imogolites. Micro-XRF elemental maps (pixel size of 104 µm) showing the distribution of S (L) and Ge (M) in DW-treated lung. (N) Scan of lung upper layer section stained with hematoxylin and eosin revealing inflammation. Squares (white or black) delimitate Ge-rich areas selected for XAS and micro-CT. (O) 2D virtual slice extracted from reconstructed micro-CT volume (voxel size of 4.24 µm). White arrows indicate brilliant areas with high X-ray attenuation attributed to the presence of Ge-imogolites. (Modified from [100].)

Chemical Distinction   The output of CT is a reconstructed 3D volume of local X-ray attenuation coefficients, which depend on both material composition and density. CT provides no direct elemental identification. In the case of single-phase material or material with number-limited phases exhibiting distinct chemistry and X-ray absorption contrast, CT provides indirect phase identification. The use of standard materials for X-ray attenuation calibration can also help for chemical distinction.

In the particular case of NOAA localization in complex and heterogeneous media (in term of chemical composition), 3D CT data generally requires coupling with 2D chemical or mineralogical mapping (e.g., micro-XRF, micro-XRD, laser-ablation–coupled ICP-MS). 3D imaging is often best considered as a part of a multiscale and multitechnique imaging strategy (this has been termed correlative tomography).

Preparation of soft tissues without staining is also crucial, as contrast-enhancement agents could mask the distribution of NOAA, especially if the linear attenuation coefficients (μ) of the contrast agents and of the nanoparticles are similar. Critical point drying of the sample represents then a good alternative to staining.

Operator Dependency, Quantitative Analysis   3D image acquisition and analysis have a strong operator-dependency. Indeed a large amount of acquisition parameters can be chosen arbitrarily all influencing the final result and visual inspection is a great qualitative tool, but strongly operator dependent. Most of the time qualitative analysis is not sufficient and quantitative results from 3D reconstructed volume are required.

In practice, the quantitative 3D analysis is often not straightforward and can be prone to several systematic errors caused by imaging artifacts (e.g., beam hardening, ring artifacts, edge-enhancement effect). The most critical operation (i.e., extremely operator dependent) in this analysis is the segmentation of the volume, which is usually based on grey-value thresholding and on analysis of the voxel intensity histogram (plot of the frequency or number of occurrences of voxels at a particular intensity/gray value). The overlapping distribution is a potential problem with intensity-based thresholding. NOAA segmentation can then be critical.

Figure 3.28 shows the reconstructed volume of a wood piece covered by a paint layer incorporating nanoparticles of CeO_2 (nano-CeO_2), and a 2D virtual slice extracted from this volume. Histogram of the total volume exhibits no distinct peaks and reveals the presence of several phases with poor X-ray absorption contrast (i.e., overlapping distribution). The histograms of selected subvolumes (white frames) help to identify the contribution of each phase (air, wood, and paint incorporating nano-CeO_2) in the total histogram. Nano-CeO_2 identification and thresholding were performed based on comparison with histogram of control paint (with no nanoparticles incorporation).

Surface Physicochemical Properties

Physical and chemical properties such as particle charge, size, chemical composition, and surface functionality are often linked. These characteristics may be modified, intentionally or unintentionally, through adsorption of species on the particle surface.

Long-Period X-Ray Standing Waves (LP-XSW)   As an element-sensitive high-resolution structural probe, the XSW technique has been used in the past to characterize a wide range of samples including surface, interface, and thin film structures. This technique has also been used for environmental studies. In particular, it has been applied to determine the distribution and speciation of metal (loid) ions, such as Zn, Pb, and Se, at the interface between single-crystals of α-Al_2O_3 and α-Fe_2O_3 and aqueous solutions or microbial biofilms [101–104]. More recently, this approach has been used to characterize the corrosion of Ag nanoparticles (Ag-NPs) at a PAA/α-Al_2O_3 interface [105].

Operating Principle   XSW methods rely on the ability to generate a coherent superposition of two beams, resulting in the formation of a standing wave with well-defined intensity field [106–108]. There are a number of mechanisms for generating an XSW. The simplest is by reflection, in which case the superposition of the reflected and incident X-ray plane waves gives rise to the standing wave. This two-beam reflection condition can be produced by (1) strong Bragg diffraction from a single crystal or from a periodically layered synthetic microstructure, (2) weak kinematical Bragg diffraction from a single-crystal thin film, or (3) total external reflection from an X-ray mirror [109]. The latter setup will be discussed in more detail in the following since it has been mainly used for environmental studies.

In the total external reflection mode (grazing incidence geometry), long-period X-ray standing waves are generated from interference of the incident and reflected beams when the incidence angle is near the

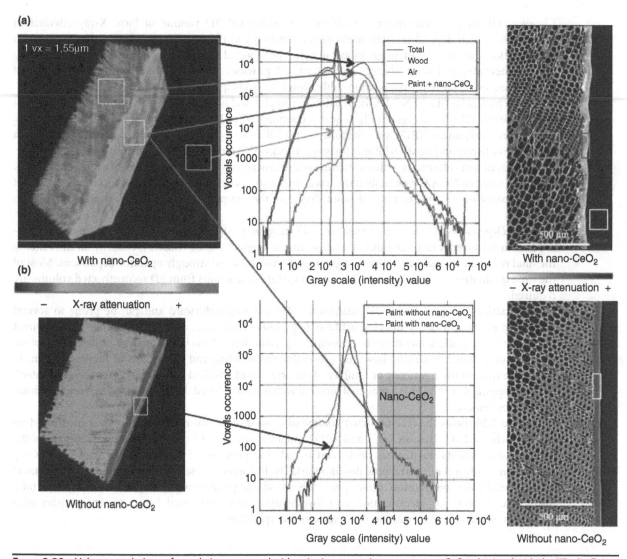

Figure 3.28 Volume renderings of wood pieces covered with paint incorporating or not nano-CeO$_2$, obtained with the XRadia Zeiss MicroXCT400 microscope (voxel size of 1.55 μm). 2D virtual slices extracted from the volume are also shown. Histograms of total volume and of selected subvolumes (white frames corresponding to air, wood, and paint with or without nano-CeO$_2$ areas) are plotted and compared for nano-CeO$_2$ identification and thresholding. (Modified from unpublished CEREGE data, in preparation.)

critical angle for total external reflection of the substrate [104]. The intensity distribution of the standing wave generated by total external reflection is perpendicular to the mirror surface. For the case of a mirror surface in contact with air or vacuum, the period varies with the incident angle α_i as, $D = \lambda/[2\sin(\theta)]$. As the incident angle is increased (Figure 3.29) the standing wave period compresses and the first antinode sweeps through the region above the mirror surface. At the critical angle, the first antinode is coincident with the surface plane and the standing wave period reaches the critical period, $D_c = \lambda/2\theta_c$ [106].

The fluorescence yield (FY) of an element residing above the substrate varies as the incident angle changes and the standing wave antinodes compress and move toward the reflecting surface [110, 111]. Furthermore, fluorescence shows a variation in intensity with angle due to the change in the position of the nodes and antinodes of the XSW. To model the florescent yield of an element that may be distributed throughout the interface region it is necessary to determine the structure of the electric field within the sample of interest as a function of incident angle. The XSW-FY data can be fitted using a general scattering model, which allows quantification of the vertical distribution of specific elements. The FY data can be modeled using the following equation [110, 112, 113]:

$$FY(\alpha_i) = \int I(z, \alpha_i) N(z) exp[-(zL_a)]dz \qquad (3.23)$$

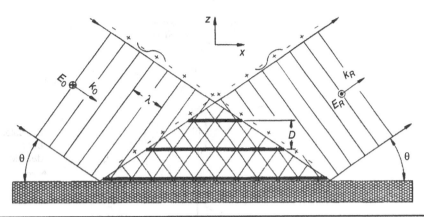

FIGURE 3.29 Illustration of the x-ray standing-wave field formed by the interference between the incident and specular-reflected plane waves above a mirror surface [106].

where $N(z)$ is the depth profile of the fluorescing element, L_a is the attenuation length of the outgoing fluorescence, and $I(z, \alpha_i)$ is the electric field intensity at depth z and angle α_i. The complete electric field intensity profile $I(z, \alpha_i)$ above and below the surface is calculated using Parratt's recursive formalism with the parameters obtained from the reflectivity fits [114–116]. A simulated fluorescent yield profile for a metal distribution in a mineral-biofilm multilayer system is shown on Figure 3.30 as an example [108].

Same principle applies for the diffraction signal of a particle residing above the surface. It is therefore possible to collect a XRD-XSW spectra at a given 2θ, characteristic of a diffraction peak of the probed crystal to get its distribution profile.

Experimental Setup Such technique requires a coherent beam with high brilliance such as synchrotron beam with specific optics. A coherent superposition of waves at all points within the sample of interest is required to generate an XSW. The loss of coherence results in the lack of interference of the field amplitudes. The experimental setup include a goniometer capable of aligning the sample such that the surface normal lays in the scattering plane defined by the incident and reflected wave vectors. It is necessary to be able to adjust the incident angle (α_i) from roughly zero to a few degrees with angular resolution on the order of 0.001°. A fluorescence detector is required to collect the fluorescence X-rays as well as an ion chamber mounted on a detector arm used to monitor the reflected beam for determination of the net reflectivity signal (Figure 3.31).

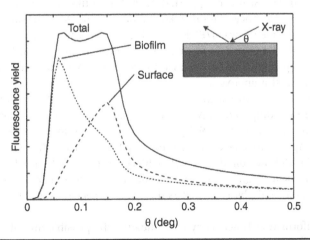

FIGURE 3.30 Simulated fluorescent yield profile for a metal distribution in a mineral-biofilm composite. The surface contribution is from metals absorbed on the mineral surface, the biofilm contribution is calculated assuming a constant concentration throughout a 1-μm-thick biofilm overlayer [108].

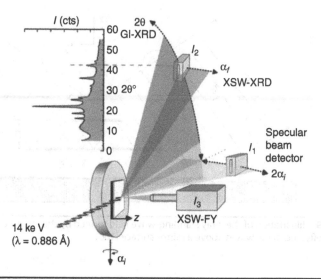

FIGURE 3.31   Typical experimental setup for the fluorescence yield and diffraction-based measurements as well as specular reflectivity using synchrotron X-rays.

The reflectivity is measured as a function of α_i and $2\alpha_i$ (°) using an ion chamber (I_1); the grazing-incidence XRD and XRD-based X-ray standing-wave profiles are measured as a function of 2θ (°) and α_f(°), respectively, using a 2D array detector (I_2). Fluorescence yield is measured orthogonal to the sample using a high-resolution fluorescence detector (I_3) [105].

Limitations   Sample preparation is probably one of the most challenging tasks as the surface roughness should be low (typically <5 nm) for the XSW to be generated at the surface of the substrate. Additionally, the sample has to be homogeneous as the projection of the beam on the sample is large in the grazing incidence geometry.

Application in the Particular Case of Nanoparticles   Recently it has been shown that the use of LP-XSW, usually dedicated to the characterization of oriented surfaces, can be successfully applied to study the reactivity of nanoparticles in contact with a complex system [105]. Corrosion processes of silver nanoparticles (Ag-NPs) has been studied in aqueous conditions and in the presence of organic matter and halide species common to many natural environments. It is of particular importance because the release of toxic Ag^+ from oxidation/dissolution of Ag-NPs may strongly impact ecosystems [117, 118]. In this context, Ag-NPs capped with polyvinolpyrrolidone (PVP) in contact with polyacrylic acid (PAA) and Cl^- in solution has been investigated. Ag-NPs were first deposited on a silicon wafer via spin-coating forming a monolayer (Figure 3.32b). The Si-wafer-supported Ag-NPs thin film layer was aged during 24 h in a 2% solution of PAA with initial Cl concentration of 0.014 mol L^{-1}. After 24 h, the sample was spin-coated leading to a dry multilayered sample (Si-wafer/PVP-capped Ag-NPs/PAA) (Figure 3.32d).A combination of synchrotron-based X-ray standing-wave fluorescence yield and X-ray diffraction-based experiments on the dry sample. Diffusion and precipitation processes at the Ag-NPs–PAA interface were characterized with a high spatial resolution using this approach.

Fitting the FY profiles using a two-box model (the first box corresponding to the Ag-NPs layer, the second corresponding to the PAA layer) indicates that the highest concentration of Ag (90.3%) is in the Ag-NPs layer. The remaining 9.7 percent of Ag occurs in the PAA film and may correspond to both suspended Ag-NPs that have migrated into the PAA film and dissolved Ag^+ that has diffused into the PAA film after dissolution from the Ag-NPs. Chloride is mainly observed in the Ag-NPs film (87%).

GI-XRD data (Figure 3.33) were collected at selected incidence angles α_i to provide additional information on solid phases as well as the spatial association of different crystalline solids that may have formed.

This information is necessary for evaluating the possible migration of Ag-NPs into the PAA film as well as the formation and spatial distribution of AgCl as a function of incidence angle. As expected, the fitting of the XRD-XSW profiles suggests a strong contribution of Ag^0 (92%) in the Ag-NPs film and a lesser contribution in the PAA film (8%). In contrast, the XRD-XSW profile based on the diffraction

Figure 3.32 Imaging of Ag-NPs thin film with and without PAA coating. (a, b) SEM and AFM (overlays) images of the PVP-capped Ag-NPs film uncoated and (c, d) coated with PAA [105].

maxima for AgCl suggests that this phase is approximately evenly distributed between the PAA film (49%) and the PVP-coated Ag-NPs film (51%). The important proportion of AgCl observed in the PAA film is consistent with the observation of new-formed particles in the PAA observed by SEM analysis [106]. The formation of AgCl during aging of the PVP-capped Ag-NPs thin film implies the following steps occurred: (1) Ag is oxidized, presumably at Ag-NPs surfaces, (2) soluble aqueous Ag^+ species diffuse into the PAA solution, (3) Ag^+ reacts with chloride ions initially present in the PAA solution resulting in the formation of the stable AgCl complex, and (4) the PAA solution becomes saturated with respect to AgCl(s) and precipitation of AgCl(s) occurs. This study demonstrates the complementary nature of XSW-FY, GI-XRD, and XSW-XRD techniques for obtaining both compositional and crystalline phase distribution information with good spatial resolution on a multilayered system. Such approach provides new insights about complex chemical processes involving nanoparticles in natural systems and could also be used in studies of a variety of multilayered systems with nanosized dimensions.

Surface Charge Characterization

Operating Principle   Surface charge is a key parameter controlling the stability of nanoparticle dispersions. In polar solvents, surfaces may have a charge of a specified density σ that can be approximated by a

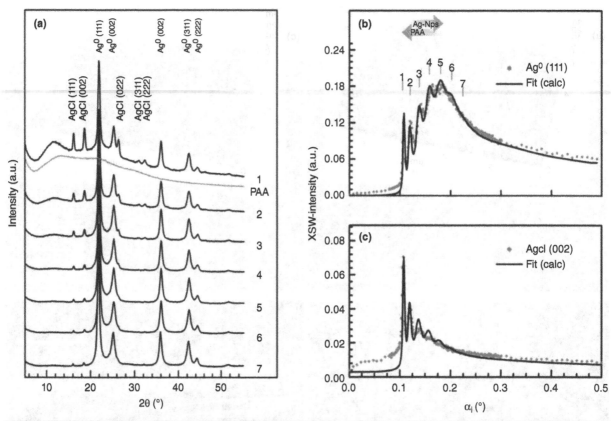

FIGURE 3.33 GI-XRD- and XRD-based XSW for AgCl and Ag°. (a) GI-XRD profiles collected at seven incidence (α_i) angles shown in (b). Diffraction maxima from crystalline phases can be indexed with the known structures of AgCl and metallic Ag. Diffuse scattering from an unreacted PAA is shown for comparison (solid grey). (b, c) XRD-XSW profiles from diffraction features corresponding to Ag-NPs (Ag°) or AgCl. The Ag° and AgCl profiles were obtained from tracking the intensities of the (111) and (002) reflections, respectively, as a function of incidence angle (α_i) [106].

potential $\tilde{\psi}$. Depending on the type of material, the surface charge can vary as a function of pH as is the case for oxide minerals. Sources of pH-dependant charge are the protonation and deprotonation of functional groups, while fixed charges result from crystal lattice defects and atomic substitution. In an aqueous medium, a diffuse electrical double layer will form at the solid-liquid interface as ionic species are attracted to the charged surface. Ions with a same charge compared to the nanoparticle surface are repelled from the surface, while those ions having an opposite charge are attracted to it. This effect decreases from the charged interface with a characteristic length κ^{-1} (the Debye screening length) depending on the solution ionic strength and composition. Various techniques can be used to determine particle surface charge. First the charge can be assessed using a proton as an atomic probe, by titrating a suspension of particles.

The sign of the charge can also be determined indirectly by determining the zeta potential (ξ) of the particles. The zeta potential is the electrical potential that exists at the "shear plane" at some distance from the particle surface. It is derived from measuring the electrophoretic mobility distribution of a dispersion of charged particles as they are subjected to an electric field. The electrophoretic mobility is defined as the velocity of a particle per electric field unit and is measured by applying an electric field to the dispersion of particles and measuring their average velocity. Depending on the concentration of ions in the solution, either the Smoluchowski (for high ionic strengths) or Hueckel (for low ionic strengths) equations are used to calculate the zeta potential from the measured mobilities. For very small particles the mobility is generally determined using laser Doppler velocimetry. It is essential to determine the modifications of the surface charge (or zeta potential) as a function of pH, ionic strength, or presence of surface coating. By doing so, it is possible to determine the point of zero charge (PZC) or isoelectric point (pH_{iep}) for a sample which partly governs the suspension stability.

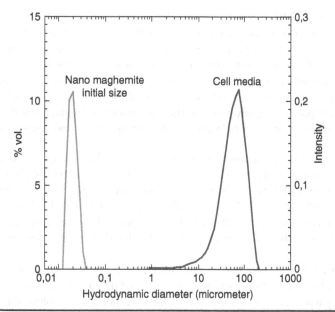

Figure 3.34 Evolution of the size of nanomaghemite nanoparticles in DMEM nutritive medium (pH = 7.4) measured by photon correlation spectroscopy (PCS). (Courtesy of M. Auffan.)

Application in the Particular Case of Nanoparticles For biological tests, knowledge of the particle PZC is crucial as it relates to the stability or propensity to aggregate and/or to more or less rapidly stick with the cell membranes. The importance of this concept may be illustrated by considering cytotoxicology experiments conducted with fibroblasts cells and nanomaghemite. Here, the PZC of the nanomaghemites is around pH 7. Adding 8 nm nanomaghemite to a DMEM nutritive medium results in aggregation of the nanomaghemite, with the particle size approaching several tens of microns (Figure 3.34).

This aggregation is due to an homoaggregation of nanomaterials (Nms) partly due to surface complexation by DMEM molecules which allows to stick Nms. In order to increase the stability of nanoparticles, their surface chemistry is often modified through functionalization or polymer encapsulation to modify their surface charge. In the case of nanomaghemites, the adsorption of DMSA (dimercaptosuccinic acid) at the maghemite surface strongly modified the surface charge leading to a stable dispersion in the DMEM nutritive medium. The DMSA is fixed through the SH group at the maghemite surface. Therefore, the external part of the DMSA-covered maghemite corresponds to the carboxylic group of the DMSA. At pH 7, the carboxylic groups are negatively charged resulting in repulsive electrostatic forces between the various particles (Figure 3.35).

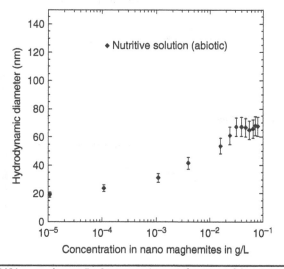

Figure 3.35 Stability of a DMSA-coated nano-Fe_2O_3 suspension as a function of their concentration in an abiotic supplemented DMEM (contact time is 48 h). (Courtesy of Mélanie Auffan.)

Single Particle ICP-MS   Inductively coupled plasma–mass spectrometry (ICP-MS) is a widely used characterization technique for determining the concentration and elemental composition of aqueous samples. The technique's inherent selectivity coupled with excellent sensitivity (low ng L^{-1} for most elements), makes it ideal for the analysis of trace amounts of metals. However, conventional ICP-MS requires that the samples be dissolved in acid prior to analysis. For nanomaterials, information pertaining to the size, shape, and morphology would be lost through the dissolution. To this end, single particle ICP-MS (spICP-MS) has been developed to take advantage of the instrument's selectivity, sensitivity and rapid data acquisition to provide information on particle size, number, composition, and concentration [119–121].

Operating Principles   As opposed to conventional ICP-MS, samples run in single-particle mode are not acidified prior to analysis. In doing so, the particle that enters the plasma is ablated as a whole, forming a cloud of ions that travels through the instrument to the detector. The breadth of the ion cloud spans approximately 500 μs [122]. As a result, the detector in spICP-MS is set to acquire data at a rapid rate (≤10 ms dwell times) such that when this ion cloud is detected, it registers as a pulse above the background (Figure 3.36). A dissolved sample on the other hand would be recorded as a constant signal as shown in Figure 3.36(a).

In order to size these particles, a calibration curve is constructed using a set of dissolved standards. This dissolved calibration curve is then converted into a mass flux curve in order to relate the intensity of the nanoparticle event to the intensity of its corresponding dissolved standard [Equation (3.24)] [123]:

$$W = t_{dwell} \times C_{STD} \times \eta_{eff} \times Q_{flow} \tag{3.24}$$

where W = mass flux (μg L^{-1})
t_{dwell} = dwell time (milliseconds)
C_{STD} = concentration of the dissolved standard (μg L^{-1})
η_{eff} = nebulization efficiency (unitless)
Q_{flow} = flow rate (mL ms^{-1})

The pulse height (intensity) of the nanoparticle event is converted into a mass using the slope of a dissolved calibration standard curve [Equation (3.25)]:

$$m_{NP} = \frac{1}{f_a} \times \frac{(I_{NP} - I_{BKGD}) - b}{m} \tag{3.25}$$

where m_{NP} = mass of nanoparticle (g)
f_a = mass fraction of particle

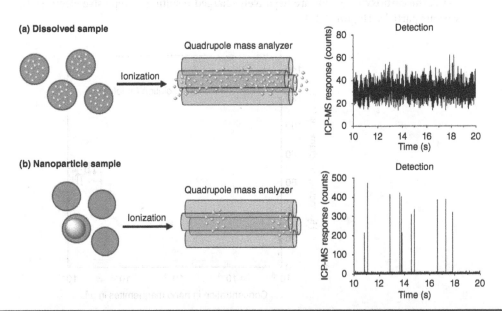

FIGURE 3.36   (a) Analysis of dissolved Au standards. (b) Analysis of 100 nm Au nanoparticles. Each pulse represents a single nanoparticle event.

I_{NP} = Intensity of the nanoparticle (counts)

I_{BKGD} = Intensity of the background (counts)

m = slope of the mass flux curve

b = intercept of the mass flux curve

From this mass, if a spherical geometry is assumed, the diameter can be calculated [Equation (3.26)]:

$$D_{NP} = \sqrt[3]{\frac{m_{NP} \times 6}{\rho_{NP} \times \pi}}$$

(3.26)

where D_{NP} = diameter of nanoparticle (nm)

ρ_{NP} = density of nanoparticle (g cm^{-3})

m_{NP} = mass of nanoparticle (g)

A typical workflow for spICP-MS is represented in Figure 3.37. An important step of the process is determining which signal is background and which results from a nanoparticle event. The intensity cutoff between nanoparticles is frequently determined by establishing an intensity threshold as the average background intensity added to three times its standard deviation ($\bar{x}_{bkgd} + 3\sigma_{bkgd}$). Pulses that exceed this threshold are considered nanoparticles and are removed from the dataset for further processing. The remaining data points are subjected to the same process iteratively until the intensity threshold converges upon a single value. The size detection limit (the smallest detectable particle) by this technique is generally considered to be the intensity at the threshold [124].

Another important parameter to consider is the transport efficiency [η_{eff} in Equation (3.24)]. In ICP-MS analysis, the aqueous sample is aspirated into the spray chamber and subsequently the plasma via a nebulizer. As a result of this design, many of the droplets that are aspirated do not reach the plasma and instead go to waste. Subsequently, ICP-MS only detects a fraction of the mass of a dissolved sample. In order to correct this, the nebulization efficiency has to be determined to accurately compare that signal from dissolved samples to the intensity of the nanoparticle event. To determine the transport efficiency value, several methods can be employed: waste collection, particle number concentration, and using a standard size particle [123, 125]. One of the most straightforward methods is to collect the waste generated from the analysis of the sample. By knowing the flow rate of the sample-introduction system, the fraction of sample that was aspirated into the torch can be determined. Transport efficiency can be calculated by analyzing a solution of known particle number concentration. This can usually be achieved using a particle of known size and mass concentration. The number of nanoparticle events detected is compared to the number of nanoparticle events expected to calculate the nebulization transport efficiency. Lastly, the nebulization efficiency can be determined by comparing the signal generated from a dissolved standard against the signal generated from a nanoparticle event. As nanoparticles of a known diameter contain the entirety of their mass in the generated signal, this can be compared to the signal from a dissolved sample that only is being generated from the fraction of droplets that are ionized in the plasma. This difference allows for the determination of the nebulizer transport efficiency.

Sample Preparation   Sample preparation for spICP-MS is relatively straightforward compared to other techniques, and many samples can be analyzed without introducing any potential preparation artifacts. In preparing samples for analysis, it is important to consider both the aggregation state of the sample

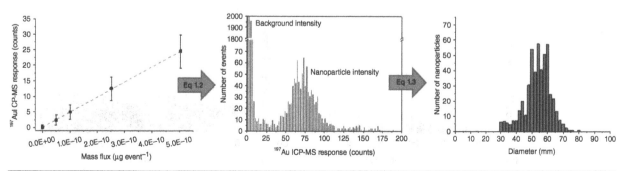

FIGURE 3.37   Workflow of spICP-MS. The mass flux curve is used to convert the nanoparticle intensity into a mass and subsequently a diameter (Montaño, unpublished data.).

as well as the particle number concentration. To reduce the potential for aggregation, many samples are dispersed prior to analysis, using either a sonic bath or a sonicator probe. It is also important to consider the particle number concentration. If a sample is too concentrated with particles, more than one particle may reach the plasma at once resulting in "coincidence" [123]. The close proximity of these particles results in the detector registering a single particle event which can lead to an overestimation of the mass (size) of the particle; as well as, a smaller particle number concentration. This problem can be alleviated by either diluting the sample further, or in some cases using faster dwell times [126–129]. In samples where there is a high background of dissolved analyte that interferes with the detection of the nanomaterial, it may be necessary to reduce or remove the dissolved analyte prior to analysis. This may be achieved by centrifuging the particle and decanting the supernatant. Suspending the particles in a new media will result in a reduction in the dissolved background signal. Reducing the background intensity has also been achieved through ion chromatography. By coupling ion chromatography to ICP-MS, the background intensity can be significantly reduced to fully resolve the nanoparticle intensity [129].

Application in the Case of Nanoparticles   Single particle ICP-MS was initially developed with the intent to analyze atmospheric particulate matter [130–133]. In the past decade, this technique has been used to analyze a variety of nanomaterials with varying compositions and morphologies in complex environmental and biological matrices. Metallic nanoparticles (e.g., gold and silver) have been studied extensively in a number of environmental systems [119, 134]. Using spICP-MS, silver nanoparticles were detected in the effluent from a wastewater treatment plant [134, 135], and was used to study the transformation of silver nanoparticles in a natural lake system [136, 137]. Nanoparticles extracted from biological tissue using tetramethylammonium hydroxide were characterized by spICP-MS, showing the applicability of this method to detect and characterize the potential uptake and bioaccumulation of nanoparticles [138].

Metal oxide nanomaterials have also been characterized using spICP-MS. Reed et al. utilized this technique to characterize titanium dioxide, cerium dioxide, and zinc oxide nanoparticles, showcasing the polydisperse nature of the nanomaterials [139]. The application of metal oxide nanoparticles in a variety of industrial processes such as chemical-mechanical polishing slurries ensures their potential entry into the environment. Using spICP-MS, these materials can be characterized both for materials characterization, as well as, detection in the environment [140].

Beyond spherical nanomaterials, this technique can be used to detect and characterize nanomaterial of unique morphologies. This technique has been applied to the detection and characterization of silver nanowires, including those extracted from the hemolymph of *Daphnia magna* [141]. Beyond metallic nanomaterials, this technique has also been applied to the detection of carbonaceous nanomaterials, specifically carbon nanotubes (CNTs). Most CNTs contain metallic impurities as a result of their synthesis. These impurities serve as vectors for detection of the CNTs (Figure 3.38), allowing for the detection and counting of CNTs in aqueous systems [142].

Limitations   Despite its utility, spICP-MS is limited in a number of ways. The most prevalent limitation is placed on the sensitivity of the ICP-MS itself. Low ion transmission through the instrument may

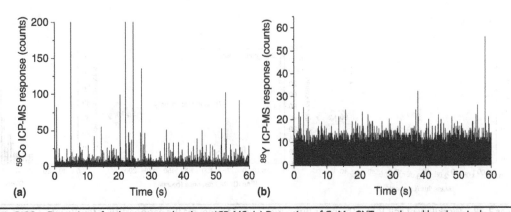

Figure 3.38 Detection of carbon nanotubes by spICP-MS. (a) Detection of CoMo CNTs produced by chemical vapor deposition. (b) Detection of CNTs synthesized by arc discharge. (Montaño, Lee, Hsu-Kim, unpublished data.)

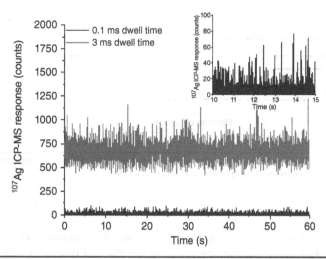

Figure 3.39   Analysis of 0.5 ppb Ag⁺ and 0.5 ppb 60 nm Ag-NP. (Inset: magnified portion of 100 μs dwell time analysis showing nanoparticle pulses above dissolved background.)

limit the number of ions that reach the detector and can subsequently be detected. For most metallic elements, the size detection ranges between 10 and 20 nm, but for some elements that are hindered by the presence of constant molecular interferences, the size detection limit can be as high as 300 nm [124]. High dissolved background concentrations and high particle number concentration may also limit the efficacy of this technique in some cases. In recent years, progress has been made in alleviating these issues by moving toward faster acquisition rates on the order of microseconds (as opposed to the conventional millisecond dwell times). The ability to analyze on the time scale of a particle event results in better resolution between particle events, and an improved signal-to-noise over a dissolved analyte background (Figure 3.39).

In quadrupole ICP-MS, only a single mass can be detected at a time, limiting the ability to analyze multiple elements within a single particle. Work is currently underway to overcome this limitation through the use of microsecond dwell times, which allow for the quadrupole to switch between two masses within a single particle event [126, 143]. Borovinskaya et al. have also achieved multielement detection by utilizing time-of-flight technology coupled to ICP-MS to detect multiple elements in a core-shell particle [144, 145].

Despite these limitations, the continued development of spICP-MS ensures that its full potential has yet to be reached. As the many previous applications of this technique have demonstrated, spICP-MS is one of the premiere technique for the analysis of inorganic nanomaterials in biological and environmental matrices where their potential ecotoxicity is a concern.

Field Flow Fractionation   The unique mechanical, optical, and electrical properties of nanoparticles are heavily governed by the size of the nanomaterials as a result of increased surface area at smaller sizes and the increased potential for quantum confinement [146]. As such, it will be important to detect and characterize engineered nanoparticles on a size basis to more accurately reflect their potential ecotoxicological behavior [147]. Furthermore, the analysis of nanoparticles in complex environmental and biological media will be hindered by several factors such as homo- and heteroaggregation, surface modification, and the ubiquitous presence of naturally occurring nano-sized colloids that interfere with the detection and characterization of the particles of interest [148]. Field flow fractionation is a chromatography-like technique capable of separating particles by a number of metrics (e.g., buoyant mass, hydrodynamic diameter, electrophoretic mobility) according to their behavior in an applied field. The power of field flow fractionation resides not only in its high resolution, but in its versatility in being coupled to a variety of detectors such as DLS and ICP-MS [149, 150].

Operating Principles
Flow-Field Flow Fractionation   One of the most common field flow fractionation techniques is flow-field flow fractionation (flow-FFF). In this method, particles are injected into a thin ribbon-like channel (75–250 μm thick) with a fluid flow applied perpendicular to the sample flow. A focusing flow acting

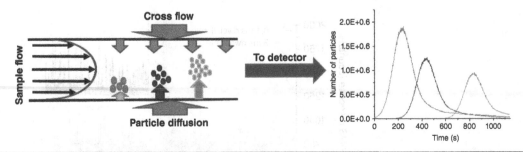

Figure 3.40   Particle separation by flow-FFF. Smaller particles diffuse further up the channel and are subject to higher flow velocities, thereby eluting faster than larger particles.

opposite the sample flow focuses the particles near the beginning of the channel while the cross flow forces particles against the accumulation wall of the channel that is covered by an ultrafiltration membrane. To begin separation, the cross flow is turned off thereby allowing particles to flow to the detector. As they flow through the channel, the smaller particles diffuse further from the accumulation wall as a result of their higher diffusion coefficient; whereas larger particles remain closer to the accumulation wall as governed by the Einstein-Stokes equation (See p. 27) [151].

Particles that diffuse farther from the accumulation wall are subject to the higher velocity flow of the parabolic flow profile from the sample flow. Subsequently, particles with smaller hydrodynamic radii will elute faster than larger particles as shown by Figure 3.40, with the retention time governed by Equation (3.27):

$$t_r = \frac{\omega^2 t_0 V_c}{6DV_0} \qquad (3.27)$$

where t_r = retention time (s)
$\quad D$ = diffusion coefficient (m^2 s^{-1})
$\quad \omega$ = channel thickness (m)
$\quad V_c$ = cross flow volumetric flow rate (mL s^{-1})
$\quad V_0$ = void volume (mL)
$\quad t_0$ = void time (s)

The most commonly applied method is asymmetrical flow-field flow fractionation (AF4), which has been successfully coupled to a number of detectors such as ICP-MS, DLS, and multiangle light scattering (MALS) for the characterization of a number of nanoparticles, colloids, and macromolecules (e.g., proteins, humics) [152–155]. Though typically for the analysis of submicrometer particles, flow-FFF can also be operated in steric mode; whereby, larger particles elute first as a results of their physical dimensions being subject to the higher velocities of the channel flow as demonstrated in Figure 3.41 [156–158].

Sedimentation-Field Flow Fractionation   Though flow-FFF is the most commonly applied fractionation technique, there has been developing interest in the capabilities of sedimentation field flow fractionation for the detection, separation, and characterization of naturally occurring and engineered nanoparticles. As opposed to using a perpendicular fluid flow, particles analyzed by sed-FFF are injected into a spinning rotor that is then accelerated at high speeds to provide a centrifugal force that forces the particles to the

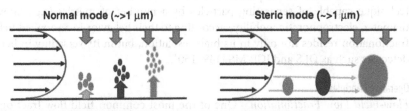

Figure 3.41   Different modes of flow-FFF operation (normal and steric).

accumulation wall. The sample flow propels the particles through the channel and particles which are denser than the carrier fluid accumulate toward the accumulation wall. Particles are separated according to their effective mass (the difference between the true mass of the particle and the mass of liquid displaced by the particle). Using centrifugal force as the applied field results in the approximate equation for retention time as shown in Equation (3.28) [149, 151, 159]:

$$t_r = \frac{\pi D_H^3 \Delta\rho G \omega t_0}{6 K_B T} \tag{3.28}$$

where D_H = diameter of particle (m)
    $\Delta\rho$ = difference in density between particle and media (g cm^{-3})
    G = gravitational acceleration (9.8 m s^{-2})
    ω = channel thickness (m)
    t_0 = void time (s)
    K_B = Boltzmann's constant (kg m^2 s^{-2} K^{-1})
    T = temperature (K)

As a result of the retention time being proportional to the diameter cubed, as opposed to diameter as in flow-FFF, sed-FFF is able to achieve greater resolution between particle sizes. In addition, separation takes place in a stainless steel channel, overcoming any potential membrane interactions that may complicate analysis by flow-FFF.

Additional FFF Techniques   Other FFF techniques that have been used for the characterization of nanoparticles include thermal-FFF and electrical-FFF. Thermal-FFF behaves similarly to flow-FFF, but uses a thermal gradient to separate particles and macromolecules according to their thermal diffusion coefficient. Particles are injected into a channel where particles are placed in between two copper plates, one which has been heated (hot wall) while the other is cooled by the carrier fluid [160, 161]. The temperature difference creates a thermal gradient where particles with a lower thermal diffusion coefficient accumulate toward a cold wall, and those with higher thermal diffusion diffuse further up into the center of the channel, thereby eluting faster. Though applicability to engineered nanoparticles has been limited, this method has been used to characterize a wide range of high-molecular-weight polymers and macromolecules [162].

Electrical-FFF has also found some application in the characterization of nanomaterials. Flow takes place in a channel where a low-voltage electrical field (0–2 V) separates particles according to electrophoretic mobility. The sensitivity of this technique to the surface charge of the particles analyzed makes it a promising platform by which to study the surface composition of nanoparticles and colloids [163].

Sample Preparation   The most important aspects of sample preparation for field flow fractionation lie in the selection of the carrier fluid. For flow-FFF, it is important to choose a carrier fluid of low viscosity as the viscosity of the medium and the cross flow velocity are directly proportional. The viscosity of the carrier fluid in sed-FFF plays a more prominent role as the difference in density between the carrier fluid and the particle density is a key aspect of determining the buoyant mass, and subsequently the effective mass of the particles analyzed [164]. In addition to viscosity, chemical compatibility is an important factor to consider. Many nanoparticles are functionalized with charged surfaces, when in contact with ionic solutions of opposite charge, may result in the bridging and aggregation of the particles intended for analysis. Many surfactants, both ionic and non-ionic, have been used in the analysis of nanomaterials such as FL-70, Triton X-100, and sodium dodecyl sulfate. Prior to analysis, it is good practice to ensure particles are dispersed to accurately reflect the size distribution of the particle population. Means of dispersing particles include sonication and vortexing; however, these techniques may introduce a number of potential analytical artifacts such as physical and thermal degradation of particles that may influence the results of the analysis.

Application in the Case of Nanoparticles   Field flow fractionation has been used to characterize a wide range of both naturally occurring and engineered nanoparticles. In particular, the coupling of FFF to an elemental detector such as ICP-MS is capable of providing information on both the size distribution as well as the elemental composition of the particles. The large specific surface area of nanomaterials makes them effective sorbents of a number of environmentally relevant contaminants in the environment. Using FFF-ICP-MS, the adsorption of uranium ions [165] and lead ions [166] onto iron oxide colloids has

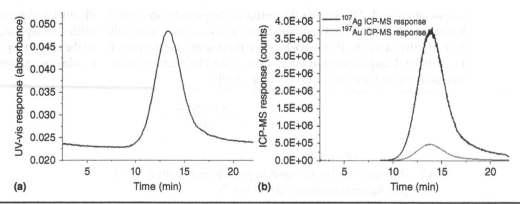

FIGURE 3.42 Analysis of gold core/silver shell nanoparticles. (a) FFF fractogram with UV-vis detection shows a single particle peak. (b) Coupling FFF with ICP-MS provides elemental composition elucidating the composition of the core (Au) and shell (Ag) of the particle.

been studied in environmental systems, providing a quantifiable assessment of the fate and transport of these contaminants in groundwater. The elemental specificity of FFF-ICP-MS is also able to differentiate a relatively broad distribution of naturally occurring nanomaterials from a narrow distribution of engineered nanomaterials comprised of elements rarely found in the natural environment (e.g., gold nanoparticles) [143]. The analysis of complex nanomaterials with core-shell structures has also been achieved by FFF techniques, demonstrating the power of these techniques for materials characterization (Figure 3.42). It has also been proposed that utilizing the high resolving power of FFF coupled with the sensitivity of ICP-MS may allow for the differentiation of naturally occurring and engineered nanomaterials by virtue of elemental ratios [143].

These techniques have also been used to characterize potential transformation of nanomaterials, such as dissolution, surface modification, and aggregation. A study of silver nanoparticles in the presence of bovine serum albumin (BSA) has demonstrated the capability of FFF to resolve primary particle size distributions, the formation of particle aggregates, in addition to the formation of a BSA-Ag$^+$ complex as a result of the dissolution of the silver nanoparticle [154]. Sed-FFF has been used in some cases to characterize silica nanoparticles that have been extracted from exposed rat lung tissue, demonstrating the capability of these analytical techniques to enhance our understanding of the potential ecotoxicological effects of nanomaterials [167].

Beyond the analysis of the nanomaterials themselves, field flow fractionation is a useful technique for separating out broad size distributions of particles. Fractions that have been separated by FFF can be analyzed by a number of techniques such as XRD, TEM, SEM, and spICP-MS [168, 169]. (See Table 3.1.)

Limitations   The main drawback of FFF arises from the potentially lengthy method development that may need to take place prior to analysis of the target nanomaterial. In order to achieve the requisite resolution required for analysis, the cross flow (or other applied field) will need to be adjusted such as to provide good separation between particles, while also allowing for particle elution within a reasonable time frame. It should be noted, that the broader the size distribution of particles to be analyzed, the longer the analysis time as a result of generating a field capable of providing good separation for a wide breadth of particle sizes. Most analyses by flow-FFF can take anywhere between 20 min to 1 h per samples depending on the conditions. Though advances in instrumentation may allow for more automated analysis and higher throughput, ideal separation may require a lengthy analysis time. Other points to consider include the selection of carrier fluid as discussed in sample preparation, but it may also be worth considering potential membrane interactions between the particle surfaces and the ultrafiltration membrane within the FFF channel. As such, it is important to assess particle recovery to ensure acceptable amounts of particle elution.

Lastly, the detection limit of these techniques will generally be a factor of the detector used to characterize the fractions that elute. For instance, a higher concentration of particles will be required to be detected by UV-vis as opposed to ICP-MS. Adjusting particle concentration to within the desired range for detection may introduce artifacts that are necessary to consider when realizing the results of these techniques.

	XRD	Raman	XAS	Mössbauer	NMR	SAXS, SANS	DLS	TEM	SEM	AFM	Zeta Potential	XPS	CT
Size	♥	♥			♥	♥♥	♥♥	♥♥	♥	♥			♥
Shape	♥					♥♥	○	♥♥	♥				♥
Surface area						♥							
Chemical composition		♥		♥	♥			♥	♥				
Speciation of elements		♥	♥♥	♥♥	♥♥							♥♥	
Redox-sate of elements			♥♥	♥♥								♥♥	
Surface chemistry			♥										
Surface charge						●					♥♥		
Crystal structure	♥♥	♥♥	♥♥	♥♥	♥	♥♥							
Agglomeration state						♥♥	○	♥	♥				♥
Heterogeneity								♥♥	♥				♥♥

♥♥ Highly applicable, gives quantitative information

♥ Applicable in some cases, gives quantitative information

● Applicable in some cases, gives quantitative information with validation with another technique

○ Applicable in some cases, gives qualitative information

TABLE 3.1 Applicability of Analytical Techniques to Providing Specific Information on Nanoparticles (Adapted from Oberdorster et al. 2005.)

References

1. Zhang, H. and J. Banfield, "Thermodynamic Analysis of Phase Stability of Nanocrystalline Titania," *Journal of Materials Chemistry,* 8(9): 2073–2076, 1998.
2. Warren, B.E., *X-Ray Diffraction,* Dover Publications, New York, 1990, 381.
3. Mamontov, E. and T. Egami, "Structural Defects in a Nano-Scale Powder of CeO2 Studied by Pulsed Neutron Diffraction," *Journal of Physics and Chemistry of Solids,* 61(8): 1345–1356, 2000.
4. Ferraro, J.R. and K. Nakamoto, *Introductory Raman Spectroscopy,* Academic Press, New York, 1994.
5. Neves, V., et al., "Cellular Localization, Accumulation and Trafficking of Double-Walled Carbon Nanotubes in Human Prostate Cancer Cells," *Nano Research,* 5(4): 223–234, 2012.
6. Robert, T., et al., "Micro-Raman Spectroscopy Study of Surface Transformations Induced by Excimer Laser Irradiation of TiO2," *Thin Solid Films,* 440(1–2): 268–277, 2003.
7. Koningsberger, D.C. and R. Prins, *X-Ray Absorption: Principles, Applications, Techniques of EXAFS, SEXAFS and XANES,* John Wiley, New York, 1988.
8. Fontaine, A., "Interactions of X-Rays with Matter: X-Ray Absorption Spectroscopy," in *Neutron and Synchrotron Radiation for Condensed Matter Studies,* Baruchel, H.J., J.-L. Itodeau, M.S. Lehmann, J.R. Regnard, and C. Schlenker, Editors. Les Editions de Physique—Springer Verlag: Les Ulis, Berlin, 1993.
9. Rehr, J.J. and R.C. Albers, "Theoretical Approaches to X-Ray Absorption Fine Structure," *Reviews of Modern Physics,* 72: 621–654, 2000.
10. Sayers, D.A., E.A. Stern, and F.W. Lytle, "New Technique for Investigating Noncrystalline Structures: Fourier Analysis of the Extended X-Ray—Absorption Fine Structure," *Physical Review Letter,* 27: 1204–1207, 1971.
11. Chemseddine, A., et al., "XAFS Study of Functionalized Nanoclusters and Nanocluster Assemblies," *Zeitschrift für Physik D Atoms, Molecules and Clusters,* 40(1–4): 566–569, 1997.
12. Auffan, M., et al., "In Vitro Interactions between DMSA-Coated Maghemite Nanoparticles and Human Fibroblasts: A Physicochemical and Cyto-Genotoxical Study," *Environmental Science and Technology,* 2006. Vol 40, 14, 4367–4374.
13. Choi, H.C., Y.M. Jung, and S.B. Kim, "Size Effects in the Raman Spectra of TiO2 Nanoparticles," *Vibrational Spectroscopy,* 37: 33–38, 2005.
14. Mulder, E., et al., "Size-Evolution Towards Metallic Behavior In Nano-Sized Gold And Platinum Clusters As Revealed By 197Au Mössbauer Spectroscopy," *NanoStructured Materials,* 7(3): 269–292, 1996.
15. Liu, M., et al., "XRD and Mössbauer Spectroscopy Investigation of Fe2O3–Al2O3 Nano-Composite," *Journal of Magnetism and Magnetic Materials,* 294: 294–297, 2005.
16. Zhu, S., et al., "Structural Studies of Iron-Doped TiO2 Nano-Composites by Moössbauer Spectroscopy, X-ray Diffraction and Transmission Microscopy," *Physica B,* 374: 199–205, 2005.
17. Harris, R.K., *Encyclopedia of Magnetic Resonance,* J. Wiley and sons, Chichester, 2011.
18. Farrar, T.C. and E.D. Becker, *Pulse and Fourier Transform NMR. Introduction to Theory and Methods,* Academic Press, New York and London, 1971.
19. Thomas, F., et al., "Aluminum(III) Speciation with Hydroxy Carboxylic Acids. ^{27}Al NMR Study," *Environmental Science & Technology,* 27(12): 2511–2516, 1993.
20. Atkins, T.M., et al., "Synthesis of Long T Silicon Nanoparticles for Hyperpolarized Si Magnetic Resonance Imaging," *ACS Nano,* 7(2): 1609–1617, 2013.
21. Hartmann, S.R. and E.L. Hahn, "Nuclear Double Resonance in the Rotating Frame," *Physical Reviews,* 128: 2042–2053, 1962.
22. Auffan, M., et al., "Surface Structural Degradation of TiO_2-Based Nanomaterial Used in Cosmetics," *Environmental Science & Technology,* 44(7): 2689–2694, 2010.
23. Denaix, L., I. Lamy, and J.Y. Bottero, "Structure and Affinity Towards Cd2+, Cu2+, Pb2+ of Synthetic Colloidal Amorphous Aluminosilicates and Their Precursors," *Colloids and Surfaces A,* 158(3): 315–325, 1999.
24. Levard, C., et al., "Synthesis of Imogolite Fibers from Decimolar Concentration at Low Temperature and Ambient Pressure: A Promising Route for Inexpensive Nanotubes," *Journal of the American Chemical Society,* 131(7): 17080–17081, 2009.
25. Levard, C., et al., "Synthesis of Large Quantities of Single-Walled Aluminogermanate Nanotube," *Journal of the American Chemical Society,* 130(18): 5862–5863, 2008.
26. Levard, C., et al., "*Formation and Growth Mechanisms of Imogolite-Like Aluminogermanate Nanotubes,*" *Chemistry of Materials,* 22(8): 2466–2473, 2010.
27. Akitt, J.W., "Multinuclear Studies of Aluminium Compounds," *Progress in Nuclear Magnetic Resonance Spectroscopy,* 21: 1–149, 1989.
28. Smith, M.E., "Application of ^{27}Al NMR Techniques to Structure Determination in Solids," *Applied Magnetic Resonance,* 4: 1–64, 1993.
29. Auffan, M., et al., "Is There a Trojan Horse Effect During Magnetic Nanoparticles and Metalloid Co-contamination of Human Dermal Fibroblasts?" *Environmental Science & Technology,* 46(19): 10789–10796, 2012.
30. Marın-Almazo, M., et al., "Synthesis and Characterization of Rhodium Nanoparticles Using HREM Techniques," *Microchemical Journal,* 81: 133–138, 2005.
31. Yan, W., et al., "Powder XRD Analysis and Catalysis Characterization of Ultra-Small Gold Nanoparticles Deposited on Titania-Modified SBA-15," *Catalysis Communications,* (6): 404–408, 2005.
32. Hiutsunomiya, S. and R. Ewing, "Application of High-Angle Annular Dark Field Scanning Transmission Electron Microscopy, Scanning Transmission Electron Microscopy-Energy Dispersive X-ray Spectrometry, and Energy-Filtered Transmission Electron Microscopy to the Characterization of Nanoparticles in the Environment," *Environmental Science & Technology,* 37: 786–791, 2003.
33. Stephan, O., et al., "Electron Energy-Loss Spectroscopy on Individual Nanotubes," *Journal of Electron Spectroscopy and Related Phenomena,* 114–116: 209–217, 2001.
33a. Brant, J.A. and Childress, A.E., "Membrane-Colloid Interactions: Comparison of Extended DLVO Predictions with AFM Force Measurements", *Environmental Engineering Science,* 19(6): 413–427, December 2002.

33b. Brant, J.A. and Childress, A.E., "Colloidal Adhesion to Hydrophilic Membrane Surfaces," *Journal of Membrane Science*, 241(2): 235–248, 2004.

34. Thill, A., S. Desert, and M. Delsanti, "Small Angle Static Light Scattering: Absolute Intensity Measurement," *The European Physical Journal, Applied Physics*, 17: 201–208, 2002.

35. Berne, B.J. and R. Pecora, *Dynamic Light Scattering*, J. Wiley, New York, 1976. pp. 376.

36. Brown, W., *Light Scattering: Principles and Development*, Clarendon Press, Oxford, 1996.

37. Glatter, O. and O. Kratky, *Small-Angle X-Ray Scattering*, Academic Press, London, 1982.

38. Higgins, J.S. and H.C. Benoit, *Polymers and Neutrons Scattering*, Clarendon Press, Oxford, 1994.

39. Guinier, A. and G. Fournet, *Small Angle Scattering of X-Rays*, Wiley, New York, 1955.

40. Glatter, O. and O. Kratky, *Small Angle X-Ray Scattering*, Academic Press, London, 1982.

41. Axelos, M., D. Tchoubar, and J.Y. Bottero, "Small-Angle X-Ray-Scattering Investigation of the Silica Water Interface—Evolution of the Structure with pH," *Langmuir*, 5(5): 1186–1190, 1989.

41a. Mandelbrot, B., 1977b. Fractals and turbulence: attractors and dispersion. Seminar on Turbulence, Berkeley 1976. Organized by Alexandre Chorin, Jerald Marsden & Stephen Smale. Edited by P. Bernard & T. Ratiu (Lecture Notes in Mathematics 615). New York: Springer, 83–93.

42. Debye, P. and A.M. Bueche, "Scattering by an Inhomogeneous Solid," *Journal of Applied Physics*, 20: 518–525, 1949.

43. Vicsek, T., *Fractal Growth Phenomena*, World Scientific, Singapore, 1989.

44. Johansson, G., "On the Crystal Structure of Some Basic Aluminum Salts," *Acta Chemica Scandinavica*, 14: 771–773, 1960.

45. Bottero, J.Y., et al., "Investigation of the Hydrolysis of Aqueous Solutions of Aluminum Chloride. 2. Nature and Structure by Small Angle X-Ray Scattering," *Journal of Physical Chemistry*, 86: 3667–3673, 1982.

46. Bottero, J.Y., et al., "Mechanism of Formation of Aluminum Trihydroxide from Keggin Al_{13} Polymers," *Journal of Colloid and Interface Science*, 117(1): 47–57, 1987.

47. Tchoubar, D., et al., Partial Hydrolysis of Ferric-Chloride Salt—Structural Investigation by Photon-Correlation Spectroscopy and Small-Angle X-Ray-Scattering," *Langmuir*, 7(2): 398–402, 1991.

48. Masion, A., et al., "Chemistry and Structure of Al(OH)/Organic Precipitates. A Small Angle X-Ray Scattering Study. 1. Numerical Procedure for Speciation from Scattering Curves," *Langmuir*, 10: 4344–4348, 1994.

49. Masion, A., et al., "Chemistry and Structure of Al(OH)/Organic Precipitates. A Small Angle X-Ray Scattering Study. 2. Speciation and Structure of the Aggregates," *Langmuir*, 10: 4349–4352, 1994.

50. Masion, A., et al., "Nucleation and Growth Mechanisms of Fe Oxyhydroxides in the Presence of PO_4 Ions. 3. Speciation of Fe by Small Angle X-Ray Scattering," *Langmuir*, 13(14): 3882–3885, 1997.

51. Maillet, P., et al., "Evidence of Double-Walled Al–Ge Imogolite-Like Nanotubes. A Cryo-TEM and SAXS Investigation," *Journal of the American Chemical Society*, 132(4): 1208–1209, 2010.

52. Fratzl, P., et al., "Position-Resolved Small-Angle X-Ray Scattering of Complex Biological Materials," *Journal of Applied Crystallography*, 30(2): 765–769, 1997.

53. Jolivet, J.-P., E. Tronc, and C. Chanéac, "Synthesis of Iron Oxide-Based Magnetic Nanomaterials and Composites," *C. R. Chimie*, 5: 659–664, 2002.

54. Richter, H., Z.P. Wang, and L. Ley, "The one phonon raman-spectrum in microcrystalline silicon", *Solid State Communication*, 39: 625, 1981.

55. Kelly, S., F.H. Pollak, and M. Tomkiewicz, "Raman Spectroscopy as a Morphological Probe for TiO2 Aerogels" *Journal of Physical Chemistry B*, 101: 2730–2734, 1997.

56. Greegor, R. and F. Lytle, "Morphology of Supported Metal-Clusters—Determination by Exafs and Chemisorption," *Journal of Catalysis*, 63(2): 476–486, 1980.

57. Jentys, A., "Estimation of Mean Size and Shape of Small Metal Particles by EXAFS," *Physical Chemistry Chemical Physics*, 1(17): 4059–4063, 1999.

58. Frenkel, A.I., C.W. Hills, and R.G. Nuzzo, "A View from the Inside: Complexity in the Atomic Scale Ordering of Supported Metal-nanoparticles: Dextran and Albumin Derivatised Iron Oxide," *The Journal of Physical Chemistry B*, 105(51): 12689–12703, 2001.

59. Fernandez-Garcia, M., et al., "Nanostructured Oxides in Chemistry: Characterization and Properties," *Chemical Review*, 104: 4063–4104, 2004.

60. Rose, J., et al., "Synthesis and Characterisation of Carboxylate-FeOOH Nanoparticles(Ferroxane) and Ferroxane-Derived Ceramics," *Chemistry of Materials*, 14: 621–628, 2002.

61. Bazin, D., et al., "Numerical Simulation of the Platinum LIII Edge White Line Relative to Nanometer Scale Clusters," *Journal of Physical Chemistry B*, 101: 5332–5336, 1997.

62. Combes, J.-M., et al., "Formation of Ferric Oxides from Aqueous Solutions: A Polyhedral Approach by X-ray Absorption Spectroscopy. 1. Hydrolysis and Formation of Ferric Gels," *Geochimica and Cosmochimica Acta*, 53(3): 583–594, 1989.

63. Bottero, J., et al., "Structure and Mechanisms of Formation of FeOOH(Cl) Polymers," *Langmuir*, 10(1): 316–319, 1994.

64. Rose, J., et al., "Nucleation and Growth Mechanisms of Fe(III) Oxy-Hydroxide in the Presence of PO4 Ions. 1. Fe-K edge EXAFS study," *Langmuir*, 12(26): 6701–6707, 1996.

65. Rose, J., et al., "Nucleation and Growth Mechanisms of Fe(III) Oxy-Hydroxide in the Presence of PO4 Ions. 2. P-K Edge EXAFS Study," *Langmuir*, 13(6): 1827–1834, 1997.

66. Aitchison, P., et al., interface—comparison of Al-doped and Zr-doped silicates with the purely siliceous sample/x-ray-absorption spectroscopy», Inorganic chemistry, 34(18), 1995, pp. 4611–4617.

67. H. Roussel, V. Briois, E. Elkaim, A. de Roy, J. P. Besse. "Cationic Order and Structure of [Zn–Cr–Cl] and [Cu–Cr–Cl] Layered Double Hydroxides: A XRD and EXAFS Study. *Journal of Physical Chemistry*. Volume 104, Issue 25 (2000).

68. Michot, L.J., E. Montargès-Pelletier, B.S. Lartiges, J.-B. d'Espinose de la Caillerie, and V. Briois, Formation mechanism of the Ga-13 keggin ion: A combined EXAFS and NMR study *Journal of American Chemical Society*, 122: 6048, 2000.

69. Chemseddine, A. and T. Moritz, Nanostructuring Titania: Control over Nanocrystal Structure, Size, Shape, and Organization *European Journal of Inorganic Chemistry*, 2: 235, 1999.

70. Turillas, X., et al., "'Hydroxyde' Precursor to Zirconia: Extended X-Ray Absorption Fine Structure Study," *Journal of Materials Chemistry*, 3(6): 583–586, 1993.

71. Helmerich, A., et al., "Structural Studies on an ORMOCER System Containing Zirconium" *Journal of Material Science*, 29: 1388–1389, 1994.

72. Peter, D., T.S. Ertel, and H. Bertagnolli, "'EXAFS Study of Zirconium as Precursor in the Sol-Gel Process: II. The Influence of the Chemical Modification," *Journal of Sol-Gel Science and Technology*, 5: 5–14, 1995.

73. Baruchel, J., et al., *X-Ray Tomography in Material Science, General Principles*, Hermes Science Publications, Paris, 2000.

74. Bossa, N., et al., "Micro- and Nano-X-Ray Computed-Tomography: A Step Forward in the Characterization of the Pore Network of a Leached Cement Paste," *Cement and Concrete Research*, 67(0): 138–147, 2015.

75. Sanchez, S., et al., "Three-Dimensional Synchrotron Virtual Paleohistology: A New Insight into the World of Fossil Bone Microstructures," *Microscopy and Microanalysis*, 18(5): 1095–1105, 2012.

76. Sleutel, S., et al., "Comparison of Different Nano- and Micro-Focus X-Ray Computed Tomography Set-ups for the Visualization of the Soil Microstructure and Soil Organic Matter," *Computers & Geosciences*, 34(8): 931–938, 2008.

77. Baker, D.R., et al., "An Introduction to the Application of X-Ray Microtomography to the Three-Dimensional Study of Igneous Rocks," *Lithos*, 148(0): 262–276, 2012.

78. Neues, F. and M. Epple, "X-Ray Microcomputer Tomography for the Study of Biomineralized Endo- and Exoskeletons of Animals," *Chemical Reviews*, 108(11): 4734–4741, 2008.

79. Salvo, L., et al., "X-Ray Micro-Tomgraphy an Attractive Characterisation Technique in Materials Science," *Nuclear Instruments and Methods in Physics Research Section B-Beam Interactions with Materials and Atoms*, 200: 273–286, 2003.

80. Stock, S.R., "Recent Advances in X-Ray Microtomography Applied to Materials," *International Materials Reviews*, 53(3): 129–181, 2008.

81. Landis, E.N. and D.T. Keane, "X-Ray Microtomography," *Materials Characterization*, 61(12): 1305–1316, 2010.

82. Maire, E. and P.J. Withers, "Quantitative X-Ray Tomography," *International Materials Reviews*, 59(1): 1–43, 2014.

83. Cnudde, V. and M.N. Boone, "High-Resolution X-Ray Computed Tomography in Geosciences: A Review of the Current Technology and Applications," *Earth-Science Reviews*, 123: 1–17, 2013.

84. Wildenschild, D. and A.P. Sheppard, "X-Ray Imaging and Analysis Techniques for Quantifying Pore-Scale Structure and Processes in Subsurface Porous Medium Systems" *Advances in Water Resources*, 51: 217–246, 2013.

85. Fusseis, F., et al., "A Brief Guide to Synchrotron Radiation-Based Microtomography in (Structural) Geology and Rock Mechanics," *Journal of Structural Geology*, 65: 1–16, 2014.

86. Mizutani, R. and Y. Suzuki, "X-ray Microtomography in Biology," *Micron*, 43(2–3): 104–115, 2012.

87. Betz, O., et al., "Imaging Applications of Synchrotron X-Ray Phase-Contrast Microtomography in Biological Morphology and Biomaterials Science. 1. General Aspects of the Technique and its Advantages in the Analysis of Millimetre-Sized Arthropod Structure," *Journal of Microscopy-Oxford*, 227(1): 51–71, 2007.

88. Wilkins, S.W., et al., "Phase-Contrast Imaging Using Polychromatic Hard X-Rays," *Nature*, 384(6607): 335–338, 1996.

89. Bronnikov, A.V., "Theory of Quantitative Phase-Contrast Computed Tomography," *Journal of the Optical Society of America A—Optics Image Science and Vision*, 19(3): 472–480, 2002.

90. Mayo, S.C., A.W. Stevenson, and S.W. Wilkins, "In-Line Phase-Contrast X-ray Imaging and Tomography for Materials Science," *Materials*, 5(5): 937–965, 2012.

91. Bech, M., et al., "Soft-Tissue Phase-Contrast Tomography with an X-Ray Tube Source," *Physics in Medicine and Biology*, 54(9): 2747–2753, 2009.

92. Lak, M., et al., "Phase Contrast X-Ray Synchrotron Imaging: Opening Access to Fossil Inclusions in Opaque Amber," *Microscopy and Microanalysis*, 14(3): 251–259, 2008.

93. Sakdinawat, A. and D. Attwood, "Nanoscale X-Ray Imaging," *Nature Photonics*, 4(12): 840–848, 2010.

94. Withers, P.J., "X-Ray Nanotomography," *Materials Today*, 10(12): 26–34, 2007.

95. Merkle, A.P. and J. Gelb, "The Ascent of 3D X-Ray Microscopy in the Laboratory," *Microscopy Today*, 21(1): 10–15, 2013.

96. Metscher, B.D., "MicroCT for Developmental Biology: A Versatile Tool for High-Contrast 3D Imaging at Histological Resolutions," *Developmental Dynamics*, 238(3): 632–640, 2009.

97. Metscher, B.D., "MicroCT for Comparative Morphology: Simple Staining Methods Allow High-Contrast 3D Imaging of Diverse Non-mineralized Animal Tissues," *BMC Physiology*, 9: 11, 2009.

98. van de Kamp, T., P. Vagovič, T. Baumbach, and A. Riedel, "A Biological Screw in a Beetle's Leg," *Science*, 333(52): 6038–6052, 2011.

99. Mielke, R.E., et al., "Differential Growth of and Nanoscale TiO2 Accumulation in *Tetrahymena thermophila* by Direct Feeding versus Trophic Transfer from *Pseudomonas aeruginosa*," *Applied and Environmental Microbiology*, 79(18): 5616–5624, 2013.

100. van den Brule, S., et al., "Nanometer-Long Ge-Imogolite Nanotubes Cause Sustained Lung Inflammation and Fibrosis in Rats," *Particle and Fibre Toxicology*, 11(1): 67, 2014.

101. Templeton, A.S., et al., "Distribution and Speciation of Metals and Metalloids at Microbe/Mineral Interfaces," *Geochimica Et Cosmochimica Acta*, 69(10): A613–A613, 2005.

102. Templeton, A.S., et al., "Selenium Speciation and Partitioning Within *Burkholderia cepacia* Biofilms Formed on a-Al2O3 Surfaces," *Geochimica Et Cosmochimica Acta*, 67(19): 3547–3557, 2003.

103. Templeton, A.S., et al., "Pb(II) Distributions at Biofilm-Metal Oxide Interfaces," *Proceedings of the National Academy of Sciences of the United States of America*, 98(21): 11897–11902, 2001.

104. Trainor, T.P., et al., "Application of the Long-Period X-Ray Standing Wave Technique to the Analysis of Surface Reactivity: Pb(II) Sorption at Alpha-Al2O3/Aqueous Solution Interfaces in the Presence and Absence of Se(VI)," *Langmuir*, 18(15): 5782–5791, 2002.

105. Wang, Y., et al., "Competitive Sorption of Pb(II) and Zn(II) on Polyacrylic Acid-Coated Hydrated Aluminum-Oxide Surfaces," *Environmental Science & Technology*, 47(21): 12131–12139, 2013.

106. Levard, C., et al., "Probing Ag Nanoparticle Surface Oxidation in Contact with (In)organics: An X-Ray Scattering and Fluorescence Yield Approach," *Journal of Synchrotron Radiation*, 18(6): 871–878, 2011.

107. Bedzyk, M.J., G.M. Bommarito, and J.S. Schildkraut, "X-Ray Standing Waves at a Reflecting Mirror Surface," *Physical Review Letters*, 62(12): 1376–1379, 1989.

108. Batterman, B.W., "Detection of Foreign Atom Sites by Their X-Ray Fluorescence Scattering," *Physical Review Letters*, 22(14): 703–705, 1969.

109. Trainor, T.P., A.S. Templeton, and P.J. Eng, "Structure and Reactivity of Environmental Interfaces: Application of Grazing Angle X-Ray Spectroscopy and Long-Period X-Ray Standing Waves," *Journal of Electron Spectroscopy and Related Phenomena*, 150(2–3): 66–85, 2006.

110. M. J. Bedzyk and L. Cheng, in Reviews in Mineralogy and Geochemistry, edited by P. Fenter et al. (Mineralogical Society of America, Washington, 2002),Vol. 49, p. 221.

111. Abruna, H.D., G.M. Bommarito, and D. Acevedo, "The Study of Solid Liquid Interfaces with X-Ray Standing Waves," *Science*, 250(4977): 69–74, 1990.

112. Bedzyk, M.J., et al., "X-Ray Standing Waves—A Molecular Yardstick for Biological Membrane," *Science*, 241(4874): 1788–1791, 1988.

113. Bommarito, G.M., J.H. White, and H.D. Abruna, "Electrosorption of Iodide on Platinum—Packing Density and Potential-Dependent Distributional Changes Observed Insitu with X-Ray Standing Waves," *Journal of Physical Chemistry*, 94(21): 8280–8288, 1990.

114. Bedzyk, M.J., et al., "Diffuse-Double Layer at a Membrane-Aqueous Interface Measured with X-ray Standing Waves," *Science*, 248(4951): 52–56, 1990.

115. Parratt, L.G., "Surface Studies of Solids by Total Reflection of X-Rays," *Physical Review*, 95(2): 359–369, 1954.

116. Krol, A., C.J. Sher, and Y.H. Kao, "X-Ray-Fluorescence of Layered Synthetic Materials with Interfacial Roughness," *Physical Review B*, 38(13): 8579–8592, 1988.

117. Deboer, D.K.G., "Glancing-Incidence X-Ray-Fluorescence of Layered Materials," *Physical Review B*, 44(2): 498–511, 1991.

118. Levard, C., et al., "Environmental Transformations of Silver Nanoparticles: Impact on Stability and Toxicity" *Environmental Science & Technology*, 46(13): 6900–6914, 2012.

119. Pace, H.E., et al., "Single Particle Inductively Coupled Plasma-Mass Spectrometry: A Performance Evaluation and Method Comparison in the Determination of Nanoparticle Size," *Environmental Science & Technology*, Volume: 46 Issue: 22 Pages: 12272–12280 (2012).

120. Laborda, F., E. Bolea, and J. Jimenez-Lamana, "Single Particle Inductively Coupled Plasma Mass Spectrometry: A Powerful Tool for Nanoanalysis," *Analytical Chemistry*, 86(5): 2270–2278, 2013.

121. Laborda, F., et al., "Critical Considerations for the Determination of Nanoparticle Number Concentrations, Size and Number Size Distributions by Single Particle ICP-MS," *Journal of Analytical Atomic Spectrometry*, 28(8): 1220–1232, 2013.

122. Olesik, J.W. and P.J. Gray, "Considerations for Measurement of Individual Nanoparticles or Microparticles by ICP-MS: Determination of the Number of Particles and the Analyte Mass in Each Particle," *Journal of Analytical Atomic Spectrometry*, 27(7): 1143–1155, 2012.

123. Pace, H.E., et al., "Determining Transport Efficiency for the Purpose of Counting and Sizing Nanoparticles via Single Particle Inductively Coupled Plasma Mass Spectrometry" *Analytical Chemistry*, 83(24): 9361–9369, 2011.

124. Lee, S., et al., "Nanoparticle Size Detection Limits by Single Particle ICP-MS for 40 Elements," *Environmental Science & Technology*, 48(17): 10291–10300, 2014.

125. Tuoriniemi, J., G. Cornelis, and M. Hasselloev, "Improving Accuracy of Single Particle ICPMS for Measurement of Size Distributions and Number Concentrations of Nanoparticles by Determining Analyte Partitioning During Nebulisation," *Journal of Analytical Atomic Spectrometry*, Volume: 29 Issue: 4 Pages: 743–752 (2014).

126. Montaño, M.D., et al., "Improvements in the Detection and Characterization of Engineered Nanoparticles Using spICP-MS with Microsecond Dwell Times," *Environmental Science: Nano*, Volume: 1 Issue: 4 Pages: 338–346 (2014).

127. Hineman, A. and C. Stephan, "Effect of Dwell Time on Single Particle Inductively Coupled Plasma Mass Spectrometry Data Acquisition Quality," *Journal of Analytical Atomic Spectrometry*, Volume: 29 Issue: 7 Pages: 1252–1257 (2014).

128. Tuoriniemi, J., G. Cornelis, and M. Hassellöv, "A New Peak Recognition Algorithm for Detection of Ultra-small Nano-Particles by Single Particle ICP-MS Using Rapid Time Resolved Data Acquisition on a Sector-Field Mass Spectrometer," *Journal of Analytical Atomic Spectrometry*, 30(8): 1723–1729, 2015.

129. Hadioui, M., C. Peyrot, and K.J. Wilkinson, "Improvements to Single Particle ICP-MS by the On-line Coupling of Ion Exchange Resins," *Analytical Chemistry*, Volume: 86 Issue: 10 Pages: 4668–4674 (2014).

130. Degueldre, C. and P.-Y. Favarger, "Colloid Analysis by Single Particle Inductively Coupled Plasma-Mass Spectroscopy: A Feasibility Study," *Colloids and Surfaces A: Physicochemical and Engineering Aspects*, 217(1): 137–142, 2003.

131. Degueldre, C. and P.-Y. Favarger, "Thorium Colloid Analysis by Single Particle Inductively Coupled Plasma-Mass Spectrometry," *Talanta*, 62(5): 1051–1054, 2004.

132. Degueldre, C., P.Y. Favarger, and C. Bitea, "Zirconia Colloid Analysis by Single Particle Inductively Coupled Plasma–Mass Spectrometry," *Analytica Chimica Acta*, 518(1–2): 137–142, 2004.

133. Degueldre, C., P.Y. Favarger, and S. Wold, "Gold Colloid Analysis by Inductively Coupled Plasma-Mass Spectrometry in a Single Particle Mode," *Analytica Chimica Acta*, 555(2): 263–268, 2006.

134. Mitrano, D.M., et al., "Detecting Nanoparticulate Silver Using Single-Particle Inductively Coupled Plasma–Mass Spectrometry," *Environmental Toxicology and Chemistry*, 31(1): 115–121, 2012.

135. Tuoriniemi, J., G. Cornelis, and M. Hasselöv, "Size Discrimination and Detection Capabilities of Single-Particle ICPMS for Environmental Analysis of Silver Nanoparticles," *Analytical Chemistry*, 84(9): 3965–3972, 2012.

136. Furtado, L., et al., "The Persistence and Transformation of Silver Nanoparticles in Littoral Lake Mesocosms Monitored Using Various Analytical Techniques," *Environmental Chemistry*, Volume: 11 Issue: 4 Pages: 419–430 (2014).

137. Mitrano, D., et al., "Tracking Dissolution of Silver Nanoparticles at Environmentally Relevant Concentrations in Laboratory, Natural, and Processed Waters Using Single Particle ICP-MS (spICP-MS)," *Environmental Science: Nano*, 1(3): 248–259, 2014.

138. Gray, E.P., et al., "Extraction and Analysis of Silver and Gold Nanoparticles from Biological Tissues Using Single Particle Inductively Coupled Plasma Mass Spectrometry," *Environmental Science & Technology*, Volume: 47 Issue: 24 Pages: 14315–14323 (2013).

139. Reed, R.B., et al., "Overcoming Challenges in Analysis of Polydisperse Metal-Containing Nanoparticles by Single Particle Inductively Coupled Plasma Mass Spectrometry," *Journal of Analytical Atomic Spectrometry*, 27(7): 1093–1100, 2012.

140. Speed, D., et al., "Physical, Chemical, and In vitro Toxicological Characterization of Nanoparticles in Chemical Mechanical Planarization Suspensions Used in the Semiconductor Industry: Towards Environmental Health and Safety Assessments," *Environmental Science: Nano*, 2(3): 227–244, 2015.

141. Scanlan, L.D., et al., "Silver Nanowire Exposure Results in Internalization and Toxicity to Daphnia Manga," *ACS Nano*, Volume: 7 Issue: 12 Pages: 10681–10694 (2013).

142. Reed, R.B., et al., "Detection of Single Walled Carbon Nanotubes by Monitoring Embedded Metals," *Environmental Science: Processes & Impacts*, 15(1): 204–213, 2013.

143. Montano, M., et al., "Current Status and Future Direction for Examining Engineered Nanoparticles in Natural Systems," *Environmental Chemistry*, 11(4): 351–366 (2014).

144. Borovinskaya, O., et al., "Simultaneous Mass Quantification of Nanoparticles of Different Composition in a Mixture by Microdroplet Generator-ICPTOFMS," *Analytical Chemistry*, 86(16): 8142–8148 (2014).

145. Borovinskaya, O., et al., "A Prototype of a New Inductively Coupled Plasma Time-of-Flight Mass Spectrometer Providing Temporally Resolved, Multi-element Detection of Short Signals Generated by Single Particles and Droplets," *Journal of Analytical Atomic Spectrometry*, 28(2): 226–233, 2013.

146. Nowack, B. and T.D. Bucheli, "Occurrence, Behavior and Effects of Nanoparticles in the Environment," *Environmental Pollution*, 150(1): 5–22, 2007.

147. Moore, M.N., "Do Nanoparticles Present Ecotoxicological Risks for the Health of the Aquatic Environment?" *Environment International*, 32(8): 967, 2006.

148. von der Kammer, F., et al., "Analysis of Engineered Nanomterials in Complex Matrices (Environment and Biota): General Considerations and Conceptual Case Studies," *Environmental Toxicology and Chemistry*, 31(1): 32–49, 2012.

149. Beckett, R., et al., "Determination of Thickness, Aspect Ratio and Size Distributions for Platey Particles Using Sedimentation Field-Flow Fractionation and Electron Microscopy," *Colloids and Surfaces A: Physicochemical and Engineering Aspects*, 120(1): 17–26, 1997.

150. Giddings, J.C., "Field-Flow Fractionation: Analysis of Macromolecular, Colloidal, and Particulate Materials," *Science*, 260(5113): 1456–1465, 1993.

151. Giddings, J.C., "Measuring Colloidal and Macromolecular Properties by FFF," *Analytical Chemistry*, 67(19): 592A–598A, 1995.

152. Baalousha, M., et al., "Size Fractionation and Characterization of Natural Colloids by Flow-Field Flow Fractionation Coupled to Multi-Angle Laser Light Scattering," *Journal of Chromatography A*, 1104(1–2): 272–281, 2006.

153. Jiménez, M.S., et al., "An Approach to the Natural and Engineered Nanoparticles Analysis in the Environment by Inductively Coupled Plasma Mass Spectrometry," *International Journal of Mass Spectrometry*, 307(1–3): 99–104, 2011.

154. Poda, A.R., et al., "Characterization of Silver Nanoparticles Using Flow-Field Flow Fractionation Interfaced to Inductively Coupled Plasma Mass Spectrometry," *Journal of Chromatography A*, 1218(27): 4219–4225, 2011.

155. Stolpe, B., et al., "High Resolution ICPMS as an On-line Detector for Flow Field-Flow Fractionation; Multi-element Determination of Colloidal Size Distributions in a Natural Water Sample," *Analytica Chimica Acta*, 535(1–2): 109–121, 2005.

156. Giddings, J.C., et al., "Fast Particle Separation by Flow/Steric Field-Flow Fractionation," *Analytical Chemistry*, 59(15): 1957–1962, 1987.

157. Giddings, J.C. and M.N. Myers, "Steric Field-Flow Fractionation: A New Method for Separating I to 100 μm Particles," *Separation Science and Technology*, 13(8): 637–645, 1978.

158. Moon, M.H. and J.C. Giddings, "Extension of Sedimentation/Steric Field-Flow Fractionation into the Submicrometer Range: Size Analysis of 0.2-15-. mu. m Metal Particles," *Analytical Chemistry*, 64(23): 3029–3037, 1992.

159. Tadjiki, S., et al., "Detection, Separation, and Quantification of Unlabeled Silica Nanoparticles in Biological Media Using Sedimentation Field-Flow Fractionation," *Journal of Nanoparticle Research*, 11(4): 981–988, 2009.

160. Pasti, L., S. Agnolet, and F. Dondi, "Thermal Field-Flow Fractionation of Charged Submicrometer Particles in Aqueous Media," *Analytical Chemistry*, 79(14): 284–5296, 2007.

161. Fraunhofer, W. and G. Winter, "The Use of Asymmetrical Flow Field-Flow Fractionation in Pharmaceutics and Biopharmaceutics," *European Journal of Pharmaceutics and Biopharmaceutics*, 58(2): 369–383, 2004.

162. Ponyik, C.A., D.T. Wu, and S.K.R. Williams, "Separation and Composition Distribution Determination of Triblock Copolymers by Thermal Field-Flow Fractionation," *Analytical and Bioanalytical Chemistry*, 405(28): 9033–9040, 2013.

163. Gigault, J., et al., "Nanoparticle Characterization by Cyclical Electrical Field-Flow Fractionation," *Analytical Chemistry*, 83(17): 6565–6572, 2011.

164. Schimpf, M.E., K. Caldwell, and J.C. Giddings, *Field-Flow Fractionation Handbook*, John Wiley & Sons, New York, 2000.

165. Lesher, E.K., J.F. Ranville, and B.D. Honeyman, "Analysis of pH Dependent Uranium(VI) Sorption to Nanoparticulate Hematite by Flow Field-Flow Fractionation—Inductively Coupled Plasma Mass Spectrometry," *Environmental Science & Technology*, 43(14): 5403–5409, 2009.

166. Hassellöv, M. and F. von der Kammer, "Iron Oxides as Geochemical Nanovectors for Metal Transport in Soil-River Systems," *Elements*, 4(6): 401–406, 2008.

167. Deering, C.E., et al., "A Novel Method to Detect Unlabeled Inorganic Nanoparticles and Submicron Particles in Tissue by Sedimentation Field-Flow Fractionation," *Part Fibre Toxicol*, 5: 18, 2008.

168. Plathe, K.L., et al., "The Role of Nanominerals and Mineral Nanoparticles in the Transport of Toxic Trace Metals: Field-Flow Fractionation and Analytical TEM Analyses After Nanoparticle Isolation and Density Separation," *Geochimica et Cosmochimica Acta*, 102: 213–225, 2013.

169. Wagner, S., et al., "First Steps Towards a Generic Sample Preparation Scheme for Inorganic Engineered Nanoparticles in a Complex Matrix for Detection, Characterization, and Quantification by Asymmetric Flow-Field Flow Fractionation Coupled to Multi-Angle Light Scattering and ICP-MS," *Journal of Analytical Atomic Spectrometry*, 30(6): 1286–1296, 2015.

169a. Oberdörster, G., Oberdörster, E., and Oberdörster, J., "Nanotoxicology: An Emerging Discipline Evolving from Studies of Ultrafine Particles," *Environ Health Perspect*, 113(7): 823–839, July 2005.

CHAPTER 4

Reactive Oxygen Species Generation on Nanoparticulate Material

Michael Hoffmann

Department of Engineering and Applied Science, California Institute of Technology, Pasadena, California, USA

Ernest (Matt) Hotze

Department of Civil & Environmental Engineering, Duke University, Durham, North Carolina, USA

Mark R. Wiesner

Department of Civil & Environmental Engineering, Duke University, Durham, North Carolina, USA

Background

Photoactive nanomaterials can be grouped roughly into two different classes: metal oxide or metal chalcogenide photocatalysts such as titanium dioxide or cadmium sulfide, respectively; and materials that can be photosensitized such as chromophores and certain types of fullerenes.

Semiconductor photochemistry behavior is governed by the bandgap that can be described as the energy difference between the valence band (fermi level highest energy electrons) and the conduction band (lowest energy unoccupied molecular orbitals). Light absorbed at wavelength less than or equal to the bandgap energy ($\lambda \leq \lambda_{bg}$) of a semiconductor will result in the promotion of an electron to the conduction band, and as a consequence, a hole or vacancy in the HOMO state of the molecular orbitals is created. Both the promoted electrons and the holes can migrate to the surface of the semiconductor and subsequently react in aqueous solution to form reactive oxygen species such as superoxide (O_2^-) and hydroxyl radical ($^{\cdot}OH$).

Photosensitizers are sensitized by light, and their electrons are excited within the molecular orbitals. Sensitized electrons can behave according to two mechanisms: type II electron transfer involving a donor molecule and type II energy transfer involving no reaction. Both of these pathways will potentially produce reactive oxygen species in solution. The net effect of both these processes is to convert light energy into oxidizing chemical energy.

In engineered systems, oxidation serves many purposes, from elimination of potential hazards to improving aesthetic qualities. Chemical oxidation may be used in a wide range of applications, such as the breakdown of organic compounds such as TCE and atrazine, the oxidation of the reduced states of metals such as iron(II) and arsenic(III), or the inactivation of pathogenic organisms in water treatment.

Oxidation is an electron transfer process in which electrons from a reductant (i.e., an electron donor) are transferred to an oxidant and electron acceptor. The thermodynamic constraints or boundary conditions for electron transfer in aqueous solution are given by the electrochemical potentials or half-cell potentials as illustrated in Equation (4.1):

$$O_2 + 4H^+ + 4e^- \rightleftharpoons 2H_2O \quad E_H^0 = +1.23 \text{ V} \tag{4.1}$$

where oxygen, O_2, is the electron acceptor in this half-reaction leading to the formation of water. The redox potential E_H, under standard conditions, is given as 1.29 V relative to the standard hydrogen electrode (NHE).

In the reduction-oxidation (i.e., redox) scale, reduction potential values (E_H^0) are positive for oxidizing species (i.e., oxidants) negative for reducing species (reductants). The redox potential for any set of conditions other than the standard conditions (i.e., concentrations are fixed at 1.0 M at a temperature of 298.15 K) can be determined with the Nernst equation as follows:

$$E_H = +1.23 - \frac{2.3RT}{nF} \log \frac{[H_2O]^2}{[O_2][H^+]^4} \tag{4.2}$$

for any set of concentration conditions where R is the universal gas constant, T is temperature, n is the number of electrons transferred, and F is the Faraday constant. For example, at pH 7 in water, the reduction potential of O_2 given as follows:

$$E_H^0(W) = +1.23 - \frac{2.3RT}{nF} \log \frac{[a_{H_2O}]^2}{P_{O_2}[10^{-7}]^4} = +0.81 \text{ V} \tag{4.3}$$

where the activity of water, $a_{H_2O} = 1$ by definition, and the partial pressure of oxygen, P_{O_2}, gas are by definition equal to 1 for standard state conditions.

Key reduction potentials for the important oxygen-containing species are given in Table 4.1.

Oxygen Species Half Reaction	E_H (pH 0)	$\Delta G/n$ (kJ mol^{-1})
$O_2 + 4H^+ + 4e^- \rightleftharpoons 2H_2O$	+1.23	−118.56
$O_2 + 2H^+ + 4e^- \rightleftharpoons H_2O_2$	+0.70	−67.47
$^3O_2 + e^- \rightleftharpoons O_2^{-\bullet}$	−0.16	+15.42
$^3O_2 + H^+ + e^- \rightleftharpoons HO_2^{\bullet}$	+0.12	−11.57
$^1O_2 + e^- \rightleftharpoons O_2^{-\bullet}$	+0.83	−80.01
$O_2^+ + e^- \rightleftharpoons O_2$	+3.20	−308.45
$^{\bullet}OH + e^- \rightleftharpoons {}^-OH$	+1.90	−183.14
$^{\bullet}OH + H^+ + e^- \rightleftharpoons H_2O$	+2.72	−262.19
$O^{-\bullet} + H^+ + e^- \rightleftharpoons HO^-$	+1.77	−170.61
$HO_2 + e^- \leftrightarrow \rightleftharpoons HO_2^-$	+0.75	−72.29
$HO_2 + H^+ + e^- \rightleftharpoons H_2O_2$	+1.50	−144.59
$H_2O_2 + 2H^+ + 2e^- \rightleftharpoons 2H_2O$	+1.77	−170.61
$H_2O_2 + e^- \rightleftharpoons {}^{\bullet}OH + H_2O$	+0.72	−69.40
$O_3 + 2H^+ + 2e^- \rightleftharpoons O_2 + H_2O$	+2.08	−200.50
$O_3 + e^- \rightleftharpoons O_3^{-\bullet}$	+1.00	−96.39
$O_3 + H^+ + e^- \rightleftharpoons O_2 + {}^{\bullet}OH$	+1.34	−129.17

$T = 298.15$ K, $P = 1.0$ atm and all concentrations and activities (by definition) are constant at 1.0 M.

TABLE 4.1   Standard State Reduction Potentials of Important Oxygen Species

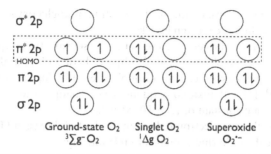

FIGURE 4.1   The 2p molecular orbitals of the dioxygen molecule. Electron spins in the HOMO states for ground-state molecular oxygen are compared to spin-paired singlet oxygen and the one-electron reduction product, superoxide.

For an overall redox reaction, coupling the half-reactions for an oxidant with a reductant, we can write the following simple equation:

$$Ox_1 + Red_2 = Red_1 + Ox_2 \tag{4.4}$$

The corresponding Nernst equation can be expressed in terms of the overall redox potential, E_{rxn}, for a given set of nonstandard conditions as follows:

$$E_{rxn} = E_H^0 - \frac{2.3RT}{nF}\log\frac{[Red_1][Ox_2]}{[Ox_1][Red_2]} \tag{4.5}$$

The highest occupied molecular orbitals (HOMO) of ground state oxygen contain unpaired electrons with parallel spins (Figure 4.1). The parallel spins are characterized by triplet signal response in an applied magnetic field, while antiparallel spins have a characteristic singlet response in a magnetic field.

As a consequence of the unpaired, parallel spins in the ground-state ($^3O_2(^3\Sigma_g^-)$) oxygen is paramagnetic. Thus, molecular oxygen (i.e., dioxygen) has a triplet spin state. However, most ground molecules in the ground electronic state are spin-paired and singlet state. Given Woodward-Hoffmann symmetry constraints, a kinetically favorable reaction occurs when the reacting molecules have the same spin state. As a consequence, triplet-state oxygen, $^3O_2$, is unlikely to be highly reactive with a singlet-state molecule such as propane, $CH_3CH_2CH_3$, due to the symmetry considerations (i.e., the reaction is spin-forbidden), even though the overall thermodynamic driving force for a four-electron transfer is highly favorable. In order to oxidize propane rapidly (i.e., rapid kinetics), ground state oxygen must undergo activation or spin-pairing, which can be achieved in a number of ways. For example, ground-state dioxygen can be activated simply by near IR light as shown in Equation (4.6). In contrast, reactive oxygen species (ROS; Table 4.1) are substantially more reactive since they have overcome symmetry restrictions.

$$O_2(^3\Sigma_g^-) + h\nu \longrightarrow O_2(^1\Delta_g)\{\lambda < 1000\text{ nm}\} \tag{4.6}$$

A primary example of the difference in reactivity between ($O_2(^1\Delta_g)$) and $O_2(^3\Sigma_g^-)$ is illustrated by the difference in rate constants for the same reductant.

$$O_2(^3\Sigma_g^-) + NO_2^- \xrightarrow{k_7} NO_2 + O_2^{-\bullet} \tag{4.7}$$

$$O_2(^1\Delta_g) + NO_2^- \xrightarrow{k_8} NO_2 + O_2^{-\bullet} \tag{4.8}$$

where in water $k_8 \gg k_7$ ($k_8 = 3.1 \times 10^6$ M^{-1}s^{-1}; $k_7 = 0.036$ M^{-1}s^{-1}).

Reactive oxygen species (ROS) include but are not limited to singlet oxygen ($O_2(^1\Delta_g)$), hydroperoxyl radical ($HO_2^\bullet$), superoxide ($O_2^{-\bullet}$), dihydrogen trioxide (H_2O_3, HO_3^-), and hydroxyl radical ($^\bullet OH$). As shown in Table 4.1, reduction potentials for most ROS species are substantially more favorable than $O_2(^3\Sigma_g^-)$.

The lifetime of singlet oxygen in water is on average 3 μs due to deactivation by collision with H_2O according to Equation (4.9).

$$O_2(^1\Delta_g) + H_2O \xrightarrow{k_d} O_2(^3\Sigma_g^-) + H_2O^* \tag{4.9}$$

The first-order decay constant for the deactivation of singlet oxygen in water is $k_d = 3.2 \times 10^5$ M^{-1}s^{-1}, where the characteristic lifetime is given by $1/k_d$.

Singlet oxygen can be generated in nature by sensitization (type II reaction) of dye compounds such as rose bengal or C_{60} molecules, and also by irradiation of naturally occurring humic acids in lakes and rivers [1, 2]. Singlet oxygen reacts with molecules of biological significance such as nucleic acids, lipids, and amino acids [3] with toxic consequences. For example, rhodopsin reacts with $O_2(^1\Delta_g)$ at pH 8.0 with a second-order rate constant of 1.1×10^9 M^{-1}s^{-1}.

Superoxide, the one-electron reduction product of dioxygen (Table 4.1), and its protonated conjugate acid, $HO_2^\bullet$, have the following equilibrium relationship:

$$HO_2 \underset{}{\overset{K_a}{\rightleftharpoons}} O_2^{-\bullet} + H^+ \tag{4.10}$$

where $pK_a = 4.8$. Superoxide is readily formed by electron transfer from sensitized dyes (type I reactions) or by sensitized oxidation of secondary alcohols. In addition, $O_2^{-\bullet}$ is formed in aqueous suspensions of semiconductor photocatalysts (e.g., ZnO or TiO_2), where oxygen is reduced by the photo-excited conduction-band electrons on the surface of the metal oxide or metal sulfide semiconductors. However, the characteristic lifetime of superoxide in aqueous solution is short due to competition from its self-reaction (i.e., dismutation) into oxygen and hydrogen peroxide.

$$HO_2^\bullet + O_2^{-\bullet} \overset{H^+}{\longrightarrow} H_2O_2 + O_2 \tag{4.11}$$

$$HO_2^\bullet + HO_2^\bullet \longrightarrow H_2O_2 + O_2 \tag{4.12}$$

$$O_2^{-\bullet} + O_2^{-\bullet} \overset{2H^+}{\longrightarrow} H_2O_2 + O_2 \tag{4.13}$$

However, there is a pronounced difference in the rates of reaction Equations (4.11) and (4.12). For example, $k_{11} = 9.7 \times 10^7$ M^{-1}s^{-1}; this can be compared to $k_{12} = 8.3 \times 10^5$ M^{-1}s^{-1} and $k_{13} < 2$ M^{-1}s^{-1}. As a consequence of the relative slowness of Equation (4.13), most living cells employ a protein "superoxide dismutase" (SOD) to catalyze the reaction under biological pH conditions (e.g., pH 7.8).

Hydroxyl radical, the three-electron reduction product of dioxygen, is the most highly reactive oxygen species in terms of redox potential and typical electron transfer rates, which are most often in the diffusion controlled regime ($k = 10^9$ to 10^{10} M^{-1}s^{-1}). Hydroxyl radical reacts by three pathways: hydrogen atom extraction, HO radical addition, or by direct electron transfer as illustrated by the following three reactions:

$$CH_3CH_2OH + {}^\bullet OH \longrightarrow CH_3\dot{C}HOH + H_2O \tag{4.14}$$

$$(4.15)$$

$$SO_3^{2-} + {}^\bullet OH \longrightarrow SO_3^{-\bullet} + H_2O \tag{4.16}$$

The array of oxygen-containing radicals known as ROS in solution is known to damage cell membranes, cellular organelles, and nucleic acids contained within RNA and DNA. Moreover, the oxidizing properties of ROS can also be generated on the surface of nanomaterials.

Reactive oxygen species are highly reactive with low selectivity. In addition, ROS species present many challenges for direct time-resolved detection due to their short lifetimes and relatively low concentrations under steady-state conditions. This limitation can be overcome by trapping the free radicals with appropriate chemical trapping agents (e.g., chemical compounds that readily react with ROS and stabilize their unpaired electrons).

Electron paramagnetic resonance (EPR/ESR) detects the small changes that an unpaired electron exerts on an applied magnetic field. The unpaired electron, in a spin state of +1/2 or −1/2, responds to a magnetic field by aligning either parallel or antiparallel to the applied field. Both spin states have distinct energy levels, which are determined by the magnetic field strength. In order to detect these energy states, the sample is exposed to electromagnetic radiation with sufficient energy to excite the electrons from the lower state to the upper state. This energy gap is given by

$$\Delta E = g\beta H \tag{4.17}$$

where ΔE is the energy gap between the $+1/2$ and $-1/2$ state, H is the applied magnetic field, β is the Böhr magneton constant, and g is the splitting factor for the free electron. This splitting factor depends on the atoms within the radical compound being detected. An adsorption spectrum can be obtained by using the applied magnetic field to detect the changes. EPR detectors take the first derivative of initial spectrum.

The spin-trapping molecules affect the signal of the unpaired electron on the free radical species that is detected; this effect is known as splitting and is most commonly caused by adjacent hydrogen atoms. Often the splitting leads to a unique pattern for the EPR signals. The combination of the hyperfine structure (number of lines), line shape, and hyperfine splitting (the distance between peaks) give a radical its unique imprint. An ideal spin-trapping molecule will react quickly and specifically with the radical species of interest and produce a characteristic signal. Two spin traps, which are most frequently used for oxygen radical detection, are 5,5-dimethyl-pyrrolin-1-oxyl (DMPO) and 2,2,6,6-tetramethylpiperidine-N-oxyl (TEMP/TEMPO). Figures 4.2, to 4.4 give the reaction of DMPO with superoxide and hydroxyl radical and TEMP with singlet oxygen.

In each case, the product of the reaction is a nitroxide compound, which is stabilized by charge delocalization between the nitrogen and oxygen atom. Figures 4.5, to 4.7 illustrate typical EPR spectra for TEMPO, DMPO-OOH, and DMPO-OH.

The TEMPO spectrum is a 1:1:1 hyperfine structure that results from the interaction of the unpaired electron with the nitrogen nucleus. Both DMPO-OH and DMPO-OOH have the same interaction, but splitting occurs due to the presence of adjacent hydrogen and oxygen atoms. Hydroxyl radical reacts with DMPO about nine orders of magnitude faster than superoxide, so the DMPO-OH signal will predominate unless hydroxyl radical is quenched (*vide infra*). As a result, superoxide detection with TEMP requires much higher concentrations than hydroxyl detection would. DMPO-OOH can also decompose to DMPO-OH giving a false positive for hydroxy radical, but there are ways to avoid this [4, 5]. Many other spin traps are available, and new ones are developed on a regular basis.

Another common option for detection of reactive oxygen is chemical reduction. Two examples are cytochrome c [6, 7] or nitroblue tetrazolium (NBT) [8–10] reduction. Both of these compounds are reduced by superoxide. Cycochrome c reacts with a rate constant of 2.6×10^5 $M^{-1}s^{-1}$ and NBT with a constant of 6×10^4 $M^{-1}s^{-1}$. While Cytochrome c has a relatively simple reaction pathway, the pathway for NBT is more complex [8]. A simplified version is depicted in Figure 4.8.

This reduction is not selective, so the presence of the reduction products does not assure superoxide activity. Thus, superoxide dismutase [6] (*vide infra*) must be added to quench superoxide and create a baseline reduction level. These reductions can be followed by simple spectrophotometry, which provides a significant experimental advantage to using these compounds to quantify and study ROS production.

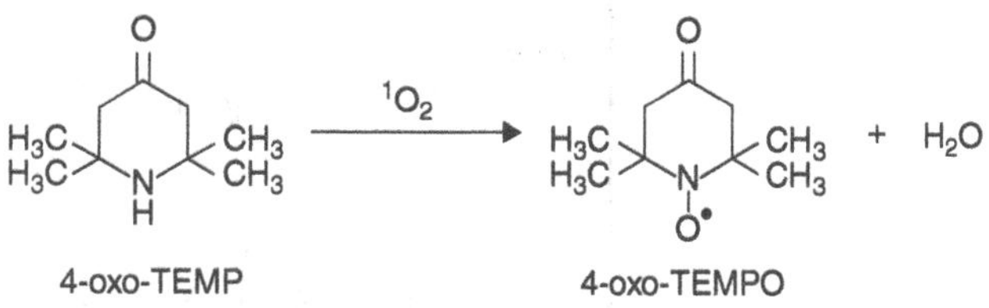

FIGURE 4.2 Reaction of TEMP with singlet oxygen.

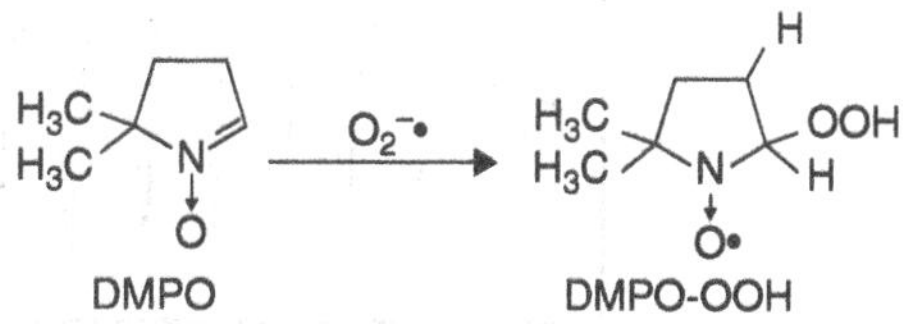

FIGURE 4.3 Reaction with DMPO with superoxide.

DMPO DMPO-OH

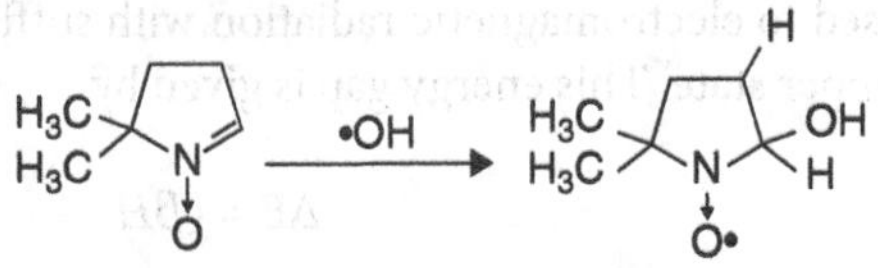

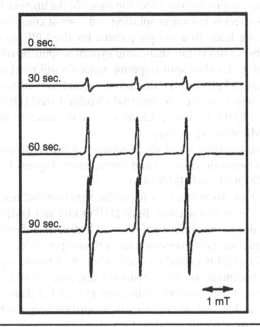

Figure 4.4 Reaction of DMPO with hydroxyl radical.

Figure 4.5 EPR signal of the TEMP-singlet oxygen adduct.

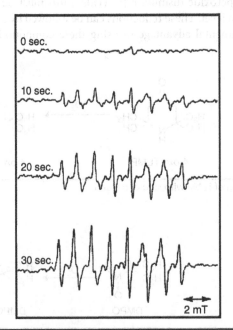

Figure 4.6 EPR signal of the DMPO-OOH adduct.

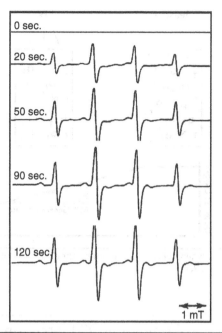

FIGURE 4.7 EPR signal of DMPO-OH adduct.

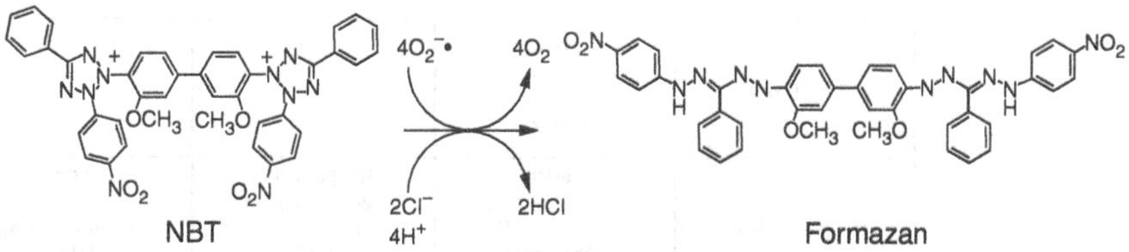

FIGURE 4.8 NBT reduction by superoxide.

Another advantage is the ability to analyze reduction data and determine rates of ROS production. However, concentration detection limits are higher than typically found using EPR methods. As in the case of EPR, there are other chemical reductants available (e.g. XTT).

In a less common approach, ROS is detected by the simple monitoring of dissolved oxygen concentration in a solution. The basic principle is shown below in Figure 4.9. A compound irreversibly traps the reactive oxygen and a drop of dissolved oxygen occurs. A simple handheld meter can be used to monitor this change.

One major advantage of the approach illustrated in Figure 4.9 is that the rate of ROS generation can be obtained from the dissolved oxygen (DO) loss by adding an excess of the trapping compound; theoretically every ROS will be trapped. However, the free radical traps must react with ROS at much higher rates than observed for ROS decay. In previous research, Zepp and coauthors used 2,5-dimethylfuran (DMF) to trap ROS [1, 2], while Haag et al. used furfuryl alcohol (FFA) as the trapping agent [11]. However, FFA reacts with singlet oxygen at relatively high rates (e.g., $k = 1.2 \times 10^8$ $M^{-1}s^{-1}$). It is often difficult to detect low concentrations of ROS when compared with EPR methods (Table 4.2).

Chemical or biochemical quenchers can also be added to target specific ROS. For example, superoxide dismutase is routinely used to quench superoxide in a reaction medium [7]. Beta carotene and azide ion (N_3^-) are known quenchers for singlet oxygen [12]. These compounds serve to eliminate the response seen from any of the above detection methods. If a response is eliminated by the addition of a quencher, it is easier to assume that the specific ROS generated the signal. However, most quenchers lack complete specificity for a single ROS component. For example, superoxide dismutase seems to have an effect on singlet oxygen production despite its supposed specificity [13] for superoxide.

Alternative methods for detection of ROS include fingerprinting of reaction products and direct chemiluminescence detection (e.g., singlet oxygen luminescence can be quantified at 1270 nm [14]).

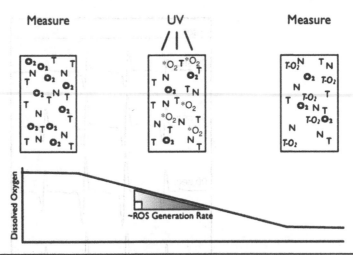

FIGURE 4.9   Dissolved oxygen ROS measurement method. O_2 = dissolved oxygen; T = ROS trap; N = nanoparticle; $^*O_2$ = ROS; T-O_2 = trapped ROS.

Common ROS Detection Methods			
Type of ROS (decreasing oxidative power)	Electron Paramagnetic Resonance (EPR)	Spectrophotometric	Dissolved Oxygen Measurement
Hydroxyl	**DMPO** [Yamakoshi 2003] [Dugan 1997] **DMSO**	N/A	N/A
Superoxide	**DMPO** [Yamakoshi 2003] [Dugan 1997] **DEPMPO** [Frejaville 1995]	**NBT** [Saito 1983] **Cytochrome C** [You 2003] **XTT** [Okado-Matsumoto 2001]	**FFA** [Haag 1984] [Pickering 2005] **DMF** [Zepp 1985]
Singlet Oxygen	**TEMP** [Wilkinson 1993]	**Dipropionic acid** [Lindig 1981]	**FFA** [Haag 1984] [Pickering 2005] **DMF** [Zepp 1985]

TABLE 4.2   A Brief Summary of ROS Detection Methods

Nanoparticulate Semiconductor Particles and ROS Generation

Semiconducting metal oxides and metal chalcogenides have been used as catalysts for a wide variety of chemical reactions in the gas-solid phase and liquid-solid reactions over a broad range of temperatures from <0 to >500°C [15]. Semiconducting oxides and sulfides can be activated by an applied electrical potential, by the absorption of photons, or by elevated temperatures. The energies required for activation of some common metal oxide and metal chalcogenide semiconductors are given in Tables 4.3 and 4.4, respectively.

Activation can occur by the input of energy (i.e., eV or nm), which is sufficient to promote an electron from the valence band to the conduction band or from energy states that lie within the bandgap (trapped or doped states), as illustrated in Figures 4.9 and 4.10. For example, absorption of photons (with $E_{hv} > E_{bg}$) by a semiconductor leads to promotion of an electron to the conduction band, e_{cb}^-, and at the same time creation of a vacancy, h_{vb}^+, in the valence band as illustrated in Figure 4.10.

The relative energy levels of the conduction band and valence band states for some of the representative semiconductors listed in Tables 4.3 and 4.4 are shown in Figure 4.11. In the specific case of TiO_2, a semiconductor used in numerous commercial products, the conduction band edge has a relative energy level at pH 0 of −0.4 V and the valence band hole has an electrochemical potential of +2.8 V. Thus, in the case of TiO_2 the conduction band electron is mildly reducing although with sufficient potential to reduce O_2 to superoxide, O_2^-, while the valence band hole at the surface of TiO_2 is a powerful oxidizing agent.

Semiconductor	E_g (eV)	E_g (nm)
TiO_2—anatase	3.2	385
TiO_2—rutile	3.0	413
ZnO	3.3	376
$SrTiO_3$	3.2	387
WO_3	2.8	443
Nb_2O_5	3.4	365
Bi_2O_3	3.2	387
CeO_2	3.4	365
In_2O_3	3.1	403
SnO_2	3.7	330
$MnTiO_3$	2.8	443
$FeTiO_3$	3.2	390
$BaTiO_3$	3.2	385
$CaTiO_3$	2.8	448
$PbTiO_3$	3.4	365
$Bi_4Ti_3O_{12}$	3.1	403
$Bi_2Ti_2O_7$	3.0	420
$Bi_{12}TiO_{20}$	3.1	386
t-ZrO_2	5.0	248
m-ZrO_2	5.2	238
Fe_2O_3	2.2	564
$YFeO_3$	2.58	481
CeO_2	3.4	365
In_2O_3	3.6, 3.1	344, 400
IrO_2	3.12	397
MoO_3	3.24	383
Bi_2WO_6	2.69	461
$K_3Ta_3Si_2O_{13}$	4.1	302
$LiTaO_3$	4.7	264
$NaTaO_3$	4.0	310
$KTaO_3$	3.3	376
$KNbO_3$	3.8	326
γ-Bi_2O_3	2.8	443
$K_4Nb_6O_{17}$	3.3	376
$CaTa_2O_6$	4.0	310
$SrTa_2O_6$	4.4	282
$BaTa_2O_6$	4.1	302
$Sr_2Ta_2O_7$	4.6	270
K_2PrTaO_{15}	3.8	326
$Sr_2Nb_2O_7$	3.9	318
$AgNbO_3$	2.9	428

*1 eV = 4.42×10^{14} Hz (s^{-1}); $\lambda = \dfrac{hc}{E_g(eV)v(Hz\ eV^{-1})}$; C = 299792458 m s^{-1}; h = 4.14×10^{-15}

TABLE 4.3   Band-Gap Energies for Metal Oxide and Mixed-Metal Oxide Semiconductors [15]

Semiconductor	E_g (eV)	E_g (nm)
$BiVO_4$	2.4	517
Ag_3VO_4	2.0	620
Bi_2WO_4	2.8	443
$La_2Ti_2O_7$	2.8	443
Bi_2YNbO_7	2.0	620
Bi_2CeNbO_7	2.10	590
Bi_2GdNbO_7	2.13	582
Bi_2SmNbO_7	2.21	561
Bi_2NdNbO_7	2.25	551
Bi_2PrNbO_7	2.26	549
Bi_2LaNbO_7	2.38	521
Ga_2O_3	4.98	249
SnO_2	3.5	354
In_2O_3	2.7	459
Cu_2O	2.6, 2.0	477, 620
$CdWO_4$	2.94	422
$CdMoO_4$	2.43	510

TABLE 4.3 Band-Gap Energies for Metal Oxide and Mixed-Metal Oxide Semiconductors [15] (*Continued*)

For example, with use of the appropriate semiconductor catalysts, it is possible to drive the photoelectrolysis of water into hydrogen and oxygen as follows:

$$2H_2O \xrightarrow{\text{UV \& visible light}} 2H_2 + O_2 \qquad (4.18)$$

The overall multielectron redox potentials for the multielectron reactions at pH 7 suggest that H_2 production with the input of UV-visible light energy is feasible.

$$H_2O \longleftrightarrow H^+ + OH^- \qquad (4.19)$$

$$2H_2O + 2\,\overline{e} \longleftrightarrow H_2 + 2\,{}^-OH \qquad (4.20)$$

$$2H_2O \longleftrightarrow O_2 + 4H^+ + 4\overline{e} \qquad (4.21)$$

The redox potential for Equation (4.21) at pH 7 is $E_H = -1.23$ V (NHE) with the corresponding half-reactions of -0.41 V [Equation (4.20)] and 0.82 V [Equation (4.21)], which gives a $\Delta G° = +237$ kJ/mole). However, with the input of light at wavelengths ≤ 1000 nm (i.e., 1.23 eV $\approx$ 1000 nm), the overall energy requirement for the photosynthetic splitting of water can be met with solar radiation in principle. On the other hand, the rate of reaction in the normal Marcus regime should depend on the overall driving force (i.e., lower wavelength irradiation is preferable kinetically) and the thermodynamics of the initial or sequential one-electron transfer processes at the semiconductor surfaces. Moreover, the one-electron transfers are much less favorable thermodynamically than the overall two-electron transfer reactions as shown below:

$$H^+ + e^-_{aq} \underset{}{\overset{E_H = -2.5 \text{ V (pH 7)}}{\rightleftharpoons}} H^{\bullet}_{aq} \qquad (4.22)$$

$$H_2O \underset{}{\overset{E_H = -2.3 \text{ V (pH 7)}}{\rightleftharpoons}} {}^{\bullet}OH + H^+ + e^-_{aq} \qquad (4.23)$$

$${}^-OH \underset{}{\overset{E_H = -1.8 \text{ V (pH 0)}}{\rightleftharpoons}} {}^{\bullet}OH + e^-_{aq} \qquad (4.24)$$

Semiconductor	E_g (eV)	E_g (nm)
ZnS	3.66	339
ZnSe	2.90	428
ZnTe	2.25	551
CdS	2.45	506
CdSe	1.74	713
CdTe	1.45	855
HgS	2.2	564
$CuAlS_2$	3.4	365
$CuAlSe_2$	2.7	459
$AgGaS_2$	2.5	496
$CuGaS_2$	2.4	517
$AgInS_2$	1.8	689
GaS_2	3.4	365
Ga_2S_3	2.8	443
GaS	2.5	496
As_2S_3	2.5	496
Gd_2S_3	2.55	486
La_2S_3	2.91	426
MnS	3.0	413
Nd_2S_3	2.7	459
Sm_2S_3	2.6	477
ZnS_2	2.7	459
ZrS_2	1.82	681
$Zn_3In_2S_6$	2.81	441
Pr_2S_3	2.4	517

TABLE 4.4 Band-Gap Energies for Metal Sulfide and Chalcogenide Semiconductors [15]

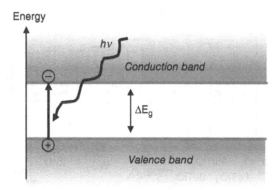

FIGURE 4.10 Schematic representation of the energy gap between the HOMO states or valence band of a semiconductor and the LUMO states or conduction band, where ΔE_g corresponds to the band-gap energy in either energy (eV) or wavelength (nm).

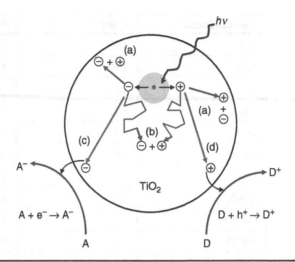

Figure 4.11   Presented above is a graphic depiction of a spherical semiconductor particle, which has absorbed a photon of sufficient energy to promote and electron from the conduction band to the valence band. After charge separation, the electron, e_{cb}^- and the hole, h_{cb}^+, may simply recombine either direct band-gap recombination (a) or indirect recombination from trapped states (b) e_{tr}^- or h_{tr}^+. Some of mobile electrons and holes can migrate to the surface of the nanoparticulate semiconductor where they can undergo electron transfer reactions (c and d). The trapped electron can be transferred to an electron acceptor, A, and the trapped hole can accept an electron from an electron donor, D.

In the presence of oxygen, the reductive process in water leading to superoxide ion, $O_2^{-\bullet}$ has the following half-reaction and a reduction potential of $E_H = -0.33$ V: at pH 7.

$$O_2 + \bar{e} \underset{pH=7}{\overset{E_H = -0.33\ V}{\rightleftharpoons}} O_2^{-\bullet} \tag{4.25}$$

On the oxidative side, the water or hydroxyl radical has corresponding potential at pH 0:

$$^-OH \underset{pH=0}{\overset{E_H = -1.8\ V}{\rightleftharpoons}} {}^\bullet OH + e_a^- \tag{4.26}$$

However, when working with hydrated metal oxide surfaces one must take into account the nature of the reactive surface sites that normally involve metal hydroxyl functionalities. In the specific case of dehydrated TiO_2 (see Figure 4.12), we must consider that when exposed to water either in a humid atmosphere or in

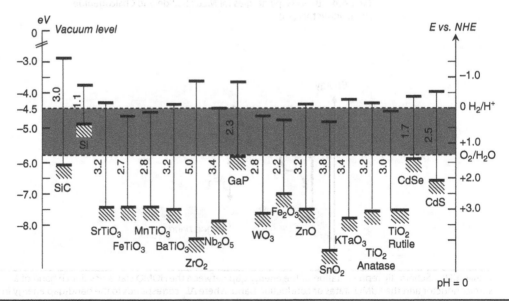

Figure 4.12   Comparison of the band-gap energies for an array of semiconductor relative to the reduction and oxidation potentials of water at pH 0.

aqueous suspension the surface titanium-oxygen bonds in the crystallites undergo hydrolysis to produce surface hydroxyl groups as follows [16–27]:

$$> Ti\text{-}O\text{-}Ti < + H_2O \underset{k_d}{\overset{k_h}{\rightleftharpoons}} 2 > TiOH \tag{4.27}$$

The metal hydroxyl surface sites (e.g., >TiOH or >FeOH) exist under ambient conditions in an open atmosphere or in an aqueous suspension. Once hydrolyzed, the surface hydroxyl groups either gain or lose a proton depending on the intrinsic acidity of the metal oxide [24, 28]. A simple formalism for characterizing the surface acidity versus pH, for quantifying the surface buffering capacity, the surface ion-exchange properties, and surface complexation capacity for cations, anions, and ligands is presented in Equations (4.28) and (4.29). The pH dependent changes in terms of the acid-base chemistry of surface hydroxyl functionalities (e.g., >MOH, >TiOH, >FeOH) can be treated as a conventional diprotic acid, although there may be more than one type of surface site undergoing protonation and deprotonation (i.e., a distribution of surface acidity constants, K_a^s.

In the case of nanoparticulate TiO_2, the titration of a colloidal suspension with NaOH gives a classical titration curve for a diprotic acid as shown in Figure 4.13. Using the titration data of Figure 4.13, the surface acidity of TiO_2 can be characterized is terms of two surface acidity constants as follows:

$$< TiOH_2^+ \underset{k_{-1}}{\overset{k_1}{\rightleftharpoons}} < TiOH + H^+$$

$$K_{a1}^s = \frac{k_1}{k_{-1}} \qquad pK_{a1}^s = 2.4 \tag{4.28}$$

$$< TiOH \underset{k_{-2}}{\overset{k_2}{\rightleftharpoons}} < TiO^- + H^+$$

$$K_{a2}^s = \frac{k_2}{k_{-2}} \qquad pK_{a2}^s = 8.0 \tag{4.29}$$

$$pH_{zpc} = \frac{\left(pK_{a1}^s + pK_{a2}^s\right)}{2} = 5.25 \tag{4.30}$$

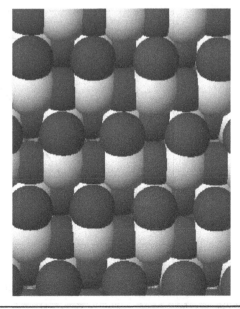

Figure 4.13 The anhydrous TiO_2 surface looking at the predominant 101″ crystalline face of TiO_2 (anatase) showing oxygen in dark grey (O^{2-}) and titanium in white (Ti^{4+}).

In the simplest case, at low ionic strength in the absence of added cations or anions, the isoelectric point or "point of zero charge" is described as follows in terms of the concentrations of the relevant surface species as a function of pH:

$$[> TiOH_2^+] = [> TiO^-] \tag{4.31}$$

At a fixed ionic strength, the surface charge on TiO_2 is a function of the solution pH as follows [24, 28]:

$$\sigma_0\big|_{[\mu]=\mathrm{const}} = \frac{\partial \sigma_0}{\partial \mathrm{pH}}\bigg|_{\mathrm{pH} \neq \mathrm{pH}_{zpc}} (\mathrm{pH} - \mathrm{pH}_{zpc}) \tag{4.32}$$

The above acidity constants and the pH of zero point of charge, pH_{zpc}, are given for quantum-sized TiO_2 in the particle size range of 1.0 to 3.0 nm. The titration data can be plotted with respects to the surface site density as shown in Figure 4.15 to get more reliable estimates to the pH_{zpc} for most semiconductors listed in Tables 4.3 and 4.4.

The number of moles of surface hydroxyl groups {>MOH} in an aqueous suspension of volume V can be estimated as follows:

$$\{>MOH\} \approx \frac{(m_{MO}/\rho_{MO} V_{MO}) A_{MO} d_{OH} (10^{18}\ nm^2/1\ m^2)}{N} \tag{4.33}$$

where A_{MO} is the surface area of the metal oxide particle (m²), d_{OH} is the site density of hydroxyl groups for MO (number per nm²), N is Avogadro's number, m_{MO} is the mass of the metal oxide, MO, in a suspension (g) of a given volume, V is the volume of a MO particle (cm³), and ρ is the density of the MO (g/cm³). The density of surficial >MOH groups, d_{OH}, ranges from 4 to 10 per nm².

Irradiation of TiO_2 (as a primary example) generates electrons and holes on a 100-femtosecond to picosecond time scale. The approximate energy level positions for TiO_2 in the anatase crystalline form are shown in Figure 4.11 with an equivalent scale of redox potentials versus NHE (normal hydrogen electrode). After photoexcitation (Figure 4.16) some of the electrons and holes migrate to the surface where they can be trapped within the nanosecond timeframe on the surface of TiO_2 in the form of >Ti(III)OH and >Ti(IV)OH•.

The surface hydration and dehydration process and photoexcitation can be followed with DRIFT (diffuse reflectance infrared Fourier transform) spectroscopy [29–31]. In Figure 4.18, evidence for the reversible hydration (Figure 4.17) and dehydration of TiO_2 is shown where in the presence of a humid atmosphere with water peak in the IR spectrum is clearly visible near 3500 cm⁻¹.

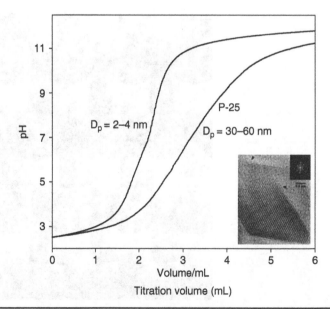

FIGURE 4.14 A typical titration curve for two different size fractions of colloidal TiO_2. This data can be used to determine the surface acidity constants and provided an estimate of the pH of zero point of charge.

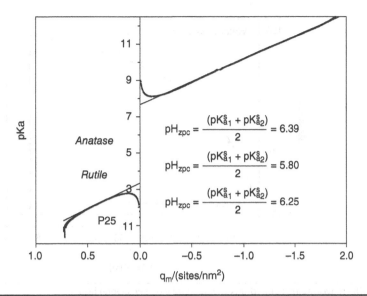

FIGURE 4.15   Using titration data similar to that presented in Figure 4.14, the individual surface acidity constants and the surface site density can be determined. The pH_{zpc} for anatase, rutile and a mixture of anatase and rutile (Degussa P25) are shown here.

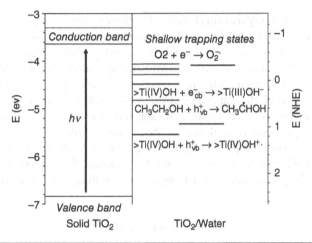

FIGURE 4.16   Energy level positions for the photoexcitation of TiO_2 ($\Delta E_g = 3.2$ eV) in the anatase form relative to the solid-solution interface redox potentials for key steps and possible electron transfer reactions. Surface trapping states within the band-gap energy domain are indicated.

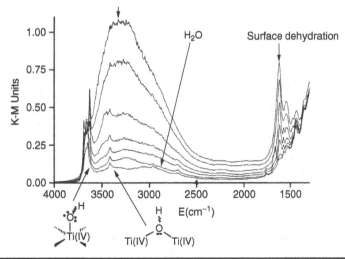

FIGURE 4.17   DRIFT spectra of TiO_2. The broad band spanning 2500 to 3900 cm^{-1} is due to >TiOH stretching vibrations in different atomic environments. With progressive dehydration, this characteristic feature disappears, and discrete stretches within 3400 to 3800 cm^{-1} arise. Complete dehydration required thermal treatment for 12 hours at 623 K under a ~1 µTorr vacuum. Dehydrated TiO_2 is reversibly rehydrated with water vapor. Surface trapping states are clearly indicated in the dehydrated spectra appear at 3716 cm^{-1}.

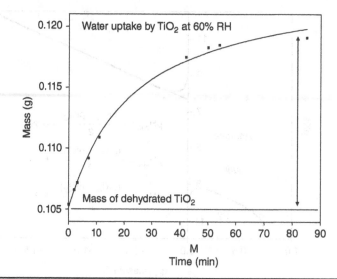

FIGURE 4.18   The rehydration of anhydrous TiO₂ takes place over several hours.

In a typical photolysis experiment (Figure 4.19), there is convincing FTIR evidence for the formation of >Ti(III)OH (e_{tr}^-) and >Ti(IV)OH• (h_{tr}^+) as the surface trapping state. The electron is trapped as >Ti(III)OH (a reduced Ti site on the surface) and the hole is trapped as surface-bound hydroxyl radical, >Ti(IV)OH•. The trapped hole, >Ti(IV)OH•, is a powerful oxidizing species that is able to undergo either direct electron transfer reactions or hydrogen abstraction reactions, depending of the chemical nature of the electron donor. On the other hand, the trapped electron (>Ti(III)OH) is a moderate to weak reductant, although it is still capable of transferring electrons to dioxygen (O₂) adsorbed on the surface. The DRIFT spectra of Figure 4.19 show a pronounced peak shift in the presence of oxygen (i.e., $O_2 + e_{tr}^- \rightarrow O_2^-$) versus the same system in a vacuum. For example, the band that appears at 3716 cm⁻¹ has been identified as the trapped electron, >Ti(III)OH, which results from the localization of conduction band electrons in a surface trapping sites. In the presence of O₂, the trapped electron disappears and surface-bound superoxide, $O_2^{-•}$, is formed.

$$O_2 + e_{tr}^- \rightleftarrows O_2^{-•} \tag{4.34}$$

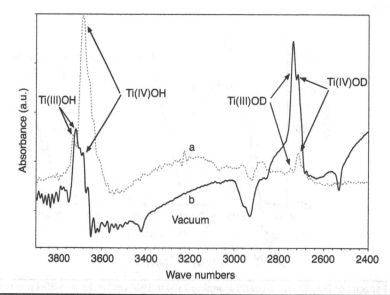

FIGURE 4.19   Surface functional groups, >Ti(III)OH and >Ti(IV)OH groups reflecting the trapping of electrons and holes after illumination under a vacuum.

As an alternative, Equation (4.34) can be written as

$$O_2 + >Ti(III)OH^- \rightleftharpoons O_2^{-\bullet} + >Ti(IV)OH \tag{4.35}$$

The trapped electron, $>Ti(III)OH^-$, is completely removed by exposure to Br_2 in the gas phase. The trapped electron has an ESR (electron spin resonance) signal with $g_\perp = 1.957$, and $g_\parallel = 1.990$. However, after irradiation the 3716 cm^{-1} band persists under a 1.0 atmosphere of dry (0 percent RH) O_2 at 300 K, whether in the dark or under illumination. Whereas the TiO_2 surface must be exposed to water vapor before the trapped electron band at 3716 cm^{-1} disappears, *ab initio* calculations of oxygen-deficient TiO_2 in the rutile form indicate that excess charge in the bulk remains spin-paired and localized at vacant oxygen sites. Inter-band-gap states on reduced TiO_2 surfaces are associated with spin-polarized Ti(III) $3d^1$ and Ti(II) $3d^2$ configurations.

The conduction band electrons in the trapped state electrons have been observed using a variety of laser-based pump-probe photolysis experiments. The so-called blue electron in either a deep or swallow state trap—that is, an internal Ti(III) site or a surface $>Ti(III)OH$ site—has a characteristic spectrum in the visible with a band peak at 600 nm. The appearance and disappearance of the "blue electron" can be followed kinetically by rapid-scan spectroscopy in order to get an estimate of the actual lifetimes of mobile electrons that are available for electron transfer.

The trapped electron and hole can recombine over a broad range of timeframes from nanoseconds to hours depending on the energy of the trapped state. In the absence of external electron donors or acceptors, the shallow trapped electrons can persist for up to 40 hours before eventual recombination as shown in Figure 4.20.

$$>Ti(IV)OH^{+\bullet} + >Ti(III)OH^- \xrightarrow{\text{recombination}} 2 >Ti(IV)OH \tag{4.36}$$

In the presence of O_2, recombination can occur through reversible electron transfer from $>Ti:O_2^{-\bullet}$ back to a bound surface hydroxyl radical as follows:

$$>Ti(IV)OH^{+\bullet} + >Ti(IV)OH:O_2^{-\bullet} \xrightarrow{\text{recombination}} 2 >Ti(IV)OH + O_2 \tag{4.37}$$

However, the majority of electrons in bandgap excited TiO_2 exist in a free state, giving rise to a broad, featureless absorption with intensity proportional to $(\lambda/\mu m)^{1.73}$. These electrons decay according to a saturation kinetic mechanism that is limited by the density of trapped states. Kinetic observations suggest that free charge carriers are relatively stable in the bulk phase, that surface charge trapping is a reversible

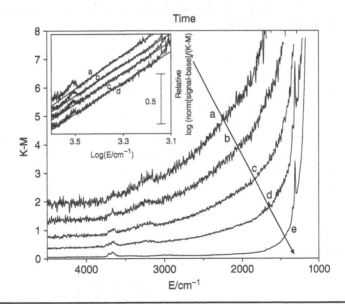

FIGURE 4.20   DRIFT spectra of TiO_2 (a) 13 minutes after UV irradiation, (b) 52 minutes after, (c) 94 minutes after, (d) 135 minutes after, and (e) following full relaxation of the excited state signal. Inset: $\log(S_{norm})$ versus $\log(\bar{v}\ (cm^{-1}))$, where S_{norm} is the background subtracted signal normalized to $S = 1$ at 2500 cm^{-1}.

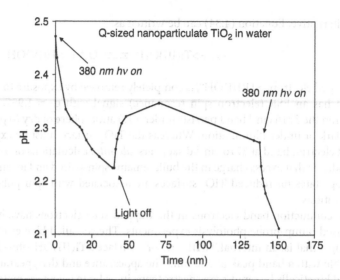

FIGURE 4.21   Irradiation of TiO$_2$ quantum-sized colloids (D_p = 2 nm) in water also produces trapped electrons and holes, which lead to shifts in the measured pH in the presence and absence of light with no added electron donor or acceptor except H$_2$O.

process, and that recombination of trapped states does not necessarily occur rapidly, even in the presence of an opposing charge acceptor. The charge trapping capacity of the surface eventually decreases following the complete recombination of a large number of free carriers. The charge carrier recombination rate increases with increasing surface hydration, indicating that surface hydroxylation assists annihilation reactions by allowing irreversible electron trapping. A surface electron trap state observed at 0.42 eV may be responsible for the mediating the annihilation mechanism.

A number of alternative spectroscopic techniques have been applied to characterize the lifetime of the mobile electrons in TiO$_2$ and other semiconductors. Martin et al. [29, 30] used microwave frequency and Herrmann et al. [31] used radio frequency spectroscopy as a probe technique after laser excitation and determined a broad range of lifetimes for the various trapping states from microseconds to milliseconds.

Most metal oxide and mixed metal oxide semiconductor surface chemistry is dominated by hydroxyl groups when in the presence of water or humid air. For example, the metal niobates, LiNbO$_3$ and KNbO$_3$, are widely used electro-optic and photorefractive materials that depend on the activation of surface protons (i.e., protons bound in hydroxyl ions, $^-$OH). The hydroxyl bound protons have activation energies in the range of 1 eV for mobility in LiNbO$_3$ and KNbO$_3$ crystals. The corresponding surface hydration in KNbO$_3$ leads to the following reactions:

$$> Nb(V)ONb(V) < + H_2O \rightleftharpoons 2 > Nb(V)OH \tag{4.38}$$

The photochemistry and photophysics of quantum-sized nanoparticles [16, 22, 24, 35] (1 nm < D_p < 10 nm) in contrast with the larger, bulk-phase particles ($D_p \approx 100$ nm), can be quite different. Many nanoparticulate semiconductors, which are often described as "quantum-sized particles" or "quantum dots" depending on their applications, exhibit a characteristic blue shift in the UV or visible absorption spectrum. Along with the blue shift in the absorption spectrum, there is a corresponding increase in the band-gap energy, ΔE_g which can be described in terms of a simple solution to the Schrödinger equation with an appropriate Hamiltonian.

$$\Delta E_g \simeq \left(\frac{\pi^2 \bar{h}^2}{2R^2} \frac{1}{\mu} \right) - \frac{1.8e^2}{\varepsilon R} \tag{4.39}$$

where R is the particle radius and μ is the reduced mass of the exciton or the electron-hole pair.

$$\frac{1}{\mu} = \left(\frac{1}{m_{e^-}^*} + \frac{1}{m_{h^+}^*} \right) \tag{4.40}$$

$m_{e^-}^*$ is the effective mass of the electron and $m_{h^+}^*$ is the effective mass of the hole, and ε is the dielectric constant of TiO$_2$, and $\bar{h}$ is Planck's constant.

According to Equation (4.39), as R decreases the band-gap energy, ΔE_g, increases (i.e., $R\downarrow = \Delta E_g\uparrow$). As an example, Kormann et al. [24] prepared Q-sized TiO_2 with a characteristic blue shift from the bulk state bandgap of 385 nm for anatase TiO_2 down to 350 nm (i.e., $\Delta E_g = 3.2$–3.35 eV). The steady-state particle size ranged from 2.0 to 2.5 nm depending on the preparation conditions and the Ti(IV) reagent used in the synthesis (e.g., $TiCl_4$ or Ti(IV)-isopropoxide). The corresponding cluster size (oligomer) for the nanoparticles ranged from 120 to 220 monomers. In an earlier study, Bahnemann et al. [16] reported that Q-ZnO exhibited bandgap increases as larger as 1 eV or $\Delta E_{g,Q\text{-}ZnO} = 4.2$ eV).

An increase in ΔE_g often enhances the reactivity of the photocatalyst by increasing its reduction/oxidation potential and thus the driving force for electron transfer in the normal Marcus regime; thereby ROS (reactive oxygen species) should also be a function of particle size.

In a subsequent study, Hoffman et al. [32] investigated the photochemical production of H_2O_2 on irradiated Q-ZnO over the wavelength range of $320 \leq \lambda \leq 370$ nm in the presence of carboxylic acids and oxygen. Steady-state concentrations up to 2 mM H_2O_2 were formed. Maximum H_2O_2 concentrations were obtained only with added electron donors (i.e., hole scavengers). The order of photochemical efficiency for H_2O_2 production with carboxylic acids as electron donors was $HCO_2^- > C_2O_4^{2-} > CH_3CO_2^- >$ citrate. Isotopic labeling of the electron acceptor, O_2, with ^{18}O verified that H_2O_2 was produced directly by the reduction of adsorbed oxygen by conduction band electrons. Quantum yields were as high as 30 percent for H_2O_2 production at low photon fluxes. At the same time, the quantum yield was shown to vary with the inverse square root of absorbed light intensity [i.e., $\phi \propto (\sqrt{I_{abs}})^{-1}$], with the wavelength of excitation $\phi \propto (\lambda)^{-1}$, and with the diameter of the Q-sized colloids (i.e., $\phi \propto D_p^{-1}$). For example, $d[H_2O_2]/dt$ is 100 to 1000 times faster on Q-sized ZnO particles ($D_p = 2$–4 nm) than with bulk-phase ZnO particles ($D_p = 100$ nm).

Hydrogen peroxide production proceeds, after initial photoactivation, by electron transfer from the conduction band to dioxygen adsorbed on the surface of the excited state metal oxide as follows:

$$2[e_{cb}^- + O_2 \longrightarrow O_2^{-\bullet}] \tag{4.41}$$

$$O_2^{-\bullet} + H^+ \underset{\longleftarrow}{\overset{pK_s = 4.8}{\longrightarrow}} HO_2^\bullet \tag{4.42}$$

$$2HO_2^\bullet \longrightarrow H_2O_2 + O_2 \tag{4.43}$$

Hoffmann and coworkers [25, 33, 34] observed a tenfold increase in the measure quantum yield for H_2O_2 production upon reduction of the mean particle diameter from 40 to 23 nm for ZnO, where O_2 was the electron acceptor and small molecular organic compounds (e.g., carboxylic acids and alcohol) the electron donor. Similar effects were reported by Hoffmann and coworkers [35, 38, 39] for photo-polymerization reactions catalyzed by Q-sized CdS, Q-ZnO, and Q-TiO_2 and for SO_2 oxidation in the aqueous phase.

In addition to ROS generated from surface hydroxyl species and from adsorbed O_2, there are other oxygen-containing free radical species that are generated on the surface of photoactivated semiconductors. For example, S(IV) ($[S(IV)] \equiv [SO_2 \times H_2O] + [HSO_3^-] + [SO_3^{2-}]$) is readily photooxidized [28] in the presence of colloidal suspensions of nanoparticulate α-Fe_2O_3.

$$O_2 + 2HSO_3^- \xrightarrow[\alpha\text{-}Fe_2O_3]{h\nu \leq 520\,nm} 2SO_4^{2-} + 2H^+ \tag{4.44}$$

Quantum yields ranged from 0.08 to 0.3 with a maximum yield found at pH 5.7. The primary initiation pathway involved irradiation at wavelengths equal to or less than the nominal bandgap of hematite, which is 2.2 eV or 560 nm. Upon band-gap illumination, conduction-band electrons and valence-band holes are separated; the trapped electrons are transferred either to surface-bound dioxygen or to Fe(III) sites on or near the surface, while the trapped holes accept electrons from adsorbed SO_3^{2-} to produce surface-bound $SO_3^{-\bullet}$. The relatively high quantum yields are attributed in part to the desorption of $SO_3^{-\bullet}$ from the α-Fe_2O_3 surface and subsequent initiation of a homogeneous aqueous-phase free radical chain oxidation of S(IV) to S(VI). The following photochemical rate expression describes the observed kinetics over a broad range of conditions:

$$-\frac{d[S(IV)]}{dt} = \phi I_0 (1 - 10^{-\varepsilon[\alpha\text{-}Fe_2O_3]\ell}) \left(\frac{K_s[HSO_3^-]}{1 + K_s[HSO_3^-]} \right) \tag{4.45}$$

where the quantum yield ϕ is defined as follows [40, 41]:

$$\phi_i(\lambda) \equiv \frac{\text{\# of molecules reacting via pathway i}}{\text{total number of photons absorbed by reacting molecule}} \tag{4.46}$$

or

$$\phi_r(\lambda)\,(\text{mol einstein}^{-1}) = \frac{\text{moles of compound transformed}}{\text{moles of photons absorbed}} \tag{4.47}$$

where

$$\sum_i \phi_i = 1 \tag{4.48}$$

A similar kinetic expression [38] was observed for the photocatalytic oxidation of S(IV) on TiO$_2$. In this case, for $\lambda \leq 385$ nm, quantum yields in excess of unity (e.g., $0.5 \leq \phi \leq 300$) were observed and attributed also to desorption of the SO$_3^{\bullet}$ radical anion from the TiO$_2$ surface leading to the initiation of homogeneous free-radical chain reactions. These chain reactions have an amplified effect on the measured quantum efficiency. Depending on the free-radical chain length, the measured ϕ values can be greater than 1. In addition, the observed quantum yields depend on the concentration and nature of free-radical inhibitors present in the heterogeneous suspension.

For SO$_2$, in water, the free-radical chain reactions involve the formation of sulfur radical species such as SO$_3^{-\bullet}$, SO$_4^{-\bullet}$, and SO$_5^{-\bullet}$ that are alternative forms of ROS with similar reactivity to superoxide and hydroxyl radicals.

Iron oxides and iron oxide polymorphs initiate the chain reaction as follows:

$$O_2 + 2HSO_3^- \xrightarrow[\alpha\text{-Fe}_2\text{O}_3]{h\nu} 2HSO_4^- \tag{4.49}$$

$$>\!FeOH + HSO_3^- \rightleftharpoons\ >\!FeSO_3^- + H_2O \tag{4.50}$$

$$>\!Fe(III)SO_3^- + h_{vb}^+ \rightleftharpoons\ >\!Fe(III) + SO_3^{-\bullet} \tag{4.51}$$

$$SO_3^{-\bullet} + O_2 \longrightarrow SO_5^{-\bullet} \tag{4.52}$$

$$SO_5^{-\bullet} + SO_3^{2-} \longrightarrow SO_4^{2-} + SO_4^{-\bullet} \tag{4.53}$$

$$SO_5^{-\bullet} + SO_3^{2-} \longrightarrow SO_5^{2-} + SO_3^{-\bullet} \tag{4.54}$$

$$SO_4^{-\bullet} + SO_3^{2-} \longrightarrow SO_4^{2-} + SO_3^{-\bullet} \tag{4.55}$$

$$SO_5^{2-} + SO_3^{2-} \longrightarrow 2SO_4^{2-} \tag{4.56}$$

The other iron(III) oxide polymorphs (e.g., γ-FeOOH, $\Delta E_g = 2.06$ eV; β-FeOOH, $\Delta E_g = 2.12$ eV; α-FeOOH, $\Delta E_g = 2.10$ eV; δ-FeOOH, $\Delta E_g = 1.94$ eV; γ-Fe$_2$O$_3$, $\Delta E_g = 2.03$ eV) are photocatalytic for certain reactions [42]. In general, the observed order of relative photochemical reactivity toward SO$_2$ oxidation [43] is γ-FeOOH $> \alpha$-Fe$_2$O$_3 > \gamma$-Fe$_2$O$_3 > \delta$-FeOOH $> \beta$-FeOOH $> \alpha$-FeOOH, while in the case of C$_2$O$_4^{2-}$ oxidation [44–46] the relative catalytic order is ferrihydrite (am-Fe(OH)$_3$, γ-Fe$_2$O$_3 > \gamma$-FeOOH $> \alpha$-Fe$_2$O$_3 > \alpha$-FeOOH $> \beta$-FeOOH.

$$2>\!FeOH + h\nu \xrightarrow{e^-/h^+}\ >\!Fe(II)OH^- +\ >\!Fe(IV)OH^+ \tag{4.57}$$

$$e_{tr}^- \qquad\qquad h_{tr}^+$$

$$>\!Fe(IV)OH^+ + C_2O_4^{2-} \longrightarrow\ >\!Fe(III)OH + CO_2 + CO_2^{-\bullet} \tag{4.58}$$

$$>\!Fe(II)OH^- + O_2 \longrightarrow\ >\!Fe(III)OH + O_2^{-\bullet} \tag{4.59}$$

Similar reactions occur with other carboxylic and dicarboxylic acids on the surface of the iron oxides and oxyhydroxides leading to superoxide and hydrogen peroxide formation. Pehkonen et al. found that the rates of H$_2$O$_2$ formation on a series of iron oxide polymorphs depended on the chemical nature of the carboxylate electron donor where HCO$_2^- >$ CH$_3$CO$_2^- >$ CH$_3$CH$_2$CH$_2$CO$_2^-$.

Metal Sulfide Surface Chemistry and Free-Radical Generation

Metal sulfide and related chalcogenide semiconductors are used in a variety of electronic applications. For example, CdS, which is an n-type semiconductor with $E_g = 2.4$ eV, has been shown to have photocatalytic activity for H_2 production under visible light irradiation under anoxic conditions in water in the presence of electron donors such as C_2H_5OH, HS^- and SO_3^{2-}.

The electronic levels of nanoparticulate or quantum-sized CdS (Q-CdS) can be tuned by changing or controlling particle size without changing the chemical composition. For example, Hoffman et al. [35] found an increase in quantum efficiency for photo-polymerization of methylmethacrylate with a corresponding decrease in particle size using Q-CdS.

The surface chemistry [47–53] of nanoparticulate CdS in the Q-size domain has some similarities, which are initiated by the hydrolysis of the surface of CdS to form surface functionalities [54] that are dominated by cadmium mercapto group, >CdSH, and cadmium hydroxyl, >CdOH, functionalities as follows:

$$\left[{}^{S^{2-}}_{Cd^{2+}} > (CdS)\right]_2 + H_2O \underset{}{\overset{K_H}{\rightleftharpoons}} {}^{S^{2-}}_{Cd^{2+}} > Cd(II)SH + {}^{S^{2-}}_{Cd^{2+}} > S(-II)Cd(II)OH \tag{4.60}$$

The variable surface charges arise from protonation and deprotonation of surface sulfhydryl and hydroxyl groups as depicted in following equations:

$$ {}^{S^{2-}}_{Cd^{2+}} > CdSH + H^+ \rightleftharpoons {}^{S^{2-}}_{Cd^{2+}} > CdSH_2^+ \tag{4.61}$$

$$ {}^{S^{2-}}_{Cd^{2+}} > CdSH_2^+ \overset{K_{a1}^s}{\rightleftharpoons} {}^{S^{2-}}_{Cd^{2+}} > CdSH + H^+ \tag{4.62}$$

$$ {}^{S^{2-}}_{Cd^{2+}} > CdSH \overset{K_{a2}^s}{\rightleftharpoons} {}^{S^{2-}}_{Cd^{2+}} > CdS^- + H^+ \tag{4.63}$$

$$ {}^{S^{2-}}_{Cd^{2+}} > CdOH + H^+ \rightleftharpoons {}^{S^{2-}}_{Cd^{2+}} > CdOH_2^+ \tag{4.64}$$

$$ {}^{S^{2-}}_{Cd^{2+}} > CdOH_2^+ \overset{K_{a1,1}^s}{\rightleftharpoons} {}^{S^{2-}}_{Cd^{2+}} > CdOH + H^+ \tag{4.65}$$

$$ {}^{S^{2-}}_{Cd^{2+}} > CdOH \overset{K_{a2,1}^s}{\rightleftharpoons} {}^{S^{2-}}_{Cd^{2+}} > CdO^- + H^+ \tag{4.66}$$

In the simplest case, at low ionic strength in the absence of added cations or anions, the isoelectric point or "point of zero charge" is described as follows in terms of the concentrations of the relevant surface species as a function of pH:

$$\left[{}^{S^{2-}}_{Cd^{2+}} > CdSH_2^+\right] + \left[{}^{S^{2-}}_{Cd^{2+}} > CdOH_2^+\right] = \left[{}^{S^{2-}}_{Cd^{2+}} > CdS^-\right] + \left[{}^{S^{2-}}_{Cd^{2+}} > CdO^-\right] \tag{4.67}$$

A likely set of surface chemical reactions that take place upon illumination of metal sulfide particles under anoxic conditions are given below for the case of an aqueous suspension of CdS in the presence of dissolved ethanol [55–60]:

$$ {}^{S^{2-}}_{Cd^{2+}} > Cd(+II)S(-II)H_2^+ \overset{h\nu}{\underset{h_{vb}^+}{\xrightarrow{\bar{e}_{cb}}}} {}^{S^{2-}}_{Cd^{2+}} > Cd(+I)S(-I)H_2^+ \tag{4.68}$$

$$ {}^{S^{2-}}_{Cd^{2+}} > Cd(+I)S(-I)H_2^+ \overset{h\nu}{\underset{h_{vb}^+}{\xrightarrow{\bar{e}_{cb}}}} {}^{S^{2-}}_{Cd^{2+}} > Cd(0)S(0)H_2^+ \tag{4.69}$$

$$ {}^{S^{2-}}_{Cd^{2+}} > Cd(0)S(0)H_2^+ \rightleftharpoons {}^{S^{2-}}_{Cd^{2+}} > Cd(+II)S(0)^+ + H_2 \tag{4.70}$$

$$ {}^{S^{2-}}_{Cd^{2+}} > Cd(+II)S(0)^+ + CH_3CH_2OH \rightleftharpoons {}^{S^{2-}}_{Cd^{2+}} > Cd(+II)S(-I)H^+ + CH_3\dot{C}HOH \tag{4.71}$$

$$ {}^{S^{2-}}_{Cd^{2+}} > Cd(+II)S(-I)H^+ + CH_3CH_2OH \rightleftharpoons {}^{S^{2-}}_{Cd^{2+}} > Cd(+II)S(-II)H_2^+ + CH_3\dot{C}HOH \tag{4.72}$$

$$ 2H^\bullet \longrightarrow H_2 \tag{4.73}$$

$$ 2\,CH_3\dot{C}HOH \longrightarrow 2\,CH_3CHO + H_2 \tag{4.74}$$

Similar photoreactions can occur at neutral >CdSH and >CdOH sites and protonated surface sites involving >CdSH$_2^+$ and >CdOH$_2^+$.

$$\underset{Cd^{2+}}{^{S^{2-}}} > Cd(+II)S(-II)H \xrightarrow{h\nu} \xrightarrow[h_{vb}^+]{\bar{e}_{cb}} \underset{Cd^{2+}}{^{S^{2-}}} > Cd(+I)S(-I)H \tag{4.75}$$

$$\underset{Cd^{2+}}{^{S^{2-}}} > Cd(+I)S(-I)H \xrightarrow{h\nu} \xrightarrow[h_{vb}^+]{\bar{e}_{cb}} \underset{Cd^{2+}}{^{S^{2-}}} > Cd(0)S(0)H \tag{4.76}$$

$$\underset{Cd^{2+}}{^{S^{2-}}} > CdOH_2^+ \xrightarrow{2\bar{e}_{cb}} \underset{Cd^{2+}}{^{S^{2-}}} > CdO^- + H_2 \tag{4.77}$$

$$\underset{Cd^{2+}}{^{S^{2-}}} > CdO^- + 2H^+ \xrightarrow{2\bar{e}_{cb}} \underset{Cd^{2+}}{^{S^{2-}}} > CdOH_2^+ \tag{4.78}$$

$$\underset{Cd^{2+}}{^{S^{2-}}} > CdOH \xrightarrow{h_{vb}^+} \underset{Cd^{2+}}{^{S^{2-}}} > CdOH^{+\cdot} \xrightarrow{CH_3CH_2OH} \underset{Cd^{2+}}{^{S^{2-}}} > CdOH_2^+ + CH_3\dot{C}HOH \tag{4.79}$$

In the presence of oxygen (i.e., under oxic conditions), O_2 would immediately react with the carbon-centered ethanolic radical, $CH_3\dot{C}HOH$, for form the corresponding peroxy radical ($RO_2^\bullet$), which is an alternative form of ROS.

Fullerene Photochemistry and ROS Generation Potential

Similar to the cases of the metal oxide and sulfide semiconductors, the photochemical properties of fullerenes can be viewed in the context of excitation of electrons across a bandgap. For example, the bandgap of pure C_{60} has been reported to be 2.3 eV, which is comparable to that of iron oxide polymorphs (Table 4.3). The bandgap for carbon nanotubes (CNTs) depends on its chirality and is inversely proportional to the diameter of the nanotube. In the case of ROS generation by fullerenes, they can act either as a photosensitizers or an electron shuttle.

Two distinct pathways are recognized for the photosensitization of fullerenes. Both pathways involve the initial excitation of the photosensitizing molecule (i.e., a fullerene). Type I sensitization involves electron transfer and depends upon the presence of a donor molecule that can reduce the triplet state of the sensitizer. The triplet state is more susceptible to electron donation than is the ground state singlet molecule. In the presence of oxygen, superoxide radical anion can be formed by direct electron transfer from this excited radical to molecular oxygen.

The type II photosensitization pathway involves the transfer of excited spin-state energy from the sensitizer to another molecule. Type II sensitization does not depend upon the presence of a donor molecule, but only requires a long-lived triplet excited state. In the presence of oxygen, this triplet state is quenched by ground state oxygen, which is transformed into singlet oxygen. As noted previously, singlet oxygen is a reactive oxygen species that can participate in reactions in solution that are spin-forbidden in the case of ground-state molecular oxygen. Singlet oxygen formation via type II photosensitization and quenching has been reviewed extensively by Wilkinson et al.

A typical photosensitizing molecule in the ground state is represented by S_o with S_1 and T_1 representing the lowest energy singlet and triplet states, respectively. Figure 4.22 is a graphical representation of their main photosensitized pathways.

Light within the absorbance range of the photosensitizer is absorbed and promotes electrons into the excited singlet state [Equation (4.80)].

$$S_o + h\nu \xrightarrow{k_w} S_1 \tag{4.80}$$

The intensity and wavelength of the light will govern the rate of S_1 formation (k_w). The singlet state then decays via at least three different pathways: fluorescence, internal conversion, or intersystem crossing [Equations (4.81) to (4.83)]. Intersystem crossing is a spontaneous nonradiative transition from the singlet to the triplet electronic state.

$$S_1 \xrightarrow{k_F} S_o + h\nu_F \tag{4.81}$$

$$S_1 \xrightarrow{k_{ic}} S_o \tag{4.82}$$

$$S_1 \xrightarrow{k_{isc}} T_1 \tag{4.83}$$

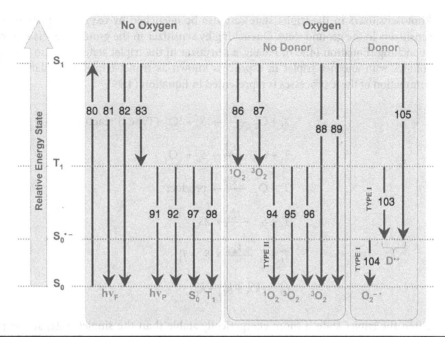

Figure 4.22   General photosensitization kinetic scheme.

The rate of singlet decay is taken as the sum of these rates of singlet decay, k_{SD}, [Equation (4.84)] and the quantum yield, ϕ_T, is defined by the fraction of the singlet state that decays to the triplet state [Equation (4.85)].

$$k_{SD} = k_F + k_{ic} + k_{isc} = \frac{1}{\tau(S_1)} \tag{4.84}$$

$$\phi_T = \frac{k_{isc}}{k_{SD}} \tag{4.85}$$

The singlet form of a photosensitizer can also be quenched by oxygen to form the triplet state of the sensitizer, generating either singlet oxygen [Equation (4.86)] or ground state oxygen [Equation (4.87)]. It can also be quenched to the ground state [Equation (4.88)] or the sensitizer may be altered through a reaction to form a new product [Equation (4.89)].

$$S_1 + O_2 \xrightarrow{k_{S\Delta}^{O_2}} T_1 + {}^1O_2^* \tag{4.86}$$

$$S_1 + O_2 \xrightarrow{k_{isc}^{O_2}} T_1 + {}^3O_2 \tag{4.87}$$

$$S_1 + O_2 \xrightarrow{k_{Sd}^{O_2}} S_o + {}^3O_2 \tag{4.88}$$

$$S_1 + O_2 \xrightarrow{k_{Sr}^{O_2}} \text{reaction product} \tag{4.89}$$

$$k_{SQ}^{O_2} = k_{S\Delta}^{O_2} + k_{isc}^{O_2} + k_{Sd}^{O_2} + k_{Sr}^{O_2} \tag{4.90}$$

The triplet state in turn decays by phosphorescence [Equation (4.91)] or intersystem crossing [Equation (4.92)]. Phosphorescence releases light energy as electrons fall to a lower energy level, and internal conversion involves the same drop in energy without the associated energy release. The rate of triplet decay is the sum of these processes [Equation (4.93)].

$$T_1 \xrightarrow{k_{Tp}} S_o + h\nu_p \tag{4.91}$$

$$T_1 \xrightarrow{k_{Td}} S_o \tag{4.92}$$

$$k_{TD} = k_{Tp} + k_{Td} = \frac{1}{\tau(T_1)} \tag{4.93}$$

Photosensitizers in the triplet state can also be quenched by oxygen [Equations (4.94) to (4.97)] or by sensitizers in the ground state. Quenching by sensitizer in the ground state is known quite simply as self-quenching [Equation (4.97)]. Finally, a sensitizer in the triplet state may also be quenched when a triplet collides with another triplet in a process known as triplet-triplet annihilation [Equation (4.98)]. The summation of these processes is represented in Equation (4.99).

$$T_1 + O_2 \xrightarrow{k_{T\Delta}^{O_2}} S_o + {}^1O_2^* \text{ (Type II reaction)} \tag{4.94}$$

$$T_1 + O_2 \xrightarrow{k_{Td}^{O_2}} S_o + {}^3O_2 \tag{4.95}$$

$$T_1 + O_2 \xrightarrow{k_{Tr}^{O_2}} \text{product} \tag{4.96}$$

$$T_1 + S_o \xrightarrow{k_{SQ}^{S_o}} 2S_o \tag{4.97}$$

$$T_1 + T_1 \xrightarrow{k_{AN}^{T_1}} S_o + T_1 \tag{4.98}$$

$$k_{TQ}^{O_2} = k_{T\Delta}^{O_2} + k_{Td}^{O_2} + k_{Tr}^{O_2} \tag{4.99}$$

Often the triplet state is more energetically stable than the singlet state, so the products of the above reactions are favored over those resulting from reactions with the singlet state in photosensitizing systems. The triplet state is quenched by oxygen by three different pathways; the fraction of triplet that participates in the type II reaction to form singlet oxygen is represented by Equation (4.100). Oxygen also competes with triplet-triplet annihilation and self-quenching as potential pathways for triplet quenching. Consequently, the proportion of the triplet state quenched with oxygen can be calculated using Equation (4.101). The quantum yield for singlet oxygen expresses how efficiently the photons are used to produce singlet oxygen [Equation (4.102)].

$$f_\Delta^{II} = \frac{k_{T\Delta}^{O_2}}{k_{TQ}^{O_2}} \tag{4.100}$$

$$P_T^{O_2} = k_{TQ}^{O_2}[O_2] \Big/ \left(k_{TD} + k_{TQ}^{O_2}[O_2] + k_{SQ}^{S_o}[S_o] + k_{AN}^{T_1}[T_1] \right) \tag{4.101}$$

$$\phi_\Delta = \phi_T f_\Delta^{II} P_T^{O_2} \tag{4.102}$$

The nonproductive pathways regarding ROS production such as triplet quenching by oxygen along with triplet-triplet annihilation and self-quenching represent inefficiencies in converting light energy to chemical energy. Increasing the lifetime of the triplet state and minimizing the effect of the nonproductive pathways will result in higher quantum yields for ROS production; these pathways are controlled by the concentration and proximity of the photosensitizer in the solution.

Type I reactions can occur in parallel with type II reactions, when the photosensitizers are in the presence of electron donors. Type I reactions are initiated by the reduction of the triplet state by an electron donor [Equation (4.103)]. The donor in this case has a reduction potential lower than either the ground state (S_o) or the excited state (T_1 or S_1) of the sensitizer [Equations (4.2) to (4.5)]. Once the reduction occurs, the sensitizer takes the form of a radical anion ($S_o^{\bullet-}$) that has the possibility of reducing oxygen to superoxide [Equation (4.104)] and subsequently returning to the ground state. Reduction of the singlet state is more thermodynamically favorable but is kinetically limited because of the short lifetime of the singlet excited state [Equation (4.105)]. As a result, singlet state reactions with the electron donor are not likely to occur.

$$T_1 + D \xrightarrow{k_{T1}^{D}} S_o^{\bullet-} + D^{\bullet+} \tag{4.103}$$

$$S_o^{\bullet-} + O_2 \xrightarrow{k_{S_o^{\bullet-}}^{O_2}} O_2^{\bullet-} + S_o \tag{4.104}$$

$$S_1 + D \xrightarrow{k_{S1}^{D}} S_o^{\bullet-} + D^{\bullet+} \tag{4.105}$$

Thus, the proportion of oxygen reacting with the triplet state sensitizer must be modified to express the reactions between the triplet state and the electron donor [Equation (4.106)].

$$P_T^{O_2} = k_{TQ}^{O_2}[O_2] \bigg/ \left(k_{TD} + k_{TQ}^{O_2}[O_2] + k_{SQ}^{S_o}[S_o] + k_{AN}^{T_1}[T_1] + k_{T1}^{D}[D] \right) \tag{4.106}$$

In order to determine the quantum yield for superoxide formation, the following assumptions are made: anion radical sensitizers react with oxygen to form superoxide (i.e., $f_{sup}^{II}=1$); anion radicals are unlikely to be formed [Equation (4.105)] due to the short lifetime of S_1; and the donor reacts only with the triplet-state molecule, T_1, of Equation (4.103).

Given these assumptions, the proportion of the triplet state molecules that react with a donor is given by Equation (4.107) with the corresponding quantum yield given by Equation (4.108).

$$P_T^{D} = k_{T1}^{D}[D] \bigg/ \left(k_{TD} + k_{TQ}^{O_2}[O_2] + k_{SQ}^{S_o}[S_o] + k_{AN}^{T_1}[T_1] + k_{T1}^{D}[D] \right) \tag{4.107}$$

$$\phi_{sup} = \phi_T P_T^{D} \tag{4.108}$$

The overall quantum yield for the production of ROS ($O_2^{-\bullet}$) in such a system can then be given as the sum of the quantum yields:

$$\phi_{ROS} = \phi_\Delta + \phi_{sup} \tag{4.109}$$

Kinetically, the triplet state is a key intermediary in photosensitizing processes. The quantum yield for ROS [Equation (4.109)] cannot be maximized unless triplet decay pathways (k_{TD}), oxygen quenching that does not lead to singlet oxygen ($k_{TQ} - k_{T\Delta}^{O_2}$), self-quenching ($k_{SQ}^{S_o}$), and triplet-triplet annihilation ($k_{AN}^{T_1}$), are minimized as potential pathways for removal of the triplet state sensitizer. Based on these kinetic limitations, the ideal photosensitizer has three properties: the absorbance of low energy light to create the singlet excited state efficiently; preferred conversion of the singlet state to the triplet state due to intersystem crossing [Equation (4.83)]; and a low occurrence of non-ROS forming triplet removal pathways. The nanomaterial class known as fullerenes holds great promise due to properties that correspond to each of these desirable traits.

ROS Production by Fullerenes

Carbon-based nanomaterials such as fullerenes have been known to be photoactive as photosensitizers from the first studies of their physical properties [61]. Fullerenes, and in particular C_{60}, have been studied intensively for applications in fields such as photodynamic therapy [62], photovoltaics [63], and materials [64].

An advantage of using fullerenes, and in particular C_{60}, as photosensitizers in an engineered system is that they are highly stable. For example, the carbon cage making up C_{60} appears to be nearly impervious to degradation by oxidation or susceptible to enzymatic attack. However, fullerenes may be modified in aqueous environments such as in the formation of epoxide derivatives of C_{60} in the presence of UV light [65] or on the surface of a metal oxides such as TiO_2 [66].

When fullerenes are illuminated under the appropriate wavelength, the electrons are excited from the ground state ($^0C_{60}$) to the singlet state [Equation (4.80)]. The singlet state ($^1C_{60}$) can decay in three main manners: fluorescence [Equation (4.81)], internal conversion [Equation (4.82)], and intersystem crossing (ISC) [Equation (4.83)]. The first two result in the ground state while the latter leads to the relaxation of singlet C_{60} to the triplet state ($^3C_{60}$). Interaction of the singlet state with oxygen can also result in the triplet state [Equations (4.86) and (4.87)]. Equation (4.86) results in the production of singlet oxygen via type II photosensitization. The triplet state, $^3C_{60}$, has a significantly longer lifetime than $^1C_{60}$ in solution, allowing it to participate in type II formation of singlet oxygen to a greater extent than does the $^1C_{60}$ [Equation (4.94)]. The triplet state is also susceptible to self-quenching [Equation (.497)] via interaction with the ground state ($^0C_{60}$) and triplet-triplet annihilation [Equation (4.98)] via interaction with another triplet ($^3C_{60}$). Type I sensitization [Equation (4.103)] occurs when the triplet state comes in contact with a donor molecule that has a more negative reduction potential than ($^3C_{60}$). The resulting radical ($C_{60}^{-\bullet}$) can then pass the electron to ground state oxygen to form superoxide [Equation (4.104)]. An illustration of these main fullerene photosensitization pathways is shown in Figure 4.23.

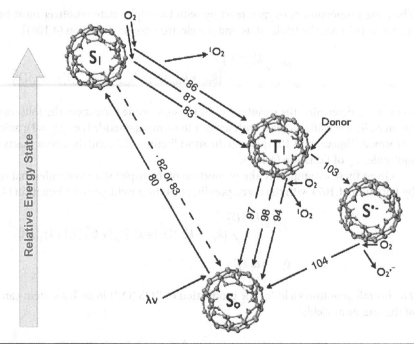

FIGURE 4.23 Major photosensitization pathways for C_{60}. Numbers *vide supra* and *vide infra* refer to the primary reaction pathways shown in this figure and Figure 4.22.

In general, conditions that lead to ROS generation can be grouped into four categories in which reaction pathways are most directly related and in many cases have been investigated together:

1. Ground state fullerene: excitation and decay [Equations (4.80) to 4.82)]
2. Intersystem crossing: fullerene triplet state formation [Equations (4.83), (4.86), (4.87)]
3. Fullerene triplet quenching: type II photosensitization [Equations (4.94), (4.97), (9.98)]
4. Fullerene triplet reduction: type I photosensitization [Equations (4.103) and (4.104)

Each of these groups of reaction pathways are affected by changes in fullerene functionality and aggregation state, such as those produced to suspend fullerene in water. In each section, we will first discuss the pathway specific properties of C_{60} found in nonpolar solvents, following with how encapsulation and functionalization of C_{60} for aqueous suspension affects that reaction group pathway.

Ground state fullerene: excitation and decay

The absorbance spectrum of free C_{60} suspended in nonpolar solvents has sharp peaks with absorbance in the UV and visible range (Figure 4.24). This has important consequences because quantum yields for the

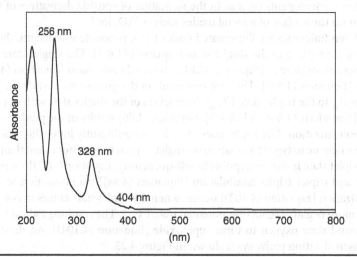

FIGURE 4.24 Typical UV/Vis absorbance of C_{60} suspended in nonpolar solvent.

photosensitized production of singlet oxygen by a suspension of C_{60} in nonpolar benzene are near unity for light in the UV and visible range [61, 67]. According to Equation (4.85), triplet quantum yield would equal singlet oxygen quantum yield under ideal energy transfer conditions, indicating that the triplet quantum yield is unity in the case of C_{60}.

As a consequence, the reaction shown in Equation (4.80) occurs efficiently when C_{60} is free and unaltered in a nonpolar solvent. Generally, C_{60} suspended in nonpolar solvent results in a broader spectrum that is shifted toward the dark grey wavelengths. This effect varies with the type of suspension [68–72], but generally the degree of broadening increases with clustering of the C_{60} within the surrounding agent. Reaction in Equation (4.80) is deleteriously affected because not all incident light can reach the surface of the C_{60}. Fluorescence [Equation (4.81)] and internal conversion [Equation (4.82)] have not been identified as important contributors to the decay of the ${}^1C_{60}$ back to ${}^0C_{60}$. However, addends decrease triplet quantum yield by promoting non-triplet forming singlet decay pathways such as florescence [Equation (4.81)] and most likely internal conversion [Equation (4.82)] [73–75]. The cage structure can be altered by various chemical reactions that perturb the extended π-bonding network and subsequently raise the energy of the LUMO electrons due to loss of conjugation [76] (Figure 4.25). This increases the energy required to excite electrons across the bandgap and into their excited states; thus requiring higher energy light [Equation (4.80)] and contributing to a reduction in quantum yield [Equation (4.85)] [73].

The photophysical properties of C_{70}, a higher order fullerene cage, are influenced by its structure, which may have an oblong shape in order to maintain the cagelike carbon structure. The singlet oxygen quantum yield [Equation (4.102)] (a measure of the lower limit of ${}^3C_{70}$ quantum yield [Equation (4.85)] was found to be around 0.81, indicating light conversion was not as efficient in this molecule [67]. This decrease is partially attributed to deactivation of ${}^1C_{70}$ by alternative pathways such as internal conversion [Equation (4.82)] that do not produce the triplet state. Thus, it appears that either increasing the size of the fullerene cage or decreasing fullerene symmetry—or both—may lead to a decrease in quantum yield [Equation (4.85)].

Intersystem Crossing: Fullerene Triplet State Formation

The characteristic rates of ISC (k_{isc}) for C_{60} and C_{70} had been determined to be 3.0×10^{10} s^{-1} and 8.7×10^9 s^{-1}, respectively [77]. Decay of the singlet excited C_{60} is predominantly ISC to the triplet state [Equation (4.83)] [61, 67]. This phenomenon can be explained in terms of small energy splitting between ${}^1C_{60}$ and ${}^3C_{60}$, low fluorescence [Equation (4.81)], and large spin-orbital interaction. The large diameter and spherical nature of C_{60} promote these properties by lowering electron repulsion and the extended π-bonding network, respectively. In C_{70}, the extended π-bonding network seems to be perturbed enough to promote internal conversion [Equation (4.82)] rather than ISC [Equation (4.83)]; this results in reduced ${}^3C_{60}$ quantum yields [Equation (4.85)]. Addends reduce the ISC rate [Equation (4.83)] in the same way by reducing the amount of π bonds on the surface of the C_{60} cage. Since the singlet state cannot be as easily relaxed, its lifetime is noticeably longer but still on the order of nanoseconds [78–80].

Fullerene Triplet Quenching: Type II Photosensitization

After excitation of C_{60} to the triplet state via ISC [Equation (4.83)], the corresponding lifetime of the triplet state is microseconds in absence of quenching by oxygen [61, 77, 81]. However, the lifetime also depends on phosphorescence [Equation (4.91)], internal conversion [Equation (4.92)], self-quenching [Equation (4.97)],

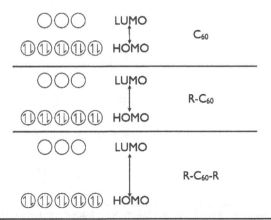

Figure 4.25 Increasing the amount of addends enlarges the gap between the HOMO and LUMO.

and triplet-triplet annihilation [Equation (4.98)]. Therefore, the triplet lifetime depends on concentration of C_{60}. By measuring triplet lifetime at various concentrations, the effect of these alternative quenching mechanisms [Equations (4.97) and (4.98)] can be eliminated from lifetime calculations, allowing for an estimation of the intrinsic triplet lifetime around 133 μs in nonpolar solvent [82]. While more generally, the triplet lifetime tends to be around 40 μs in most studies involving nonpolar solvents because a single concentration is used to measure the triplet lifetime. These triplet lifetimes are exceptional, but along with type II photosensitization rates lifetimes change dramatically when C_{60} is suspended in the aqueous environment.

Encapsulating agents such as γ-cyclodextrin (γ-CD), Triton-X, and poly(vinylpyrrolidine) (PVP) increase the lifetime of the triplet state [Equation (4.93)] up to 130 μs regardless of concentration [68–70, 83]. This is likely due to the encapsulating agent's ability to reduce contact between fullerenes making self-quenching [Equation (4.97)] and triplet-triplet annihilation [Equation (4.98)] less frequent (i.e., γ-CD encapsulation reduced triplet-triplet annihilation by four times as compared with free C_{60} in toluene [69]). Photosensitization rates benefit from the increased lifetime of the triplet state, but the same encapsulation effects reduce the rate of type II singlet oxygen formation [Equation (4.94)]. In the case of γ-CD, the triplet quenching by oxygen was determined to be half that of free C_{60} in toluene after correction for oxygen diffusion rates [69]. In addition, illuminated PVP/C_{60} was monitored for the characteristic singlet oxygen emissions band at 1270 nm. However, the IR emission was not observed [68] and thus, it was concluded that type II sensitization [Equation (4.94)] was not occurring or that it was taking place at a very low rate.

A more specific example, "mechanically entrapped C_{60}," recently has been developed for C_{60} suspension is [84]. The carbon cage is entrapped in an all-silica zeolite Y supercage (Figure 4.26).

In the absence of O_2, the triplet lifetime [Equation (4.93)] is extended into the order of minutes. Presumably, this is due to the complete lack of quenching mechanism activity and indicates the C_{60} molecules are more than likely entrapped individually and not as clusters. Otherwise, triplet-triplet annihilation [Equation (4.98)] and self-quenching [Equation (4.97)] would lower the triplet lifetime significantly. Despite this, in the presence of oxygen, the type II [Equation (4.94)] mechanism occurs effectively but at a slight rate decrease from the diffusion controlled quenching of free C_{60}.

Increasing the number of addends on the C_{60} cage decreases the quantum yield of the triplet state [Equation (4.85)] and as a result the quantum yield of singlet oxygen [Equation (4.102)] [85]. The fused core area can be correlated to the triplet C_{60} and singlet oxygen quantum yields (Figures 4.27 and 4.28) [86].

However, since the ratio of quantum yields (ϕ_Δ/ϕ_T) is approximately unity for all different functional groups, the nature of the addend does not hinder the ability of oxygen to quench the $^3C_{60}$ via a type II mechanism [Equation (4.94)] [85, 86].

All the photosensitization properties of functionalized fullerene were first measured in nonpolar solvent in order to accurately gauge the induced changes with increasing addends. Nonetheless, these modifications were originally intended for increased water solubility so it was only a matter of time before photosensitizing properties were tested in the aqueous environment. An immediate consequence of placing

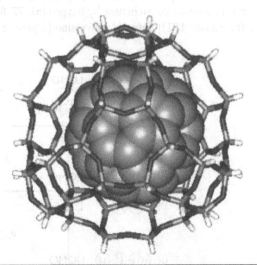

Figure 4.26 A molecular model representation of C_{60} trapped in a zeolite cage [84].

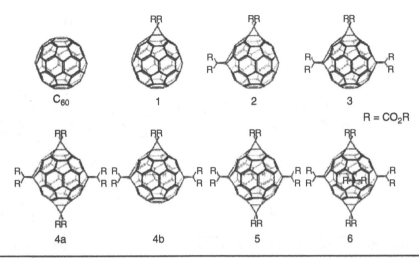

FIGURE 4.27 Common examples of increasingly functionalized C_{60} [86].

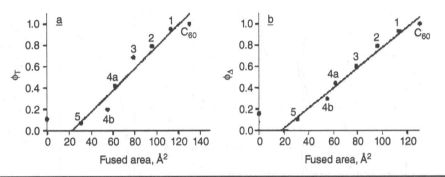

FIGURE 4.28 $^3C_{60}$ quantum yield and singlet oxygen quantum yield correlated with functionalized cage area [86].

functionalized C_{60} in the aqueous system is clustering. Monofunctionalized malonic acid C_{60} ($C_{60}C(COOH)_2$) is a good example. The nonpolar ends of these molecules are thought to group together facing the negatively charged carboxyl groups out into the polar environment [71] (Figure 4.29).

Clustering reduces the lifetime of the triplet state two orders of magnitude [71] by promoting triplet-triplet annihilation [Equation (4.98)] and self-quenching [Equation (4.97)]. Therefore, the triplet excited state does not survive long enough to participate in photosensitization reactions. In order to alleviate this problem γ-CD can be added to cap the exposed nonpolar ends and reduce clustering. This increases triplet lifetime comparable to that observed for γ-CD encapsulated C_{60} [69].

Type II quenching by oxygen [Equation (4.94)] for the encapsulated monofunctionalized compounds has a rate constant on the same order of magnitude as that of γ-CD encapsulated C_{60} [71]. However, bi-functionalized, tri-functionalized, and poly-functionalized compounds can be suspended in the aqueous environment without the use of an encapsulating agent (Figure 4.27 provides examples of multi-functionalized compounds). Bi-functional and tri-functional C_{60} singlet oxygen formation rates can be compared with free C_{60} and they show only a slight slowdown in singlet oxygen formation, whereas the poly-functionalized fullerol ($C_{60}(OH)_x$) is about an order of magnitude slower. The slowdown is attributed to the extremely perturbed π bonding system caused by addition of hydroxyl groups [71].

Suspensions of fullerol have been observed to exhibit two distinct triplet lifetimes. A shorter time constant for triplet decay appears to be concentration dependent, while triplet lifetime simultaneously decays with a longer time constant that is concentration independent [87]. Annihilation [Equation (4.98)] and self-quenching [Equation (4.97)] play an important role in the shorter decaying component, while the longer decaying component is probably associated with the presence of individually suspended fullerol. EPR spin-trapping methods can be used to monitor the type II formation of singlet oxygen [Equation (4.94)], and the singlet oxygen signal is both time and concentration dependent in the presence of UV illumination (Figure 4.30) [3, 88].

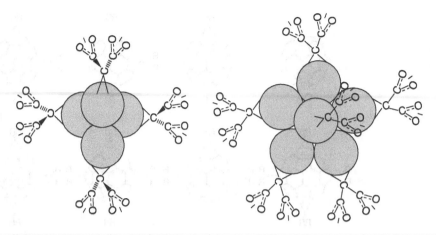

FIGURE 4.29 Drawing of possible cluster formations of monofunctionalized malonic acid C_{60} derivatives in the aqueous environment [71].

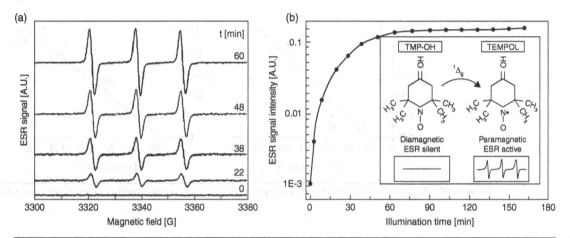

FIGURE 4.30 EPR signal growth of the TEMP-singlet oxygen adduct in the presence of fullerol suspended in an aqueous system.

Fullerol exhibits lower singlet oxygen quantum yields due to its perturbed π-bond system, but the hydroxyl groups increase the triplet lifetime by reducing cage contact so that the molecule can participate in type II reactions in water.

This phenomenon, of increasing overall reactivity with increased surface functionalization can be generalized to some non-fullerene that photocatalytically generate reactive oxygen species on the nanoparticle surface. Such particles include semiconducting nanoparticles such as TiO_2 as well as the family of caged structures that include graphenes, carbon nanotubes as well as C_{60}, C_{70} and others. When such nanoparticles form clusters, their surfaces may interact to quench ROS producing reactions. In nanomaterials such as graphene, CNTs and fullerene, this involves triplet-triplet annihilation as described above. In the case of semiconductors, this electron holes generated on surfaces of one particle are in increased proximity with electrons generated at the surface of another nanoparticles within the cluster[83]. When functionalization of the nanoparticle surface increases nanoparticle stability and therefore decreases aggregation (see Chapter 6), but also reduces the efficiency of the intrinsic reaction (reduced quantum yield), then an optimum in ROS generation may occur with respect to overall functionalization (Figure 4.31).

However functionalization will also affect the affinity of the nanoparticle for any surface where the ROS may be delivered influencing the impact of reactions such as microbial inactivation or oxidation. Badireddy and coworkers [84] found that inactivation rates for MS2 viruses by several fullerenes followed the order: aqu-nC_{60} < $C_{60}(OH)_6$ ≈ $C_{60}(OH)_{24}$ < $C_{60}(NH_2)_6$ which was explained by differences in $^1O_2$ production rates arising from changes in the quantum yield linked to aggregate structure and differences in proximity between the fullerenes and the MS2 in suspension.

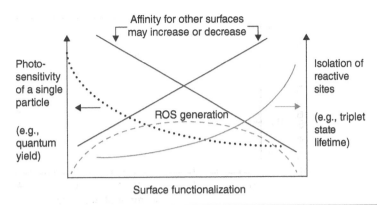

FIGURE 4.31 Interplay between nanoparticle surface functionalization, single particle photosensitivity (e.g., quantum yield), and proximity of surfaces that may affect quenching and, in the case of fullerenes, CNTs and graphenes, triplet state life lifetime. In the case of fullerenes, increasing derivatization disrupts the pi-electron shell decreasing the quantum yield and $^1O_2$ generation. However, derivatization also leads to smaller, more open aggregates, decreasing the likelihood of triplet-triplet annihilation and self-quenching and thereby increasing the lifetime of the excited triplet state, resulting in greater $^1O_2$ production. Depending on the surface functionalization, the affinity for a nanoparticle for surface where reactive oxygen may be delivered may either increase or decrease. An optimal derivatization maximizing ROS generation as well as increasing affinity for the target surface will favor delivery of ROS to the surface. (Adapted from [84])

Fullerene Triplet Reduction: Type I Photosensitization

In the presence of appropriate donor compounds, the high electron affinity of C_{60} results in type I reactions. C_{60} can be reduced with up to five electrons in benzonitrile solvent with progressive reduction potentials of (−0.36, −0.83, −1.42, −2.01, −2.60 V versus SCE) [89, 90]. This affinity is due to the extended π-bonding that can spread the extra electrons across the surface. In addition to its electron affinity, C_{60} also has a stable triplet state that is about 1.56 eV higher than $^0C_{60}$. The higher energy $^3C_{60}$ is more easily reduced because the reduction potential is raised by this energy (1.56V-0.42V = 1.14V vs. SCE) [61]. As a consequence, when an electron donor of lower reduction potential than $^3C_{60}$ is present, excitation to the triplet state plays an important role in Type I reactions [Equation (4.103)].

The electron transfer capabilities of γ-CD encapsulated C_{60} as opposed to free C_{60} in propan-2-ol can be compared in terms of their bi-molecular rate constants [Equation (4.103)]. Interestingly enough, the rate constant is about a factor of 2 slower in the encapsulating agent [72]. This is consistent with the rate of oxygen quenching by γ-CD/C_{60} [69]. Similar C_{60} micellular suspensions formed with the non-ionic surfactant Triton-X 100 form monomeric or colloidal suspensions of C_{60} depending on the preparation method. The bi-molecular rate constant for reduction by a donor [Equation (4.103)] is three orders of magnitude less than the free C_{60} in toluene [83]. Independent measurement confirms that triton X encapsulation slows reduction by one order of magnitude compared with γ-CD [70]. The inability of the donor molecule to approach the surface of C_{60} is likely due to steric and charge repulsion effects [72]. Poly(vinylpyrrolidone) (PVP) is another encapsulating agent that has been used extensively to suspend C_{60} in aqueous solution at concentrations of up to 400 mg/L [68]. In the presence of adenosine 5'-(trihydrogen diphosphate) {NADH}, a C_{60} suspension has been shown to damage DNA; concurrently EPR and NBT detection confirms superoxide formation via Type I reaction [Equation (4.104)] but at reduced rates from free C_{60} [9, 14, 88]. As noted previously for Type II reactions, encapsulation represents a tradeoff between triplet lifetime [Equation (4.93)] and quantum yield of Type I [Equation (4.108)] reactions.

As discussed earlier, the LUMO increases with the addition of addends, and because C_{60} is fully occupied in the HOMO, a reducing electron must jump a larger and larger gap in order to complete the reduction. This translates into increasingly more negative reduction potentials that drop about 0.1 to 0.15 V for each additional addend (Figure 4.32) [76, 78–90]. Concurrently, the triplet energy increases with the addition of addends (Figure 4.33) [86]. However, this energy increase is not as dramatic as the decrease in reduction potential, and upon the summation of these two effects a net decrease in reduction potential for the triplet state of the increasingly functionalized C_{60} cage occurs. As $^3C_{60}$ is increasingly functionalized it takes on electrons less readily than nonfunctionalized $^3C_{60}$ (Figure 4.33). However, the energy of $^3C_{60}$ is approximately 1.5 V higher than C_{60} and as a result all functionalized triplet states remain easier to reduce than ground state C_{60} despite the negative trend in their reduction potentials.

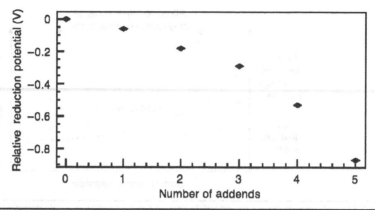

FIGURE 4.32 Reduction potential of increasing number of functional groups relative to the ground state C_{60}.

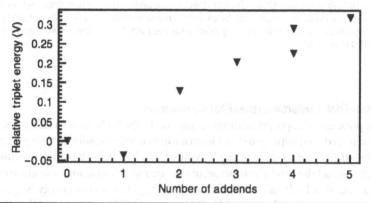

FIGURE 4.33 Relative triplet energy with increasingly functionalized C_{60} cage.

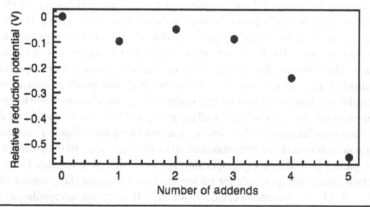

FIGURE 4.34 Relative reduction potential of increasingly functionalized $^3C_{60}$.

As a result, the change in reduction potential for monofunctionalized/encapsulated forms of C_{60} have lower rates of reduction than the unfunctionalized/encapsulated C_{60} [80]. The type of addend does seem to make a significant difference. Cages with positively charged amino functional groups are more susceptible to reduction than the negatively charged carboxyl groups.

In the case of fullerol, the reduction potential is estimated between −0.358 and −0.465 V vs. NHE (−0.600 and −0.707 vs. SCE), indicating the reduction potential decrease with addition of addends (Figure 4.34) has a limit. Fullerol has been shown to produce type I [Equation (4.103)] superoxide in the

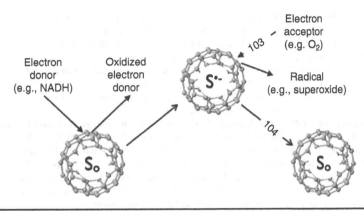

FIGURE 4.35 Dark "type I" reaction reduction pathway.

presence of UV light by selectively quenching with azide. Reduction of triplet state fullerol is possible in neutral aqueous solution because electron approach is not hampered by the presence of hydroxyl groups, but acidity may influence the photochemical reaction via increased donor access to the surface at low pH due to protonation of surface hydroxyl groups [13].

Dark "Type I" Pathway

In some cases, a donor may be present in solution with ground state C60 or a derivative thereof. If the donor has a lower reduction potential than the C60 (−0.42 V vs. SCE), there is no need for light to excite the fullerene into the higher energy triplet state. This can be categorized as a dark Type I reaction (Figure 4.35). In this case, the donor can reduce ground state C60. Furthermore, the product $C_{60}^{\bullet-}$ radical anion can transfer its electron to molecular oxygen with the formation of superoxide.

The electron transfer and photosensitization reactions described here for C_{60} have also been observed to apply for carbon nanotubes and graphenes. All of these materials have been observed to have antimicrobial properties to some extent, even in the absence of light. Indeed, these "dark reactions" appear to be the primary pathway toward inactivation of microorganisms. In this case, the ability to inactive microorganisms appears to stem from a variation on the dark reaction described above, where the fullerene or similar material acts as an electron acceptor, and oxidizes materials on the microbe surface [90].

References

1. Zepp, R.G., A.M. Braun, J. Hoigne, and J.A. Leenheer, "Photoproduction of Hydrated Electrons from Natural Organic Solutes in Aquatic Environments." *Environmental Science & Technology*, 1987; **21**(5): 485–490.
2. Zepp, R.G., P.F. Schlotzhauer, and R.M. Sink, "Photosensitized Transformations Involving Electronic-Energy Transfer in Natural-Waters—Role of Humic Substances." *Environmental Science & Technology*, 1985; **19**(1): 74–81.
3. Vileno, B., et al., "Singlet Oxygen ((1)Delta(g))-Mediated Oxidation of Cellular and Subcellular Components: ESR and AFM Assays." *Journal of Physics-Condensed Matter*, 2005; **17**(18): S1471–S1482.
4. Halliwell, B. and J.M.C. Gutteridge, *Free Radicals in Biology and Medicine*. Vol. 3rd. Ed. 1999, Oxford, England: Oxford University Press.
5. Frejaville, C., et al., "5-(Diethoxyphosphoryl)-5-Methyl-1-Pyrroline N-Oxide—a New Efficient Phosphorylated Nitrone for the In-Vitro and In-Vivo Spin-Trapping of Oxygen-Centered Radicals." *Journal of Medicinal Chemistry*, 1995; **38**(2): 258–265.
6. McCord, J.M. and I. Fridovich, "Reduction of Cytochrome C by Milk Xanthine Oxidase." *Journal of Biological Chemistry*, 1968; **243**(21): 5753.
7. McCord, J.M. and I. Fridovich, "Utility of Superoxide Dismutase in Studying Free Radical Reactions. 2. Mechanism of Mediation of Cytochrome-C Reduction by a Variety of Electron Carriers." *Journal of Biological Chemistry*, 1970; **245**(6): 1374.
8. Bielski, B.H.J., G.G. Shiue, and S. Bajuk, "Reduction of Nitro Blue Tetrazolium by CO_2^- and O_2^- Radicals." *Journal of Physical Chemistry*, 1980; **84**(8): 830–833.
9. Yamakoshi, Y., N. Umezawa, A. Ryu, K. Arakane, N. Miyata, Y. Goda, T. Masumizu, et al., "Active Oxygen Species Generated from Photoexcited Fullerene (C_{60}) as Potential Medicines: $O_2^{-\bullet}$ versus $^1O_2$." *Journal of the American Chemical Society*, 2003; **125**(42): 12803–12809.
10. Saito, I., T. Matsuura, and K. Inoue, "Formation of Superoxide Ion via One-Electron Transfer from Electron-Donors to Singlet Oxygen." *Journal of the American Chemical Society*, 1983; **105**(10): 3200–3206.
11. Haag, W.R. and J. Hoigne, "Singlet Oxygen in Surface Waters .3. Photochemical Formation and Steady-State Concentrations in Various Types of Waters." *Environmental Science & Technology*, 1986; **20**(4): 341–348.

12. Schmidt, R., "Deactivation of $O_2(^1\Delta_g)$ Singlet Oxygen by Carotenoids: Internal Conversion of Excited Encounter Complexes." *Journal of Physical Chemistry A*, 2004; **108**(26): 5509–5513.

13. Pickering, K.D. and M.R. Wiesner, "Fullerol-Sensitized Production of Reactive Oxygen Species in Aqueous Solution." *Environmental Science & Technology*, 2005; **39**(5): 1359–1365.

14. Nakanishi, I., et al., "Direct Detection of Superoxide Anion Generated in C_{60}-Photosensitized Oxidation of NADH and an Analogue by Molecular Oxygen." *Journal of the Chemical Society-Perkin Transactions 2*, 2002; (11): 1829–1833.

15. Hoffmann, M.R., S.T. Martin, W.Y. Choi, and D.W. Bahnemann, "Environmental Applications of Semiconductor Photocatalysis." *Chemical Reviews*, 1995; **95**(1): 69–96.

16. Bahnemann, D.W., C. Kormann, and M.R. Hoffmann, "Preparation and Characterization of Quantum Size Zinc-Oxide—a Detailed Spectroscopic Study." *Journal of Physical Chemistry*, 1987; **91**(14): 3789–3798.

17. Carraway, E.R., A.J. Hoffman, and M.R. Hoffmann, "Photocatalytic Oxidation of Organic-Acids on Quantum-Sized Semiconductor Colloids." *Environmental Science & Technology*, 1994; **28**(5): 786–793.

18. Choi, W., A. Termin, and M.R. Hoffmann, "Effects of Metal-Ion Dopants on the Photocatlytic Reactivity of Quantum-Sized TiO_2 Particles." *Angewandte Chemie-International Edition in English*, 1994; **33**: 1091–1092.

19. Choi, W.Y. and M.R. Hoffmann, "Photoreductive Mechanism of CCl_4 Degradation on TiO_2 Particles and Effects of Electron-Donors." *Environmental Science & Technology*, 1995; **29**(6): 1646–1654.

20. Choi, W.Y. and M.R. Hoffmann, "Kinetics and Mechanism of CCl_4 Photoreductive Degradation on TiO_2: The Role of Trichloromethyl Radical and Dichlorocarbene." *Journal of Physical Chemistry*, 1996; **100**(6): 2161–2169.

21. Choi, W.Y. and M.R. Hoffmann, "Novel Photocatalytic Mechanisms for $CHCl_3$, $CHBr_3$, and $CCl_3CO_2^-$ Degradation and the Fate of Photogenerated Trihalomethyl Radicals on TiO_2." *Environmental Science & Technology*, 1997; **31**(1): 89–95.

22. Choi, W.Y., A. Termin, and M.R. Hoffmann, "The Role of Metal-Ion Dopants in Quantum-Sized TiO_2 Correlation between Photoreactivity and Charge-Carrier Recombination Dynamics." *Journal of Physical Chemistry*, 1994; **98**(51): 13669–13679.

23. Choi, W.Y., A. Termin, and M.R. Hoffmann, "Effects of Metal-Ion Dopants on the Photocatalytic Reactivity of Quantum-Sized TiO_2 Particles." *Angewandte Chemie-International Edition in English*, 1994; **33**(10): 1091–1092.

24. Kormann, C., D.W. Bahnemann, and M.R. Hoffmann, "Preparation and Characterization of Quantum-Size Titanium-Dioxide." *Journal of Physical Chemistry*, 1988; **92**(18): 5196–5201.

25. Kormann, C., D.W. Bahnemann, and M.R. Hoffmann, "Photocatalytic Production of H_2O_2 and Organic Peroxides in Aqueous Suspensions of TiO_2, Zno, and Desert Sand." *Environmental Science & Technology*, 1988; **22**(7): 798–806.

26. Kormann, C., D.W. Bahnemann, and M.R. Hoffmann, "Environmental Photochemistry—Is Iron-Oxide (Hematite) an Active Photocatalyst—a Comparative Study—α-Fe_2O_3, ZnO, TiO_2." *Journal of Photochemistry and Photobiology a Chemistry*, 1989; **48**(1): 161–169.

27. Kormann, C., D.W. Bahnemann, and M.R. Hoffmann, "Photolysis of Chloroform and Other Organic-Molecules in Aqueous TiO_2 Suspensions." *Environmental Science & Technology*, 1991; **25**(3): 494–500.

28. Faust, B.C., M.R. Hoffmann, and D.W. Bahnemann, "Photocatalytic Oxidation of Sulfur-Dioxide in Aqueous Suspensions of α-Fe_2O_3." *Journal of Physical Chemistry*, 1989; **93**(17): 6371–6381.

29. Martin, S.T., H. Herrmann, W.Y. Choi, and M.R. Hoffmann, "Time-Resolved Microwave Conductivity .1. TiO_2 Photoreactivity and Size Quantization." *Journal of the Chemical Society-Faraday Transactions*, 1994; **90**(21): 3315–3322.

30. Martin, S.T., H. Herrmann, and M.R. Hoffmann, "Time-Resolved Microwave Conductivity .2. Quantum-Sized TiO_2 and the Effect of Adsorbates and Light-Intensity on Charge-Carrier Dynamics." *Journal of the Chemical Society-Faraday Transactions*, 1994; **90**(21): 3323–3330.

31. Herrmann, H., S.T. Martin, and M.R. Hoffmann, "Time-Resolved Radio-Frequency Conductivity (Trrfc) Studies of Charge-Carrier Dynamics in Aqueous Semiconductor Suspensions." *Journal of Physical Chemistry*, 1995; **99**(45): 16641–16645.

32. Hoffman, A.J., G. Mills, H. Yee, and M.R. Hoffmann, "Q-Sized CdS: Synthesis, Characterization, and Efficiency of Photoinitiation of Polymerization of Several Vinylic Monomers." *Journal of Physical Chemistry*, 1992; **96**(13): 5546–5552.

33. Hoffman, A.J., E.R. Carraway, and M.R. Hoffmann, "Photocatalytic Production of H_2O_2 and Organic Peroxides on Quantum-Sized Semiconductor Colloids." *Environ. Sci. Technol.*, 1994; **28**(5): 776–785.

34. Hong, A.P., D.W. Bahnemann, and M.R. Hoffmann, "Cobalt(Ii) Tetrasulfophthalocyanine on Titanium-Dioxide—a New Efficient Electron Relay for the Photocatalytic Formation and Depletion of Hydrogen-Peroxide in Aqueous Suspensions." *Journal of Physical Chemistry*, 1987; **91**(8): 2109–2117.

35. Hong, A.P., D.W. Bahnemann, and M.R. Hoffmann, "Cobalt(Ii) Tetrasulfophthalocyanine on Titanium-Dioxide .2. Kinetics and Mechanisms of the Photocatalytic Oxidation of Aqueous Sulfur-Dioxide." *Journal of Physical Chemistry*, 1987; **91**(24): 6245–6251.

36. Hoffman, A.J., H. Yee, G. Mills, and M.R. Hoffmann, "Photoinitiated Polymerization of Methyl-Methacrylate Using Q-Sized Zno Colloids." *Journal of Physical Chemistry*, 1992; **96**(13): 5540–5546.

37. Cornu, C.J.G., A.J. Colussi, and M.R. Hoffmann, "Quantum Yields of the Photocatalytic Oxidation of Formate in Aqueous TiO_2 Suspensions under Continuous and Periodic illumination." *Journal of Physical Chemistry B*, 2001; **105**(7): 1351–1354.

38. Cornu, C.J.G., A.J. Colussi, and M.R. Hoffmann, "Time Scales and pH Dependences of the Redox Processes Determining the Photocatalytic Efficiency of TiO_2 Nanoparticles from Periodic Illumination Experiments in the Stochastic Regime." *Journal of Physical Chemistry B*, 2003; **107**(14): 3156–3160.

39. Frank, S.N. and A.J. Bard, "Heterogeneous Photocatalytic Oxidation of Cyanide and Sulfite in Aqueous-Solutions at Semiconductor Powders." *Journal of Physical Chemistry*, 1977; **81**(15): 1484–1488.

40. Leland, J.K. and A.J. Bard, "Photochemistry of Colloidal Semiconducting Iron Oxide Polymorphs." *Journal of Physical Chemistry*, 1987; **91**(19): 5076–5083.

41. Siefert, R.L., S.O. Pehkonen, Y. Erel, and M.R. Hoffmann, "Iron Photochemistry of Aqueous Suspensions of Ambient Aerosol with Added Organic-Acids." *Geochimica Et Cosmochimica Acta*, 1994; **58**(15): 3271–3279.

42. Erel, Y., S.O. Pehkonen, and M.R. Hoffmann, "Redox Chemistry of Iron in Fog and Stratus Clouds." *Journal of Geophysical Research-Atmospheres*, 1993; **98**(D10): 18423–18434.

43. Siefert, R.L., S.O. Pehkonen, Y. Erel, and M.R. Hoffmann, "Photoreduction of Iron Oxyhydroxides in the Presence of Important Atmospheric Organic-Compounds." *Environmental Science & Technology*, 1993; **27**(10): 2056–2062.

44. Davis, A.P. and C.P. Huang, "Adsorption of Some Substituted Phenols onto Hydrous Cds(S)." *Langmuir*, 1990; **6**(4): 857–862.

45. Davis, A.P. and C.P. Huang, "The Photocatalytic Oxidation of Sulfur-Containing Organic-Compounds Using Cadmium-Sulfide and the Effect on Cds Photocorrosion." *Water Research*, 1991; **25**(10): 1273–1278.

46. Davis, A.P. and C.P. Huang, "Effect of Cadmium-Sulfide Characteristics on the Photocatalytic Oxidation of Thioacet-amide." *Langmuir*, 1991; **7**(4): 709–713.

47. Davis, A.P., Y.H. Hsieh, and C.P. Huang, "Photooxidative Dissolution of Cds(S)—the Effect of Cu(Ii) Ions." *Chemosphere*, 1994; **28**(4): 663–674.

48. Hsieh, Y.H. and C.P. Huang, "Photooxidative Dissolution of Cds (S) .1. Important Factors and Mechanistic Aspects." *Colloids and Surfaces*, 1991; **53**(3–4): 275–295.

49. Hsieh, Y.H., C.P. Huang, and A.P. Davis, "Photooxidative Dissolution of Cds(S) in the Presence of Heavy-Metal Ions." *Chemosphere*, 1992; **24**(3): 281–290.

50. Hsieh, Y.H., C.P. Huang, and A.P. Davis, "Photooxidative Dissolution of Cds(S)—the Effect of Pb(Ii) Ions." *Chemosphere*, 1993; **27**(5): 721–732.

51. Park, S.W. and C.P. Huang, "The Surface-Acidity of Hydrous Cds(S)." *Journal of Colloid and Interface Science*, 1987; **117**(2): 431–441.

52. Borgarello, E., K. Kalyanasundaram, M. Graetzel, and E. Pelizzetti, "Visible Light Induced Generation of Hydrogen from Hydrogen Sulfide in Cadmium Sulfide Dispersions with Hole Transfer Catalysis by Ruthenium(IV) Oxide. *Helvetica Chimica Acta*, 1982; **65**(1): 243–248.

53. Buehler, N., K. Meier, and J.F. Reber, "Photochemical Hydrogen Production with Cadmium Sulfide Suspensions." *Journal of Physical Chemistry*, 1984; **88**(15): 3261–3268.

54. Darwent, J.R. and G. Porter, "Photochemical Hydrogen Production Using Cadmium Sulfide Suspensions in Aerated Water." *Journal of the Chemical Society, Chemical Communications*, 1981(4): 145–146.

55. De, G.C., A.M. Roy, and S.S. Bhattacharya, "Photocatalytic Production of Hydrogen and Concomitant Cleavage of Industrial-Waste Hydrogen-Sulfide." *International Journal of Hydrogen Energy*, 1995; **20**(2): 127–131.

56. De, G.C., A.M. Roy, and S.S. Bhattacharya, "Photocatalytic Production of Hydrogen and Concomitant Cleavage of Industrial Waste Hydrogen Sulfide." *International Journal of Hydrogen Energy*, 1995; **20**(2): 127–131.

57. De, G.C. and A.M. Roy, "Photocatalytic Hydrogen Production with CdS and CdS/ZnS Modified by Different Electron Hole Transfer Additives Using Visible Light. *Journal of Surface Science and Technology*, 1999; **15**(3–4): 147–158.

58. Arbogast, J.W., et al., "Photophysical Properties of C_{60}." *Journal of Physical Chemistry*, 1991; **95**(1): 11–12.

59. Wang, S.Z., R.M. Gao, F.M. Zhou, and M. Selke, "Nanomaterials and Singlet Oxygen Photosensitizers: Potential Applications in Photodynamic Therapy." *Journal of Materials Chemistry*, 2004; **14**(4): 487–493.

60. Kamat, P.V., M. Haria, and S. Hotchandani, "C-60 Cluster as an Electron Shuttle in a Ru(II)-Polypyridyl Sensitizer-Based Photochemical Solar Cell." *Journal of Physical Chemistry B*, 2004; **108**(17): 5166–5170.

61. Echegoyen, L. and L.E. Echegoyen, "Electrochemistry of Fullerenes and Their Derivatives." *Accounts of Chemical Research*, 1998; **31**(9): 593–601.

62. Creegan, K.M., et al., "Synthesis and Characterization of (C60)O, the 1st Fullerene Epoxide." *Journal of the American Chemical Society*, 1992; **114**(3): 1103–1105.

63. Ziolkowski, L., K. Vinodgopal, and P.V. Kamat, "Photostabilization of Organic Dyes on Poly(Styrenesulfonate)-Capped TiO_2 Nanoparticles." *Langmuir*, 1997; **13**(12): 3124–3128.

64. Arbogast, J.W. and C.S. Foote, "Photophysical Properties of C-70." *Journal of the American Chemical Society*, 1991; **113**(23): 8886–8889.

65. Yamakoshi, Y.N., T. Yagami, K. Fukuhara, S. Sueyoshi, and N. Miyata, "Solubilization of Fullerenes into Water with Polyvinylpyrrolidone Applicable to Biological Tests." *Journal of the Chemical Society-Chemical Communications*, 1994(4): 517–518.

66. Andersson, T., K. Nilsson, M. Sundahl, G. Westman, and O. Wennerstrom, "C-60 Embedded in Gamma-Cyclodextrin—a Water-Soluble Fullerene." *Journal of the Chemical Society-Chemical Communications*, 1992(8): 604–606.

67. Guldi, D.M., R.E. Huie, P. Neta, H. Hungerbuhler, and K.D. Asmus, "Excitation of C-60, Solubilized in Water by Triton X-100 and Gamma-Cyclodextrin, and Subsequent Charge Separation via Reductive Quenching." *Chemical Physics Letters*, 1994; **223**(5–6): 511–516.

68. Guldi, D.M., H. Hungerbuhler, and K.D. Asmus, "Unusual Redox Behavior of a Water-Soluble Malonic-Acid Derivative of C-60—Evidence for Possible Cluster Formation. *Journal of Physical Chemistry*, 1995; **99**(36): 13487–13493.

69. Hungerbuhler, H., D.M. Guldi, and K.D. Asmus, "Incorporation of C-60 into Artificial Lipid-Membranes." *Journal of the American Chemical Society*, 1993; **115**(8): 3386–3387.

70. Anderson, J.L., Y.Z. An, Y. Rubin, and C.S. Foote, "Photophysical Characterization and Singlet Oxygen Yield of a Dihydrofullerene." *Journal of the American Chemical Society*, 1994; **116**(21): 9763–9764.

71. Sayes, C.M., et al., "The Differential Cytotoxicity of Water-Soluble Fullerenes." *Nano Letters*, 2004; **4**(10): 1881–1887.

72. Bensasson, R.V., et al., "Photophysical Properties of 3 Hydrofullerenes." *Chemical Physics Letters*, 1995; **245**(6): 566–570.

73. Boudon, C., J.P. Gisselbrecht, M. Gross, L. Isaacs, H.L. Anderson, R. Faust, and F. Diederich, "Electrochemistry of Mono-Adducts through Hexakis-Adducts of C-60." *Helvetica Chimica Acta*, 1995; **78**(5): 1334–1344.

74. Wasielewski, M.R., M.P. Oneil, K.R. Lykke, M.J. Pellin, and D.M. Gruen, "Triplet-States of Fullerenes C60 and C70—Electron-Paramagnetic Resonance-Spectra, Photophysics, and Electronic-Structures." *Journal of the American Chemical Society*, 1991; **113**(7): 2774–2776.

75. Guldi, D.M., "Capped Fullerenes: Stabilization of Water-Soluble Fullerene Monomers as Studied by Flash Photolysis and Pulse Radiolysis." *Journal of Physical Chemistry A*, 1997; **101**(21): 3895–3900.

76. Guldi, D.M. and M. Maggini, "Synthesis and Photophysical Properties of Electro- and Photoactive Fulleropyrrolidines." *Gazzetta Chimica Italiana*, 1997; **127**(12): 779–785.

77. Guldi, D.M. and K.D. Asmus, "Photophysical Properties of Mono- and Multiply-Functionalized Fullerene Derivatives." *Journal of Physical Chemistry A*, 1997; **101**(8): 1472–1481.

78. Fraelich, M.R. and R.B. Weisman, "Triplet-States of C-60 and C-70 in Solution—Long Intrinsic Lifetimes and Energy Pooling." *Journal of Physical Chemistry*, 1993; **97**(43): 11145–11147.

79. Hamano, T., et al., "Singlet Oxygen Production from Fullerene Derivatives: Effect of Sequential Functionalization of the Fullerene Core." *Chemical Communications*, 1997(1): 21–22.

80. Prat, F., R. Stackow, R. Bernstein, W.Y. Qian, Y. Rubin, and C.S. Foote, "Triplet-State Properties and Singlet Oxygen Generation in a Homologous Series of Functionalized Fullerene Derivatives." *Journal of Physical Chemistry A*, 1999; **103**(36): 7230–7235.

81. Mohan, H., D.K. Palit, J.P. Mittal, L.Y. Chiang, K.D. Asmus, and D.M. Guldi, "Excited States and Electron Transfer Reactions of C-60(OH)(18) in Aqueous Solution." *Journal of the Chemical Society-Faraday Transactions*, 1998; **94**(3): 359–363.

82. Vileno, B., P.R. Marcoux, M. Lekka, A. Sienkiewicz, I. Feher, and L. Forro, "Spectroscopic and Photophysical Properties of a Highly Derivatized C-60 Fullerol." *Advanced Functional Materials*, 2006; **16**(1): 120–128.

83. Jassby, D., J. Farner-Budarz, and M. Wiesner, "Impact of Aggregate Size and Structure on the Photocatalytic Properties of TiO$_2$ and ZnO Nanoparticles." *Abstracts of Papers of the American Chemical Society*, 2011; **242**: 1.

84. Badireddy, A.R., J.F. Budarz, S. Chellam, M.R. Wiesner, "Bacteriophage Inactivation by UV-A Illuminated Fullerenes: Role of Nanoparticle-Virus Association and Biological Targets." *Environmental Science & Technology*, 2012; **46**(11): 5963–5970.

85. Dubois, D., K.M. Kadish, S. Flanagan, R.E. Haufler, L.P.F. Chibante, L.J. Wilson, "Spectroelectrochemical Study of the C60 and C70 Fullerenes and Their Monoanions, Dianions, Trianions, and Tetraanions." *Journal of the American Chemical Society*, 1991; **113**(11): 4364–4366.

86. Dubois, D., K.M. Kadish, S. Flanagan, and L.J. Wilson, "Electrochemical Detection of Fulleronium and Highly Reduced Fulleride (C-60(5-)) Ions in Solution." *Journal of the American Chemical Society*, 1991; **113**(20): 7773–7774.

87. Eastoe, J., E.R. Crooks, A. Beeby, and R.K. Heenan, "Structure and Photophysics in C-60-Micellar Solutions." *Chemical Physics Letters*, 1995; **245**(6): 571–577.

88. Yamakoshi, Y., S. Sueyoshi, K. Fukuhara, and N. Miyata, "Center Dot OH and O-2(Center Dot-) Generation in Aqueous C-60 and C-70 Solutions by Photoirradiation: An EPR Study." *Journal of the American Chemical Society*, 1998; **120**(47): 12363–12364.

89. Guldi, D.M. and M. Prato, "Excited-State Properties of C-60 Fullerene Derivatives." *Accounts of Chemical Research*, 2000; **33**(10): 695–703.

90. Lyon, D.Y., L. Brunet, G.W. Hinkal, M.R. Wiesner, and P.J.J. Alvarez, "Antibacterial Activity of Fullerene Water Suspensions (nC(60)) Is Not due to ROS-Mediated Damage." *Nano Letters*, 2008; **8**(5): 1539–1543.

CHAPTER 5

Ecotoxicology Principles for Manufactured Nanomaterials

Catherine Mouneyrac

MMS Laboratory, L'UNAM, Université Catholique de l'Ouest, Angers, France

Jason M. Unrine

Department of Plant and Soil Sciences, University of Kentucky, Lexington, Kentucky, USA

Laure Giamberini

LIEC, UMR 7360, CNRS and Université de Lorraine, Lorraine, France

Olga V. Tsyusko

Department of Plant and Soil Sciences, University of Kentucky, Lexington, Kentucky, USA

Catherine Santaella

LEMIRE-BIAM, UMR 7265 CEA–CNRS–Aix-Marseille Université, Aix-en-Provence, France

Richard T. Di Giulio

Nicholas School of the Environment, Duke University, Durham, North Carolina, USA

Fabienne Schwab

CEREGE–CNRS–Aix-Marseille Université–IRD–Collège de France, Aix-en-Provence, France

Introduction

At the inception of nanoecotoxicology, in the early 2000s, it was frequently hypothesized that nanomaterials may exert toxicity through mechanisms, which were distinct from the mechanisms of toxicity for molecules or dissolved ions [1, 2]. Such effects are often termed nano-specific or particle-specific effects. In some cases, particle-specific effects may include differences in the mode of delivery of dissolved species [3]. This led to fears that there could be unpredicted adverse effects of nanomaterials [4]. Over the past decade or so of research, many of these fears have been allayed; however, there are still many examples in the literature of nano-specific effects. The purpose of this chapter is to give a brief explanation of the basic principles of ecotoxicology and ask the question whether it is necessary to take into account such nano-specific effects and if so to highlight those differences. For a more complete introduction to ecotoxicology we would refer the reader to previous works [5, 6].

Ecotoxicology has been defined as the scientific discipline that investigates the adverse effects of contaminants on the biosphere and its constituents [5]. It is often approached through a hierarchy of biological organization proceeding from the molecular level to the ecosystem level (Figure 5.1) [7]. One of the central axioms of toxicology is that all toxic effects are initiated at the molecular level. Such initiating actions may range from fairly non-specific chemical reactions or highly specific receptor mediated effects. On the other hand, ecotoxicology is primarily concerned with predicting adverse ecological outcomes on the level of individuals, populations, communities, and ecosystems. The majority of ecological risk assessments and environmental regulations are concerned with effects on the level of the population and above. In the

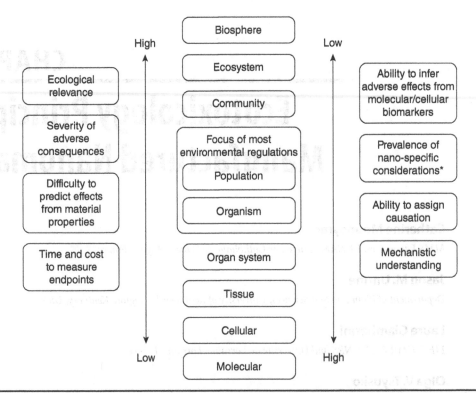

FIGURE 5.1   Ecotoxicological effects originate at the molecular level and propagate through different levels of biological organization. The arrows indicate how considerations for nanoecotoxicology scale across these levels of biological organization (adapted and modified from Newman (2015)) [5]. *Although the prevalence of nano-specific considerations appears to be greater at lower levels of biological organization, there are examples of nano-specific considerations at the ecosystem level.

United States, environmental laws, such as the Clean Water Act, protect populations of aquatic organisms, not individual organisms [8]. However, a large focus of ecotoxicological research is at the level of individual organisms. This is because it is often assumed that measurement of parameters such as growth, metabolism, reproduction, and mortality are of the greatest value in predicting population level effects and above, but at the same time are more tractable and repeatable than studies on higher levels of biological organization. On the other hand, factors influencing toxicity may propagate from the higher levels of biological organization, which most often exhibit stochastic dynamics, to lower, more highly regulated, levels of organization [9]. Furthermore, even extrapolation from individual level to population level endpoints is most often not predictive of population dynamics due to complex interactions among life history traits [10]. However, for practical, rather than scientific reasons, the majority of ecotoxicology studies have been conducted at the level of the organism and below. Additionally, current demands on regulatory bodies to test ever larger numbers of individual chemicals for a greater array of endpoints makes it less than feasible to perform whole animal studies for every new chemical on a representative set of species, let alone population level studies, a problem that is particularly germane to nanomaterials. This has led to a push to try and predict adverse ecological effects from cellular or molecular level responses. Recent attempts have been made to establish causal frameworks, such as adverse outcome pathways, to link molecular level endpoints to population level [11]. An additional argument for molecular level biomarkers is that they may serve as an early warning of adverse effects before they are evident at the population level and may be useful for biomonitoring of populations [12].

In the subsequent sections, proceeding from the molecular level to the ecosystem level, we will give a brief and general introduction to how toxicants interact with biological systems on that level. Then we will highlight key studies that illustrate nano-specific effects and evaluate whether this nano-specificity likely operates at that level of organization. This discussion will include both toxicokinetic (absorption, distribution, metabolism and excretion of a toxicant) and toxicodynamic (interaction of the toxicant with its target and subsequent effects) considerations; however, we do not discuss bioaccumulation, trophic transfer or maternal transfer in detail.

Molecular Level Impacts

Toxicity is initiated at the molecular level when the structure, function, or quantity of a biomolecule is altered by a toxicant. While it is not possible to review every possible molecular mechanism of toxicity, there are some emerging themes. One theme involves binding of the toxicant to the active site of a macromolecule preventing binding of the intended substrate of that macromolecule. This is one of the most specific types of molecular mechanisms as it requires that the conformation of the toxicant align in a specific way with the binding site of the target. A classic example of this is binding of carbon monoxide (CO) to hemoglobin and the structurally related cytochromes. CO binds to hemoglobin with 200 to 300 times the affinity of O_2, so very small partial pressures of CO in air can reduce the oxygen carrying capacity of blood in organisms that use hemoglobin as an oxygen carrier. However, this only partially explains CO toxicity. Dissolved CO also enters mitochondria and binds to heme within cytochromes, ultimately inhibiting oxidative phosphorylation [13]. Another example would be binding of an organophosphate compound, such as the insecticide malathion, to the active site of acetylcholinesterase (AChE). This binding occurs when the organophosphate forms a phosphoester bond with the serine residue in the "esteric" site of AChE. Formation of this phosphoester involves coordination of a leaving group on the organophosphate with the "anionic site" on AChE. Hydrolysis of a second leaving group results in aging or fixation of the phosphoester bond preventing its hydrolysis [14]. Binding of the organophosphate compound prevents the binding of the neurotransmitter acetylcholine so that it can be broken down. This leads to accumulation of acetylcholine in cholinergic synapses and neuromuscular junctions and causes inappropriate firing of neurons and contraction of muscles. These molecular level interactions are so specific that it is possible to tune the toxicity of an organophosphate by altering the nature and arrangement of these leaving groups. Such highly specific interactions are often termed "key and lock" mechanisms and they usually involve mimicry of the structure of an endogenous biochemical substance. Inorganic substances can engage in biological mimicry as well. For example, strontium can substitute for calcium in bone greatly reducing bone mineralization rates [15]. Similarly, selenium can substitute for sulfur in amino acids altering their redox properties [16]. However, it is not always the case that this involves a loss of function, or antagonism. For example, several substances have been shown to mimic estradiol and act as either an agonist for the estrogen receptor (although antagonism is also possible) [17].

Dichlorodiphenyltrichloroethane (DDT), an organochlorine pesticide, can bind to voltage gated sodium channels in neurons causing them to remain in the open position resulting in uncontrolled nerve impulses [17].

There are very few examples where a nanomaterial has been shown to act through such a mechanism. There are a couple of reasons for this. First, nanomaterials are typically much larger than binding sites of macromolecules and as such cannot act as a ligand (Figure 5.2). In most cases, nanomaterials range from similar in size to much larger than proteins. The binding sites of enzymes and other macromolecules typically measure a few Å across. One nanometer is equal to 10 Å, so only the particles that are less than 1 nm in diameter have any possibility of fitting into these active sites. Secondly, the binding of substrates to the binding sites requires that the substrate have very precise geometry so that functional groups on the substrate can coordinate or form bonds with the complimentary functional groups in the binding site within a precise coordination geometry. Metal clusters are a grey area between molecules and nanoparticles and typically have a defined structure. To the extent that one might consider a cluster to be a nanomaterial, there are some examples of this type of binding. For example, Au55 clusters (clusters of exactly 55 Au atoms) have been shown to bind to the major grove of B-DNA [18]. These clusters had a precise structure with an Au core size of 14 Å and a ligand shell of 4 Å. Interestingly, the clusters were degraded to Au13 when the conformation of the DNA transitioned during dehydration from B-DNA, which has a larger major grove, to A-DNA that has a smaller major groove. It is important to keep in mind that it is possible for nanomaterials to dissolve and release ions or other molecules that may then bind to these active sites.

There are also much less specific molecular mechanisms of toxicity. Proteins can be damaged in nonspecific ways that lead to denaturing, or changing of the tertiary structure of the proteins. This inevitably leads to a change in the function of the protein and it has to either be re-folded into the proper confirmation with the help of molecular chaperones or targeted for degradation in the proteasome [19]. There are examples in the literature that suggest that this may be a mechanism of toxicity for nanomaterials. For example, 4 nm gold nanoparticles, which are highly insoluble, caused unfolded protein responses in the nematode *Caenorhabditis elegans* [20]. This may be due to the intramolecular forces that maintain the conformation of proteins being disrupted by contact with the nanomaterials [21]. Forces of attraction and repulsion between the particle surface and the peptide chain may disrupt such forces within the peptide

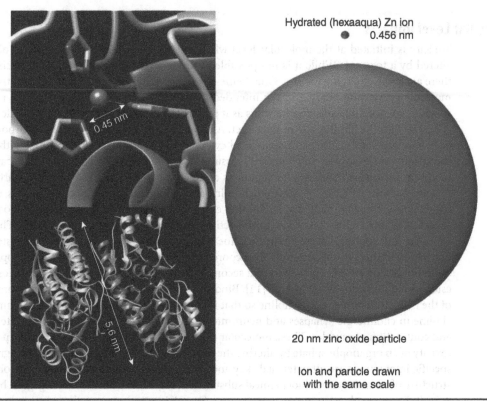

Figure 5.2 Illustration of interaction distances involved with ions and particles. On the right, a 20 nm ZnO nanoparticle and hydrated Zn ion are drawn to the same scale. On the left, the relative size of a Zn ion and Zn binding protein are shown at the bottom and at the top the size and geometry of the Zn binding site are shown. This protein is the high affinity Zn binding protein ZnuA from *Escherichia coli*. Protein models were generated and the distances were measured from a 1.9 Å resolution X-ray crystal structure using Chimera version 1.10.1 and protein data bank file 1PQ4 [35]. In the top left panel, zinc (purple sphere) is coordinated by three histidine residues (nitrogens in blue) and one water molecule (not visible because it is behind the zinc atom).

and result in protein denaturation [22]. If the extent of damage to proteins is sufficient this ultimately leads to cell death either through alteration in functions provided by the proteins or due to accumulation of protein aggregates within the cell [23]. Direct contact between nanomaterials and phospholipid bilayers has also been reported to cause structural changes and damage [24].

Damage to macromolecules can also occur indirectly through oxidative stress, which occurs when antioxidant defenses become overwhelmed by pro-oxidant species. Generation of reactive oxygen or nitrogen species (ROS or RNS) in excess of the ability of the cell to neutralize these species or repair damage from them can lead to adduct formation or damage of biomolecules. For example, cerium dioxide nanomaterials have been demonstrated to cause oxidative damage to proteins as quantified in rats by assaying protein carbonyls [25]. When proteins accumulate a sufficient number of carbonyls or other types of irreversible oxidative damage, they can be denatured and lose their function [26]. This is by no means an effect that is specific to nanomaterials. A wide array of toxicants ranging from metals to organic compounds are known to cause oxidative stress [27, 28]. Despite the fact that oxidative stress has received a great deal of attention in nanotoxicology, it is not specific to nanomaterials.

DNA damage is a change in chemical structure of DNA (oxidized DNA, removal of DNA base, formation of a bulky adduct, and single or double strand breakage) and it can be caused by both exogenous agents as well as endogenous processes. Common types of DNA damage are oxidative and nitrosative DNA damage which occurs via interaction with ROS/RNS generated during normal cell metabolism and in response to toxicant's exposure [29, 30]. For instance, reaction of DNA with superoxide, hydrogen peroxide, or hydroxyl radical can result in oxidized DNA bases as well as single or double DNA strand breaks [30]. A frequently measured product of oxidatively damaged DNA is 8-oxo-2′-deoxyguanosine, which has been shown to increase in response to tobacco smoking, exposure to heavy metals, asbestos fibers [31]. Another type of DNA damage is hydrolytic where loss of DNA bases can occur due to a labile N-glycosidic bond (between a pentose sugar and nucleobase) and such sites are called apurinic/apyrimidinic sites [32]. A type of DNA damage that is often not recognized is DNA-protein crosslink, which is formed through covalent binding of proteins to DNA. Ultraviolet and ionizing radiation as well as some metals such as arsenic, chromium, and nickel have been demonstrated to cause DNA-protein crosslinks [33]. If the DNA damage cannot be repaired or is not repaired correctly, this can interfere with replication, transcription and possibly lead to mutations and chromosomal aberrations [34]. If mutations occur in the germline, there is a possibility of their transfer to offspring [35].

Multiple studies have shown that several different types of nanomaterials can cause DNA damage [36–38]. In many instances similarly to other mutagens, the DNA damage observed was induced by ROS generated after exposure to nanomaterials. For instance, exposure to TiO_2 nanomaterials resulted in increased ROS and increased DNA damage in HepG2 cells [37]. An interesting finding was that overexpression of toll-like receptors (TLRs), transmembrane proteins responsible for activating cells' innate and adaptive immunity in response to bacteria and viruses, increased or decreased DNA damage depending on TLRs location (on the surface or endosome, respectively). It was also correlated with both increases and decreases in ROS, induction of caspase-3 and apoptosis [37, 39]. Also, increase in the product of oxidatively damaged DNA by ROS, 8-oxo-7,8-dihydro-2′-deoxyguanosine (8-oxodG) has been shown in rats in lung and/or liver tissue after exposure to carbon black, fullerene, single wall nanotubes, titanium dioxide, and copper oxide nanoparticles [36].

In addition to ROS mediated DNA damage, direct binding of nanomaterials to DNA can result in DNA conformational changes, as it has been observed in a study of Au nanomaterials with functionalized surfaces of different hydrophobicity [40] and also with Ag nanoparticles that were grown on plasmid DNA [41]. There is also a report of DNA damage by nanoparticles observed in vitro in human fibroblasts across cellular barriers without actually crossing that barrier but through intercellular signaling [42]. This "bystander" genotoxic effect has been documented in response to low doses of ionizing radiation where the effect was observed in cells that were not directly exposed to radiation [43]. Bystander signals have been shown to induce genomic instability and further in vivo studies are necessary to determine whether nanoparticles indeed can induce DNA damage in this way. Multiple signaling pathways are associated with DNA damage and are involved in DNA damage response (DDR). Cell cycle arrest, DNA repair, and apoptosis are all parts of the DDR. The deficiency in the DDR are associated with neurodegenerative diseases, aging, and cancer [34]. Among several main players that have been identified in responses to metal nanomaterials (cobalt and titanium dioxide) were *Ataxia telangiectasia mutated* (ATM), a protein kinase that serves as a mediator of DDR and is activated by double DNA strand breaks [44].

Molecular level considerations for toxicokinetics of nanomaterials are different than for molecular or ionic substances. Molecular and ionic substances can cross biological membranes in several different ways including both active and passive transport mechanisms. The active (energy dependent) mechanisms include active transport through ion pumps and endocytosis. Because of the size differences between ions and nanomaterials as previously highlighted, there is little evidence or reason to believe that ion pumps or molecular transporters play a role in nanomaterial uptake. There is however, abundant evidence that endocytosis, pinocytosis and phagocytosis play important roles in uptake of nanomaterials [45]. For ionic contaminants, binding to proteins normally taken up by cells via endocytosis can occur. For example, plutonium can bind to transferrin, a protein that normally facilitates iron uptake, and be taken into the cell by endocytosis of this protein [46]. There is evidence that the protein corona of nanomaterials, or the layer of adsorbed proteins on the particle surface, plays a critical role in their endocytosis [47, 48]. These surface adsorbed proteins, or other opsonins, interact with cell surface receptors thereby activating endocytosis. Endocytosis and subsequent intracellular trafficking will be further discussed in the following section on cellular mechanisms of toxicity.

Among the passive uptake mechanisms, diffusion through ion channels (either gated or ungated), aquaporins, and carrier molecules facilitated that diffusion plays a role for ions and molecules, but as of yet there little evidence that these mechanisms are important for nanomaterials. The molecular geometry of the molecules involved makes it unlikely. For example, the pore size of aquaporin is around 0.28 nm [49]. Passive diffusion of nanomaterials across cell membranes can also occur [50].

Metabolism of organic chemicals typically occurs in a series of three phases (although not all three phases are necessarily applicable to a particular toxicant). Phase one adds a functional group to the molecule, often a hydroxyl group. Phase one metabolism is often mediated by enzymes in the cytochrome P450 oxidase family. Enzymes involved in phase two add conjugates to the molecule such as glutathione or glucuronic acid in order to increase the water solubility of the compound to aid in elimination from the body. Some toxicants undergo a third phase of metabolism where the conjugates added in phase two are further modified to act as affinity tags for removal from cells by multidrug resistance proteins [51]. Metal cations, such as Zn^{2+}, can be extracted directly by transporter proteins [52] or detoxified by incorporation into insoluble minerals or binding to metal binding proteins such as metallothionein [53]. Like metals, nanomaterials are not metabolized in the same sense as organic chemicals; however, they can be biotransformed [54]. Redox transformation (reduction) of ceria nanomaterials has been observed in vivo in nematodes for example [55]. Dissolution of nanomaterials is an important biotransformation that might facilitate their elimination [56]. One study showed that biotransformation of Au nanomaterials might actually increase their bioavailability, perhaps due to opsonization of the particles [57]. In vivo transformations of nanomaterials is an area that is ripe for further study.

Cellular and Tissue Level Considerations

Histopathology studies the changes in cells and tissues as a result of disease, or in this case exposure to toxicants. Histopathology is an important tool in ecotoxicology because it provides a clear visualization of toxicant damage to an organism. For example, examination of the ultrastructure of fish gill tissues can provide a clear indication of toxicant induced changes, which can adversely affect respiration. Gill ultrastructural changes have been shown in response to mercury exposure in eastern mosquitofish (*Gambusia holbrooki*) [58]. Other examples include observations of neoplasms (tumors) in the liver of fish exposed to polycyclic aromatic hydrocarbons [59], or observation of female reproductive cells in male reproductive organs (intersex) exposed to wastewater effluent [60]. Although cells have frequently been examined under an electron microscope to verify the intracellular localization of nanomaterials, there has been relatively little work examining histopathological changes due to nanomaterial exposure outside of studies of lung histopathology [61].

One type of lesion that may be important for nanomaterials, particularly insoluble nanomaterials, is granuloma. Granulomas are collections of immune cells that form around foreign materials in tissues when the immune system perceives that material is foreign but cannot be eliminated. It is an attempt by the body to wall off this material to prevent damage to the surrounding tissue. This would be an unusual lesion for traditional molecular or ionic toxicants as they typically form around pathogenic organisms or foreign objects in the tissue. For example, individuals who have spent some of their life in the Ohio and Mississippi river valleys of the United States commonly have granulomas in the lungs as the result of a fungal infection called histoplasmosis [62]. Granulomas may be an important histopathological endpoint for recalcitrant, insoluble nanomaterials. For example, granulomas were observed in liver tissue of rats exposed intravenously to ceria nanomaterials and aggregates of ceria were observed within the granulomas [63]. Ceria is very resistant to dissolution and was very poorly eliminated by the rats via the urine or feces [63]. These granulomas were likely a response of the inability of the rat body to eliminate the nanomaterials.

The ultimate manifestation of cellular toxicity is cell death. Cell death can either be programmed cell death, or apoptosis, which minimizes damage to surrounding cells, or necrosis, which is unregulated cell death. Necrotic cell death causes damage to surrounding cells when proteolytic enzymes and other substances are released from subcellular compartments of the dying cell. This leads to the formation of necrotic lesions in tissues and generally causes inflammation. Necrosis has limited value as a biomarker of toxicity because many different disease processes other than toxicity can cause necrosis, such as infections, and it occurs at relatively high exposure concentrations [5]. In contrast, apoptosis is often the focus of toxicity studies and it occurs at lower toxicant exposure levels than necrosis. Apoptosis is a highly regulated process and its initiation is mediated by activation a class of proteases called caspases and disruption of the mitochondrial membrane potential [64]. It can be initiated by a variety of molecular insults. There have been numerous studies that demonstrate that nanomaterials can induce apoptosis similarly to ordinary molecular/ionic toxicants [65]. While the molecular machinery regulating apoptosis is presumably the same regardless of the nature of the toxicant, there are some examples that show somewhat unique aspects of the mechanisms for nanomaterials. For example, gold nanoparticles were conjugated with a peptide that caused them to be trafficked to the nucleus of cancer cells, where DNA strand breakage occurred [66]. As a result the cell cycle was arrested during cell division (cytokinesis) and apoptosis was induced.

The pathways for the uptake and intracellular trafficking of the nanoparticles and not the apoptosis pathways, as cited in the gold nanoparticle example above, represents the nano-specific portion of the toxicity pathway.

Bioavailability

The term "bioavailability" eludes a consensus definition among chemists, pharmacologists, toxicologists, and engineers. Bioavailability has been defined as "the extent to which a toxic contaminant is available for biologically mediated transformations and/or biological actions in the environment" [67]. According to the definition of the International Union of Pure and Applied Chemistry (IUPAC), bioavailability is "the extent of absorption of a substance by a living organism, as compared to a standard system [68]. In aquatic ecotoxicology, the biological availability has been defined as the extent to which a contaminant is free for uptake by an organism and to which it can cause an effect at the site of action [69]. Each of these definitions describes the activity of a compound that has the potential to interact with biological systems [67].

Bioaccessibility is defined as the potential for a substance to interact with and be absorbed by an organism. Bioaccessibility is related to bioavailability but is more encompassing. Nanomaterials may be occluded from the organism, hidden in soil organic matter or in exopolymers produced by bacterial cells for instance, or bioavailable after a period of time as for slowly reversible sorbed compounds [70]. Finally, they can occupy a different spatial range of the environment than the organism, as in soil matrix, with micro- and macropores that are habitats for bacteria and worms, respectively. Nanoparticles can become available within the order of seconds from these locations and hence are bioavailable, following release from labile or reversible pools; or, the organism can move into contact with the contaminant. Alternatively, release may occur over much longer timescales, and render the nanomaterials bioaccessible. Bioaccessibility includes all of these cases. In other words, bioaccessibility includes what is actually bioavailable and what is potentially bioavailable [70]. It is clear that predicting hazard from nanomaterials in some sense begins with a prediction of when nanomaterials are bioaccessible. For example, Xiao and Wiesner [71] have shown that hydrophobicity can be an indicator of the bioaccessibility of some nanoparticles to bacterial cells. The attachment of hydrophobic nanoparticles to the bacterial surface, planktonic or attached (i.e., biofilm), with or without exopolymers was greater than for hydrophilic nanoparticles. However, our knowledge of nanoparticle-organism interactions is only sufficient at the present time to predict bioaccessibility of nanomaterials from laboratory measurements in a limited number of cases.

In the environmental context, exposure to Engineered Nano Materials (ENMs) refers to the process of release of NMs adsorbed or agglomerated, diffusion and transport, adsorption and attachment to the organism and potential uptake. A key issue in nanotoxicology is that the amount of nanomaterial that eventually interacts with the organism, is far different from the amount added to the system, due to processes such as homo- and heteroaggregation of NMs, dissolution, and chemical or biological transformations (Chapters 6 and 7). Bioavailability of NMs can be modulated by environmental factors or by features that control cell and NM surface properties. For this reason, surface affinity (Chapter 6) is hypothesized to be a predictor of bioavailability.

Bioavailability and bioaccumulation therefore address processes that connect environmental exposure to organismal effects. Hou et al. [72] have reviewed the available literature (over 65 studies) on the biological accumulation of ENMs in different aquatic compartments (water, soil, or sediment). They identified fish, daphnids, and earthworms as the three most commonly used organisms, and attributed this primarily) to their recommended use as test organisms in standard test guidelines. Based on values of the bioconcentration factor (BCF), *calculated as the ratio of tissue concentration to water concentration* of the bioaccumulated material, biofilms, clams, and daphnids present a high possibility to accumulate ENM through adsorption or ingestion, which represent a major route for ENMs uptake. This is consistent with the feeding behavior of these filter-feeding organisms. In contrast, the crustacean *Gammarus roeseli* when exposed to CeO_2 NPs exhibited a fourfold higher BCF than that observed for the zebra mussel, *Dreissena polymorpha*. This was explained by the NPs behavior in the water media and the life traits of these organisms [73]. BCF values for fish were 1 to 2 orders of magnitude smaller than those of daphnids and for earthworms, the bioaccumulated sediment accumulation factor (BSAF) was less than 1. However, the approaches used in these studies present some limitations. Specifically, ENMs taken up by daphnids may remain in the gut with limited translocations across gut walls [74] and whole-body BCFs values in fish are almost always smaller compared to whole-body concentrations in small invertebrates but this does imply that fish are less sensitive to a given material than are other animals. Luoma et al. [75] have reviewed the applicability of bioaccumulation models available for metals to metal-based ENMs. They concluded that the Biological Ligand Model could be employed provided that metal-based ENM dissolution is known to account the fraction of metal-based ENM uptake driven by environmental dissolution of metal-based ENMs. The biodynamic modeling approach [75] measuring trace metal accumulation in aquatic organisms by characterizing the individual processes of uptake and elimination (that collectively result in steady-state tissue concentrations) has been applied successfully to NPs [76]. In addition, the application of this model allow one to discriminate between the relative importance of dietary and waterborne exposure routes. Experiments conducted with the freshwater snail *Lymnaea stagnalis* suggested that dietary exposure was the primary route of bioaccumulation of Ag-ENMs [77].

While nanomaterials may dissolve or produce reactive species in solution, greater proximity of a nanomaterial to a biological receptor can be expected to increase any effects that nanoparticle reactivity may engender. For the most part, attachment is a prerequisite for interaction of NMs with cells and potential internalization. Several types of intermolecular forces control the interaction between particles in water: electrostatic forces arising the surface charge and chemical interactions such as hydrogen bonding and receptor–ligand interactions, van der Waals attractive forces, steric interaction and structural forces that result from the ordering of water molecules, such as repulsive hydration forces or attractive hydrophobic interaction. The relationship between forces acting on particles and interfaces, including biological interface can be captured quantitatively by measurements of attachment efficiency or surface affinity as described in Chapter 6.

Nanoparticle Uptake into Cells

Different mechanisms permit ENMs to cross the semipermeable membrane that surrounds all living cells and enter the interior of the cell (Figure 5.3). The transmembrane particle transport mechanism referred to as endocytosis corresponds to the mechanism involving the invagination of the plasma membrane to form intracellular vesicles containing the endocytosed material. The forms of endocytosis (size of formed vesicles and molecular process) may be organism and cell-type specific. Three major endocytotic pathways have been identified in in vitro studies according to their size specificity [75, 78]:

1. Clathrin—mediated endocytosis is a receptor-mediated process that may be important for particles up to approximately 120 nm in size. The internalized particle is encapsulated within endosomes (intracellular vesicles) with the clathrin coat protein. The endosomes fuse with acidic lysosomes containing enzymes to break down foreign bodies or macromolecules. Nanomedicine has exploited this pathway to release therapeutic payloads from particles taken up by this mechanism. For metal-based NPs, this process may be responsible for the uptake of metals via a "Trojan horse" effect. The release of toxic metal ions from metal-based NPs occurs in lysosomes. For example, CdSe quantum dots were observed to be taken into the lysosomes of hepatopancreas cells of oysters (*Crassostrea virginica*) [79] and lysosomal uptake of Ag-ENMs into oyster embryos have been observed [80]. The release of metal ions into the cytosol occurs after lysosomal destabilization.

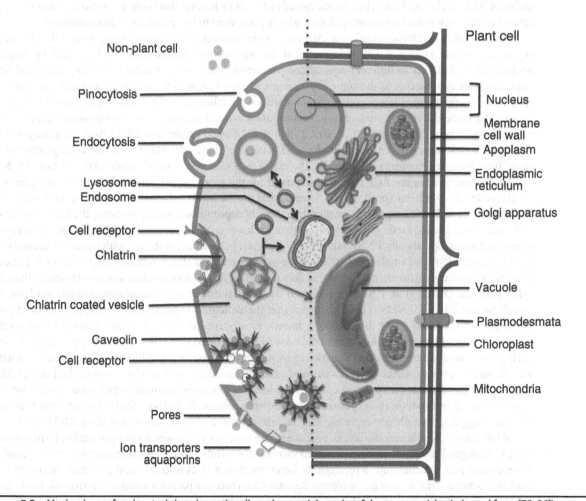

FIGURE 5.3 Mechanisms of endocytosis in eukaryotic cells and potential uptake of the nanoparticles (adapted from [75, 86]).

2. Clathrin-independent endocytosis involves internalization pathways such as caveolar endocytosis (particles up to 60 nm) that do not require a receptor. In contrast with clathrin-coated endosomes, caveosomes are unlikely to fuse with lysosomes.

3. Phagocytosis (particles >1 mm) and macro-pinocytosis (<1 mm) concern uptake of fluids and solutes. ENMs within macropinosomes are not exposed to lysosomal degradation and may remain in the cell in particulate form.

In addition to endocytotic uptake, NPs may diffuse through or disrupt the plasma membrane, and metal nanoparticles may dissolve releasing ions, which via protein-mediated transport may be taken up over channels or transporters (including metal specific transporters, such as the Cu transporter).

Once taken up by endocytosis, nanoparticles undergo intracellular trafficking. After phagocytosis, engulfed materials are contained within intracellular vesicles known as phagosomes. These phagosomes merge with lysosomes, which are intracellular vesicles that contain a wide variety of hydrolytic enzymes that degrade the contained material. Phagosomes may also produce a wide variety of reactive oxygen species, such as hydrogen peroxide, superoxide anion, and hypochlorite as a mechanism to kill pathogenic organisms after they are ingested [81]. These substances may also be important for degradation of non-living materials such as nanomaterials. A similar process occurs for endocytosis where the endocytosed vesicles merge with early endosomes. These early endosomes mature to late endosomes and eventually acquire hydrolytic enzymes to lysosomes [82]. The pH of lysosomes is typically around 5 and the hydrolytic enzymes have peak activity around this pH [82]. In plants and fungi the vacuoles perform a similar function as endosomes [83]. These conditions may be sufficient to degrade certain relatively soluble nanomaterials such as silver or zinc oxide, but not more recalcitrant materials such as titanium dioxide, cerium dioxide or gold, that are resistant to oxidation, enzymatic degradation, and pH dependent dissolution.

The importance of endocytosis in cellular uptake of nanomaterials may partially help to explain their distribution among organ systems in animals. The mononuclear phagocyte system (MPS or reticuloendothelial system) is part of the immune system that consists of phagocytic cells, such as monocytes, macrophages, and Kupffer cells. These cells accumulate in certain organs in vertebrates, particularly the lymph nodes, spleen, and liver. Studies of insoluble nanomaterials such as gold and cerium dioxide in vertebrates have shown that the liver and spleen are major sites for accumulation presumably because this is where there is a high level of phagocytic activity [48, 57, 63]. However, very small nanomaterials (5 nm or smaller) may not be recognized by the MPS [84] and coating of nanomaterials with substances like polyethylene glycol to prevent adsorption of proteins and other opsonins may also help nanomaterials to evade the MPS [85].

Nanoparticle Uptake by Plants

A growing number of studies report the uptake of nanoparticles (NPs) in a broad variety of plants. The uptake mechanisms, pathways, and physical barriers for nanoparticle uptake by plants have been reviewed in detail [87]. Unlike animal cells, plants possess exterior to their cell membranes an additional rigid cell wall that selectively limits the uptake of external agents including NPs [88]. This cell wall is made of loosely arranged cellulose fibers to resist tension, and a tight coextensive structural protein network (pectin) to resist compression and shear forces. The cell wall can be further reinforced by complex organic polymers such as lignin, suberin, or callose, and by inorganic compounds such as biomineralized silica and calcium or calcium carbonate deposits. Reinforcement occurs particularly as a response to cell wall injuries (callose, lignin), and in the outermost cells of the plants (epidermis), the passage to the vasculature (Casparian strip), the vasculature itself (endodermis, phloem, and xylem), and cells strengthening the cells such as those of the pericycle. The thickness, composition, and thus permeability of NPs to cell walls are highly heterogeneous and temporally dynamic. The porosity of the cell wall can range from 4 to 100 nm. Occasionally particles up to approximately 150 nm have been found in the cell wall [87]. In line with this, Larue et al. reported in an uptake study on nondissolving nano-TiO_2 in wheat an uptake threshold for NPs less than 140 nm in diameter into roots, and a threshold of less than or equal to 36 nm for translocation into aboveground tissue [89].

In addition to cell walls, plants possess two other important but often overlooked and less intensively studied barriers that can block NPs from being internalized. First, a hydrophobic cuticle covers most air-exposed tissue of plants including above- and below-ground tissue, mainly to prevent water evaporation and attachment of dirt. Second, mucilage or exudates cover most other regions of the plant that are not covered by a cuticle. These coating facilitate local uptake of nutrients and water. The release of exudates among other things serves to mobilize nutrients and trap bacteria. Capture of large amounts of NPs by mucilage or exudates on the outside of roots has been observed, for example, for cucumber plants [90]. The cell wall is the first barrier for NP plant uptake in very specific plant parts only, such as in root hairs, trichomes, lenticels, hydathodes; or in submerged aquatic macrophytes that use their leaves for nutrient uptake [91]. Because of the crucial role of root hairs for nutrient and water uptake, the actual permeability of the cell wall for NPs is nevertheless an important research gap.

NP Uptake across the Plant Cell Membrane

Like mammalian cells, plant cells can undergo endocytosis. This has been demonstrated using nanobeads, 20 nm in diameter, which were observed to be partially internalized in plant cells through receptor-independent mechanism such as fluidic phase endocytosis [92]. But the major endocytic mechanism is clathrin mediated [92]. Particles, 100 nm in size, were not internalized by endocytosis. In plant cells, lysosomes are usually absent in favor of a large central vacuole, dedicated to molecular degradation and storage. However in some cells, low pH lytic compartments resembling lysosome structure coexist with the vacuole.

Endocytosis is, to date, the most frequently reported uptake mechanism for NPs across the plant cell membranes. Other suggested uptake pathways without much solid evidence until now are diffusion through pores by mediated transport, binding to carrier proteins, through porins, ion channels, aquaporins, binding to nutrients or plant exudates, or by altering the membrane and creating new pores. Uptake of NPs into plants is likely to be achieved upon wounding as consequence of below-ground herbivores and mechanical injuries. Such pathways, in addition to lateral root junctions, are known entry routes for bacteria and certainly are open for efficient particle uptake.

NP Cell-to-Cell Transport

Once the cell wall or membrane has been crossed, NMs can be transported through the plasmodesmata (symplastic pathway) or through the apoplasm (apoplastic pathway) and translocated to the shoots. Alternatively, NMs can enter leaves from the cuticular and stomatal pathways. NM entry in plant cells varies according to the plant species, culture growth conditions and NM properties [93]. Microscopic channels of about 50 to 60 nm at the widest midpoint, the plasmodesmata, connect plant cell walls, enabling transport of NPs of less than 3 to 4 nm in diameter [94], and communication between cells. The symplasm is the inner side of the plasma membrane, while the apoplasm consists of cell walls and spaces between cells. Inside both of them, water and low-molecular-weight solutes can freely diffuse.

Representative Studies Reporting NP Uptake in Higher Plants

Lin et al. [95] proposed that individual fullerene C_{70} nanoparticles enter rice roots through osmotic pressure, capillary forces, pores on cell walls, and intercellular plasmodesmata or via the symplastic route. Once in the plant roots and stems, C_{70} nanoparticles were proposed to share the vascular system with water and nutrients and may be transported via the evaporation of water from the plant leaves. Fullerene surface properties play an important role in determining their uptake by cells. Small sized and hydrophilic fullerol particles ($C_{60}(OH)_{20}$) readily permeated through the plant cell wall, driven by concentration gradient but were mostly excluded by the amphiphilic plasma membrane due to their hydrophilicity, mutual electrostatic repulsion, and hydrogen bonding with water. Capillary and van der Waals forces allowed fullerol nanoparticles to accumulate and to eventually protrude through the fluidic plasma membrane, helped by low-level steady-state endocytosis and loss of membrane integrity. Hydrophobic moieties and larger C_{70} fullerenes coated with natural organic matter were mostly retained by the porous cell wall due to their hydrophobic interaction with cellulose. Single-walled carbon nanotubes can be internalized in intact plant cells (i.e., with cell walls) through fluidic-phase endocytosis [96]. Multi-walled carbon nanotubes (MWCNTs) could enter cells by creating pores [97]. Long MWCNTs (>200 nm) accumulated in subcellular organelles while the shorter ones (30–100 nm) were found into vacuoles, nucleus and plastids [98]. Metal NPs (e.g., Au, Ag, and Pt) and metal oxide NPs such as TiO_2, CeO_2, Fe_3O_4, and CuO also enter plant cells (for a review see [93]). Most of them contain metallic elements for which plants have acknowledged ion transporters. The presence of inorganic NPs inside the plant roots could be due to direct entering of NPs or uptake of dissolved Ce and possible re-formation inside the plants. Using Zr/CeO_x-NP, Schwabe et al. [87] have shown that uptake of dissolved Ce(III), due to plant root activity, followed by re-precipitation is as an important pathway in CeO_2-NP uptake by plants.

NP Uptake by Green Algae

Algae protect themselves against NP intrusion by cell walls and mucilage like higher plants, and are capable of endocytosis. Some reviews [99, 100] provide examples where ENMs either adsorb on the cell wall (e.g., QDs on the green alga *Chlamydomonas* sp., [95]) without any internalization or, contrastively, were taken up by algae (e.g., CdTe/CdS QDs and *Chlamydomonas reinhardtii* [101]). Macro-pinocytosis was the main route for internalization of thioglycolic acid stabilized CdTe quantum dots by the freshwater alga *Ochromonas danica* [102]. Even if endocytosis can be an uptake pathway in algae, the predominant route of ENMs entry into algae is most of the time undetermined. Finally, analyzing the internalization process requires a reliable measure of internalization that differentiates true uptake for surface attachment. After exposure to Ag NPs, resuspension and centrifugation of the green algae *C. reinhardtii* in presence of a strong ligand of Ag such as cysteine, decreased the intracellular Ag by more than 95 percent, showing that Ag NPs internalization across the cell membrane was limited [103].

NP Uptake by Cyanobacteria and Bacteria

Bacteria are classified as gram negative or gram positive based on their membrane structures. The structural differences lie in the organization of a key component of the membrane, peptidoglycan. Gram-negative bacteria exhibit only a thin peptidoglycan layer (about 2–3 nm) between the cytoplasmic membrane and the outer membrane. In contrast, gram-positive bacteria lack the outer membrane but have a peptidoglycan layer of about 30 nm thick.

Up until recently, bacteria were thought to lack the endocytosis mechanism. In *Escherichia coli*, protein uptake involves transport through the cell membrane via pore-like structures in the 10 to 16 nm size range for gram-negative bacteria and 2.9 to 5.5 nm in gram-positive bacteria such as *Bacillus subtilis*. However, in some bacteria, as in eukaryotes, proteins can be internalized via the formation of endosome-like structures.

As reported for algae, the internalization of ENMs in bacterial cells has been variously described [104] and is subjected to contradictory conclusions. Behra et al. [99] review recent literature on Ag ENMs entry into gram-negative and gram-positive bacteria. They report that most studies do not unequivocally prove the entrance of ENMs into cells, a process that is made more difficult to substantiate due to potential artifacts of Transmission Electron Microscopy (TEM) observation and the absence of combined elemental analysis in studies to date. Although some researchers have suggested that Ag ENMs may enter bacteria to a limited extent, most evidence indicates that Ag NPs do not readily enter either gram-negative or gram-positive bacteria. Kloepfer et al. [105] showed that adenine- and AMP-conjugated QDs of less than 5 nm could be internalized in *B. subtilis* and *E. coli* through a purine metabolism dependent process. However, the QDs did not cross the membrane through purine permeases. While binding of QD was specific, entry appeared to occur through transient membrane damages due to reactive oxygen species produced upon light exposure, triggering electron transfer between the QD and the surrounding medium. QDs particles larger than 5 nm were not internalized possibly because of the lessened reactivity of these QDs rather than their size.

The cyanobacterium *Microcystis aeruginosa* internalizes CuO ENMs, a main form of intracellular NPs being Cu_2O, with no hazard to membrane integrity. Organic matter enhances the uptake, and the main transmembrane pathway for CuO NPs seems to occur via endocytosis, based on comparisons of uptake observed with and without inhibitors of endocytosis [106].

Bradford et al. [107] studied the effect of triplicate doses of Ag-NPs, which have well-known antimicrobial activity, on natural microbial communities of estuarine sediments. They show low or no impact on estuarine sediment bacterial diversity, possibly because the chloride ions in estuary water affected the chemistry, behavior and availability of Ag-NPs. Extracellular polymeric substances (EPS) secreted a the surface of bacteria can protect the cells, decreasing NMs bioavailability, adsorption and impact [108, 109]. The possible role of surface affinity (Chapter 6) as a predictor of bioavailability is illustrated by the conclusion in one study that natural organic matter (NOM) strongly modifies the bioavailability of nanoscale zero valent iron NMs (nZVI) and consequently their toxicity by preventing the adhesion of NPs to bacterial cells [110]. If exposure to and bioavailability of nZVI in soil is not reduced to any large extent by soil, the observed toxicities in soils with a higher clay content have been observed to be less than those obtained from a sandy loam soil.

Most of studies to date have been conducted in relatively well-controlled conditions at high exposure concentrations (>ppm) to trigger measurable effects and to detect accumulation above background. A major challenge in understanding the environmental implications of nanotechnology implies evaluating ENM uptake in organisms at environmentally realistic exposure concentrations. Application of tracer techniques such as radioisotopes or enriched stable isotopes are well suited to distinguish the tracer from the background naturally present in environmental compartments (water, sediment) and biota. Another challenge is related to the characterization of the chemistry and morphology of ENMs in biota. Internal distribution of ENMs in the different organs and their cellular localization can be assessed using advanced microscopic and imaging techniques such as X-ray absorption spectroscopy and synchrotron X-ray fluorescence (SXRF) microscopy, a promising tool to quantify elements in plant and animal cells. ICP-MS coupled to laser ablation can be used in addition to X-ray based 2D and 3D images to enhance the limit of detection.

Molecular and Cellular Responses and Biomarkers

Specific assays are used to evaluate how cells respond to environmental stressors. These assays may probe for complex reactions such as oxidative stress, or general effects such as genotoxicity.

Oxidative Stress Biomarkers

Among changes in biological activities induced by NPs, reactive oxygen species (ROS) generation is one of the most frequently reported NP-associated toxicities. Various inorganic ENMs (CeO_2, TiO_2, ZnO, Cu, Ag) have been shown to induce ROS generation in bacteria and algae (for a review see [100]).

The cellular metabolism in aerobic organisms is one of the most efficient forms of energy metabolism involving the reduction of molecular oxygen (O_2). However, O_2 partial reduction in endogenous reactions gives rise to the formation of highly toxic ROS including superoxide anion radical (O_2^-), hydroxyl radical ($OH^\bullet$), peroxyl radical ($ROO^\bullet$), alkoxy radical ($RO^\bullet$), hydroperoxyl radical ($HO_2^\bullet$) and nitric oxide ($NO^\bullet$), and nonradical forms such as hydrogen peroxide (H_2O_2), singlet oxygen ($^1O_2$), ozone (O_3), hypochlorous acid (HOCl) and peroxynitrite (OOHNO). ENMs can induce intra- and extracellular ROS formation in organisms via different mechanisms that may depend on (1) NPs inherent redox-active properties or composition of surface properties (chemical reactivity), as well as of impurities present in particle preparation; (2) physical interaction of NPs with cellular and subcellular components involved in the catalysis of redox processes; and (3) NPs persistence in biological systems that can lead to continuous availability over time (dissolution) inducing site-specific ROS formation.

Oxidative stress is a parameter that is convenient to measure because cells respond to oxidative stress by induction of protective or damage responses that can easily be measured as enzymatic and nonenzymatic responses. Aerobic organisms have evolved antioxidant defense mechanisms whose main function is to block off and to deactivate ROS. These systems are composed of cytosolic enzymes (mainly superoxide dismutases: SODs, peroxidases, catalases), reducing molecules of low molecular weight (glutathione, ascorbates, urates) and several liposoluble vitamins (α-tocopherol, β-carotene). SODs correspond to a metallo-enzyme family (containing Cu, Zn, Fe, or Mn) capable of converting the superoxide anion to oxygen or hydrogen peroxide, H_2O_2. Among peroxidases, glutathione peroxidases (GPx), uses reduced glutathione (GSH) to reduce different types of peroxides. Its enzymatic activity generates GSH from the oxidized form of glutathione (GSSG). Catalases are hemoproteins occurring in peroxisomes and act by decomposing H_2O_2 into H_2O and O_2. Among these biochemical tools, glutathione S-transferase (GST) and catalase (CAT) were the most sensitive revealing the enhancement of antioxidant defenses in the bivalve *S. plana* and the ragworm *N. diversicolor* exposed to sublethal concentrations of various metal-based ENPs [111].

In plant cells, the ascorbate-glutathione cycle is major hydrogen peroxide detoxifying system [112]. The dehydroascorbate reductase (DHAR) and glutathione reductase (GR) enzymes generate the ascorbate needed by ascorbate peroxidase (APX) to catalyze the reduction of H_2O_2. APX enzymes catalyze the conversion of H_2O_2 into H_2O, using ascorbate as a specific electron donor. APXs are specific to plants and algae and are key players to protect chloroplasts from damage form H_2O_2 and HO°. Rico et al. [86] showed that in rice seedlings, CeO_2 NPs (8 nm, 500 $mg \cdot L^{-1}$) increased the H_2O_2 content in roots by 162 percent relative to the control, however with no increase in lipid peroxidation. CeO_2 NPs also enhanced the antioxidant enzyme activities of GR, and guaiacol peroxidase and APX that controlled the oxidative stress.

Metallothioneins (MTs) are low molecular weight proteins with high cysteine content (30%) and heat-stability, that promoted cellular metal homeostasis by binding and detoxifying metals such as Zn, Cu, Cd, and Ag. MTs are also known to have a protective role against oxidative damage caused by ROS by binding and sequestering transition metals or scavenging oxyradicals as hydrogen peroxide. Consequently, this protein may have a valuable role as a biomarker of NP exposure, particularly in the case of metal-based NPs. The increase of MT in response to a variety of metal-based NPs has been reported in several organisms. Gold NPs (0.1 mg/L) induced MT increase in the endobenthic clam *Scrobicularia plana* [113]. Several studies have focused on the effects of CuO NPs and an increase of MT proteins has been documented in at least two mollusk species, *Mytilus galloprovincialis* and *S. plana* [114, 115]. Since no dissolution of Cu from CuO NPs occurred in the exposure medium of *S. plana*, MT induction could be explained by intracellular Cu dissolution to ionic forms; oxyl-radical formation; or a combination of both. In the ragworm *Nereis diversicolor* exposed to nano-Ag in sediments, no increase in MT-like proteins was observed whereas soluble Ag caused MT induction [116]. Nano-Ag increased MT gene expression in embryo and adult hepatopancreas cells of *Crassostrea virginica* in the nanomolar range [80].

Metallothionein gene showed differential expression in wheat roots exposed to silver as different forms. The MT gene was induced in wheat roots exposed to silver as NPs (10 nm 2.5 mg·kg⁻¹) or a high concentration of Ag ion (2.5 mg·kg⁻¹) but was not expressed in plants exposed to bulk Ag (44 μm) and low Ag ion (63 μg·kg⁻¹) [117].

Nevertheless, when pro-oxidant forces overcome antioxidant defenses, oxidative stress occurs. The cellular response to the presence of ROS in excess can lead to an increased cell proliferation, an increased production of peroxides and free radicals that can damage all cellular components, as proteins, lipids and DNA and to a further extent to a disruption in normal mechanisms of cellular signaling and cellular death. The production of damaging ROS is also associated to inflammation and immunity responses.

Lipid Peroxidation

The most widely studied of cellular damage induced by ROS is the peroxidation of lipids (LPO) capable of inducing structural and chemical alterations of both cellular and mitochondria membranes [118]. The process of lipid peroxidation involves a chain of reactions leading to the breakdown of polyunsaturated fatty acids since they are sensitive to oxidative reactions. Their degradation induces the formation of various toxic compounds (lipid alkoxy radicals, ketones, alkanes, epoxides, and aldehydes). Among them, malondialdehyde (MDA) is the most important and the most studied and ROS mediated membrane lipid oxidation of *E. coli* treated with bare or capped ZnO NPs (7–9 nm) has been observed by detection and spectrophotometric measurement of MDA by TBARS (thiobarbituric acid-reactive species) assay [119]. ZnO and TiO_2 ENPs appear to induce oxidative stress leading to genotoxicity and cytotoxicity in *E. coli*. The potent antibacterial activity of CuO nanoparticles may be due to ROS-generation a lipid peroxidation by the nanoparticles attached to the bacterial cells, which in turn provokes an enhancement of the intracellular oxidative stress [120] and TiO_2 NPs may also produce transitory growth inhibition related to lipid peroxidation. Chen et al. [121] reported a transient production of MDA in *C. reinhardtii* exposed to nano-TiO_2 (21 nm, 0.1, 1, 10, 20, 100 mg·L⁻¹ of TiO_2 for a cell density of 10^4 cells·mL⁻¹ in exponential growth phase) MDA reached maximum values after 8 h exposure and then decreased to a moderately low level at 72 h. These studies highlight the need to envision lipid peroxidation as an event that triggers kinetic cell response and to perform time dependent analyses.

Ghosh et al. [122] detected increased level of malondialdehyde (MDA) in *Allium cepa* roots at exposed to 4 mM (0.9 μM) of TiO_2 NPs. Zhao et al. [123] germinated and grown corn plants (*Zea mays*) in soil treated with CeO_2 NPs (10 ± 1 nm) at 400 or 800 mg·kg⁻¹. CeO_2 NP treatments increased accumulation of H_2O_2, up to day 15, in phloem, xylem, bundle sheath cells and epidermal cells of shoots. The activities of catalase and ascorbate peroxidase, two antioxidant enzymes, were also increased in the corn shoot, concomitant with the H_2O_2 levels. Both 400 and 800 mg·kg⁻¹ CeO_2 NPs triggered the up-regulation of the HSP70 in roots, indicating a systemic stress response. However, none of the CeO_2 NPs increased the level of TBA reacting substances, indicating that no lipid peroxidation occurred. CAT, APX, and HSP70 might help the plants defend against CeO_2 NP-induced oxidative injury and survive NP exposure.

In mussels (*M. edulis*) exposed to Au NPs (15.6 nm; 750 μg·L⁻¹) increase of LPO levels was observed in the digestive gland [124]. Ringwood and coworkers [80] found no oxidative damage in the digestive gland of oysters (*C. virginica*) exposed to C60 fullerene (10–100 nm; 1–500 μg·L⁻¹) in gills of mussels (*Mytilus* spp.) after in vitro exposure to Fe_2O_3 NPs (5–90 nm; 1 mg–L⁻¹) [125] and in *S. plana* and *N. diversicolor* [111] exposed to various metal-based NPs (Ag, CuO, ZnO, CdS) suggesting that antioxidant defenses were sufficiently effective to prevent oxidative damage.

Genotoxicity

Consistent with the involvement of the genome throughout the maintenance of cellular homeostasis and the transmission of genetic information between generations, genotoxicity assessment in organisms exposed to NPs is of great interest. The well-known sensitivity of DNA to oxidizing agents, suggests that genotoxicity may be a potential pathway of toxicity of NP exposure. Moreover, since NPs can penetrate the nucleus via nuclear pore complexes due to their small size (1–100 nm) direct NPs genotoxicity can be due by direct interactions of the particles and DNA due to their physic-chemical properties (e.g., surface charge), by intracellular release of metal ions or with cellular components associated to NPs. The use of a battery of biomarkers has been proved to be efficient tools for evaluating the genotoxic effect of NPs: (1) the alkaline comet assay revealing DNA strand breaks (single and double), which represents one of the major types of oxidative damage to DNA via oxidative stress, (2) the RAPD assay evaluating genome instability, (3) the micronucleus test and the nuclear abnormality assay determining chromosomal damage induced by clastogenic (DNA breakage) or aneugenic (abnormal segregation) effects.

In bacteria, genotoxicity assays are based on genetically engineered microorganisms. The principle of the mutagenicity assay is a reversion of a growth inhibited bacterial strain, back to a fast growing phenotype (Ames test) or promoter-reporter fusions that generate a quantifiable dose-dependent signal in the presence of potential DNA damaging compounds and the induction of repair mechanisms. An adaptation of the Ames test proved necessary to improve the NP-cell interactions and revealed mutagenic potential of NPs. Preexposure of bacteria to NPs in a low ionic strength solution (NaCl 10 mM) at a pH below the nanoparticle isoelectric points (pH 5.5) improved the accuracy of the test [126]. Genotoxicity has been reported for TiO_2 NPs (25 nm anatase/rutile forms of 80/20, and 5–30 nm 100% anatase) at concentrations between 10 and 100 mg·L^{-1} in *Salmonella typhimurium* strains [126]. The genotoxicity was not related to the internalization of NPs by the cells, but rather to lipid peroxidation and MDA production.

Numerous studies reported genotoxic effects of NPs in different species including TiO_2 NP (<25 nm, 99.7% anatase) induced chromosomal aberrations and micronucleus formation in *Allium cepa* (the onion) root tips due to the internalization of the particles and the resultant oxidative stress. Allium micronuclei and chromosomal aberrations correlated with the reduction in root growth. TiO_2 NPs (100 nm, 2–10 mM concentration range) induced genotoxicity in *Allium cepa* and *Nicotiana tabacum* [122]. In *Allium*, micronuclei and chromosomal aberrations correlated with the reduction in root growth. Lipid peroxidation, measured by increased level of malondialdehyde (MDA) concentration, was proposed as one of the mechanism leading to DNA damage.

Exposure of *Vicia faba* root tips to Ag NPs (60 nm, 12.5–100 mg·L^{-1}) significantly increased the number of chromosomal aberrations, micronuclei, and decreased the MI in exposed groups compared to control [127]. CeO_2 NPs (8 nm) demonstrated genotoxicity at high doses in soybean roots. Random amplified polymorphic DNA assay was applied to detect DNA damage and mutations caused by NPs. Exposure of soybean plants to CeO_2 NPs results in the appearance of four new bands at 2 g·L^{-1} and three new bands at 4 g·L^{-1} treatment [128] traducing damages to DNA.

In coelomocytes of the annelid polychaete *N. diversicolor* DNA damage has been observed after 10-day sediment exposure to AgNPs (PVP-capped, 20–200 nm), micron, and ionic Ag [129]. In all cases NPs were most toxic than the other forms (micron, ionic) suggesting a nanoscale effect. In the lugworm *Arenicola marina* exposed to TiO_2 NPs (32 nm spheres) contaminated sediment, DNA damage of coelomocytes has been also reported [130]. In bivalves genotoxic effects of NPs have been reported. In mussels (*M. galloprovincialis*) exposed in natural seawater to CuO NPs (31 nm; 10 µg·L^{-1}) and Ag NPs (42 nm; 10 µg·L^{-1}) genotoxicity was mediated by oxidative stress and both NPs show lower genotoxic effects than their soluble forms [131]. In contrast in the clam *Macoma balthica* no genotoxic effects were observed in after exposure to sediment spiked with Ag NPs (20, 80 nm) and CuO NPs (<100 nm) [132]. In an outdoor mesocosm study, similar genotoxicity of CuO NPs (29.5 nm) and soluble Cu was observed in *S. plana* after 21 days of exposure at 10 µg·L^{-1} [114] whereas later work indicated that genotoxicity was induced more severely by Ag NPs than soluble Ag. Cd-based QDs induced DNA damage in marine and freshwater mussels. An additive effect of genotoxic potential produced by C60 (100–200 nm; 0.1–1 mg·L^{-1}) and polycyclic aromatic hydrocarbons (PAHs; fluoranthene) was observed in *Mytilus* hemocytes. In fish, DNA damage was reported in the gonadal cells of rainbow trout exposed to TiO_2 NPs in the presence of UV light in the skin cells of goldfish (GFSK-S1) in the absence of UV light. Long-term exposure to TiO_2 NPs has been observed to induce chromosomal instability and cell transformation in in vitro and in vivo. In the marine fish *Trachinotus carolinus*, TiO_2 NPs caused genotoxicity leading to DNA damage and resulting in the formation of micronuclei and other erythrocyte nuclear abnormalities.

Cell death by apoptosis is a general biomarker, which is expressed in multiple cell systems, conserved phylogenetically, modulated by environmentally relevant levels of chemical contaminants, and indicates a state of stress of organisms. Analyzing the morphological, biochemical, and molecular changes that take place during this universal biological process can identify cells that undergo apoptosis. The best recognized biochemical hallmark of both early and late stages of apoptosis is the activation of cysteine proteases (caspases). The detection of active caspase-3 in cells and tissues, a key enzyme in execution stage of apoptotic pathway, is an important method used to assess the induction of apoptosis. For example, DNA breaks in *S. plana* were higher following exposure to Ag NPs compared to soluble form. Thus, at least in this bivalve, exposure to Ag NPs probably caused DNA damage in cells too severe to be repaired. Consequently, bivalve exposure to Ag NP aggregates might lead to an earlier induction of apoptosis compared to soluble silver.

Hallmarks of programmed cell death in eukaryotic models are transferable to prokaryotic models. Bacterial cell death induced by treatment with different bactericidal antibiotics is accompanied by several biochemical markers of apoptosis, including DNA fragmentation, chromosomal condensation, exposure of phosphatidylserine (PS) to the outer leaflet of the plasma membrane, and dissipation of membrane potential. Bao et al. [133] reported the first induction of apoptosis in *E. coli* induced by AgNPs. AgNPs (5–10 nm; 5 and 10 mg·L^{-1}) influenced the number of *E. coli* by decreasing the growth rate. Exposure of PS on the outer leaflet of the plasma membrane of cells and membrane damage unrelated to necrosis but to apoptosis were shown together with the inhibition of newborn DNA synthesis.

Organismal Level Considerations

General Measures

Measurement of endpoints at the level of the whole organism has been historically important in ecotoxicology as it provides stronger inference of effects at the population level than molecular or cellular effects. Perhaps the most well-known measurement is that for the Median Lethal Dose or LD50, which quantifies the dose of a contaminant (expressed in mass of contaminant per mass of organism) that leads to mortality of 50 percent of the organisms. This is therefore a measure of the acute toxicity of a material. In addition to lethality, there are many sublethal and/or chronic effects that can have important ecological consequences. While sublethal endpoints have been measured for a wide variety of nanomaterials, no sublethal endpoint appears to be specific to nanomaterials and there are therefore to our knowledge not any nano-specific considerations at this level of organization. For example, there are a variety of endpoints related to growth and reproduction that have implications for population level impacts. Such endpoints might include, number of offspring, size of offspring, quality or performance of offspring, time to first reproduction, time to last reproduction, and so on. These are measures essentially related to the Darwinian fitness of an organism, or its potential to contribute its genes to the next generation. Growth is also a critical organismal endpoint which may also be reflected in reproduction, as larger individuals may tend to have more and larger offspring for example. It may be that nanomaterials cause sublethal effects in ways that are not explained simply by the release of ions or degradation products. For example, reproductive toxicity was observed in *C. elegans* exposed to Ag nanomaterials (either pristine or sulfidized) and while the effect itself differed only in the dose required to elicit an effect, not all of the reproductive toxicity could be explained by the release of free ions in the exposure medium [134]. In this example, it was likely that damage of the cuticle by the silver sulfide nanoparticles was responsible for the observed toxicity. However, the measured organismal endpoint is not specific to nanomaterials and the nano-specific aspect of the toxicity pathway is likely a molecular level aspect.

Reproductive Effects

Most toxicants have the potential to impact reproduction either directly or indirectly. Blickley et al. [135] have shown in the reproductively active adult fish (*Fundulus heteroclitus*) dietary exposed to CdSe/ZnS QDs declined trends in fecundity suggesting that chronic QD exposure could have consequences at the population level.

Repair of toxicant damage is an energy dependent process. An organism must allocate energy between growth, maintenance, reproduction and activity (e.g., foraging or social activity). Energy used for repair of damage from toxicants has to be taken from one of these categories assuming that caloric intake is limited [136]. Measurement of basal metabolic rates and adenylate energy charge are endpoints that provide an indication of the energetic costs of toxicant exposure [136, 137]. Additionally, measurement of nonpolar lipid content can provide an indication of energetic reserves [138]. Energetic endpoints can be integrated within a dynamic energy budget (DEB) model and used to predict population level impacts. Recently a DEB was parameterized to describe the effects of ZnO nanomaterials on the energy budget of the mussel *M. galloprovincialis* [139] and in the marine diatom *Thalassiosira pseudonana* [140].

Developmental Effects

Many toxicants disrupt the normal growth and development of organisms. Developmental toxicity might include embryo lethality, teratogenesis (malformations), or even subtle effects such as fluctuating asymmetry. There are several standard assays for measurement of developmental toxicity such as the Frog Embryo Teratogenesis Assay (FETAX), which utilizes African clawed frog (*Xenopus laevis*) embryos/larvae. Teratogenicity testing with zebrafish (*Danio rerio*) has also been a very important model system in recent decades. Numerous studies have demonstrated malformations as a result of nanomaterial exposure [140]. One interesting study in *D. rerio* showed that silica nanowires with an aspect ratio greater than 1 caused teratogenesis while silica materials with a smaller aspect ratio did not [141].

Behavior is another set of integrated organismal functions that is amenable to toxicological testing. These behaviors can range from predator avoidance to mating behavior and they can be directly correlated to the Darwinian fitness of an organism. The advantage of behavioral endpoints is that they can be extremely sensitive and they integrate the physiology of an organism or group of organisms with their operational environment. The endpoints thus provide a natural link to population, community and ecosystem level endpoints. In one study, earthworms (*Eisenia fetida*) avoided soils containing silver nanoparticles and silver ions at similar concentrations, although the avoidance of ions was immediate, while there was a delay in avoidance of the nanoparticles [142]. The delay in avoidance was later related to a delay in the intracellular dissolution of the nanoparticles suggesting that the proximate mechanism of the avoidance involved silver ions in both cases [3].

Neuroendocrine Effects

Other organismal level effects in vertebrates include effects on the neuroendocrine system, such as the hypothalamic-pituitary-thyroid axis that regulates development and metabolism, the hypothalamic-pituitary-gonadal axis that regulates reproduction, and the adrenocortical axis that regulates metabolism, the immune system, and osmoregulation. The hypothalamic-pituitary-adrenal axis is particularly sensitive to stress and is involved in both the fight-or-flight response (adrenaline mediated) and longer-term adaptation to stress (corticosteroid mediated). Neuroendocrine endpoints do seem to be susceptible to nanomaterial disruption. For example, silver nanomaterials have been shown to disrupt the thyroid axis in *X. laevis* [143].

With the exception of growth, mortality, reproduction and to some extent developmental toxicity, the organismal endpoints are not well explored in nanotoxicology. However, there is no compelling reason to believe that there would be nano-specific endpoints or effects on the organismal level other than those that have been manifested as such on lower levels of biological organization. One area that does require attention, as was mentioned previously with reference to the MPS, is immunotoxicology. There is evidence that nanomaterials elicit proinflammatory responses and may act as an adjuvant enhancing immune responses [144]. It is possible that nanoscale phenomena are responsible for this.

As mentioned in the discussion of the organ system level, there may be some special considerations for bioaccumulation of nanomaterials. While the bioavailability of nanomaterials is generally fairly low, nanomaterials that are not able to be dissolved within tissues may be extremely poorly eliminated. In vertebrates the particles larger than about 5 nm may not be eliminated from the kidney. For this reason ceria nanomaterials larger than 5 nm have been shown to have negligible elimination in rats [63]. It was also shown that even though the bioavailability of gold nanoparticles was low in earthworms (*E. fetida*), the elimination rate was essentially zero, indicating that the particles should bioaccumulate over time [57].

Immunotoxicity

The function of the immune system is to fight foreign pathogens and to prevent the growth of malignant cells by a variety of effector cells, molecules, and chemical messengers. Protection against foreign materials occurs through two different components of the immune system; the innate immunity, which is immediate, produces a nonspecific response and the acquired immunity, which is on the long term (immunological memory) produces a specific response. In aquatic vertebrates such as teleost fish the immune system has a high degree of similarity with that of mammals. In contrast, in lower vertebrates and in invertebrate species such as mollusks, crustaceans, sponges, and oligochaetes, the complexity of the immune system is less and relies mainly on nonspecific immunity. Conservation of the general mechanisms of innate immunity from invertebrates to mammals allows understanding common biological responses to environmental contaminants. This nonspecific immune response includes phagocytosis, cytotoxic activity, inflammatory reactions, and soluble mediators (enzymes, antimicrobial peptides, and cytokine-like molecules). Increasing evidence supports the hypothesis that the immune system of animals represents a significant target of ENMs [145]. Since different types of NPs may induce immunostimulation or immunosuppression in different experimental models, immunotoxicity tests have been widely applied to define robust assays that can be used for effective screening of NP induced immunomodulatory effects. Suspension-feeding animals, such as bivalves, are particularly at risk to NP exposure, since they have extremely developed systems for uptake of particles (nano and micro) integral to intracellular digestion and cellular immunity. The blue mussel *Mytilus* has been the species most utilized so far for studying the effects and mechanisms of action of ENMs on innate immunity [146]. Granular hemocytes representing the dominant hemocyte cell type of these bivalves are involved in cell-mediated immunity through high phagocytosis activity and various cytotoxic reactions. In these cells, the immune function is modulated by conserved components of kinase-mediated cell signaling. Immunocytotoxicity can be also mediated by lysosome damage generally evaluated using the Neutral Red Retention Time (NRRT) assay. Moderate but significant decreases in lysosomal membrane stability (LMS) were observed in bivalves exposed to ZnO, TiO_2, SiO_2, and CeO_2. In the sea urchin, a successful marine model, globally distributed in almost all depths, latitudes, temperatures, and environments in the sea, intracellular uptake of metal-oxide NPs induces lysosomal perturbations [147]. Interestingly, the in vivo effects of TiO_2 NPs on mussel (*M. galloprovincialis*) immune parameters were observed at concentrations (1–10 µg/L) much lower than those used in classical ecotoxicity tests on aquatic species, and closer to predicted environmental concentrations [148]. Not surprisingly, numerous studies have observed that cadmium-containing QDs induce significant effects on hemocyte immune parameters of marine and freshwater mussels (*M. edulis*, *M. galloprovincialis*, *Elliptio complanata*). In oysters (*Crassostrea virginica*) C60 fullerene agglomerates may concentrate in lysosomes and induce lysosomal membrane destabilization.

Both TiO_2 and ZnO NPs significantly increased phagocytosis of Neutral Red-conjugated zymosan particles at the lowest concentration tested (1 µg·mL^{-1}), whereas higher (5 µg·L^{-1}, 10 µg·mL^{-1}) concentrations induced a dramatic decrease in phagocytic activity. SiO_2 NPs usually did not affect phagocytosis whereas CeO_2 NPs inhibited phagocytosis at all the concentrations tested (1, 5, 10 µg·mL^{-1}) [145].

Plant immunity relies on preformed mechanisms based on structure (cell walls) and compounds (e.g., enzymes and antimicrobial metabolites) and infection-triggered responses that regulated in part by signaling hormones (e.g., salicylic acid, methyl jasmonate, and ethylene). Exposure of *Arabidopsis thaliana* to 5 mg·L^{-1} AgNPs (20 nm) for 10 days resulted in down-regulation of genes associated with response to pathogens and hormonal stimuli such as auxin-regulated genes, ethylene signaling pathway, and systemic acquired resistance against fungi and bacteria [149]. Interestingly, three highly up-regulated genes in the presence of Ag NPs, but not Ag$^+$, belong to the thalianol biosynthetic pathway, which could be involved in the plant defense system. Chu et al. [150] showed that nanosized Ag–silica hybrid complex (1–10 ppm) triggered plant defense responses in *A. thaliana*, inducing the expression of pathogenesis-related genes that are implicated in systemic acquired resistance.

Behavioral Biomarkers and Neurotoxicity

Although biochemical biomarkers are useful determining the mechanistic basis of toxic effects of NPs on biological systems, their lack of ecological relevance is frequently underlined and their use in ecological assessment questioned. Thus, the use of behavioral biomarkers, such as burrowing, feeding rate, are preconized since their disturbances may have ecological consequences. Indeed, burrowing changes favor predation and impairment of inter and intra-specific competition whereas decreased feeding rate limits energy acquisition necessary for survival, growth, and reproduction. The use of these behavioral biomarkers was shown as important tools to assess the impact of metal-based NP at concentrations far below lethal effects in bivalves and polychaetes [111]. For example, large Au NPs (40 nm) induce stronger inhibition of burrowing kinetics of *S. plana* when compared to small ones (5–15 nm; 100 µg·L^{-1}) [112]. Burrowing of *S. plana* was also modified after exposure to CuO (10–100 nm; 10 g·L^{-1}) and ^{67}ZnO NPs (21–34 nm; 3 mg·kg^{-1} sediment) [151, 152]. Furthermore, the exposure route is an important approach to evaluate behavioral responses in bivalves exposed to NPs. Dietary exposure decreases the feeding rate in *S. plana* more than the waterborne exposure to Ag NPs (40–50 nm; 10 µg·L^{-1}) [152].

The nervous system is also vulnerable to oxidative damage (high content of unsaturated fatty acids, high oxygen consumption rate). Acetylcholinesterase (AChE) is a crucial enzyme in the nervous system since involved in the correct transmission of nerve impulses by hydrolyzing acetylcholine (neurotransmitter) into choline and acetic acid in cholinergic synapses. NPs may bind to AChE and affect its activity. The association between behavioral changes and neurotoxicity effects (AChE inhibition) is still controversial for bivalves exposed to NPs. Usually, exposure to different metal-based NPs (CuO, ZnO, Ag) did not change AChE activity in *S. plana* [111], and no obvious neurotoxic effects were observed in *Mytilus* spp. after in vitro exposure to Fe NPs [125]. In contrast, Au NPs increased AChE activity in the clam *S. plana* [113], and was probably associated with a phenomenon of overcompensation. In the mussel *M. galloprovincialis* exposed to CuO NPs Gomes et al. [114] showed a significant AChE inhibition but behavioral biomarkers were not investigated in this study.

Probing Organismal Responses Using "Omics"

ENM effects on cells and organisms have been conducted using methods that seek characterize sweeping biomolecular responses that include monitoring genes (genomics), mRNA (transcriptomics), proteins (proteomics), and metabolites (metabolomics) in a specific biological sample in a nontargeted and nonbiased manner. Genomics provides information on changes at the heart of the cellular system, transcriptomics informs the manner in which cells formulate a response, proteomics is the response that the cell eventually makes, and metabolomics encompasses the whole and complex set of metabolic outcomes produced by the cell. Genomics (transcriptomics, proteomics, and metabolomics) have been integrated into ecotoxicology. Such "ecotoxicogenomics" is defined as the study of gene and protein expression in nontarget organisms that is important in responses to environmental toxicant exposures. "Omic" technologies are capable of producing vast amounts of data that may be used to assist both hypothesis-driven research and data-driven research, typically accompanied by machine learning, to discern patterns that may lead to the formulation of hypotheses. In the following sections, we provide examples where "omics" methods have been used to track changes in organisms following exposure to ENMs.

Omics and Bacteria

Transcriptomics/Bacteria   Microarrays have been used to analyze the impact of bactericidal copper oxide NMs on the bacterium *Legionella pneumophila*, the causative agent of the Legionnaires' pulmonary disease [154]. CuO-NP-exposed *L. pneumophila* cells (160 µg/mL for 3 hours) displayed changes in gene expression involving metabolism, nucleotide replication, transcription, and DNA repair. Together with the repression of recQ, which encodes a protein that plays a critical role in maintaining genomic stability, the results were consistent with previous reports that copper ions may bind to DNA molecules, disrupting biochemical processes, or cause damage to the helical structure by cross-linking within and between the nucleic acid strands. Another significant response observed was the induction of oxidative stress genes, *dnaK* and *katG*, suggesting toxicity to CuO NPs to *L. pneumophila*. A number of virulence genes were highly induced. These virulence gene responses may represent a kind of defense mechanisms to mitigate the toxic effect of CuO NPs. In drinking water, the main environmental stresses faced by environmental bacteria are disinfectants and competition. The analysis of the cell response could be used to design activity assays for in situ detection of disinfection efficacy. Two genes (*ceg29* and *rtxA*) were significantly induced in a dose- and time-dependent manner and could provide qRT-PCR targets, for improved surveillance of *Legionella* in drinking water.

For metal-based nanoparticles, transcriptomics has been used to discriminate the organism response to metals in ionic- or NP-form and provided insights into the molecular mechanisms of the response of organisms specific to NPs. Pelletier et al. [155] deciphered the genetics-based response mechanisms of *E. coli* after exposure to CeO_2 nanoparticles or Ce^{3+}, using whole-genome microarray. The transcriptomic results were indicative of a mild, general stress response and a highly interrelated, complex, and coordinated gene regulatory network response. The expression of a set of deregulated genes (*cydAB, NirD,* terminal oxidase cytochrome *bd-I,* succinate dehydrogenase, cytochrome *b* terminal oxidase and genes related to the respiration) in response to cerium, was consistent with iron limitation, oxidative stress exposure and alteration of respiration, when cerium was introduced as either NPs or as ions. Interestingly, eight genes (*rnt, thiS, cysW, yciW, cysI, ilvG, cysN, pyrB*) were significantly differentially expressed between the CeO_2 NPs and salt treatments. The majority of these genes are involved in sulfur metabolism but, like other related genes, such as *cysD* and *cysJ*, responded also to other treatments.

McQuillan et al. [156] report the whole-transcriptome response of *E. coli* to acute treatment with silver nanoparticles (Ag NPs) or silver ions (Ag(I)) as silver nitrate using gene expression microarrays. Both forms of silver induced significant regulation of heat shock response genes in line with protein denaturation associated with protein structure vulnerability indicating Ag(I)-labile –SH bonds. Disruption to iron-sulfur clusters led to the positive regulation of iron-sulfur assembly systems and the expression of genes for iron and sulfate homeostasis. Ag ions induced a redox stress response associated with an up-regulation of the *E. coli soxS* transcriptional regulator gene. The electronic homology of Ag(I) and Cu(I), explains why genes associated with copper homeostasis were positively regulated indicating Ag(I)-activation of copper signaling. Differential gene expression was observed for the silver nitrate and Ag NP silver delivery. In total, 379 genes were differentially regulated: Ag NPs only regulated 309 genes and silver nitrate regulated 70 genes. There was clear evidence that Ag causes redox stress but the greatest response, in terms of the number of genes regulated, was the unfolded proteins, reflecting the pan-metabolism action of Ag(I) on protein structure and function. Accordingly, this process is the primary mechanism of Ag toxicity to *E. coli*.

Proteomics and Bacteria   Proteomics is the next step following transcriptomics in the study of biological systems. Lok et al. [157] investigated the mode of antibacterial action of silver nanoparticles (nano-Ag) of against *E. coli* by proteomic approaches (2-DE and MS identification), in parallel to analyses involving solutions of Ag^+ ions. The proteomic data revealed that a short exposure of *E. coli* cells to nano-Ag resulted in an accumulation of envelope protein precursors, indicative of the dissipation of proton motive force. Consistent with these proteomic findings, nano-Ag was shown to destabilize the outer membrane, collapse the plasma membrane potential and deplete the levels of intracellular ATP. The mode of action of nano-Ag was also found to be similar to that of Ag^+ ions; however, the effective concentrations of nano-Ag and Ag^+ ions were at nanomolar and micromolar levels, respectively. Mirzajani et al. [158] deciphered the impact of silver nanoparticles against *Bacillus thuringiensis* by a proteomic approach. An exposure of *B. thuringiensis* cells to AgNPs resulted in an accumulation of envelope protein precursors, indicative of the dissipation of a proton motive force. The proteins identified were involved in oxidative stress tolerance, metal detoxification, transcription and elongation processes, protein degradation, cytoskeleton remodeling and cell division. These two studies show some common responses of gram-negative and gram-positive bacteria to nano-Ag.

Although in vitro studies of zinc oxide nanoparticles suggest that toxicity to the soil bacterium *Cupriavidus necator* is exerted in a similar manner to zinc acetate, Neal et al. [159] found no free Zn ion is associated with nanoparticle suspensions. The proteome of cells subjected to equal concentrations of either the nanoparticle or zinc acetate suggest that the mode of toxicity is quite different for the two forms of Zn, with a number of membrane-associated proteins up-expressed in response to nanoparticle exposure. Their data suggests that nanoparticles act to interrupt cell membranes thereby causing cell death rather than exerting a strictly toxic effect. We also identify potentially useful genes to serve as biomarkers of membrane disruption in toxicogenomic studies with nanoparticles or to engineer biosensor organisms.

Metabolomics and Bacteria   mRNA gene expression data and proteomic analyses reveal the set of gene products being produced in the cell, data that represents one aspect of cellular function. Metabolic profiling can give an instantaneous snapshot of the physiology of that cell. Metabolomics is defined as the quantitative measurement of the dynamic multiparametric metabolic response of living systems to pathophysiological stimuli or genetic modification. Metabolomics is useful to provide biomarkers or main biochemical pathways related to the response of an organism to some perturbations.

Organic salts of bismuth are currently used as antimicrobial agents against *Helicobacter pylori*. The effect of Bi NPs on metabolomic footprinting of *H. pylori* was further evaluated by ^{1}H NMR spectroscopy [160]. Exposure of *H. pylori* to an inhibitory concentration of Bi NPs (100 µg/mL) led to release of some metabolites such as acetate, formic acid, glutamate, valine, glycine, and uracil from bacteria into their supernatant. These findings confirm that these nanoparticles interfere with Krebs cycle, nucleotide, and amino acid metabolism and shows anti-*H. pylori* activity.

Combinations of Omics Approaches   Omics measures may be combined to better evaluate the molecular impact of a cell stressor. Sohm et al. [161] used transcriptomic and proteomic global approaches to examine the response of *E. coli* to TiO_2 in the dark (23 ± 4.9 nm, 84:16 ratio of anatase and rutile, 10 mg/L). They showed a cellular response related to osmotic stress, metabolism of cell envelope components and uptake/metabolism of endogenous and exogenous compounds. This primary mechanism of bacterial NP TiO_2 toxicity is supported by the observed massive cell leakage of K^+/Mg^{2+} concomitant with the entrance of extracellular Na^+, and by the depletion of intracellular ATP level.

However, the time-response of mRNA is not always correlated with protein abundance due to differential regulation and to differences in the lifetime of molecules. Fajardo et al. [162] focused on zero-valent iron (nZVI) effects on a common soil bacterium *Bacillus cereus*. No changes at the transcriptional level were detected in genes of particular relevance in cellular activity (*narG, nirS, pykA, gyrA,* and *katB*), the proteomic approach highlights differentially expressed proteins in *B. cereus* under nZVI exposure. Proteins involved in oxidative stress-response and tricarboxylic acid cycle (TCA) modulation were overexpressed and proteins involved in motility and wall biosynthesis were repressed. This study enabled to detect a molecular-level response as early warning signal, providing new insight into first line defense response of a soil bacterium after nZVI exposure.

Omics and Algae

In addition to microarrays, the transcriptome can be examined by RNA-sequencing using the capabilities of next-generation sequencing to reveal a snapshot of RNA presence and quantity from a genome at a given moment in time. Simon et al. [163] used RNA-seq to evaluate the effects of exposure to four different metal-based NPs, nano-Ag, nano-TiO$_2$, nano-ZnO, and CdTe/CdS QDs, in the eukaryotic green alga *C. reinhardtii* a sentinel freshwater species. nZnO caused nonspecific global stress to the cells under environmentally relevant conditions. Genes with photosynthesis-related functions were decreased drastically during exposure to TiO$_2$ NPs and slightly during exposures to the other NP types. This pattern suggests either toxicological effects in the chloroplast or effects that mimic a transition from low to high light. Ag NPs exposure dramatically elevated the levels of transcripts encoding known or predicted components of the cell wall and the flagella, suggesting that it damages structures exposed to the external medium. Exposures to TiO$_2$ and ZnO NPs, and QDs, elevated the levels of transcripts encoding subunits of the proteasome, suggesting proteasome inhibition.

Taylor et al [164] used unbiased transcriptomics and metabolomics approaches to investigate the potential toxicity of 4 to 5 nm ceria NPs to the unicellular green alga, *C. reinhardtii*. Molecular perturbations were detected at supra-environmental ceria NP-concentrations, primarily down-regulation of photosynthesis and carbon fixation with associated effects on energy metabolism. For acute exposures to small mono-dispersed particles, it can be concluded there should be little concern regarding their dispersal into the environment for this trophic level.

Omics and Plants

Many genes differentially expressed following exposure to Ag NPs or Ag$^+$ were found to be involved in the response of plants to various stresses. Kaveh et al. [165] reported changes in gene expression in *Arabidopsis thaliana* exposed to polyvinylpyrrolidone-coated Ag NPs and silver ions (Ag$^+$). Up-regulated genes were primarily associated with the response to metals and oxidative stress (e.g., vacuolar cation/proton exchanger, superoxide dismutase, cytochrome P450-dependent oxidase, and peroxidase), while down-regulated genes were more associated with response to pathogens and hormonal stimuli [e.g., auxin-regulated gene involved in organ size (ARGOS), ethylene signaling pathway, and systemic acquired resistance (SAR) against fungi and bacteria]. A significant overlap was observed between genes differentially expressed in response to Ag NPs and Ag$^+$ (13% and 21% of total up- and down-regulated genes, respectively), suggesting that Ag NP-induced stress originates partly from silver toxicity and partly from nanoparticle-specific effects. Three highly up-regulated genes in the presence of Ag NPs, but not Ag$^+$, belong to the thalianol biosynthetic pathway, which is thought to be involved in the plant defense system. Results from this study provide insights into the molecular mechanisms of the response of plants to Ag NPs and Ag$^+$.

Taylor et al. [166] used a microarray-based study to monitor the expression of candidate genes involved in metal uptake and transport in *Arabidopsis* upon gold exposure. There was up-regulation of genes involved in plant stress response such as glutathione transferases, cytochromes P450, glucosyl transferases, and peroxidases. In parallel, data showed a significant down-regulation of a discreet number of genes encoding proteins involved in the transport of copper, cadmium, iron, and nickel ions, along with aquaporins, which bind to gold. With physiological data, their results suggest that the plant predominantly takes up gold as an ionic form, and that they respond to gold exposure by up-regulating genes for plant stress and down-regulating specific metal transporters to reduce gold uptake.

The effect of exposure to 100 mg/L zinc oxide (ZnO NP), fullerene soot (FS) or titanium dioxide (TiO$_2$ NPs) nanoparticles on gene expression in *A. thaliana* roots was also studied using microarrays [167]. The genes induced by ZnO NPs and FS include mainly ontology groups annotated as stress responsive, including both abiotic (oxidative, salt, water deprivation) and biotic (wounding and defense to pathogens) stimuli. The down-regulated genes upon ZnO NP exposure were involved in cell organization and biogenesis, including translation, nucleosome assembly and microtubule-based process. FS largely repressed the transcription of genes involved in electron transport and energy pathways. Only mild changes in gene expression were observed upon TiO$_2$ NP exposure, which resulted in up- and down-regulation of genes involved mainly in responses to biotic and abiotic stimuli. The data clearly indicate that the mechanisms of phytotoxicity are highly nanoparticle dependent despite a limited overlap in gene expression response.

Proteomic analysis [168] revealed that an exposure of *Oryza sativa* L. roots to different concentrations of Ag NPs resulted in an accumulation of protein precursors, indicative of the dissipation of a proton motive force. The proteins identified are involved in oxidative stress tolerance, Ca^{2+} regulation and signaling, transcription and protein degradation, cell wall and DNA/RNA/protein direct damage, and cell division and apoptosis.

Proteomic changes have also been reported to be induced in *Eruca sativa*, commonly called rocket, in response to Ag NPs or $AgNO_3$. The low level of overlap of differentially expressed proteins indicates that Ag NPs and $AgNO_3$ caused different plant responses. Both Ag treatments triggered changes in proteins involved in the redox regulation and in the sulfur metabolism, with a potential role to maintain cellular homeostasis. Only the Ag NP exposure caused the alteration of some proteins related to the endoplasmic reticulum and vacuole indicating these two organelles as targets of the Ag NPs action. These data add further evidences that the effects of Ag NPs are not simply due to the release of Ag ions.

Omics and Bivalves

Because they are sessile and filter feeders, bivalves are able to accumulate high levels of contaminants and so are able to reflect spatial and temporal variations in aquatic pollution. For this reason they have been used as sentinel organisms in the biomonitoring programs of aquatic ecosystems mainly for marine pollution. A wide range of exposure and effect biomarkers have been developed at different levels of the biological organization in numerous marine but also in freshwater species in order to assess the impact of pollution. However, in aquatic ecotoxicology the integration of genomic and proteomic tools have been done primarily with marine microorganisms or other model species with full genome sequencing. Bivalves, even ecologically relevant, have not yet had their genome completely sequenced.

During this last decade, these species were the subject of omic studies to support the environmental toxicology. Using bivalve species, environmental proteomics can be successfully applied to assess the impacts of a wide range of anthropogenic mineral and organic contaminants including metals, persistent organic pollutants, endocrine disrupting chemicals, and natural toxins, under both laboratory and field conditions (see for example [168] and references cited within). But to date, extremely few molecular studies have been performed with emerging contaminants such as manufactured nanoparticles.

Transcriptomic approaches have been used in *M. galloprovincialis* exposed to different concentrations of nano-TiO_2 suspensions and multiple responses were evaluated in the digestive gland, gills and/or immune cells [170]. Transcription of selected genes related to (1) metal exposure (metallothionein, isoforms MT20, MT10), (2) antioxidant defense (GST, catalase), (3) stress response and apoptosis (HSP70, p53), (4) reproduction (Mytilus Estrogen Receptor genes MeERs), and (5) immune system (lysozyme, toll-like receptor isoform i) were measured by RT-QPCR as possible targets for NPs alone or in mixture with other metal or organic toxics. In the digestive gland, TiO_2 NPs decreased transcription of antioxidant and immune-related genes highlighting a general down-regulation of immune-related transcriptional responses. Transcriptomic methods were also reported only in a freshwater bivalve, *Corbicula fluminea* exposed to gold NPs [169] Using real-time RT-. PCR, these authors focused on the expression of six genes that encode proteins involved in (1) metal detoxification (metallothionein), (2) antitoxic and antioxidative defense (glutathione S transferase, catalase, superoxide-dismutase), (3) mitochondrial concentration and respiratory chain (RNA12s, subunit 1 of the cytochrome-C-oxidase) and oxidative stress.

More generally, proteomics is a relevant complement to transcriptomic approaches and can lead to the identification of new biomarkers of contaminant toxicity. Tedesco and coworkers [123] revealed in *Mytilus edulis* following gold NPs exposure, increased ubiquitination in digestive gland and also carbonylation in the gill. The same NPs decreased thiol-containing proteins inducing oxidative stress in bivalves but without causing lipid peroxidation. In the same species exposed to CuO NPs, Hu et al. [171] used redox proteomic approach with two-dimensional gel electrophoresis separation and peptide mass fingerprinting identification. In gill tissues, actin and triosephosphate isomerase showed decreased protein thiols while alpha-tubulin, tropomyosin and Cu–Zn superoxide dismutase showed increased carbonylation indicating protein oxidation of cytoskeleton and enzymes in response to increasing concentrations of CuO NPs.

In another marine bivalve, *M. galloprovincialis* the same molecular approaches yielded evidence of tissue and metal-dependent responses with Ag NPs affecting similar cell pathways as the ionic form in cytoskeleton and cell structure (catchin, myosin heavy chain), stress response (heat shock protein 70), oxidative stress (glutathione s-transferase), transcriptional regulation (nuclear receptor subfamily 1G), adhesion and mobility (procollagen-P) and energy metabolism (ATP synthase, NADH dehydrogenase). Moreover, NPs alone induced proteomic-specific responses in relation to stress response and to cytoskeleton and cell structure (paramyosin) [172]. The same research group later showed that copper NPs up-regulate proteins in the gills and down-regulate them in the digestive gland, while Cu^{2+} induced the opposite trends with distinct expression of proteins that could be common or specific to each Cu form and tissue, reflecting different mechanisms involved in their toxicity.

Omics and Earthworms

In terrestrial ecosystems, earthworms are considered as bio-indicators of soil quality and have been widely used for ecotoxicological studies in order to assess the biological effects (from the molecular to the individual levels) of diverse soil contaminants including recently NPs. The nematode *Caenorhabditis elegans*, is mainly found in the liquid phase of soil and leaf-litter environments. During the last decade, this worm species became a very attractive model in the field of environmental toxicology involving omic tools because of its completely sequenced genome. In most published studies, a transcriptomic approach was conducted using microarray and/or quantitative real-time PCR (qRT-PCR). Single-walled carbon nanotubes (SWCNTs) induced deregulation of genes involved in chemo-perception, signaling, growth, cell cycle, lifespan, and reproduction but also down-regulation of biological pathways involved in MAPK signaling cascades, the citrate cycle and endocytosis, which seems to be responsible of the toxicological effects in exposed worms. The same pathway is affected when this species is exposed to Au-NPs associated with endoplasmic reticulum alteration and Ca activation leading first to cell necrosis and then to mortality [173]. The metallothioneins (mtl-1, mtl-2), the phytochelatin synthase (pcs-1) and an apoptotic marker (cep-1) were transcriptionally activated when *C. elegans* was exposed to ZnO NPs [174] and the cytochrome pathway (cyp35a2) was affected under CeO_2 or TiO_2 NPs exposure [175]. The functional genomics analyses evidenced the deregulation of known stress-response genes *sod-3* (superoxide dismutase) and *daf-12* (abnormal dauer formation proteins) explaining reproductive failure and highlighting the important role of oxidative stress involved in Ag NPs induced toxicity. Moreover, using gene and enzyme responses in another earthworm species, *Eisenia fetida* exposed to the same metallic NPs, Tsyusko et al. [176] confirmed this toxicity mechanism.

Ratnasekhar et al. [177] used gas chromatography mass spectrometry (GC-MS)-based metabolomics approach to understand the toxicity of sublethal concentrations (7.7 and $38.5\,\mu g \cdot ml^{-1}$) of TiO_2 NP (<25 nm) in the soil nematode *Caenorhabditis elegans*. The biological pathways affected due to the exposure of TiO_2 NP were mainly associated with the tricarboxylic acid (TCA) cycle, arachidonic acid metabolism and glyoxylate dicarboxylate metabolism. The manifestation of differential metabolic profile in organism exposed to TiO_2 (NP or bulk particle) was witnessed as an effect on reproduction.

Omics Studies of ENPs and Fish

Zebrafish (*Danio rerio*) at different stages have been used as a model in toxicological studies for some time, to evaluate the effects of different classes of environmental contaminants including nanomaterials. Recently, omics approaches have been conducted using this species mainly dealing with transcriptomics.

Sequencing-based transcription-profiling conducted on embryos have shown that the toxicity of AgNP is mainly associated with bioavailable Ag^+ explained by the inhibition of the oxidative phosphorylation pathway. A gene response unique to Ag NPs was identified that may have resulted from alterations in light availability to the developing embryos [178]. Chronic exposure to the same metallic NPs exposed to adult fish produced a distinct transcriptomic response with increasing exposure concentrations. Microarray analyses showed that both extracellular and intracellular processes were affected, indicating that Ag NPs exerted effects both outside and inside the cell associated with DNA damage, possibly through production of ROS [179]. The same research group found that, in zebra fish exposed to nano-Ag, Cu, and Ti or their soluble forms, a distinct gene expression profile was induced suggesting a metal specific biological mechanism response.

In rainbow trout using a DNA microarray associated to quantitative polymerase chain reaction on a selection of genes, the toxicity of Ag NP and ionic Ag was concluded to pass by different mechanisms of action with the ionic form involving the mobilization of metals and oxidative stress and Ag NPs, inducing protein denaturation and inflammation processes that were also observed with quantum dots [180]. The studies discussed here, conducted with toxicogenomic approaches using bivalves, worm or fish species, have shown that functional genomics technologies, proteomics, and the other omic tools are particularly powerful. They provide a nonbiased and non–a priori assessment of the molecular responses of organisms to emergent contaminants, such as nanomaterials, for which the target and biological processes are largely unpredictable.

Population Level

Paradoxically, ecological impacts on populations, while the focus of environmental assessments and regulations, are not frequently studied [181]. The goal of many environmental laws is to maintain persistent healthy populations of organisms. Although many endpoints used to assess population level impacts are measured on the level of individuals (e.g., mortality, reproduction, growth), the dynamics of real populations are extremely complex and may be difficult to extrapolate from individual level measurements [10]. Populations of individuals are structured with respect to the age of individuals, the size of individuals, the reproductive status of individuals (demographics), and they are also structured within time and space. Individual subpopulations may also interact in patchy landscapes as meta-populations. All of these factors must be taken into account within the ecological context of a population to assess its health. Simple models to examine toxicant effects on population growth rate in age structured populations include the life table approach [182] and the Leslie matrix [183]. These models account for age specific mortality and natality and make predictions of population growth rate. While not well explored, contamination often has a spatial structure across a landscape. Within a meta-population, individual populations within a contaminated patch may act as a sink population while uncontaminated patches act as source populations [184].

Populations also have a genetic structure and all genes (alleles) in a population represent the gene pool. Change in allele frequencies over time within a population is microevolution. Some of the foundational work in ecotoxicology involved changes in the genetic structure of populations caused by contamination. For instance air pollution in the United Kingdom caused industrial melanism, or increase in the frequency of alleles causing dark coloration, in the peppered moth (*Biston betularia*) [185]. A study of natural plant cattail (*Typha* spp.) populations near the Chernobyl nuclear site in Ukraine revealed an increase in the genetic diversity in the populations closest to the reactor. However, this increase was not due to radiation-induced mutations but due to changes in the plant preferences for sexual reproduction versus clonal reproduction, highlighting the complexities of understanding changes to the genetic structure of a population [186]. These examples demonstrate that contamination can drive microevolution, with the evolved genotype/phenotype having a competitive advantage over the unaltered structure. This has recently been demonstrated in a population of the teleost *Fundulus heteroclitus* inhabiting a hydrocarbon-contaminated estuary in Virginia [187]. However, that population also incurred important fitness costs. As nanomaterials become more abundant in the environment, evolutionary effects may warrant consideration.

Effects from multigenerational exposures are frequently understudied but can be predictive of the population level effects. One of the potential mechanisms for these responses are epigenetic, such as changes in DNA methylation, histone methylation, and microRNA [188]. For example, a transgenerational study of the endocrine disruptor, vinclozolin, in rats showed increase in sperm phenotypic abnormalities for four generations. The mechanism appeared to be epigenetic and as these abnormalities were correlated with changes in DNA methylation patterns [189]. Transgenerational and mutigenerational effects have been also observed for nanoparticles. For instance, exposure of *C. elegans* to gold nanoparticles resulted in decreased reproduction in the second generation but recovery occurred after that in the third and fourth generations [190]. Similarly, continuous exposure of *Drosophila melanogaster* to silver nanoparticles also showed decrease in reproduction in second generation and slow recovery by generation five and eight [191].

It is also important to note here that population density itself a factor that can influence toxicity. High population density can cause physiological stress that exacerbates the effects of a toxicant. For example, when more submissive rainbow trout (*Oncorhynchus mykiss*) are coexposed with more dominant individuals, copper toxicity is exacerbated due to stress induced impairment of osmoregulation [192].

Because there has not yet been widespread release of nanomaterials to the environment at high concentrations, there have not been any studies of population level impacts in real-world populations. There is also no compelling reason to believe that nanomaterials would have any nano-specific effects at the population level. There have been a few laboratory experiments aimed at examining population level impacts of nanomaterials. For example, titanium dioxide nanoparticles were shown to decrease population growth rate in 5/10 species of algae and increase it in 5/10 species [193]. This is an unexpected result given that it is expected that the titanium dioxide would have caused phototoxicity to individual algal cells and highlights the complexities of extrapolating from individual effects to population dynamics.

Community and Ecosystem Effects

Groups of interacting populations of different species form a community. This community forms within an abiotic environment. The whole of the abiotic environment and the community of organisms inhabiting it is an ecosystem. There are different approaches to assessing the potential for community level effects.

One approach assumes that if the most sensitive species do not experience toxicity that the community as a whole will not. For example, the distribution of no observed effect concentration (NOEC) for a variety of species can be analyzed to determine the lower 95 percent confidence interval for the mean NOEC. This concentration is defined as a community level NOEC and assumes that only 5 percent of species in a community would be affected [194]. There are a few criticisms of this approach. First of all it assumes that the sample of species tested is representative of the species in the community of interest. Second, it assumes that among the 5 percent of species affected, there is no keystone species that plays a key role in maintaining the integrity of the community. This approach also assumes that the species respond the same in single species tests as they do within the context of a community. It is known that interactions among species or among individuals within a population can affect toxicity. For example, predator cues from bluegill sunfish (*Lepomis macrochirus*) increased toxicity of fipronil to the cladoceran *Ceriodaphnia dubia* [195]. In the case of silver nanomaterials, exudates from aquatic macrophytes (*Egeria densa* and *Potamogeton diversifolia*) reduced the toxicity of polyvinylpyrrolidone (PVP) and gum arabic–coated silver nanoparticles to zebrafish embryos [196, 197]. In the latter case nano-specific effects were observed. The mechanism of detoxification depended on the coating of the particles. The exudates changed the surface chemistry of the PVP-coated particles, decreasing their attachment efficiency, while they enhanced the dissolution of the gum arabic–coated particles and bound the released Ag ions to sulfhydryl groups on the exudates. In this sense the transformation of the particle and subsequent influence on toxicity could be considered as a nano-specific effect. Nanomaterials were also shown to be more effective at inhibiting the symbiotic relationship between rhizobia and legumes than bulk/dissolved metals in biosolids amended soils [198, 199].

General indices such as species richness, evenness and diversity can be used to quantify toxicant induced changes on community structure. One would expect the species indices to decrease as a consequence of contaminant exposure. A meta-analysis of effects of common contaminants on community indices in marine environments found that species richness was the most sensitive [200]. In freshwater streams the index of biotic integrity (IBI) is often used to measure contaminant impacts. The IBI takes into account the number of species within specific taxonomic and functional/trophic groups as well as their condition [201]. However, to our knowledge, these approaches to assessing the effects of contaminants have not been applied to nanomaterials, with the exception of microbes, discussed below.

Recent advances in DNA sequencing technology have enabled detailed investigations of microbial community structure opening a new frontier in community ecotoxicology. While there are many examples of contaminants altering the composition of microbial communities, functional redundancy within the community may mean that the actual function of the community is resilient. For example, a recent study showed that while metal exposure changed microbial community composition, it did not decrease the functional diversity of forest soils from a polluted region of Poland [202]. There is some indication that nanomaterials may change microbial community composition in soils differently than dissolved or bulk chemicals. For example, a recent study showed that soils amended with biosolids from a wastewater treatment plant receiving nanomaterials had a different microbial community structure than soils receiving biosolids from a wastewater treatment plant dosed with dissolved/bulk metals [197]. This was also shown for silver nanoparticles as compared to silver nitrate in terrestrial mesocosms amended with biosolids [203].

Ultimately when entire ecosystems are stressed the flow of energy and nutrients may be disrupted. Ecosystem responses to contaminants predicted by Odum included increased turnover and decreased cycling of nutrients as well as decreased productivity and increased respiration [204]. These ecosystem services are perhaps a more tangible than changes to microbial community structure. Only a handful of studies have examined ecosystem function with respect to nanomaterials. A study of forest soil showed that iron and tin oxide nanoparticles increased the C/N ratio in soil, perhaps reflecting changes in nutrient cycling [205]. The previously mentioned terrestrial mesocosms that were dosed with silver nanomaterial containing biosolids experienced increased flux of nitrous oxide relative to silver nitrated dosed mesocosms [203]. It is unclear if the silver nitrate and sliver nanoparticles affected the microorganisms via different mechanisms or if it was the fate and transport of the materials within the soil that caused different outcomes. For example, different species of microorganisms may live at different depths within the soil profile and may therefore be differentially exposed to nanomaterials versus dissolved metals if the distribution of the toxicants is different within the soil profile. This difference in the spatial and temporal aspects of exposure may lead to differences in toxicity between nanomaterials and ordinary chemicals and may lead to effects that are different between nanomaterials and ordinary chemicals.

Summary

The basic principles of ecotoxicology are applicable to both nanomaterials and ordinary chemicals with few additional considerations for nanomaterials. Perhaps the greatest differences between nanomaterials and ordinary chemicals have been demonstrated on the molecular and cellular levels. The difference in scale and forces of interaction between nanomaterials and ordinary chemicals mean that very different mechanisms of uptake and intracellular transport are involved. Although some ordinary toxicants are taken up by endocytosis, it appears to be of primary importance for nanomaterials. Certain histopathological lesions, such as granulomas, may occur as a result of the inability to degrade nanomaterials within tissues. On the molecular, cellular, and tissue level much of the same molecular machinery involved in dealing with pathogens is also used to process nanomaterials. In the case of nanomaterials, damage to biomolecules tends to occur through non-specific mechanisms, such as generation of ROS/RNS or through denaturation of proteins. From the organismal to the community level, there appear to be few nano-specific considerations and there are no endpoints that are specific to nanomaterials.

The emergence of nanomaterials has made a preexisting problem even larger—the need to test an ever-expanding list of chemicals for ecotoxicity. It is not possible to test every chemical and every variation of nanomaterial for every endpoint of importance at every level of biological organization. One approach to combat this problem is to identify common toxicity pathways and screen chemicals using a biological pathway approach [206]. Under this approach, these toxicity pathways are causally linked to higher order effects for a few well-studied chemicals and larger numbers of chemicals can be screened on the molecular level for involvement in these toxicity pathways. It still remains to be seen if such approaches will be successful even for ordinary chemicals given the complicating influence of top-down ecological effects, but such an approach can be envisioned for nanomaterials (Figure 5.4). Also, it is now well known that the properties of nanomaterials are dependent on their surroundings and environmental histories, complicating this approach even further [207]. Training these models using pristine nanomaterials may not yield accurate predictions given the transformations of nanomaterials and properties that are dependent on the environment. To address this complicating factor, an approach that identifies intermediate measures which integrate a wide array of nanomaterial properties within specific environmental systems would be predictive of toxicity under different exposure scenarios may be required. Such an approach, termed the functional assay approach, has been recently described [208].

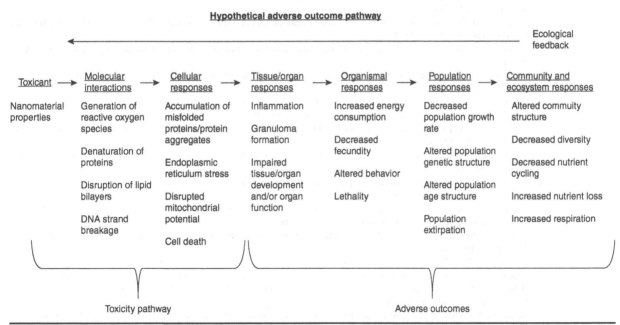

FIGURE 5.4   Hypothetical generic adverse outcome pathway for nanomaterials illustrating examples of endpoints that would be relevant at different levels of biological organization. Figure based on a generic adverse outcome pathway presented by Ankley et al. [11]. In contrast to the model of Ankley et al., we add a community and ecosystem level and show that ecological feedbacks can propagate effects from the top.

References

1. Nel, A., T. Xia, and N. Li, "Toxic Potential of Materials at the Nanolevel." *Science*, 2006; **311**: 622–627.
2. Unrine, J., P. Bertsch, and S. Hunyadi, "Bioavailability, Trophic Transfer, and Toxicity of Manufactured Metal and Metal Oxide Nanoparticles in Terrestrial Environments," in *Nanoscience and Nanotechnology: Environmental and Health Impacts*, V. Grassian, Editor. 2008, Hoboken, NJ: John Wiley and Sons, Inc., pp. 345–366.
3. Tsyusko, O.V., et al., "Short-Term Molecular-Level Effects of Silver Nanoparticle Exposure on the Earthworm, *Eisenia fetida*." *Environmental Pollution*, 2012; **171**: 249–255.
4. Service, R.F., "Nanotoxicology: Nanotechnology Grows Up." *Science*, 2004; **304**(5678): 1732–1734.
5. Newman, M., *Fundamentals of Ecotoxicology: The Science of Pollution*. 4th ed. 2015, Boca Raton, FL: CRC Press.
6. Di Giulio, R.T. and M.C. Newman, "Ecotoxicology," in *Casarett & Doull's Toxicology: The Basic Science of Poisons*. 2013, New York: McGraw-Hill.
7. Newman, M. and C. Jagoe, *Ecotoxicology: A Hierarchical Treatment*. 1996, Boca Raton, FL: CRC press.
8. Borm, P.J.A., "Particle Toxicology: From Coal Mining to Nanotechnology." *Inhalation Toxicology*, 2002; **14**(3): 311–324.
9. Beketov, M.A. and M. Liess, "Ecotoxicology and Macroecology—Time for Integration." *Environmental Pollution*, 2012; **162**: 247–254.
10. Forbes, V.E. and P. Calow, "Is the per Capita Rate of Increase a Good Measure of Population-Level Effects in Ecotoxicology?" *Environmental Toxicology and Chemistry*, 1999; **18**(7): 1544–1556.
11. Ankley, G.T., et al., "Adverse Outcome Pathways: A Conceptual Framework to Support Ecotoxicology Research and Risk Assessment." *Environmental Toxicology and Chemistry*, 2010; **29**(3): 730–741.
12. Fossi, M.C., "Biomarkers as Diagnostic and Prognostic Tools for Wildlife Risk Assessment: Integrating Endocrine-Disrupting Chemicals." *Toxicology and Industrial Health*, 1998; **14**(1–2): 291–309.
13. Roderique, J.D., et al., "A Modern Literature Review of Carbon Monoxide Poisoning Theories, Therapies, and Potential Targets for Therapy Advancement. *Toxicology*, 2015; **334**: 45–58.
14. Čolović, M.B., et al., "Acetylcholinesterase Inhibitors: Pharmacology and Toxicology." *Current Neuropharmacology*, 2013; **11**(3): 315–335.
15. Wornham, D.P., et al., "Strontium Potently Inhibits Mineralisation in Bone-Forming Primary Rat Osteoblast Cultures and Reduces Numbers of Osteoclasts in Mouse Marrow Cultures." *Osteoporosis International*, 2014; **25**(10): 2477–2484.
16. Janz, D., et al., "Selenium Toxicity to Aquatic Organisms," in *Ecological Assessment of Selenium in the Aquatic Environment*, W.J. Adams, et al., Editors. 2010, Pensacola, FL: SETAC Press.
17. Sonnenschein, C. and A.M. Soto, "An Updated Review of Environmental Estrogen and Androgen Mimics and Antagonists." *Journal of Steroid Biochemistry and Molecular Biology*, 1998; **65**(1–6): 143–150.
18. Ariyasu, S., et al., "Alignment of Gold Clusters on DNA via a DNA-Recognizing Zinc Finger-Metallothionein Fusion Protein." *Bioconjugate Chemistry*, 2009; **20**(12): 2278–2285.
19. Vallentine, P., et al., "The Ubiquitin—Proteasome Pathway Protects Chlamydomonas Reinhardtii against Selenite Toxicity, but Is Impaired as Reactive Oxygen Species Accumulate." *AoB Plants*, 2014; **6**: 1–11.
20. Tsyusko, O.V., et al., "Toxicogenomic Responses of the Model Organism Caenorhabditis elegans to Gold Nanoparticles." *Environmental Science & Technology*, 2012; **46**(7): 4115–4124.
21. Fei, L. and S. Perrett, "Effect of Nanoparticles on Protein Folding and Fibrillogenesis." *International Journal of Molecular Sciences*, 2009; **10**(2): 646–655.
22. Chen, P., et al., "Contrasting Effects of Nanoparticle Binding on Protein Denaturation." *The Journal of Physical Chemistry C*, 2014; **118**(38): 22069–22078.
23. Nakamura, T. and S. Lipton, "Cell Death: Protein Misfolding and Neurodegenerative Diseases." *Apoptosis*, 2009; **14**(4): 455–468.
24. Leroueil, P.R., et al., "Wide Varieties of Cationic Nanoparticles Induce Defects in Supported Lipid Bilayers." *Nano Letters*, 2008; **8**(2): 420–424.
25. Yokel, R.A., et al., "Biodistribution and Oxidative Stress Effects of a Systemically-Introduced Commercial Ceria Engineered Nanomaterial." *Nanotoxicology*, 2009; **3**(3): 234–248.
26. Berlett, B.S. and E.R. Stadtman, "Protein Oxidation in Aging, Disease, and Oxidative Stress." *Journal of Biological Chemistry*, 1997; **272**(33): 20313–20316.
27. Valko, M., H. Morris, and M.T. Cronin, "Metals, Toxicity and Oxidative Stress." *Current Medicinal Chemistry*, 2005; **12**(10): 1161–1208.
28. Di Giulio, R.T., et al., "Biochemical Responses in Aquatic Animals: A Review of Determinants of Oxidative Stress." *Environmental Toxicology and Chemistry*, 1989; **8**(12): 1103–1123.
29. Jones, L.E., et al., "Differential Effects of Reactive Nitrogen Species on DNA Base Excision Repair Initiated by the Alkyladenine DNA Glycosylase." *Carcinogenesis*, 2009; **30**(12): 2123–2129.
30. De Bont, R. and N. van Larebeke, "Endogenous DNA Damage in Humans: A Review of Quantitative Data." *Mutagenesis*, 2004; **19**(3): 169–185.
31. Valavanidis, A., T. Vlachogianni, and C. Fiotakis, "8-Hydroxy-2′-Deoxyguanosine (8-OHdG): A Critical Biomarker of Oxidative Stress and Carcinogenesis." *Journal of Environmental Science and Health, Part C*, 2009; **27**(2): 120–139.
32. Helleday, T., S. Eshtad, and S. Nik-Zainal, "Mechanisms Underlying Mutational Signatures in Human Cancers." *Nature Reviews Genetics*, 2014; **15**(9): 585–598.
33. Barker, S., M. Weinfeld, and D. Murray, "DNA–Protein Crosslinks: Their Induction, Repair, and Biological Consequences." *Mutation Research/Reviews in Mutation Research*, 2005; **589**(2): 111–135.
34. Polo, S.E. and S.P. Jackson, "Dynamics of DNA Damage Response Proteins at DNA Breaks: A Focus on Protein Modifications." *Genes & Development*, 2011; **25**(5): 409–433.
35. Banerjee, S., et al., "Structural Determinants of Metal Specificity in the Zinc Transport Protein ZnuA from Synechocystis 6803." *Journal of Molecular Biology*, 2003; **333**(5): 1061–1069.

36. Moller, P., et al., "Measurement of Oxidative Damage to DNA in Nanomaterial Exposed Cells and Animals." *Environmental and Molecular Mutagenesis*, 2015; **56**(2): 97–110.

37. El-Said, K.S., et al., "Molecular Mechanism of DNA Damage Induced by Titanium Dioxide Nanoparticles in Toll-Like Receptor 3 or 4 Expressing Human Hepatocarcinoma Cell Lines." *Journal of Nanobiotechnology*, 2014; **12**: 48.

38. Wan, R., et al., "DNA Damage Caused by Metal Nanoparticles: Involvement of Oxidative Stress and Activation of ATM." *Chemical Research in Toxicology*, 2012; **25**(7): 1402–1411.

39. El-Said, K.S., et al., "Effects of Toll-like Receptors 3 and 4 Induced by Titanium Dioxide Nanoparticles in DNA Damage-Detecting Sensor Cells." *Journal of Biosensors & Bioelectronics*, 2013; **4**(5): 1–5.

40. Goodman, C.M., et al., "DNA-Binding by Functionalized Gold Nanoparticles: Mechanism and Structural Requirements." *Chemical Biology & Drug Design*, 2006; **67**(4): 297–304.

41. Roy, S., S. Basak, and A.K. Dasgupta, "Nanoparticle Induced Conformational Change in DNA and Chirality of Silver Nanoclusters." *Journal of Nanoscience and Nanotechnology*, 2010; **10**(2): 819–825.

42. Bhabra, G., et al., "Nanoparticles Can Cause DNA Damage across a Cellular Barrier." *Nature Nanotechnology*, 2009; **4**(12): 876–883.

43. Mothersill, C. and C. Seymour, "Radiation-Induced Bystander Effects, Carcinogenesis and Models." *Oncogene*, 2003; **22**(45): 7028–7033.

44. Khalil, H.S., et al., "Targeting ATM Pathway for Therapeutic Intervention in Cancer." *BioDiscovery*, 2012; **1**(3): 1–13.

45. Sahay, G., D.Y. Alakhova, and A.V. Kabanov, "Endocytosis of Nanomedicines." *Journal of Controlled Release*, 2010; **145**(3): 182–195.

46. Jensen, M.P., et al., "An Iron-Dependent and Transferrin-Mediated Cellular Uptake Pathway for Plutonium." *Nature Chemical Biology*, 2011; **7**(8): 560–565.

47. Chithrani, B.D., A.A. Ghazani, and W.C.W. Chan, "Determining the Size and Shape Dependence of Gold Nanoparticle Uptake into Mammalian Cells." *Nano Letters*, 2006; **6**(4): 662–668.

48. Storm, G., et al., "Surface Modification of Nanoparticles to Oppose Uptake by the Mononuclear Phagocyte System." *Advanced Drug Delivery Reviews*, 1995; **17**(1): 31–48.

49. Agre, P., "The Aquaporin Water Channels." *Proceedings of the American Thoracic Society*, 2006; **3**(1): 5–13.

50. Nel, A.E., et al., "Understanding Biophysicochemical Interactions at the Nano-Bio Interface." *Nature Materials*, 2009; **8**: 543–557.

51. Xu, C., C.Y. Li, and A.N. Kong, "Induction of Phase I, II, and III Drug Metabolism/Transport by Xenobiotics." *Archives of Pharmacal Research*, 2005; **28**(3): 249–268.

52. Sekler, I., et al., "Mechanism and Regulation of Cellular Zinc Transport." *Molecular Medicine*, 2007; **13**(7–8): 337–343.

53. Wallace, W.G. and S.N. Luoma, "Subcellular Compartmentalization of Cd and Zn in Two Bivalves. II. Significance of Trophically Available Metal (TAM)." *Marine Ecology Progress Series*, 2003; **257**: 125–137.

54. Graham, U.M., et al., "In Vivo Processing of Ceria Nanoparticles inside Liver: Impact on Free-Radical Scavenging Activity and Oxidative Stress." *ChemPlusChem*, 2014, **79**(8): 1083–1088.

55. Collin, B., et al., "Influence of Natural Organic Matter and Surface Charge on the Toxicity and Bioaccumulation of Functionalized Ceria Nanoparticles in *Caenorhabditis elegans*." *Environmental Science & Technology*, 2014; **48**(2): 1280–1289.

56. Utembe, W., et al., "Dissolution and Biodurability: Important Parameters Needed for Risk Assessment of Nanomaterials." *Part Fibre Toxicology*, 2015; **12**: 11.

57. Unrine, J.M., et al., "Trophic Transfer of Au Nanoparticles from Soil along a Simulated Terrestrial Food Chain." *Environmental Science & Technology*, 2012; **46**(17): 9753–9760.

58. Jagoe, C.H., A. Faivre, and M.C. Newman, "Morphological and Morphometric Changes in the Gills of Mosquitofish (*Gambusia holbrooki*) after Exposure to Mercury (II)." *Aquatic Toxicology*, 1995; **34**: 163–183.

59. Vogelbein, W.K., et al., "Hepatic Neoplasms in the Mummichog Fundulus Heteroclitus from a Creosote-Contaminated Site." *Cancer Research*, 1990; **50**(18): 5978–5986.

60. Nolan, M., et al., "A Histological Description of Intersexuality in the Roach." *Journal of Fish Biology*, 2001; **58**(1): 160–176.

61. Roda, E., et al., "Comparative Pulmonary Toxicity Assessment of Pristine and Functionalized Multi-Walled Carbon Nanotubes Intratracheally Instilled in Rats: Morphohistochemical Evaluations." *Histology and Histopathology*, 2011; **26**(3): 357–367.

62. Kibria, R., et al., "'Ohio River Valley Fever' Presenting as Isolated Granulomatous Hepatitis: A Case Report." *Southern Medical Journal*, 2009; **102**(6): 656–658.

63. Yokel, R., et al., "Distribution, Elimination and Biopersistence to 90 Days of a Systemically-Introduced 30 m Ceria Engineered Nanomaterial in Rats." *Toxicological Sciences*, 2012. In Press.

64. Elmore, S., "Apoptosis: A Review of Programmed Cell Death." *Toxicologic Pathology*, 2007; **35**(4): 495–516.

65. Lapied, E., et al., "Silver Nanoparticle Exposure Causes Apoptotic Response in the Earthworm Lumbricus Terrestris (Oligochaeta)." *Nanomedicine*, 2010; **5**(6): 975–984.

66. Kang, B., M.A. Mackey, and M.A. El-Sayed, "Nuclear Targeting of Gold Nanoparticles in Cancer Cells Induces DNA Damage, Causing Cytokinesis Arrest and Apoptosis." *Journal of the American Chemical Society*, 2010; **132**(5): 1517–1519.

67. Hamelink, J., et al., *Bioavailability: Physical, Chemical, and Biological Interactions.* 1994, CRC Press, Boca Raton, FL.

68. Nordberg, M., J.H. Duffus, and D.M. Templeton, "Explanatory Dictionary of Key Terms in Toxicology: Part II (IUPAC Recommendations 2010). *Pure and Applied Chemistry*, 2010; **82**(3): 679–751.

69. Newman, M., *Fundamentals of Ecotoxicology.* 3rd ed. Boca Raton, FL: CRC Press.

70. Semple, K.T., et al., "Peer Reviewed: Defining Bioavailability and Bioaccessibility of Contaminated Soil and Sediment Is Complicated." *Environmental Science & Technology*, 2004; **38**(12): 228A–231A.

71. Xiao, Y. and M.R. Wiesner, "Characterization of Surface Hydrophobicity of Engineered Nanoparticles." *Journal of Hazardous Materials*, 2012; **215**: 146–151.

72. Hou, W.-C., P. Westerhoff, and J.D. Posner, "Biological Accumulation of Engineered Nanomaterials: A Review of Current Knowledge." *Environmental Science: Processes & Impacts*, 2013; **15**(1): 103–122.

73. Garaud, M., et al., "Multibiomarker Assessment of Cerium Dioxide Nanoparticle (nCeO2) Sublethal Effects on Two Freshwater Invertebrates, *Dreissena polymorpha* and *Gammarus roeseli*." *Aquatic Toxicology*, 2015; **158**: 63–74.

74. Zhu, X., Y. Chang, and Y. Chen, "Toxicity and Bioaccumulation of TiO_2 Nanoparticle Aggregates in *Daphnia magna*. *Chemosphere*, 2010; **78**(3): 209–215.

75. Luoma, S.N., F.R. Khan, and M.-N.l. Croteau, "Bioavailability and Bioaccumulation of Metal Based Engineered Nano-materials (Me-ENMs) in Aquatic Environments: Concepts and Processes," in *Frontiers of Nanoscience*, J. Lead and E. Valsami-Jones, Editors. Amsterdam, Netherlands. 2014; **7**: 157–193.

76. Khan, F.R., et al., "Bioaccumulation Dynamics and Modeling in an Estuarine Invertebrate Following Aqueous Exposure to Nanosized and Dissolved Silver." *Environmental Science & Technology*, 2012; **46**(14): 7621–7628.

77. Croteau, M.-N.l., et al., "Silver Bioaccumulation Dynamics in a Freshwater Invertebrate after Aqueous and Dietary Exposures to Nanosized and Ionic Ag." *Environmental Science & Technology*, 2011; **45**(15): 6600–6607.

78. Moore, M.N., "Do Nanoparticles Present Ecotoxicological Risks for the Health of the Aquatic Environment?" *Environment International*, 2006; **32**(8): 967–976.

79. Ringwood, A.H., et al., "Characterization, Imaging and Degradation Studies of Quantum Dots in Aquatic Organisms." *MRS Online Proceedings Library*, 2011; 895–910.

80. Ringwood, A.H., et al., "The Effects of Silver Nanoparticles on Oyster Embryos." *Marine Environmental Research*, 2010; **69**(Suppl 1): S49–S51.

81. Hampton, M.B., A.J. Kettle, and C.C. Winterbourn, *Inside the Neutrophil Phagosome: Oxidants, Myeloperoxidase, and Bacterial Killing*. 1998; **92**: 3007–3017.

82. Alberts, B., et al., *Molecular Biology of the Cell*. 2002, New York: Garland.

83. Marty, F., "Plant Vacuoles." *The Plant Cell*, 1999; **11**(4): 587–599.

84. Dan, M., et al., "Ceria-Engineered Nanomaterial Distribution in, and Clearance from, Blood: Size matters." *Nanomedicine*, 2012; **7**(1): 95–110.

85. Verrecchia, T., et al., "Non-Stealth (Poly(Lactic acid/albumin)) and Stealth (Poly(Lactic Acid-Polyethylene Glycol)) nanoparticles as injectable drug carriers." *Journal of Controlled Release*, 1995; **36**(1–2): 49–61.

86. Rico, C.M., et al., "Cerium Oxide Nanoparticles Modify the Antioxidative Stress Enzyme Activities and Macromolecule Composition in Rice Seedlings." *Environmental Science & Technology*, 2013; **47**(24): 14110–14118.

87. Schwab, F., et al., "Barriers, Pathways and Processes for Uptake, Translocation and Accumulation of Nanomaterials in Plants—Critical Review." *Nanotoxicology*, 2015; **11**: 1–22.

88. Nair, R., et al., "Nanoparticulate Material Delivery to Plants." *Plant Science*, 2010; **179**(3): 154–163.

89. Larue, C., et al., "Accumulation, Translocation and Impact of TiO_2 Nanoparticles in Wheat (*Triticum aestivum* spp.): Influence of Diameter and Crystal Phase." *Science of The Total Environment*, 2012; **431**: 197–208.

90. Zhang, Z.Y., et al., "Uptake and Distribution of Ceria Nanoparticles in Cucumber Plants." *Metallomics*, 2011; **3**(8): 816–822.

91. Schwab, F., et al., "Dissolved Cerium Contributes to Uptake of Ce in the Presence of Differently Sized CeO_2-Nanoparticles by Three Crop Plants." *Metallomics*, 2015; **7**(3): 466–477.

92. Bandmann, V., et al., "Uptake of Fluorescent Nano Beads into BY2-Cells Involves Clathrin-Dependent and Clathrin-Independent Endocytosis." *FEBS Letters*, 2012; **586**(20): 3626–3632.

93. Rico, C., et al., "Interaction of Nanoparticles with Edible Plants and Their Possible Implications in the Food Chain." *Journal of Agricultural Food Chemistry*, 2011; **59**(8): 3485–3498.

94. Dietz, K.-J. and S. Herth, "Plant Nanotoxicology." *Trends in Plant Science*, 2011; **16**(11): 582–589.

95. Lin, C., et al., "Studies on Toxicity of Multi-Walled Carbon Nanotubes on Arabidopsis T87 Suspension Cells." *Journal of Hazardous Materials*, 2009; **170**: 578–583.

96. Liu, Q., et al., "Carbon Nanotubes as Molecular Transporters for Walled Plant Cells." *Nano Letters*, 2009; **9**(3): 1007–1010.

97. Wild, E. and K. Jones, "Novel Method for the Direct Visualization of In Vivo Nanomaterials and Chemical Interactions in Plants." *Environmental Science & Technology*, 2009; **43**: 5290–5294.

98. Serag, M., et al., "Trafficking and Subcellular Localization of Multiwalled Carbon Nanotubes in Plant Cells." *ACS Nano*, 2011; **5**: 493–499.

99. Behra, R., et al., "Bioavailability of Silver Nanoparticles and Ions: From a Chemical and Biochemical Perspective." *Journal of the Royal Society Interface*, 2013; **10**(87): 20130396.

100. von Moos, N. and V.I. Slaveykova, "Oxidative Stress Induced by Inorganic Nanoparticles in Bacteria and Aquatic Microalgae-State of the Art and Knowledge Gaps." *Nanotoxicology*, 2013; **8**(6): 605–630.

101. Domingos, R.F., et al., "Bioaccumulation and Effects of CdTe/CdS Quantum Dots on *Chlamydomonas reinhardtii*—Nanoparticles or the Free Ions?" *Environmental Science & Technology*, 2011; **45**(18): 7664–7669.

102. Wang, Y., et al., "Bioaccumulation of CdTe Quantum Dots in a Freshwater Alga *Ochromonas danica*: A Kinetics Study." *Environmental Science & Technology*, 2013; **47**(18): 10601–10610.

103. Piccapietra, F., et al., "Intracellular Silver Accumulation in *Chlamydomonas reinhardtii* upon Exposure to Carbonate Coated Silver Nanoparticles and Silver Nitrate." *Environmental Science & Technology*, 2012; **46**(13): 7390–7397.

104. von Moos, N., P. Bowen, and V.I. Slaveykova, "Bioavailability of Inorganic Nanoparticles to Planktonic Bacteria and Aquatic Microalgae in Freshwater." *Environmental Science: Nano*, 2014; **1**(3): 214–232.

105. Kloepfer, J., R. Mielke, and J. Nadeau, "Uptake of CdSe and CdSe/ZnS Quantum Dots into Bacteria via Purine-Dependent Mechanisms." *Applied and Environmental Microbiology*, 2005; **71**(5): 2548–2557.

106. Wang, Z., et al., "Toxicity and Internalization of CuO Nanoparticles to Prokaryotic Alga *Microcystis aeruginosa* as Affected by Dissolved Organic Matter." *Environmental Science & Technology*, 2011; **45**(14): 6032–6040.

107. Bradford, A., et al., "Impact of Silver Nanoparticle Contamination on the Genetic Diversity of Natural Bacterial Assemblages in Estuarine Sediments." *Environmental Science & Technology*, 2009; **43**(12): 4530–4536.

108. Zeyons, O.l., et al., "Direct and Indirect CeO_2 Nanoparticles Toxicity for *Escherichia coli* and *Synechocystis*." *Nanotoxicology*, 2009; **3**(4): 284–295.

109. Planchon, M., et al., "Exopolysaccharides Protect *Synechocystis* against the Deleterious Effects of Titanium Dioxide Nanoparticles in Natural and Artificial Waters." *Journal of Colloid and Interface Science*, 2013; **405**: 35–43.

110. Li, Z., et al., "Adsorbed Polymer and NOM Limits Adhesion and Toxicity of Nano Scale Zerovalent Iron to *E. coli*." *Environmental Science & Technology*, 2010; **44**(9): 3462–3467.

111. Mouneyrac, C., et al., "Fate and Effects of Metal-Based Nanoparticles in Two Marine Invertebrates, the Bivalve Mollusc *Scrobicularia plana* and the Annelid Polychaete *Hediste diversicolor*." *Environmental Science and Pollution Research*, 2014; **21**(13): 7899–7912.

112. Caverzan, A., et al., "Plant Responses to Stresses: Role of Ascorbate Peroxidase in the Antioxidant Protection." *Genetics and Molecular Biology*, 2012 Dec; **35**(4 Suppl): 1011–1019.

113. Pan, J.-F., et al., "Size Dependent Bioaccumulation and Ecotoxicity of Gold Nanoparticles in an Endobenthic Invertebrate: The Tellinid Clam *Scrobicularia plana*." *Environmental Pollution*, 2012; **168**: 37–43.

114. Gomes, T.N., et al., "Effects of Copper Nanoparticles Exposure in the Mussel Mytilus galloprovincialis." *Environmental Science & Technology*, 2011; **45**(21): 9356–9362.

115. Buffet, P.-E., et al., "A Mesocosm Study of Fate and Effects of CuO Nanoparticles on Endobenthic Species (*Scrobicularia plana, Hediste diversicolor*)." *Environmental Science & Technology*, 2013; **47**(3): 1620–1628.

116. Garcia-Alonso, J., et al., "Cellular Internalization of Silver Nanoparticles in Gut Epithelia of the Estuarine Polychaete *Nereis diversicolor*." *Environmental Science & Technology*, 2011: **45**(10): 4630–4636.

117. Dimkpa, C.O., et al., "Silver Nanoparticles Disrupt Wheat (*Triticum aestivum* L.) Growth in a Sand Matrix." *Environmental Science & Technology*, 2013; **47**(2): 1082–1090.

118. Roméo, M. and L. Giamberini, "Ecological Biomarkers," in *Ecological Biomarkers Indicators of Ecotoxicological Effects*, C. Amiard-Triquet, J.C. Amiard, and P.S. Rainbow, Editors. 2013, Boca Raton, FL: CRC Press, pp. 15–44.

119. Dutta, R.K., B.P. Nenavathu, and M.K. Gangishetty, "Correlation between Defects in Capped ZnO Nanoparticles and Their Antibacterial Activity." *Journal of Photochemistry and Photobiology B: Biology*, 2013; **126**: 105–111.

120. Applerot, G., et al., "Understanding the Antibacterial Mechanism of CuO Nanoparticles: Revealing the Route of Induced Oxidative Stress." *Small*, 2012; **8**(21): 3326–3337.

121. Chen, L., et al., "Toxicological Effects of Nanometer Titanium Dioxide (Nano-TiO_2) on *Chlamydomonas reinhardtii*." *Ecotoxicology and Environmental Safety*, 2012; **84**: 155–162.

122. Ghosh, M., M. Bandyopadhyay, and A. Mukherjee, "Genotoxicity of Titanium Dioxide (TiO_2) Nanoparticles at Two Trophic Levels: Plant and Human Lymphocytes." *Chemosphere*, 2010; **81**(10): 1253–1262.

123. Zhao, L., et al., "Stress Response and Tolerance of Zea Mays to CeO(2) Nanoparticles: Cross Talk among H(2)O(2), Heat Shock Protein and Lipid Peroxidation." *ACS Nano*, 2012; **6**(11): 9615–9622.

124. Tedesco, S., et al., "Oxidative Stress and Toxicity of Gold Nanoparticles in *Mytilus edulis*. *Aquatic Toxicology*, 2010; **100**(2): 178–186.

125. Kádár, E., et al., "Stabilization of Engineered Zero-Valent Nanoiron with Na-Acrylic Copolymer Enhances Spermiotoxicity." *Environmetal Science & Technology*, 2011; **45**(8): 3245–3251.

126. Jomini, S.P., et al., "2012Modifications of the Bacterial Reverse Mutation Test Reveals Mutagenicity of TiO_2 Nanoparticles and Byproducts from a Sunscreen TiO_2-Based Nanocomposite." *Toxicology Letters*, **215**(1): 54–61.

127. Patlolla, A.K., et al., "Genotoxicity of Silver Nanoparticles in Vicia faba: A Pilot Study on the Environmental Monitoring of Nanoparticles." *International Journal of Environmental Research and Public Health*, 2012; **9**(5): 1649–1662.

128. Lopez-Moreno, M.L., et al., "Evidence of the Differential Biotransformation and Genotoxicity of ZnO and CeO_2 Nanoparticles on Soybean (Glycine max) Plants. *Environmental Science & Technology*, 2010; **44**(19): 7315–7320.

129. Cong, Y., et al., "Toxic Effects and Bioaccumulation of Nano-, Micron- and Ionic-Ag in the polychaete, *Nereis diversicolor*. *Aquatic Toxicology*, 2011; **105**(3-4): 403–411.

130. Galloway, T., et al., "Sublethal Toxicity of Nano-Titanium Dioxide and Carbon Nanotubes in a Sediment Dwelling Marine Polychaete." *Environmental Pollution*, 2010; **158**(5): 1748–1755.

131. Gomes, T., et al., "Genotoxicity of Copper Oxide and Silver Nanoparticles in the Mussel *Mytilus galloprovincialis*." *Marine Environmental Research*, 2013; **84**: 51–59.

132. Dai, L., et al., "Effects, Uptake, and Depuration Kinetics of Silver Oxide and Copper Oxide Nanoparticles in a Marine Deposit Feeder, *Macoma balthica*." *ACS Sustainable Chemistry & Engineering*, 2013; **1**(7): 760–767.

133. Bao, H., et al., "New Toxicity Mechanism of Silver Nanoparticles: Promoting Apoptosis and Inhibiting Proliferation." *PLoS One*, 2015; **10**(3): e0122535.

134. Starnes, D.L., et al., "Impact of Sulfidation on the Bioavailability and Toxicity of Silver Nanoparticles to *Caenorhabditis elegans*." *Environmental Pollution*, 2015; **196**: 239–246.

135. Blickley, T., et al., "Effects of Dietary CdSe/ZnS Quantum Dot Exposure in Estuarine Fish: Bioavailability, Oxidative Stress Responses, Reproduction, and Maternal Transfer." *Aquatic Toxicology*, 2014; **148**: 27–49.

136. Congdon, J.D., et al., "Resource Allocation-Based Life Histories: A Conceptual Basis for Studies of Ecological Toxicology." *Environmental Toxicology and Chemistry*, 2001; **20**(8): 1698–1703.

137. Rowe, C.L., et al., "Elevated Maintenance Costs in an Anuran (Rana Catesbeiana) Exposed to a Mixture of Trace Elements during the Embryonic and Early Larval Periods." *Physiological Zoology*, 1998; **71**(1): 27–35.

138. Din, Z. and J. Brooks, "Use of Adenylate Energy Charge as a Physiological Indicator in Toxicity Experiments." *Bulletin of Environmental Contamination and Toxicology*, 1986; **36**(1): 1–8.

139. Muller, E.B., et al., "Impact of Engineered Zinc Oxide Nanoparticles on the Energy Budgets of Mytilus Galloprovincialis." *Journal of Sea Research*, 2014; **94**: 29–36.

140. Miller, R.J., et al., "Impacts of Metal Oxide Nanoparticles on Marine Phytoplankton." *Environmental Science & Technology*, 2010; **44**(19): 7329–7334.

141. Nelson, S.M., et al., "Toxic and Teratogenic Silica Nanowires in Developing Vertebrate Embryos." *Nanomedicine: Nanotechnology, Biology, and Medicine*, 2010; **6**(1): 93–102.

142. Shoults-Wilson, W.A., et al., "Evidence for Avoidance of Ag Nanoparticles by Earthworms (*Eisenia fetida*)." *Ecotoxicology*, 2011; **20**(2): 385–396.

143. Carew, A.C., et al., "Chronic Sublethal Exposure to Silver Nanoparticles Disrupts Thyroid Hormone Signaling during *Xenopus laevis* Metamorphosis. *Aquatic Toxicology*, 2015; **159**: 99–108.

144. Di Gioacchino, M., et al., "Immunotoxicity of Nanoparticles." *International Journal of Immunopathology and Pharmacology*, 2011; **24**(1 Suppl): 65s–71s.

145. Ciacci, C., et al., "Immunomodulation by Different Types of N-Oxides in the Hemocytes of the Marine Bivalve *Mytilus galloprovincialis*." *PLoS One*, 2012; **7**(5): e36937.

146. Canesi, L., et al., "Bivalve *molluscs* as a Unique Target Group for Nanoparticle Toxicity." *Marine Environmental Research*, 2012; **76**: 16–21.

147. Falugi, C., et al., "Toxicity of Metal Oxide Nanoparticles in Immune Cells of the Sea Urchin." *Marine Environmental Research*, 2012; **76**: 114–121.

148. Barmo, C., et al., "In Vivo Effects of n-TiO$_2$ on Digestive Gland and Immune Function of the Marine Bivalve *Mytilus galloprovincialis*." *Aquatic toxicology*, 2013; **132–133**: 9–18.

149. Kaveh, R., et al., "Changes in *Arabidopsis thaliana* Gene Expression in Response to Silver Nanoparticles and Silver Ions." *Environmental Science & Technology*, 2013; **47**(18): 10637–10644.

150. Chu, H., et al., "A Nanosized Ag-Silica Hybrid Complex Prepared by γ-Irradiation Activates the Defense Response in Arabidopsis." *Radiation Physics and Chemistry*, 2012; **81**(2): 180–184.

151. Buffet, P.-E., et al., "Behavioural and Biochemical Responses of Two Marine Invertebrates *Scrobicularia plana* and *Hediste diversicolor* to Copper Oxide Nanoparticles." *Chemosphere*, 2011; **84**(1): 166–174.

152. Buffet, P.-E., et al., "Fate of Isotopically Labeled Zinc Oxide Nanoparticles in Sediment and Effects on Two Endobenthic Species, the Clam *Scrobicularia plana* and the Ragworm *Hediste diversicolor*." *Ecotoxicology and Environmental Safety*, 2012; **84**: 191–198.

153. Buffet, P.-E., et al., "A Mesocosm Study of Fate and Effects of CuO Nanoparticles on Endobenthic Species (*Scrobicularia plana, Hediste diversicolor*)." *Environmental Science & Technology*, 2013; **47**(3): 1620–1628.

154. Lu, J., et al., "*Legionella pneumophila* Transcriptional Response following Exposure to CuO Nanoparticles." *Applied and Environmental Microbiology*, 2013; **79**(8): 2713–2720.

155. Pelletier, D.A., et al., "Effects of Engineered Cerium Oxide Nanoparticles on Bacterial Growth and Viability." *Applied and Environmental Microbiology*, 2010; **76**(24): 7981–7989.

156. McQuillan, J.S. and A.M. Shaw, "Differential Gene Regulation in the Ag Nanoparticle and Ag+-Induced Silver Stress Response in *Escherichia coli*: A Full Transcriptomic Profile." *Nanotoxicology*, 2014; **8**(S1): 177–184.

157. Lok, C.-N., et al., "Proteomic Analysis of the Mode of Antibacterial Action of Silver Nanoparticles." *Journal of Proteome Research*, 2006; **5**(4): 916–924.

158. Mirzajani, F., et al., "Proteomics Study of Silver Nanoparticles Toxicity on Oryza sativa L." *Ecotoxicology and Environmental Safety*, 2014; **108**: 335–339.

159. Neal, A.L., et al., "Can the Soil Bacterium *Cupriavidus necator* Sense ZnO Nanomaterials and Aqueous Zn2+ Differentially?" *Nanotoxicology*, 2012; **6**(4): 371–380.

160. Nazari, P., et al., "The Antimicrobial Effects and Metabolomic Footprinting of Carboxyl-Capped Bismuth Nanoparticles against *Helicobacter pylori*." *Applied Biochemistry and Biotechnology*, 2014; **172**(2): 570–579.

161. Sohm, B., et al., "Insight into the Primary Mode of Action of TiO$_2$ Nanoparticles on *Escherichia coli* in the Dark." *Proteomics*, 2015; **15**(1): 98–113.

162. Fajardo, C., et al., "Transcriptional and Proteomic Stress Responses of a Soil Bacterium *Bacillus cereus* to Nanosized Zero-Valent Iron (nZVI) Particles." *Chemosphere*, 2013; **93**(6): 1077–1083.

163. Simon, D.F., et al., "Transcriptome Sequencing (RNA-seq) Analysis of the Effects of Metal Nanoparticle Exposure on the Transcriptome of *Chlamydomonas reinhardtii*." *Applied and Environmental Microbiology*, 2013; **79**(16): 4774–4785.

164. Taylor, N.S., et al., "Molecular Toxicity of Cerium Oxide Nanoparticles to the Freshwater Alga *Chlamydomonas reinhardtii* Is Associated with Supra-Environmental Exposure Concentrations." *Nanotoxicology*, 2015; **10**(1): 32–41.

165. Kaveh, R., et al., "Changes in *Arabidopsis thaliana* Gene Expression in Response to Silver Nanoparticles and Silver Ions." *Environmental Science & Technology*, 2013; **47**(18): 10637–10644.

166. Taylor, A.F., et al., "Investigating the Toxicity, Uptake, Nanoparticle Formation and Genetic Response of Plants to Gold." *PLoS One*, 2014; **9**(4): e93793.

167. Landa, P., et al., "Nanoparticle-Specific Changes in *Arabidopsis thaliana* Gene Expression after Exposure to ZnO, TiO$_2$, and Fullerene Soot." *Journal of Hazardous Materials*, 2012; **241–242**: 55–62.

168. Campos, A., et al., "Proteomic Research in Bivalves: Towards the Identification of Molecular Markers of Aquatic Pollution." *Journal of Proteomics*, 2012; **75**(14): 4346–4359.

169. Renault, S., et al., "Impacts of Gold Nanoparticles Exposure on Two Freshwater Species: A Phytoplanctonic Alga (*Scenedesmus subspicatus*) and a Benthic Bivalve (*Corbucula fluminea*)." *Gold Bulletin*, 2008; 41: 116–126.

170. Balbi, T., et al., "In Vitro Effects of Combined Exposure to n-TiO$_2$ and Cd 2+ in Mussel Cells." *Comparative Biochemistry and Physiology, Part A*, 2012; **163**: S41.

171. Hu, W., et al., "Toxicity of Copper Oxide Nanoparticles in the Blue Mussel, *Mytilus edulis*: A Redox Proteomic Investigation." *Chemosphere*, 2014; **108**: 289–299.

172. Gomes, T., et al., "Differential Protein Expression in Mussels *Mytilus galloprovincialis* Exposed to Nano and Ionic Ag." *Aquatic Toxicology*, 2013; **136–137**: 79–90.

173. Tsyusko, O.V., et al., "Toxicogenomic Responses of the Model Organism *Caenorhabditis elegans* to Gold Nanoparticles." *Environmental Science Technology*, 2012; **46**(7): 4115–4124.

174. Polak, N., et al., "Metalloproteins and Phytochelatin Synthase May Confer Protection against Zinc Oxide Nanoparticle Induced Toxicity in *Caenorhabditis elegans." Comparative Biochemistry and Physiology Part C: Toxicology & Pharmacology*, 2014; **160**: 75–85.

175. Roh, J.-Y., et al., "Ecotoxicological Investigation of CeO$_2$ and TiO$_2$ Nanoparticles on the Soil Nematode *Caenorhabditis elegans* Using Gene Expression, Growth, Fertility, and Survival as Endpoints." *Environmental Toxicology and Pharmacology*, 2010; **29**(2): 167–172.

176. Tsyusko, O.V., et al., "Short-Term Molecular-Level Effects of Silver Nanoparticle Exposure on the Earthworm, *Eisenia fetida." Environmental Pollution*, 2012; **171**: 249–255.

177. Ratnasekhar, C., et al., "Metabolomics Reveals the Perturbations in the Metabolome of *Caenorhabditis elegans* Exposed to Titanium Dioxide Nanoparticles." *Nanotoxicology*, 2015; **9**(8): 994–1004.

178. van Aerle, R., et al., "Molecular Mechanisms of Toxicity of Silver Nanoparticles in Zebrafish Embryos." *Environmental Science & Technology*, 2013; **47**(14): 8005–8014.

179. Griffitt, R.J., et al., "Chronic Nanoparticulate Silver Exposure Results in Tissue Accumulation and Transcriptomic Changes in Zebrafish." *Aquatic Toxicology*, 2013; **130**: 192–200.

180. Gagné, F., et al., "Immunocompetence and Alterations in Hepatic Gene Expression in Rainbow Trout Exposed to CdS/CdTe Quantum Dots." *Journal of Environmental Monitoring*, 2010; **12**(8): 1556–1565.

181. EPA, *Summary Report on Issues in Ecological Risk Assessment*. 1991, United States Environmental Protection Agency: Washington, DC.

182. Ramírez-Pérez, T., S.S.S. Sarma, and S. Nandini, "Effects of Mercury on the Life Table Demography of the Rotifer *Brachionus calyciflorus* Pallas (Rotifera)." *Ecotoxicology*, 2004; **13**(6): 535–544.

183. Chandler, G.T., et al., "Population Consequences of Fipronil and Degradates to Copepods at Field Concentrations: An Integration of Life Cycle Testing with Leslie Matrix Population Modeling." *Environmental Science & Technology*, 2004; **38**(23): 6407–6414.

184. Rowe, C.L., W.A. Hopkins, and V.R. Coffman, "Failed Recruitment of Southern Toads (*Bufo terrestris*) in a Trace Element-Contaminated Breeding Habitat: Direct and Indirect Effects That May Lead to a Local Population Sink." *Archives of Environmental Contamination and Toxicology*, 2001; **40**(3): 399–405.

185. Kettlewell, H.D., "Phenomenon of Industrial Melanism in Lepidoptera." *Annual Review of Entomology*, 1961; **6**: 245–262.

186. Tsyusko, O.V., et al., "Genetics of Cattails in Radioactively Contaminated Areas around Chornobyl." *Molecular Ecology*, 2006; **15**(9): 2611–2625.

187. Di Giulio, R.T. and B.W. Clark, "The Elizabeth River Story: A Case Study in Evolutionary Toxicology." *Journal of Toxicology and Environmental Health, Part B- Critical Reviews*, 2015. In Press.

188. Baccarelli, A. and V. Bollati, "Epigenetics and Environmental Chemicals." *Current Opinion in Pediatrics*, 2009; **21**(2): 243–251.

189. Anway, M.D., et al., "Epigenetic Transgenerational Actions of Endocrine Disruptors and Male Fertility." *Science*, 2005; **308**(5727): 1466–1469.

190. Kim, S.W., J.I. Kwak, and Y.-J. An, "Multigenerational Study of Gold Nanoparticles in *Caenorhabditis elegans*: Transgenerational Effect of Maternal Exposure." *Environmental Science & Technology*, 2013; **47**(10): 5393–5399.

191. Panacek, A., et al., "Acute and Chronic Toxicity Effects of Silver Nanoparticles (NPs) on *Drosophila melanogaster." Environmental Science & Technology*, 2011; **45**(11): 4974–4979.

192. Sloman, K.A., et al., "Social Interactions Affect Physiological Consequences of Sublethal Copper Exposure in Rainbow Trout, *Oncorhynchus mykiss." Environmental Toxicology and Chemistry*, 2002; **21**(6): 1255–1263.

193. Kulacki, K.J. and B.J. Cardinale, "Effects of Nano-Titanium Dioxide on Freshwater Algal Population Dynamics." *PLoS One*, 2012; **7**(10): e47130.

194. Rand, G.M., ed. *Fundamentals of Aquatic Toxicology: Effects, Environmental Fate, and Risk Assessment*. 1995, Philadelphia, PA: Taylor & Francis.

195. Qin, G., et al., "Effects of Predator Cues on Pesticide Toxicity: Toward an Understanding of the Mechanism of the Interaction." *Environmental Toxicology and Chemistry*, 2011; **30**(8): 1926–1934.

196. Bone, A.J., et al., "Biotic and Abiotic Interactions in Aquatic Microcosms Determine Fate and Toxicity of Ag Nanoparticles: Part 2–Toxicity and Ag Speciation." *Environmental Science & Technology*, 2012; **46**(13): 6925–6933.

197. Unrine, J.M., et al., "Biotic and Abiotic Interactions in Aquatic Microcosms Determine Fate and Toxicity of Ag Nanoparticles. Part 1. Aggregation and Dissolution." *Environmental Science & Technology*, 2012; **46**(13): 6915–6924.

198. Chen, C., et al., "Toxicogenomic Responses of the Model Legume *Medicago truncatula* to Aged Biosolids Containing a Mixture of Nanomaterials (TiO$_2$, Ag and ZnO) from a Pilot Wastewater Treatment Plant." *Environmental Science & Technology*, 2015; **49**(14): 8759–8769.

199. Judy, J.D., et al., "Nanomaterials in Biosolids Inhibit Nodulation, Shift Microbial Community Composition, and Result in Increased Metal Uptake Relative to Bulk/Dissolved Metals." *Environmental Science & Technology*, 2015; **49**(14): 8751–8758.

200. Johnston, E.L. and D.A. Roberts, "Contaminants Reduce the Richness and Evenness of Marine Communities: A Review and Meta-Analysis." *Environmental Pollution*, 2009; **157**(6): 1745–1752.

201. Fausch, K.D., J.R. Karr, and P.R. Yant, "Regional Application of an Index of Biotic Integrity Based on Stream Fish Communities." *Transactions of the American Fisheries Society*, 1984; **113**(1): 39–55.

202. Azarbad, H., et al., *Microbial Community Composition and Functions Are Resilient to Metal Pollution along Two Forest Soil Gradients*, ed. D.C. Nakatsu. 2015; **91**: 1–11.

203. Colman, B.P., et al., "Low Concentrations of Silver Nanoparticles in Biosolids Cause Adverse Ecosystem Responses under Realistic Field Scenario." *PLoS One*, 2013; **8**(2): e57189.

204. Odum, E.P., "Trends Expected in Stressed Ecosystems." *BioScience*, 1985; **35**(7): 419–422.

205. Vittori Antisari, L., et al., "Toxicity of Metal Oxide (CeO$_2$, Fe$_3$O$_4$, SnO$_2$) Engineered Nanoparticles on Soil Microbial Biomass and Their Distribution in Soil." *Soil Biology and Biochemistry*, 2013; **60**: 87–94.

206. McPartland, J., H.C. Dantzker, and C.J. Portier, "Building a Robust 21st Century Chemical Testing Program at the US Environmental Protection Agency: Recommendations for Strengthening Scientific Engagement." *Environmental Health Perspectives*, 2015: **123**(1): 1–5.

207. Lowry, G.V., et al., "Transformations of Nanomaterials in the Environment." *Environmental Science & Technology*, 2012; **46**(13): 6893–6899.

208. Hendren, C.O., et al., "A Functional Assay-Based Strategy for Nanomaterial Risk Forecasting." *Science of the Total Environment*, 2015; **536**: 1029–1037.

CHAPTER 6

Physical Chemical Properties: Transport, Aggregation, and Deposition

Jonathan Brant

Department of Civil & Architectural Engineering, University of Wyoming, Laramie, Wyoming, USA

Jérôme Labille

*CEREGE UMR 7330 CNRS–Aix-Marseille University, Marseille, France, and
iCEINT CNRS–Duke University, Aix-en-Provence, France*

Jean-Yves Bottero

*CEREGE UMR 7330 CNRS–Aix-Marseille University, Marseille, France, and
iCEINT CNRS–Duke University, Aix-en-Provence, France*

Mark R. Wiesner

Department of Civil & Environmental Engineering, Duke University, Durham, North Carolina, USA

Introduction

This chapter explores physical chemical factors that govern key processes controlling the transport and fate of nanomaterials in aquatic environments, including aggregation and deposition. There is a well-developed body of work addressing the behavior of particles in water. In particular, we consider applications of the principles of colloid chemistry to specific nanomaterials and the insight that theory provides for the case of very small colloidal particles. The quantities of nanomaterials produced, the formats they manner they are integrated into products (e.g., as pure materials, hybrids, or composites), and the characteristics of the value chain of production, use, and disposal that may serve as a basis for initial release of nanomaterials to the environment are addressed in Chapter 14. Once released, these materials may undergo a variety of transformations (Chapter 7) that will be closely linked to the processes of aggregation and deposition that control the initial distribution of nanomaterials in natural and engineered systems.

Aggregation and deposition are analogous processes involving particle-surface interactions occurring in sequential steps of transport and attachment. In aggregation, the surface may be that of another particle or a growing aggregate. In deposition, the surface is an immobile "collector," where particles accumulate. Deposition typically (but not always) involves interactions between surfaces with different surface chemistries. Aggregation may occur between particles of identical composition (referred to as auto- or homogeneous aggregation) or with other materials (heteroaggregation). Heteroaggregation therefore resembles deposition in the sense that both usually involve interactions between surfaces having different properties.

Physical chemical processes controlling nanoparticle attachment or "surface affinity" for each other, as well as for abiotic or biotic surfaces such as bacteria, algae, clays, and the tissues of higher organisms, will play a key role in determining whether nanoparticles associate with each other (autoaggregation) or with larger "background" particles (heteroaggregation) that accumulate in sediments, whether nanoparticles adhere to gills or skin (deposition), or whether they stick to the media in ground water aquifers or water treatment filters (deposition). Aggregation and deposition are therefore likely to impact transport, bioavailability and bio-uptake of nanoparticles. Combined with information on nanoparticle transport, measurement of the affinity of nanoparticles for different surfaces (surface affinity) may therefore provide a means predicting important dimensions of where nanoparticles will go in the environment and how fast they will get there as mediated by the processes of particle aggregation and deposition.

Physicochemical Interactions

In aqueous environments, the nature of particle surfaces is intimately linked with the solution conditions (pH, ionic strength, temperature, etc.). Characterization of nanoparticles is therefore relevant largely within the context of the characteristic of the solution in which the nanoparticle is suspended. Those properties of particular interest for environmental analyses include surface charge, the presence of surface functional groups, Hamaker constant, and interaction energy with water (wettability). However, the characterization of materials in the nanometer regime is particularly challenging due to the fact that materials in this size range fall into a gray area where they may, in many cases, be considered as either small particles, or large solutes.

Brownian Motion

Particles, molecules and ions in fluid environments, regardless of size, possess a Brownian energy (BR) equal to $1.5\,kT$ [2], where k is the Boltzmann constant (1.38×10^{-23} J) and T is the absolute temperature in degrees Kelvin. Brownian potential energy decays exponentially from this value ($1.5\,kT$) with distance, with a decay length equal to the respective particle's radius of gyration, R_g. (The radius of gyration is the root mean square of mass weighted distances of all subvolumes in a particle from its center of mass. It has special interest in particle science because it can be applied to irregularly shaped particles.) The energy associated with Brownian motion may be compared with potential energies of interaction between particles and nearby surfaces and gives us an indication of the whether particles may "jump" out of a local energy well or over an energy barrier. These latter interactions are often considered in the context a theory that combines descriptions of forces that attract particles to surfaces and forces that may repel particles known as DLVO theory.

DLVO Theory

A benchmark context for understanding particle interactions in aqueous environments is the classical model developed referred to as Derjaguin-Landau-Verwey-Overbeek (DLVO) theory [3]. DLVO theory expresses the total interaction energy between two surfaces as the sum of (usually) attractive Lifshitz-van der Waals (LW) and repulsive electrostatic (EL) interactions originating from particle surface charge. Other forces, defined as non-DLVO forces, have also been found to be significant for surfaces in aqueous environments [2, 4] and have thus been included in the form of an extended DLVO (XDLVO) approach. Here, the total interaction energy between two surfaces in water may be written as:

$$U_{123}^{XDLVO} = U_{123}^{LW} + U_{123}^{EL} + U_{123}^{AB} + U_{123}^{BO}$$

(6.1)

Here, U^{XDLVO} is the total interaction energy between two surfaces immersed in water, U^{LW} is the Lifshitz-van der Waals interaction term; U^{EL} is the electrostatic interaction term; U^{AB} is the acid-base interaction term; and U^{BO} is the interaction energy due to Born repulsion. The subscripts 1, 2, and 3 correspond to surfaces 1 and 3 separated by an aqueous medium 2. Other interactions, such as steric interactions, are also likely and should be considered, though they are not included in the energy balance presented here. Steric interactions generally result from the adsorption of polymers, or other long chained molecules and can act to either stabilize or destabilize a particle suspension. This topic is addressed later in this chapter.

When plotted as a function of separation distance, the total interaction energy shows the evolution of the magnitude and type of interaction (repulsive or attractive) that occurs as two surfaces approach each other (Figure 6.1). Three cases arise, depending on the relative magnitudes of the attractive (negative energy potential) and repulsive (expressed as a positive quantity) phenomena involved. In the first case, the interaction between surfaces is attractive at all separation distances. In the second case, the interaction is repulsive as the surfaces approach one another until a repulsive energy barrier is overcome, at which point the interactions become attractive. In the third case (depicted in Figure 6.1), interactions are at first attractive, then repulsive, and finally attractive once again (as in the second case) as the approach distances became smaller.

The features of the potential energy curve in this last case are designated as an attractive secondary minimum (I), a repulsive barrier (II), and an attractive primary minimum (III). As separation distance

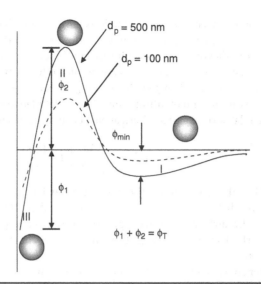

Figure 6.1   Interaction energy profile for two particle sizes illustrating the total interfacial energy (ϕ_T) that must be overcome in order for a particle to detach from a collector surface and become resuspended in the bulk suspension.

between two surfaces increases the different components of the interaction energies diminish from their corresponding values at near-contact following a unique decay pattern, which affects the shape of the resulting energy plot [3]. For instance, the secondary minimum develops because repulsive electrostatic interactions decay exponentially with separation distance while attractive van der Waals interactions decay somewhat more slowly over a longer range. The magnitudes of different interaction energies depend on particle size and thus, so does the overall interaction energy curve. For example, as particle size decreases the height of the energy barrier, if present, also decreases; similarly, the secondary minimum becomes more shallow with decreasing particle size. This is particularly evident at separation distances greater than 5 nm. Below this separation distance, shorter-range interactions tend to be less impacted by particle size. Nevertheless, size effects on interaction energies have consequences with regards to the stability of particle dispersions and particle mobility [5].

The components of interaction energy that are accounted for in the DLVO and other similar theories are discussed in more detail in the following sections. These discussions, however, are not meant to be exhaustive reviews of these subjects; the reader is instead referred to other references (e.g., [3, 6, 7]) for such detailed overviews. The purpose here is to consider the effect of diminishing particle size on those interaction energies as they may be relevant in determining nanoparticle fate and transport in the environment.

van der Waals Interactions

van der Waals interactions are in most cases attractive and act universally between surfaces in aqueous media (Table 6.1). These interactions incorporate three different electrodynamic forces: dispersion, induction, and orientation [3].

Type	Nature	Expressions
Debye interaction	(Induction) Permanent dipole versus induced dipole	$\Phi_D = -\dfrac{\alpha_1\mu_2^2 + \alpha_2\mu_1^2}{(4\pi\varepsilon_0)^2}x^{-6}$
Keesome interaction	(Orientation) Permanent dipole versus permanent dipole	$\Phi_K = -\dfrac{2}{3}\dfrac{\mu_1^2\mu_2^2}{k_B T(4\pi\varepsilon_0)^2}x^{-6}$
London interaction	(Dispersion) Induced dipole versus induced dipole	$\Phi_L = -\dfrac{3h}{2}\dfrac{v_1 v_2}{v_1+v_2}\dfrac{\alpha_{0,1}\alpha_{0,2}}{(4\pi\varepsilon_0)^2}x^{-6}$

Table 6.1   Expressions for Different Types of van der Waals Interactions between Molecules

These three potential energies of interactions have a dependence on separation distance to the power of six and may be summed together to describe the total van der Waals interfacial energy term. When integrated over the bodies of two approaching infinite blocks (an idealization of two flat plates) the resulting Lifshitz-van der Waals interaction energies are predicted to decay according to L^{-2}, where L represents the distance between the two infinitely long flat plates [2]. Using the so-called Derjaguin approximation of the spherical surface as an integral of infinite ring-shaped flat plates of infinitesimal width, the LW interaction energy between a flat surface and a spherical particle may be calculated as follows:

$$U_{123}^{LW} = -\left(\frac{Aa_p}{6h}\right)\left(1+\frac{14h}{\lambda}\right)^{-1}$$

(6.2)

Here, A is the Hamaker constant for the interacting surfaces across the medium; λ is the characteristic wavelength of the dielectric, usually taken to be equal to 100 nm; a_p is the radius of the spherical particle; and h is the surface to surface separation distance. As particle size decreases the van der Waals attraction similarly decreases in magnitude and acts of shorter separation distances, consistent with Equation (6.2) [8].

The Hamaker constant illustrates the importance of simultaneously considering the intrinsic properties of the nanoparticle and the properties of the system in which the nanoparticle is immersed. For example, to describe a nanoparticle suspended in water that aggregates with another nanoparticle or with a clay particle, the pertinent Hamaker constants in these interactions reflect the intrinsic property of the material making up nanoparticle as well as the properties of the system, which in this simple case consists of clay particles and water. Interaction between two nanoparticles of the same material (2) in a vacuum are described by the Hamaker constant A_{22}. When dispersed in a medium of Hamaker constant A_{11}, the Hamaker constant for van der Waals interactions between these two particles can be written as [9]:

$$A_{212} = A_{11} + A_{22} - 2A_{12}$$

(6.3)

With the approximation that $A_{12} \approx \sqrt{A_{11}A_{22}}$, Equation (6.3) can be rewritten as:

$$A_{212} = (\sqrt{A_{11}} - \sqrt{A_{22}})^2$$

(6.4)

If the two interacting bodies are of different materials as in the case of a nanoparticle (2) heteroaggregating with a clay particle (3) in water (1), the Hamaker constant A_{312} can be written as:

$$A_{312} = A_{21} + A_{23} - A_{12} - A_{13}$$

(6.5)

Applying the approximation $A_{ij} = \sqrt{A_{ii}A_{jj}}$ leads to the expression:

$$A_{312} = (\sqrt{A_{11}} - \sqrt{A_{22}})(\sqrt{A_{11}} - \sqrt{A_{33}})$$

(6.6)

Electrostatic Interactions

When a particle is immersed in an electrolyte, the surface charge σ, leads to an enrichment of ions near the particle surface of opposite charge (counterions) referred to as a double layer. This counterion-enriched "cloud" near a charged surface balances the surface charge. As a particle approaches another surface (another particle or an immobile surface) the ionic environments of the two approaching surfaces interact, generally producing a repulsion arising from the increasing osmotic pressure of the overlapping double layers. This electrostatic repulsion tends to decrease with decreasing particle size (Figure 6.2). Hogg et al. [10] derived an expression for the electrostatic interaction energy between two surfaces under the assumption of constant surface potential. When modeling electrostatic interactions between surfaces a theoretical constraint of constant surface potential or constant surface charge is generally made [6]. In reality, the actual condition lies between these two experimental extremes. In practical terms, surface potentials are often approximated by quantities calculated from measurements of electrokinetic phenomena such as electrophoretic mobility or streaming potential. In fact, these measurements probe the electrical potential at a small distance out from the surface, referred to as the shear plane. The potential at the shear plane calculated from electrokinetic measurements is referred to as the zeta potential, ζ. The zeta potential

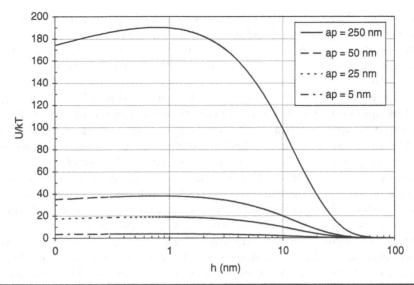

FIGURE 6.2   Electrostatic repulsion between a spherical particle and a flat surface plotted for several different particle sizes as a function of separation distance ($\zeta_1 = -30$ mV; $\zeta_2 = -25$ mV; pH = 7; $T = 20°C$; $I = 1$ mM NaCl).

can be used to estimate the potential energy of interacting double layers, U^{EL}. For example, the electrostatic interaction energy per unit area between a spherical particle and a flat surface decays with separation distance according to:

$$U_{123}^{EL}(h) = \pi \varepsilon_r \varepsilon_0 a_p \left(2\zeta_1 \zeta_3 \ln \left(\frac{1+e^{-\kappa h}}{1-e^{-\kappa h}} \right) + \left(\zeta_1^2 + \zeta_3^2 \right) \ln(1 - e^{-2\kappa h}) \right) \qquad (6.7)$$

Here, $\varepsilon_0 \varepsilon_r$ is the dielectric permittivity of the suspending fluid; κ is the Debye-Huckel parameter; and ζ_1 and ζ_2 are the zeta potentials of the interacting surfaces. Hogg et al. also derived an expression for the EDL interaction potential between two dissimilar spheres (in both size and surface potential) by approximating the interacting spherical surfaces as an integral of infinite ring-shaped flat plates of infinitesimal width. This so-called Derjaguin approximation [11] (which we also encountered in considering van der Waals interactions) is reasonable when the interaction range is small compared to the particle size ($\kappa h \gg 1$). However, for particles in the nanometric range, κh will typically be small and solutions employing the Derjaguin approximation will not be accurate. Exact analytical solutions that avoid the Derjaguin approximation as appropriate for nanoparticles have been recently derived [12]. Despite their theoretical inaccuracy for nanoscale particles, the relatively simple analytical solutions such as Equation (6.3) that employ the Derjaguin approximation yield some intuitive notion of the role of particle size and surface potential in particle-surface interactions. Another simplifying assumption used to obtain analytical solutions concerns the nature of the ions in a solution. In particular, one often makes the assumption that an electrolyte (salt) solution is a "z:z" electrolyte where cations and anions have equal valencies such as NaCl or $CaSO_4$, where cations and anions have charges, respectively, of +1 and −1 in the first case and +2 and −2 in the second case.

The Debye constant, κ, is determined according to the following equation for z:z electrolytes:

$$\kappa = \sqrt{\frac{e^2 \sum n_i z_i^2}{\varepsilon_r \varepsilon_0 kT}} \qquad (6.8)$$

Here, e is the electron charge; n_i is the number concentration of ion i in the bulk solution; and z_i is the valence of ion i.

According to Equation (6.7), the electrostatic interaction energies decay exponentially with distance, and are a function of both the separation distance (h) and the Debye length ($1/\kappa$) [6, 13]. The variability of $1/\kappa$ is dependent on the ionic strength of the solution and the valency of the ionic species present [14]. For instance, in a 100-mM NaCl solution $1/\kappa$ is equal to roughly 1.0 nm, while for a 0.01-mM NaCl solution $1/\kappa$ is approximately 100 nm. Thus, the distance over which electrostatic interactions exert their influence changes based on the ionic strength and types of ions in the solution. Therefore, it becomes important when considering EL interactions in describing nanoparticle behavior in water to consider the salt concentration and valence of the ions present. Furthermore, the thickness of the diffuse double layer has specific consequences unique to nanoparticles as they may be present at similar scales (1 to 100 nm). This is important as many models are based on the assumption that κh is much less than the diameter of the particle.

Nanoparticles have large surface area to volume ratios and potentially high sorption capacities for other aqueous species, such as ionic materials and natural organic matter [15], that would tend to favor complexation processes that are often at the origin of surface charge. For example, the high electron-affinity of fullerenes has been shown to facilitate covalent, charge transfer, and donor-acceptor interactions with other compounds [16]. Additionally, because a significant fraction of atoms are exposed at the nanoparticle surface, rather than contained in the bulk interior, nanoparticle surface chemistry can be significantly altered by surface complexation processes [15, 17]. Such processes can have a dramatic effect on nanoparticle surface charge characteristics and in turn the electrostatic interactions with other surfaces. Adsorption of ionic species can impart a charge to an otherwise uncharged particle. For example, adsorption of water and subsequent deprotonation to form hydroxyl groups has previously been observed for hydrophobic oil droplets in water, which was concluded to be the source of the measured electrophoretic mobility for these particles [18]. Similarly, the adsorption of hydroxyl groups and charge transfer interactions with solvents have been proposed as likely charging mechanisms for fullerene C_{60} nanoclusters [19, 20]. These processes are then particularly significant for fullerene nanoparticles, which might otherwise have little affinity for the aqueous phase and thus decreased stability and mobility. Furthermore, the adsorption of ionic species may lead to charge reversal for some nanoparticles [21].

Born Repulsion

Born repulsion results from the strong repulsive forces between atoms as their electron shells begin to overlap. This is a short-range interaction that acts over a distance of up to several nanometers. Prieve and Ruckenstein [22] derived an expression for the Born repulsion interaction energy between a spherical particle and a flat surface assuming pairwise additivity of the atomic Lennard-Jones potential:

$$U_{123}^{BO} = \frac{A\sigma_B^6}{7560}\left[\frac{8a_p + h}{(2a_p + h)^7} + \frac{6a_p - h}{h^7}\right] \qquad (6.9)$$

Here, σ_B is the Born collision diameter. A value for σ_B of 0.5 nm is often assumed [5]. The separation distance at which this force becomes important is predicted to decrease with particle size. Because repulsive Born interactions act over such a short distance they may not affect nanoparticle interactions on approach in aqueous media [13]. However, the Born repulsion does significantly affect the depth of the primary minimum and may possibly affect the reversibility of nanoparticle attachment relative to larger particles.

Acid-Base Interactions and the Hydrophobic Effect

Acid-base (AB) interactions characterize the hydrogen bonding properties of a surface or interacting surfaces and thus describe how that surface interacts with water. Water molecules interact with one another and structure themselves through hydrogen bonding [3, 23]. This structuring tendency can result in either attractive hydrophobic or repulsive hydrophilic interactions between particles in water. Functionality on a particle's surface results in the coordination of water molecules on the surface, which subsequently provides a preferential orientation of water through hydrogen bonding. The adsorption of water at functional sites on a surface may be associated with the net release of a proton or hydroxide to the solution. Thus, the surface may act as an acid or a base. Layer(s) of relatively ordered water on surfaces may impeded particle attachment or aggregation due to the additional energy required to "squeeze" water out as separation distances become small. In contrast, hydrophobic surfaces do not interact with water and are essentially pushed toward one another as water molecules preferentially bond with one another.

This latter phenomenon can be interpreted as an attractive interaction referred to as the hydrophobic effect [2, 23]. The interactions between particles and the bulk water are a function of the size of the particles or hydrophobic surfaces [23]. Hydrophobic nanoparticles will perturb water structuring to a much smaller degree than will larger particles with a similar chemistry [8]. In the case of small nanoparticles or other hydrophobic molecules, water may form a cage structure around the nanoparticle or molecular core resulting a relatively stable configuration referred to as a clathrate. Methane clathrates in deep ocean waters are thought to represent an enormous reservoir of hydrocarbons on our planet. Clathrate formation has been suggested as a mechanism behind the hydration of otherwise hydrophobic

C_{60} nanoparticles (which have a diameter of a little over 1 nm) and may explain how these fullerenes acquire a charge and interact with water to form stable colloidal suspensions of C_{60} aggregates.

As particle size becomes larger than 2 nm, an energetically unfavorable condition results. To compensate for a larger "disruption" between hydrogen bonds, such as hydrophobic particle, the surrounding water molecules must still restructure themselves to maintain their hydrogen-bonding network with other water molecules. If the particle is too large, water must form appropriate sized cavities with a cage-like structure. This water depletion or drying between two interacting surfaces results a net attraction between hydrophobic surfaces. It is important to bear in mind that the surfaces themselves do not actually attract each other; in fact the water simply "likes" itself too much to allow the surface to remain exposed. The balance between particle-solvent and solvent-solvent interactions and hydrogen bonding energies ultimately determines whether cavitation or strict microscopic dewetting will occur. Therefore, for all but the smallest size fraction of nanoparticles (1 nm < d < 2 nm) hydrophobic interactions will be significant and will tend to favor particle aggregation.

In contrast with hydrophobic surfaces, hydrophilic surfaces possess surface groups that may coordinate water molecules through hydration [2, 3]. Subsequent layers of water molecules may hydrogen bond with the hydrated water, resulting in several layers of relatively ordered water extending from the surface. The kinetic energy of the water molecules will tend to break hydrogen bonds. Thus, lower temperatures should favor extension of ordered water further into the bulk. Although there is some controversy regarding the extent to which water may extend into the bulk, it appears that at least two to three layers of ordered water are likely present at most hydrated surfaces. As two hydrated surfaces approach one another, dehydration must occur before the underlying surfaces are in direct contact. The additional free energy required for dehydration represents a repulsive barrier between the two approaching surfaces. Hydration forces act over a shorter range (decay length (λ) = 0.2–1.1 nm) than attractive hydrophobic interactions and decay exponentially with separation distance.

The acid-base interaction energy (U^{AB}) between a sphere and a flat surface as a function is predicted to decay exponentially with separation distance according to the following expression:

$$U_{mlc}^{AB}(h) = 2\pi a_p \lambda \Delta G_{y_0}^{AB} \exp\left[\frac{y_0 - h}{\lambda}\right] \tag{6.10}$$

Here, $\Delta G_{y_0}^{AB}$ is the acid-base free energy of interaction at contact; λ is the characteristic decay length of AB interactions in water, whose value is between 0.2 and 1.0 nm, a value of 0.6 nm is commonly used [2, 3]. Acid-base interactions decrease greatly with decreasing particle size but can be nonetheless comparable to electrostatic interactions even when particle size is below several tens of nanometers.

Definition of the Stability Ratio Theoretical consideration of the attractive and repulsive forces that particles experience as they approach a surface can be used to estimate the "stickiness" of a particle, or its affinity for a surface. The stickiness coefficient, α is the reciprocal of the stability ratio, W, can be calculated from theory as an integral measure of the interaction potentials between two particles (or a particle approaching any other surface). For the case of two particles:

$$W = 2r \int_{2r}^{\infty} \frac{\exp[V(R)/kT]}{R^2 G(R)} dR \tag{6.11}$$

Here, R is the center-to-center separation distance between two particles; $V(R)$ is the interaction potential between two particles at distance R; and $G(R)$ is a dimensionless hydrodynamic resistance function. The coefficient $G(R)$ accounts for the additional resistance caused by the squeezing of the fluid molecules between two approaching particles, thus as particle size decreases the resistance imposed by the fluid molecules also decreases (imagine a basketball versus a golf ball traveling through water). The hydrodynamic resistance function is close to unity for nanoparticles and may therefore generally be neglected in these cases. Unfortunately, in many cases observed stability ratios do not typically compare quantitatively with those calculated from theory, which is attributed to an incomplete assessment of the interfacial energy conditions between interacting surfaces in water and geometrical considerations. Calculations of the stability ratio have typically relied on extended DLVO theory to describe the interaction potential. Although calculated values of the stability ratio provide insight into conditions that favor aggregation and those that do not, there is not quantitative agreement between calculated and experimentally observed values of the stability ratio.

Implications of the Continuum Approximation for Nanoparticles

The foregoing discussion of the configuration of water on surfaces underscores a key limitation in our ability to describe many interactions at the nanometeric scale. DLVO and other theories describing particle behavior in aqueous media typically treat the intervening fluid (water) as a uniform, structureless medium that is well described in terms of its bulk properties [3] such as density, viscosity, and dielectric constant. As illustrated in the case of ordered water near surfaces, a molecular view of interactions between particles, surfaces and fluid molecules may be required to adequately describe phenomena that affect nanoparticle aggregation and deposition. A primary challenge in this regard lies in bridging phenomena that apply at the atomic or molecular scale with phenomena with those observed in the larger scale system. Given the size and complex composition of any real system, it is not possible to simply calculate and sum all of the interactions that occur at the molecular scale. The problem remains computationally intractable. Approaches for bridging this gap in length scales include averaging across many molecular interactions at a given scale, or using bulk properties as boundary conditions for performing detailed calculations at a given location at the molecular scale.

Limitations on theory that assume that particles and ions exist in a fluid described as a continuum are particularly apparent when separation distances between two surfaces approaches 5 nm or less, when considering particles with dimensions similar to that of ions, molecular interactions, such as steric repulsion, become significant. Similar limitations exist in describing particle surfaces where it may not be possible to average over many functional groups on a surface as is typically done in surface complexation modeling.

Aggregation

While we tend to think of nanoparticles as unusually small objects, in fact they will often be present in aqueous systems as larger clusters of the primary nanoparticles or heteroaggregated material, even in the absence of any potential destabilizing agents (e.g., salts, polymers, organic materials). Aggregation may occurs when attractive forces such as van der Waal's attraction are greater than the repulsive forces a particle experiences as it approaches another particle (or stationary surface in the case of deposition).

The propensity of nanoparticles to aggregate, particularly in natural systems, is an important consideration in determining not only their mobility, fate and persistence, but also their toxicity. Particles in the size range of an individual nanoparticles will have negligible settling rates. However aggregation may result in a growth in mean particle size to the extent that settling rates increase. Transport and attachment processes determine the morphology of particle aggregates and deposits, which in turn affects the nature of subsequent transport. The persistence of the aggregated nanomaterial in suspension may decrease as aggregates settle or flow toward surfaces where they may deposit, effectively reducing the ambient concentrations to which organisms will be exposed. Reductions in active surface area that occur during aggregation and an increased proximity of particle surface area within aggregates may also fundamentally reduce reactivity compared with nanoparticles in an unaggregated state. Some aggregated nanoparticles experience a decreased ability to produce reactive oxygen species [24–26] or dissolution rate [27] with implications regarding their toxicity [28, 29]. Aggregation may also alter the bioavailability of the nanomaterial if larger materials are less capable of entering cells. Conversely, aggregation may be at the heart of the toxic response as in the case of carbon nanotube agglomeration within lungs, which has been observed to lead to suffocation in laboratory animals where these materials were intentionally introduced [30, 31]. It is possible that heteroaggregation of nanoparticles with organic molecules or other materials may essential imbed the nanomaterial in another functionality or may alter the availability of the nanomaterial due to changes in size or chemistry.

Aggregation will be equally important in environmental engineering applications of nanomaterials. The use of nanoparticles for ground water remediation is one example where nanoparticle aggregation may affect the mobility of the nanomaterial and therefore the ability to deliver nanoparticles to a desired location in the subsurface.

Definitions

Particle dispersions are thermodynamically unstable if the total free energy of the systems may be lowered through a reduction in interfacial area via aggregation. Aggregation involves the formation and growth of clusters and is controlled by both the reaction conditions and interfacial chemical interactions [13, 15, 32].

Clusters of particles may be referred to as either aggregates or agglomerates, terms that some scientists use to differentiate the relative strength of the forces binding particles together in a cluster. Since the difference between particles that are strongly bound together in an aggregate and those that are more loosely bound in an agglomerate is somewhat arbitrary and operationally defined, here we will refer to clusters, agglomerates, and aggregates synonymously, generally giving preference to the term "aggregate" employed by colloid chemists.

The aggregation state of nanoparticles in suspension is indicated by the letter "n" which indicates the number of nanoparticles present in an aggregate of a given size. Thus, a suspension of "nC_{60}" or "$nTiO_2$" is a suspension of aggregates of C_{60} nanoparticles or TiO_2 nanoparticles, where in each case aggregates contain "n" nanoparticles. An aggregation state of n = 1 refers to a suspension of distinct nanoparticles while larger values of "n" indicate clusters containing more nanoparticles. In practice, a given suspension will contain nanoparticles of variable aggregation states corresponding to a distribution of aggregate sizes. In some cases, as illustrated by C_{60}, the size of the primary particle is rigorously defined by the chemical composition of the nanomaterial. In other cases, such as TiO_2, the size of the primary particle will be determined approximately by a quantity and structure of material making up the particle, as for example occurs in a crystal lattice of titanium and oxygen of a given mass. Continuing with the example of TiO_2, a single nanoparticle having dimensions roughly in the range of 1 to 100 nm would be referred to as a TiO_2 nanoparticle, or in some cases simply as nano-TiO_2. (Note that some authors mix the use of "n" and "nano" in designating nanomaterials so that it may not be clear if they are referring to a suspension of individual nanoparticles or partially aggregated nanoparticles.) In addition to primary particle size and aggregation state, more information may be needed to completely specify a nanoparticle and its aggregates, as in the case of TiO_2 that may exist in several crystal forms including anatase and rutile, or may be present as a hybrid particle such as alumina oxide-coated TiO_2 used in some sun screens.

Some Observations of Aggregation

Light scattering and TEM analyses of several common nanoparticles reveal that the spontaneous formation of nanoparticle aggregates occurs with many nanomaterials dispersed in water (Figure 6.3). This may be attributed in part to a decreasing electrostatic repulsion between charged particles with decreasing size for those particles that develop a charge in water. This topic was covered earlier in this chapter showing the influence of particle size on the repulsive interfacial energy for particles in water. The size, structure, and chemical properties of these clusters are dependent on the characteristics of the constituent nanoparticles [33] and the process by which the particles are put into suspension [33, 34]. Unmodified titania nanoparticles form stable clusters with a narrow size distribution. Cluster size measured by light scattering is confirmed by TEM imagery, and the primary TiO_2 particles within the cluster are evident. A striking characteristic of the nC_{60}, shown in Figure 6.3, is the presence of well-defined facets and hexagonal shape in 2-D projection. In this case, the nC_{60} is formed by solvent exchange, where the C_{60} is initially dissolved in an organic solvent and then mixed with water. The images of clusters produced in this case suggest that aggregate formation may resemble a process of crystal growth rather than undirected particle aggregation. Similar observations have been made for nC_{60} produced through other solvent-exchange techniques for dispersing these materials in water [34]. When nC_{60} is formed by extended mixing in water without the intermediary use of organic solvents, the resulting aggregates are much less organized.

It is also interesting to note that modification of the nanoparticle surface chemistry to enhance the solubility of the nanoparticle may not in fact result in a molecular dispersion [35]. In other words, cluster formation may occur even when the particle surface has been modified (e.g., functionalization or surfactant wrapping) to enhance stability. For instance, hydroxylation of the C_{60} molecule to make fullerol does indeed increase the rate at which it goes into suspension. This is because hydroxylation of the C_{60} molecule to form fullerol produces a generally hydrophilic and charged particle. However, contrary to some initial assumptions concerning the state of fullerol in water, these molecules do not exist as discrete entities (Figure 6.3). Instead, as shown in the TEM of a fullerol cluster, they tend to form spherical clusters composed of many $C_{60}OH_{20-24}$ [36]. For fullerol, cluster formation is likely due to the distribution of OH groups across the C_{60} surface resulting in heterogeneous interfacial interactions. The distribution of —OH groups on the C_{60} results in the formation of distributed hydrophobic/hydrophilic regions on the C_{60} molecule [35, 37]. Agglomeration of the $C_{60}OH_{20-24}$ may occur through attractive patch-patch interactions between the hydrophobic regions. Graphite similarly shows a strong tendency to aggregate and forms highly fractal aggregates in water [38].

(a)

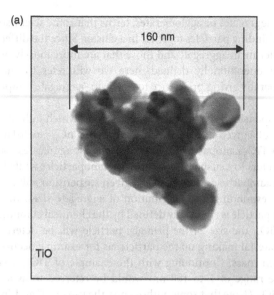

(b)

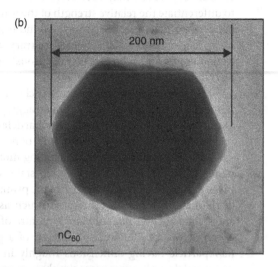

(c)

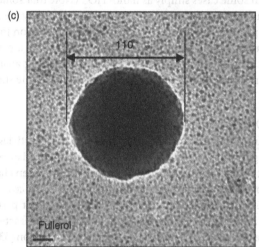

FIGURE 6.3 TEM images of titanium dioxide, nC_{60}, and fullerol nanoclusters formed after introduction into water.

Aggregation Kinetics and Particle Stability

The stability of particle dispersions (i.e., their tendency not to aggregate with themselves) may be determined experimentally by comparing the aggregation rates of "sticky" and less sticky particles. Here, the stickiness is accounted for in terms of an attachment efficiency [39] or affinity coefficient, α. The attachment efficiency largely reflects chemical factors controlled by surface functionality and solution chemistry. In the case where particles have a high affinity for one another (perfectly sticky particles), all collisions between particles result in an aggregation event and the attachment (or affinity) coefficient, α, has a value of unity. The ratio of the aggregation rates between a perfectly sticky (fast) and less sticky (slow) case under conditions that are otherwise identical (concentration, mixing, etc.) is characterized in terms of a stability ratio, W, which may be written as:

$$W = \frac{k_{11fast}}{k_{11slow}}$$

(6.12)

Here, k_{11fast} and k_{11slow} are the rate constants for the early stages of aggregation and the attachment coefficient, α is simply the reciprocal of the stability ratio. These fast ($\alpha = 1$) and slow ($\alpha < 1$) regimes of aggregation correspond respectively to cases where either transport is limiting (diffusion-limited aggregation, DLA) or where attachment is limiting (reaction-limited aggregation, RLA). For particles that are stabilized primarily by electrostatic repulsion (as described subsequently), there is a critical ionic strength (which varies as a function of electrolyte valence) referred to as the critical coagulation concentration (CCC), above which aggregation is assumed to be transport-limited and the value of W is taken as unity. At ionic strengths below the CCC, repulsive electrostatic interactions become significant and W is greater than 1.

Autoaggregation rates can be measured using static or dynamic light scattering (depending on particle size) of mean particle diameter. In the case of particles stabilized by electrostatic repulsion, the stability ratio is determined by taking the ratio of rates of change of aggregate average diameter observed at ionic strengths below and well above the CCC. Evaluation of the affinity coefficient for heteroaggregation between a suspension of nanoparticles and another population of particles such as bacteria, is somewhat more complex. As an alternative to extracting initial aggregation rates from measurements of mean aggregate diameter over time, measurements of the particle size distribution over time can be compared with that calculated from a particle population model. In this case, W is treated as a fitting coefficient. Such an approach is well suited to polydisperse suspensions and has been used to track nanoparticle aggregation [17].

Kinetic expressions for aggregation first proposed by Von Smoluchowski [40] can be extended to describe the evolution of aggregate sizes in a population of suspended particles over time. The kinetics of aggregation governing the evolution of the particle size distribution over time is written as a system of differential equations [41]. The assumption that aggregation occurs in sequential steps of transport and attachment is described by multiplying rate constants that describe particle transport producing collisions between particles or aggregates of different sizes, by affinity coefficients that describe attachment probability following collision. In the case of an irreversible aggregation, the evolution of the number concentration, n_k, of a particle in size class k can be written as follows:

$$\frac{dn_k}{dt} = \frac{1}{2} \sum_{i+j \to k} \alpha_{ij} \beta_{ij} n_i n_j - n_k \sum_{i=1}^{\infty} \alpha_{ik} \beta_{ik} n_i \qquad (6.13)$$

Here, β_{ij} is the collision frequency (transport) between two aggregates containing i and j particles and the $\alpha_{i,j}$ are the attachment efficiencies between particles or aggregates in different size classes. The first summation term in Equation (6.8) describes the generation of aggregates of aggregation state "k" (or size k aggregates) from the aggregation of smaller size aggregates, while the second term corresponds to the loss of size k aggregates due to their attachment to any other size aggregate. The population balance described by Equation (6.13) can be extended to include additional processes such as particle sedimentation, particle breakup, particle dissolution, particle formation, or biouptake.

Collision Mechanisms

The collision frequency kernel β_{ij} is typically composed of three principal collision mechanisms: Brownian diffusion (β_{br}), shear-induced collisions (β_{sh}), and differential settling (β_{ds}). The relative importance of collision mechanisms changes as a function of particle size. This can be seen by considering the case of a spherical particle much larger than the solvent molecules. In this case, the Brownian diffusion coefficient for the particle can be calculated from the Stokes-Einstein equation:

$$D = \frac{k_B T}{6 \pi \eta r} \qquad (6.14)$$

Here, k_B is the Boltzmann constant, T the temperature, η the fluid viscosity and r the particle hydrodynamic size. Equation (6.14) shows that the Brownian motion as a transport process becomes more significant with decreasing particle size. In contrast, forces such as gravity or those originating from fluid flow (e.g., drag forces), increase with increasing particle size. For example, relative settling velocities scale with diameter square, and so transport by differential settling will likely be more important for nanoparticle aggregates and for interactions between nanoparticles and larger background particles rather than for interactions between individual nanoscale particles. Similarly, shear-induced collisions are also predicted to increase with particle diameter, making these interactions of increasing importance for heteroaggregation.

Calculation of the Collision Rate Kernel

Two ideal cases have been presented in the literature for calculating particle collisions by each of these mechanisms. In one case, the rectilinear model, particles are assumed to have no effect on the fluid they are suspended in, as though all fluid passes through the aggregate or particle as it diffuses, settles, or is carried by the fluid. In contrast, the curvilinear model [42, 43] considers the effect of creeping flow around particles and compressed streamlines as two particles approach one another. The curvilinear model predicts many fewer collisions between particles than does the rectilinear model since fluid must be displaced before particle contact occurs. The curvilinear model with particle transport defined primarily by Brownian diffusion would appear to be appropriate for describing fullerene transport early during the aggregation process; particularly if the resulting aggregates are compact and highly ordered. However, if aggregation leads to the formation of porous aggregates, neither the rectilinear nor the curvilinear model is satisfactory. An intermediate model [44, 45] is required that considers the entire range of cases, where the effect of fluid flow around each aggregate as well as the potential for fluid to pass through the aggregate is taken into account. Closed-form approximations have been proposed [44] for calculating the collision frequency kernel as a function of the ratio of drag force on a permeable aggregate to that on an impermeable sphere, Ω, and the fraction of fluid that passes through an isolated aggregate, η:

$$\beta_{Br}(i,j)=\frac{2kT}{3\mu}\left(\frac{1}{\Omega_i r_i}+\frac{1}{\Omega_j r_j}\right)(r_i+r_j) \qquad (6.15)$$

$$\beta_{sh}(i,j)=\frac{4}{3}\left(\sqrt{\eta_i}\,r_i+\sqrt{\eta_i}\,r_i\right)^3 G \qquad (6.16)$$

$$\beta_{ds}(i,j)=\pi\left(\sqrt{\eta_i}\,r_i+\sqrt{\eta_i}\,r_i\right)^2\left|u_i^*-u_j^*\right| \qquad (6.17)$$

Here, r is the radius of the aggregate or particle, μ is the dispersing medium viscosity (0.000890 kg/m s for water), G is the mean shear rate (s^{-1}), and u^* is the settling velocity of the particle or aggregate. For the case where all streamlines pass through the aggregate (the fraction of fluid that passes through an isolated aggregate, $\eta = 1$) and the drag on the particles is equal to that of an impermeable sphere ($\Omega = 1$), Equations (6.15) to (6.17) reduce to the rectilinear model. Assuming that the transport step of aggregation can be adequately expressed by calculated values for the collision kernel, β_{ij}, then the stability ratio can be determined by adjusting its reciprocal, α, so that calculations of particles size distributions over time correspond with the observed particle size distributions.

For a suspension of nanoparticles all having the same size, assuming that collisions between particles j leads to irreversible attachment with an efficiency of α, the decrease in number concentration of unit particles of size j in suspension can be estimated as:

$$\frac{dn_j}{dt}=-\alpha\beta_{jj}\left(\frac{n_j}{2}\right)^2 \qquad (6.18)$$

Here, β_{jj} is the specific particle collision frequency, and n_j is the number concentration of particles j. If all particle contacts result in aggregation, then $\alpha = (1/W) = 1$ and $\frac{dn_j}{dt}=-\beta_{ii}\left(\frac{n_j}{2}\right)^2 = k_{jj,\text{fast}}$. Compared to micron-sized particles, nanoparticles will have considerably smaller inertia and approach velocities prior to impact with another particle [17]. This results in extended times in the interaction fields of the other particle meaning that nanoparticles are more likely to be deflected away as a result of repulsive forces. This would also suggest that nanoparticles do not "collide" in the literal sense, but rather slow down to a velocity of zero as they approach one another. If we describe aggregation as occurring between two groups of identically-sized particles, each with an initial concentration equal to one-half of the total number concentration n_j the solution to Equation (6.18) for the time t_a required to reduce the initial particle concentration to a fraction $(1 - n_j/n_{j0})$ of its initial value through aggregation is:

$$t_a=\frac{4}{\alpha\beta_{jj}n_{j0}}\left(\frac{n_{j0}}{n_j}-1\right) \qquad (6.19)$$

From Equation (6.19) it becomes evident that t_a varies inversely with the collision frequency β_{ij} and with the initial particle concentration n_{j0}.

Heteroaggregation

Another simplifying case to Equation (6.13) is the case of heteroaggregation between particles in two size classes representing, on one hand, a concentration of nanoparticles, n, and on the other, a concentration, B, of background particles such as clays, or bacteria that are assumed to remain constant. During the early stages of heteroaggregation the affinity coefficient can be expressed as a function of the distribution constant $\gamma(t)$, which describes the concentration of unheteroaggregated nanoparticles (nanoparticles not yet aggregated with background particles) to the concentration of heteroaggregated nanoparticles per mass of background particles:

$$\ln(\gamma(t)B + 1) = \alpha\beta(n, B)Bt \tag{6.20}$$

If an initial concentration n_0 of nanoparticles is allowed to heteroaggregate with background particles of concentration B for a prescribed series of mixing times, and then allowed to settle leaving a concentration "$n(t)$" in suspension, a plot of $\ln(\gamma(t)B + 1)$, or equivalently $\ln(n_0/n(t))$, versus the aggregation (mixing) time is predicted to yield a linear relationship if all of the background particles are removed by settling. The slope of this plot, $\alpha\beta B$ is a measure of the relative affinity of the nanoparticles for the background particles [46]. The slope of a best fit line to these data plotted as described by Equation (6.20) yields an estimate of $\alpha\beta B$ which can be normalized to measurements obtained when $\alpha = 1$ to obtain values of the affinity coefficient between nanoparticles and background particles.

More complex cases of heteroaggregation involve tracking many size classes of particles and following the association of nanoparticles with each of these size classes. The more complicated case of tracking the multitude of interactions between aggregates composed of two or more types of particles, such as occurs in heteroaggregation between nano- and background particles, requires solving the Smoluchowski equations while accounting for the nature and proportions of each of the particle types constituting each class of aggregates. The fractal nature of the resulting aggregates is tracked, and collision rate kernel is calculated using the intermediate model described by Equations (6.15) to (6.17). Three different types of interactions may result from the collision between two aggregate units, depending on their respective composition and orientation. Nanoparticle-nanoparticle, background-background, or nanoparticle-background interaction may be induced, each related to a respective sticking efficiency α_{NN}, α_{BB}, and α_{NB}. The following expression for the overall sticking efficiency α_{global} in a two components system was proposed by Therezien et al.:

$$\alpha_{global}(f_i, f_j) = f_i f_j \alpha_{NN} + (1-f_i)(1-f_j)\alpha_{BB} + ((1-f_i)f_i + (1-f_j)f_j)\alpha_{NB} \tag{6.21}$$

Here, f_i and f_j are the ratios of nanoparticles to background in the heteroaggregates i and j, respectively.

Simulations such as these for a range of conditions that might be anticipated in aquatic environments from freshwater wetlands to wastewater treatment plants predict that in most cases nanoparticles will rapidly associate through heteroaggregation with background particles, such that the equivalent "half life" of a free nanoparticle in these systems is likely to be on the order of minutes to days [47, 48]. This is due to the fact that concentrations of background particles are likely to be orders of magnitude higher than that in nanoparticles in these aqueous systems. Heteroaggregation can be observed experimentally as in the case where micron-sized background particles (silica or clay) were mixed with TiO_2 nanoparticles, both components being initially monodisperse and very distinct from each other in size. Rather than measuring the affinity coefficient, α_{NB}, directly, this parameter was treated as a fitting coefficient. The evolution in the particle size distribution was monitored using laser diffraction and physicochemical conditions (pH, ionic strength, and ion composition) were varied. A model encompassing Equations (6.13) and (6.15) to (6.17) [49, 50] was solved [51, 52] treating α_{NB} as a fitting coefficient. It was shown that the initial heteroaggregation kinetics correlates to the surface area ratio of nanoparticles attached to the surface of the background particles, confirming the multiple collision scenarios expected from Equation (6.21).

Heteroaggregation as Related to Nanoparticle Reactivity

Heteroaggregation is often the initial step that leads to a nanoparticle reacting with a surface. Examples of reactions include silver nanoparticle dissolution, leading to plant toxicity [53], photosensitization of fullerene leading to singlet oxygen generation and viral inactivation [54], reduction of CeO_2 nanoparticles on contact with human dermal fibroblasts in vitro leading to DNA damage [55], or titanium dioxide photocatalysis that leads to hydroxyl radical formation and subsequent reactions on surfaces [26]. Surface affinity is related to nanoparticle reactivity when nanoparticles promote a reaction on the surfaces they attach to following transport and attachment (Figure 6.4). Transport is described by the collision rate kernel, β, and attachment is described by the relative affinity, α_{NB} of nanoparticles of concentration, n, for background particles (such as bacteria, viruses, algae, or fibroblasts) of concentration, B. Under this assumption that NPs must first attach to background particles before the pertinent reaction can occur, the apparent rate constant, $k_{apparent}$ for the overall reaction (e.g., rate of bacterial inactivation

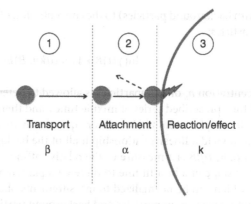

FIGURE 6.4   The role of heteroaggregation in nanoparticle reactivity.

or rate of cerium reduction) is related to the rates of heteroaggregation and the intrinsic reaction rate constant, k_{in} as:

$$k_{apparent} = \frac{k_{in}\alpha_{NB}\beta Bn}{k_{in} + \alpha_{NB}\beta Bn} \qquad (6.22)$$

Thus, by Equation (6.22) when heteroaggregation is not limiting ($k_{in} \ll \alpha_{NB}\beta Bn$), $k_{apparent} \cong k_{in}$. Conversely, when heteroaggregation is limiting, due, for example, to a small affinity between nanoparticles and background particles, then $k_{in} \geq \alpha_{NB}\beta Bn$ and the overall reaction rate constant, $k_{apparent}$, observed experimentally is predicted to be proportional to α_{NB}. Evaluation of surface affinity in conjunction with data interpreted using Equation (6.22) allows nanoparticle reactivity to assessed and potentially controlled by altering surface affinity to achieve a desired overall reaction rate.

We can conclude that the transport and fate (apart from speciation) of nanomaterials in these environments will be largely determined by heteroaggregation rates and the transport of "background" particles. The rate of heteroaggregation will in turn depend largely on the affinity of nanoparticles for these background surfaces. Reactivity and the possible effects of nanoparticles on the organisms they come in contact with are also predicted to depend on the surface affinity. The affinity coefficient will depend on the chemical nature of interactions between nanoaparticles and the surfaces they encounter as well as the solution chemistry of the systems in which they are immersed. The following section surveys some of the factors affecting nanoparticle surface chemistry and attachment (surface affinity).

Surface Chemistry of Nanoparticles as It Affects Attachment

Nanoparticles are often engineered with surface properties that reduce their affinity for auto- or heteroaggregation, or conversely, to enhance their affinity for specific kinds of surfaces. A variety of techniques have been developed for dispersing particles into suspension and reducing their affinity for one another, including: solvent exchange for buckminsterfullerene [56–58], wrapping or grafting molecules, such as surfactants, that impart hydrophilicity to the nanoparticles [59] or direct functionalization of the nanoparticle to produce hydrophilic groups. C_{60} cluster formation can be controlled according to the functionality given to the nanoparticle surface [35, 36, 60, 61]. In addition to the functionalization intentionally engineered into a given nanoparticle, nanoparticles may also acquire charge through interactions with solutes. Indeed, in aqueous environments, the nature of particle surfaces is intimately linked with the solution conditions such as pH, ionic strength, and ionic composition. In addition to ionic species, the surface chemistry of nanoparticles will be influenced by the sorption of macromolecules such as proteins in blood or humic materials in water or soil. Characterization of nanoparticles is therefore relevant largely within the context of the characteristics of the system into which the nanoparticle is introduced. In aqueous systems, for example, a large number of extrinsic (system-dependent) properties of nanoparticles such as surface charge, surface potential, hydrophobicity, and macromolecule conformation are determined by interactions between the particle surface and the solution chemistry. The list of intrinsic properties of nanoparticles is much shorter and includes composition and structure. Thus, a consideration of nanoparticle affinity for surfaces involves a consideration of the interactions between intrinsic and extrinsic nanoparticle properties as well as the properties of the surfaces (e.g., sand, lung tissue, algae) in the system of interest.

Origin of Particle Surface Charge

Surface charge leads to the electrostatic interactions between particles and the surfaces they encounter that may greatly affect their tendency to aggregate or deposit. Particles in water are typically charged due to the mechanisms of surface dissociation, specific adsorption of nonconstituting ions, preferential adsorption of constituting ions, and/or crystal lattice defects [6]. Surface dissociation requires the presence of functional groups such as the hydroxyl group on a metal oxide surface. These functional groups may protonate or deprotonated with changes in pH yielding a net surface charge that may be positive, neutral, or negative [62]. Functional groups on nanoparticles may form over time due to particle aging or interactions with the solvent. For example, as previously noted extended stirring of the fullerene C_{60} in water leads to the formation of a stable colloidal suspension of C_{60} clusters (nC_{60}) as shown in Figure 6.3 [21, 63]. The stability of these clusters is interesting in itself since the initial C_{60} is virtually insoluble in water [64–66]. The fact that cluster formation occurs is suggestive of changes in the surface chemistry of the C_{60} with time and exposure to water. This hypothesis is supported by adsorption isotherms for C_{60}, which illustrate the affinity of these materials for water (Figure 6.5). Here, the sample mass is measured as a function of varying the water vapor pressure ($P_{(H_2O)}/P_0$) from 0 to 1 (adsorption isotherm) and back to 0 (desorption isotherm). The isotherm can be decomposed in two steps: (1) at $P/P_0 < 0.7$, the low slope of the adsorption isotherm indicates a weak affinity for water, and (2) at $P/P_0 = 0.7$, the step in the slope indicates a new surface chemistry resulting in an exponential increase of the isotherm and the multilayer adsorption of water molecules at higher vapor pressures. The C_{60} is thus hydrophilic for $P/P_0 > 0.7$. Moreover, this modification in the C_{60} surface appears to be irreversible since a significant hysteresis persists between the adsorption/desorption steps over the entire vapor pressure range. After desorption, one monolayer of water remains adsorbed on the C_{60} surface, perhaps due to clathrate formation [20, 56] but more likely due to surface hydroxylation around $P/P_0 = 7$ during the adsorption step [67]. The resulting nC_{60} aggregates have a negative surface charge as measured by zeta potential.

Surface functional groups may also specifically adsorb ions from solution such as when Ca^{2+} adsorbs to a γ-Al_2O_3 surface, imparting a positive charge. Other common examples of specific adsorption include the adsorption of citrate onto gold or silver nanoparticles or of ligands and macromolecules such as proteins onto metals, metal oxides [69]. Preferential adsorption of constituting ions is illustrated in the classic case of silver iodide (AgI) particles which have a surface potential determined by the relative abundance of Ag^+ and I^- ions as described by the Nernst equation. Finally, defects arising from the substitution of ionic species in a crystal lattice by other ionic species of unequal charge may also result in net particle surface charge. The distinction between these mechanisms may be blurred. For example, adsorbed ligands containing dissociable functional groups may themselves undergo acid-base reactions as the pH of the

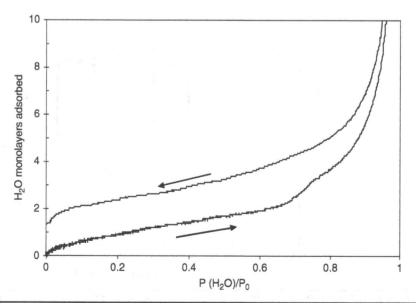

FIGURE 6.5   Adsorption/desorption isotherm of water vapor onto pristine C_{60} powder obtained from gravimetric analysis.

solution changes while ligands may interact with a variety of anions and cations that will affect the overall surface charge of the particle. Thus macromolecules such as proteins and humic materials that sorb to nanoparticles are often observed to affect particle charge.

pH Effects

The effects of surface charge on heteroaggregation can be illustrated for the case of TiO_2 nanoparticles and larger sand particles. The TiO_2 nanoparticles have relatively slow settling rates and tend to remain in suspension in the absence of aggregation. But after heteroaggregating with the larger sand particles, the nanoparticles may settle from suspension. The tendency to heteroaggregate in this case is largely determined by the relative charge of the TiO_2 and sand. When a mixture of these particles is shaken for one hour, then left to settle [68], the fraction of nanoparticles remaining in suspension is observed to decrease as a function of pH. The sand is negatively charged (deprotonated) over the whole pH range studied, whereas TiO_2 tends to reverse charge at a pH value of about 5.5, which is referred to as the pH point of zero charge (pH_{PZC}). When the surface charges have the same sign, that is, when pH > 5:5 (II), there is no change in the concentration of nanoparticles in suspension after introducing the sand, hence no attractive interaction. On the other hand, when the surface charges have opposite signs, that is, when pH < 5:5 (I), the concentration of nanoparticles in suspension falls sharply, indicating a high level of nanoparticle adsorption onto the sand. The influence of surface modification on nanoparticle aggregation over variable pH is further illustrated in the case of maghemite (Fe_2O_3) nanoparticles. At a pH of 7, maghemite nanoparticles aggregate to form large clusters [1]. When the pH is reduced to a value of 3, protonation of the maghemite surface promotes nanoparticle dispersion.

Ionic Strength Effects

In many cases, nanoparticle aggregation follows the qualitatively expectations of the DLVO-type models, that is, particles tend to aggregate more quickly at higher ionic strengths and/or at pH values near the point of zero charge [70]. Both of these changes in solution chemistry reduce repulsive electrostatic interactions. Adherence of nanoparticle stability to classical colloidal models is illustrated by the case of nC_{60} aggregation in two different electrolytes, NaCl and $CaCl_2$, of variable concentration (Figure 6.6).

The average diffusion coefficient (D^{nC60}) of nC_{60} clusters and higher order aggregates is inversely correlated to the average size of the aggregates via the Stokes-Einstein relation [Equation (6.14)].

When the salt concentration increases above a critical value, the critical coagulation concentration (CCC), D decreases abruptly, corresponding to an increase in average size due to aggregation. The respective CCCs measured with $CaCl_2$ and NaCl for the nC_{60} in this case are 2 ± 0.4 mM and 100 ± 0.4 mM. These values conform to the dependence of the CCC with the sixth power of the counterion valence, as predicted by the Schulze-Hardy rule for ideal systems [71].

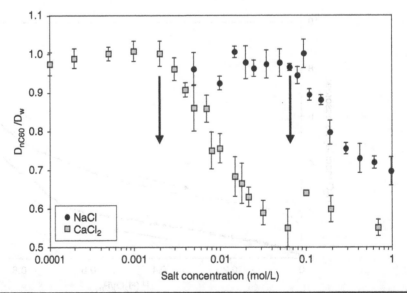

FIGURE 6.6 Average diffusion coefficient of C_{60} clusters in aqueous dispersion as a function of salt concentration and valence; $CaCl_2$ (squares) and NaCl (solid circles). Data are normalized to the reference diffusion coefficient in absence of salt D_w. The arrows point the respective critical coagulation concentrations, above which the dispersion is destabilized.

Electrophoretic mobility measurements (taken as an indicator of surface charge) for a variety of different nanoparticles as a function of pH, reveal a classic curve of increasingly negative electrophoretic mobility as solution pH becomes more basic, while mobility approaches zero with increasing ionic strength [3, 13, 72, 73]. DLVO calculations suggest that as ionic strength increases, there is a reduction in the energy barrier between nanoparticles due to compression of the electric double layer, which allows for the attractive van der Waals interactions to dominate, leading to the formation of larger aggregates.

Differences in aggregation behavior may occur amongst different nanoparticle size fractions. In other words, nanomaterials cannot be treated as a monotonic class of particles, where a 10-nm particle is treated the same as that of a 100-nm particle. In fact, a range of behaviors may be observed for particles that are characterized by sizes on two different ends of the nano-spectrum (1–100 nm).

Effects of Adsorbing Macromolecules

Environmental systems present a number of naturally occurring macromolecules that may interact with nanoparticle surfaces including bacterial exudates, polysaccharides, and numerous compounds resulting from the degradation of organic matter of animal or plant origin. Such natural organic matter (NOM) represents a broad class of organic macromolecules of highly variable composition, ubiquitous to nearly all soil and water systems. While the focus here will be on NOM, proteins and lipids play an analogous role in physiological systems in modifying nanoparticle surfaces. The formation of a protein "corona" on particles [74] has been recognized as a means of stabilizing colloidal suspensions for several decades, although protein adsorption may also act to reduce dispersal in environmental systems [75].

Particles in natural waters, regardless of size, may associate with organic macromolecules that subsequently alter the particle's interfacial and physical characteristics (e.g., charge, reactivity, size) [76]. For example, association of NOM with nanoparticles can alter their effective charge, reactivity, and hydrodynamic radius [77]. Macromolecules, in turn, are affected by virtually every other element in solution, such as ionic strength and pH, and these interactions may alter macomolecular conformation, and adsorption. The high specific surface area of nanoparticles also implies that these particles will be likely targets for macromolecule adsorption, producing particularly pronounced changes in surface properties.

The size of nanoparticles may be small, comparable to, or large, relative to the macromolecules present in environmental and physiological systems. Humic substances include materials of irregular morphology ranging in molecular weight from 100 to over 10^6 daltons, corresponding to molecular sizes (radii of gyration) of less than 1 nanometer to hundreds of nanometers. Higher ionic strength and/or low pH will tend to reduce the size of charged macromolecules due to reduced charge and charge screening. Polysaccharides will tend to be more linear, while proteins may be more spherical. Smaller nanoparticles may associate with polysaccharides (or larger NOM molecules) in a "string of pearls" configuration, where several nanoparticles attach along various portions of the polysaccharide or within a larger organic matrix Figure 6.7. In this scenario, the organic material acts as the adsorbing surface and thus, mobility may be

Figure 6.7   Illustration of nanoparticles entangled in a collection of organic macromolecules.

dominated by the properties of the macromolecule rather than the nanoparticle. Smaller proteins or humic molecules may in contrast accumulate on the surface of larger nanoparticles, in which case the overall size is dominated by the nanoparticle dimension but the surface chemistry of the nanoparticle is altered by the adsorbed material.

Where nanoparticles present a clear interface for adsorption, NOM or ions may form direct chemical bonds on the particle surface (inner sphere adsorption). Conversely, in outer sphere adsorption, no direct chemical bonds are formed and instead the adsorbate is held at the particle surface through electrostatic and/or hydrogen bonding forces. Adsorption of NOM to a nanoparticle surfaces is a function of the solution chemistry (pH and ionic strength) and the characteristics of the organic molecule, such as molecular size and charge [77]. NOM has been shown to adsorb to negatively charged particles in aqueous media (e.g., iron oxide colloids). Under environmentally relevant conditions, humic acids readily adsorb to other negatively charged surfaces and in turn modify the physicochemical characteristics (e.g., charge) of the solid-liquid interface. The amount of adsorbed material generally increases with increasing ionic strength [76, 77] and with increasing concentration of the organic macromolecule [76], but individual molecules may have a more collapsed compact configuration. Adsorption of NOM on colloidal iron particles appears to be favored by neutral pH values (pH in the range of 4–6) and ionic strengths of 1 to 10 mM NaCl. Similar conditions may favor association with other nanoscale particles.

NOM adsorption can also alter aggregate structure [77]. This is because aggregate structure is generally determined by both aggregation kinetics and short-range chemical interactions between the agglomerating particles. As NOM adsorption alters these interactions it will ultimately influence aggregate structure. For instance attractive van der Waals interactions between adsorbed NOM may result in the formation of a tighter aggregate structure.

Although the majority of surfaces in aqueous environments carry a net negative charge, the magnitude of this charge can vary considerably with differences in functionality and charge density. Adsorption of NOM to a surface will alter the charge properties of that surface, in most cases making it more negatively charged due to the functional groups (typically carboxylic or phenolic groups) [76]. Modification of the particle surface charge will depend on a variety of factors including the charge characteristics of the NOM (e.g., charge density and functionality), the mode of adsorption (inner versus outer sphere) [78], particle surface chemistry, and the solution chemistry.

Conversely, nanoparticle surface charge will affect the propensity to adsorb NOM as illustrated by the case of nC_{60} and clusters of hydroxylated C_{60}. Recalling that these clusters are very similar with the exception of their surface chemistries a clear difference exists in their interaction with the NOM. Clusters of the fullerol (hydroxylated C_{60}) readily adsorb the tannic acid as indicated by an increasingly negative surface potential with higher NOM concentration. On the other hand, very little change in surface charge is observed for the SON/nC_{60} suggesting that there is less adsorption of the NOM. Beyond merely altering particle surface charge, adsorption of NOM may also produce steric stabilization of particles. If the entropy of sorbed molecules would decrease as layers overlap, a repulsive interaction is produced. The combined effects of charge on the macromolecules and steric repulsion typically play a role in the electro-steric repulsion produced by NOM. However, the relative thickness of the sorbed layers to the diffuse layer thickness must also be considered. For example, under conditions of high salinity, macromolecules may sorb to a greater extent, individual molecules will have a more compact configuration, and diffuse layers will be small. If the thickness of the adsorbed organic layer exceeds the thickness of the diffuse double layer, surface interactions will be dominated by the macromolecules [76]. As ionic strength decreases and the thickness of the diffuse double layer increases, there is the potential for interfacial interactions to be dominated by EL interactions. Steric interactions may also increase when the ionic strength is low as the organic macromolecules may expand under such conditions as a result of increased electrostatic repulsion between charged functional groups on the adsorbed macromolecule.

NOM adsorption therefore affects nanoparticle stability in two principle ways. First, it modifies the effective interfacial charge and thus the magnitude and possibly sign of the electrostatic interactions between surfaces. Second, short-range steric barriers and hydration forces are produced that prevent interparticle approach and diminish the impact of attractive van der Waals interactions [78]. The degree to which NOM adsorption alters nanoparticle stability is dependent on the solution chemistry and the characteristics of the nanoparticle and the organic macromolecules. As an example, steric interactions are a strong function of the molecular weight (an indicator of size) of the NOM; larger macromolecules result in more substantial steric repulsions. For solutions composed of monovalent cations like Na^+ and K^+ NOM adsorption will generally result in more stable nanoparticle suspensions. For example, Mylon et al. [77] found that the CCC for NOM-coated hematite nanoparticles increased from 30 to 107 mM NaCl. This is due to an increase in the nanoparticle surface charge, as was previously discussed. In this case, adsorption of NOM to the nanoparticle surface enhances its stability in the presence of increasing monovalent cation concentrations.

Macromolecular coatings (engineered or acquired) may not always decrease particle interactions with surfaces [79]. Consider the case when macromolecules protrude from the particle surface beyond the diffuse layer, and encounter a surface without a similar macromolecule coating. In this case, macromolecules can serve to bridge across the region of EL repulsion, and allow particles to attach more easily to the uncoated surface than they would in the absence of the macromolecules [80].

Small concentrations of NOM may also favor aggregation, particularly at higher ionic strength. This is illustrated by the case of the dispersion stability of TiO_2 nanoparticles used in sunscreens for a variety of NOM types (polysaccharide, humic acid, tannic acid) [81]. Measuring the extent of aggregation by the turbidity of suspensions remaining after a given mixing and settling period, some types of NOM are observed to stabilize while other destabilize suspensions of the TiO_2 nanoparticles. Destabilization occurs for low NOM/nanoparticle ratios, while NOM at higher concentrations produces a stabilizing effect. The addition of NaCl tends to enhance aggregation as is consistent with DLVO theory.

Deposition

Particle Deposition in Porous Media

The deposition on stationary surfaces is a key factor in determining nanoparticle mobility in the environment, the capability of water treatment systems to remove these particles, and potentially bioavailability. The mobility of submicron and micron-sized particles in porous media has been considered for nearly a century [5, 82–86]. Analysis of the transport and deposition of particles in porous media typically begins with the convective-diffusion equation, which under steady-state conditions is generally written as [13]:

$$\frac{\partial n_j}{\partial t} + \nabla \cdot (\mathbf{u}n_j) = \nabla \cdot (\mathbf{D} \cdot \nabla n_j) - \nabla \left(\frac{\mathbf{D} \cdot \mathbf{F}}{kT} n_j \right) + \frac{\partial n_j}{\partial t}_{RXN} \tag{6.23}$$

Here, n_j is the number concentration of particles of size or type j in the suspension; $\mathbf{D}$ is the particle diffusion tensor; $\mathbf{u}$ is the particle velocity vector induced by the fluid flow; and $\mathbf{F}$ is the external force vector, and $\frac{\partial n_j}{\partial t}_{RXN}$ is the reaction of these particles with a surface (deposition) or with other particles (aggregation). The external force vector may be used to account for forces such as those arising from gravity and interfacial chemical interactions typically accounted for in the extended DLVO model (see Section 6.2).

As in the case of aggregation, particle deposition can be considered as a sequence of particle transport (in this case to an immobile collector) followed by attachment to the surface [87]. In a porous medium such as a filter or groundwater aquifer, fluid flow can be described by the Happel sphere in cell model [88] in which grains within the porous medium are assumed to be spherical collectors of radius a_c, surrounded by an imaginary outer fluid sphere of radius b with a free surface (Figure 6.8). The imaginary fluid envelope contains the same amount of fluid as the relative volume of fluid to the collector volume in the entire medium (i.e., $(a_c/b)(1 - \varepsilon)^{1/3}$), where ε is the porosity of the given medium. Particles can be transported to the collector surface through a combination of interception, gravitational settling, and diffusion transport mechanisms. Interception occurs when fluid streamlines pass sufficiently close to the collector such that contact results. Contact through gravitational settling and diffusion results from the particles crossing the fluid streamlines via the respective mechanisms within a critical region around the collector and contacting the surface.

Analytical solutions for particle transport due to Brownian diffusion have been combined with particle trajectory calculations to yield a closed-form solution for the transport of particles to the surface of spherical collectors expressed as the theoretical single collector contact efficiency (η_0) [89, 90]:

$$\eta_0 = \eta_D + \eta_I + \eta_G \tag{6.24}$$

Here, η_D is the single collector contact efficiency for transport by diffusion; η_I is the single collector contact efficiency for transport by interception; and η_G is the single collector contact efficiency for transport by gravity. Equation (6.24) is based on the additivity assumption, where η_0 is determined by summation of the three independently determined transport mechanisms.

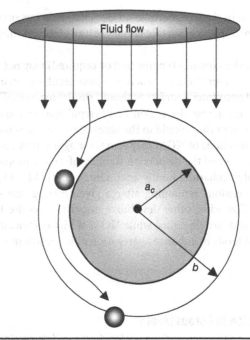

FIGURE 6.8 Illustration of the Happel sphere in cell model b in which the spherical collector having a radius a_c is surrounded by a fluid envelope having a radius b.

It is instructive to look at the components of η_0 plotted as a function of particle size (Figure 6.9). For particles in the micron size range transport to a collector surface is largely governed by interception and gravitational settling mechanisms. However, the transport of unaggregated nanoparticles to a collector will typically be dominated by diffusion [89]. It is evident that as particle size approaches that of the nanoscale ($d < 100$ nm) the interception and gravity terms begin to approach zero. On the other hand, once particle size is smaller than approximately 300 nm, the particle contact efficiency is largely controlled by η_D. The diffusion component η_D is a function of the porosity of the transporting medium, the aspect ratio between the collector and particle sizes, the particle approach velocity to the collector surface, and the Hamaker constant for the interacting surfaces. With the high-diffusion component for nanoparticles their mobility may be especially low in porous media characterized by low flow rates or Peclet numbers, such as groundwater aquifers.

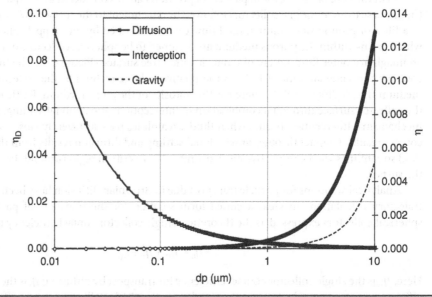

FIGURE 6.9 Respective contributions to the total single collector contact efficiency from each of three transport mechanisms: diffusion (η_D), interception (η_I), and gravitational settling (η_G). The magnitude of each mechanism is plotted as a function of particle size while holding other particle characteristics constant.

We obtain a similar picture of particle mobility when the efficiency of single collectors is integrated across a length of the porous medium that they comprise. In Figure 6.10, particle removal, represented as the mass concentration of particles removed, $1 - C/C_0$, is plotted as a function of particle size on a semilog scale. Increasing the fluid flow rate or flow velocity serves to increase particle mobility. Particles in the nanosize range (d_p = 1–100 nm) are relatively immobile, which is indicated by their high removal values. As particle size increases, mobility increases until a relative maximum in mobility is reached at around 1.5 μm. For particles larger than approximately 1.5 μm mobility begins to decrease as the collector efficiency increases due to a larger contribution by interception and gravity. Holding all other variables constant, nanoparticles should be relatively immobile under a given set of conditions in comparison to other micron-sized particles of similar surface chemistry.

Quantifying Nanoparticle Deposition and Mobility

It is somewhat nonintuitive that the very small size of nanoparticles may actually lead to less mobility (a higher deposition rate) in a porous medium such as a ground water aquifer or a filter in comparison with larger particles. Observations of particle deposition in porous media over a wide range of particle sizes suggest that the current description of particle transport to collectors is reasonably complete [87, 89, 92]. However, when particle attachment is not favorable only a fraction of the particles transported to the collector surface will attach; therefore, the single collector efficiency must be modified.

Indeed, transport is only part of the story. The surface chemistry of nanoparticles rather than their size is likely to be factor that determines mobility. Just as aggregation was described mathematically to be the product of factors transport (β) and attachment (α), probability that a particle approaching a single collector deposits on the collector can be written [93] as the product of the collector contact efficiency and the attachment efficiency:

$$\eta = \eta_0 \alpha \tag{6.25}$$

When particle removal is calculated for a lower value of the attachment or stickiness coefficient, particle removal decreases, corresponding to decreased affinity for the porous medium and an increased mobility (Figure 6.10).

The ratio of the rate of particle deposition on a collector to the rate of collisions with that collector is the attachment efficiency factor, α and is analogous to the stickiness coefficient or 1/W. Theoretical predictions of the attachment efficiency in this case are identical to those for particle aggregation, typically based on DLVO-type calculations (e.g., Figure 6.1) that consider the balance of forces arising from interactions at very small separation distances between particles and the collector surface. These phenomena may be important over length scales that are large by comparison with nanoparticle dimensions. Similar to

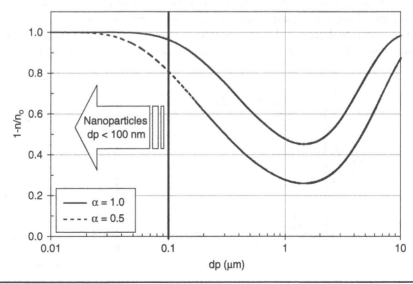

FIGURE 6.10 Particle removal assuming two different attachment efficiencies as a function of particle size calculated using equation 4.4 (v = 0.14 cm/s, ε = 0.43, a_c = 355 μm, L = 9.25 cm).

the case where the stickiness coefficient is treated as a fitting variable in the particle population models, the attachment efficiency can be treated as an empirical parameter that captures all aspects of particle deposition not described by the more extensively validated particle transport models. The empirically determined attachment efficiency should vary with changes in surface and solution chemistry.

Measurements of particle removal across a length (L) of a homogeneous porous medium composed of spherical grains of radius (r_c) and porosity ε can be combined with calculations of particle transport to yield estimates of the attachment efficiency factor α:

$$\alpha = \frac{4a_c}{3(1-\varepsilon)\eta_0 L}\ln\left(\frac{n_j}{n_{jo}}\right) \tag{6.26}$$

Here, n_j and n_{jo} are respectively the particle number concentrations present at distance L in the column effluent and influent to the column; η_0 is the clean bed single collector contact efficiency, which describes the particle transport to an individual collector and can be calculated as a function of the Darcy velocity, porous medium grain size, porosity, and temperature among other variables using Equation (6.26). Using experimentally measured n_j/n_{jo} values (fraction of influent particles remaining after passing through the porous medium) and theoretical η_0 values [9], values of α can be calculated for a given particle suspension.

In such column experiments, nanoparticles are introduced to the column as a step input and the nanoparticle concentration n_j is measured at the column outlet as time over time. The resulting breakthrough curve (e.g., Figure 6.11 which shows mass concentration of nanoparticles in the outlet normalized by the inlet concentration over time) rises to a plateau value. It is this plateau value that is used to obtain n/n_0 (or C/C_0) for the purposes of calculating the affinity coefficient as given in Equation (6.26). When nanoparticle feed at the influent is stopped, the nanoparticle concentration at the outlet drops. Column experiments can be conducted under conditions or varying pH, natural organic matter, and other variations in solution chemistry that may affect surface affinity (attachment) as described previously.

Particle deposition may also take place in the attractive secondary minimum if it is present [93]. The ability of particles to deposit in a secondary minimum is dependent on their size and on the kinetic energy imparted onto them by the fluid flow. As particle size decreases the secondary minimum becomes shallower, while the energy barrier decreases in height. In this way a transition between deposition in the secondary and primary minimum will exist according to particle size. For example, Petit et al. [1973] found that for selenium sols this transition from primary to secondary minima deposition occurred at a particle size of around 55 nm. In other words, particles bigger than 55 nm are deposited in the secondary minimum, while those smaller than 55 nm are deposited in the primary minimum. This value of course will vary as a function of solution chemistry and particle surface chemistry.

In most cases relatively good agreement exists between model predictions and experimental results when favorable deposition conditions exist while there is more disagreement when unfavorable conditions are present. Deposition in the secondary minimum may resolve some of the discrepancy between theoretical and observed values of the attachment efficiency.

When electrostatic repulsion is a primary source of particle stability, the value of α, may be manipulated through changes in solution chemistry (ionic strength and pH). Consistent with DLVO theory, experimentally determined values of α for fullerene nanoclusters passing through a packed column of silicate glass beads are observed to increase with increasing ionic strength and valence.

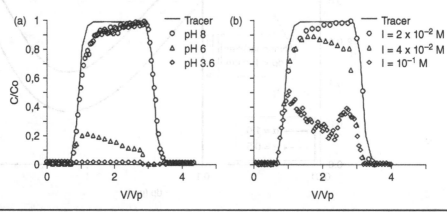

FIGURE 6.11   Breakthrough curves of the TiO$_2$ nanoparticles within sand porous medium as a function of pH at $I = 10^{-3}$ M (a), of salt concentration at pH = 8 (b) [68].

	dp (nm)	U (µm cm/ s⁻¹V⁻¹)	C/C₀	ϕ_{max} (kT)	α	Distance to 99.9% Deposition (m)
TTA/nC₆₀	44	−0.86	0.56	10	0.0001	14
C₆₀OH₂₀₋₂₄	106	−1.50	0.99	75	0.001	9.8
SWNT	16	−3.88	0.94	37	0.008	2.4
Silica ZL	106	−3.20	0.68	297	0.039	0.6
Silica OL	57	−1.95	0.97	135	0.169	0.2
Anatase	142	−0.60	0.56	50	0.298	0.1
FeRT	51	1.23	0.30	0	0.336	0.1
AlRT	24	1.72	0.85	0	0.895	0.06

Here U is electrophoretic mobility, which is an indicator of surface charge and ϕ_{max} is the height of the energy barrier between the two surfaces. The distance to reduce the initial particle concentration by 99.9 percent can be taken as a qualitative measure of nanoparticle mobility.

TABLE 6.2 Measured and Calculated Transport Characteristics of Selected Fullerene and Metal Oxide Nanoparticles in a Column Packed with Silicate Glass Beads

Theory suggests that as particle size decreases, the attachment efficiency should increase [94] due to a reduction in the height of the energy barrier as discussed earlier. However, experiments with model systems looking at both particle stability and mobility have found that the attachment efficiency is surprisingly insensitive to particle size. This discrepancy between theory and reality is attributed to an incomplete assessment of interfacial interaction energies and to changes in the nature of interactions with decreasing particle size. It is therefore unclear whether or not nanoparticles should be inherently less stable or more prone to depositing onto surfaces under favorable conditions than are larger particles.

Attachment efficiencies as calculated from experiments of particle deposition in a well-characterized medium [95] show wide variability as a function of composition and size (Table 6.2).

Information on nanoparticle transport and surface affinity can be combined to calculate an index of particle mobility in porous media based on the characteristic distance for removal from an initial source. The reciprocal of this distance, commonly referred to as the filter coefficient, is calculated as a function of the single collector efficiency and surface affinity as:

$$\lambda = \frac{3}{2}\frac{(1-\varepsilon)}{d_c}\alpha\eta_0 \tag{6.27}$$

Here λ has units of inverse distance. [When $(1/\lambda)$ is multiplied by $\ln n/n_0$ one obtains an estimate of the distance required to reduce the initial particle concentration n_0 to a concentration of n.] Plotting reciprocal λ as a function of particle size (Figure 6.12) it is evident that even for nanoparticles with low attachment efficiencies, the characteristic distance to which they will travel in a homogeneous media is low. Real systems present numerous complexities that may increase or decrease the true mobility of nanoparticles. For example, physical inhomogeneities such as fissures may increase mobility by providing preferential flow paths. Heterogeneities in media size may decrease porosity, thereby increasing deposition. Heterogeneities in surface chemistry may decrease the effective surface area available for deposition.

However, particle size also interacts with heterogeneities to determine the effective surface of the collectors that may be "visible" to a particle. While larger particles may see an average surface that is unfavorable to deposition, nanoparticles may be able to sample the surface at a finer scale, seeking-out more favorable attachment sites. The ability to access more of the surface, combined with a higher rate of transport to the surface due to Brownian diffusion may combine to reduce nanoparticle mobility. In one study by Schrick et al. [96] observed that the mobility of iron nanoparticles in soil columns was less than that of larger iron particles (Figure 6.12).

This observation has direct implications for the use of nanopartices in groundwater remediation applications and removal by granular media filters. Highly reactive nanoparticles, such as nano-iron, have been proposed as a possible remediation tool for contaminants susceptible to reduction by Fe°. However, both theory and experiments suggest that these materials should have relatively low mobilities in such applications. Adjustments in particle size and surface chemistry to balance reactivity with mobility may resolve this issue. Low nanoparticle mobilities in porous media also suggest that it may be possible to remove many of these materials from water using granular media filters.

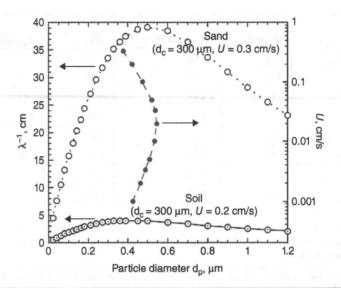

FIGURE 6.12   Open circles: Calculated filtration length (λ^{-1}) as a function of iron particle diameter, dp. Upper curve shows a calculation for Ottawa sand ($\varepsilon = 0.42$, $d_c = 300$ m, Darcy flow velocity $U = 0.3$ cm/s). Lower curve shows a calculation for an average soil ($\theta = 0.42$, $d_c = 100$ m, $U = 0.2$ cm/s). Filled circles: Particle diameter with maximum calculated value of -1 as a function of Darcy flow velocity for soil ($\theta = 0.42$, $d_c = 100$ m). The shaded region highlights particle diameters associated with the greatest predicted mobilities. (Adapted from [96].)

Detachment

Once a particle attaches to a collector surface it is subjected to a number of forces that simultaneously act to retain or displace it (Figure 6.13) [97].

Like attachment, particle detachment from a collector is also dependent on the nature of the interaction between the two surfaces. In the absence of an energy barrier the rate of particle detachment from the collector surface will be controlled by the ability of the particle to diffuse across the diffusion boundary layer [97]. When an energy barrier is present however, the deposited particle must overcome an energy (ϕ_T) that is equivalent to the depth of the primary minimum (ϕ_1) plus the height of the energy barrier (ϕ_2) (Figure 6.1) in order to go back into the bulk suspension. Similarly, nanoparticles deposited in a secondary minimum must overcome an energy that is equivalent to its depth (ϕ_{min}). The height of primary and secondary minima decreases with decreasing particle size. With this in mind nanoparticle remobilization should be more sensitive to changes in solution chemistry than larger particles.

The balance of forces acting on the particle determines whether the deposited particles may become remobilized through changes in solution chemistry and/or hydrodynamic conditions. The three principal forces acting on a deposited particle in a porous medium are typically taken to be the fluid drag force (F_D), the lift force (F_L), and the adhesive force (F_A). The adhesive forces act to retain the particle on the collector

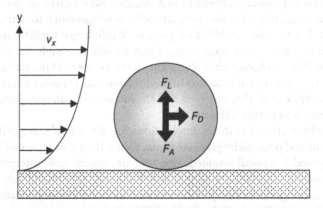

FIGURE 6.13   Illustration showing the fluid velocity gradient and forces acting on a particle once it has deposited onto a surface.

while the hydrodynamic forces (F_D and F_L) act to favor particle detachment. The fluid drag force acting on a retained particle on a collector surface is calculated according to [97, 98]:

$$F_D = (1.7005)6\pi\mu v_p a_p \qquad (6.28)$$

Here, the leading coefficient (1.7005) accounts for wall effects near the collector surface; a_p is the radius of the retained particle; u is the viscosity of the fluid; and v_p is the fluid velocity at the center of the retained particle. The fluid velocity at the center point of the retained particle is calculated using the following relationship, which is derived using a representative pore structure such as the constricted tube model [98].

$$v_p = \frac{Q/N_{pore}}{(\pi/4)d_z^2}\frac{4(d_z/2 - a_p)}{(d_z/2)^2} \qquad (6.29)$$

Here, Q is the volumetric flowrate through the porous medium; N_{pore} is the number of pores in a cross section of the packed column; and d_z is the diameter of the pore space in between the collectors. In this case, the pore space is comprised of a series of parabolic constrictions with the diameter being a function of the distance along the pore (z).

$$d_z = 2\left\{\frac{d_{max}}{2} + \left[4\left(\frac{d_c}{2} - \frac{d_{max}}{2}\right)\left(0.5 - \frac{z}{h}\right)^2\right]\right\} \qquad (6.30)$$

Here, d_z is the constriction diameter at a distance z along the pore; d_c is the equivalent diameter of the constriction; d_{max} is the maximum pore diameter, and h is the pore length. For less well-defined and more complex flow geometries, accurately modeling the hydrodynamic torque will be difficult. The lift force on a spherical particle attached to a collector surface may be approximated as follows:

$$F_L = \frac{81.2a_p^2\mu\omega^{0.5}v_p}{v^{0.5}} \qquad (6.31)$$

Here, ω is the velocity gradient at the collector surface and v is the kinematic viscosity of the fluid. The lift force arises from the different pressure forces acting on the top and bottom of the particle due to the velocity gradient. For nanoparticles, F_L is negligible in comparison to the adhesive force and may likely be omitted [97]. The result is a torque exerted on the particles as the result of the hydrodynamic drag given by:

$$T_D = 1.399a_p F_D \qquad (6.32)$$

Here, the leading coefficient (1.399) indicates that the drag force acts on the deposited particle at an effective distance of $1.399a_p$ from the surface [97]. In other words, the hydrodynamic torque acts over a lever arm of $l_y = 0.339a_p$.

The adhesive forces are a result of physicochemical interactions occurring between the two interacting bodies. The favorability or adhesion strength may be quantified in terms of the free energy of adhesion (W_A) using a XDLVO type approach for determining the free energy at contact. The adhesion force (F_A) may then be determined using different scaling models, such as the Johnson-Kendell-Roberts (JKR) [99] and Derjaguin-Mullen-Toporov (DMT) [100] models. This adhesive force resists the torque exerted by the drag force, and is manifested as a torque acting over a specified lever arm of l_x:

$$T_A = F_A l_x \qquad (6.33)$$

The value of l_x is taken to be equal to the radius of the contact area between the collector and the retained nanoparticle. The size of the contact area is determined by a number of different parameters, the most significant of which are the elasticity of the interacting bodies, particle size, and surface roughness.

Elasticity describes the stiffness or malleability of a material. As elasticity increases the contact area will also increase in size as the surfaces can deform and wrap around each other [101]. The contact area also decreases both on approach and at contact with decreasing particle size. Nanoparticles interact with a smaller region of the collector surface as they approach and ultimately make contact with the collector surface. Surface roughness may result in either an increase or a decrease in the contact area depending on particle size and the density of asperities on the collector surface. Roughness is likely to result in an increased contact area between the collector and the nanoparticle, resulting in an increase in the adhesive force.

Typically, fluid drag or hydrodynamic shearing forces are significant only for particles larger than several hundred nanometers when deposition occurs in the primary minimum [5]. This is due to several factors, such as surface roughness effects and the relative insignificance of hydrodynamic interactions in relation to thermodynamic ones at the collector interface. However, drag forces play a larger role in nanoparticle detachment when deposition occurs in the secondary minimum where the thermodynamic interactions are much weaker. The importance of hydrodynamic interactions relative to thermodynamic ones for nanoparticles can be understood by comparing the relative magnitudes of each. The mean kinetic energy (U_{KE}) imparted to a particle in a fluid flowing through a packed bed is determined according to [5]:

$$U_{KE} = \frac{1}{2} m_p \left(\frac{U}{\varepsilon} \right)^2 \tag{6.34}$$

Here, U is the superficial pore velocity and m_p is the particle mass. This assumes that the particle velocity does not lag the fluid velocity, an assumption that should be valid for nanoparticles. The relationship between particle size and kinetic energy in a porous media is illustrated in Figure 6.14. Calculations of kinetic energy are plotted for two different fluid velocities that are representative of typical groundwater Darcy velocities. Particles smaller than 100 nm are the least impacted by changes in fluid velocity and have a relatively weak kinetic energy associated with them. The low kinetic energy associated with nanoparticles suggests that deposition into even shallow secondary minima may be possible. This may explain a lack of dependence of particle mobility on fluid flow velocity if the kinetic energy does not exceed the depth of a secondary minima or the height of the energy barrier.

Effect of Surface Roughness and Aggregate Morphology

Surface heterogeneities are commonly cited as the principle reasons for discrepancies between theoretical predictions and experimental results for surface controlled processes and have received considerable attention in the research literature [34, 102–106]. These heterogeneities may be physical (e.g., roughness) or chemically (e.g., charge distribution) [3, 13]. Heterogeneity tends to become more apparent at smaller length-scales. Because the interaction area decreases with decreasing particle size nanoparticles will be more affected by surface heterogeneities than larger ones [107].

One form of chemical heterogeneity is that of an uneven distribution of charge resulting from the uneven distribution of surface functional groups and crystalline structure defects, and the presence of surface impurities or contaminants, such as ferric, aluminum, and manganese oxides. The distribution of these heterogeneities may be thought of in terms of patches having different charge properties [108].

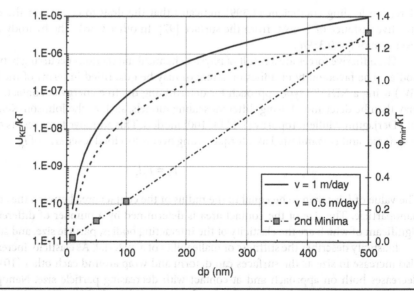

FIGURE 6.14 Kinetic energy and the depth of the secondary interaction energy minima as a function of particle size ($\zeta_p = -30$ mV; $\zeta_c = -20$ mV; $H_A = 10^{-20}$ J; $\varepsilon = 0.4$).

These patch-patch interactions appear to explain some of the observed variability in particle transport in chemically heterogeneous systems [109].

Greater surface roughness has been observed to increase deposition rate coefficients observed for some nanoparticles [98]. Similar observations have also been made for colloid deposition onto membrane surfaces [110, 111]. Enhanced particle deposition rates on rough surfaces are attributed to a combination of a reduced energy barrier on approach and physical trapping of the nanoparticles by surface features. For ideally smooth surfaces, interactions occur normal to the interacting surfaces; however, for a particle approaching a rough surface it experiences both normal and tangential forces that can capture it in surface depressions. Furthermore, the variable height of the surface topography means that the approaching particle is experiencing a range of interactions as some are more prominent than others at different separation distances. For this reason it is difficult to fully describe a true surface interaction between heterogeneous surfaces in terms of a single energy curve. Surface roughness can also alter the magnitude and type (repulsive or attractive) of interaction between two surfaces [110]. This results from a reduction in the relative interaction area between the particle and the collector surface. One approach is to model surface asperities as a series of small hemispheres instead of a flat surface. The resulting calculations yield smaller interaction energies as the interaction energy is determined by the radii of curvature of the protrusions rather than that of the interacting bodies. Consequently, roughness reduces the height of the energy barrier between two surfaces in aqueous media, thus making particle deposition more favorable [105, 106]. For instance, Suresh and Walz [105] found that the van der Waals interaction energy significantly increased when the separation distance between two surfaces approached the height of the surface asperities. The number of asperities per unit area on the surface had a smaller influence on the magnitude of the interaction energy than did asperity size. A similar conclusion was reached by these authors regarding the electrostatic interactions; electrostatic repulsion occurs at a larger separation distance than would be expected for a smooth surface. At small separation distances, surface roughness has a more profound impact on van der Waals interactions (i.e., makes the attraction stronger). Hence, at shorter separations the height of the energy barrier is substantially reduced. In summary, surface roughness tends to reduce the depth of a secondary minimum and shifts it to longer separation distances while the repulsive energy barrier is decreased and the primary minimum is moved to larger separation distances.

Roughness features may also physically trap nanoparticles on the surface once contact has been made. This occurs through a combination of enhanced surface adhesion and modification of the surface fluid velocity pattern. Upon deposition, particles are principally subject to shearing forces that would cause them to "roll" across a surface. This rolling motion is resisted by the adhesion force acting between the surfaces as discussed earlier in this section. Surface asperities act to increase the particle's adhesion lever arm while also reducing the shearing forces acting on the deposited nanoparticle. The impact of surface roughness on the adhesion between two interacting bodies has been explored at length [105–107, 109–111]. Even small asperities (<1 nm) on a collector surface may dramatically increase the contact area, between the nanoparticle and collector surfaces and, in turn, the adhesive lever arm for the nanoparticle (Figure 6.15) [98]. This is in contrast with case of larger particles where the adhesion is reduced through a reduction in the contact area due to the asperities [107, 112, 113]. For larger colloids, roughness may decrease the actual contact area leading to higher mobilization rates. However, the opposite scenario is likely for nanoparticles as their size will allow them to fit in between even tightly spaced surface asperities. In other words, if the particle diameter is greater than the distance between any two asperities, then adhesion will tend to be low, as the contact area is low [111]. However, if the particle diameter is less than this distance, then the adhesion will be higher, due to a higher contact area. Thus, roughness should tend to increase the contact area between the collector and the depositing nanoparticle [109] and resuspension should be reduced.

Similarly, irregularities on the surface of an aggregate of nanoparticles also appear to dominate surface interactions during deposition of the aggregate [114]. The strength of colloidal interaction for the aggregates of nanoparticle is calculated to be of the same order of magnitude as that for the primary particles making up the aggregate. Thus, analogous to irregularities in surface roughness, it is the size of the primary particles, not that of the aggregates, that determines the strength of the colloidal interaction between the aggregate and an environmental surface.

NOM Adsorption and Particle Transport

As discussed earlier, adsorption of organic macromolecules can enhance nanoparticle stability and mobility [76, 116, 117] in environmental systems by altering their surface chemistry, and in turn the respective interfacial interactions between other surfaces. By associating with nanoparticles organic compounds alter

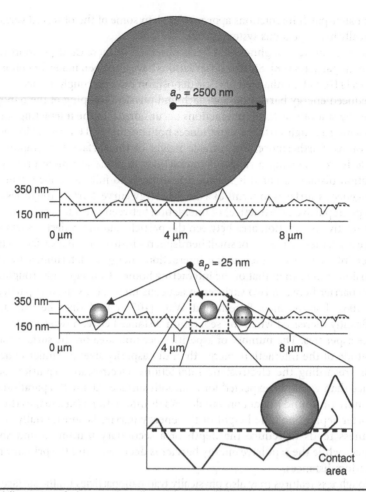

FIGURE 6.15   Scaled illustration of a large and small particle(s) on a AFM generated height profile of a rough surface. The surface had an average roughness of 65 nm. The contact area between one of the small ($a_p = 25$ nm) particles with the rough surface is highlighted.

the particle surface charge (making them more negatively charged), and possibly other interfacial properties, and inhibit particle deposition by generating short-range barriers (e.g., hydration forces and steric interactions) to surface contact [117]. Organic matter may also act to mobilize otherwise immobile materials, as has been observed for organic acids in promoting colloidal transport [118, 119]. Finally, organics may also piggy back on particles in a process referred to as facilitated transport, as has been observed for mercury and for radionucleotides transported by colloids [117].

Nanoparticle Transport in Matrices Such as Gels and Biofilms

Nanoparticles may be present in complex structures such as porous flocs and biofilms. For example, as a result of heteroaggregation, highly random structures may be formed that include nanoparticles within the structure. In many environmental systems, surfaces exposed to nonsterile and humid conditions can be colonized by microorganisms that encage themselves in hydrated exopolymeric substances (EPS), which contain mixtures of polysaccharides, proteins, and nucleic acids. This process is collectively called biofilm formation [120]. The bioavailability of nanoparticles to aquatic organisms, and in turn their access to the chain food, is likely to depend on the filtering properties of the biofilms that surround various microbes. Attractive interactions between the nanoparticles and biofilms are expected to favor retention [121–123], while repulsive interactions will prevent nanoparticle diffusion in the organic network.

Diffusion in Biofilms and Gel-Like Structures

In porous and disordered media like soil, biofilms or bacterial flocs, the random movement of diffusing particles is constrained. Diffusion in these disordered gel-like systems is often modeled as transport through a fractal structure [124]. In uniform Euclidean systems, for every number of dimensions, the random diffusion motion of a nanoparticle denoted by superscript A is described as [125]:

$$\overline{x^2(t)}^A = 2dD^A t \tag{6.35}$$

Here $\overline{x^2(t)}^A$ is the mean square displacement of the diffusing solute at time t; D is the nanoparticle diffusion coefficient; and d is the dimensionality of space. The diffusion properties of small particles in gels has been related to three distinct parameters [126–128] that affect the distribution of particles in a gel saturated with water described by a global partition coefficient Φ [126]:

$$\Phi = \vartheta \alpha \pi = \frac{[A]_g}{[A]_w} \tag{6.36}$$

Here, $[A]_g$ and $[A]_w$ are the nanoparticle concentrations in the gel and in the bulk solution, respectively; and ϑ, α, and π are the respective contributions of purely steric, chemical, and electrostatic effects.

Steric Repulsion

Steric interactions arise in disordered systems when the nanoparticle approaches the size of the interconnected pores in the media. (This is different than the steric interactions affecting particle stability discussed earlier.) The potential for collisions with the pore walls in the media reduce the degrees of freedom for particle movement with a resulting reducing in the macroscopic diffusion coefficient of the particle in the disordered media compared with unconstrained diffusion in the bulk fluid. These obstructions may be characterized in terms of a steric obstruction coefficient [129]:

$$\vartheta = (1 - \phi) \times \left(1 - \frac{R_A}{R_P}\right)^2 \tag{6.37}$$

Here, R_A and R_p are the respective diameters of the particles and the pores; ϕ is the gel volume fraction. This equation is based on a number of limiting assumptions such as the particles being perfect spheres and that the particle and pore sizes being described each by a single value. Although the theoretical constraints of this equation will rarely be met in natural gels and biofilms, it provides some insight into the role of particle and pore size as they affect the steric contribution.

Steric effects on diffusion can also be viewed in terms of decrease in mean squared displacement, $\overline{x^2(t)}^A$, that occurs as the particle moves through a fractal structure [124, 130–134]:

$$\overline{x^2(t)}^A = \Gamma t^{2/dw} \tag{6.38}$$

Here, d_w is the fractal dimension of diffusion and Γ is proportional to the constrained diffusion coefficient. In normal random diffusion, d_w is equal to 2 [Equation (6.35)]. In anomalous diffusion, d_w is greater than 2 corresponds to the slowing down of the transport caused by the delay of the diffusing particles in the disordered structure. By measuring the respective characteristic times $t_c(x)^A$ required for the nanoparticle to cover several distances x in the disordered structure, one can calculate d_w by plotting Equation 6.38 and solving for the inverse slope of a log/log plot of $t_c(x)$ versus x (e.g., $\log(x) \sim \frac{1}{d_w}\log(t)$).

Due to this steric contribution, particles above a critical size are not allowed to diffuse within a disordered structure. This critical size is of the order of that of the interconnected pores in the structured network. In bacterial biofilms, Lacroix-Gueu et al. [134] showed the enabled anomalous diffusion of latexes and bacteriophages with 55 nm radius. In 1.5 to 2 wt% agarose gels, the sterric obstruction appear with solute sizes above 10 nm and a critical size is measured around 70 nm [127, 137]. However, the critical size is logically inversely correlated to the gel density. Labille et al. [136] showed with the same agarose gel that in a gel/solution interface, the fiber density is locally increased from 1.5 to 5 wt% in a 100-μm-thickness interphase layer, through which the maximum size of diffusing particles is decreased to around 50 nm radius. The diffusion motion of macromolecules has also been measured in intra-cellular cytoplasm, where dextran with molecular weight up to 2106 Da (~45 nm radius) undergoes impaired diffusion [138].

Electrostatic Interactions

Most gels, flocs, and biofilms encountered in aquatic and soil environments are mainly composed of polysaccharides and humic substances. Both of these highly reactive components are negatively charged in a large pH range due to the presence of acidic groups in their chemical structure [139, 140]. This induces a negative surface charge to the global diffusing media, which is characterized as follows by the Donnan potential Ψ, which is the average difference in potential between gel and bulk water:

$$\psi = \frac{kT}{zF} \sinh^{-1} \frac{\rho}{2zFc} \tag{6.39}$$

Here, ρ is the charge density of the gel; c and z are, respectively, the molar concentration and charge of the electrolyte in the external bulk solution; and F is the Faraday constant. The diffusion motion in the gel of charged nanoparticles is thus affected by an electrostatic contribution π that can be described by a Boltzmann distribution:

$$\pi = \frac{[A]_P}{[A]_W} = \exp\left(-\frac{Z_A F \psi}{kT}\right) \tag{6.40}$$

Here, $[A]_P$ is the concentration of particles A in gel pores exclusively controlled by electrostatic interactions and Z_A is electrical charge of the particles.

Specific Chemical Bonds

When the particle surface presents a specific chemical affinity for sites S on the gel fibers, an adsorption reaction occurs to form a complex SA according to the following intrinsic equilibrium constant K_A^{int}:

$$K_A^{int} = \frac{[SA]}{[A]_P[S]} \tag{6.41}$$

Here, $[SA]$ and $[S]$ are the respective concentrations of complexes SA and free sites S. Then, when $\Psi \neq 0$, by combining Equations (6.36), (6.39), and (6.40), one obtains the following expression for Φ:

$$\Phi = \frac{[SA]+[A]_P}{[A]_W} = \exp\left(-\frac{Z_A F \Psi}{kT}\right)\left(1 + K_A^{int}[S]\right) = \pi\alpha \tag{6.42}$$

The existence of specific chemical interactions has been put in evidence for example in the case of silica nanoparticles (ludox HS_{30}) and amine- or carboxylate-terminated dendrimers, diffusing in an agarose gel [126, 127], where the diffusion coefficients measured by FCS showed more reduced diffusivity, compared to pure steric effect expected.

Airborne Nanoparticles

A detailed consideration of the origins, transport, and characteristics of airborne nanoparticles is beyond the scope of this chapter. However, these materials are of major concern for human health and occupational safety [141]. In fact, nanoscale particles are perhaps more ubiquitous in the atmosphere than in any other environment. Clinical studies have suggested a strong link between particulate air pollution and respiratory disease. Many of the principles governing nanoparticle transport in aqueous systems apply to atmospheric systems as well. In particular, the framework for analyzing the kinetics of particle aggregation and deposition are directly applicable to atmospheric particles. However, the origins and characteristics of particles in air and water may differ. Due to the importance of combustion as a source of exposure to nanoscale particles, and the potential similarity between soot and fullerenes, we briefly address this aspect.

Atmospheric nanoparticles may be detected and characterized by simultaneously measuring the light scattering, the photoelectric charging, and the diffusion charging [142]. These methods may also be used to determine the likely source of the nanoparticles and differentiate between those from combustion processes and background nanoparticles. Particles carrying polycyclic aromatic hydrocarbons are detected by their large photoelectric charging signature whereas particles from other sources only exhibit light scattering and diffusion charging. Although carbon nanotubes have not yet been detected as unintentional combustion products, C_{60} may be present in some cases.

Nanoscale particles may be produced and dispersed in the atmosphere through both natural processes and human activities. A significant source of man-made nanoscale particles is the burning of hydrocarbon fuels such as diesel, gasoline, and propane [141, 143]. Industrial processes are another source of atmospheric nanoparticles, though their contribution is far less than that of vehicle exhaust. Thus, relatively high concentrations of these nanoparticles occur along roadsides and other areas immediately surrounding combustion sources [141]. Nanoscale particles formed from the burning of these fuels generally fall in a

	Diesel Exhaust, %	Gasoline Exhaust, %
Carbon black	68	32
Organic carbon	31	61
Other materials	1	7

Particles in diesel and gasoline exhaust at the point of emission are generally smaller than 50 nm, with diesel exhaust also containing some larger particles up to 1000 nm in diameter.

TABLE 6.3 General Composition of Nanoparticles Emitted from Diesel and Gasoline Engines [143]

size range of 10 to 60 nm [141, 144] and are primarily composed of unburned oils, polycyclic aromatic hydrocarbons, inorganic compounds and sulfates [143, 144]. The composition and properties of the airborne nanoparticles will vary according to the type of fuel and engine that is used (Table 6.3) [143, 144]. For instance, gasoline engines tend to produce smaller sized nanoparticles while diesel engines emit larger agglomerates in a size range of 50 to 1000 nm. Similar to nanoparticles in aqueous media the behavior and properties of atmospheric nanoparticles are dynamic [141, 144].

Conventional diesel exhaust (in the absence of after treatment) emits 10 to 100 times more particle mass and up to 10^5 times more particle number than gas engines. For instance, soot produced from the burning of diesel fuel has been characterized as consisting of agglomerated spherical particles with a mean diameter between 20 nm to several micrometers. The primary particles here are homogeneous small particles with a mean size distribution of 25 nm. Soot had a surface area of 175 m^2/g, compared to 11 m^2/g for commercial carbon black. The core of the soot was characterized as containing disordered polycyclic aromatic hydrocarbons which have been reported to act as nuclei for soot formation [145]. The small primary particle size and high surface area, and presence of adsorbed hydrocarbons all contribute to the high reactivity of these particles [143]. Furthermore, absorbed material in diesel exhaust particulate matter is specifically responsible for adverse health effects; particles in the smaller size fraction of the particulate matter may have a larger fraction of absorbed material.

The size distribution of combustion-derived atmospheric nanoparticles evolves as they are dispersed from the point source and is a function of the characteristics of the system and the particles [144]. System properties that have been found to be particularly significant include the meteorological conditions (wind speed, wind direction, atmospheric temperature, and relative humidity), particle concentration, presence, and concentration of trace gases (e.g., NO_x), and the concentration of materials that may induce coagulation, condensation, and evaporation processes [141]. The variation in particle size depends on the balance between growth through coagulation and condensation and shrinkage by evaporation. Nanoparticles may aggregate either with each other (self-coagulation) or with other larger background particles, a process known as heteroaggregation [144]. Aggregation of atmospheric nanoparticles is typically a rapid process [144] and accounts for the more substantial concentrations of atmospheric nanoparticles in the immediate vicinity of combustion or other generation sources.

Given the dependence of particle size distribution on metrological conditions, or actually the "solution chemistry" it is reasonable to expect that the characteristics of the nanoparticles will have seasonal variations. Minoura et al. [140] observed that nanoparticles had on average a larger peak diameter in the summer months compared to those in the winter ones. This difference may be attributed to a variety of factors as has been discussed here. Additionally, though it was found that nanoparticles in the winter were determined to come from a point source while those in the summer were thought to come from both a point source and through photochemical nucleation, which takes place primarily in summer. The more favorable formation and growth conditions (e.g., higher humidity and particle concentrations) in summer months also result in higher particle aggregation or growth rates [140]. For example, higher particle concentrations result in higher particle growth rates (60 nm/day in the winter and 103 nm/day in the summer) as a result of higher collision frequencies.

Summary

For many nanoparticles larger than several tens of nanometers, many traditional relationships and models used for colloidal systems may be used for describing nanoparticle behavior in aqueous systems. However, when particles are smaller than approximately 20 nm, particle behavior increasingly resembles that of a molecular solute and intermolecular forces play a greater role in determining the transport, aggregation and deposition of these materials. Heterogeneities of the surfaces with which nanoscale particles may interact will also play an increasing important role, and the characterization of these surfaces is therefore important in predicting nanoparticle behavior.

Nanoparticle transport at the mesoscopic scale in aqueous systems is dominated by their characteristically high diffusion coefficients as a result of their small size. While this confers a high mobility to nanoparticles in a liquid or gas it also results in them having high contact efficiencies with potential collector surfaces, making them relatively immobile even when they possess low attachment efficiencies (α < 0.1). The stability of nanoparticle dispersions will largely determine how long nanoparticles are likely to remain in the nano domain. Nanoparticles often aggregate to form clusters both with and without the presence of destabilizing agents or changing chemical conditions. It is therefore necessary to consider the transport of nanomaterials both as nanoparticles and as materials that may transition into the colloidal domain or larger, where they may be subject to transport mechanisms such as gravitational settling. Aggregation will therefore likely reduce the persistence of these materials in the environment and possibly their bioavailability.

The deposition of nanoparticles on surfaces ranging from sand grains in aquifers to leaves on trees will also reduce their persistence. Nanoparticles can be entrapped by myriad media ranging from mineral formations to biofilms. The interaction between nanoparticles and surfaces may be assessed within the context of various established interfacial models, such as the DLVO model discussed here.

According to the DLVO theory and its extensions, the stability of particle dispersions, results from the sum of repulsive and attractive forces at their interface with neighboring solid interfaces. Generally, repulsive electrostatic interactions between like charged particles control dispersion stability, though the nature of these and other interactions changes with particle size. It is therefore imperative that size effects on interfacial interaction energies be considered, particularly given that some interactions may become significant for nanoparticles that may otherwise not be so for colloidal sized materials of the same composition (e.g., hydration interactions and silica particles).

The surface reactivity of suspended particles is a key parameter that controls their interaction with other materials that may be present in suspension, and thus their behavior in the given dispersion; even more so for nanoparticles, for which their large surface area to volume ratio induces a high sorption capacity for foreign species. The nature of these interactions is always specific to the chemical characteristics of the nanoparticles, and their affinity for the foreign species present in the system. In many cases adsorption occurs through charge mediated interactions, or via the formation of specific chemical bonds. In most cases, bond formation between nanoparticles and other materials results in a modification of the surface reactivity of both, tending to lower the system free energy. This favors intraparticle attractions and aggregation. One obvious example of this phenomenon is the absorption and embedding of nanoparticles into organo-mineral flocs, which cloaks the properties of the nanoparticles and producing transport and fate behavior that is determined by the floc. Similarly, adsorption or reaction with other materials will affect nanoparticles fate and transport.

References

1. Wiesner, M.R., et al., "Decreasing Uncertainties in Assessing Environmental Exposure, Risk, and Ecological Implications of Nanomaterials." *Environmental Science & Technology*, 2009; **43**(17): 6458–6462.
2. van Oss, C.J., "Acid-Base Interfacial Interactions in Aqueous Media." *Colloids and Surfaces A*, 1993; **78**: 1.
3. Israelachvili, J.N., *Intermolecular and Surface Forces*, 2nd ed. 1992, London: Academic Press Harcourt Brace Jovanovich. p. 450.
4. Brant, J.A. and A.E. Childress, "Assessing Short-Range Membrane-Colloid Interactions Using Surface Energetics." *Journal of Membrane Science*, 2002; **203**: 257–273.
5. Hahn, M.W., D. Abadzic, and C.R. O'Melia, "Aquasols: On the Role of the Secondary Minima." *Environmental Science & Technology*, **38**(22): 5915–5924, 2004.
6. Hunter, R.J., *Foundations of Colloid Science*. Vol. 2. 1989, Oxford: Clarendon Press.
7. van Oss, C.J., *Interfacial Forces in Aqueous Media*. 1994, New York: Marcel Dekker. p. 179.
8. Choudhury, N. and B.M. Pettitt, "On the Mechanism of Hydrophobic Association of Nanoscopic Solutes." *Journal of the American Chemical Society*, **127**(10): 3556–3567, 2005.
9. Gregory, J., "The Calculation of Hamaker Constants." *Advan. Colloid Interface Sci.*, 1969; **2**: 396–417.
10. Hogg, R.I., T.W. Healy, and D.W. Fuerstenau, "Mutual Coagulation of Colloidal Dispersions." *Transactions of the Faraday Society*, 1966; **62**: 1638.
11. Derjaguin, B.V., "Friction and Adhesion IV. The Theory of Adhesion of Small Particles." *Kolloidnyi Zhurnal*, 1934; **69**: 155–164.

12. Lin, S.H. and M.R. Wiesner, "Exact Analytical Expressions for the Potential of Electrical Double Layer Interactions for a Sphere-Plate System." *Langmuir*, 2010; **26**(22): 16638–16641.
13. Elimelech, M., et al., *Particle Deposition and Aggregation: Measurement, Modeling, and Simulation*. 1995, Oxford, England: Butterworth-Heinemann.
14. van Oss, C.J., "Hydrophobic, Hydrophilic and Other Interactions in Epitope-Paratope Binding." *Molecular Immunology*, 1995; **32**(3): 199–211.
15. Fukushi, K. and T. Sato, "Using a Surface Complexation Model to Predict the Nature and Stability of Nanoparticles." *Environmental Science & Technology*, 2005; **39**(5): 1250–1256.
16. Andrievsky, G.V., et al., "On the Production of an Aqueous Colloidal Solution of Fullerenes." *Chemical Communications*, 1995; **8**(12): 1281–1282.
17. Schwarzer, H.-C. and W. Peukert, "Prediction of Aggregation Kinetics Based on Surface Properties of Nanoparticles." *Chemical Engineering Science*, 2005; **60**(1): 11–25.
18. Marinova, K.G., et al., "Charging of Oil-Water Interfaces due to Spontaneous Adsorption of Hydroxyl Ions. *Langmuir*, 1996; **12**: 2045–2051.
19. Brant, J., et al., "Comparison of Electrokinetic Properties of Colloidal Fullerenes (n-C-60) Formed Using Two Procedures." *Environmental Science & Technology*, 2005; **39**(17): 6343–6351.
20. Andrievsky, G.V., et al., "Comparative Analysis of Two Aqueous-Colloidal Solutions of C60 Fullerene with Help of FTIR Reflectance and UV-Vis Spectroscopy. *Chemical Physics Letter*, 2002; **364**: 8–17.
21. Brant, J., H. Lecoanet, and M.R. Wiesner, "Aggregation and Deposition Characteristics of Fullerene Nanoparticles in Aqueous Systems." *Journal of Nanoparticle Research*, 2005; **7**(4–5): 545–553.
22. Prieve, D.C. and E. Ruckenstein, "Rates of Deposition of Brownian Particles Calculated by Lumping Interaction Forces into a Boundary Condition." *Journal of Colloid and Interface Science*, 1976; **57**(3): 547–550.
23. Yaminsky, V.V.V. and E.A. Vogler, "Hydrophobic Hydration." *Current Opinion in Colloid & Interface Science*, 2001; **6**: 342–349.
24. Guldi, D.M. and M. Prato, "Excited-State Properties of C_{60} Fullerene Derivatives." *Accounts of Chemical Research*, 2000; **33**(10): 695–703.
25. Hotze, E.M., J.Y. Bottero, and M.R. Wiesner, "Theoretical Framework for Nanoparticle Reactivity as a Function of Aggregation State. *Langmuir*, 2010; **26**(13): 11170–11175.
26. Jassby, D., J.F. Budarz, and M. Wiesner, "Impact of Aggregate Size and Structure on the Photocatalytic Properties of TiO_2 and ZnO Nanoparticles." *Environmental Science & Technology*, 2012; **46**(13): 6934–6941.
27. Elzey, S. and V.H. Grassian, "Agglomeration, Isolation and Dissolution of Commercially Manufactured Silver Nanoparticles in Aqueous Environments." *Journal of Nanoparticle Research*, 2010; **12**(5): 1945–1958.
28. Brunner, T.J., et al., "In Vitro Cytotoxicity of Oxide Nanoparticles: Comparison to Asbestos, Silica, and the Effect of Particle Solubility." *Environmental Science & Technology*, 2006; **40**(14): 4374–4381.
29. Cheng, Y.W., et al., "Toxicity Reduction of Polymer-Stabilized Silver Nanoparticles by Sunlight." *Journal of Physical Chemistry C*, 2011; **115**(11): 4425–4432.
30. Lam, C.W., et al., "Pulmonary Toxicity of Single-Wall Carbon Nanotubes in Mice 7 and 90 Days after Intratracheal Installation." *Toxicological Sciences*, 2004; **77**: 126–134.
31. Warheit, D.B., et al., "Comparative Pulmonary Toxicity Assessment of Single-Wall Carbon Nanotubes in Rates." *Toxicological Sciences*, 2004; **77**: 117–125.
32. Kobayashi, M., et al., "Aggregation and Charging of Colloidal Silica Particles: Effect of Particle Size. *Langmuir*, 2005; **21**(13): 5761–5769.
33. Barnard, A.S. and L.A. Curtiss, "Prediction of TiO_2 Nanoparticle Phase and Shape Transitions Controlled by Surface Chemistry." *Nano Letters*, 2005; **5**(7): 1261–1266.
34. Brant, J.A., et al., "Characterizing the Impact of Preparation Method on Fullerene Cluster Structure and Chemistry." *Langmuir*, 2006; **22**(8): 3878–3885.
35. Georgakilas, V., et al., "Supramolecular Self-Assembled Fullerene Nanostructures." *Proceedings of the National Academy of Sciences*, 2002; **99**(8): 5075–5080.
36. Guo, Z.-X., et al., "Nanoscale Aggregation of Fullerene in Nafion Membrane." *Langmuir*, 2002; **18**: 9017–9021.
37. Natalini, B., et al., "Chromatographic Separation and Evaluation of the Lipophilicity by Reversed-Phase High Performance Liquid Chromotography of Fullerene-C_{60} Derivatives." *Journal of Chromatography A*, 1999; **847**: 339–343.
38. Moraru, V., N. Lebovka, and D. Shevchenko, "Structural Transitions in Aqueous Suspensions of Natural Graphite." *Colloids and Surfaces A: Physicochemical and Engineering Aspects*, 2004; **242**(1–3): 181–187.
39. O'Melia, C.R., "Aquasols: The Behavior of Small Particles in Aquatic Systems." *Environmental Science and Technology*, 1980; **14**(9): 1052–1060.
40. Smoluchowski, M., "Versuch einer Mathematischen Theorie der Koagulations- Kinetik Kolloider Losungen." *Zeitschrift fur Physikalische Chemie*, 1917; **92**: 129.
41. O'Melia, C.R., "Coagulation and Flocculation," in *Physicochemical Processes-for Water Quality Control*, W.J. Weber, Editor. 1972, John Wiley & Sons: New York. pp. 61–107.
42. Adler, P.M., "Interaction of Unequal Spheres I. Hydrodynamic Interaction: Colloidal Forces." *Journal of Colloid and Interface Science*, 1981; **84**(2): 461–474.
43. Han, M. and D.F. Lawler, "Interactions of Two Settling Spheres: Settling Rates and Collision Efficiency." *Journal of Hydraulic Engineering-ASCE*, 1991; **117**(10): 1269–1289.
44. Veerapaneni, S. and M.R. Wiesner, "Hydrodynamics of Fractal Aggregates with Radially Varying Permeability." *Journal of Colloid and Interface Science*, 1996; **177**: 45–57.
45. Adler, P.M., "Streamlines in and Around Porous Particles." *Journal of Colloid and Interface Science*, 1981; **81**(2): 531–535.

46. Barton, L.E., et al., "Theory and Methodology for Determining Nanoparticle Affinity for Heteroaggregation in Environmental Matrices Using Batch Measurements." *Environmental Engineering Science*, 2014; **31**(7): 421–427.

47. Therezien, M., A. Thill, and M.R. Wiesner, "Importance of Heterogeneous Aggregation for NP Fate in Natural and Engineered Systems." *Science of the Total Environment*, 2014; **485–486**: 309–318.

48. Sani-Kast, N., et al., "Addressing the Complexity of Water Chemistry in Environmental Fate Modeling for Engineered Nanoparticles." *Science of the Total Environment*, 2015; **535**: 150–159.

49. Wiesner, M.R., "Kinetics of Aggregation Formation in Rapid Mix." *Water Resources*, 1992; **26**(3): 379–387.

50. Thill, A., et al., "Flocs Restructuring during Aggregation: Experimental Evidence and Numerical Simulation." *Journal of Colloid and Interface Science*, 2001; **243**(1): 171–182.

51. Labille, J., et al., "Heteroaggregation of Titanium Dioxide Nanoparticles with Natural Clay Colloids." *Environmental Science & Technology*, 2015; **49**(11): 6608–6616.

52. Praetorius, A., et al., "Heteroaggregation of Titanium Dioxide Nanoparticles with Model Natural Colloids under Environmentally Relevant Conditions." *Environmental Science & Technology*, 2014; **48**(18): 10690–10698.

53. Yin, L.Y., et al., "More than the Ions: The Effects of Silver Nanoparticles on Lolium multiflorum." *Environmental Science & Technology*, 2011; **45**(6): 2360–2367.

54. Hotze, E.M., et al., "Mechanisms of Bacteriophage Inactivation via Singlet Oxygen Generation in UV Illuminated Fullerol Suspensions." *Environmental Science & Technology*, 2009; **43**(17): 6639–6645.

55. Auffan, M., et al., "CeO$_2$ Nanoparticles Induce DNA Damage towards Human Dermal Fibroblasts In Vitro." *Nanotoxicology*, 2009. in press, corrected proof.

56. Andrievsky, G.V., et al. "Are Fullerene Soluble in Water?" in *Fullerenes: Recent Advances in the Chemistry and Physics of Fullerenes and Related Materials*. 1995. Reno, Nevada: The Electrochemical Society.

57. Deguchi, S., R.G. Alargova, and K. Tsujii, "Stable dispersions of fullerenes, C-60 and C-70, in water. Preparation and characterization." *Langmuir*, 2001; **17**(19): 6013–6017.

58. Scrivens, W.A. and J.M. Tour, "Synthesis of ^{14}C-Labeled C$_{60}$, Its Suspension in Water, and Its Uptake by Human keratinocytes." *Journal of the American Chemical Society*, 1994; **116**: 4517–4518.

59. Saleh, N., et al., "Absorbed Triblock Copolymers Deliver Reactive Iron Nanoparticles to the Oil/Water Interface." *Nano Letters*, 2005; **5**(12): 2489–2494.

60. Angelini, G., et al., "Study of the Aggregation Properties of a Novel Amphilic C$_{60}$ Fullerene Derivative." *Langmuir*, 2001; **17**: 6404–6407.

61. Guldi, D.M., et al., "Ordering Fullerene Materials at Nanometer Dimensions." *Accounts of Chemical Research*, 2005; **38**(1): 38–43.

62. Stumm, W., *Chemistry of the Solid-Water Interface*. 1992, New York, NY: Wiley.

63. Cheng, X., A.T. Kan, and M.B. Tomson, "Naphthalene Adsorption and Desoption from Aqueous C$_{60}$ Fullerene." *Journal of Chemical and Engineering Data*, 2004. ACS ASAP Article.

64. Ruoff, R.S., et al., *Solubility of C$_{60}$ in a variety of solvents*. Journal of Physical Chemistry, 1993. **97**: p. 3379–3383.

65. Marcus, Y., et al., "Solubility of C$_{60}$ fullerene." *Journal of Physical Chemistry B*, 2001; **105**: 2499–2506.

66. Nakamura, E. and H. Isobe, "Functionalized Fullerenes in Water. The First 10 Years of Their Chemistry, Biology, and Nanoscience." *Accounts of Chemical Research*, 2003; **36**(11): 807–815.

67. Labille, J., et al., "Hydration and Dispersion of C-60 in Aqueous Systems: The Nature of Water-Fullerene Interactions." *Langmuir*, 2009; **25**(19): 11232–11235.

68. Solovitch, N., et al., "Concurrent Aggregation and Deposition of TiO(2) Nanoparticles in a Sandy Porous Media." *Environmental Science & Technology*, 2010; **44**(13): 4897–4902.

69. Fauconnier, N., et al., "Thiolation of Maghemite Nanoparticles by Dimercaptosuccinic Acid." *Journal of Colloid and Interface Science*, 1997; **194**(2): 427–433.

70. Bellona, C., et al., "Factors Affecting the Rejection of Organic Solutes during NF/RO Treatment—a Literature Review." *Water Research*, 2004; **38**: 2795–2809.

71. Hsu, J.-P. and Y.-C. Kuo, "An Algorithm for the Calculation of the Electrostatic Potential Distribution of Ion-Penetrable Membranes Carrying Fixed Charges." *Journal of Colloid and Interface Science*, 1995; **171**: 483–489.

72. Mchedlov-Petrossyan, N.O., V.K. Klochkov, and G.V. Andrievsky, "Colloidal Dispersions of Fullerene C$_{60}$ in Water: Some Properties and Regularities of Coagulation by Electrolytes." *Faraday Transactions*, 1997; **93**(24): 4343–4346.

73. Hunter, R.J., *Zeta Potential in Colloid Science: Principles and Applications*. 1981, London: Academic Press.

74. Cedervall, T., et al., "Understanding the Nanoparticle-Protein Corona Using Methods to Quantify Exchange Rates and Affinities of Proteins for Nanoparticles." *Proceedings of the National Academy of Science USA*, 2007; **104**(7): 2050–2055.

75. Moreau, J.W., et al., "Extracellular Proteins Limit the Dispersal of Biogenic Nanoparticles." *Science*, 2007; **316**: 1600–1603.

76. Franchi, A. and C.R. O'Melia, "Effects of Natural Organic Matter and Solution Chemistry on the Deposition and Reentrainment of Colloids in Porous Media." *Environmental Science & Technology*, 2003; **37**: 1122–1129.

77. Mylon, S.E., K.L. Chen, and M. Elimelech, "Influence of Natural Organic Matter and Ionic Composition on the Kinetics and Structure of Hematite Colloid Aggregation: Implications to Iron Depletion in Estuaries." *Langmuir*, 2004; **20**(21): 9000–9006.

78. Johnson, S.B., et al., "Adsorption of Organic Matter at Mineral/Water Interfaces. 6. Effect of Inner-Sphere versus Outer-Sphere Adsorption on Colloid Stability." *Langmuir*, 2005; **21**(14): 6356–6365.

79. Labille, J., et al., "Flocculation of Colloidal Clay by Bacterial Polysaccharides: Effect of Macromolecule Charge and Structure." *Journal of Colloid and Interface Science*, 2005; **284**(1): 149–156.

80. Lin, S.H., et al., "Polymeric Coatings on Nanoparticles Prevent Auto-Aggregation But Enhance Attachment to Uncoated Surfaces." *Abstracts of Papers of the American Chemical Society*, 2011; **242**: 2.

81. Labille, J., et al., "Aging of TiO$_2$ Nanocomposites Used in Sunscreen. Dispersion and Fate of the Degradation Products in Aqueous Environment. *Journal of Environment and Pollution*, 2010; **158**(12): 3482–3489.

82. Iwasaki, T., "Some Notes on Sand Filtration." *Journal American Water Works Association*, 1937; **29**: 1597–1602.

83. Ives, K.J., "Mathematical Models of Deep Bed Filtration," in *The Scientific Basis of Filtration*, K.J. Ives, Editor. 1975, Noordhoff International Publishing, division of A.W. Sijthoff International Publishing Company: Netherlands. pp. 203–224.

84. O'Melia, C.R. and W. Stumm, "Theory of Water Filtration." *Journal American Water Works Association*, 1967; **59**: 1393–1412.

85. Rajagopalan, R. and C. Tien, "The Theory of Deep Bed Filtration," in *Progress in Filtration and Separation*, R.J. Wakeman, Editor. 1979, Elsevier: New York. pp. 179–269.

86. Elimelech, M. and C.R. O'Melia, "Kinetics of Deposition of Colloidal Particles in Porous Media. *Environmental Science & Technology*, 1990; **24**: 1528–1536.

87. O'Melia, C.R., "Particle-Particle Interactions in Aquatic Systems." *Colloids and Surfaces*, 1989; **39**: 255.

88. Happel, J., "Viscous Flow in Multiparticle Systems: Slow Motion of Fluids Relative to Beds of Spherical Particles." *Journal of American Institute of Chemical Engineers*, 1958; **4**(2): 197–201.

89. Rajagopalan, R. and C. Tien, "Single Collector Analysis of Collection Mechanisms in Water Filtration. *Canadian Journal of Chemical Engineering*, 1977; **55**: 246–255.

90. Tufenkji, N. and M. Elimelech, "Correlation Equation for Predicting Single-Collector Efficiency in Physiochemical Filtration in Saturated Porous Media." *Environmental Science & Technology*, 2004; **38**: 529–536.

91. Derjaguin, B.V. and L. Landau, "Theory of the Stability of Strongly Charged Lyophobic Sols and of the Adhesion of Strongly Charged Particlesin Solutions of Electrolytes." *Acta Physicochimica URSS*, 1941; **14**: 633–662.

92. Yao, K., M. Habibian, and C. O'Melia, "Water and Waste Water Filtration: Concepts and Applications." *Environmental Science & Technology*, 1971; **5**(11): 1105–1112.

93. Hahn, M.W. and C.R. O'Melia, "Deposition and Reentrainment of Brownian Particles in Porous Media under Unfavorable Chemical Conditions: Some Concepts and Applications." *Environmental Science & Technology*, 2004; **38**: 210–220.

94. O'Melia, C.R., *Kinetics of Colloid Chemical Processes in Aquatic Systems*. W. Stumm, Editor. 1990, John Wiley & Sons, Inc.: New York.

95. Lecoanet, H.F., J.Y. Bottero, and M.R. Wiesner, "Laboratory Assessment of the Mobility of Nanomaterials in Porous Media. *Environmental Science & Technology*, 2004; **38**(19): 5164–5169.

96. Schrick, B., et al., "Delivery Vehicles for Zerovalent Metal Nanoparticles in Soil and Groundwater." *Chemistry of Materials*, 2004; **16**(11): 2187–2193.

97. Ryan, J. and M. Elimelech, "Colloid Mobilization and Transport in Groundwater." *Colloids and Surfaces*, 1996; **107**: 1–56.

98. Li, X., et al., "Role of Hydrodynamic Drag on Microsphere Deposition and Re-entrainment in Porous Media under Unfavorable Conditions." *Environmental Science & Technology*, 2006; **39**(11): 4012–4020.

99. Johnson, K.L., K. Kendall, and A.D. Roberts, "Surface Energy and the Contact of Elastic Solids." *Proceedings of the Royal Society of London Series A Mathematical and Physical Sciences*, 1971; **324**(1558): 301–313.

100. Derjaguin, B.V., J.J. Muller, and Y.P. Toporov, "Effect of Contact Deformations on the Adhesion of Particles." *Journal of Colloid and Interface Science*, 1975; **53**: 314–326.

101. Johnson, K.L., "Mechanics of Adhesion." *Tribology International*, 1998; **31**(8): 413–418.

102. Elimelech, M., et al., "Relative Insignificance of Mineral Grain Zeta Potential to Colloid Transport in Geochemically Heterogeneous Porous Media." *Environmental Science & Technology*, 2000; **34**: 2143–2148.

103. Johnson, P.R., N. Sun, and M. Elimelech, "Colloid Transport in Geochemically Heterogeneous Porous Media: Modeling and Measurements." *Environmental Science & Technology*, 1996; **30**: 3284–3293.

104. Song, L., P.R. Johnson, and M. Elimelech, "Kinetics of Colloid Deposition onto Heterogeneously Charged Surfaces in Porous Media. *Environmental Science & Technology*, 1994; **28**(6): 1164–1171.

105. Suresh, L. and J.Y. Walz, "Effect of Surface Roughness on the Interaction Energy between a Colloidal Sphere and a Flat Plate." *Journal of Colloid and Interface Science*, 1996; **183**: 199–213.

106. Suresh, L.W. and J.Y. Walz, "Direct Measurement of the Effect of Surface Roughness on the Colloidal Forces between a Particle and Flat Plate." *Journal of Colloid and Interface Science*, 1997; **196**: 177–190.

107. Brant, J.A. and A.E. Childress, "Colloidal Adhesion to Hydrophilic Membrane Surfaces." *Journal of Membrane Science*, 2004; **241**(2): 235–248.

108. Rajasekar V., "Double Layer Calculations for the Attachment of a Colloidal Particle with a Charged Surface Patch onto a Substrate," *Separations Technology*, 1992; **2**: 98–103.

109. Song, L. and M. Elimelech, "Transient Deposition of Colloidal Particles in Heterogeneous Porous Media." *Journal of Colloid and Interface Science*, 1994; **167**: 301–313.

110. Elimelech, M., et al., "Role of Membrane Surface Morphology in Colloidal Fouling of Cellulose Acetate and Composite Aromatic Polyamide Reverse Osmosis Membranes." *Journal of Membrane Science*, 1997; **127**(1): 101–109.

111. Vrijenhoek, E.M., S. Hong, and M. Elimelech, "Influence of Membrane Surface Properties on Initial Rate of Colloidal Fouling of Reverse Osmosis and Nanofiltration Membranes." *Journal of Membrane Science*, 2001; **188**: 115–128.

112. Cappella, B. and G. Dietler, "Force-Distance Curves by Atomic Force Microscopy." *Surface Science Reports*, 1999; **34**: 1–104.

113. Bowen, W.R., et al., "An Atomic Force Microscopy Study of the Adhesion of a Silica Sphere to a Silica Surface-Effects of Surface Cleaning." *Colloids and Surfaces A: Physicochemical and Engineering Aspects*, 1999b; **157**: 117–125.

114. Bowen, W.R. and T.A. Doneva, "Atomic Force Microscopy Studies of Membranes: Effect of Surface Roughness on Double-Layer Interactions and Particle Adhesion." *Journal of Colloid and Interface Science*, 2000; **229**: 544–549.

115. Lin, S.H. and M.R. Wiesner, "Deposition of Aggregated Nanoparticles—A Theoretical and Experimental Study on the Effect of Aggregation State on the Affinity between Nanoparticles and a Collector Surface." *Environmental Science & Technology*, 2012; **46**(24): 13270–13277.

116. Thoral, S., et al., "XAS Study of Iron and Arsenic Speciation during Fe(II) Oxidation in the Presence of As(III)." *Environmental Science & Technology*, 2005; **39**(24): 9478–9485.

117. Slowey, A.J., et al., "Role of Organic Acids in Promoting Colloidal Transport or Mercury from Mine Tailings." *Environmental Science & Technology*, 2005; **39**: 7869–7874.

118. Lowry, G.V., et al., "Macroscopic and Microscopic Observations of Particle-Facilitated Mercury Transport from New Idria and Sulfur Bank Mercury Mine Tailings." *Environmental Science & Technology*, 2004; **38**: 5101–5111.

119. Slowey, A.J., J.J. Rytuba, and G.E.J. Brown, "Speciation of Mercury and Mode of Transport from Placer Gold Mine Tailings." *Environmental Science & Technology*, 2005; **39**(6): 1547–1554.

120. Costerton, J.W., et al., "Bacterial Biofilms in Nature and Disease." *Annual Review of Microbiology*, 1987; **41**: 435–464.

121. Ballance, S., et al., "Influence of Sediment Biofilm on the Behaviour of Aluminum and Its Bioavailability to the Snail Lymnaea Stagnalis in Neutral Freshwater." *Canadian Journal of Fisheries and Aquatic Sciences*, 2001; **58**(9): 1708–1715.

122. Battin, T.J., et al., "Contributions of Microbial Biofilms to Ecosystem Processes in Stream Mesocosms." *Nature*, 2003; **426**(6965): 439–442.

123. Jordan, R.N., "Evaluating Biofilm Activity in Response to Mass Transfer-Limited Bioavailability of Sorbed Nutrients." *Biofilms*, 1999; **310**: 393–402.

124. Havlin, S. and D. Ben-Avraham, "Diffusion in Disordered Media." *Advances in Physics*, 2002; **51**(1): 187–292.

125. Havlin, S. and D. Ben-Avraham, "Diffusion in Disordered Media." *Advances in Physics*, 1987; **36**(695–798).

126. Fatin-Rouge, N., et al., "Diffusion and Partitioning of Solutes in Agarose Hydrogels: The Relative Influence of Electrostatic and Specific Interactions." *Journal of Physical Chemistry B*, 2003; **107**(44): 12126–12137.

127. Fatin-Rouge, N., K. Starchev, and J. Buffle, "Size Effects on Diffusion Processes within Agarose Gels." *Biophysical Journal*, 2004; **86**: 2710–2719.

128. Fatin-Rouge, N., et al., "A Global Approach of Diffusion in Agarose Gles from SANS and FCS Experiments." To be submitted.

129. Giddings, J.C., et al., "Statistical Theory for the Equilibrium Distribution of Rigid Molecules in Inert Porous Networks. Exclusion Chromatography." *Journal of Physical Chemistry*, 1968; **72**(13): 4397–4408.

130. Alexander, S. and R. Orbach, "Density of States on Fractals—Fractons. *Journal De Physique Lettres*, 1982; **43**(17): L625–L631.

131. Webman, I., "Effective-Medium Approximation for Diffusion on a Random Lattice." *Physical Review Letters*, 1981; **47**(21): 1496–1499.

132. Benavraham, D. and S. Havlin, "Diffusion on Percolation Clusters at Criticality." *Journal of Physics a-Mathematical and General*, 1982; **15**(12): L691–L697.

133. Gefen, Y., A. Aharony, and S. Alexander, "Anomalous Diffusion on Percolating Clusters." *Physical Review Letters*, 1983; **50**(1): 77–80.

134. Rammal, R. and G. Toulouse, "Random-Walks on Fractal Structures and Percolation Clusters." *Journal De Physique Lettres*, 1983; **44**(1): L13–L22.

135. Lacroix-Gueu, P., et al., "In Situ Measurements of Viral Particles Diffusion Inside Mucoid Biofilms." *Comptes Rendus Biologies*, 2005; **328**(12): 1065–1072.

136. Labille, J., N. Fatin-Rouge, and J. Buffle, "Local and Average Diffusion of Nanosolutes in Agarose Gel: The Effect of the Gel/Solution Interface Structure." *Langmuir*, 2007; **23**(4): 2083–2090.

137. Pluen, A., et al., "Diffusion of Macromolecules in Agarose Gels: Comparison of Linear and Globular Configurations." *Biophysical Journal*, 1999; **77**(1): 542–552.

138. Verkman, A.S., "Solute and Macromolecule Diffusion in Cellular Aqueous Compartments." *Trends in Biochemical Sciences*, 2002; **27**(1): 27–33.

139. Rinaudo, M., et al., "Electrostatic Interactions in Aqueous Solutions of Ionic Polysaccharides." *International Journal of Polymer Analysis and Characterization*, 1997; **4**(1): 57–69.

140. Labille, J., et al., "Flocculation of Colloidal Clay by Bacterial Polysaccharides: Effect of Macromolecule Charge and Structure." *Journal of Colloid and Interface Science*, 2005; **284**: 149–156.

141. Minoura, H. and H. Takekawa, "Observation of Number Concentrations of Atmospheric Aerosols Andanalysis of Nanoparticle Behavior at an Urban Background Area in Japan." *Atmospheric Environment*, 2005; **39**: 5806–5816.

142. Siegmann, K. and H.C. Siegmann, "Fast and Reliable 'In Situ' Evaluation of Particles and Their Surfaces with Special Reference to Diesel Exhaust." SAE Technical Paper 2000-01-1995, 2000, doi:10.4271/2000-01-1995.

143. Muller, J.-O., et al., "Diesel Engine Exhaust Emission: Oxidative Behavior and Microstructure of Black Smoke Soot Particulate." *Environmental Science & Technology*, 40(4): 1231–1236, 2006.

144. Jacobson, M.Z. and J.H. Seinfield, "Evolution of Nanoparticle Size and Mixing State Near the Point of Emission." *Atmospheric Environment*, 2004; **38**: 1839–1850.

145. Ishiguro, I., Y. Takitori, and K. Akihara, "Microstructure of Diesel Soot Particles Probed by Electron Microscopy: First Observation of Inner Core and Outer Shell." *Combust Flame*, 1997; **108**: 231–234.

CHAPTER 7

Environmental Transformations of Nanomaterials and Nano-Enabled Products

Clément Levard

*CEREGE UMR 7330 CNRS–Aix-Marseille University, Marseille, France, and
iCEINT CNRS–Duke University, Aix-en-Provence, France*

Mélanie Auffan

*CEREGE UMR 7330 CNRS–Aix-Marseille University, Marseille, France, and
iCEINT CNRS–Duke University, Aix-en-Provence, France*

Gregory V. Lowry

*Center for the Environmental Implications of Nanotechnology and Department of Civil & Environmental Engineering,
Carnegie Mellon University, Pittsburgh, Pennsylvania, USA*

Introduction

Assessing the risks of engineered nanomaterials (NMs) to the environment and to human health requires an understanding of the potential exposure routes and toxicological effects from acute and chronic exposures. Significant knowledge has been generated regarding the fate, transport, and toxic properties of pristine "as manufactured" engineered NMs. However, many NMs are highly reactive and will transform when released into the environment or upon entering an organism. These transformations will affect their fate, transport, and toxic properties, but proportionally less effort has gone into assessing the fate and effects of transformed NMs as compared with "as manufactured" materials. We require more knowledge of the types, rates, and extent of transformations expected for NMs in environmental and biological systems, and need to better assess the impact of those transformations on the fate, transport, and toxicity of NMs. Without such understanding, our conclusions about NM toxicity may be biased, and it will be challenging to develop and parameterize models for environmental fate and effects because we do not know the properties of the materials we intend to model, that is, the transformed NM.

This chapter presents the current understanding of NM transformations in biological and environmental media, and discusses the implications of these transformations on their fate and effects. It presents laboratory and field data on known NM transformations. It also presents research on NM transformations in environmentally realistic mesocosm studies and through case studies using real NM-enabled products in order to extend our understanding to complex environments and products.

Nanomaterial Chemical Transformations

One element of nanotechnology research is to correlate the properties of an NM with its behavior (e.g., reactivity or toxicity). For many applications of NMs this is a tractable goal because the NM can be isolated and protected from interactions with air or water, for example, photovoltaic NMs encapsulated in a solar panel. However, for environmental releases of nanomaterials, the interactions of the NM with the environment, for example, air, water, organic matter greatly complicates the ability to correlate the NM properties with its behavior because that NM will transform to a new material, often with very different properties than the parent NM. This is now a well-accepted reality for nano Environment, Health and Safety (EHS) research, and the recent research framework proposed by the U.S. National Research Council recommends that nano EHS research focus on identifying "critical elements of nanomaterial interactions" affecting exposure, hazards, and hence risks posed by engineered NMs [1]. This includes redox reactions, dissolution and ligation, and biologically mediated degradation of nanomaterials that ultimately lead to changes in the physical and chemical properties of NMs, and their persistence, bioavailability or biouptake, reactivity, and toxicity.

A variety of consequences resulting from redox transformations affect both exposure and toxicity. The oxidation of metal NMs can result in the formation of an oxide shell or enhanced dissolution. The oxidation or reduction of a metal oxide NM can alter the surface charge of the particle, or can change the crystal phase of the material (e.g., oxidation of magnetite to maghemite [2]). Oxidation may result in the accumulation of a relatively insoluble oxide surface coating on the nanoparticle (NP) that passivates the surface and reduces subsequent oxidation. This can lead to highly persistent core-shell NMs that may increase the potential for exposure due to their persistence. However, the oxide shell may also greatly decrease the release of toxic metal ions, or the redox activity of the particles, which can significantly decrease their toxicity [3–5]. The metal oxide phases formed on the NM exterior may have a high capacity for binding ions from solution. This can make them useful for applications that remove contaminants from water [6] but can also affect the bioavailability of the adsorbed toxic ions; that is, it may increase availability through the "Trojan horse" effect [7], or decrease availability by removing them from solution [8]. For many zerovalent metal NPs (e.g., Ag NPs or NZVI), the oxidation of the metal to a +I, +II, or +III state is required to obtain appreciable dissolution. For example, oxidation of Ag^0 to $Ag(I)$ is a prerequisite to dissolution and release of bactericidal Ag^+ [9].

This section reviews the known chemical transformations that NMs undergo in biological and environmental media, and describes how these transformations affect exposure and toxicity potential of engineered NMs.

Redox Reactions

Reduction and oxidation reactions (redox reactions) are key reactions in environmental and biological systems. Redox reactions involve the transfer of electrons to and from different chemical species. In a redox reaction one species is oxidized (loss of one or more electrons) and the other species is reduced (gain of one or more electrons).

$$
\begin{array}{lll}
M \rightarrow M^+ + e^- & E_1^\circ & \text{(oxidation)} \\
N^+ + e^- \rightarrow N & E_2^\circ & \text{(reduction)} \\
\hline
M + N^+ \rightarrow N + M^+ & E_1^\circ + E_2^\circ & \text{(net reaction)}
\end{array}
$$

Such reactions are the basis of energy conversion (e.g., oxidation of methane to CO_2 and H_2O to produce energy) and metabolism of all living things on the planet. Many NMs are made from elements that can exist in multiple oxidation states (e.g., Ce, Ag, Fe, Mn) and therefore can be redox active. Whether or not the metal atoms in NPs will become more reduced or more oxidized depends on the standard potential for the redox couple for the reaction between the NM and the compound of interest ($E_1^\circ + E_2^\circ$) after being adjusted for environmental conditions, that is, using the actual concentrations of the species rather than 1mol assumed for standard conditions. Thermodynamically, any positive value of ($E_1^\circ + E_2^\circ$) results in a negative Gibbs free energy and therefore is thermodynamically favorable. However, the availability of a suitable oxidant or reductant is necessary. Moreover, the rate of the transformation must be rapid enough to be relevant over the time scales of interest.

The redox state of the metal atoms in a crystalline NM (or at least at the surface) will depend on the crystal structure, redox conditions of the environment, the availability of oxidants and reductants, and the standard oxidation potential for the metal. The standard oxidation potentials for some of the most common metals used in engineered NMs is provided in Table 7.1.

Environmental redox conditions can swing from highly oxidizing (+0.7 V versus SHE) to highly reducing (−0.4 V versus SHE) at neutral pH depending on the environmental conditions. Clean surface waters and most aerated soils will be largely oxidizing environments, with the redox potential being controlled by the availability of oxygen. Molecular oxygen is a good oxidant, with a standard oxidation potential of +0.695 V (in water). Therefore, thermodynamically the metal atoms in many NMs will be

Metal	Potential Oxidation States	Standard Reduction Potential (V)
Au	0, +1	1.68
Pt	0, +2	1.2
Ag	0, +1	0.8
Fe	0, +2, +3	−0.04 [for Fe(III)] −0.41 [for Fe(II)]
Cu	0, +1, +2	0.34 [for Cu(II)] 0.18 [for Cu(I)]
Sn	+2, +4	−0.14 [for Sn(II)] 0.15 [for Sn(IV)]
Pb	0, +2	−0.13
Zn	+2	−0.76
Al	+3	−1.66

TABLE 7.1 Oxidation States Standard Reduction Potentials in Water versus Standard Hydrogen Electrode (SHE) for Solvated Metal Ions That Are Common in Many Engineered NMs

oxidized in these environments, at least at the surface. In carbon-rich subaquatic sediments and ground water the environment may be depleted of oxygen. In these "reducing" environments the redox potential is controlled chemical species other than oxygen, for example, HS^-. In these environments, the metal atoms in NM may be present in their reduced states. Some environments have dynamic redox conditions, for example, wetting and drying soils, and the potential for cycling of NMs between different redox states exists. The implications of these redox swings on NM fate and toxicity have been largely unexplored.

Many NMs have been shown to participate in redox reactions in aquatic and terrestrial environments. Metal NMs made from metal atoms with appropriate redox potentials (e.g., silver [9, 10], copper [11], and iron [12, 13]) become oxidized in oxicnatural waters. Nanoscale zerovalent iron (NZVI) is engineered to be oxidized by selected environmental contaminants such as chromium or chlorinated solvents [12]. However, they are also oxidized by water in anoxic waters resulting in H_2 generation [14]. Cerium oxide (ceria) NPs show particularly interesting redox behaviors in soils and water [15], and they can contain both Ce(III) and Ce(IV) in their structure simultaneously [16]. Macromolecules in the environment can lead to additional oxidation and reduction, altering the ratio of Ce(III) to Ce(IV) on the NP surfaces [16]. This indicates that in particular, the surface atoms on the cerium oxide NMs actively participate in redox reactions under normal environmental conditions. The Cu in CuO NMs can also undergo redox transformations in sediments. The speciation of copper between Cu(0), Cu(I), and Cu(II) compounds will depend on the redox conditions in the sediment [17].

In addition to the redox active metal elements in an NM, other elements may be susceptible to oxidation. For example, the sulfur and selenium in some metal sulfide and metal selenide NMs, major components of quantum dots, are susceptible to oxidation. The oxidation of reduced sulfur to elemental sulfur, sulfite, or sulfate, results in the release of soluble, toxic metal ions such as Cd^{2+} [18]. Such reactions cannot be overlooked in assessing the fate, persistence, and toxicity of NMs, especially in systems where oxidants are added, for example, ozone treatment of drinking waters, or when NMs are expected to move between anoxic and aerobic zones, for example, resuspension of subaquatic sediment from a river or lake.

Sunlight-catalyzed redox reactions (photo-oxidation and photo-reduction) may prove to be a very important transformation process affecting NM coatings, oxidation state, generation of reactive oxygen species (ROS), and persistence in the environment. This may be particularly important for carbonaceous NMs, such as fullerene and fullerene-like NMs [19]. Exposure of aqueous fullerene suspensions to sunlight resulted in the oxidation of carbon in the structure, functionalizing the surfaces with hydroxyl and carboxyl groups. Prolonged exposure to sunlight leads to partial mineralization of the fullerene to CO_2. Sunlight exposure can also degrade the polymeric coatings on the NM. For example, exposure to natural light caused the degradation of gum arabic coatings on Ag NPs and induced aggregation and sedimentation of the Ag NPs from solution [20].

Dissolution and Ligation

Two processes that greatly influence NM behavior and toxicity include dissolution and ligation. Dissolution is the process by which the bonds holding the metal atoms within the NM obtain sufficient energy to release those atoms into solution (often water). For many nanomaterials, they are small enough in size that the surface energy (pressure) is increased due to curvature and the resulting stress and strain on the crystal structure. This can sometimes result in enhanced dissolution relative to larger, "non-nano" (i.e., bulk) forms of the materials, or this higher energy can lead to a rearrangement of surface atoms that ultimately decreases the surface energy and thereby decrease the dissolution rate and extent. For Ag NMs, there was no apparent stress or strain on the NPs all the way down to 6 nm in diameter, however, the solubility increased with decreasing size due to the curvature (Kelvin effect) [21]. Ligation is the process whereby a ligand, such as sulfide or functional groups on natural organic matter, can bind with the NM surface to alter its properties. Both ligation and dissolution are affected by the solvent properties (e.g., presence of salt), and/or by the NM properties (e.g., size).

Dissolution and ligation are especially relevant for NMs made from soft metal cations (e.g., Ag, Zn, and Cu) because they form slightly to moderately soluble metal oxides, and because they often have a strong affinity for inorganic and organic sulfide ligands. Other ligands of environmental and biological importance include chloride and phosphate. For example, in the absence of sulfide, Ag NPs will oxidize and readily react with dissolved chloride to form $AgCl(s)/Ag^0$ core-shell particles [22–24]. The low solubility of the AgCl shell prevents further oxidation and dissolution of the Ag NPs [23]. In seawater, the Ag/Cl ratio is such that soluble $AgCl_x^{1-x}$ complexes form instead, increasing the solubility of the Ag NPs [23]. For ZnO NPs in a wastewater treatment plant, both sulfide and phosphate are important ligands. ZnO NPs entering the wastewater treatment plant are dissolved and transformed to ZnS, $Zn_3(PO_4)_2$, and Zn-ferrihydrite particles [25–27] due to the strong ligation of Zn by these competing ligands.

Dissolution will influence the persistence, exposure potential, and the toxicity of NMs. Dissolution will decrease the persistence of many NPs released to the environment. Based on modeling, the lifetime of Ag NPs in an oxic and oligotrophic subaquatic sediment is estimated to be several decades due to the dissolution of Ag NPs, but is anticipated to be centuries or more in a eutrophic lake where sulfidation of the Ag NPs decreases their solubility [28]. Since many NMs (especially Ag NPs, Zn-based NPs, and Cu-based NPs) express toxicity through dissolution and release of toxic cations, their dissolution may also serve to increase their toxicity. It has been demonstrated that ZnO and Ag NP dissolution increased its toxicity to plants in model soils, [29] and to microorganisms [30] A summary of the relevant dissolution and ligation processes are given in Figure 7.1. These are discussed in this section.

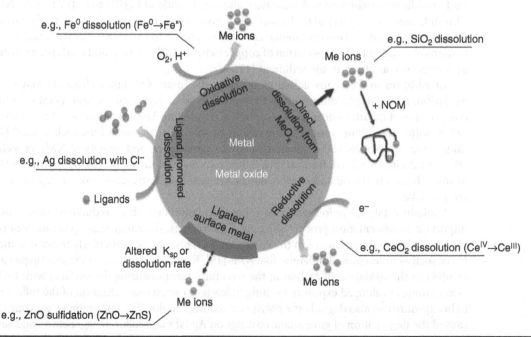

FIGURE 7.1 Schematic illustration of various metal (Me) and metal oxide (MeOx) NP dissolution and ligation processes.

Ligation

Nanomaterials have a high surface-to-volume ratio, making them highly surface active toward ligands in the environment. There is a myriad of inorganic (e.g., sulfide, chlorine, phosphate, carbonate) and organic ligands (e.g., natural organic matter, plant exudates and extracellular polymeric substances EPS, proteins) present in the environment. This combination of ligands and high surface active materials leads to a high rate of ligation for most NMs. Ligation of NMs by these species can affect the NM charge, hydrodynamic diameter, solubility, affinity for other surfaces, identification and uptake by organisms, toxicity, and overall fate in the environment. The NM ligands can generally fall into two classes: organic ligands, for example, macromolecules, and inorganic ligands, for example, phosphate. Each of these ligands is discussed.

Small organic macromolecules such as carboxylic acids and thiols can have a strong affinity for the NM surface. Similarly, larger organic macromolecules, for example, polymers used to stabilize nanomaterials against aggregation [31] or natural organic matter [32] can ligate NMs. This ligation can occur through specific chemical interactions with the metal or metal oxide surface as is the case with thiols and phosphate-based ligands, or with carboxylic acids groups on the organic molecules. This type of chemical reaction with the NM surface is often termed a *specific* interaction and results in a relatively strong covalent or electrostatic bond between the ligand and NM. In the absence of a specific interaction, there still can be association of the macromolecule with the NM surface. This can occur through nonspecific van der Walls interactions or through hydrophobic interaction between the NM and the macromolecule. In general, these interactions are weaker than for specific interactions. However, the strength of interaction tends to increase with increasing molecular weight of the macromolecule, and for large macromolecules (molecular weight >1000 g/mol) these interactions are often considered to be irreversible, or reversible on very long time scales [33].

The ligation of NMs by organic molecules can greatly affect properties of the NMs that are important determinants of fate and effects. The charge of an NM can be greatly influenced by the adsorption of ligands. The charge on an NM can increase when the macromolecule is a polyelectrolyte and therefore brings more charge to the NM surface. The charge of an NM can be reversed from positive to negative with the adsorption of a negatively charged macromolecule such as natural organic matter. However, the adsorption of an uncharged macromolecule or one with low charge density can decrease the electrophoretic mobility of an NM rather than increase it. This happens because the macromolecule increases the drag on an NM in the electric field more than its charge, thereby lowering its mobility in the electric field [32]. The presence of organic ligands can also affect the hydrodynamic diameter of an NM. This is most apparent for larger macromolecules on the NM surfaces because the large tails of the macromolecules extend from the NM, increasing drag on the particle and slowing its diffusion rate. The hydrodynamic diameter is determined from the diffusion rate in light-scattering measurements. Macromolecules and organic ligands can also affect the dissolution rate of the NM. For example, the effect of small organic acids such as oxalate and succinate affect the dissolution rate of iron oxides nanoparticles as ferrihydrite [34]. Another key property on NMs affected by the ligation by organic macromolecules is the affinity of the NM for other surfaces, also known as the attachment affinity. The adsorption of macromolecules and small molecules to NM surfaces give rise to electrostatic and steric repulsive forces that prevent attachment (aggregation and deposition) [35]. Finally, the ligation of NMs by organic ligands can alter the particle's "identity" and therefore uptake by organisms, bioavailability, and toxicity potential, as well as overall fate in the environment.

Ligation by Inorganic Ligands

Many metal and metal oxide nanoparticles are made from metal atoms that form strong complexes with common inorganic dissolved species such as phosphate, carbonate, and sulfide. The ability of these ligands to form complexes with both the free metal atoms or metal atoms at the NM surface can have several consequences. First, complexation of the released metal atoms can decrease the metal atom concentration in fluid surrounding the NM, leading to a greater driving force for dissolution and therefore a higher rate of dissolution. Second is the potential to form core-shell NMs with the outer shell being comprised of the insoluble metal-ligand complex, for example, metal sulfide shell formation for NMs made from class B soft metal cations (e.g., Ag, Cd, Pb, Cu). This shell formation can lead to a decrease in the rate of dissolution of the NMs, thereby increasing the persistence of the metal or metal oxide (core) [28].

Sulfidation of metal NMs made from chalcophilic metals are a special case of strong ligation by sulfide. The effect of such ligation on the solubility of the resulting NMs depends on the process of sulfidation. In some cases, partial dissolution and strong ligation of the metal may lead to the formation of a metal ligand shell on the particles (Figure 7.1), which can greatly decrease the rate of further oxidation and dissolution

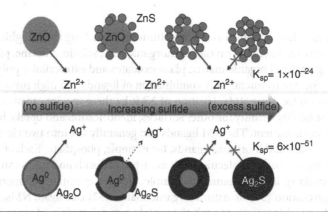

FIGURE 7.2   Potential sulfidation processes for Ag^0 and ZnO NPs, and resulting NP morphology and effect on aqueous solubility. Arrows indicate dissolution of the NP; the dashed arrow indicates dissolution from a partially sulfidized Ag NP with an incoherent sulfide shell, and crossed arrows indicate hindered dissolution due to the decreased solubility of Ag_2S and ZnS transformation products. (Reprinted with permission from [41].)

(e.g., formation of a Ag_2S shell on Ag NPs) [36]. Formation of this relatively insoluble metal sulfide shell on the Ag NP particle surface altered the surface charge and induced aggregation [36]. For ZnO, weaker ligation by sulfide or faster dissolution led to the precipitation of 4 to 5 nm ZnS NPs rather than the formation of a coherent shell (Figure 7.2) [37]. The effect of sulfide on the rate of dissolution of ZnO NPs was much less than for Ag NPs due to the formation of the incoherent shell on ZnO compared to the coherent shell formed on Ag NPs. These cases are end-member examples of shell formation, but the formation of a coherent metal sulfide shell will be dependent on the NM properties as well as the conditions for sulfidation. For example, a close observation of the sulfide shell on a sulfidized Ag NP from a wastewater treatment plant revealed that the shell was less coherent [38].

The formation of a metal sulfide shell (or metal phosphate or metal carbonate shell) can affect the dissolution rate and persistence of NMs. However, the overall stability of the metal-ligand NPs that form is not yet determined for most species. For example, the metal sulfides may dissolve if they are moved into an oxic environment (e.g., the water column). Thermodynamic predictions suggest this will be the case [39], but the rate at which this occurs has not been determined. ZnS and Zn adsorbed to ferrihydrite in wastewater treatment plant biosolids (with ZnO NPs as the primary source of Zn) were relatively labile during swings in redox conditions that occurred upon wetting and drying. There was more ZnS observed under anoxic conditions and more Zn adsorbed to ferrihydrite present under oxic conditions when the biosolids were cycled between these conditions [27]. The $Zn_3(PO_4)_2$ formed was relatively stable under all redox conditions. Additional work is needed to determine the particle properties (e.g., particle size, capping agent) and environmental conditions (redox state and availability of free sulfide or other ligands) that affect their dissolution and/or ligation rates [40].

Biologically Mediated Transformations

Many NMs have sizes similar to biological macromolecules, for example, proteins [42]. Or they are made from metals that are part of the normal biogeochemical cycling of the elements, for example, Fe or Cu. Therefore, biological transformations of NMs are highly likely to occur in living tissues (both intracellular and extracellular) and environmental media (e.g., soils). Redox reactions take place in the cytoplasm, cell wall, cell membrane, and extracellularly via redox-labile enzymes and cytochromes or through intracellular ROS production. Overall, intercellular ligation with sulfur species or acid-promoted dissolution appears to be the most commonly observed transformations of NMs *in vitro* and *in vivo*. There is limited characterization (but increasing) of the types of biological transformations that may occur within and by organisms, and the effect of these transformations on NM properties and toxicity [43]. Biologically mediated transformations of both the NM core, inorganic shell, or organic coatings are possible. These transformations can affect the behavior of the NMs including surface charge, aggregation state, and reactivity, which affects transport, bioavailability, and toxicity. Studies on biological transformations have primarily been conducted on cell lines *in vitro*, but also *in vivo* for bacteria and fungi, and plants. The current state of the art for transformations of NMs in each of these systems is discussed.

Living organisms and especially eukaryotic cells metabolism cannot support high pH variation. For most eukaryotic cells pH is buffered in a 7 to 8 pH range. In the case of Eh, oxidative phosphorylation in mitochondria is based on a series of redox reactions at near-circumneutral pH for which potentials are in a range of -0.32 (NAD+/NADH) to 0.29 V (cytochromes) where NAD = **Nicotinamide Adénine Dinucléotide**. Besides oxidative phosphorylation (respiration) extracellular Eh is generally controlled by thiol/disulfide redox systems (mainly GSH/GSSH and Cys/CySS) for which Eh vary in a range of -0.140 to -0.08 V. In such conditions many elements can be redox unstable that lead to electron exchange between NP surface and surrounding media. This could be the starting point of disequilibrium of the redox balance leading to toxic effect. In the specific case of CeO_2 NMs, several studies found that the toxicity was either exerted by direct contact with cells [44–47], membrane damages [48], cell disruption [46], release of free Ce(3+) [45, 46], or DNA damage [47, 49]. Even if free Ce(3+) is toxic, release of Ce(3+) from the CeO_2 nanoparticles did not explain by itself the toxicity observed in these studies. However, the reduction of the Ce(4+) into Ce(3+) at the surface of the CeO_2 nanoparticles correlates with the toxicity and oxidative stress observed. Using XANES at Ce L_3-edge, Thill et al. and Auffan et al. showed that the cytotoxicity/genotoxicity of CeO_2 could be related to the reduction of surface Ce(4+) atoms to Ce(3+) [44, 45, 50]. But, further research is needed to find out whether the oxidative activity of ceria could be responsible.

Another example regards iron-based NMs. In an effort to understand the long-term fate and effects of magnetic resonance imaging (MRI) contrast agents, Mejias et al. demonstrated that superparamagnetic dimercaptosuccinic acid–capped magnetite NPs are transformed to non-superparamagnetic iron oxides *in vitro* [51]. Similar behavior was also observed using different magnetite NPs and organisms, suggesting that this is a feature of the magnetite NPs regardless of capping agent [3, 52]. Biological transformation of NMs have been shown to occur in lysosomes where the intracellular pH can be low (pH = 4–5), thereby promoting transformation of NMs. Another study reported that the oxidation of iron in contact to bacteria can be at the origin of an oxidative stress. The potential of the redox couple characteristic of Fe^0 nanoparticles at pH = 7 is in a Eh $Fe^{2+}/Fe^0 = -0.40/-0.7$ V as a function of the Fe concentration which is lower than the typical Eh in biological media. This allowed the oxidation of $Fe^0(s)$ in contact with *E. coli*. This oxidation is directly responsible of toxic effects through the generation of an oxidative stress, as demonstrated by an enhanced toxicity of these Fe nanoparticles toward a mutant strain of *E. Coli* deprived of defense mechanism against oxidative stress [3]. Both microorganisms and fungi can promote the environmental transformations of NMs. The redox reactions between bacteria and naturally occurring, nanoscale iron oxide are well understood [54]. For example, bacteria such as *Geobacter* and *Shewanella* spp. can respire using nanoscale ferrihydrite as an electron acceptor, producing nanoscale magnetite.

CuO NPs are also readily dissolved in lysosomes at pH = 4.5, leading to toxicity from the release of Cu^{2+} ions, but intercellular sulfide species can act to transform CuO NPs to much less soluble CuS NPs [53]. However, intracellular H_2O_2 promoted the oxidative dissolution of the CuS NPs [53].

The biodegradation of carbon nanotubes by bacteria was confirmed using ^{14}C-labeled carbon nanotubes, and the factors influencing biodegradation of carbon nanotubes by bacteria have been reviewed [55]. Biotransformation of polymer coatings used on many NMs for biomedical applications is also likely to occur in the environment. Covalently bound polyethylene glycol (PEG) coatings on engineered NMs were shown to be bioavailable to microorganisms isolated from an urban stream [56]. This degradation of the coating led to aggregation of the NMs. Fungi can also transform a variety of NP types. This was first reported by Allen et al., who demonstrated the oxidation and carboxylation of CNTs by OH radicals produced from the horseradish peroxidase enzyme [57]. This oxidation increased the surface charge of the CNTs and stability against aggregation while decreasing hydrophobicity. It also decreased the toxicity potential of CNTs [58]. More recently, Khan and Ahmad showed that fungi could transform 150- to 200-nm TiO_2 (anatase) to 5- to 28-nm TiO_2 (brookite), leading to disaggregation of the initial TiO_2 particles [37]. Fungi were also shown to promote the crystallization of amorphous silica into crystalline SiO_2 NPs [59]. Thus, biotransformation of NPs by bacteria and fungi can lead to phase changes, oxidation, and alteration of their coatings.

A commonly invoked exposure pathway for NMs is the use of biosolids from wastewater treatment used on crop lands. This has led to a number of recent studies on the transformation of NMs in wastewater treatment plants and on the uptake and transformation of NMs by different plant species. For instance, two varieties of nanoceria, pristine and citrate-functionalized, were followed in an aerobic bioreactor simulating wastewater treatment by conventionally activated sludge [60]. This study indicates that the majority of nanoceria (>90%) was associated with the solid phase where a reduction of the Ce(IV) NPs to Ce(III) occurred. After 5 weeks in the reactor, 44% ± 4% reduction was observed for the pristine nanoceria and 31% ± 3% for the citrate-functionalized nanoceria, illustrating surface functionality dependence.

The authors suggest that the likely Ce(III) phase generated would be Ce_2S_3. At maximum, 10% of the CeO_2 will remain in the effluent and be discharged as CeO_2, a Ce(IV) phase.

Once in contact with plant, common transformation mechanisms include dissolution and various crystal phase transformations within the plant roots, shoots, and leaves for some nanomaterials. For example, *Elsholtzia splendens* took up CuO NPs largely intact [61]. This is likely due to the relative low solubility of CuO NPs in soil at neutral pH. Conversely, Zhang et al. demonstrated uptake and transformation of CeO_2 NPs to $CePO_4$ and Ce carboxylates in cucumber plants despite the very low solubility of CeO_2 [62]. However, CeO_2 NPs were not transformed in cilantro after being taken up [63]. For La_2O_3 NP uptake into cucumber plants, the La_2O_3 particles were dissolved and transformed to $LaPO_4$ needle-like nanoclusters. Recently, Stegemeier and coworkers demonstrated that Ag_2S was taken up by alfalfa plants differently that for Ag^0, predominantly due to differences in solubility between the particle types [64]. These studies indicate that dissolution and phase transformation of NMs in plants can occur, but no general trends in NP types or plant types and their ability to be transformed have emerged.

Biological transformations of both carbon-based and metal oxides NMs, as well as their organic coatings, may ultimately act to attenuate their concentrations in the environment. However, there are too few studies at this point in time to make general conclusions about the types and rates of these processes and their importance in real systems. More research is needed to fully catalog the nature and extent of biological transformations expected in the environment, but doing so is vital to assessing the persistence and long-term risks that these materials may pose to the environment.

Mesocosms: Evaluating Nanomaterial Transformations in Complex Environmental Systems

The complexity of natural systems and the unpredictable behaviors of many nanomaterials in those systems makes predicting NM fate challenging. Therefore, transformations and long-term fate of engineered NMs must be measured in realistic complex natural systems to accurately assess the risks that they may pose. Not intended to be an exhaustive list of results, this section presents two studies performed in freshwater mesocosms highlighting the chemical transformations of two redox-sensitive NPs (Ag and CeO_2) in complex and realistic environmental systems.

Ag NPs Oxidation in a Simulated Large-Scale Freshwater Emergent Wetland

In 2012, Lowry et al. were the first to report the long-term environmental fate and transformation of a commercially available Ag NPs applied to a large-scale mesocosm simulating a freshwater emergent wetland [65]. These mesocosms were designed to allow for the natural processes affecting Ag NP behavior on a scale large enough to capture the complexity and dynamic nature of real environments.

These mesocosms have a sloped bed that allows for the existence of both an aquatic/subaquatic environment and a terrestrial environment. The terrestrial environment had an unsaturated vadose zone and was largely oxic, whereas the subaquatic sediment was consistently flooded and predominantly anoxic. The Ag NPs dosed in these mesocosms were reported by the manufacturer to be 10 nm in diameter with a Poly Vinyl Pyrolidone (PVP)coating (10,000 g/mol). The mesocosms were dosed in either the terrestrial compartment (an aqueous suspension containing 4.2 g Ag NPs containing approximately 2.9 g Ag was sprayed onto the soil) or the aquatic compartment (the same mass of Ag NPs in aqueous suspension was sprayed onto the water surface to provide approximately 25 mg/L Ag NPs containing 16.6 mg/L Ag in the water column). The speciation of Ag present in the mesocosms was determined using X-ray absorption spectroscopy (XAS) at the Ag K-edge (Figure 7.3).

Eighteen months after dosing, the Ag NPs added to the system were partially oxidized and sulfidized, with the degree of sulfidation dependent on the compartment where they resided. Ag NPs recovered from terrestrial soils contained 52% ± 1% Ag_2S (acanthite), and 47% ± 1% Ag^0 (Figure 7.3). Ag NPs recovered from the subaquatic sediment were more oxidized than those added to the terrestrial compartment; ~18% of the original Ag^0 character remained after aging. Oxidized silver species were more variable than for the terrestrial case and included a match to reference spectra for $Ag_2S = 55\% \pm 6\%$ and Ag-cysteine = 27% ± 6% (Figure 7.3) along with the initial Ag^0. While it is unlikely that Ag existed solely as Ag-cysteine in the samples, the need to incorporate the reference spectra for this compound suggests that a proportion of the silver was coordinated to sulfhydryl-containing organic ligands, likely natural organic matter.

The presence of Ag_2S and Ag-sulfhydryl species (e.g., Ag complexed with reduced S in organic matter) as transformation by-products is consistent with the fact that these species are known to be the most thermodynamically stable, low temperature, geochemical phases for monovalent Ag. Moreover, the

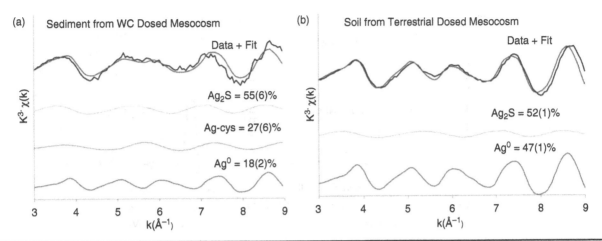

Figure 7.3. Speciation of Ag NPs aged in the mesocosms for 18 months determined from EXAFS analysis and linear combination fitting with spectra from reference materials (Ag$_2$S, Ag-cysteine) complexes, and Ag^0 foil. (a) Determined in terrestrial soils for Ag NPs dosed onto the terrestrial soils. (b) For Ag NPs in subaqueous surficial sediment taken from the mesocosm that had been dosed with Ag NPs in the water column. (Adapted from [65].)

lower extent of oxidation and sulfidation of Ag NPs aged in the terrestrial environment was consistent with the drier and more oxic conditions in this compartment compared to sediments, which were always anoxic. There was also less acid volatile sulfide (AVS) in the terrestrial environment available to the particles.

Lowry et al. [65] found that the level of sulfidation in Ag NPs in terrestrial soils was greater than might be expected for purely oxic soils, but likely occurred during periods of prolonged rainfall when the terrestrial soils were flooded for extended periods, allowing the soils to go anoxic or anaerobic through microbial action, resulting in partial sulfidation of the Ag NPs. These results suggest that Ag NPs that are continuously exposed to anoxic conditions and submerged in sediment will be more quickly oxidized and sulfidized than those residing in relatively dryer and oxic terrestrial environments. Moreover, they observed that the transformation of Ag NPs to Ag$_2$S remained incomplete even after 18 months. Thus, sulfidation occurred much slower than had been observed in laboratory studies using Na$_2$S as the source of sulfide [66]. The slower sulfidation of the Ag NPs in the sediments compared to laboratory sulfidation may be a result of the relatively low AVS/Ag molar ratio (0.5:1 in the terrestrial soil and sediment) and the presence of other metals as Fe that would compete for the sulfide. The slower sulfidation could also be due to a limited amount of oxidant (e.g., dissolved oxygen) that would be required to oxidize the Ag NPs during the sulfidation process [67]. It was also noted in the study that the Ag NPs used in that mesocosm study consisted of large (hydrodynamic diameters of several hundreds of nanometers) aggregates when they were dosed into the mesocosm. This aggregation may have also slowed the rate of oxidation and sulfidation. Subsequent studies in similar mesocosms using highly dispersed Ag NPs were completely sulfidized in less than a month.

CeO$_2$ NPs Reduction in Aquatic Mesocosms Simulating a Pond Ecosystem

In 2014, Tella et al. determined the transfer and redox transformation of bare CeO$_2$ NPs (pristine NPs) and citrate-coated CeO$_2$ NPs (nanocomposite used in wood stain) exposed in indoor mesocosms to organisms part of a pond trophic system [68, 69]. Contrary to the mesocosm study described above where a pulse of Ag NPs was introduced, this exposure represented a long-term and multiple dosing exposure scenario (10 × 100 µg/L of CeO$_2$) that is more consistent with expectations in natural systems with continuous low-level inputs of NPs.

Seven mesocosms (750 x 200 x 600 mm) were filled with artificial sediment and 46 L of water with pH and conductivity close to the natural pond water [pH = 7, 11.7°C, 413 µS cm^{-2}, (O$_2$) = 9.5 mg L^{-1}]. The sediment was largely anoxic, while the water column remained oxic during the experiment. The CeO$_2$ NPs dosed in these mesocosms were reported by the manufacturer to be crystallites of cerianite (3–4 nm) with average hydrodynamic diameters in the stock suspension of ~8 nm. The citrate-coated CeO$_2$ NPs stock suspension consist of 2.2 × 10^5 mg L^{-1} of Ce in equilibrium with 5.1 ± 0.3 × 10^3 mg L^{-1} of citrate [70]. The speciation of Ce present in the mesocosms was determined using XAS at the Ce L$_3$-edge.

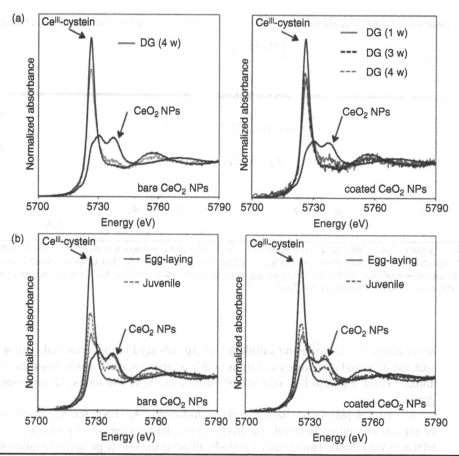

FIGURE 7.4 Ce L_3-edge XANES spectra in *P. corneus*. (a) Digestive gland from adult *P. corneus* exposed to bare and coated CeO$_2$ NPs at different periods of time in mesocosms. (b) Egg-laying (sampled after 1 week) and juvenile (sampled after 4 weeks). (1 w = 1 week; 3 w = 3 weeks; 4 w = 4 weeks.)

Over time, CeO$_2$ NPs tend to accumulate on the surficial sediment. After 4 weeks Ce was observed in the digestive gland of adult mollusks (*Planorbarius corneus*), and also associated with juvenile and egg-laying organisms. Adult digestive glands, whole juveniles, and egg-laying (eggs + organic matrix) were analyzed by XANES (Figure 7.4). Both bare and coated CeO$_2$ NPs are initially composed by Ce(IV). This was confirmed by their XANES spectra characterized by two absorption peaks at 5730.0 eV and 5738.0 eV while Ce(III) reference compounds only present one absorption edge at 5726.2 eV (Figure 7.4). Linear combination fit of XANES spectra indicated that in the surficial sediments, Ce(IV) was the major phase present after 4 weeks with minor contribution of Ce(III). However, 79% ± 15% and 77% ± 15% of Ce(IV) was in the form of Ce(III) in the digestive gland of adult organisms after 4 weeks for coated and bare CeO$_2$ NPs, respectively. This reduction into Ce(III) occurred less than a week after exposure. For juvenile and egg-laying *P. corneus*, Ce(IV) reduction was lower. Only 47% ± 7% of the Ce interacting with juveniles was reduced into Ce(III) after 4 weeks and 22% ± 1% of the Ce was reduced into Ce(III) in egg-laying ones. No significant difference in Ce speciation was observed as a function of the type of NPs. Based on the strong aggregation and sedimentation of the NPs after 4 weeks, on the habitat and feeding regime of the snail in the mesocosms, and on the slight Ce(IV) reduction at the surface of the sediment, they hypothesized that the strong Ce(III) percentage observed in the digestive gland occurred after ingestion by snails [68, 69].

Case Study on the Chemical Transformation of Ag NPs: From Socks to Crops

In addition to measuring NM transformations in realistic environmental systems, another gap is to assess NM transformations during the entire life cycle of the NMs as they may be transformed well before entering the environment. The case study presented below illustrates the importance of considering all the different stages from manufacturing of the NMs up to the release of the NMs in the environment.

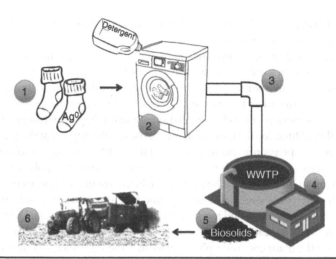

FIGURE 7.5   From socks to crops: different steps that have to be considered when assessing the transformation of Ag NPs initially present in textile and released into the environment upon washing.

Among the NPs that may undergo significant chemical transformations, the case of Ag NPs is of particular interest for the following reasons: (1) It is incorporated in many consumer products. (2) Ag was shown to be released from Ag-containing products in large amount during their use. (3) Ag is the focus of a vast amount of studies because of its known ecotoxicity to a variety of organisms in its pristine (metallic Ag) form. (4) Ag is redox sensitive and will therefore not be stable under most uses and environmental conditions [39]. Once oxidized, Ag strongly reacts with both organic and inorganic ligands that are ubiquitous in the environment as discussed earlier in this chapter (O_2, chloride, sulfurs).

The use of Ag NPs in textile for antimicrobial and antiodor functionality is one of the most often mentioned uses of Ag NPs. A substantial amount of research literature has been published on this topic revealing the importance of characterizing the speciation of Ag during the lifetime of the product and after release [71–75]. In this regard, we have chosen this specific scenario to illustrate why considering environmental transformations is essential for accurately predicting the fate of NPs in the environment. This will include transformations of Ag within the textile, during the release upon washing, in the sewer system and in the wastewater treatment plant (WWTP) that are believed to be a sink for manufactured NPs and finally in the environment via the use of digested sludge as fertilizer (Figure 7.5).

Ag NP

Transformations During the Use of Textiles [Figure 7.5 (1)]

During the use of the Ag-containing textile, three different processes can potentially cause NP corrosion. The first one consists in air corrosion during the use of the textile or when being stored between uses. To the best of our knowledge, no studies have focused on the transformation of Ag NPs within textiles when exposed to air only. However, slight sulfidation (also call tarnishing) has been suspected and observed [72]. The two other processes involve water-induced corrosion due to sweating or during washing cycles. Hedberg et al. [72] have investigated the release of ionic silver from Ag NPs exposed in a synthetic sweat media, but characterization of the Ag speciation was not explored. Transformations and release behavior during washing is strongly dependent on the initial Ag-bearing textile and the compositions of the detergents. The speciation of Ag after washing varies depending upon the different detergent products used. As expected, based on the detergent composition and the high affinity of Ag(I) for halides, Ag partially transforms into AgCl in most cases [71, 75]. As an example, Impellitteri et al. have shown that a significant fraction (up to 50%) of the metallic silver present in Ag NP–containing socks was converted to AgCl when exposed to an hypochlorite/detergent solution [71]. The extent of transformation reported and the nature of the transformed products differ significantly between studies. Such discrepancies are certainly due to the high variability in the physico-chemical properties of the initial Ag that was used to manufacture the textile.

Speciation and Quantification of Released Ag upon Washing [Figure 7.5 (2)]

Few studies have investigated the release of Ag from commercially available textiles under relevant conditions. As an example, Geranio et al. have shown that in most cases, more than 75% of the Ag in the textile was released as particles bigger than 450 nm with only 5% to 15% of smaller particles and very low amount of dissolved species were measured [73].

Recently, Hedberg et al. investigated a more realistic exposure scenario [72]. They studied Ag release in sequential contact with synthetic sweat, laundry detergent, and freshwater, simulating a possible transport path to the environment. Interestingly they found that the amount of Ag released in the sequential exposure was diminished by a factor of 2 compared to the sum of each separate exposure.

Release amount of Ag and its speciation are strongly product-dependent. This can easily be explained by the different functionalization strategies of the textiles applied at the industrial scale (dipping the product in an Ag NPs suspension *versus* immersion of the textile in an Ag salt solution followed by *in situ* reduction). Other manufacturing methods incorporate silver wires into the polymer fabric (X-static). The methodologies of incorporating Ag or Ag NPs into the fabrics will in part govern the lifetime of the Ag within the textile. Such differences could also be explained by the initial Ag state in the textile. Unfortunately, this information is proprietary and rarely reported by the manufacturers. Despite the complexity of the different systems, it clearly appears that Ag is released in wastewaters upon washing. Partial transformation into AgCl seems to be the main corrosion phenomena that the Ag undergoes during washing although a significant fraction of metallic Ag seems to persist and trace amounts of Ag_3PO_4 and Ag_2S were also observed.

Ag Speciation in Sewers [Figure 7.5 (3)]

Urban wastewater systems are expected to be a major sink for nanoparticles released from textiles. As discussed above a significant fraction of Ag contained in textile is released upon washing and will ultimately end up in the sewer systems. To the best of our knowledge only one study has focused on the fate of Ag NPs in sewers before it reaches WWTP. Kaegi et al. have shown that the Ag NPs were efficiently transported to the WWTP without substantial losses in mass. However, they concluded that Ag NP discharged to the wastewater stream will become sulfidized to various degrees in the sewer system before reaching WWTP [38].

Wastewater Treatment Plant [Figure 7.5 (4)]

The WWTP is also an important stage to consider when assessing the transformations of Ag NPs on its way to the environment. Given the high affinity of Ag for sulfur, any unsulfidized Ag entering the WWTP is expected to become sulfidized in those facilities. In particular, plant operation such as anaerobic digesters where lot of sulfide is produced by microorganisms. The finding of Ag_2S nanoparticles in the final sewage sludge of a full-scale WWTP confirmed that Ag NPs are likely to become fully sulfidized in WWTPs [76]. Kaegi et al. experimentally showed almost complete sulfidation of spiked Ag NPs in a pilot WWTP. The facility used in this study was fed with municipal wastewater and consisted in a nonaerated and an aerated tank with a secondary clarifier [77]. Additional batch experiments suggested that sulfidation occurred in the nonaerated tank within less than 2 hour. Sulfidation was found to be size dependent. They have also shown that Ag NPs are efficiently retained by activated sludge, irrespective of size and coating (>98.9% in activated-sludge batch experiment) [38]. The same lines of evidence were shown by Kent et al. who observed sulfidation almost exclusively in anaerobic zones of the WWTP for particles fabricated by nanosphere lithography and placed in both aerated and nonaerated tanks [78]. The sulfidation rate in the nonaerated tank was about 11 to 14 nm of Ag transformed into Ag_2S per day. Almost complete sulfidation was observed for samples after 10 days. Similar results were obtained for NPs that were added to the influent of a pilot WWTP [27]. Ag speciation performed on the biosolids shows that Ag was entirely transformed into Ag_2S.

3.5 Sludge Processing [Figure 7.5 (5)]

Waste sludge from the WWTP anaerobic digester undergoes a series of treatments before it is applied to agricultural fields. The sludges are typically dewatered and dried to 20 to 50 percent solids by weight content. The dewatered biosolids are then further dried, subjected to lime addition and heat treatment, and/or composted. These different processes can potentially affect Ag speciation because of the change of the redox conditions and pH. Although Ag_2S is only very poorly water soluble ($Ksp = 10^{-51}$), the behavior of nano-Ag_2S or amorphous Ag_2S as well as Ag bound to thiols in contact with bacteria and Organic Matter (OM) is unknown.

Final sewage sludge is usually composted prior to their use. This involves wetting and drying cycles and resulting swings in redox conditions. The effects of composting (over 3 weeks) with different redox conditions and moisture content, as well as the effect of lime addition and heating, were investigated in laboratory reactors. Both treatments did not affect the speciation of Ag confirming the very high stability of Ag_2S even in oxic environment. Similarly, others have shown that Ag_2S formed in sewage sludge resists oxidation even after 6 months of simulated stockpiling and composting [79]. These authors have also shown that Ag sulfidized regardless of the initial form of Ag ($AgNO_3$, Ag NPs, AgCl NPs), or coating for the pristine Ag NPs (citrate, polyvinyl sulfonate, and mercaptosuccinic acid).

3.6 Land Application and Bioavailability [Figure 7.5 (6)]

Behavior of Ag NPs in soil pore water (dissolution and aggregation) was investigated in the context of sewage sludge amendment to soils [80]. Interestingly, when particles were aged in sewage sludge before incubation in soils, initial coating does not affect their behavior in soils. As expected, silver sulfide is the main transformed product. Conversely, when the particles are incubated without sewage sludge, coating (citrate *versus* PVP) was shown to strongly affect the behavior of Ag NPs in term of aggregation state and partitioning to pore water.

Ag_2S NP bioavailability resulting from the sulfidation of Ag NPs has been poorly studied compared to Ag NPs bioavailability although it seems much more environmentally relevant considering the transformations that Ag NPs will undergo before reaching environmental compartments.

Conclusions and Identified Gaps

Ag NPs used in textiles start to transform very early in their life cycle. The extent and nature of the initial transformations upon washing are strongly product dependent because of the different strategies used to incorporate the Ag in the textiles but also because of the variety of the initial Ag form used. The story seems to be much more consistent once the particles are released into the urban sewer systems. All studies agree on the fact that Ag NPs will undergo rapid sulfidation. Such transformation occurs regardless of initial size and coating. Sulfidation is a key transformation that strongly affects Ag solubility and toxicity as discussed earlier. Silver sulfide is probably the most relevant form of silver to consider assessing Ag ecotoxicity for this specific scenario. Strangely, the majority of toxicology and ecotoxicity studies focus on the pristine material rather than the transformed product. Other studies have looked at the toxicity of Ag leached from socks and have shown that the leachate, mainly consisting of dissolved Ag species, was highly toxic to zebrafish embryos [81]. Even though efforts are produced to better mimic realistic exposure scenarios, important gaps still need to be filled to better predict the overall environmental risk that Ag represents.

- Sequential studies along the product value chain are needed to assess the behavior of Ag NPs from textile manufacturing to the soil for a more realistic understanding of Ag fate in the environment. An interesting approach was proposed by Hedberg et al. [72] but should be extended to capture the entire life cycle. Such studies will likely have to also consider multiple-use patterns and exposure scenarios to capture the breadth of uses for these products, for example, soaps versus textiles versus water filters.

- Ecotoxicity studies should consider more the sulfidized form of Ag rather than the pristine Ag NPs to account for transformations that consistently occur before reaching the environment.

- Another important gap is the long-term stability of the sulfidized product. It seems that part of the sulfidized Ag is amorphous or bound to natural organic matter through interaction with thiol groups. One can hypothesize that the mobility and bioavailability of these transformed phases will be enhanced compared to crystalline silver sulfide.

References

1. National Research Council, *Research Progress on Environmental, Health, and Safety Aspects of Engineered Nanomaterials*; National Research Council, 2013.
2. Reinsch, B. C.; Forsberg, B.; Penn, R. L.; Kim, C. S.; Lowry, G. V., Chemical Transformations during Aging of Zerovalent Iron Nanoparticles in the Presence of Common Groundwater Dissolved Constituents. *Environmental Science & Technology* 2010,*44*, (9), 3455–3461.
3. Auffan, M.; Achouak, W.; Rose, J.; Roncato, M.-A.; Chanéac, C.; Waite, D. T.; Masion, A.; Woicik, J. C.; Wiesner, M. R.; Bottero, J.-Y., Relation between the Redox State of Iron-Based Nanoparticles and Their Cytotoxicity toward Escherichia coli. *Environmental Science & Technology* 2008,*42*, (17), 6730–6735.
4. Nel, A.; Xia, T.; Meng, H.; Wang, X.; Lin, S.; Ji, Z.; Zhang, H., Nanomaterial Toxicity Testing in the 21st Century: Use of a Predictive Toxicological Approach and High-Throughput Screening. *Accounts of Chemical Research* 2013,*46*, (3), 607–621.

5. Phenrat, T.; Long, T. C.; Lowry, G. V.; Veronesi, B., Partial Oxidation ("Aging") and Surface Modification Decrease the Toxicity of Nanosized Zerovalent Iron. *Environmental Science & Technology* 2009,*43*, (1), 195–200.

6. Yavuz, C. T.; Mayo, J. T.; Yu, W. W.; Prakash, A.; Falkner, J. C.; Yean, S.; Cong, L.; et al., Low-Field Magnetic Separation of Monodisperse Fe3O4 Nanocrystals. *Science* 2006,*314*, (5801), 964–967.

7. Studer, A. M.; Limbach, L. K.; Van Duc, L.; Krumeich, F.; Athanassiou, E. K.; Gerber, L. C.; Moch, H.; Stark, W. J., Nanoparticle Cytotoxicity Depends on Intracellular Solubility: Comparison of Stabilized Copper Metal and Degradable Copper Oxide Nanoparticles. *Toxicology Letters* 2010,*197*, (3), 169–174.

8. Plathe, K. L.; von der Kammer, F.; Hassellov, M.; Moore, J. N.; Murayama, M.; Hofmann, T.; Hochella, M. F., The Role of Nanominerals and Mineral Nanoparticles in the Transport of Toxic Trace Metals: Field-Flow Fractionation and Analytical TEM Analyses after Nanoparticle Isolation and Density Separation. *Geochimica Et Cosmochimica Acta* 2013,*102*, 213–225.

9. Lok, C. N.; Ho, C. M.; Chen, R.; He, Q. Y.; Yu, W. Y.; Sun, H.; Tam, P. K. H.; Chiu, J. F.; Che, C. M., Silver Nanoparticles: Partial Oxidation and Antibacterial Activities. *Journal of Biological Inorganic Chemistry* 2007,*12*, (4), 527–534.

10. Henglein, A., Colloidal Silver Nanoparticles: Photochemical Preparation and Interaction with O-2, CCl4, and Some Metal Ions. *Chemistry of Materials* 1998,*10*, (1), 444–450.

11. Mudunkotuwa, I. A.; Pettibone, J. M.; Grassian, V. H., Environmental Implications of Nanoparticle Aging in the Processing and Fate of Copper-Based Nanomaterials. *Environmental Science & Technology* 2012,*46*, (13), 7001–7010.

12. Liu, Y. Q.; Majetich, S. A.; Tilton, R. D.; Sholl, D. S.; Lowry, G. V., TCE Dechlorination Rates, Pathways, and Efficiency of Nanoscale Iron Particles with Different Properties. *Environmental Science & Technology* 2005,*39*, (5), 1338–1345.

13. Auffan, M.; Achouak, W.; Rose, J.; Roncato, M.-A.; Chaneac, C.; Waite, D. T.; Masion, A.; Woicik, J. C.; Wiesner, M. R.; Bottero, J.-Y., Relation between the Redox State of Iron-Based Nanoparticles and Their Cytotoxicity toward *Escherichia coli*. *Environmental Science & Technology* 2008,*42*, (17), 6730–6735.

14. Liu, Y.; Phenrat, T.; Lowry, G. V., Effect of TCE Concentration and Dissolved Groundwater Solutes on NZVI-Promoted TCE Dechlorination and H2 Evolution. *Environmental Science & Technology* 2007,*41*, (22), 7881–7887.

15. Karakoti, A.; Singh, S.; Dowding, J. M.; Seal, S.; Self, W. T., Redox-Active Radical Scavenging Nanomaterials. *Chemical Society Reviews* 2010,*39*, (11), 4422–4432.

16. Baalousha, M.; Le Coustumer, P.; Jones, I.; Lead, J. R., Characterisation of Structural and Surface Speciation of Representative Commercially Available Cerium Oxide Nanoparticles. *Environmental Chemistry* 2010,*7*, (4), 377–385.

17. Ma, R.; Stegemeier, J.; Levard, C.; Dale, J. G.; Noack, C. W.; Yang, T.; Brown, G. E.; Lowry, G. V., Sulfidation of Copper Oxide Nanoparticles and Properties of Resulting Copper Sulfide. *Environmental Science: Nano* 2014,*1*, (4), 347–357.

18. Derfus, A. M.; Chan, W. C. W.; Bhatia, S. N., Probing the Cytotoxicity of Semiconductor Quantum Dots. *Nano Letters* 2004,*4*, (1), 11–18.

19. Hou, W. C.; Jafvert, C. T., Photochemical Transformation of Aqueous C(60) Clusters in Sunlight. *Environmental Science & Technology* 2009,*43*, (2), 362–367.

20. Cheng, Y. W.; Yin, L. Y.; Lin, S. H.; Wiesner, M.; Bernhardt, E.; Liu, J., Toxicity Reduction of Polymer-Stabilized Silver Nanoparticles by Sunlight. *Journal of Physical Chemistry C* 2011,*115*, (11), 4425–4432.

21. Ma, R.; Levard, C.; Marinakos, S. M.; Cheng, Y. W.; Liu, J.; Michel, F. M.; Brown, G. E.; Lowry, G. V., Size-Controlled Dissolution of Organic-Coated Silver Nanoparticles. *Environmental Science & Technology* 2012,*46*, (2), 752–759.

22. Levard, C.; Michel, F. M.; Wang, Y. G.; Choi, Y.; Eng, P.; Brown, G. E., Probing Ag Nanoparticle Surface Oxidation in Contact with (In)Organics: An X-Ray Scattering and Fluorescence Yield Approach. *Journal of Synchrotron Radiation* 2011,*18*, (6), 871–878.

23. Levard, C.; Mitra, S.; Yang, T.; Jew, A. D.; Badireddy, A. R.; Lowry, G. V.; Brown, G. E., Effect of Chloride on the Dissolution Rate of Silver Nanoparticles and Toxicity to *E. coli*. *Environmental Science & Technology* 2013,*47*, (11), 5738–5745.

24. Li, X. A.; Lenhart, J. J.; Walker, H. W., Dissolution-Accompanied Aggregation Kinetics of Silver Nanoparticles. *Langmuir* 2010,*26*, (22), 16690–16698.

25. Lombi, E.; Donner, E.; Taheri, S.; Tavakkoli, E.; Jamting, A. K.; McClure, S.; Naidu, R.; Miller, B. W.; Scheckel, K. G.; Vasilev, K., Transformation of Four Silver/Silver Chloride Nanoparticles During Anaerobic Treatment of Wastewater and Post-processing of Sewage Sludge. *Environmental Pollution* 2013,*176*, 193–197.

26. Lombi, E.; Donner, E.; Tavakkoli, E.; Turney, T. W.; Naidu, R.; Miller, B. W.; Scheckel, K. G., Fate of Zinc Oxide Nanoparticles during Anaerobic Digestion of Wastewater and Post-treatment Processing of Sewage Sludge. *Environmental Science & Technology* 2012,*46*, (16), 9089–9096.

27. Ma, R.; Levard, C.; Judy, J. D.; Unrine, J. M.; Durenkamp, M.; Martin, B.; Jefferson, B.; Lowry, G. V., Fate of Zinc Oxide and Silver Nanoparticles in a Pilot Wastewater Treatment Plant and in Processed Biosolids. *Environmental Science & Technology* 2014,*48*, (1), 104–112.

28. Dale, A. L.; Lowry, G. V.; Casman, E. A., Modeling Nanosilver Transformations in Freshwater Sediments. *Environmental Science & Technology* 2013,*47*, (22), 12920–12928.

29. Priester, J. H.; Ge, Y.; Mielke, R. E.; Horst, A. M.; Moritz, S. C.; Espinosa, K.; Gelb, J.; et al., Soybean Susceptibility to Manufactured Nanomaterials with Evidence for Food Quality and Soil Fertility Interruption. *Proceedings of the National Academy of Sciences of the United States of America* 2012,*109*, (37), E2451–E2456.

30. Xiu, Z.-m.; Zhang, Q.-b.; Puppala, H. L.; Colvin, V. L.; Alvarez, P. J. J., Negligible Particle-Specific Antibacterial Activity of Silver Nanoparticles. *Nano Letters* 2012,*12*, (8), 4271–4275.

31. Phenrat, T.; Song, J. E.; Cisneros, C. M.; Schoenfelder, D. P.; Tilton, R. D.; Lowry, G. V., Estimating Attachment of Nano- and Submicrometer-Particles Coated with Organic Macromolecules in Porous Media: Development of an Empirical Model. *Environmental Science & Technology* 2010,*44*, (12), 4531–4538.

32. Louie, S. M.; Spielman-Sun, E. R.; Small, M. J.; Tilton, R. D.; Lowry, G. V., Correlation of the Physicochemical Properties of Natural Organic Matter Samples from Different Sources to Their Effects on Gold Nanoparticle Aggregation in Monovalent Electrolyte. *Environmental Science & Technology* 2015,*49*, (4), 2188–2198.

33. Kim, H.-J.; Phenrat, T.; Tilton, R. D.; Lowry, G. V., Fe0 Nanoparticles Remain Mobile in Porous Media after Aging Due to Slow Desorption of Polymeric Surface Modifiers. *Environmental Science & Technology* 2009,*43*, (10), 3824–3830.

34. Cornell, R. M.; Schindle, P. W., Photochemical Dissolution of Goethite in Acid/Oxalate Solution. *Clays and CLay Minerals* 1987,*35*, (5), 347–352.
35. Phenrat, T.; Saleh, N.; Sirk, K.; Kim, H.-J.; Tilton, R.; Lowry, G., Stabilization of Aqueous Nanoscale Zerovalent Iron Dispersions by Anionic Polyelectrolytes: Adsorbed Anionic Polyelectrolyte Layer Properties and Their Effect on Aggregation and Sedimentation. *Journal of Nanoparticle Research* 2008,*10*, (5), 795–814.
36. Levard, C.; Reinsch, B. C.; Michel, F. M.; Oumahi, C.; Lowry, G. V.; Brown, G. E., Sulfidation Processes of PVP-Coated Silver Nanoparticles in Aqueous Solution: Impact on Dissolution Rate. *Environmental Science & Technology* 2011,*45*, (12), 5260–5266.
37. Khan, S. A.; Ahmad, A., Phase, Size and Shape Transformation by Fungal Biotransformation of Bulk TiO2. *Chemical Engineering Journal* 2013,*230*, 367–371.
38. Kaegi, R.; Voegelin, A.; Ort, C.; Sinnet, B.; Thalmann, B.; Krismer, J.; Hagendorfer, H.; Elumelu, M.; Mueller, E., Fate and Transformation of Silver Nanoparticles in Urban Wastewater Systems. *Water Research* 2013,*47*, (12), 3866–3877.
39. Levard, C.; Hotze, E. M.; Lowry, G. V.; Brown, G. E., Environmental Transformations of Silver Nanoparticles: Impact on Stability and Toxicity. *Environmental Science & Technology* 2012,*46*, (13), 6900–6914.
40. Ma, R.; Levard, C.; Marinakos, S. M.; Cheng, Y. W.; Liu, J.; Michel, F. M.; Brown, G. E.; Lowry, G. V., Size-Controlled Dissolution of Organic-Coated Silver Nanoparticles. *Environmental Science & Technology* 2012,*46*, (2), 752–759.
41. Ma, R.; Levard, C.; Michel, F. M.; Brown, G. E.; Lowry, G. V., Sulfidation Mechanism for Zinc Oxide Nanoparticles and the Effect of Sulfidation on Their Solubility. *Environmental Science & Technology* 2013,*47*, (6), 2527–2534.
42. Liu, W.; Chaurand, P.; Di Giorgio, C.; De Meo, M.; Thill, A.; Auffan, M.; Masion, A.; et al., Influence of the Length of Imogolite-Like Nanotubes on Their Cytotoxicity and Genotoxicity toward Human Dermal Cells. *Chemical Research in Toxicology* 2012,*25*, (11), 2513–2522.
43. Auffan, M.; Rose, J.; Wiesner, M. R.; Bottero, J.-Y., Chemical Stability of Metallic Nanoparticles: A Parameter Controlling Their Potential Cellular oxicity in vitro. *Environmental Pollution* 2009,*157*, (4), 1127-1133.
44. Thill, A.; Zeyons, O.; Spalla, O.; Chauvat, F.; Rose, J.; Auffan, M.; Flank, A. M., Cytotoxicity of CeO2 Nanoparticles for *Escherichia coli*. Physico-Chemical Insight of the Cytotoxicity Mechanism. *Environmental Science & Technology* 2006,*40*, (19), 6151–6156.
45. Zeyons, O.; Thill, A.; Chauvat, F.; Menguy, N.; Cassier-Chauvat, C.; Orear, C.; Daraspe, J.; Auffan, M.; Rose, J.; Spalla, O., Direct and Indirect CeO2 Nanoparticles Toxicity for *E.coli* and Synechocystis. *Nanotoxicology* 2009,*3*, (4), 284–295.
46. Rodea-Palomares, I.; Boltes, K.; Fernandez-Pias, F.; Leganas, F.; Garcia-Calvo, E.; Santiago, J.; Rosal, R., Physicochemical Characterization and Ecotoxicological Assessment of CeO2 Nanoparticles Using Two Aquatic Microorganisms. *Toxicological Sciences* 2011,*119*, (1), 135–145.
47. Auffan, M.; Rose, J.; Orsiere, T.; De Meo, M.; Thill, A.; Zeyons, O.; Proux, O.; et al., CeO2 Nanoparticles Induce DNA Damage towards Human Dermal Fibroblasts In vitro. *Nanotoxicology* 2009,*3*, (2), 161–U115.
48. Fang, X. H.; Yu, R.; Li, B. Q.; Somasundaran, P.; Chandran, K., Stresses Exerted by ZnO, CeO2 and Anatase TiO2 Nanoparticles on the *Nitrosomonas europaea*. *Journal of Colloid and Interface Science* 2010,*348*, (2), 329–334.
49. Benameur, L.; Auffan, M.; Cassien, M.; Liu, W.; Culcasi, M.; Rahmouni, H.; Stocker, P.; et al., DNA Damage and Oxidative Stress Induced by CeO2 Nanoparticles in Human Dermal Fibroblasts: Evidence of a Clastogenic Effect as a Mechanism of Genotoxicity. *Nanotoxicology* 2015,*9*, 696–705.
50. Auffan, M.; Rose, J.; Orsiere, T.; De Meo, M.; Thill, A.; Zeyons, O.; Proux, O.; et al., CeO2 Nanoparticles Induce DNA Damage towards Human Dermal Fibroblasts In vitro. *Nanotoxicology* 2009,*3*, (22), 161–171.
51. Mejias, R.; Gutierrez, L.; Salas, G.; Perez-Yague, S.; Zotes, T. M.; Lazaro, F. J.; Morales, M. P.; Barber, D. F., Long Term Biotransformation and Toxicity of Dimercaptosuccinic Acid-Coated Magnetic Nanoparticles Support Their Use in Biomedical Applications. *Journal of Controlled Release* 2013,*171*, (2), 225–233.
52. Levy, M.; Luciani, N.; Alloyeau, D.; Elgrabli, D.; Deveaux, V.; Pechoux, C.; Chat, S.; et al., Long Term In vivo Biotransformation of Iron Oxide Nanoparticles. *Biomaterials* 2011,*32*, (16), 3988–3999.
53. Wang, Z. Y.; von dem Bussche, A.; Kabadi, P. K.; Kane, A. B.; Hurt, R. H., Biological and Environmental Transformations of Copper-Based Nanomaterials. *Acs Nano* 2013,*7*, (10), 8715–8727.
54. Richter, K.; Schicklberger, M.; Gescher, J., Dissimilatory Reduction of Extracellular Electron Acceptors in Anaerobic Respiration. *Applied and Environmental Microbiology* 2012,*78*, (4), 913–921.
55. Zhang, L. W.; Petersen, E. J.; Habteselassie, M. Y.; Mao, L.; Huang, Q. G., Degradation of Multiwall Carbon Nanotubes by Bacteria. *Environmental Pollution* 2013,*181*, 335–339.
56. Kirschling, T. L.; Golas, P. L.; Unrine, J. M.; Matyjaszewski, K.; Gregory, K. B.; Lowry, G. V.; Tilton, R. D., Microbial Bioavailability of Covalently Bound Polymer Coatings on Model Engineered Nanomaterials. *Environmental Science & Technology* 2011,*45*, (12), 5253–5259.
57. Allen, B. L.; Kichambare, P. D.; Gou, P.; Vlasova, II; Kapralov, A. A.; Konduru, N.; Kagan, V. E.; Star, A., Biodegradation of Single-Walled Carbon Nanotubes through Enzymatic Catalysis. *Nano Letters* 2008,*8*, (11), 3899–3903.
58. Kang, S.; Mauter, M. S.; Elimelech, M., Physicochemical Determinants of Multiwalled Carbon Nanotube Bacterial Cytotoxicity. *Environmental Science & Technology* 2008,*42*, (19), 7528–7534.
59. Bansal, V.; Ahmad, A.; Sastry, M., Fungus-Mediated Biotransformation of Amorphous Silica in Rice Husk to Nanocrystalline Silica. *Journal of the American Chemical Society* 2006,*128*, (43), 14059–14066.
60. Barton, L. E.; Auffan, M.; Bertrand, M.; Barakat, M.; Santaella, C.; Masion, A.; Borschneck, D.; et al., Transformation of Pristine and Citrate-Functionalized CeO2 Nanoparticles in a Laboratory-Scale Activated Sludge Reactor. *Environmental Science & Technology* 2014,*48*, (13), 7289–7296.
61. Shi, J. Y.; Peng, C.; Yang, Y. Q.; Yang, J. J.; Zhang, H.; Yuan, X. F.; Chen, Y. X.; Hu, T. D., Phytotoxicity and Accumulation of Copper Oxide Nanoparticles to the Cu-Tolerant Plant *Elsholtzia splendens*. *Nanotoxicology* 2014,*8*, (2), 179–188.
62. Zhang, P.; Ma, Y. H.; Zhang, Z. Y.; He, X.; Zhang, J.; Guo, Z.; Tai, R. Z.; Zhao, Y. L.; Chai, Z. F., Biotransformation of Ceria Nanoparticles in Cucumber Plants. *Acs Nano* 2012,*6*, (11), 9943–9950.

63. Morales, M. I.; Rico, C. M.; Hernandez-Viezcas, J. A.; Nunez, J. E.; Barrios, A. C.; Tafoya, A.; Flores-Marges, J. P.; Peralta-Videa, J. R.; Gardea-Torresdey, J. L., Toxicity Assessment of Cerium Oxide Nanoparticles in Cilantro (*Coriandrum sativum L.*) Plants Grown in Organic Soil. *Journal of Agricultural and Food Chemistry* 2013,*61*, (26), 6224–6230.

64. Stegemeier, J. P.; Schwab, F.; Colman, B. P.; Webb, S. M.; Newville, M.; Lanzirotti, A.; Winkler, C.; Wiesner, M. R.; Lowry, G. V., Speciation Matters: Bioavailability of Silver and Silver Sulfide Nanoparticles to Alfalfa (Medicago sativa). *Environmental Science & Technology* 2015,*49*, (14), 8451–8460.

65. Lowry, G. V.; Espinasse, B. P.; Badireddy, A. R.; Richardson, C. J.; Reinsch, B. C.; Bryant, L. D.; Bone, A. J.; et al., Long-Term Transformation and Fate of Manufactured Ag Nanoparticles in a Simulated Large Scale Freshwater Emergent Wetland. *Environmental Science & Technology* 2012,*46*, (13), 7027–7036.

66. Levard, C.; Reinsch, B. C.; Michel, F. M.; Oumahi, C.; Lowry, G. V.; Brown, G. E., Sulfidation Processes of PVP-Coated Silver Nanoparticles in Aqueous Solution: Impact on Dissolution Rate. *Environmental Science & Technology* 2011,*45*, (12), 5260–5266.

67. Liu, J. Y.; Hurt, R. H., Ion Release Kinetics and Particle Persistence in Aqueous Nano-Silver Colloids. *Environmental Science & Technology* 2010,*44*, 2169–2175.

68. Auffan, M.; Tella, M.; Santaella, C.; Brousset, L.; Pailles, C.; Barakat, M.; Espinasse, B.; et al., An Adaptable Mesocosm Platform for Performing Integrated Assessments of Nanomaterial Risk in Complex Environmental Systems. *Scientific reports* 2014,*4*, 5608.

69. Tella, M.; Auffan, M.; Brousset, L.; Issartel, J.; Kieffer, I.; Pailles, C.; et al., Transfer, Transformation and Impacts of Ceria Nanomaterials in Aquatic Mesocosms Simulating a Pond Ecosystem. *Environmental Science & Technology* 2014,*48*, (16), 9004–9013.

70. Auffan, M.; Masion, A.; Labille, J.; Diot, M. A.; Liu, W.; Olivi, L.; Proux, O.; et al., Long-Term Aging of a CeO2 Based Nanocomposite Used for Wood Protection. *Environmental Pollution* 2014,*188*, 1–7.

71. Impellitteri, C. A.; Tolaymat, T. M.; Scheckel, K. G., The Speciation of Silver Nanoparticles in Antimicrobial Fabric Before and After Exposure to a Hypochlorite/Detergent Solution. *Journal of Environmental Quality* 2009,*38*, (4), 1528–1530.

72. Hedberg, J.; Skoglund, S.; Karlsson, M.-E.; Wold, S.; Wallinder, I. O.; Hedberg, Y., Sequential Studies of Silver Released from Silver Nanoparticles in Aqueous Media Simulating Sweat, Laundry Detergent Solutions and Surface Water. *Environmental Science & Technology* 2014,*48*, (13), 7314–7322.

73. Geranio, L.; Heuberger, M.; Nowack, B., The Behavior of Silver Nanotextiles during Washing. *Environmental Science & Technology* 2009,*43*, (21), 8113–8118.

74. Benn, T. M.; Westerhoff, P., Nanoparticle Silver Released into Water from Commercially Available Sock Fabrics. *Environmental Science & Technology* 2008,*42*, (11), 4133–4139.

75. Lombi, E.; Donner, E.; Scheckel, K. G.; Sekine, R.; Lorenz, C.; Von Goetz, N.; Nowack, B., Silver Speciation and Release in Commercial Antimicrobial Textiles as Influenced by Washing. *Chemosphere* 2014,*111*, 352–358.

76. Kim, B.; Park, C. S.; Murayama, M.; Hochella Jr., M. F., Discovery and Characterization of Silver Sulfide Nanoparticles in Final Sewage Sludge Products. *Environmental Science & Technology* 2010,*44*, (19), 7509–7514.

77. Kaegi, R.; Voegelin, A.; Sinnet, B.; Zuleeg, S.; Hagendorfer, H.; Burkhardt, M.; Siegrist, H., Behavior of Metallic Silver Nanoparticles in a Pilot Wastewater Treatment Plant. *Environmental Science & Technology* 2011,*45*, (9), 3902–3908.

78. Kent, R. D.; Oser, J. G.; Vikesland, P. J., Controlled Evaluation of Silver Nanoparticle Sulfidation in a Full-Scale Wastewater Treatment Plant. *Environmental Science & Technology* 2014,*48*, (15), 8564–8572.

79. Lombi, E.; Donner, E.; Taheri, S.; Tavakkoli, E.; Jaemting, A. K.; McClure, S.; Naidu, R.; Miller, B. W.; Scheckel, K. G.; Vasilev, K., Transformation of Four Silver/Silver Chloride Nanoparticles during Anaerobic Treatment of Wastewater and Post-processing of Sewage Sludge. *Environmental Pollution* 2013,*176*, 193–197.

80. Whitley, A. R.; Levard, C.; Oostveen, E.; Bertsch, P. M.; Matocha, C. J.; Kammer, F. v. d.; Unrine, J. M., Behavior of Ag Nanoparticles in Soil: Effects of Particle Surface Coating, Aging and Sewage Sludge Amendment. *Environmental Pollution* 2013,*182*, 141–149.

81. Gao, J.; Sepúlveda, M. S.; Klinkhamer, C.; Wei, A.; Gao, Y.; Mahapatra, C. T., Nanosilver-Coated Socks and Their Toxicity to Zebrafish (*Danio rerio*) Embryos. *Chemosphere* 2015,*119*, (0), 948–952.

PART 3

Environmental Applications of Nanomaterials

CHAPTER 8

Nanomaterials for Groundwater Remediation

Gregory V. Lowry

Carnegie Mellon University, Pittsburgh, Pennsylvania, USA

Introduction

Contamination of subsurface soil and groundwater by organic and inorganic contaminants is an extensive and vexing environmental problem that stands to benefit from nanotechnology. The EPA reports that contamination by chlorinated organic pollutants such as trichloroethylene (TCE), and heavy metals such as lead and hexavalent chromium are primary concerns at Superfund National Priorities List sites, which are the most contaminated sites in the United States. Associated health risks have led to an extensive remediation effort for the past 30 years. Remediation is costly and poses significant technical challenges. For example, life-cycle treatment costs are estimated to exceed $2 billion for approximately 3000 contaminated Department of Defense sites (Stroo et al. 2003).

Remediation efforts aimed at removing deep subsurface contamination (e.g., pump and treat) have had limited success because most pollutants are not highly mobile in the subsurface. For example, many organic contaminants are only weakly water-soluble and tend to remain as a separate nonaqueous phase liquid (NAPL) in the subsurface. Many organic contaminants are denser than water (DNAPL) and migrate downward in the aquifer. As sketched in Figure 8.1, DNAPL residual ganglia and pooled DNAPL are trapped in the porous soil. Heavy metals such as Pb(II) or Cr(VI) also tend to be concentrated in areas near their release point. The residual saturation acts as a long-term source for contaminant leaching to the groundwater, resulting in large plumes of dissolved contaminants and decades-long remediation times. Often the contaminant source may be difficult to locate or is too deep for excavation to be a cost-effective solution, and the prevailing "pump-and-treat" technologies cannot meet cleanup targets in a reasonable amount of time in most cases (Water Science Technology Board 2004). "Pump-and-treat" attempts to remove the contamination by pumping groundwater from the contaminated plume to the surface to remove the contaminants. "Pump-and-treat" and other technologies that address primarily the plume tend to fail because they do not address or remove the source. Accordingly, the Department of Energy recently advocated the development of novel *in situ* technologies to accelerate the rate at which contaminated sites are restored back to an acceptable condition (U.S. DOE 2005).

Nanotechnology has the potential to create novel and effective *in situ* treatment technologies for groundwater contaminant source zones. Rapid advances in nanotechnology have led to the creation of novel nanoparticles with unique and tunable physical and chemical properties. Their properties can be adjusted to make them highly reactive with common organic pollutants, and to minimize the formation of unwanted toxic by-products. Highly reactive nanoparticles such as nanoscale zerovalent iron ("nanoiron" or NZVI) (Liu et al. 2005a; Liu et al. 2005b), nanocatalysts (e.g., Au/Pd bimetallic nanoparticles [Nutt et al. 2005]), or nanosized sorbents (Tungittiplakorn et al. 2005; Tungittiplakorn et al. 2004) have been developed specifically to remediate contamination by organic and inorganic contaminants. In principle, their small size (10–500 nm) also provides an opportunity to deliver these remedial agents to subsurface contaminants *in situ*, and provides access to contamination trapped in the smallest pores in an aquifer matrix. Highly mobile nanoparticles are needed that can transport in the subsurface to where contaminants

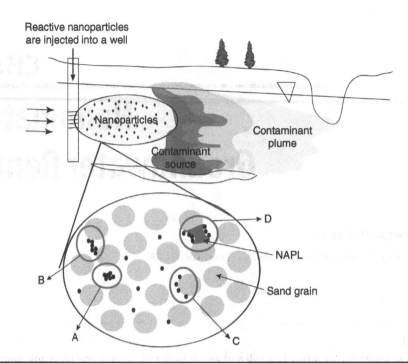

FIGURE 8.1   The macroscale problem. Illustration of DNAPL distribution as residual saturation (sources) and a plume of dissolved contaminants in an aquifer. Nanotechnology offers the potential to effectively target the chemical treatment to the residual saturation zone *in situ*. Reactive nanoparticles are injected into the aquifer using a well. The particles are transported to the contaminant source where they can degrade the contaminant. Nanoparticles can aggregate (A) and be filtered from solution via straining (B) or attachment to aquifer grains (C). Methods to target the nanoparticles to the contaminant (D) could improve the efficiency of the technology. Factors affecting the mobility of nanoparticles in the subsurface and the ability to target contaminants are discussed later in this chapter.

are located, which increases the potential for offsite migration and thus the potential for any unwanted ecotoxicity or human health risks. The high reactivity, and the potential for facile delivery directly to the contaminant source, suggests that nanoparticles can accelerate the degradation rate of contaminants in the source zone, and decrease the time and cost of remediation relative to traditional treatment technologies that address the plume.

Here, we focus on recent advances in the use of nanoparticles for *in situ* remediation of contaminated groundwater. In particular, the focus is on the use of nanoscale zerovalent iron and bimetallics (e.g., Fe^0/Pd) for the rapid *in situ* degradation of chlorinated organic compounds and reduction of heavy metals in contaminant source zones. *In situ* source zone treatment using nanoiron is one of the early adopters of environmental nanotechnology. There are already many documented applications of nanoiron ranging from pilot to full-scale, and early data suggest that nanoiron can be effective and can lower the costs of contaminant source zone remediation in groundwater (Gavaskar et al. 2005). Despite this rush to market, there are critical aspects of nanoparticle-based remediation strategies that have not yet been fully addressed. The delivery of nanomaterials in the subsurface is analogous to filtration in porous media. Particle slurries injected into the subsurface must migrate from the point of injection through the subsurface to the source of the contamination. Natural geochemical conditions (e.g., pH and ionic strength) can destabilize the nanoparticles and allows aggregation (labeled A in Figure 8.1). Particle aggregates may then be removed by straining and pore plugging (labeled B in Figure 8.1). These same conditions will also increase nanoparticle deposition onto media grains (labeled C in Figure 8.1). High deposition rates onto media grains or straining and pore plugging will limit nanoparticle transport. While maximizing transportability, ideally the nanomaterials will have an affinity for the target contaminant. For example, nanomaterials with hydrophobic character may partition to the DNAPL/water interface thereby delivering the particles to where they are most needed (labeled D in Figure 8.1). A fundamental understanding of the particle properties controlling the reactivity (rates and products/intermediates) and lifetime of iron nanoparticles, optimal methods of nanoiron injection/transport to the contaminant source zone, and methods for targeting specific subsurface contaminants or source areas to increase their effectiveness are needed.

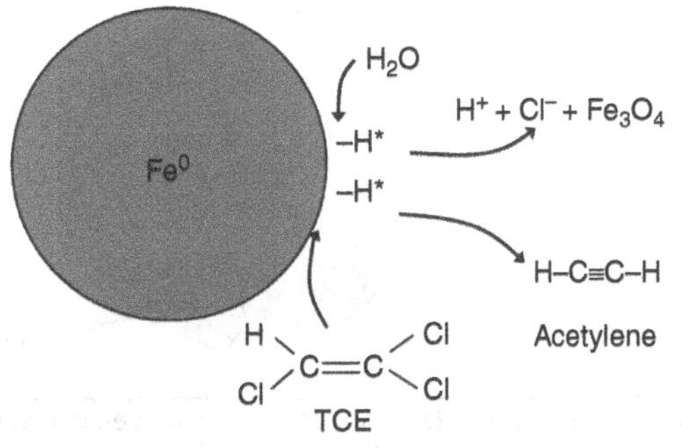

FIGURE 8.2A **Reactive nanoparticles.** TCE dechlorination by nanoiron. Fe^0 oxidation provides electrons for the reduction of TCE. The Fe^0 core shrinks while the Fe_3O_4 oxide shell grows. The particles are no longer active once all of the Fe^0 is oxidized.

Reactivity, Fate, and Lifetime

Nanoparticles can sequester groundwater contaminants (via adsorption or complexation), making them immobile, or can degrade or transform them to innocuous compounds. Contaminant transformations by nanoiron, which is a strong reductant, are typically redox reactions. When the oxidant or reductant is the nanoparticle itself, it is considered a reactive nanoparticle (Figure 8.2). A good example of a reactive nanoparticle is TCE dechlorination by nanoiron. Oxidation of Fe^0 in the particle provides electrons for the reduction of TCE, which acts as the oxidant. The Fe^0 core shrinks and is used up in the reaction. The particles are no longer active once all of the Fe^0 is oxidized (Figure 8.2a). Nanoparticles that *catalyze* redox reactions but are not themselves transformed are catalytic nanoparticles, which requires an additional reagent that serves as the reductant or oxidant (e.g., Pd nanoparticles require H_2 as a reductant [Lowry and Reinhard 1999]). H_2 (the reductant) is activated by the Pd to form adsorbed reactive H species. TCE adsorbs to the Pd surface where it is reduced by the reactive H species on the Pd surface (Figure 8.2b). In principle, the catalyst can repeat this indefinitely so long as reductant is continually supplied. In practice, precipitation of minerals or natural organic matter on the Pd surface or adsorption of reduced sulfur species deactivate the catalyst and it has a finite lifetime (Lowry and Reinhard 2000). Some nanomaterials are engineered to strongly sequester contaminants (Figure 8.2c). The high affinity for the contaminant allows the nanoparticle to significantly lower the aqueous phase concentrations, to out-compete natural geosorbents such as organic carbon, and serves to concentrate the contaminants onto the particles. Once concentrated onto the nanoparticles, the contaminants can be removed along with the nanoparticles. This can be highly effective for hydrophobic organic contaminants such as PCBs and PAHs and for heavy metals.

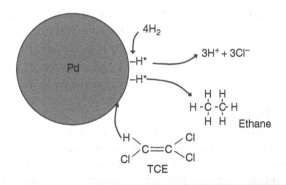

FIGURE 8.2B **Catalytic nanoparticles.** TCE dechlorination by catalytic Pd particles. H_2 is supplied as the reductant for TCE dechlorination. In principle, the Pd catalyst is not altered by the reaction and can remain active as long as H_2 is supplied. In practice, catalyst deactivation occurs and the particle lifetime is finite. Catalyst regeneration may extend the life of the particle.

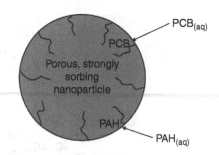

FIGURE 8.2C   Adsorbent nanoparticles. Nanoparticles engineered as very strong sorbents can be used to strongly sequester organic or inorganic contaminants. Once adsorbed, the contaminants are no longer bioavailable.

For *in situ* remediation with any type of nanoparticle, it is important to know which groundwater contaminants will respond to the treatment and which will not. It is also important to know how long the reactive or catalytic particles will remain active as this will determine important operation decisions such as how much to inject and when reinjection may be necessary.

Degradation of halogenated hydrocarbons, particularly chlorinated solvents, occurs via a reductive process. The Fe^0 in the nanoiron is oxidized by the chlorinated solvent, which is subsequently reduced. For chlorinated hydrocarbons, the reduction typically results in the replacement of a chlorine atom with a hydrogen atom. For heavy metals, the metal, such as Pb(II) or Cr(VI), is reduced to its zerovalent form on the nanoiron surface, or forms mixed (Fe-Metal) precipitates that are highly insoluble (Ponder et al. 2000). The general half-reactions for the oxidation of iron and the reduction of chlorinated organic compounds (COC) or heavy metals are given in Equations (8.1) to (8.3), where Me is a metal ion of charge a.

$$Fe^0 \rightarrow Fe^{2+} + 2 \cdot e^- \tag{8.1}$$

$$COC + n \cdot e^- + m \cdot H^+ \rightarrow products + 3Cl^- \tag{8.2}$$

$$Me^{a+} + b \cdot e^- \rightarrow Me^{a-b} \tag{8.3}$$

In the case of nanoiron, or Fe^0-based bimetallics, the reduction of the contaminant is surface-mediated, and the particle itself is the reductant. The attractiveness of nanoiron is that the particles have a high surface-to-volume ratio and therefore have high reactivity with the target contaminants. The following generalizations can be made about the reactivity and lifetime of all nanoparticulate remedial agents that are themselves the reactive material—that is, not true catalysts according to the formal definition of a catalyst:

- Any process that affects the surface properties of the particles (e.g., formation of an Fe oxide on the surface) can affect their *reactivity*.

- Any oxidant (e.g., O_2 or NO_3^-) competing with the target contaminant will utilize electrons and may lower the rate and efficiency of the nanoiron treatment for the target contaminants.

- Reactive nanoparticles that serve as a reactant rather than a catalyst will have a finite lifetime, the length of which depends on the concentration of the target contaminant, the presence of competing oxidants, and the selectivity of the particles for the desired reaction.

Nanoiron Reactivity

Since 1997, there have been many laboratory studies conducted to determine the range of contaminants that are amenable to reductive dechlorination by nanoiron or bimetallics such as Fe^0/Pd nanoparticles. A summary of the reactive surface area normalized pseudo-first-order reaction rate constants for each type of contaminant and iron is given in Table 8.1. For some forms of nanoiron, the reported surface area-normalized reaction rate constants are similar to those reported for iron filings, which have been used successfully as *in situ* reactive barriers for groundwater remediation in shallow aquifers (Wilkin et al. 2005), suggesting that there is not a quantum nanosize effect for these particles (Liu et al. 2005b; Nurmi et al. 2005). Rather, the

Compound	Iron Type	Pseudo-First-Order k_{rxn} ($\times 10^{-3}$ L/h $\cdot$ m²)	Ref.
Chlorinated Methanes			
CCl₄	Fe⁰	0.5–240	(Lien and Zhang 1999; Nurmi et al. 2005)
CHCl₃	Pd/Fe	0.08	(Lien and Zhang 1999)
Chlorinated Ethanes			
HCA	Fe⁰	770 ± 70	(Song and Carraway 2005)
PCA		796 ± 63	
1,1,2,2-TeCA		538 ± 56	
1,1,1,2-TeCA		30.3 ± 4.2	
1,1,2-TCA		151 ± 13	
1,1,1-TCA		2.31 ± 0.34	
1,2-DCA		0	
1,1-DCA		0.02 ± 0.004	
HCA			(Lien and Zhang 2005)
PCA	Fe/Pd	20	
1,1,1,2-TeCA		27	
1,1,2,2-TeCA		21	
1,1,1-TCA		9	
Chlorinated Ethenes		**6**	
TCE	Fe⁰	1–14	(Liu et al. 2005a)
			(Liu et al. 2005b)
			(He and Zhao 2005)
	Fe⁰-starch stabilized	20	(He and Zhao 2005)
	Fe/Pd	18	(Wang and Zhang 1997)
	Fe/Pd-starch stabilized	67	(He and Zhao 2005)
PCBs	Fe⁰	$10^{-6} - 5 \times 10^{-5}$ L/yr $\cdot$ m²	(Lowry and Johnson 2004)
	Fe⁰	0	(He and Zhao 2005)
	Fe/Pd	0.053	
	Fe/Pd-starch stabilized	0.31	
Inorganics/Metals			
Perchlorate	Fe⁰	0.013 mg Cl/g Fe/hr	(Cao et al. 2005)
Nitrate	Fe⁰	70	(Caoe et al. 2000)
Selenate	Fe⁰	2.5	(Mondal et al. 2004)
	Fe/Ni	0.9	
Arsenic(III)	Fe⁰	200–600	(Kanel et al. 2005)
Cr(VI)	Fe⁰	110–230	(Ponder et al. 2000)
Pb(II)	Fe⁰	130	

TABLE 8.1 Organic Contaminants and Their Reaction Rates in Reactions with Nanoscale Iron or Bimetallic Iron Particles

high surface area of nanoiron (10–50 m²/g) compared to iron filings (0.01–1 m²/g) makes them attractive for *in situ* remediation of groundwater because of the higher reaction rates that they can provide. This is exemplified in the reaction with perchlorate (Cao et al. 2005); perchlorate reduction by nanoiron was rapid enough to make it useful for *in situ* perchlorate reduction, whereas iron filings reduced perchlorate far too slowly to be useful (Moore et al. 2003). There are instances where nanosize effects may be evident. Nanoiron made from the borohydride reduction of dissolved Fe has exhibited behavior that may be attributable to small particle size (i.e., a nanosize effect). First, nanoiron made from reduction of dissolved Fe by sodium borohydride has the ability to activate H₂ gas and use it in the reduction of chlorinated ethenes (Liu et al. 2005a, 2005b). This ability was attributed to the amorphous structure and the nanosized Fe⁰ crystallites (~1 nm) in these particles. Micron-scale iron filings and other forms of nanoiron with larger crystallites cannot active H₂. Second, this nanoiron that does not have any noble metal catalyst can dechlorinate polychlorinated biphenyl compounds (PCBs) at ambient temperature and pressure (Lowry and Johnson 2004; Wang and Zhang 1997). Iron filings and other types of nanoiron without an added catalyst cannot dechlorinate PCBs under these conditions. Novel nanomaterials that leverage the nanosize effects may degrade highly recalcitrant compounds that are not amenable to degradation by currently available materials.

Laboratory investigations (Table 8.1) indicate that the reactivity afforded by nanoiron and Fe^0/Pd bimetallic nanoparticles is sufficient to make it useful for *in situ* remediation of groundwater. It is important to note, however, that most of these studies use fresh nanoiron particles and do not observe reactivity over the lifetime of the particles so they should be considered as initial rates. There have been few investigations on the effect of aging (oxidation) or precipitation of Fe-containing minerals (e.g., $FeCO_3$) on nanoiron reactivity throughout its lifetime. It is likely that many of the conclusions regarding geochemical effects on the reactivity of iron filings will be applicable to nanoiron systems, but this has not yet been demonstrated. If reactivity of nanoiron is found to be insufficient, addition of Pd or some other noble metal catalyst increases the reactivity, but this also increases the cost, which currently ranges from $50 to $110/kg for nanoiron without added catalyst.

Reaction Products, Intermediates, and Efficiency

There are two reasons why *in situ* remediation using reactive nanoparticles must consider the reaction products formed. First, a reaction product may potentially be more toxic or mobile than the parent compound, and thus increases the risks posed by the site rather than decreasing them. Second, the products formed can strongly influence the effectiveness and costs of remediation. Developers of reactive nanoparticles for remediation need to consider the production and potential negative consequences of reaction intermediates and products. TCE offers a good example of how *in situ* remediation may increase the risks at a site rather than decrease them. Under reducing conditions afforded by nanoiron, TCE can be sequentially dechlorinated from TCE to dichloroethene (DCE), and to vinyl chloride (VC), and then to ethene, and under some conditions to ethane (Figure 8.3). VC is classified as a known human carcinogen, while TCE is classified as a suspected human carcinogen. Thus, conversion of TCE to VC may in fact increase the risk at a site due to VC's higher toxicity. In the case of nanoiron, most laboratory investigations have shown negligible production of chlorinated intermediates for most chlorinated solvents, which is optimal. The exception is carbon tetrachloride (CCl_4), where reduction by nanoiron results in chloroform ($CHCl_3$) as a long-lived reactive intermediate or product. Chloroform is more mobile and toxic than carbon tetrachloride, and reactive particles that can promote the complete conversion of carbon tetrachloride to methane (CH_4) or to carbon dioxide (CO_2) is most desirable.

For reactive nanoparticles, where the particle itself is the reductant and therefore has a finite reducing power, it is important to understand the factors influencing the Fe^0 utilization efficiency. The iron utilization efficiency is defined as the mass of target contaminant degraded per unit mass of nanoiron added.

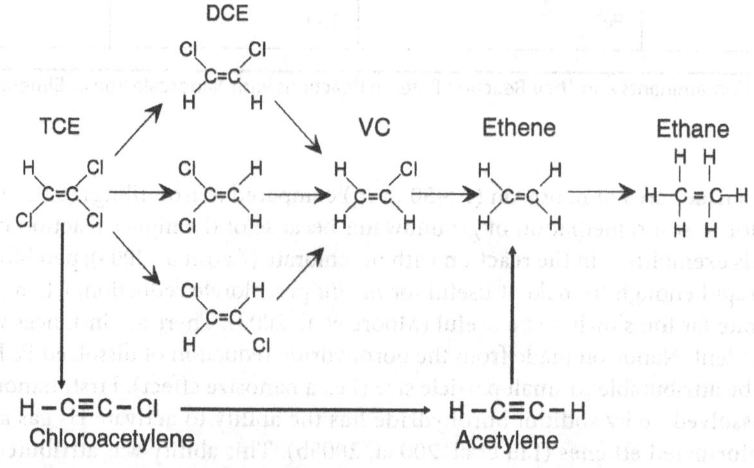

Figure 8.3 Dechlorination pathways of TCE. Partial dechlorination may lead to production of dichloroethene isomers and vinyl chloride, which is more toxic (known carcinogen) than the parent compound TCE (suspected carcinogen).

The degradation products formed can strongly influence the nanoiron mass required to degrade a given mass of contaminant to innocuous products. For example, using Fe^0 as the reductant TCE can dechlorinate to partially or fully saturated dechlorination products [Equation (8.4)], where the values of the coefficients a and b are determined by the products formed (e.g., $a = 4$, $b = 3$ for ethane, and $a = 2$, $b = 3$ for acetylene). Equation (8.4) is the net reaction derived from the two half-reactions shown in Equations (8.1) and (8.2).

$$aFe^0 + TCE + (2a-b)H^+ \rightarrow prod + aFe^{2+} + bCl^- \tag{8.4}$$

For TCE, removing all of the chlorines to form the corresponding hydrocarbon makes it nontoxic. For nanoiron made from borohydride reduction of dissolved iron, the dominant reaction product is ethane (C_2H_6) (Liu et al. 2005a). For another type of nanoiron, the dominant reaction product is acetylene (C_2H_2) (Liu et al. 2005b). Both are nonchlorinated and are equally nontoxic, however, the carbon in ethane is highly reduced (C average oxidation state is –III), while the carbon in acetylene is not as reduced (C average oxidation state in acetylene is –I). The average oxidation state of C in the parent compound (TCE) is +I, so reduction of TCE to ethane requires twice as much Fe^0 as for reduction of TCE to acetylene. This is also evident in the value of a for ethane ($a = 4$) versus acetylene ($a = 2$). Reactive nanoparticles can potentially be designed for optimal efficiency by selecting for products that are nontoxic but require the smallest possible redox swing.

Fate and Lifetime

Nanoparticles have great potential to benefit the environment by improving groundwater remediation, but before their widespread release it is prudent to evaluate the potential risks associated with their use. This requires a fundamental understanding of their long-term fate and lifetime. Most nanoiron is comprised of primary particles ranging in size from ~40 to 100 nm that have an Fe^0 core and Fe-oxide shell that is primarily magnetite and maghemite (Figure 8.4). Magnetite (Fe_3O_4) is the dominant Fe oxide at the Fe^0/Fe-oxide interface, while maghemite (Fe_2O_3) is the dominant Fe oxide at the Fe-oxide/water interface. As the particles oxidize (and the contaminants reduce), the Fe^0 core shrinks and ultimately the particles become Fe oxide. For some nanoiron, the particles undergo reductive dissolution, followed by precipitation of the dissolved Fe as another Fe oxide form (Liu et al. 2005a).

Nanoiron is a unique nanoparticle because its corrosion by water produces H_2 gas [Equation (8.5)] (Liu et al. 2005a, 2005b), which can be used as an energy source by hydrogenotrophic bacteria (Sorel et al. 2001;

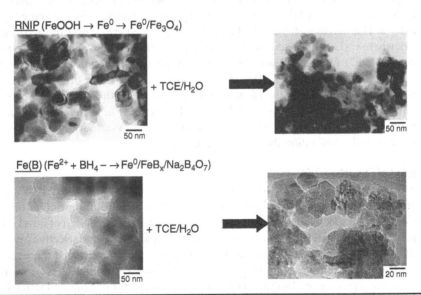

FIGURE 8.4   TEM images of fresh nanoiron samples showing before reaction and after reaction with TCE in water. Both particle types have an apparent core-shell morphology. RNIP, made from gas phase reduction of Fe oxides in H_2, appear to have a shrinking core during reaction in water. Fe(B), made from reduction of Fe^{2+} in a water/methanol solution using sodium borohydride appear to undergo oxidative dissolution followed by precipitation of the dissolved Fe to hematite (Liu et al. 2005b). The shell on Fe(B) is predominantly borate (Nurmi et al. 2005).

Vikesland et al. 2003) that also remove surficial H_2 (i.e., cathodic depolarization) (Hamilton 1999) to enhance or sustain iron surface reactivity.

$$Fe + H^+ + e^- \xrightarrow{k_1} Fe-H \xrightarrow{k_2} \frac{1}{2}H_2 \qquad (8.5)$$

Further, dissimilatory iron reducing bacteria (DIRB) such as *Geobacter* or *Shewanella* spp. could maintain active Fe^0 surfaces by reductively dissolving Fe oxides or Fe oxyhydroxides formed at the water/particle interface (Gerlach et al. 2000), or by generating reactive surface-associated Fe(II) species (Williams et al. 2005). Homoacetogens are another group that could enhance TCE removal—either directly through cometabolism (using H_2 as primary substrate), or indirectly by stimulating heterotrophic activity through acetate production ($4Fe^0 + 2CO_2 + 5H_2O \rightarrow CH_3COO^- + 4Fe^{+2} + 7OH^-$) (Oh and Alvarez 2002). Microbially induced corrosion could also ensure the localized dissolution of the iron nanoparticles, thereby eliminating possible concerns from off-site migration and risk. The data collected thus far on the fate of nanoiron have been largely collected in the laboratory in deionized water. Porewater constituents such as carbonate, sulfate, and chloride may inhibit or promote nanoiron corrosion (Agrawal and Tratnyek 1996; Bonin et al. 1998; Phillips et al. 2000; Vogan et al. 1999). It is critical to know what the end product of nanoiron oxidation is under real environmental conditions, but a well controlled *in situ* study of the fate of nanoiron in the subsurface has yet to be conducted.

Effect of pH

The role of pH is an important one for nanoiron lifetime. As seen in Equation (8.5), decreasing the pH should increase the rate of H_2 evolution from nanoiron if the reduction of H^+ at the iron surface is the rate controlling step. According to Equation (8.5), in the absence of any other oxidants and assuming that the reaction is first order with respect to the specific surface area of the iron, the rate of H_2 evolution is given by Equation (8.6).

$$\frac{dH_2}{dt} = k_{H_2} \cdot [SA] \cdot [H^+]^b = k_{H_2 obs}[H^+]^b \qquad (8.6)$$

k_{H_2} is the first-order reaction rate constant for H_2 evolution and $[SA]$ is the specific surface area for the reaction, $k_{H_2 obs}$, is the observed pseudo-first-order reaction rate constant. The H_2 evolution from nanoiron in solution at pH ranging from 6.5 to 8.9, the equilibrium pH for an iron/water system (Pourbaix 1966), is shown in Figure 8.5a. Indeed, as the pH decreases, the rate of H_2 evolution increases. For pH ranging from 6.5 to 8.0, there is a log linear relationship between the rates, indicating that b in Equation (8.6) is ≈ 1 (Figure 8.5b). Thus H_2 evolution from RNIP is pseudo-first-order with respect to H^+. At pH greater than 8, the mechanism for Fe^0 oxidation by H^+ appears to change, although some H_2 evolution is occurring at pH > 8. This agrees with published reports for hydrogen evolution from zerovalent metals at near neutral pH, where k_1 [Equation (8.5)] has been shown to be the rate controlling step (Wang and Farrell 2003, Bockris et al. 1987). Most importantly is the effect of pH on the lifetime of RNIP. Using the data in Figure 8.5a, at pH > 8, the lifetime of RNIP is expected to be on the order of 9 to 12 months, whereas at pH near neutral (typical for soils) the lifetime of RNIP will be on the order of 2 weeks. This only considers the unintended competing reaction of RNIP with water/H+. In the presence of additional oxidants such as the target contaminant (e.g., TCE), the lifetime could be even shorter because this adds another oxidant to the system that will utilize the Fe^0 in the particles.

Effect of Competing Oxidants

Any *in situ* groundwater remediation technology that employs nanomaterials will have to consider the cost of those materials, as material costs are likely to be a significant portion of the total cost of remediation. Contaminant specificity should therefore be considered in the design of novel reactive (or sorptive) nanomaterials. The concept of contaminant specificity is known as the selectivity ratio. At many contaminated sites a mixture of contaminants may be present and nanoiron injected there may have several competing oxidants. For example, a site may contain TCE along with nitrate (NO_3^-), which are both amenable to reduction by nanoiron (Yang and Lee 2005). In addition, the oxidation of water/H^+ to produce H_2 is always operable, which provides parallel competing oxidation pathways for the nanoiron (Figure 8.6).

If the reaction with TCE is the desired reaction, nanoiron designed to give selectivity ratios of $k_2[TCE]/k_3[NO_3^-]$ and $k_2[TCE]/k_1[H^+]$ as large as possible is desirable. For first-order reactions such as those provided by RNIP, the relative concentrations of each oxidant are also important as the rate is proportional to the

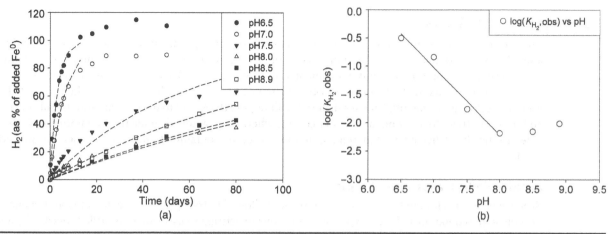

FIGURE 8.5 H_2 production from RNIP. (a) H_2 production from 500 mg/L RNIP with an Fe^0 content of 27 wt% at different pHs. pH was buffered using 50 mM HEPES buffer. Data are from duplicate reactors. The curve represents the first-order data fit (k_{obs}, H_2 = 0.0067 1/day, r^2 = 0.996). (b) k_{obs}, H_2 versus pH production from 500 mg/L RNIP with an Fe^0 content of 27 wt%. Lines are interpolated (not fit) and only meant to guide the eye.

concentration of contaminant (e.g., rate = k_2[TCE]). Any application of reactive nanomaterials for groundwater remediation will require extensive site characterization to determine the suitability of the materials for that particular site. Reactive or sorptive nanomaterials may not be suitable at a site where the concentrations of competing oxidants at a given site are high, especially if the nanomaterial cannot be designed with a selectivity ratio that minimizes the effect of competing oxidants. There is considerable room for improvement regarding the selectivity ratio of nanomaterials, and controlling the reactivity of these materials at the nanoscale is an exciting research avenue that will continue to be pursued.

The economic feasibility of reactive nanoiron for TCE DNAPL source zone remediation highlights the need to understand the products formed and the potential competing reactions. The goal is to reductively dechlorinate the contaminant (TCE) to nontoxic products using electrons derived from the nanoiron. General equations describing the transformation of trichloroethylene (TCE) were given in Equation (8.4) (Liu et al. 2005a). The two electrons from Fe^0 oxidation can be used to dechlorinate TCE Equation (8.4), or can be used to produce H_2 [Equation (8.5)]. The mass of TCE dechlorinated per mass of Fe^0 will depend on the relative rates of each of these reactions, and on the value of n in Equation (8.4), which is a function of the dechlorination products formed. The products of TCE degradation using nanoiron vary from acetylene (using RNIP) to ethane (using iron made from the borohydride reduction of dissolved iron). Acetylene is the best- case scenario as this pathway only requires 4 moles of electrons per mole of TCE ($n = 4$). It requires twice as many to transform TCE to ethane ($n = 8$). Assuming nanoiron costs \$50/kg and that all injected zerovalent iron is used to transform TCE, the cost of treatment ranges from \$44/kg TCE (for acetylene) to \$88/kg TCE (for ethane). If oxidants other than TCE were present (e.g., NO_3^- or dissolved oxygen), the process efficiency may decrease and the cost for iron would increase. For example, if only 10 percent of the electron from iron oxidation were used to transform TCE, the cost would increase by a factor of 10. Careful consideration of the site geochemistry and appropriate bench scale feasibility testing are necessary to assess the economic feasibility of reactive nanoparticles at a particular site.

$$+ \quad TCE \quad \xrightarrow{k_2} \quad \text{Nontoxic dechlorination products}$$

$$Fe^0 \quad + \quad NO_3^- \quad \xrightarrow{k_3} \quad NO_2^-, NH_4^+, N_2$$

$$+ \quad H^+ \quad \xrightarrow{k_1} \quad H_2, OH^-$$

FIGURE 8.6 Parallel reaction pathways for competing oxidants in a contaminated aquifer.

Delivery and Transport Issues

The use of reactive nanomaterials for groundwater remediation is promising considering the availability and effectiveness of many types of nanomaterials for degrading or sequestering environmental contaminants of concern. The attractiveness of nanomaterials is their potential to be used *in situ*, directly degrading the contaminants in the subsurface without the need to excavate them or pump them out of the ground. Realizing the potential of nanomaterials will require the ability to inject them into the subsurface and transport them to the contaminant source zone where they rapidly degrade the contaminants. If particles cannot be delivered to and remain at the source of contamination, they will not be utilized efficiently.

Injection Methods and Delivery Vehicles

Reactive nanoparticles can be injected into a well and allowed to transport down gradient from the injection site to the contaminated area. Typically this is done in existing wells, or in wells drilled specifically for the remediation process. Drilling and packing a well, even a small diameter (3-inch) well is quite expensive. Direct push wells (Figure 8.7) are a lower cost alternative to drilled wells, and are the most often used delivery tool for remediation with nanoiron. These are hydraulic devices mounted on a truck or tractor. They can create wells that are 2 to 3 inches in diameter that can then be used for injecting reactive nanoparticles. A nanoparticle slurry can be injected along part of all of the vertical range of the probe (Figure 8.8) to provide treatment to specific regions in the aquifer. This helps to target the reactive nanoparticles to the contaminated regions of the subsurface where they are needed.

Transport

The effectiveness of remediation, as well as the potential for unwanted exposure of humans and other biota to these reactive nanoparticles, depends on how easily these materials transport in porous media. Uncertainties regarding the fate, transport, and potential toxicity of engineered nanomaterials have prompted investigations on the fate and transport of various nanoparticles (very fine colloids) ranging in size from 1.2 to 300 nm (Lecoanet et al. 2004; Lecoanet and Wiesner 2004; Royal Academy of Engineering 2004;

Figure 8.7   Truck-mounted Geo-Probe® for creating direct push wells for nanoiron injection. Photo is courtesy of GeoProbe® Systems, Inc., Salina, Kansas.

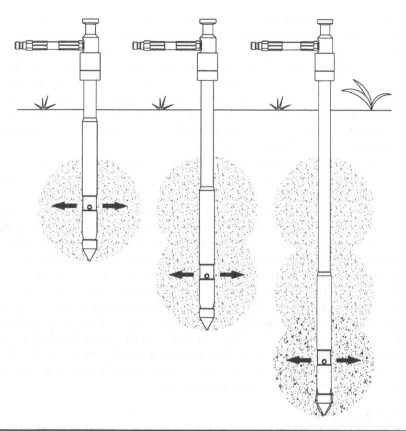

Figure 8.8   GeoProbe® injection of nanoparticles into the subsurface. Multiple injection points along the probe can provide a zone of influence over a portion of or the entire vertical section of the contaminated area. Photo is courtesy of GeoProbe® Systems, Inc., Salina, Kansas.

Saleh et al. 2007; Schrick et al. 2004). These studies have demonstrated that nanoiron and many other types of engineered nanomaterials do not transport easily in saturated porous media. Transport distances range from a few centimeters for nanoiron (Saleh et al. 2007; Schrick et al. 2004) to a few meters for carbon nano-materials under typical groundwater conditions (Lecoanet et al. 2004; Lecoanet and Wiesner 2004). This has prompted the use of supports such as hydrophilic carbon or polyacrylic acid to enhance the transport of nanoiron (Schrick et al. 2004), the use of surfactant micelles to deliver nanoiron directly to entrapped NAPL (Quinn et al. 2005), or adsorbed polymers to improve transport distances and target entrapped DNAPL (Saleh et al. 2007). The reasons for limited transport and the need for advanced delivery vehicles are addressed here.

Investigations of the transport and fate of colloids (micron and submicron particles) in the environment has been extensively studied for decades, as these processes are critical to understanding deep bed filtration (Yao et al. 1971) and the movement of colloids in natural systems (Elimelech et al. 1995; Ryan and Elimelech 1996). Typically this has focused on particles on the order of 1 micrometer, but particles that are a few hundred nanometers have also been studied. As previously discussed, the fate and transport of nanoparticles in porous media can be considered a filtration problem (Figure 8.1). Filtration theory indicates that the magnitude and rate of aggregation or deposition will depend on *physical* properties, including the nanoparticle size, pore size distribution of the media, and the flow velocity, and by the *chemical* properties such as the pH, ionic strength, and ionic composition, which control the magnitude and polarity of the attractive and repulsive forces between the nanoparticles and between the nanoparticles and mineral grains. Thus, the *hydrogeochemistry* of the system, along with the properties of the nanoparticles, will determine the transportability of nanoparticles at a specific site. Increasing ionic strength or the presence of divalent cations such as Ca^{2+} and Mg^{2+} can destabilize the particles by decreasing electrostatic double layer (EDL) repulsions between particles and allow aggregation (labeled A in Figure 8.1). This will also increase nanoparticle deposition onto media grains (labeled C in Figure 8.1). Greater attachment efficiency to media grains (or membrane surfaces in a membrane filter) will limit nanoparticle transport. Alternatively, aggregation of nanoparticles to larger-sized aggregates may potentially serve to increase particle mobility. The diffusion rate of the larger aggregates will be slower than for the larger particles, which may decrease the rate of particle-media interactions. Nanoparticle aggregation or the presence of high concentrations of particles could also lead to straining (labeled B in Figure 8.1), which will limit or retard transport. The flow velocity also plays a significant role. At a high porewater velocity, the residence time of the nanoparticles at the collector surface may be too short to allow for attachment to occur. Low attachment efficiency will result in longer transport distances. For nanoparticles ($d_p < 100$ nm) in environmental media (e.g., soil, sediments), it is difficult to predict a priori how the various interactions among the particles (aggregation) and media grains will affect their transport.

There are limitations to applying standard deep-bed (or clean-bed) filtration models in natural systems, and especially for remediation applications where concentrated suspensions of particles will be injected. First, typical filtration models have to make many simplifying assumptions, such as homogeneous media, monodisperse particle size distribution of nanomaterials, constant porewater velocity, and uniform surface properties of nanoparticles and filter media. These are typically not applicable in real environmental systems, which are physically and chemically heterogeneous. Another assumption of filtration models is that the bed is clean—that is, only particle-media grain interactions are explicitly accounted for. This may not be the case for groundwater remediation using engineered nanoparticles, where high concentration slurries of polydisperse particles will be injected to achieve a 0.1 wt% to 0.5 wt% concentration of particles in the aquifer. Under these conditions, particle-particle interactions are important, and filter-ripening models may be more appropriate than clean bed filtration models. Regardless of the model type (clean bed or filter ripening), model inputs will include particle size, attachment coefficients, flow velocity, ionic strength, and ionic composition as discussed below. Aggregation and surface modification, and how these affect the mobility or engineered nanomaterials and the ability to target specific regions in the subsurface, are discussed.

Effect of Aggregation

If aggregation is rapid, and creates particles that are greater than a few microns in diameter, there is potential for this aggregation to limit their transportability. For many nanoparticles, aggregation will likely be a mechanism that significantly limits their transport in the environment, and makes it possible to remove them using standard water-treatment practices such as flocculation/sedimentation or membrane filtration. The rate of particle aggregation and the size and morphology of the aggregates formed depends on both the collision frequency (transport) and the collision/attachment efficiency (i.e., the magnitude of the attractive and repulsive forces between the particles).

Experience shows that many nanoparticle suspensions are typically highly unstable and rapidly aggregate (Saleh et al. 2005a), making them difficult to handle. The reason for the rapid aggregation is that their small size (~100 nm) gives them high diffusion coefficients (transport rates) and an exceptionally high number of collisions, so even dilute suspensions of nanoparticles with relatively low sticking coefficients would rapidly aggregate and not remain as individual nanoparticles in suspension under normal environmental conditions. Moreover, the presence of cations and anions, particularly divalent cations such as Mg^{2+} and Ca^{2+}, further destabilizes nanoparticles and increases their rates of aggregation. For nanoiron, 100-nm particles present at a volume fraction (Φ) as low as $\Phi = 10^{-5}$ rapidly aggregate into 5-micron particles in just a few minutes (Figure 8.9). This problem is exacerbated for nanoparticles such as nanoiron that are magnetic and therefore subject to non-DLVO magnetic attractive forces. The time scale of colloid dispersion stability is determined by the magnitude of the energy barrier between particles. According to classical DLVO theory, the major attractive energy is van der Waals energy (V_{vdW}) while the major repulsive energy is electrostatic interaction energy (V_{ES}) (de Vicente 2000; Elimelech et al. 1995; Evans 1999; Heimenz 1997; Plaza 2001; Strenge 1993). The V_{vdW} attractive force between spherical particles can be expressed as (de Vicente 2000)

$$V_{vdW} = \frac{-A}{6} \left[\frac{2r^2}{s(4r+s)} + \frac{2r^2}{(2r+s)^2} + \ln s \frac{(4r+s)}{(2r+s)^2} \right] \tag{8.7}$$

where A is the Hamaker constant, which is 10^{-19} N $\cdot$ m for Fe, γ-Fe_2O_3 and Fe_3O_4 (Rosensweig 1985). r(m) is the radius of particles, and s(m) is distance between surfaces of two interacting particles. Electrostatic repulsion between two identical particles, V_{ES} can be expressed as (de Vicente 2000)

$$V_{ES} = 2\pi \varepsilon_r \varepsilon_0 r \zeta^2 \ln \left[1 + e^{-\kappa s} \right] \tag{8.8}$$

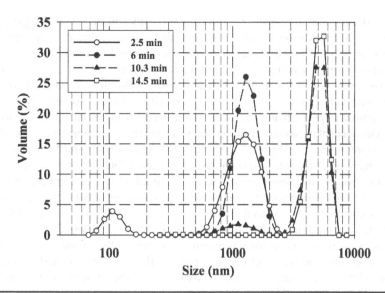

FIGURE 8.9   DLS indicates that nanoiron particles (100 nm) present at 80 mg/L (volume fraction ~10-5) rapidly flocculate to form 5-micron size aggregates.

where ε_r is the relative dielectric constant of the liquid, ε_0 is the permittivity of the vacuum, ζ is electrokinetic or zeta potential of diffuse layer of charged particles, and κ is the reciprocal Debye length. Applying classical DLVO theory, an energy barrier of RNIP is predicted to be 7.0 $k_B T$ (Figure 8.10). This energy barrier is sufficient to prevent rapid aggregation, suggesting that dispersions of these particles should be stable. This contrasts the observed behavior.

Iron nanoparticles that behave as a single domain magnetic particle have an intrinsic permanent magnetic dipole moment $\mu = (4\pi/3)r^3 M_s$ even in the absence of an applied magnetic field (Butter 2003a; Butter 2003b; de Gennes 1970; McCurrie 1994; Neto 2005; Rosensweig 1985). When particle dipoles are oriented in head-to-tail configuration, the maximum magnetic attraction energy (V_M) can be expressed as (de Vicente 2000)

$$V_M = \frac{-8\pi\mu_0 M_s^2 r^3}{9\left(\dfrac{s}{r}+2\right)^3} \tag{8.9}$$

where μ_0 is the permeability of the vacuum. The potential energy of interaction for RNIP that includes this magnetic attraction is also shown in Figure 8.10. For magnetic nanoparticles like RNIP and magnetite, the

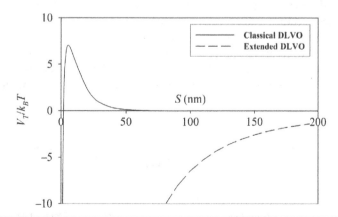

FIGURE 8.10   Classical DLVO simulations including EDL repulsions and van der Waals attractive forces predict an energy barrier to aggregation of ~7.0 $k_B T$, which should limit aggregation. Including magnetic attractive forces (dashed line) indicates no energy barrier, and that attractive forces may be as long range as a few hundred nanometers. S is the separation distance between the particles in nanometers. V_T is the sum of the attractive and repulsive forces acting on the particles.

magnetic attraction dominates the interaction energy and there is no longer a predicted energy barrier to aggregation. In fact, extended DLVO suggests that there are relatively long-range attractive forces (~250 nm) for nanoiron particles. This is in agreement with the rapid aggregation observed for RNIP.

Rapid aggregation makes it difficult to predict their transport in porous media since the rates of these transport processes are influenced by particle size. It also makes it difficult to predict the potential toxicity associated with these particles—it is not known if 5-micron-sized aggregates of nanometer-sized particles illicit a toxic response that is the same as or different from a concentrated suspension of the 100-nm-sized particles of equal mass concentration. Rapid aggregation, however, may have advantages. Because flocculation/aggregation of fine particles followed by sedimentation and filtration are commonly used to treat drinking water (Viessman and Hammer 1998), the tendency of nanoparticles to rapidly aggregate suggests that they will be easily removed in traditionally used water-treatment systems.

The rapid aggregation of most types of nanoparticles has led to the use of coatings to modify their surface chemistry in a way that minimizes aggregation and increases the stability of aqueous dispersions. Surfactants and natural and synthetic polymers have been proposed as a means to stabilize nanoiron suspensions in order to make them transport effectively in the subsurface (Saleh et al. 2005, 2007). This will affect their fate and transport characteristics in the environment, and may also affect their ability to be removed from water supplies or during treatment, so the impact of surface coatings on the environmental fate and transport should be considered in the design process for the coatings. For example, coatings could be used to deliver particles in the subsurface, but then biodegrade, transform, or desorb after some time so that the particles are no longer mobile and the potential for exposure is minimized.

Surface Modification

Effective use of many of the new nanoparticles being developed will require that particle aggregation be limited. For example, metal-containing nanoparticles being considered as MRI contrast agents will have to be designed to be dispersible in aqueous environments with high osmolarity such as blood (Sitharaman et al. 2004; Veiseh et al. 2005). Surface functionalization is a common tool for minimizing or controlling nanoparticle aggregation. This niche of nanotechnology has the potential to create nanoparticles that could be widely distributed and mobile in the environment, easily assimilated into people or other biological organisms, and difficult to remove by remediation and treatment if needed. The uncertainties surrounding the effect of surface coatings on the fate, transport, and potential toxicity of engineered nanomaterials makes this a rich area for current and future research on nanotechnology. Nanoparticles used for *in situ* groundwater remediation will require surface coatings for their intended application (Saleh et al. 2005a; Saleh et al. 2007). There are three classes of typical surface coatings used, including polymers, polyelectrolytes, and surfactants. These coatings can impart charge to the particles (positive or negative) and can provide three modes of colloidal stabilization that make them mobile in the environment: electrostatic, steric, or electrosteric (Figure 8.11). Both natural and synthetic varieties of each type of modifier are widely available and used (Table 8.2).

In general, high molecular weight polymers (synthetic or natural) provide steric repulsions that may limit nanoparticle-bacteria interactions (Figure 8.11). Polyelectrolytes are large polymers containing charged functional groups (anionic or cationic) and provide electrosteric repulsions. Surfactants can provide electrostatic repulsive or attractive forces (depending on their charge), but are less effective than polymers or polyelectrolytes because their small size is unsuitable for steric repulsions. Polymers adsorb strongly (effectively irreversible) to nanoparticles, while surfactant adsorption is more easily reversible (Braem et al. 2003; Fleer et al. 1993; Holmberg et al. 2003; Velegol and Tilton 2001). Electrostatic, steric, or electrosteric repulsions decrease nanoparticle interactions with mineral grains (Saleh et al. 2007) and

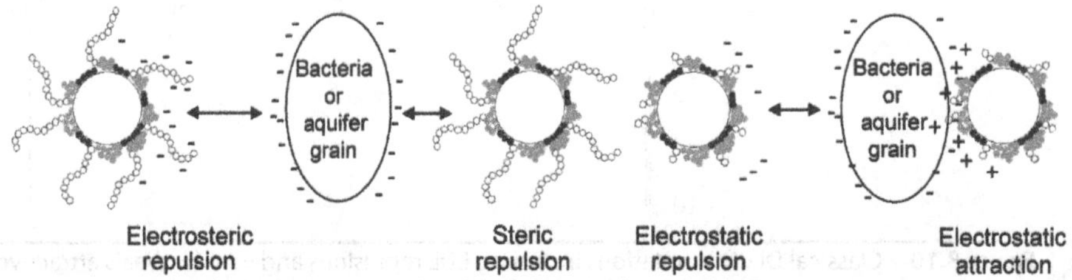

| Electrosteric repulsion | Steric repulsion | Electrostatic repulsion | Electrostatic attraction |

FIGURE 8.11 Surface coatings may prevent interactions between nanoparticles and bacteria or mineral grains. Polyelectrolytes provide electrosteric repulsive forces, while polymers provide steric repulsive forces. Short chain anionic or cationic surfactants can provide electrostatic repulsive or attractive forces.

Coating	Charge	Stabilization Type	Relevance
Polymers			Common nontoxic polymer used to stabilize NP dispersions.
Polyethylene glycol (PEG)	Nonionic	Steric	Adsorbed PVA reduces bacterial adsorption to surfaces (Koziarz
Polyvinyl alcohol (PVA)	Nonionic	Steric	and Yamazaki 1999). Cellulose is used to prepare stabilized
Carboxymethyl cellulose	Nonionic	Steric	nanoiron dispersions (He and Zhao 2005). Inexpensive and
Guargum	Nonionic	Steric	biodegradable.
Polyelectrolytes			Used to prepare stabile nanoiron dispersions that partition to
Triblock copolymers (PMAA$_x$-PMMA$_y$-PSS$_z$)	Anionic	Electrosteric	the TCE/water interface (Saleh et al. 2005a). PSS homopolymer is a common inexpensive alternative used to stabilize NP dispersions. Polylysine is a biodegradable polypeptide and
Polystyrene sulfonate	Anionic	Electrosteric	bactericidal (Roddick-Lanzilotta and McQuillan 1999).
Poly(aspartic acid)	Anioinc	Electrostatic	Poly(aspartic acid) should be biodegradable and less bactericidal.
Surfactants			SDBS is a common anionic surfactant shown to enhance
SDBS	Anionic	Electrosteric	nanoiron mobility (Saleh et al. 2007) but resists biodegradation.
Alkyl polyglucosides	Nonionic	Steric/Hydration	Alkyl polyglucosides are biodegradable (Matsson et al. 2004).

TABLE 8.2   Coatings from Each Class to Be Evaluated

potentially also with soil bacteria, which may decrease the observed bactericidal properties of nano-particles. For example, Goodman et al. (2004) found that gold nanoparticles with positively charged side chains were toxic to *E. coli*, but negatively charged particles were not. Rose et al. (2005) found that positively charged CeO$_2$ nanoparticles adsorbed to *E. coli* and were bactericidal, showing a clear dose-response.

Poly(methacrylic acid)-b-poly(methylmethacrylate)-b-poly(styrene sulfonate) triblock copolymers (PMAA-PMMA-PSS), PSS homopolymer polyelectrolyte, sodium dodecylbenzene sulfonate (SDBS), polyethylene glycol (PEG), carboxymethyl cellulose (CMC), and guar gum have been shown to adsorb to and stabilize dispersions of nanoiron and Fe oxide nanoparticles to improve particle mobility in the environment due to steric, electrosteric, or electrostatic repulsions between the particles and between the particles and soil grains (He and Zhao 2005; Saleh et al. 2007). Poly(aspartic acid), an anionic poly-peptide, also adsorbs to iron oxide surfaces via acid-base interactions through the carboxyl groups (Chibowski and Wisniewska 2002; Drzymala and Fuerstenau 1987; Nakamae et al. 1989) and can be an effective stabilizer for metal-oxide nanoparticles. Alkyl polyglucosides are an emerging class of sur-factant synthesized from renewable raw materials and are nontoxic, biocompatible, and biodegradable. They adsorb to metal oxide surfaces (Matsson et al. 2004). They are desirable due to their low cost and "green" nature and widely used as detergents in manufacturing. A systematic investigation of the ability of these types of surface modifiers to enhance nanoiron mobility by minimizing particle-particle interac-tions as well as particle-media grain interactions is underway.

Polyelectrolytes have been used to provide functionality to nanoiron [e.g., affinity for the DNAPL/water interface (Saleh et al. 2005a)] as well as to enhance their transport in laboratory columns. For example, transport of triblock copolymer-modified nanoiron is significantly enhanced in a 10-cm sand column rela-tive to the unmodified nanoiron, especially at the high nanoiron concentration that is needed to be cost-effective (Figure 8.12) (Saleh et al. 2007). In this case, enhanced transport is due to an adsorbed layer of a strong polyelectrolyte that provides electrosteric repulsions, which limit the particle-particle interactions as well as the particle sand-grain interactions. Enhanced mobility comes at a price, however, as modified particles were four to nine times less reactive with TCE than for unmodified particles (Saleh et al. 2007). Despite the lower reactivity, the particles are sufficiently reactive with TCE to be effective *in situ* ground-water remediation agents.

Effect of Modifier Type on Transport

Different types of modifiers will provide different modes of stabilizing nanoparticles against aggregation and attachment to aquifer media grains. These differences lead to different elutability and transport distances. For example, the eluted masses of bare MRNIP (a polyaspartate-modified nanoiron), a triblock copolymer-modified nanoiron, and a surfactant-modified (SDBS) nanoiron through a 12.5-cm sand-filled column are

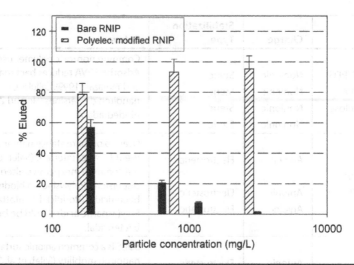

Figure 8.12   Surface modification by strong polyelectrolyte polymers enhances transport of nanoiron (RNIP) transport through saturated porous media. Polymer-modified iron (striped bars) is highly transportable through water-saturated sand columns, while unmodified nanoiron (black bars) does not transport well through the column, particularly at high concentrations (3 g/L).

shown in Figure 8.13. At 3 g/L, bare RNIP has very low transportability (1.4 ± 3% mass elution) through a saturated sand column at low ionic strength. Retrieving the sand from the column and analyzing for iron revealed that most particles were trapped within the first 1 to 2 cm of the column. MRNIP-, polymer-, and SDBS-modified RNIP elution was much higher, with the triblock copolymer and MRNIP elution at 95 percent and 98 percent, respectively. SDBS was not as effective as the polymer but still improved RNIP elution to approximately 50 percent. These results indicate that surface modification is essential for reasonable transport, even at low ionic strength. These differences could be used to synthesize particles with specific transport distances that can then be used for controlled delivery of nanoparticles to specific regions in the subsurface.

Geochemical Effects on Transport (pH, Ionic Strength, and Ionic Composition)

Each of the modified particles also responds differently to changes in ionic strength and to ionic composition. Both Na^+ and K^+ cations and Ca^{2+} and Mg^{2+} cations are prevalent in natural water systems. The presence of these monovalent and divalent cations tends to shield EDL repulsions between particles and between particles and aquifer grains. Shielding these repulsive forces increases the attachment efficiency of the particles to the sand grains and decreases the distance that they can travel through saturated porous media before being filtered from solution. Divalent cations are much more efficient at shielding EDL

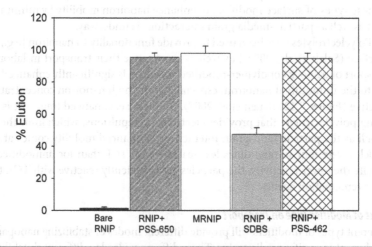

Figure 8.13   Percent mass of bare and modified RNIP eluted through a 12.5-cm silica sand column with porosity of 0.33. Modifying agents, i.e., PMAA-PMMA-PSS polymer or SDBS, were added at 2 g/L concentration in each case. Polyaspartic acid (MRNIP) was added at a 6:1 mass ratio. The approach velocity was 93 m/d.

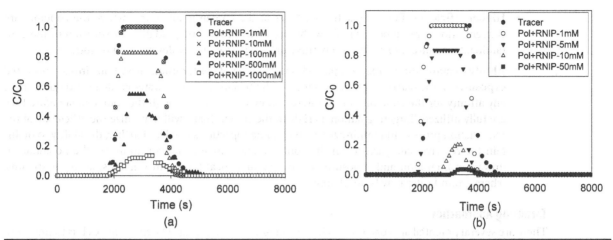

FIGURE 8.14   (a) Effect of ionic strength (NaCl) and (b) cation type (Ca^{21}) on transport of polyelectrolyte-modified nanoiron through a 21-cm water-saturated sand column.

repulsions than are monovalent cations due to their higher charge density. Surface modifications that provide strong EDL repulsions as well as steric hindrances, termed electrosteric repulsions, should in principle do the best job of minimizing attachment of particles to sand grains. Saleh et al. (2007) measured the elution of each surface modified particle under varying ionic strength conditions and determined that large molecular weight polyelectrolytes indeed provided the best elutability. This was attributed to the ability of the large molecular weight polymers to provide strong electrosteric repulsions compared to short chain polyelectrolytes such as polyaspartic acid, or short chain surfactants such as SDBS. These results demonstrate that the site-specific geochemistry must be taken into account when developing dispersants to enhance nanoiron delivery.

Conditions such as high ionic strength and the presence of divalent cations in even small quantities tend to increase retention of nanoparticles by porous media. Since groundwater aquifers and surface waters typically have ionic strengths in excess of 10^{-4} M and frequently have significant concentrations of calcium or magnesium, conditions should tend to favor nanomaterial deposition. Even the electrosteric repulsions provided by surface modifiers such as polymers or polyelectrolytes can be overcome at high ionic strength and in the presence of divalent cations such as Ca^{2+} and Mg^{2+}. For example, the approximately 100 percent transport of PMAA-PMMA-PSS modified RNIP through a 21-cm sand column up to 10 mM of a monovalent cations (NaCl) (Figure 8.14a), decreases at ionic strengths of greater than 100 mM. In the presence of divalent cations, such as Ca^{2+} which are much more effective at screening the charge on particles, transport is reduced dramatically at concentrations of Ca^{2+} greater than 5 mM (Figure 8.14b). The electrosteric repulsions provided by the polyelectrolyte used here are much more effective than for surfactants that rely solely on electrosteric repulsions, or natural polymers or surfactants (e.g., alginate), and should be considered a best-case scenario (i.e., most transportable). Even under these conditions, only transport distances of a few 10s of meters are expected since these cations are present in most surface and subsurface waters. Thus even nanomaterials that are engineered to be highly mobile in the subsurface are not expected to be exceptionally mobile in the subsurface. This poses challenges from a standpoint of delivering the particles *in situ*, but comes as a relief from the standpoint of potential risks posed by releasing these particles into the environment. Further, this implies that the incidental release of engineered nanomaterials that inadvertently contaminant our current water-treatment infrastructure should be easily removed as these processes employ ionic strength increases and addition of divalent cations to destabilize particles to remove them.

Targeting

The ability to target specific contaminants is important in any *in situ* groundwater remediation scheme. Specificity is advantageous for several reasons:

- **Higher contaminant transformation rates.** Since most reactions and sorption processes are first order with respect to the contaminant of interest, placing reactive or sorptive particles in the sources area where contaminant concentrations are highest will provide the most rapid transformations.

- **Higher efficiency.** Placing reactive particles at the contaminant source where the contaminant concentrations are highest will allow the preferred process (degradation or sorption of the contaminant) to out-compete the competitive processes (e.g., H_2 evolution from nanoiron).

- **Ability to maintain the reactive particles in the area of contamination and minimize unwanted exposures.** If the reactive particles migrate with the natural groundwater gradient and do not have any affinity for the contaminant of interest, they may move through the source area before they are fully utilized. Targeting nanomaterials to the contaminant will maximize the efficiency of the remediation process and minimize the mass of nanoparticles needed. Further, the ability to maintain the reactive nanoparticles in the contaminant source zone will minimize the potential for unwanted migration and exposure of sensitive biological targets in wetlands, lakes, or streams, where groundwater may be discharging.

Targeting Approaches

There are several potential approaches to achieve contaminant targeting. The first approach is to impart to the particles some affinity for the specific contaminant of interest. This targeted delivery approach is inspired by targeted drug delivery technology—localizing the remediation agents (the "drug") at the contaminant source (the "diseased tissue") via "smart" thermodynamic interactions between the particles and the contaminant. This approach has been recently evaluated in the laboratory (Saleh et al. 2005a, 2005b, 2007). Saleh et al. (2005) used multifunctional triblock copolymers (Figure 8.15) adsorbed to the nanoiron particles to disperse nanoiron into water for good aquifer transportability, minimize undesirable adhesion to mineral and natural organic matter (NOM) surfaces, and preferentially anchor nanoiron at the DNAPL/water interface for source-zone accumulation (Figure 8.16) (Saleh et al. 2005a). The multifunctional polymer assemblies adsorbed on the nanoiron surfaces present hydrophilic blocks that form stable nanoparticle suspensions in water, a requirement for eventual subsurface delivery. The hydrophilic blocks are strong polyanions, so they repel the predominantly negatively charged surfaces of minerals, NOM, and NOM-coated minerals that would be encountered in soils (Beckett and Le 1990; Day et al. 1994). When in contact with DNAPL, hydrophobic polymer blocks respond by anchoring the nanoparticle at the DNAPL/aqueous interface (Figure 8.16).

The affinity of the triblock copolymer modified nanoiron has been examined in the laboratory. RNIP interfacial anchorage by the same polymers that promoted their transport through sand columns was demonstrated by the ability of modified nanoiron to stabilize TCE-in-water emulsions (Saleh et al. 2005a). Emulsification of immiscible fluids is proof of adsorption at the fluid interface, as bare interfaces are unstable to droplet coalescence and macroscopic phase separation. Figure 8.17 shows a micrograph of a TCE-in-water emulsion that was stable for over 3 months. It was stabilized by RNIP particles that had been modified by irreversible adsorption of $PMAA_{42}$-$PMMA_{26}$-PSS_{466} followed by several cycles of centrifugation and washing. For comparison, TCE and water could not be emulsified by unmodified RNIP particles, proving that the polymer was essential for anchorage at the interface. Likewise, no emulsion was formed using the supernatant of a centrifuged polymer-modified RNIP sample, proving that the interface was stabilized by adsorption of polymer-modified RNIP, and not by adsorption of free polymer that may have been present in solution with the nanoiron. Furthermore, RNIP modified by PSS homopolymers was unable to

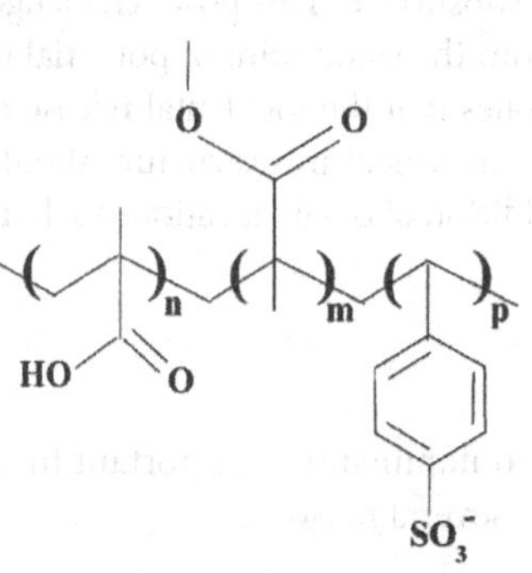

Figure 8.15 Hydrophobic-hydrophilic triblock copolymers containing a short PMAA anchoring group (left), PMMA hydrophobic block (middle), and a hydrophilic sulfonated polystyrene block (right).

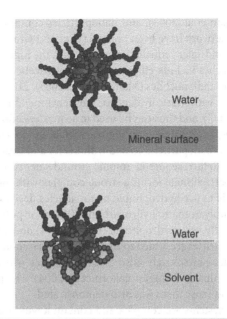

FIGURE 8.16 The targeting mechanism. Adsorbed block copolymers contain a polyelectrolyte (dark gray) block that has an affinity for water and a hydrophobic (light gray) block that has an affinity for DNAPL. In water, the polyelectrolyte block swells and the hydrophobic block collapses. The reverse happens in the DNAPL phase. This amphiphilicity anchors the particle at the DNAPL/water interface. The polyelectrolyte block is large enough to stably suspend particles in water without aggregating. The strong negative charge in the polyelectrolyte block minimizes particle adhesion to negatively charged mineral or natural organic matter surfaces in the soil before reaching the DNAPL.

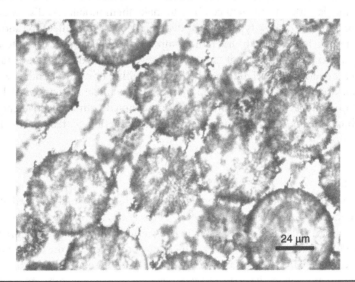

FIGURE 8.17 Optical micrograph of TCE-in-water emulsion stabilized by PMAA-PMMA-PSS-modified RNIP particles, proving that the modified particles adsorb at the TCE/water interface. Nanoiron did not partition into the TCE droplets.

emulsify TCE and water. This demonstrates that the type and composition of the surface modifiers are critical design variables for interfacial targeting.

There are many other approaches that could be used to achieve significant targeting of contaminants. For example, enzymes-based approaches may be possible. Enzymes are proteins or conjugated proteins produced by living organisms and functioning as biochemical catalysts. They typically have complex active sites that are highly selective for specific compounds. This selectivity may be leveraged to tailor reactive nanoparticles for specific compounds. Enzyme-coated carbon nanotubes have been used as single molecule biosensors (Besteman et al. 2003), and nanoparticles containing a single enzyme protected by a porous inorganic/organic network have been created (Kim and Grate 2003). These approaches may eventually be used to develop groundwater remediation agents with the highest specificity for target compounds possible. Less specific approaches include ethylenediaminetetraacetic acid (EDTA) coatings on TiO_2 nanoparticles designed to sequester radionuclides (Mattigod et al. 2005). EDTA is a strong chelating agent that forms coordination compounds with most divalent (or trivalent) metal ions, such as calcium (Ca^{2+}) and magnesium (Mg^{2+}) or copper (Cu^{2+}), and thus can be used to sequester cationic groundwater contaminants such as Cu^{2+} or Pb^{2+}. Once strongly sequestered, these toxic metals are no longer bioavailable and pose less or no risk to biota. Anatase (TiO_2) nanoparticles coated with EDTA and treated with Cu^{2+} to form an EDTA/Cu(II) complex have been used to then sequester anionic groundwater contaminants such as pertechtinate (TcO_4^-). It is proposed that the pertechtinate forms a strong complex with the bound EDTA/Cu(II) on the TiO_2 surface. Another approach is to use hydrophobic nanoparticles designed to strongly sequester hydrophobic contaminants such as polyaromatic hydrocarbons (PAHs) or polychlorinated biphenyls (PCBs). These contaminants are typically very hydrophobic (log $K_{OW} > 4$) and strongly adsorb to soil and sediment. These hydrophobic nanoparticles are added to the PCB- or PAH-contaminated soil or sediment, where the hydrophobic contaminants can strongly adsorb to them, after which, the nanoparticles are removed, thereby removing the contaminants (Tungittiplakorn et al. 2004). The potential to remove the contaminants from the nanoparticles and reuse them was also demonstrated.

Environmental systems are complex and contain a wide variety of organic and inorganic constituents that can compete for reactive sites regardless of the contaminant of choice. For example, polymeric hydrophobic nanoparticles designed to sequester PAHs or PCBs might also sequester hydrophobic soil organic matter and block access to the contaminants. In the case of DNAPL targeting, the available surface area of DNAPL could be small in comparison to the total mass (e.g., pools), thereby limiting the mass of nanoiron that could be delivered to the DNAPL/water interface. It may be sufficient to use less specific methods to control the travel distance such that reactive groundwater reagents can simply be placed near the vicinity of the source and remain there. This avoids the need for highly complex surface chemistry to provide selectivity, which will likely be expensive. Most metal-oxide nanoparticles will be relatively immobile in saturated porous media without sufficient coatings to make them mobile. As discussed previously, the transport distance for a specific coating type will vary depending on the modifier type and on the geochemical conditions at the site (pH, ionic strength, and ionic composition). This implies the potential to match the modifier type to the geochemical conditions at the site to achieve a specified transport distance. With this approach, remedial agents could be delivered to specific regions within or near the source zone. This could also be done by using surface coatings that desorb at a specific rate such that the nanoparticles travel a specific distance and then stop. Nanoparticle surface modifications, coupled with innovative engineered delivery schemes, will undoubtedly be necessary to achieve adequate targeting of nanoparticulate reagents for groundwater remediation.

Challenges

Subsurface heterogeneity and complex NAPL architecture make remediation difficult (Dai et al. 2001; Daus et al. 2001; Illangasekare et al. 1995). These issues also pose significant challenges to accurately delivering nanoparticulate remedial agents to subsurface contaminants. Nanoparticles will tend to travel along zones of high hydraulic conductivity, which may or may not be the desired placement area. For nanoparticles to reach the NAPL/water interface, particles may be required to diffuse across flow lines as they travel through the porous media. Even 10-nm particles may have prohibitively low diffusion coefficients. For example, a micromodel study by Baumann et al. (2005) investigated the targeting ability of a triblock copolymer-modified nanoiron that had been shown to partition to the NAPL/water interface ex situ (Saleh et al. 2007). At approach velocities of 2 to 13 m/d (residence time of 1 to 10 seconds), particles tended to flow past entrapped TCE rather than migrate to the interface (Figure 8.18).

Even though emulsification occurs under high shear conditions that of course are far removed from aquifer conditions, preliminary experiments in sand-packed columns and dodecane-coated sand-packed columns indicate some potential for *in situ* targeting as long as adequate time is available for nanoparticles to diffuse to the NAPL/water interface. Sand column transport studies conducted with NAPL-coated sand, under flow conditions similar to the clean sand experiments described above, indicated a 10 percent reduction in elution for $PMAA_{42}$-$PMMA_{26}$-PSS_{466}-coated RNIP compared to the clean sand column (Saleh et al. 2007). To achieve this, however, the flow had to be stopped for 24 hours to allow time for the particles to transport to the interface. Surface modifications that impart more hydophobicity to the particle should

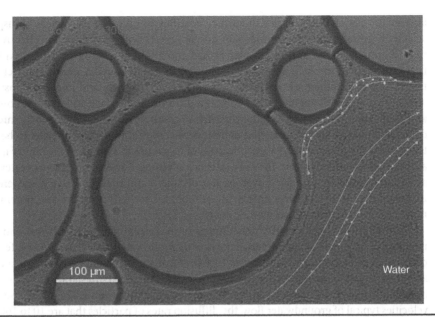

FIGURE 8.18   Trajectories of polymer-coated nanoiron in an etched silica micromodel containing water and partial TCE saturation. Particles tended to migrate toward the TCE/water interface as the approach velocity decreased. Trajectory points are 300 ms apart (Baumann et al. 2005).

provide better NAPL targeting. Using a higher hydrophobe/hydrophile ratio, or changing the middle hydrophobic block from methyl-methacrylate to butyl-methacrylate to lower the glass transition temperature and thus promote swelling of the hydrophobe in contact with NAPL may further enhance targeting. These targeting experiments indicate potential for *in situ* targeting, but additional research to optimize the block size and type and hydrophile/hydrophobe ratio are needed.

In principle, nanoparticles that have a strong affinity and selectivity for the target contaminant could be injected into the subsurface and allowed to migrate nonspecifically toward the source area where they will locate and degrade or sequester the target contaminant. This is analogous to targeted drug delivery where drugs are introduced into the body and migrate toward the target (diseased tissue). If this is not achievable, delivering nanoparticles near the source area may be acceptable. This will, however, require that the source area be very well characterized in terms of the contaminant masses and locations, and that the hydrogeology at the site be well known. Without this characterization, it will be difficult to engineer and implement an effective delivery system. Source zone characterization is difficult, however, so techniques to effectively characterize these areas are needed. Partitioning tracers (e.g., alcohols) and interfacial tracers (e.g., surfactants) have shown some promise for characterizing DNAPL source zones (Annable et al. 1998). Methods to characterize sources of heavy metals such as Cr(VI) are less well developed.

Summary and Research Needs

Unique and inexpensive reactive or highly sorptive engineered nanomaterials are becoming commercially available. The use of these novel engineered nanomaterials for improved *in situ* remediation of groundwater contaminants is promising, but engineering nanomaterials that are highly selective for the reactions of interest and that have long lifetimes such that they are cost-effective for *in situ* remediation of groundwater contaminants remain active areas of research. Potential concerns about the exposure and human health effects of these engineered nanomaterials must also be addressed to ensure the safe deployment of these materials. Future research developing novel nanomaterials for groundwater remediation will have to consider the reactivity, selectivity, and longevity of materials as well and the potential exposure and human health effects of the materials.

Even though relatively inexpensive commercially produced reactive nanomaterials for *in situ* groundwater remediation (e.g., nanoiron) are available, methods to deliver these materials to subsurface contaminants remain a challenge. Intuitively, the assumption is often made that nanoparticles will be more mobile in porous media due to their small size. However, all other factors being equal, smaller particles are less mobile due to their relatively large diffusivity that produces more frequent contacts with the surfaces of aquifer porous media. The use of surface coatings to enhance subsurface transport and to encourage selectivity toward specific groundwater contaminants will be needed for effective deployment of engineered nanomaterials in groundwater remediation. These coatings can be used to control aggregation (particle-particle interactions) and attachment to aquifer media grains (particle-media grain interactions). The geochemistry at each site (e.g., pH, ionic strength, and ionic composition) will dramatically affect the mobility of engineered nanomaterials in the subsurface and must be considered in their design and application. The ability to tailor the surface coatings to site-specific geochemical conditions for well-controlled placement of engineered nanomaterials in the subsurface appears to be obtainable in the near term and will enhance the cost-effectiveness of this remedial approach. A higher mobility of nanomaterials in the environment due to the use of surface coatings, however, implies a greater potential for exposure as nanomaterials are dispersed over greater distances and their effective persistence in the environment increases. The trade-off between enhanced mobility and effectiveness and potential exposure and risks must be considered on a case-by-case basis.

The ability to target specific contaminants or to concentrate the reactive nanomaterials in the contaminant source zone will be required to make *in situ* groundwater remediation cost-effective. The use of surface coatings to enhance target specificity is very promising and remains an active area of research. Despite the ability to develop nanomaterials with target specificity, there remain significant challenges for *in situ* delivery of nanomaterials in the subsurface due to unfavorable hydrodynamics in many cases. Even at low approach velocities typical of groundwater flow, the diffusion rates of particles that are 10 to 100 nm in diameter across flow lines to adsorbed contaminants may be prohibitively slow to allow them to diffuse across flow lines to reach their targets. Methods to tailor the surface chemistry of nanomaterials to concentrate reactive materials in specific regions in the subsurface appear promising, and may be a more obtainable near-term goal to achieve well-controlled placement of nanomaterials in contaminant source zones.

Acronyms and Symbols

aq: aqueous

CMC: carboxy methylcellulose

COC: chlorinated organic compounds

DCA: dichloroethane

DCE: dichloroethene

DIRB: dissimilatory iron reducing bacteria

DLVO: Deraguin-Landau-Verwey-Overbeek

DNAPL: dense nonaqueous phase liquid

d_p: particle diameter (L)

e^-: electron

EDL: electrostatic double layer

EDTA: ethylenediaminetetraacetic acid

Fe(B): nanoiron synthesized from reduction of dissolved iron using sodium borohydride

H^*: adsorbed hydrogen

HCA: hexachloroethane

k_B: Boltzman's constant (J/K)

k_{H_2obs}: observed rate constant for H_2 evolution from nanoiron corrosion in water (t^{-1})

K_{OW}: octanol-water partition coefficient, $M_{i,o}/M_{i,w}$ (–)

k_{rxn}: surface area normalized pseudo-first-order reaction rate constant (L hr^{-1} m^{-2})

Me^{a+}: metal ion of charge a

MRNIP: modified RNIP (modified with polyaspartic acid)

NAPL: nonaqueous phase liquid

NOM: natural organic matter

PAH: polyaromatic hydrocarbon

PCA: pentachloroethane

PCB: polychlorinated biphenyl

PEG: polyethylene glycol

$PMAA_x$: poly(methacrylic) acid; subscript represents the degree of polymerization

$PMMA_y$: poly(methylmethacrylate); subscript represents the degree of polymerization

PSS_z: poly(styrene sulfonate); subscript represents the degree of polymerization

PVA: polyvinyl alcohol

RNIP: reactive nanoscale iron particles

SA: surface area

SDBS: sodium dodecylbenzene sulfonate

T: temperature (K)

TCA: trichloroethane

TCE: trichloroethylene

TeCA: tetrachloroethane

VC: vinyl chloride

V_T: potential (J)

Φ: volume fraction; volume of particles/total volume (–)

References

1. Agrawal, A., and Tratnyek, P. G. (1996). "Reduction of Nitro Aromatic Compounds by Zero-Valent Iron Metal." *Environmental Science & Technology*, 30(1), 153–160.
2. Annable, M. D., Jawitz, J. W., Rao, P. S. C., Dai, D. P., Kim, H., and Wood, A. L. (1998). "Field Evaluation of Interfacial and Partitioning Tracers for Characterization of Effective NAPL-Water Contact Areas." *Ground Water*, 36, 495–503.
3. Baumann, T., Keller, A. A., Auset-Vallejo, M., and Lowry, G. V. "Micromodel Study of Transport Issues during TCE Dechlorination by ZVI Colloids." *American Geophysical Union Fall Meeting*, San Francisco, CA.
4. Beckett, R., and Le, N. P. (1990). "The Role of Organic Matter and Ionic Composition in Determining the Surface Charge of Suspended Particles in Natural Waters." *Colloids and Surfaces*, 44, 35–49.
5. Besteman, K., Lee, J.-O., Wiertz, F. G. M., Heering, H. A., and Dekker, C. (2003). "Enzyme-Coated Carbon Nanotubes as Single-Molecule Biosensors." *Nano Letters*, 3(6), 727–730.
6. Bonin, P. M., Odziemkowski, M. S., and Gilham, R. W. (1998). "Influence of Chlorinated Solvents on Polarization and Corrosion Behaviour of Iron in Borate Buffer." *Corrosion Science*, 40, 1391–1409.
7. Braem, A. D., Biggs, S., Prieve, D. C., and Tilton, R. D. (2003). "Control of Persistent Nonequilibrium Adsorbed Polymer Layer Structure by Transient Exposure to Surfactants." *Langmuir*, 19, 2736–2744.
8. Butter, K., Bomans P. H., Frederik, P.M., Vroege, G.J., Philipse, A.P. (2003a). "Direct Observation of Dipolar Chains in Iron Ferrofluids by Cryogenic Electron Microscopy." *Nature Materials*, 2, 88–91.
9. Butter, K., Bomans P. H., Frederik, P.M., Vroege, G.J., and Philipse, A.P. (2003b). "Direct Observation of Dipolar Chains in Iron Ferrofluids in Zero Field Using Cryogenic Electron Microscopy." *Journal of Physics: Condensed Matter*, 15, 1415–1470.
10. Cao, J., Elliott, D., and Zhang, W.-X. (2005). "Perchlorate Reduction by Nanoscale Iron Particles." *Journal of Nanoparticle Research*, 7, 499–506.
11. Chibowski, S., and Wisniewska, M. (2002). "Study of Electrokinetic Properties and Structure of Adsorbed Layers of Polyacrylic Acid and Polyacrylamide at Fe_2O_3-Polymer Solution Interface." *Colloids and Surfaces A*, 208, 131–145.
12. Choe, S., Chang, Y.-Y., Hwang, K.-Y., and Khim, J. (2000). "Kinetics of Reductive Denitrification by Nanoscale Zero-Valent Iron." *Chemosphere*, 41, 1307–1311.
13. Dai, D., Barranco, F. T. J., and Illangasekare, T. H. (2001). "Partitioning and Interfacial Tracers for Differentiating NAPL Entrapment Configuration: Column-Scale Investigation." *Environmental Science & Technology*, 35, 4894–4899.
14. Daus, D. A., Kent, B., and Mosquera, G. C. B. (2001). "A Case Study of DNAPL Remediation in Northwestern Brazil." *Journal of Environmental Science & Health*, 36, 1505–1513.
15. Day, G. M., Hart, B. T., McKelvie, I. D., and Beckett, R. (1994). "Adsorption of Natural Organic Matter onto Goethite." *Colloids and Surfaces A*, 89, 1–13.
16. de Gennes, P.-G. P., P.A. (1970). "Pair Correlations in a Ferromagnetic Colloid." *Phys. Kondens. Mater.*, 11, 189–198.
17. de Vicente, J., Delgado, A. V., Plaza, R. C., Durán, J. D. G., and González-Caballero, F. (2000). "Stability of Cobalt Ferrite Colloidal Particles: Effect of pH and Applied Magnetic Fields." *Langmuir*, 212, 14–23.
18. Drzymala, J., and Fuerstenau, D. W. (1987). "Adsorption of Polyacrylamide, Partially Hydrolyzed Polyacrylamide and Polyacrylic Acid on Ferric Oxide and Silica." *Polymer Process Technology*, 4, 45–60.

19. Elimelech, M., Gregory, J., Jia, X., and Williams, R. (1995). *Particle Deposition and Aggregation: Measurement, Modeling, and Simulation*, Butterworth-Heinemann, Oxford, England.

20. Evans, D. F. W., H. (1999). *The Colloidal Domain: Where Physics, Chemistry, Biology, and Technology Meet*, Wiley-VCH, New York.

21. Fleer, G. J., Cohen, Stuart, M. A., Scheutjens, J. M., Cosgrove, H. M., and Vincent, T. B. (1993). *Polymers at Interfaces*, Chapman and Hall, New York.

22. Gavaskar, A., Tatar, L., and Condit, W. (2005). "Cost and Performance Report: Nanoscale Zero-Valent Ion Technologies for Source Remediation." *CR-05-007-ENV*, Naval Facilities Engineering Command, Port Hueneme.

23. Gerlach, R., Cunningham, A. B., and Caccavo, F. (2000). "Dissimilatory Iron-Reducing Bacteria Can Influence the Reduction of Carbon Tetrachloride by Iron Metal." *Environmental Science & Technology*, 34(12), 2461–2464.

24. Goodman, C. M., McCusker, C. D., Yilmaz, T., and Rotello, V. M. (2004). "Toxicity of Gold Nanoparticles Functionalized with Cationic and Anionic Side Chains." *Bioconjugate Chemistry*, 15, 897–900.

25. Hamilton, W. A. "Microbially Influenced Corrosion in the Context of Metal Microbe Interactions." *Microbial Corrosion: Proceedings of the 4th International EFC Workshop*, Sequeira, Portugal, 3–17.

26. He, F., and Zhao, D. (2005). "Preparation and Characterization of a New Class of Starch-Stabilized Bimetallic Nanoparticles for Degradation of Chlorinated Hydrocarbons in Water." *Environmental Science & Technology*, 39, 3314–3320.

27. Heimenz, P. C. R. (1997). *Principles of Colloid and Surface Chemistry*, Marcel Dekker, New York.

28. Holmberg, K., Jönsson, B., Kronberg, B., and Lindman, B. (2003). *Surfactants and Polymers in Aqueous Solution*, John Wiley & Sons, West Sussex.

29. Illangasekare, T. H., Yates, D., N., and Armbruster, E. J. (1995). "Effect of Heterogeneity on Transport and Entrapment on Nonaqueous Phase Waste Products in Aquifers: An Experimental Study." *Journal of Environmental Engineering*, 121, 572–579.

30. Kanel, S. R., Manning, B., Charlet, L., and Choi, H. (2005). "Removal of Arsenic(III) from Groundwater by Nanoscale Zero-Valent Iron." *Environmental Science & Technology*, 39, 1291–1298.

31. Kim, J., and Grate, J. W. (2003). "Single-Enzyme Nanoparticles Armored by a Nanometer-Scale Organic/Inorganic Network." *Nano Letters*, 3(9), 1219–1222.

32. Koziarz, J., and Yamazaki, H. (1999). "Stabilization of Polyvinyl Alcohol Coating of Polyester Cloth for Reduction of Bacterial Adhesion." *Biotechnology Techniques*, 13(4), 221–225.

33. Lecoanet, H. F., Bottero, J.-Y., and Wiesner, M. R. (2004). "Laboratory Assessment of the Mobility of Nanomaterials in Porous Media." *Environmental Science & Technology*, 38(19), 5164–5169.

34. Lecoanet, H. F., and Wiesner, M. R. (2004). "Velocity Effects on Fullerene and Oxide Nanoparticle Deposition in Porous Media." *Environmental Science & Technology*, 38(16), 4377–4382.

35. Lien, H. L., and Zhang, W. X. (1999). "Transformation of Chlorinated Methanes by Nanoscale Iron Particles." *Journal of Environmental Engineering*, 125, 1042–1047.

36. Lien, H.-L., and Zhang, W.-X. (2005). "Hydrodechlorination of Chlorinated Ethanes by Nanoscale Pd/Fe Bimetallic Particles." *Journal of Environmental Engineering*, 131, 4–10.

37. Liu, Y., Choi, H., Dionysiou, D., and Lowry, G. V. (2005a). "Trichloroethene Hydrodechlorination in Water by Highly Disordered Monometallic Nanoiron." *Chemistry of Materials*, 17(21), 5315–5322.

38. Liu, Y., Majetich, S. A., Tilton, R. D., Sholl, D. S., and Lowry, G. V. (2005b). "TCE Dechlorination Rates, Pathways, and Efficiency of Nanoscale Iron Particles with Different Properties." *Environmental Science & Technology*, 39(5), 1338–1345.

39. Lowry, G. V., and Johnson, K. M. (2004). "Congener Specific Dechlorination of Dissolved PCBs by Microscale and Nanoscale Zero-Valent Iron in a Water/Methanol Solution." *Environmental Science & Technology*, 38(19) 5208–5216.

40. Lowry, G. V., and Reinhard, M. (1999). "Hydrodehalogenation of 1- to 3-Carbon Halogenated Organic Compounds in Water Using Palladium Catalyst and Hydrogen Gas." *Environmental Science & Technology*, 33(11), 1905–1911.

41. Lowry, G. V., and Reinhard, M. (2000). "Pd-catalyzed TCE Dechlorination in Groundwater: Solute Effects, Biological Control, and Oxidative Catalyst Regeneration." *Environmental Science & Technology*, 34(15), 3217–3223.

42. Matsson, M. K., Kronberg, B., and Claesson, P. M. (2004). "Adsorption of Alkyl Polyglucosides on the Solid/Water Interface: Equilibrium Effects of Alkyl Chain Length and Head Group Polymerization." *Langmuir*, 20(10), 4051–4058.

43. Mattigod, S. V., Fryxell, G. E., Alford, K., Gilmore, T., Parker, K., Serne, J., and Engelhard, M. (2005). "Functionalized TiO_2 Nanoparticles for Use for in Situ Anion Immobilization." *Environmental Science & Technology*, 39(18), 7306–7310.

44. McCurrie, R., A. (1994). *Ferromagnetic Materials: Structure and Properties*, Academic Press, London.

45. Mondal, K., Jegadeesan, G., and Lalvani, S. B. (2004). "Removal of Selenate by Fe and NiFe Nanosized Particles." *Industrial & Engineering Chemistry Research*, 43, 4922–4934.

46. Moore, A. M., De Leon, C. H., and Young, T. M. (2003). "Rate and Extent of Aqueous Perchlorate Removal by Iron Surfaces." *Environmental Science & Technology*, 37(14), 3189–3198.

47. Nakamae, K., Tanigawa, S., Nakano, S., and Sumiya, K. (1989). "The Effect of Molecular Weight and Hydrophilic Groups on the Adsorption Behavior of Polymers onto Magnetic Particles." *Colloids and Surfaces A*, 37, 379.

48. Neto, C. B., Bonini, M., and Baglioni, P. (2005). "Self-Assembly of Magnetic Nanoparticles into Complex Superstructures: Spokes and Spirals." *Colloids and Surfaces A*, 269, 96–100.

49. Nurmi, J. T., Tratnyek, P. G., Sarathy, V., Baer, D. R., Amonette, J. E., Pecher, K., Wang, C., et al. (2005). "Characterization and Properties of Metallic Iron Nanoparticles: Spectroscopy, Electrochemistry, and Kinetics." *Environmental Science & Technology*, 39(5), 1221–1230.

50. Nutt, M. O., Hughes, J. B., and Wong, M. S. (2005). "Designing Pd-on-Au Bimetallic Nanoparticle Catalysts for Trichloroethene Hydrodechlorination." *Environmental Science & Technology*, 39(5), 1346–1353.

51. Oh, B. T., and Alvarez, P. J. J. (2002). "Hexahydro-1,3,5-Trinitro-1,3,5-Triazine (RDX) Degradation in Biologically Active Iron Columns." *Water Air and Soil Pollution*, 141(1–4), 325–335.

52. Phillips, D. H., Gu, B., Watson, D. B., Roh, Y., Liang, L., and Lee, S. Y. (2000). "Performance Evaluation of a Zerovalent Iron Reactive Barrier: Mineralogical Characteristics." *Environmental Science & Technology*, 34(19), 4169–4176.

53. Plaza, R. C., de Vicente, J. Gómez-Lopera, S., and Delgado, A. V. (2001). "Stability of Dispersions of Colloidal Nickel Ferrite Spheres." *Journal of Colloid Interface Science*, 242, 306–313.
54. Ponder, S., Darab, J., and Mallouk, T. (2000). "Remediation of Cr(IV) and Pb(II) Aqueous Solutions Using Supported Nanoscale Zero-valent Iron." *Environmental Science & Technology*, 34(12), 2564–2569.
55. Pourbaix, M. J. N. (1966). *Atlas of Electrochemical Equilibria in Aqueous Solutions*, Pergamon Press, New York.
56. Quinn, J., Geiger, C., Clausen, C., Brooks, K., Coon, C., O'Hara, S., Krug, T., et al. (2005). "Field Demonstration of DNAPL Dehalogenation Using Emulsified Zero-Valent Iron." *Environmental Science & Technology*, 39(5), 1309–1318.
57. Roddick-Lanzilotta, A., and McQuillan, J. (1999). "An in Situ Infrared Spectroscopic Investigation of Lysine Peptide and Polylysine Adsorption to TiO_2 from Aqueous Solutions." *Journal of Colloid Interface Science*, 217, 194–202.
58. Rose, J., Auffan, M., Thill, A., Decome, L., Masion, A., Orsiere, T., Botta, A., et al. "Nanoparticles as Adsorbents (for Liquid Waste and Water Treatment)." *Franco-American Conference on Nanotechnologies for a Sustainable Environment*, Rice University.
59. Rosensweig, R. E. (1985). *Ferrohydrodynamics*, Cambridge University Press, New York.
60. Royal Academy of Engineering, T. R. S. (2004). "Nanoscience and Nanotechnologies: Opportunities and Uncertainties." Royal Academy of Engineering, London.
61. Ryan, J., and Elimelech, M. (1996). "Colloid Mobilization and Transport in Groundwater." *Colloids and Surfaces*, 107, 1–56.
62. Saleh, N., Phenrat, T., Sirk, K., Dufour, B., Ok, J., Sarbu, T., Matyjaszewski, K., et al. (2005a). "Adsorbed Triblock Copolymers Deliver Reactive Iron Nanoparticles to the Oil/Water Interface." *Nano Letters*, 5(12), 2489–2494.
63. Saleh, N., Sarbu, T., Sirk, K., Lowry, G. V., Matyjaszewski, K., and Tilton, R. D. (2005b). "Oil-in-Water Emulsions Stabilized by Highly Charged Polyelectrolyte-Grafted Silica Nanoparticles." *Langmuir*, 21(22), 9873–9878.
64. Saleh, N., Sirk, K., Phenrat, T., Dufour, B., Matyjaszewski, K., Tilton, R. D., and Lowry, G. V. (2007). "Surface Modifications Enhance Nanoiron Transport and NAPL Targeting in Saturated Porous Media." *Environmental Engineering Science*, 24(1), 45–57.
65. Schrick, B., Hydutsky, B., Blough, J., and Mallouk, T. (2004). "Delivery Vehicles for Zerovalent Metal Nanoparticles in Soil and Groundwater." *Chemistry of Materials*, 16(11), 2187–2193.
66. Sitharaman, B., Bolskar, R. D., Rusakova, I., and Wilson, L. J. (2004). "Gd@C60[C (COOH)2]10 and Gd@C60(OH)x: Nanoscale Aggregation Studies of Two Metallofullerene MRI Contrast Agents in Aqueous Solution." *Nano Letters*, 4(12), 2373–2378.
67. Song, H., and Carraway, E. R. (2005). "Reductive of Chlorinated Ethanes by Nanosized Zero-Valent Iron: Kinetics, Pathways, and Effects of Reaction Conditions." *Environmental Science & Technology*, 39, 6237–6245.
68. Sorel, D., Warner, S. D., Longino, B. L., Honniball, J. H., and Hamilton, L. A. "Dissolved Hydrogen Measurements at a Permeable Zero-Valent Iron Reactive Barrier." *American Chemical Society Annual Conference*, San Diego, CA.
69. Strenge, K. (1993). "Structure Formation in Disperse Systems." *Coagulation and Flocculation: Theory and Applications*, B. Dobiáš, ed., Marcel Dekker, New York.
70. Stroo, H., Unger, M., Ward, C. H., Kavanaugh, M., Vogel, C., Leeson, A., Marquesee, J., et al. (2003). "Remediating Chlorinated Solvent Source Zones." *Environmental Science & Technology*, 37(11), 193a–232a.
71. Tungittiplakorn, W., Cohen, C., and Lion, L. W. (2005). "Engineered Polymeric Nanoparticles for Bioremediation of Hydrophobic Contaminants." *Environmental Science & Technology*, 39(5), 1354–1358.
72. Tungittiplakorn, W., Lion, L., Cohen, C., and Kim, J. (2004). "Engineered Polymeric Nanoparticles for Soil Remediation." *Environmental Science & Technology*, 38(5), 1605–1610.
73. U.S.DOE. (2005). "Guidance for Optimizing Ground Water Response Actions at Department of Energy Sites."
74. Veiseh, O., Sun, C., Gunn, J., Kohler, N., Gabikian, P., Lee, D., Bhattarai, N., et al. (2005). "Optical and MRI Multifunctional Nanoprobe for Targeting Gliomas." *Nano Letters*, 5(6), 1003–1008.
75. Velegol, S. B., and Tilton, R. D. (2001). "A Connection between Interfacial Self-Assembly and the Inhibition of Hexadecyltrimethylammonium Bromide Adsorption on Silica by Poly-L-lysine." *Langmuir*, 17(1), 219–227.
76. Viessman, W., and Hammer, M. (1998). *Water Supply and Pollution Control*, Addison Wesley Longman, Inc., Menlo Park, CA.
77. Vikesland, P. J., Klausen, J., Zimmermann, H., Roberts, A. L., and Ball, W. P. (2003). "Longevity of Granular Iron in Groundwater Treatment Processes: Changes in Solute Transport Properties Over Time." *Journal of Contaminant Hydrology*, 64(1–2), 3–33.
78. Vogan, J. L., Focht, R. M., Clark, D. K., and Graham, S. L. (1999). "Performance Evaluation of a Permeable Reactive Barrier for Remediation of Dissolved Chlorinated Solvents in Groundwater." *Journal of Hazardous Materials*, 68(1–2), 97–108.
79. Wang, C. B., and Zhang, W. X. (1997). "Synthesizing Nanoscale Iron Particles for Rapid and Complete Dechlorination of TCE and PCBs." *Environmental Science & Technology*, 31(7), 2154–2156.
80. Water Science Technology Board, N. (2004). *Contaminants in the Subsurface: Source Zone Assessment and Remediation*, National Academies Press, Washington, D.C.
81. Wilkin, R. T., Su, C., Ford, R. G., and Paul, C. J. (2005). "Chromium-Removal Processes during Groundwater Remediation by a Zerovalent Iron Permeable Reactive Barrier." *Environmental Science & Technology*, 39(12), 4599–4605.
82. Williams, A. G. B., Gregory, K. B., Parkin, G. F., and Scherer, M. M. (2005). "Hexahydro-1,3,5-Trinitro-1,3,5-Triazine Transformation by Biologically Reduced Ferrihydrite: Evolution of Fe Mineralogy, Surface Area, and Reaction Rates." *Environmental Science & Technology*, 39(14), 5183–5189.
83. Yang, G. C., and Lee, H. L. (2005). "Chemical Reduction of Nitrate by Nanosized Iron: Kinetics and Pathways." *Water Research*, 39(5), 884–894.
84. Yao, K. M., Habibian, M. T., and O'Melia, C. R. (1971). "Water and Wastewater Filtration: Concepts and Applications." *Environmental Science & Technology*, 5, 1105–1102.

CHAPTER 9

Membrane Processes

Mark R. Wiesner

Department of Civil & Environmental Engineering, Duke University, Durham, North Carolina, USA

Andrew R. Barron

College of Engineering, Swansea University, Wales, United Kingdom

Jérôme Rose

*CEREGE UMR 7330 CNRS–Aix-Marseille University, Marseille, France, and
iCEINT CNRS–Duke University, Aix-en-Provence, France*

O ver half a century since the invention of synthetic polymers and the asymmetric membrane, refinements and new developments in membrane technologies are active areas of research that are rapidly expanding our capabilities to restructure production processes, protect the environment and public health, and provide technologies for sustainable growth. Today, membrane technologies are playing an increasingly important role as unit operations for environmental quality control, resource recovery, pollution prevention, energy production, and environmental monitoring. Membranes are also key technologies at the heart of fuel cells and bioseparations.

The nanotechnology-inspired approach, of building objects from the molecular scale, and the new materials that this approach inspires have great potential for stimulating innovation in membrane science. Nanoscale control of membrane architecture may yield membranes of greater selectivity and lower cost in both water treatment and water fabrication. Nanomaterials may serve as the basis for new membrane processes based on passive diffusion or active transport. New materials may also be used in conjunction with membranes to create new nanomaterial/membrane reactors (NMRs). Multifunctional membranes that simultaneously separate and react or detect are possible. Nanotechnology is likely to improve the reliability, and efficiency of membrane processes while broadening the range of applications.

In this chapter, we review the fundamentals of membrane processes as a basis for understanding where nanomaterials are likely to kindle innovations in membrane processes used to protect our environment and public health. We then examine several examples of nanomaterial-based approaches to improved membrane technologies.

Overview of Membrane Processes

A membrane, or more properly, a semipermeable membrane, is a thin layer of material that is capable of separating materials based on their physical or chemical properties when a driving force is applied across the membrane (Figure 9.1). Materials to be separated are introduced to the membrane on the feed or "concentrate" side where the portion of the materials rejected by the membrane accumulate. The concentration of rejected materials is typically highest near the membrane, setting up a concentration gradient for diffusion away from the membrane and back into the bulk concentrate. The "permeate" side of the membrane is enriched in materials that are able to move through the membrane more easily. The efficiency of membrane rejection, R, for a given component (also referred to as the efficiency of separation) is generally defined as 1.0 minus the ratio of the concentrations of that component in the permeate and feed:

$$R = 1 - \frac{c_{\text{permeate}}}{c_{\text{feed}}}$$

(9.1)

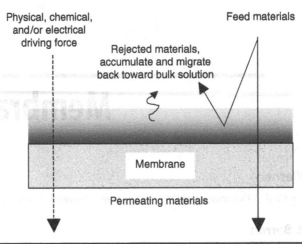

FIGURE 9.1   Separation of material by a semipermeable membrane under a driving force.

Water filtration presents a very simplified case where particles are rejected by the membrane, potentially accumulating as a cake on the concentrate side, while water passes through the membrane as permeate. The accumulation of materials in, on, or near the membrane may have undesirable consequences for transport across the membrane resulting in a condition referred to as membrane fouling. For example, in water filtration the formation of a cake on the membrane may impede the permeation of water under the applied driving force of pressure. The issue of membrane fouling will be taken up later in this chapter.

Membranes can be fabricated in many shapes and sizes. *Symmetric membranes* have an approximately uniform composition throughout their entire thickness so that a slice through any layer of the membrane parallel to its surface would look essentially identical. By contrast, *asymmetric membranes* consist of a thin membrane skin that is responsible for the separation of permeate from rejected species. The skin of an asymmetric is supported by a layer (typically much thicker) that offers little resistance to transport and does not play a role in membrane selectivity. The surface chemistry and composition of different membranes can be quite variable depending on the application. Hydrophobic, uncharged membrane material may be desirable for some membrane distillation applications, while high charge density membranes are characteristic of polymer electrolyte membranes (PEMs) used in fuel cells. Dense SiO_2 membranes have been used for gas separations while porous alumina membranes can be used to treat high temperature brines emanating from oil and gas wells.

Membranes are packaged as several *elements* grouped into units that are referred to as *modules*, *vessels*, or *stacks*, depending on the type of membrane and its application. The most frequently encountered element geometries are as flat sheets, capillary fibers, hollow fibers, tubes, or spiral wound. The geometry of the membrane element is critical in determining the economics of the membrane process since it determines how much membrane area can be fit into a given volume. For example, both hollow fiber and tubular membranes share a cylindrical geometry, but tubular membranes have a much larger diameter. Thus, the area of membrane available for mass transport per volume of module (referred to as the *packing density*) will be less for tubular membrane (Table 9.1). The cost of a membrane system tends to decrease as the packing density of the membrane module increases as the costs of module housing, instrumentation and hook-up surrounding the module, and space for the installation are spread out over more membrane area. However, as the packing density increases, less space is available within the module to allow for other functions. For example, in water filtration space must be provided in the module to allow for rejected materials to circulate freely without obstructing flow, or the feed stream to the membrane must be pretreated to remove these materials.

Transport Principles for Membrane Processes

The driving force for transport across membrane may be physical (as in the case of pressure), chemical, or electrical; acting individually or in concert. In theory, a gradient across the membrane in any parameter that affects the chemical potential of a compound can be used as a driving force for transport

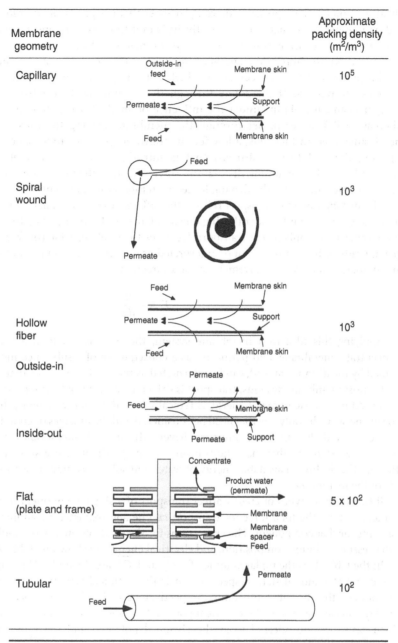

Membrane geometry		Approximate packing density (m²/m³)
Capillary		10^5
Spiral wound		10^3
Hollow fiber		10^3
Flat (plate and frame)		5×10^2
Tubular		10^2

TABLE 9.1   Approximate Packing Densities for a Number of Membrane Geometries

across the membrane. Examples of driving forces and some corresponding membrane processes are listed in Table 9.2.

Pressure-driven membrane processes have been developed at large scale for the treatment of water (both potable and wastewater) and other fluids. In water filtration, pressure-driven membrane processes are typically differentiated based on the range of materials rejected. Microfiltration (MF) and

Driving Force	Examples of Membrane Processes
Temperature gradient	Membrane distillation
Concentration gradient	Dialysis, pervaporation
Pressure gradient	Reverse osmosis, ultrafiltration
Electrical potential	Electrodialysis, electro-osmosis

TABLE 9.2   Categorization of Membrane Processes by Driving Force

ultrafiltration (UF) are pressure-driven processes that use *porous membranes* to separate micron-sized and nanometer-sized materials, respectively. Nanofiltration (NF) and reverse osmosis (RO) use *dense membranes* to separate solutes ranging from macromolecules, nanoparticles, and larger ions in the case of NF to simple salts and very low-molecular-weight organic compounds in the case of RO. Unlike pressure-driven processes in which solvent passes through the membrane, electrodialysis involves the passage of the solute through the membrane. In electrodialysis, ions pass through a semipermeable membrane under the influence of an electrical potential and leave the water behind, whereas in RO, water passes through the membrane leaving the ions behind. Polymer electrolyte membranes used in fuel cells, allow for the transport protons under a concentration gradient, while rejecting the fuel (e.g., hydrogen) that is introduced to a catalyst at the "concentrate" side of the membrane. In membrane distillation, solvent (e.g., water) evaporates and is transported across a porous membrane under the driving force of a temperature gradient.

The driving force for transport reflects the differences in the available energy on the two sides of the membrane. Analogous to water flowing down hill, material is transported from one side of the membrane to the other if and only if that transport decreases the total available (or free) energy of the system. Any spatial gradient in energy, E, can be interpreted as a force, F. The force is in the direction of decreasing energy so, for example, for movement in the x direction:

$$F_x = -\frac{dE}{dx} \tag{9.2}$$

Applying this idea to a membrane system, the decrease in total available energy of the system (concentrate, membrane, and permeate) as a consequence of a substance moving across the membrane divided by the distance moved, can be interpreted as the driving force for that movement.

In most membrane systems, transport is driven by an externally imposed gradient in a single type of energy. Although each type of energy is typically linked, in many systems, the imposed gradient in one type of energy is the only one that need be considered and the linkages or coupling with the other gradients can be ignored. For instance, pressure-driven MF and UF membranes do not reject solutes to an appreciable extent. In this case, the transmembrane gradient in pressure is the only significant factor affecting the decline in available energy of water and solutes as they cross from the feed to the permeate side of the membrane.

If two or more types of energy affect transport of a single component in the feed, then at any single moment in time the total force for transport can be approximated by adding the corresponding energy gradients for that component. The expressions for the available energy per mole of a substance associated with pressure, solution composition, and electrical energy are shown in Table 9.3.

In the table, V_i is the molar volume of i; $\overline{G}_i$ and $\overline{G}_i^o$ are the molar Gibbs free energy of i in the given system and at standard state, respectively; a_i is the chemical activity of i; R and T are the universal gas constant and the absolute temperature, respectively; z_i is the charge on species i (including sign); F is the Faraday constant; and Ψ is the electrical potential. $\overline{G}_i$ and $\overline{G}_i^o$ are also commonly written as μ_i and μ_i^o, in which case they are referred to as the chemical potential and the standard chemical potential of i, respectively. The sum of chemical and electrical potentials ($\overline{E}_{chem,i} + \overline{E}_{elec,i}$) is called the electrochemical potential.

Type of Energy	Expression for Energy/Mol of i	
Mechanical (pressure-based)	$\overline{E}_{p,i} = V_i P$	(9.3)
Chemical (concentration-based)	$\overline{E}_{chem,i} = \overline{G}_i = \overline{G}_i^o + RT \ln a_i$	(9.4)
Electrical	$\overline{E}_{elec,i} = z_i F \Psi$	(9.5)
Total available energy(a)	$\overline{E}_i = \overline{E}_{p,i} + \overline{E}_{chem,i} + \overline{E}_{elec,i}$	(9.6)
(a)In a typical membrane system; in other systems, other types of energy would have to be included.		

TABLE 9.3 Expressions of Available Energy per Mole of Chemical Species

Substituting Equations (9.3) through (9.5) into Equation (9.6), and noting that the standard molar Gibbs energy of any species i is independent of the system conditions, the overall energy change accompanying transport of species i across a membrane per unit mass of i transported can be computed as:

$$\Delta \overline{E}_i = V_i \Delta P + \Delta(\overline{G}_i^o + RT \ln a_i) + z_i F \Delta \Psi$$
$$= V_i \Delta P + RT \Delta \ln a_i + z_i F \Delta \Psi \tag{9.7}$$

where the Δ's are taken as the permeate value minus the feed value and are assumed to be small. As noted previously, a spatial gradient in energy can be interpreted as a force. Therefore, approximating ΔP, Δa_i, and $\Delta \Psi$ as the changes in those parameters across the thickness δ_m of the membrane, the driving force for transport of i across the membrane from feed to permeate, per mole of i, is

$$F_i = \frac{\Delta \overline{E}_i}{\delta_m} = V_i \frac{\Delta P}{\delta_m} + RT \frac{\Delta \ln a_i}{\delta_m} + z_i F \frac{\Delta \Psi}{\delta_m} \tag{9.8}$$

Simplifications to Equation (9.8) in which one driving force dominates in the transport of a single component of a system give rise to several frequently encountered expressions such as Fick's law (concentration gradient), Ohm's law (only electrical potential), and D'Arcy's law (pressure gradient). For example, a single component ($i = 1$) where transport is driven by differences in concentration alone ($\Delta \psi = \Delta P = 0$). The driving force per molecule for transport is obtained by dividing Equation (9.8) by Avogadro's number, N_A. If we further assume that the molecules rapidly achieve a steady, average diffusion velocity, and that the molecules experience a force of resistance to movement, F_{resist}, which is proportional to this average velocity, v_{diff}, the driving force for diffusion must be balanced by this resistive force:

$$F_{\text{resist}} = v_{\text{diff}} f = F_{\text{tot},1} = \frac{kT}{\delta_m} \frac{\Delta c_1}{c_1} \tag{9.9}$$

where the Boltzmann constant, k, is equal to R/N_A and f is the proportionality factor relating resistive force to velocity. The diffusive flux is the product of the average diffusion velocity and the concentration of the diffusing component. Thus, rearranging Equation (9.9), the diffusive flux can be expressed as:

$$J_1 = v_{\text{diff}} c_1 = \frac{kT}{f} \frac{\Delta c_1}{\delta_m} \tag{9.10}$$

Equation (9.10) is a restatement of Fick's law, where the concentration gradient is approximated by the difference in concentration across the membrane divided by the thickness of the membrane, and the diffusion coefficient, D, is equal to kT/f. Similarly, when a difference in pressure is the sole driving force for moving a fluid through a network of pores and this force is balanced by a resistance force that is proportional to fluid velocity ($F_{\Delta P} = F_{\text{resist}}$), then we obtain an expression similar to D'Arcy's law:

$$v_{\text{fluid}} = J_{\text{fluid}} = \frac{V_i}{f_p} \frac{\Delta P}{\delta_m} \tag{9.11}$$

where f_p is the friction coefficient for the pores. In each of these cases, the flux of one component in the system is shown to be proportional to a driving force.

However, in many systems of practical interest, energy gradients and the transport of multiple components do not act in isolation. The transport of one component in the system may affect the transport of another. One example is the flow of fluid through a membrane pore carrying a charge, which gives rise to a flux of current known as the streaming current. Desalination by reverse osmosis is a second example. In RO desalination, a pressure gradient is applied to transport water across a semipermeable membrane that rejects much of the salts in solution. Calculation of the flux of either salt or water requires that gradients in both pressure and concentration be considered. Phenomena such as these are examples of coupled transport.

Irreversible thermodynamics addresses coupled transport in a generalization of expressions such as Fick's law that describe the flux of a component of a system as being proportional to a driving force. We generalize the relationship for flux as a function of driving force by considering the flux of any one component, i, as a function of a vector of driving forces, X. A Taylor series expansion from the point at which the driving forces disappear ($X = 0$ at thermodynamic equilibrium where the fluxes will also be equal to zero) to X yields the following result:

$$J_i(X) = J_i(0) + \sum_j \frac{\partial J_i}{\partial X_j}\bigg|_{X=0} \cdot X_j + \sum_{j,k} \frac{1}{2!} \frac{\partial^2 J_j}{\partial X_j \partial X_k} : X_j X_k + O(X^3) \tag{9.12}$$

If the system is near equilibrium, then we can ignore the higher order terms and truncate the Taylor's series after the linear terms. This implies that the partial derivatives evaluated at $X = 0$ can be treated as constants. These constants are referred to as phenomenological coefficients, that relate the flux of component i to driving force j.

The result is the second postulate of irreversible thermodynamics referred to as the Onsager principle, that states that for small deviations from equilibrium, the fluxes of materials across a membrane, J_i, can be expressed as linear combinations of all the pertinent driving forces, X_j [1, 2]. Thus, for the case of several driving forces, the flux of component i, J_i is calculated as:

$$J_i = \sum_i L_{ij} X_j \tag{9.13}$$

where the L_{ij}, are the phenomenological coefficients.

Consider the case of transport across a membrane involving two molecular species (i and j) such as might occur in a diffusion dialysis separation. In this case, the two conjugate driving forces (X_i and X_j) are simply the gradient of the chemical potential for each of these species across the membrane. In the absence of gradients in pressure or electrical potential, and for dilute solutions, these driving forces are proportional to the concentration gradients for each species. The phenomenological coefficients for $i \neq j$ account for the possibility that the flux of i (J_i) may affect the flux of j (J_j), that is to say that these fluxes are *coupled*.

In the case of our two molecules transported by diffusion:

$$J_i = L_{ii} X_i + L_{ij} X_j$$
$$J_j = L_{ji} X_i + L_{jj} X_j$$

The matrix of phenomenological coefficients will be a square matrix since the flux of each species "i" is described as a linear sum of the gradients of the chemical potentials of all other species "j." We can begin to better appreciate the meaning of the phenomenological coefficients by rewriting Equation (9.13) such that terms with $j = i$ are removed from the summation:

$$J_i = L_{ii} X_i + \sum_{k \neq i} L_{ik} X_k \tag{9.14}$$

The first term in Equation (9.14) describes the flux of component i due to a driving force such as a concentration gradient in i. In the case of Fick's law, L_{ii} is proportional to the diffusion coefficient. For the case of Darcy's law, L_{ii} can be interpreted as the membrane hydraulic permeability. The second term in Equation (9.14) describes the impact that the transport of other components may have on the flux of component i. The phenomenological coefficients in the summation are referred to as *coupling coefficients* since they relate the impact of a driving force for one component (and hence the transport of that component) on the transport of a second component. The Onsager reciprocal relationship further states that coupling effects are symmetric and thus:

$$L_{ik} = L_{ki} \tag{9.15}$$

Application to Dense Membranes: Reverse Osmosis

Reverse osmosis membranes are permeable to water while rejecting, to a large extent, solutes present in the feed. If, in the absence of any applied pressure, pure water is placed on one side of such a membrane (1) and an aqueous salt solution on the other (2), water will is pass from side 1 to side 2 by osmosis resulting in a decrease in the concentration of salt in compartment 2. If the compartments separated by the membrane are open to the atmosphere and allow for the height of water to adjust in each compartment (as in Figure 9.2), water will diffuse until the driving force of an increasing pressure from a greater height of water in compartment 2 comes to equilibrium with the decreasing driving force for diffusion from 1 to 2.

At equilibrium, the total available energy per mole of water, $\bar{E}_w$, is equal on the two sides of the membrane, that is, $\Delta \bar{E}_w = 0$. In the absence of an electrical potential, and further noting that water is uncharged ($z_w = 0$), according to Equation (9.6), this condition is met when:

$$0 = V_w \Delta P + RT \Delta \ln a_w$$

$$V_w (P_2 - P_1) = -RT \ln \frac{a_{w,2}}{a_{w,1}} \tag{9.16}$$

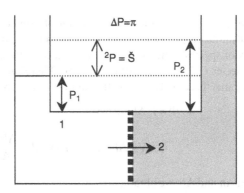

FIGURE 9.2 Pure water (1) diffuses across a salt-rejecting membrane resulting in a dilution of (2) and difference in pressure between the two compartments.

where the subscripts 1 and 2 refer to the dilute and concentrated solutions, respectively. Equation (9.15) indicates that, if the activity of water is greater on one side of the membrane, the system can be in equilibrium only if the hydrostatic pressure is greater on the other side.

If we take the activity of the pure water to be unity, and note that the pressure differential $(P_2 - P_{\text{pure water}})$ is known as the *osmotic pressure* of the solution, commonly designated as π, we obtain an expression for the osmotic pressure:

$$\pi = \frac{RT}{V_w} \ln a_{w,2} \tag{9.17a}$$

$$\pi \approx RT \sum c_s \tag{9.17b}$$

where π and a_w are the osmotic pressure and activity of water in the solution of interest and c_s is the molar concentration of salt. Equations (9.17b) applies if the solution is dilute where we can approximate the activity of water as being 1 minus the mole fraction of the salt in compartment 2 and note that $\ln(1 - x) \approx -x$. High-molecular-weight solutes produce less osmotic pressure than do low-molecular-weight compounds. This can be illustrated by considering the case of a dilute solution of molecules with molecular weight $\bar{M}$ at a mass concentration C. The osmotic pressure can be approximated as:

$$\pi \cong \frac{C}{\bar{M}} RT \tag{9.18}$$

Equations (9.17a), (9.17b), and (9.18) are all ways of writing what is known as the van't Hoff equation. Note that Equation (9.18) predicts that the osmotic pressure will decrease as the molecular weight of the species increases. For this reason, the osmotic pressure exerted by larger particles and macromolecules can typically be ignored. In contrast, small nanoparticles, while exerting an osmotic pressure much less than that of an ionic solute at equal mass concentration, may nonetheless result in a significant osmotic pressure.

Substituting Equation (9.17a) into Equation (9.6), we obtain the following result for the difference in available energy of the water on the two sides of the membrane in the absence of an electrical potential:

$$\Delta \overline{E}_w = V_w \Delta P + RT \Delta \ln a_w = V_w (\Delta P - \Delta \pi) \tag{9.19}$$

In other words, a pressure ΔP greater than the osmotic pressure must be applied across the membrane to create a driving force sufficient to move water across the membrane.

We now write the corresponding expression for the solute:

$$\Delta \overline{E}_s = V_s \Delta P + RT \Delta \ln a_s \tag{9.20}$$

Based on Equation (9.17), for this system, $\Delta \pi \approx RT \Delta c_s$, so $\Delta \pi / RT \Delta c_s = 1$. Substituting that relationship into Equation (9.20), and assuming that activity is approximated by concentration, we obtain:

$$\Delta \overline{E}_s = V_s \Delta P + \frac{\Delta \pi}{RT \Delta c_s} RT \Delta \ln c_s \tag{9.21}$$

Equations (9.19) and (9.21) indicate the amount of available energy that is released (dissipated) per mole of water and solute, respectively, which cross from the feed to the permeate side of the membrane. If J_i is the molar flux of species i, the product $J_i\Delta\bar{E}_i$ indicates the rate of energy dissipation associated with the transport of species i per unit area of membrane. Therefore, the overall rate of energy dissipation per unit area of membrane, Φ, considering the flux of both water and solute is

$$\Phi = J_w\Delta\bar{E}_w + J_s\Delta\bar{E}_s$$

$$= J_wV_w(\Delta P - \Delta\pi) + J_s\left(V_s\Delta P + \frac{\Delta\ln c_s}{\Delta c_s}\Delta\pi\right) \tag{9.22}$$

Grouping terms in $\Delta\pi$ and ΔP, we obtain:

$$\Phi = (J_wV_w + J_sV_s)\Delta P + \left(\frac{J_s\Delta\ln c_s}{\Delta c_s} - J_wV_w\right)\Delta\pi \tag{9.23}$$

$$= J_V\Delta P + J_D\Delta\pi$$

Equation (9.23) describes the rate of energy dissipation in our simple system as the sum of two terms. The first term on the right-hand side of Equation (9.23) is the product of the total volume flux, J_V, of the components (water and solute) and a driving force, the difference in mechanical pressure across the membrane (ΔP). The second term is the product of the diffusive flux of solute relative to the flux of water, J_D, and a second driving force for transport, the osmotic pressure. Thus, in this two-component system, the rate of energy dissipation is described by two fluxes and their two corresponding conjugate driving forces. This is an important result since it reveals the appropriate fluxes and conjugate driving forces to be substituted into Equation (9.13) yielding the result for reverse osmosis performance [3]:

$$J_V = L_V\Delta_P + L_{VD}\Delta\pi \tag{9.24a}$$

$$J_D = L_{DV}\Delta_P + L_D\Delta\pi \tag{9.24b}$$

where L_V corresponds to L_{ii}, and so on. Since we have two components, we end up with two equations and four phenomenological coefficients, however, by the Onsager reciprocal relationship [Equation (9.25)], $L_{VD} = L_{DV}$. In a three-component system, applying the Onsager reciprocal relationship we would have three equations and six coefficients, and so on. Numerous variations, expansions, and simplifications on Equations (9.24a) and (9.24b) have been developed in the literature. However, they virtually all share the feature that volume flux is proportional to a net pressure drop ($\Delta P - \Delta\pi$) while solute flux is proportional to the concentration difference across the membrane, which in turn is proportional to $\Delta\pi$.

For example, assuming dilute solutions, Equations (9.24a) and (9.24b), known as the Kedem and Katchalsky model [3], can be expressed as:

$$J_V = L_V(\Delta_P + \sigma\Delta\pi)$$

$$J_s = \omega\Delta\pi + (1 - \sigma)c_{lm}J_V \tag{9.25}$$

where $\sigma = \left.\dfrac{\Delta p}{\Delta\pi}\right|_{J_v=0} = \dfrac{L_{VD}}{L_V}$ is the reflection coefficient, c_{lm} is the log mean average concentration across the membrane and, $\omega = (L_VL_D - L_{VD}^2)/L_V$.

The reflection coefficient can be shown to be a product of an equilibrium term expressing relative affinity (or exclusion) of salt with respect to the membrane, and a kinetic term that expresses the relative mobilities of water and salt and their potentially coupled migration [4].

The parameter permeability, P, of a membrane to a given molecule also reflects a multiplicative relationship between affinity and mobility. In the solution-diffusion model [5], permeability is defined as the product of the molecule's solubility in the membrane, K_s, and its diffusivity, D:

$$P = K_sD \tag{9.26}$$

This leads to an equation for water flux that is again similar to that obtained from irreversible thermodynamics:

$$J_w = A_w(\Delta_P - \Delta\pi) \tag{9.27}$$

where A_w is the hydraulic permeability, $A_w = D_w c_{m,1}^w \bar{V}_w / RT\delta_m$ and $D_w, c_{m,1}^w$ are the diffusivity and concentration of water in the membrane, respectively, and δ_m is the thickness of the membrane. For solutes, the solution-diffusion model yields following expression for the flux of solute, J_s:

$$J_s = D_s K_s \frac{(C_{\text{feed}} - C_{\text{permeate}})}{\delta_m} \tag{9.28}$$

where K_s is the solubility coefficient for the solute.

Application to Porous Membranes: Fluid Filtration

Materials with high diffusivities are not typically removed from fluids by porous membranes such as MF and UF membranes operated under a pressure differential. As a result, coupled transport of mass can typically be ignored. (However, the development of a streaming current when flow moves through a charged membrane may entail coupling, particularly for smaller pore sizes.) As a result, we need only consider the first term in Equation (9.13), which we have shown reduces to a form resembling D'Arcy's law where fluid flux $J_{f,\text{vol}}$, (volume of fluid per unit area of membrane per unit time) is proportional to the difference in pressure across the membrane. This is typically expressed as:

$$J_{f,\text{vol}} = \frac{\Delta p}{\mu R_m} \tag{9.29}$$

where Δ_p is the pressure drop across the membrane [the transmembrane pressure (TMP)], μ is the absolute viscosity of the fluid, and R_m is the hydraulic resistance of the membrane, with dimensions of reciprocal length. Membrane performance is often expressed as the ratio of permeate flux, J, to the pressure drop across the membrane, Δ_p. This quantity is called the specific permeate flux, with an initial value equal to $1/\mu R_m$.

If each pore is modeled as a capillary, permeate flux can be represented as Poiseuille flow through a large number of these capillaries in parallel. In each pore, the velocity of the fluid is assumed to be zero at the wall of the pore (termed the "no-slip condition"), and at a maximum value in the center of the pore. The no-slip condition at the pore wall is ultimately a consequence of an affinity between fluid molecules and with those of the membrane pore and leads to a parabolic velocity profile.

Using the Hagen-Poiseuille equation to describe flow through cylindrical membrane pores (idealized as such, or perhaps truly cylindrical) the following expression is obtained for the permeate flux through a membrane characterized by an effective pore radius of r_{pore}:

$$J = \frac{A_{\text{pore}} r_{\text{pore}}^2 \Delta P}{A_{\text{membrane}} 8\mu\theta\delta_m} = \frac{\Delta P}{\mu R_m} \tag{9.30}$$

where $R_m = \frac{A_{\text{membrane}} 8\theta\delta_m}{A_{\text{pore}} r_{\text{pore}}^2}$, A_{pore}/A_m is the ratio of the open pore area (A_{pore}) to the entire area of the membrane surface (A_m), θ is the pore tortuosity factor, and δ_m is the effective thickness of the membrane. Note that Equation (9.30) predicts that flux should decrease with the square of decreasing pore size. If the assumptions of the Hagen-Poisseuille model hold, very high pressures would be required to induce flow through membranes with nanometer-sized pores.

Polarization Phenomena and Membrane Fouling

The rejection of materials by a membrane leads to the accumulation of these materials near, on, or sometimes within the membrane. This can lead to a decrease in membrane performance. For example, ion exchange membranes used in electrodialysis have a polymeric support structure with fixed charged sites and water filled passages. Charged functional groups on these membranes attract ions of opposite charge (counterions). This is accompanied by a deficit of like-charged ions (co-ions) in the membrane and results in a so-called Donnan potential and the exclusion of ions from ion exchange membranes with like-charged functional groups. When an electrical potential is applied across these membranes, ions migrate to the electrode of opposite charge. However, the ion exchange membrane rejects co-ions, resulting in boundary layers on either side of the membrane (referred to as *concentration polarization* layers) that are either enriched in co-ions (the feed side) of the membrane or have a deficit of these ions (the permeate side or "dialysate"). Because there are fewer ions on the dialysate side, there is an increase in electrical resistance that leads to an increase in power consumption to achieve separation.

A similar phenomenon occurs in RO where salts are rejected by the membrane, leading to a concentration polarization layer near the membrane. The concentration polarization layer increases the local osmotic pressure, resulting in the need for a higher pressure to overcome this osmotic pressure, as well as a lower rejection of salt by the membrane. The concentration profile of rejected species can be calculated from a mass balance on solute in a differential volume in the concentration polarization layer. For a simplified mass balance around the concentration polarization layer, the advective flux of solutes toward the membrane is balanced by diffusive back transport of solute:

$$v_w c = -D_B \frac{\partial c}{\partial y} \tag{9.31}$$

where v_w is the fluid velocity in the y direction (perpendicular to the membrane), and D_B is the Brownian diffusion coefficient of the solute. This expression can be integrated to yield an expression for the limiting permeate flux as a function of the bulk concentration c_{bulk}, the limiting wall concentration, c_{wall}, the diffusion coefficient for the solute, and the concentration-polarization layer thickness, δ_{cp},

$$J_{lim} = v_{w,lim} = \frac{D}{\delta_{cp}} \ln\left(\frac{c_{wall,lim}}{c_{bulk}}\right) \tag{9.32}$$

Equation (9.32) predicts that the limiting permeate flux should decrease with decreasing D. Because the diffusion coefficient increases as particle size decreases, we can expect that when membranes are used to separate nanoparticles, the limiting permeate flux for membrane operation should be higher for these species in comparison with larger colloidal species, if there is no resistant cake deposited on the membrane. However, if a layer of nanoparticles deposits on the membrane, this may lead to a decrease in flux due to membrane fouling.

In pressure-driven processes, fouling may be manifest as either an increase in the pressure drop, ΔP, across the membrane, (called the TMP) required to maintain a constant flux or, for a constant TMP, by a decrease in the permeate flux. This is illustrated in Figure 9.3 where the specific permeate flux (J/TMP) of a laboratory membrane system is seen to decline quickly over time due to fouling under conditions when the feed contains many foulants. When the membrane is taken out of service and washed with water, a portion of the fouling is reversed (reversible fouling). However, after multiple cycles of operation and washing, the amount of permeate flux that is recovered by washing decreases (irreversible fouling). At some point, membranes must be cleaned more aggressively with chemicals to at least partially reverse the "irreversible" fouling. Fouling is not reserved to pressure-driven membrane processes. The deposition of material on a membrane may change its functionality and therefore its ability to effectively separate compounds. Additional layers of material may impede mass transport originating from any driving force.

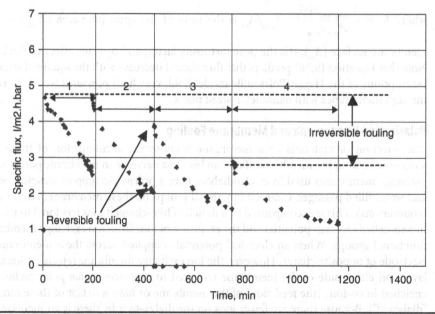

Figure 9.3 Decline in specific flux over time. Cleaning events reverse some, but not all, of the fouling.

The choice of chemicals used to clean a membrane is dependent on the chemical nature of the foulant. Dissolved organic materials are the primary foulants in many water-based membrane applications. Even when mass concentrations of particle exceed those of organic matter, organic matter may dominate fouling. Organic matter may form a gel layer on the surface of the membrane or adsorb to the surface or within the membrane matrix. In gas separations, gas-phase impurities may foul membranes by adsorption. The possibility of precipitative fouling is a serious concern in the operation of membranes designed specifically to remove scale-forming species (e.g., RO, membrane distillation, and electrodialysis). Common foulants of concern for precipitation include calcium, magnesium, and iron (primarily ferrous), any of which might precipitate as a hydroxide, carbonate, or sulfate solid. Colloids of all kinds may accumulate on or near membranes, forming a cake. Biofouling of membranes is a key concern in applications in which membranes are in contact with an aqueous environment. Bacteria may colonize membranes forming biofilms that consist of the bacteria and compounds secreted by the bacteria.

In pressure-driven membranes, these potential causes of fouling can be represented mathematically by modifying Equation (9.29) to include resistance terms for each of the causes of fouling, including those due to the changes in membrane permeability over time, $R_m(t)$, and the formation of a cake or biofilm, $R_c(t)$. One approach to describing the resistance of the cake, R_c, is to assume that the cake has a uniform structure with a specific resistance, $\hat{R}_c$, and thickness, δ_c (that may change over time). An expression for hydraulic permeability such as the Kozeny equation, can then be used to predict the specific resistance, assuming that the cake is incompressible and is composed of uniform particles:

$$\hat{R}_c = \frac{180(1-\varepsilon_c)^2}{d_p^2 \varepsilon_c^3} \tag{9.33}$$

where ε_c is the porosity of the cake, and d_p is the diameter of the particles that it comprises. This expression predicts that resistance to permeation by a deposited cake should increase as one over the square of the diameter of particles comprising the cake. Thus, in comparison with larger colloidal species, an accumulation of nanoparticles on the membrane should form a cake of relatively high specific resistance.

The effect of concentration polarization on permeate flux is primarily due to the accumulation of solutes near the membrane and the resultant osmotic pressure effect. It will typically be negligible for MF and UF that reject only particles or macromolecules. The concentration polarization term for RO and NF applications can be incorporated into an osmotic pressure term in the numerator, and the following expression is obtained:

$$J = \frac{(\Delta p - \sigma \Delta \Pi)}{\mu(R_m(t) + R_c(\delta_c(t), \ldots))} \tag{9.34}$$

Concentration polarization may be indirectly related to irreversible fouling through its effects on adsorption, cake formation, and precipitation. However, reductions in permeate flux (or increases in TMP) due directly to concentration polarization are completely reversible, making this resistance term quite different from the others.

Membranes in Fuel Cell Applications

A polymer electrolyte membrane fuel cell (PEMFC) consists of a proton exchange membrane sandwiched between two layers of catalyst material. Hydrogen or an alternative fuel such as methanol reacts catalytically at the anode to form electrons and protons (Figure 9.4). The proton exchange membrane (PEM) selectively

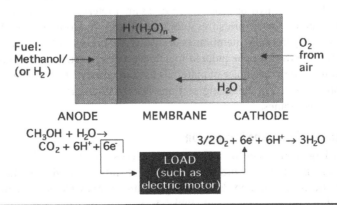

FIGURE 9.4 Diagram of a fuel cell, in this case using methanol.

transports protons to the cathode where they react catalytically with oxygen to form water. The membrane must have a high rejection for the fuel on the anode or "feed" side of the membrane, and for the oxygen on the cathode side of the membrane. The most widely used membranes in fuel cells are perfluorosulfonate polymeric membranes, most widely marketed by Dow as the Nafion™ membrane. The perfluorosulfonic polymers are strongly hydrophobic with hydrophilic groups that can adsorb large quantities of water. They are highly conductive while fully hydrated at low temperatures. Perfluorosulfonic polymers are both electron insulators and excellent proton conductors. The weak attraction to the SO_3^{3-} group allows the H^+ ions in the sulfonic groups to move upon hydration [6]. At high relative humidity (> 80%), conductivity in the range of 3×10^{-2} S/cm [7] up to nearly 9×10^{-2} S/cm is typical for materials such as of Nafion 117 at room temperature [8]. Increasing operating temperature produces limited improvements in the conductivity of Nafion due to progressive dehydration [9] and a resultant decrease in the more mobile bulk water. Indeed, these membranes become nonconductive at operating temperatures above approximately 80°C. In addition, these perfluorosulfonate membranes are expensive, representing approximately 46 percent of the cost of current fuel cell stacks.

Other polymeric membranes have been investigated for their suitability in fuel cell applications. Fluorine-free, hydrocarbon-based membrane materials are less expensive and are commercially available. They contain polar groups with high water uptake. A hydrocarbon-based membrane, however, does not provide long-term thermal stability for fuel cell applications [10]. Polymeric materials such as polyetherketone [11], sulfonated polyaromatic polysulfones [12], grafted fluorinated polymers by radiation[13], and polyvinylidene fluoride and sulfonated polystyrene-co-divinylbenzene have all shown promise. However, these polymers also exhibit thermal, chemical, or mechanical instability; and/or they do not provide a protonic conductivity comparable to that of Nafion.

Although fuel cells are typically regarded as technologies for producing electricity, the prospects for improving fuel cell operation as water supply technology are particularly intriguing. In contrast with the conventional approach to obtaining high-quality water is to *treat* the water either by removing contaminants from the water (e.g., filtration) or by removing the water from the contaminants (e.g., distillation), water production by fuel cells occurs in a manner that exemplifies the nanotechnology paradigm; *water is fabricated* rather. Cogeneration of electricity and water may in some instances be a cost-effective option, particularly when sources of high-quality water are scarce or when the final quality of water required is demanding. A typical U.S. household consumes approximately 4200 kWh of electricity per year. The theoretical yield of water from a fuel cell is based on the stoichiometry of the reaction ($2H_2 + O_2 = 2H_2O$) and the heat of formation [14]. The current generation of fuel cells is capable of producing approximately 1 L/kWh of electricity, equivalent to some 10 L of high-purity water per day for a typical U.S. household without limitations on water quality degradation in the distribution system. However, one factor that limits the use of fuel cells for water production is the need to use much of the water produced by the fuel cell to keep the membranes hydrated and sustain proton conductivity.

Membrane Fabrication

Membranes are fabricated from a rich variety of materials that range from inorganic minerals, to organic polymers, to mixtures of organic and inorganic materials. While both porous and dense membranes may be made from virtually all of these materials, the practicality of a given membrane formulation is determined by the intended use and the driving forces involved. For example, inorganic ceramic membranes may be particularly well-suited for treatment applications that will entail aggressive feed streams and/or cleaning mixtures. Polymeric hydrophobic membranes are preferred in membrane distillation applications due to the higher liquid entry pressures (compared with hydrophilic membranes) and their low thermal conductivity. Hydrophilic membranes are generally preferred for RO and forward osmosis applications. Membranes can therefore be tailored to achieve a desired structure and functionality appropriate to each application through the choice of materials used to fabricate the membranes and control of the membrane formation process. The following sections briefly summarize the most common cases of ceramic and polymeric membranes.

Conventional Membrane Fabrication

The two largest grouping of membrane materials are those of inorganic and polymeric membranes. Inorganic membranes may include sintered steel and porous glass bodies. However, ceramic membranes tend to be the most common class of inorganic membranes. Ceramic membranes are typically asymmetric, composed of a porous support onto which is deposited one or more films of metal oxide particles that are then sintered at high temperature onto the support to form the rejecting membrane skin. While simple porous ceramics have been used for centuries as filters, commercialized ceramic membranes for large-scale applications have been produced employing nanochemistry methods for several decades as presented subsequently in the discussion on membrane fabrication using nanomaterials.

The richness of polymer chemistry has been harnessed to create a large family of membranes. Porous polymeric membranes (e.g., MF and UF) are most commonly made using the phase-inversion process. The phase-inversion process begins with a dispersion of a given polymer material in a compatible solvent. When the polymer dispersion is immersed in a nonsolvent, the nonsolvent (coagulant) is exchanged with the solvent, and the polymers transition to a solid phase. The membrane may be cast on a supporting fabric, as self-supporting asymmetric membranes or as symmetric or asymmetric hollow fibers. The process can be applied using a wide range of polymers including cellulosic formulations, polysufones, polyethersulfones, polyvinylidine fluoride, polytetrafluoroethylene, polypropylene, and polyacrylonitrile among the most commonly used.

Thin film composite membranes consist of a more porous supporting layer—typically formed by phase inversion, onto which a more dense film is cast by interfacial polymerization. These membranes are the mainstay of RO membranes, usually incorporating a polyamide thin film, although polyuria, polyether-amide, and other formulations have also been developed.

Membrane Fabrication Using Nanomaterials

The performance of membranes is intimately linked to the materials they are made from. The composition of the membrane will determine important properties such as rejection (selectivity), propensity to foul, mechanical strength, and reactivity. Membrane composition may even determine the element geometries that are (or are not) possible and, of course, the cost of the membrane. It is therefore not surprising that the creation of new nanomaterials opens the door for many new approaches to fabricating and improving membranes. We will consider three cases. In the first case, we will consider examples where nanomaterials are used to make membranes. In the second case, we will examine enhancements in existing membranes materials achieved through the introduction of nanomaterials to create new composites. Finally, we will take a look at the use of nanomaterials to "mold" membranes in a process known as nanomaterial templating.

Membranes Made from Nanomaterials

Ceramic Membranes Derived from Mineral Nanoparticles   Mineral membranes have been made from a variety of mineral nanoparticle precursors. Commercially available ceramic membranes are typically made from metal oxides such as Al_2O_3, ZrO_2, and TiO_2 [15]. However, membranes can be made from many other nanomaterials ranging from gold [16, 17] to SiO_2 [18]. In most cases, nanoparticles are deposited on a support surface and then calcined to create the membrane. Processes differ in the manner in which nanoparticle precursors are prepared. One common procedure for producing nanoparticle precurors to these membranes is to precipitate particles under controlled conditions creating a suspension or *sol* of nanoparticles that are deposited on a surface and dried to form a *gel*. This procedure is known as *sol-gel*.

Sol-gel involves a four-stage process: dispersion, gelation, drying, and firing. A stable liquid dispersion or *sol* of the colloidal ceramic precursor is initially formed in a solvent with appropriate additives. In the case of alumina membranes, this first step may be carried out with 2-butanol or *iso*-propanol. By removing the alcohol, the polymerization of aluminum monomers occurs, leading to a precipitate. This material is acidified, typically using nitric acid, to produce a colloidal suspension. By controlling the extent of aggregation in the colloidal sol, a gel of desired properties can be produced. The aggregation of colloidal particles in the sol may be controlled by adjusting the solution chemistry to influence the diffuse layer interactions between particles, by adding stabilizing agents such as surfactants, or through ultrasonification. Knowing that the properties of the gel will influence the permeability of the future membrane, it is clear that the gelation step is extremely important. This gel is then deposited, typically by a slip-cast procedure, on an underlying porous support. In variations on this procedure, functionalized surfaces may also be used to achieve a more ordered deposit or micelles can be used to direct film formation in specific geometries through self-assembly. In conventional sol-gel, the excess liquid is removed by drying and the final ceramic is formed by firing the resulting gel at higher temperatures. Drying and firing conditions have been shown to be very important in the structural development of the membranes, with higher drying rates resulting in more dense membrane films.

The sol-gel approach of reacting small inorganic molecules to form oligomeric and polymeric nanoparticles has several limitations such as difficulties in controlling the reaction conditions, and the stoichiometries, solubility, and processability of the resulting gel. It would thus be desirable to prepare nanoparticles in a one-pot bench-top synthesis from readily available, and commercially viable, starting materials, which would provide control over the products. One strategy for producing nanostructured membranes involves an environmentally benign alternative to the sol-gel process for ceramic membrane formation. Metal nanoparticles such as alumoxanes [19] and ferroxanes [20] can be produced based upon the reaction of boehmite, $[Al(O)(OH)]_n$, (or lepidicrocite in the case of the ferroxanes) with carboxylic acids [21]. The physical properties of such metal-oxanes are highly dependent on the identity of the alkyl substituents, R, and range from insoluble crystalline powders to powders that readily form solutions or gels in hydrocarbon solvents and/or water. Thus, a high degree of control over the nanoparticle precursors is possible. Metal-oxanes have been found to be stable over periods of at least years. Whereas the choice of solvents in sol-gel synthesis is limited, the solubility of the carboxylate metal-oxanes is dependent on the identity of the carboxylic acid residue, which is almost unrestricted. The solubility of the metal-oxanes may therefore be readily controlled so as to make them compatible with a coreactant. Furthermore, the incorporation of metals into the metal-oxane core structure allows for atomic scale mixing of metals and formation of meta-stable phases. In the case of the aluminum-based alumoxanes, the low price of boehmite ($ 0.5 kg^{-1}$) and the availability of an almost infinite range of carboxylic acids, make these species ideal as precursors for ternary and doped aluminum oxides.

Application of a metal-oxane–based approach to creating ceramic membranes reduces the use of toxic solvents and energy consumption. By-products formed from the combustion of plasticizers and binders are minimized, and the use of strong acids eliminated. Moreover, the use of tailored nanoparticles and their deposition on a suitable substrate presents an extremely high degree of control over the nanostructure of the resulting sintered film. The versatility of the process can be used to tightly control pore-size distributions with the potential for fabricating ceramic membranes with enhanced specificity. The Molecular Weight Cut Off (MWCO) of the first generation of alumoxane-derived membranes is approximately 40,000 daltons [22], which is in the ultrafiltration range. Table 9.4 shows a comparison of the ceramic and sol-gel methods with that of the carboxylate alumoxanes for the synthesis of alumina and ternary aluminum oxides. The ease of modification of the alumoxanes suggests that a single basic coating system can be modified and optimized for use with a range of substrates.

There has been interest in using ceramics as electrolyte materials for proton exchange membrane fuel cells because of their thermal, chemical, and mechanical stability, and their lower material costs [23]. However, traditionally ceramic membranes have exhibited comparatively small proton conductivities. The conductivities of silica glasses fired at 400 to 800°C is in the order of 10^{-6} to 10^{-3} S/cm [24]. The conductivities of silica, alumina, and titania sintered at 300 and 400°C are in the range of 10^{-7} to 10^{-3} S/cm [25].

However, recent work suggests that membranes derived from ferroxane nanoparticles may be attractive alternatives for such proton exchange membranes (Figure 9.4). With a conductivity of approximately 10^{-2} S/cm the ferroxane-derived membrane represents a large improvement over other ceramic materials prepared by the traditional sol-gel method, with conductivities close to that of Nafion (Table 9.5).

The protonic conductivity of these membranes varies as a function of the temperature at which they are sintered. For example, when ferroxane films are sintered at 300°C the resulting membranes display a higher conductivity (~0.03 S/cm) at all values of relative humidity compared with the green body and with the ferroxane sintered at 400°C (Figure 9.5). Sintering at higher temperatures likely results in the sacrifice of the small pores in the ceramic as it forms and a resulting decrease in conductivity. Under the conditions examined to date, proton conductivity of these membranes appears to be relatively insensitive to the relative humidity within the membrane. A membrane with high proton conductivity at a low humidity would be highly desirable, allowing the fuel cell to operate at higher temperatures and with less difficulty in keeping

	Alumoxane	Sol-Gel
Methodology	Simple	Complex
Atomic mixing	Yes	Yes
Meta-stable phases	Yes	Yes
Stability	Excellent	Fair
Solubility	Readily controlled	Difficult to control
Processability	Good	Good
Time	<8 h	>20 h
Cost	Low	Med–high

TABLE 9.4 Comparison of the Alumoxane and Sol-Gel Synthesis Methods

Material/Study	Conductivity (S/cm)
Nafion 117	0.06
(Sumner, 1998)	0.09
(Sone, 1997)	0.09
(Kopitzke, 2000)	
SiO_2 glasses	10^{-6}–10^{-3}
(Nogami, 1998)	
$(SiO_2$-$P_2)_5$ glass	9.43E-3
(Tung and Hwang, 2004)	
Sol-gel Al_2O_3	6.0E-4
(Vichi, 1999)	
Alumoxane-derived Al_2O_3	6.7E-4
(Tsui et al.)	
Ferroxane-derived membranes	0.03
(Tsui et al.)	

TABLE 9.5 Representative Conductivities of Oxide Membranes and Nafion® Compared with Preliminary Results for Ferroxane-Derived Membrane; Conductivity Is Reported at 100 percent Humidity and at 20°C

the membrane hydrated. Similar to the Nafion membrane, the green body ferroxane material shows a strong dependence of proton conductivity on humidity. This dependence on humidity implies a structure diffusion mechanism, while the transport mechanism in the case of the ferroxane-derived ceramics is less clear. However, it is likely to involve proton "hopping" between hydrogen-bonded water or hydroxyl groups along the oxide structure. By comparison, Nafion conductivity may decrease by 200 percent or more over the range of 100 percent RH to 80 percent RH [7, 9]. The structure diffusion mechanism that typically dominates proton transport in Nafion membranes is typically orders of magnitude larger than the hopping mechanism. Transport dominated by a hopping mechanism in the ferroxane-derived membranes would imply a potential for low methanol permeability, which has been confirmed, but may not explain the very high protonic conductivities that have been observed.

Another alternative to the sol-gel method for creating membrane skins on ceramic membranes is to deposit a layer of zeolites on the porous ceramic support. Zeolites are aluminosilicates framework that contains cavities, which allow for the selective transport of liquids or gases. Naturally occurring, they can be synthesized to produce materials with a range of Si/Al ratios that exhibit increasing hydrophobicity and decreasing charge as this ratio increases. Zeolites can be deposited on the ceramic support by forming them

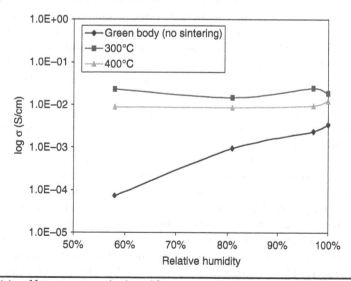

FIGURE 9.5 Conductivity of ferroxane green body and ferroxane-derived ceramics as a function of humidity.

in situ on support by hydrothermal synthesis. Due to the dimension of pores within zeolites, they are used to create membranes with rejections in the range of RO or NF membranes.

Fullerene-Based Membranes   Fullerenes have unique properties of strength, ability to tailor size, flexibility in modifying functionality, and electron affinity that have create much excitement around their potential for new membrane-based technologies. For example, the ability of fullerenes to act as electron shuttles has been considered as a possible basis for creating light-harvesting membranes using C_{60} or C_{70} contained in lipid bilayers [26] or incorporated into porous polymers [27]. The photocurrent density obtained from the C_{70}-bilayer system was observed to be about 40 times higher than that of the artificial system previously observed to be the most efficient [28]. In addition to photovoltaics, fullerenes may find uses in fuel cells. Fullerenes share some of the properties of the perfluorosulfonic polymers typically used in Polymer Electrolyte Fuel Cells (PEFCs) in that they are quite stable, anhydrous, and yet modifiable in a wide variety of manners through the introduction of proton-binding functions on the fullerene surface. These features make them interesting candidates for proton exchange membranes in fuel cells [29]. In addition to the unique properties that make fullerenes such as carbon nanotubes (CNTs) interesting materials for creating new membranes for fuel cells, fullerenes have also drawn interest as the basis for new pressure-driven membranes, particularly for the treatment of water. The small and controllable diameter of fullerene nanotubes suggests that membranes made from these materials in a fashion where fluid flows through the center of the CNT, might be highly selective.

However, Equation (9.30) predicts that the resistance to flow through a membrane composed of nanometer-sized pores should be very high and potentially prohibitive for practical applications. Surprisingly, molecular modeling first suggested that flow through pores composed of CNTs might not have the same limitations as those observed for other nanometer-sized pores [30]. Simulations indicate that water should be able to flow much faster through hydrophobic CNTs due to the formation of ordered hydrogen bonds. In the confined space of a nanotube, water is present in ordered crystalline domains. When the nanotube wall interacts significantly with the water, such as in the case of a small silica channel where silinol groups may anchor water molecules to the wall of the channel, the ordered water in the pore is thought to be less mobile. In contrast, the hydrophobic surface in the interior of a defect-free carbon nanotube appears to allow for a nearly frictionless flow that has been compared with the flow through the protein channel aquaporin-1 [30]. Visualization of water within CNTs confirms the lack of interaction between water molecules and the interior surface of CNTs [31]. While both CNT and aquaporin-based membranes have been proposed as low energy membranes for desalination, production methods will ultimately determine whether these membranes will be competitive with conventional RO membranes which already operate at full-scale near the thermodynamic limit for water-salt separation.

There are many challenges to be overcome in aligning CNTs and fabricating such a membrane. The problem of aligning membranes was first approached by filtering suspensions of Single Wall Carbon NanoTubes (SWNTs) in a strong magnetic field [32]. A more promising approach is to grow arrays of CNTs on a substrate where nanoparticles catalysts for CNT growth have been arranged in a distinct pattern that defines the number and spacing of the resulting CNTs. The diameter of the nanotube is controlled by the size of the nanoparticle catalyst initially arranged on the substrate [33]. A working membrane of aligned CNTs requires that the spaces between CNTs be filled with a material that seals the membrane to flow between CNTs, allowing flow only through the interior of the CNTs. Among the approaches taken to accomplish this has been spin-coating the CNT arrays with a polymer solution [34] or filling the interstices of the aligned CNT with silcon nitride [35]. The permeate flux of water measured across a membrane of aligned multiwall CNTs with 7-nm-diameter pores imbedded in a polystyrene matrix has been reported to be 4 to 5 orders of magnitude greater than that predicted by Equation (9.30) [36]. However, smaller-diameter CNTs may exhibit a bamboo-like structure [37] that impedes fluid flow [35]. Similar to these CNT-based membranes, fullerene-based membranes have also been made by grafting C_{60} onto the surface of track-etched membranes [38].

These fullerene-based membranes can be thought of composite membranes in that they are composed of the fullerenes and at least one other material. The properties of these composite membranes reflect the sum of the properties of the components of the membrane. As such, they resemble conventional thin film composite membranes such as those used in RO. RO composite membranes rely on a thin layer of material (typically a polyamide) on the surface or skin of the membrane to provide the essential rejection characteristics of the membrane. The thicker, underlying layer (often polysulfone) serves as a support. The overall property of the membrane is approximated by the rejection characteristics of the skin plus the mechanical characteristics of the support. Similarly, in the current generation of aligned CNT membranes, the CNTs determine the transport properties of the membrane, while the support material envelopes rather than underlies the CNTs.

Nanocomposites: Modifications to Existing Materials with Nanoparticles

The inherent limitations of temperature and water retention by fuel cells membranes made from perfluorosulfonic polymers (typically Nafion) has stimulated much research to develop nanocomposites that display high proton conductivity at high temperatures and low humidity. One approach has been to add nanoparticles designed to promote proton conductivity to polymer matrices with greater resistance to temperature than Nafion. The modification of polysulfonated membranes with solid acids in the form of silica [39] or zirconium phosphate [40] nanoparticles has resulted in membranes that can operate at higher temperatures, but still, with a lower conductivity than that of Nafion [41]. The brittleness of the ferroxane-derived membrane described previously is a key limitation to the development of this technology. Zhang et al. [42] attempted to address this limitation by preparing proton-conductive composite membranes derived from the combination of ferroxane nanoparticles and polyvinyl alcohol (PVA), where the high protonic conductivity of ferroxane nanoparticles was paired with good mechanical properties of PVA. The ferroxane-PVA membrane has a PVA skeleton with ferroxane nanoparticles distributed throughout the composite matrix. This membrane exhibits proton conductivities comparable to Nafion at moderate relative humidities and excellent mechanical properties. However, the proton conductivity of this composite membrane remains low when the relative humidity is less than 60 percent.

The electrical conductivity of several polymer-CNT blends has been evaluated [43]. While these materials may have some promise as electrode materials in fuel cells, their potential as fuel cell membranes is yet to be demonstrated. For example, poly(methyl methacrylate) (PMMA) nanocomposites containing MWNTs were found to have electric conductivities on the order of 10^{-4} to 10^{-2} S/cm [44].

Much consideration has also been given to improvements in the catalyst/membrane support materials used in fuel cells through the incorporation of fullerenes into these electrode/supports. SWNTs have been used to replace carbon black in fuel cell electrodes yielding an order of magnitude lower resistance to charge-transfer [45]. These electrodes can then be used as supports for the PEM. More efficient use of catalyst through the formation of nanoparticles with high ratio of surface area to volume has been an important element in reducing the costs of fuel cells. Nanoparticles of Pd [46] or Pt [47] catalyst assembled on a Nafion membrane have also been reported to increase methanol rejection by the membrane (reduced crossover) in direct methanol fuel cells.

Electrically conductive membranes have also been developed with the goal of reducing biofouling [48, 49] in pressure-driven membranes. It has been demonstrated that applying an electrical bias to a surface prevents the formation of biofilms [50–52]. An electrically conductive CNT-polyamide thin film, cast on a polyethersulfone (PES) support to yield a thin film composite membrane, showed near-complete recovery of permeate flux following a surface wash when a current was applied.

Many other examples of mixed-matrix membranes have been produced using zeolites, silica nanoparticles, carbon nanotubes, C_{60}, graphene oxide, metals such as silver, bismuth, or gold, and metal oxides such as TiO_2.

For example, the strength of CNTs, coupled with reported antibacterial properties, suggest that CNT-polymer composites may find use in creating membranes that resist breakage or inhibit biofouling. CNT-polymer composite membranes show significant improvement in tensile strength [53].The incorporation of C_{60} into polymeric membranes has been observed to effect membrane structure and rejection [54].

Polymer-TiO_2 nanoparticle composites have been created with the objective of creating antifouling membranes. These membranes are conceived to exploit the photocatalytic properties of TiO_2 to produce hydroxyl radicals that would then, in turn, oxidize organic foulants depositing on the membrane surface. Membranes decorated with TiO_2 nanoparticles have been fabricated in both UF [55] and RO [56] formats. These membranes are formed by self-assembly of the TiO_2 particles at functional sites (such as sulfone or carboxylate groups) on the membrane surface. Alternatively, the TiO_2 nanoparticles can be immobilized within the membrane matrix by introducing these materials as a mix during the process of membrane casting [57]. While both formats (decorated and immobilized) appear to mitigate fouling by bacterial growth, the decorated format appears to be more effective due to the higher amount of TiO_2 on the membrane surface [57].

An inherent conflict in membrane design must be overcome in developing a photocatalytic system for reducing membrane fouling. The economics of membrane module design dictate that a maximum of membrane area be contained within a given volume (high packing density). However, this design objective is in conflict with the objective of providing adequate illumination of the membrane surface to promote photocatalysis. Similarly, the use gold nanoparticles to heat water on the surface of membranes used in membrane distillation when exposed to sunlight suffers from the same limitations; the membrane surface must be exposed to light. There may be niche applications for such photocatalytic and light-activated membranes where the high cost associated with these low packing density systems is not a consideration, for example, due to a lack of alternative treatment options. However, the larger-scale application of these membranes will not be cost-effective until new strategies for delivering light to membranes in a high packing density configuration are developed.

A wide variety of hybrid naomaterials have also been used in mix-matrix membranes. Hybrid nanomaterials are composed of two integrated nanocomponents such as TiO_2 nanoparticles decorating graphene sheets, or mixed metal catalysts. Membranes may be modified by depositing particles on the surface for the purpose of simultaneous catalytic degradation, sensing, or simply improved selectivity. Nanoparticles with molecular imprints of a specific compound can be created and then attached to conventional membranes to impart a high specificity of separation for the imprinted molecule [58]. Catalytic nanoparticles (such as TiO_2 or iron) can be attached to membrane surfaces with the objective of degrading some compounds while physically separating others. The objective of creating a reactive membrane is fraught with conflicts in the efficiency of the overall system. Reaction rate considerations tend to favor a membrane with a slow permeation rate (high residence time) to allow for sufficient conversion of the desired compound(s) as the fluid they are contained within flows across the membrane. On the other hand, the efficiency in fluid separation pushes the system toward a higher permeation rate and shorter residence time. The conflict between reaction rate and residence time is resolved for the case of reactive membranes that work on materials sorbed to the surface of the membrane. This might be the case for either the destruction of sorbed foulants or the accumulation and subsequent destruction of a contaminant for which the membrane is designed to have an enhanced adsorptive affinity.

Numerous researchers have exploited the antimicrobial properties of silver nanoparticles [59] by incorporating these particles into or on polymeric or ceramic membranes to produce biofouling-resistant membranes. Similarly nano-bismuth may be used to the same effect [60]. While membranes such as these have been commercialized for niche point-of-use applications, the sacrificial nature of the Ag or Bi applied to these membranes excludes their use in large-scale applications.

Nanomaterial Templating

Nanomaterials can also be used as "molds" or templates for membranes. In this case, the initial nanomaterials are no longer present in the membrane structure, but rather have imparted their structure to another material. Nanoparticles can be deposited on a substrate to form the initial template for a porous solid that will take on the mirror image of the initial deposit [61].

Particle size, stability, and/or depositional trajectories can be controlled to engineer template morphology and yield membranes with a desired structure. When particles deposit on a surface following very predictable trajectories, such as occurs when their transport is dominated by gravity, an electrical field, or laminar flow, their depositional trajectory is said to be ballistic. The templates formed by depositing particles on the surface via ballistic trajectories tend to be compact. In contrast, when particles follow more random, diffusive trajectories to deposition they tend to form more open deposits that resemble dendrites or objects resembling small trees. Particle surface chemistry can also be altered to control template morphology. When particle-particle interactions are favorable ("sticky" particles) the resulting template tends to be dendritic. When particles are not sticky, compact deposits form. Electrostatic forces, dispersion forces and capillary forces are likely to dominate particle "stickiness" during template formation. For example, if templates are formed by dip-coating a substrate in a suspension of particles and evaporating the suspending fluid, the deformation of the liquid-fluid interface due to trapped colloidal particles gives rise to capillary forces exerted on the particles. These forces are usually attractive [62]. Thus, by controlling solution chemistry (e.g., ionic strength) and the conditions of template drying, the morphology of the template can be controlled. The self-assembly technique by capillary forces provides precise control of the thickness of the film through sphere size and concentration in solution. Control of template morphology is illustrated in SEM images of deposits of silica nanoparticles of 244 nm average diameter (Figure 9.6). Case (a) is a template formed from particles suspended in ethanol. In the case of ethanol, the particles experience a net repulsive electrostatic force in the bulk as they approach the glass surface on which they were deposited. Thus, these "nonsticky" particles form a compact deposit. In contrast, particles suspended in an aqueous solution of 1.5 M ionic strength (b) form a dendritic template.

The voids in these resulting templates are filled with a polymeric or inorganic material and upon etching (or buring) of the particles, a porous material with a three-dimensional structure is formed

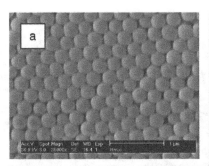

FIGURE 9.6 Templates of silica particles with an average diameter of 244 nm formed from suspensions (a) ethanol (nonsticky) and (b) an aqueous solution 1.5 M NaCl (sticky).

(Figure 9.7). The use of nanoparticles in the templating process allows for a high degree of control over chamber and pore size. The interior of the templated object can be subsequently functionalized or functionality can be introduced through the choice of material used to make the membrane. A high degree of control over both the structure and internal functionality of these membranes might be exploited to perform highly controlled reactions, with each chamber of the membrane serving as a reactor. Membrane selectivity can be modified with respect to both size exclusion and chemical affinity.

Alternatively, asymmetric templates can be formed via the sequential deposition of particle layers of nanoparticles from Langmuir-Blodgett films. A layer of small particles is first deposited onto the support, followed by a layer of larger particles. After casting the membrane around this template, a membrane with a more complex structure composed of distinct layers is formed (Figure 9.8).

CNTs have also been used as membrane templates. Beginning with an array of aligned carbon nanotubes, spaces are filled between the tubes as described earlier for the case of aligned CNT membranes. Using silicon nitride as the fill material, the CNT can then be subsequently oxidized, leaving behind a silcon nitride membrane with CNT-size pores [35].

Nanoparticle Membrane Reactors

Nanomaterials may also be used in conjunction with membranes as a nanomaterial/membrane reactor. In this case, the membrane serves only as a separation process to recover a nanomaterial that is introduced upstream for the purposes of adsorption, photocatalysis, disinfection, or some other function for which the nanomaterial is particularly well-suited. The residence time of nanomaterials in the system is controlled to optimize their effectiveness. As nanomaterials lose their effectiveness (e.g., adsorption capacity is exhausted) they are removed from the system and regenerated.

The separation of nanoparticles used in such membrane reactors may present special challenges for separation due to nanoparticle size. Nanosized materials are likely to have relatively small diffusivities compared with conventional solutes, but large diffusivities in comparison with larger colloids. While nanoparticles will therefore be more susceptible to concentration polarization [Equation (9.32)] compared

FIGURE 9.7 Cross-section of a templated membrane formed from nonsticky particles.

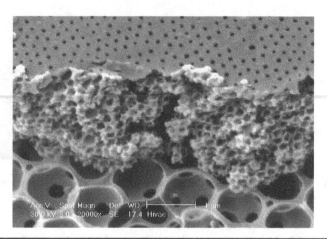

FIGURE 9.8   Asymmetric templated membrane produced from the Langmuir-Blodgett process.

with solutes, their osmotic pressure will be less than, for example, an equivalent mass concentration of ionic solutes due to their relatively high molecular weight [Equation (9.18)]. Unlike larger micron-sized particles that may form cakes on MF or UF membranes, the specific resistance of a cake formed by nanometer-sized particles will present a high specific resistance [Equation (9.33)].

Also, nanoparticles with a high degree of functionality per surface area may create cakes with a relatively high charge density. Coupling effects associated with the deformation of diffuse layers within these cakes and the flow of permeate across these cakes may be manifest as a significant electroviscous effect in which the viscosity of the fluid appears to be greater than the bulk viscosity due to the back-migration of ions.

Active Membrane Systems

Future convergence between nanochemistry and membrane science will likely yield a generation of active membrane systems. Nanomaterials might be used to develop membranes in the future with the capability to simultaneously sense and separate contaminants in a fashion that allows membranes to vary their selectivity as a function of the conditions in the feed stream. For example, self-regulating membranes might allow membranes to operate in a high permeability/low energy mode during periods where high rejections of low-molecular-weight materials are not required. Seasonal peaks in concentration of a given contaminant (e.g., a target pesticide) would be detected by the membranes and trigger an increase in the membrane molecular weight cutoff.

Nanomaterials might also be incorporated into membranes to impart properties that are activated by an electrical or chemical signal. For example, a membrane composite that is capable of producing reactive oxygen in the presence of an electron donor might be activated by the introduction of such a compound with the purpose of periodically cleaning the membrane. Membranes might also be engineered to allow for local heating of the membrane skin with the purpose of promoting membrane distillation.

Living organisms are the ultimate nanotechnology. The ability of cell membranes to selectively transport materials, often against concentration gradient, and to avoid fouling is impressive. As the field of nanochemistry advances, engineered biomimetic systems based on selective transport or rejuvenating layers of self-organizing materials may be developed for performing critical separations in energy and environmental applications.

References

1. Onsager, L., *Reciprocal relations in irreversible processes I.* Physics Review, 1931a. 37: p. 405–426.
2. Onsager, L., *Reciprocal relations in irreversible processes II.* Physics Review, 1931b. 38: p. 2265–2279.
3. Kedem, O. and A. Katchalsky, *Thermodynamic analysis of the permeability of biological membranes in non-electrolytes.* Biochem. Biophys. Acta, 1958. 27: p. 229.
4. Spiegler, K.S. and O. Kedem, *Thermodynamics of hyperfiltration (reverse osmosis): criteria for efficient membranes.* Desalination, 1966. 1: p. 311.
5. Lonsdale, H.K., U. Merten, and R.L. Riley, *Transport properties of cellulose acetate osmotic membranes.* J. Appl. Polym. Sci., 1965. 9: p. 1341–1362.

6. Eikerling, M., A.A. Kornyshev, and U. Stimming, *Electrophysical properties of polymer electrolyte membranes: a random network model.* J, Phys, Chem. B, 1997. 101(50): p. 10807–10820.

7. Sumner, J.J., et al., *Proton Conductivity in Nafion117 and in a novel bis[(perfluoroalkyl)sulfonyl]imide ionomer membrane.* J. Electrochem. Soc., 1998. 148(1): p. 107–110.

8. Zawodzinski Jr., T.A., et al., *Determination of water diffusion coefficients in perfluorosulfonate ionomeric membranes.* J. Phys. Chem. B, 1991. 95: p. 6040–6044.

9. Sone, Y., P. Ekdunge, and D. Simonsson, *Proton conductivity of Nafion 117 as measured by a four-electrode AC impedance method.* J. Electrochem. Soc., 1996. 143(4): p. 1254–1259.

10. Brandon, N.P., S. Skinner, and B.C.H. Steele, *Recent advances in materials for fuel cells.* Annu. Rev. Mater. Res., 2003. 33(1): p. 183–214.

11. Bailly, C., et al., *The sodium salts of sulfonated poly(aryl-ether-ether-ketone) (PEEK): preparation and characterization.* Polymer, 1987. 28(6): p. 1009–1016.

12. Nolte, R., et al., *Partially sulfonated poly(arylene ether sulfone)—A versatile proton conducting membrane material for modern energy conversion technologies.* J. Membr. Sci., 1993. 83(2): p. 211–220.

13. Gupta, B., F.N. Büchi, and G.G. Scherer, *Cation exchange membranes by pre-irradiation grafting of styrene into FEP films. I. Influence of synthesis conditions.* J. Polym. Sci. A Polym. Chem., 1994. 32(10): p. 1931–1938.

14. Wang, C.Y., *Fundamental models for fuel cell engineering.* Chem. Rev., 2004. 104(10): p. 4727.

15. Bhave, R.R., *Inorganic Membranes: Synthesis, Characteristics and Applications.* 1991, New York: Van Nostrand Reinhold.

16. Cai, X.M., et al., *Porous metallic films fabricated by self-assembly of gold nanoparticles.* Thin Solid Films, 2005. 491(1–2): p. 66–70.

17. Hu, X.G., et al., *Fabrication, characterization, and application in SERS of self-assembled polyelectrolyte-gold nanorod multilayered films.* J. Phys. Chem. B, 2005. 109(41): p. 19385–19389.

18. Fujii, T., et al., *The sol-gel preparation and characterization of nanoporous silica membrane with controlled pore size.* J. Membr. Sci., 2001. 187(1–2): p. 171–180.

19. Callender, R.L., et al., *Aqueous synthesis of water soluble alumoxanes: environmentally benign precursors to alumina and aluminum-based ceramics.* Chem. Mater., 1997. 9: p. 2418–2433.

20. Rose, J., et al., *Synthesis and characterization of carboxylate-FeOOH nanoparticles (ferroxanes) and ferroxane-derived ceramics.* Chem. Mater., 2002. 14(2): p. 621–628.

21. Landry, C.C. and A.R. Barron, *From minerals to materials: synthesis of alumoxanes from the reactions of boehmite with carboxylic acids.* J. Mater. Chem., 1995. 5(2): p. 331–341.

22. Cortalezzi, M.M., et al., *Characteristics of ceramic membranes derived from alumoxane nanoparticles.* J. Membr. Sci., 2002. 205: p. 33–43.

23. Grahl, C.L., *Ceramic opportunities in fuel cells.* Ceramic Industry, 2002. 152(6): p. 35–39.

24. Nogami, M., R. Nagao, and C. Wong, *Proton conduction in porous silica glasses with high water content.* J. Phys. Chem. B, 1998. 102: p. 5772–5775.

25. Vichi, F.M., M.T. Colomer, and M.A. Anderson, *Nanopore ceramic membranes as novel electrolytes for proton exchange membranes.* Electrochem. Solid State Lett., 1999. 2(7): p. 313–316.

26. Bensasson, R.V., et al., *Transmembrane electron-transport mediated by photoexcited fullerenes.* Chem. Phys. Lett., 1993. 210(1–3): p. 141–148.

27. Garaud, J.L., et al., *Photoinduced electron-transfer properties of porous polymer membranes doped with the fullerene C(60) associated with phospholipids.* J. Membr. Sci., 1994. 91(3): p. 259–264.

28. Hwang, K.C. and D. Mauzerall, *Photoinduced electron-transport across a lipid bilayer mediated by C70.* Nature, 1993. 361(6408): p. 138–140.

29. Miyatake, K. and M. Watanabe, *Recent progress in proton conducting membranes for PEFCs.* Electrochemistry, 2005. 73(1): p. 12–19.

30. Hummer, G., J.C. Rasaih, and J.P. Noworyta, *Water conduction through the hydrophobic channel of a carbon nanotube.* Nature, 2001. 414: p. 188–190.

31. Naguib, N., et al., *Observation of water confined in nanometer channels of closed carbon nanotubes.* Nano Lett., 2004. 4(11): p. 2237–2243.

32. Walters, D.A., et al., *In-plane-aligned membranes of carbon nanotubes.* Chem. Phys. Lett., 2001. 338(1): p. 14–20.

33. Kanzow, H., C. Lenski, and A. Ding, *Single-wall carbon nanotube diameter distributions calculated from experimental parameters.* Phys. Rev. B, 2001. 6312: p. 5402.

34. Hinds, B.J., et al., *Aligned multiwalled carbon nanotube membranes.* Science, 2004. 303(5654): p. 62–65.

35. Holt, J.K., et al., *Fabrication of a carbon nanotube-embedded silicon nitride membrane for studies of nanometer-scale mass transport.* Nano Lett., 2004. 4(11): p. 2245–2250.

36. Majumder, M., et al., *Nanoscale hydrodynamics: enhanced flow in carbon nanotubes.* Nature, 2005. 438(7070): p. 930.

37. Wang, Y.Y., et al., *Hollow to bamboolike internal structure transition observed in carbon nanotube films.* J. Appl. Phys., 2005. 98: p. 014312.

38. Biryulin, Y.F., et al., *Fullerene-modified dacron track membranes and adsorption of nitroxyl radicals on these membranes.* Technical Physics Letters, 2005. 31(6): p. 506–508.

39. Antonucci, P.L., et al., *Investigation of a direct methanol fuel cell based on a composite Nafion-silica electrolyte for high temperature operation.* Solid State Ionics, 1998. 125(1–4): p. 431–437.

40. Costamagna, P., et al., *Nafion 115/zirconium phosphate composite membranes for operation of PEMFCs above 100°C.* Electrochim. Acta, 2002. 47(7): p. 1023–1033.

41. Boysen, D.A., et al., *Polymer solid acid composite membranes for fuel-cell applications.* J. Electrochem. Soc., 2000. 147(10): p. 3610–3613.

42. Zhang, L.W., et al., *Proton-Conducting Composite Membranes Derived from Ferroxane-Polyvinyl Alcohol Complex.* Environmental Engineering Science, 2012. **29**(2): p. 124–132.

43. Wu, M. and L. Shaw, *Electrical and mechanical behaviors of carbon nanotube-filled polymer blends.* Journal of Applied Polymer Science, 2006. **99**(2): p. 477–488.

44. Sung, J.H., et al., *Nanofibrous membranes prepared by multiwalled carbon nanotube/poly(methyl methacrylate) composites.* Macromolecules, 2004. 37(26): p. 9899–9902.

45. Girishkumar, G., et al., *Single-wall carbon nanotube-based proton exchange membrane assembly for hydrogen fuel cells.* Langmuir, 2005. **21**(18): p. 8487–8494.

46. Tang, H.L., et al., *Self-assembling multi-layer Pd nanoparticles onto Nafion(TM) membrane to reduce methanol crossover.* Colloids and Surfaces a-Physicochemical and Engineering Aspects, 2005. **262**(1–3): p. 65–70.

47. Jiang, S.P., et al., *Self-assembly of PDDA-Pt nanoparticle/nafion membranes for direct methanol fuel cells.* Electrochemical and Solid State Letters, 2005. **8**(11): p. A574–A576.

48. de Lannoy, C.-F., et al., *Aquatic Biofouling Prevention by Electrically Charged Nanocomposite Polymer Thin Film Membranes.* Environmental Science & Technology, 2013. **47**(6): p. 2760–2768.

49. de Lannoy, C.F., et al., *A highly electrically conductive polymer-multiwalled carbon nanotube nanocomposite membrane.* Journal of Membrane Science, 2012. **415**: p. 718–724.

50. Hong, S.H., et al., *Effect of electric currents on bacterial detachment and inactivation.* Biotechnology and bioengineering, 2008. **100**(2): p. 379–386.

51. Nakasono, S., et al., *Electrochemical prevention of marine biofouling with a carbon-chloroprene sheet.* Applied and environmental microbiology, 1993. **59**(11): p. 3757–3762.

52. Wake, H., et al., *Development of an electrochemical antifouling system for seawater cooling pipelines of power plants using titanium.* Biotechnology and bioengineering, 2006. **95**(3): p. 468–473.

53. de Lannoy, C.F., E. Soyer, and M.R. Wiesner, *Optimizing carbon nanotube-reinforced polysulfone ultrafiltration membranes through carboxylic acid functionalization.* Journal of Membrane Science, 2013. **447**: p. 395–402.

54. Polotskaya, G., Y. Biryulin, and V. Rozanov, *Asymmetric membranes based on fullerene-containing polyphenylene oxide.* Fullerenes Nanotubes and Carbon Nanostructures, 2004. **12**(1–2): p. 371–376.

55. Luo, M.L., et al., *Hydrophilic modification of poly(ether sulfone) ultrafiltration membrane surface by self-assembly of TiO2 nanoparticles.* Applied Surface Science, 2005. **249**(1–4): p. 76–84.

56. Kwak, S.Y., S.H. Kim, and S.S. Kim, *Hybrid organic/inorganic reverse osmosis (RO) membrane for bactericidal anti-fouling. 1. Preparation and characterization of TiO2 nanoparticle self-assembled aromatic polyamide thin-film-composite (TFC) membrane.* Environmental Science & Technology, 2001. **35**(11): p. 2388–2394.

57. Bae, T.H. and T.M. Tak, *Effect of TiO2 nanoparticles on fouling mitigation of ultrafiltration membranes for activated sludge filtration.* Journal of Membrane Science, 2005. **249**(1-2): p. 1–8.

58. Lehmam, M., H. Brunner, and G.E.M. Tovar, *Selective separations and hydrodynamic studies: a new approach using molecularly imprinted nanosphere composite membranes.* Desalination, 2002. **149**(1-3): p. 315–321.

59. Sondi, I. and B. Salopek-Sondi, *Silver nanoparticles as antimicrobial agent: a case study on E-coli as a model for Gram-negative bacteria.* Journal of Colloid and Interface Science, 2004. **275**(1): p. 177–182.

60. Badireddy, A.R., et al., *Lipophilic nano-bismuth inhibits bacterial growth, attachment, and biofilm formation.* Surface Innovations, 2013. **1**(SI3): p. 181–189.

61. Cortalezzi, M.M., C. V., and M.R. Wiesner, *Controlling nanoparticle template morphology: effect of solvent chemistry.* Colloid and Interface Science, 2005. **283**: p. 366–372.

62. Kralchevsky, P.A., et al., *Capillary meniscus interaction between colloidal particles attached to a liquid-fluid interface.* Journal of Colloid and Interface Science, 1992. **151**(1): p. 79–94.

CHAPTER 10

Nanomaterial-Enabled Sensors

Peter Vikesland

Institute for Critical Technology and Applied Science (ICTAS) Center for Sustainable Nanotechnology (VTSuN) and NSF-EPA Center for the Environmental Implications of Nanotechnology (CEINT), Virginia Polytechnic Institute and State University, Blacksburg, Virginia, USA

Haoran Wei

Department of Civil and Environmental Engineering and NSF-EPA Center for the Environmental Implications of Nanotechnology (CEINT), Virginia Polytechnic Institute and State University, Blacksburg, Virginia, USA

Introduction

Nanomaterial-enabled sensors (nanosensors) are increasingly attracting attention due to their potential to either match or outperform existing detection methods as well as their capacity to facilitate analyte detection in challenging or previously inaccessible locations [e.g., within cells and cellular compartments (Dowd et al. 2014)]. The aim of this introductory chapter is to summarize the essential components of a nanosensor, discuss recent progress in nanosensor application, and illustrate the challenges and opportunities facing nanosensor deployment for real-world applications. A number of excellent review articles on nanosensors further illustrate the extreme breadth of this field and the capacity for nanosensors to detect a broad array of environmentally relevant analytes such as heavy metals, organic pollutants, and pathogens (Swierczewska et al. 2012, Pandey and Mishra 2014, Wujcik et al. 2014, Singh 2015). The reader is referred to these articles to gain further insights to this fascinating field.

Nanomaterial-Enabled Biosensors

At its simplest a nanosensor generally consists of (1) a nanomaterial, (2) a recognition element, and (3) a mechanism for signal transduction (Figure 10.1; Vikesland and Wigginton 2010). Analytes interact with the recognition element and induce a change in the nanosensor that is detectable (i.e., the difference between the input and output signals in Figure 10.1). When engineered properly a nanosensor exhibits both, a high degree of specificity as well as extreme sensitivity. Specificity is achieved either by detecting the intrinsic signal of the analyte or by relying on highly specific recognition elements that ideally bind only to the target (Figure 10.1). The extreme sensitivity of many nanosensors is a result of the properties of the nanomaterial itself, as well as the signal transduction method used to generate a detectable signal following exposure to an analyte. As discussed below, antibodies and aptamers are the most frequently used recognition elements, while optical and electrochemical methods are the most commonly used signal transduction mechanisms (Sharma et al. 2015). We note that sample pretreatment (i.e., purification or concentration) may be required for optimal nanosensor deployment. Such steps, which are also often required for the use of many standard analysis methods (e.g., gas or liquid chromatography), are outside the scope of this chapter.

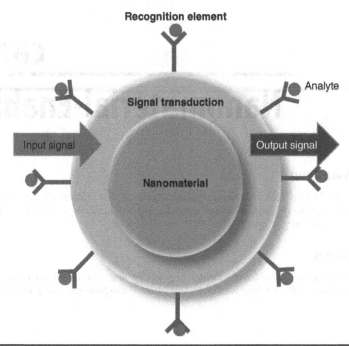

FIGURE 10.1 Schematic of the three components of nanomaterial-enabled biosensors.

Recognition Elements

The specificity of the recognition element dictates the selectivity of a nanosensor. Accordingly, the choice of recognition element is vitally important to nanosensor design. Environmental contaminants such as heavy metals, organic chemicals, and pathogens differ significantly with respect to their physical and chemical properties, and thus the recognition elements chosen for targets generally differ from one another. Antibodies are the most widely applied type of recognition element for pathogens due to the well-established literature describing antigen-antibody interactions and their long and successful history of use. In contrast, for heavy metals, chemicals that form complexes or chelates with the metal of interest are frequently used. We note that for some special cases where the intrinsic signal from the analyte can be sensitively detected (e.g., via surface-enhanced Raman spectroscopy, SERS), it is unnecessary to use a recognition element. The following section briefly introduces three classes of biomolecules that are commonly applied as recognition elements: antibodies (Dou et al. 1997, Abdel-Hamid et al. 1999, Cui et al. 2006, Campbell and Mutharasan 2008), aptamers (Sharma et al. 2015), and carbohydrates (De la Fuente and Penades 2006, El-Boubbou et al. 2007, Shen et al. 2007).

Antibodies

Antibodies are natural proteins generated as part of the immune response to infectious agents. Antibodies exhibit highly specific binding to a target antigen. This binding arises due to the combined activity of the Fab (fragment, antigen-binding) region and the Fc (fragment, constant) region of the antibody. The Fab provides antigen specificity, while the Fc region enables conjugation to the nanoparticle surface in a manner that does not disrupt the recognition process. Three categories of antibodies are often employed in nanosensors: polyclonal, monoclonal, and engineered Fab fragments (Desphande 1996, Iqbal et al. 2000). Of these three types, most nanosensors rely on monoclonal antibodies (mAbs) because they are less likely to be subject to nonspecific binding (i.e., binding to nonantigenic materials), are less likely to cross react with debris, and generally exhibit a high degree of specificity (Desphande 1996, Iqbal et al. 2000). Monoclonal antibodies can be too specific, however, in which case they may not detect species variants that lack a particular surface epitope. Recently, engineered antibody fragments that retain the specificity of antibodies, but offer improvements in production cost and substrate coverage densities have also been used within nanosensors.

Due to their widespread commercial availability and ease of use antibodies are commonly employed for pathogen detection (Iqbal et al. 2000, Leonard et al. 2003). In many applications, antibody-based nanosensors utilize a sandwich assay in which antibodies arrayed across a substrate capture target pathogens. Subsequently, antibody-functionalized nanoparticles are incubated with the pathogens to generate a detectable signal

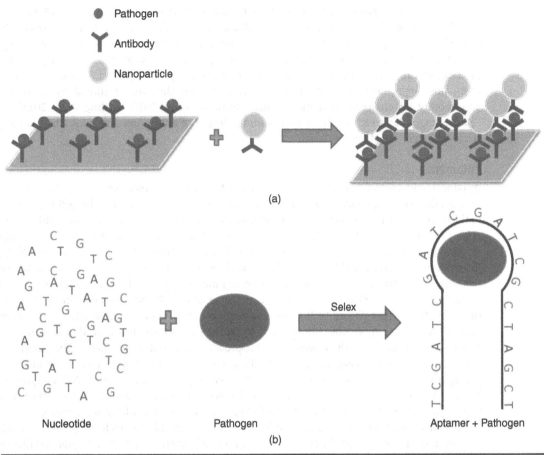

FIGURE 10.2 (a) Schematic of sandwich assay using antibody as the recognition element. (b) Schematic of screening of aptamer for a target analyte by SELEX.

(Figure 10.2a). Such an approach was recently used to detect *Escherichia coli* O157:H7 in beef and water samples (Hassan et al. 2015). In this effort, the authors utilized antibody-functionalized magnetic beads to isolate *E. coli* and then applied antibody-functionalized gold nanoparticles to complete the sandwich (Figure 10.3). The *E. coli* concentration was then determined via a novel hydrogen evolution reaction catalyzed by the gold nanoparticles and detected by chronoamperometry. Limits of detection (LODs) of 148, 457, and 309 CFU/mL were obtained in buffer, minced beef, and tap water samples, respectively.

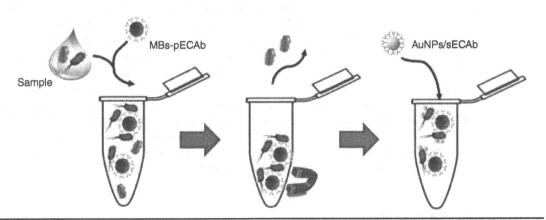

FIGURE 10.3 Schematic of a nanosensor designed to isolate and detect *E. coli* in water and food samples. Left: *E. coli* O157:H7 is captured by magnetic beads functionalized with anti–*E. coli* antibody (MBs-pECAb). Center: Unbound nontarget bacteria and debris are removed. Right: Gold nanoparticles functionalized with antibodies (AuNPs/sECAb) are added to complete the sandwich. The *E. coli* concentration is then determined by chronoamperometry. *Reprinted with permission from Hassan et al. 2015. Copyright 2014 Elsevier B.V.*

While antibody-based nanosensors are well-established in the literature (Dou et al. 1997, Abdel-Hamid et al. 1999, Cui et al. 2006, Campbell and Mutharasan 2008), there remain a number of disadvantages that limit their broad application. First of all, antibodies can only be produced *in vivo* and are highly sensitive to changes in their external environment. Large fluctuations of temperature, pH, or the presence of interferents can make antibodies lose their activity (Jayasena 1999). Furthermore, the high preparation costs, poor reproducibility, and short shelf lives of antibodies are additional concerns that limit their applicability for environmental monitoring (Bordeaux et al. 2010, Trilling et al. 2013). Despite these disadvantages the great specificity and binding affinity of antibodies still make them amongst the most reliable recognition elements.

Aptamers

Aptamers—single-stranded oligonucleotides of DNA or RNA—are emerging recognition elements with some unique advantages. Aptamers bind a broad range of targets with potentially higher affinity than antibodies due to the three-dimensional structure that the oligonucleotide develops upon interacting with an analyte (Jenison et al. 1994, Win et al. 2006). Aptamers can be selected and then produced at large scale using an artificial synthetic process known as *systematic evolution of ligands by exponential enrichment* (SELEX) (Ellington and Szostak 1990, Robertson and Joyce 1990, Tuerk and Gold 1990). In this process, a given analyte target is exposed to a large (>10^{14} strands) pool of oligonucleotides of random sequence. Following multiple cycles of separation and amplification, the oligonucleotide sequences that exhibit the highest affinity to the analyte are selected and then amplified further (Figure 10.2b). Once the sequence of the aptamer oligonucleotide has been determined via SELEX or other approaches, the capacity to synthesize aptamers based on that sequence is considerably simpler and cheaper than the biology-based approaches required to produce antibodies (Low et al. 2009). Furthermore, because aptamers can theoretically be raised against any target, their application is not limited to pathogens or other materials that raise an antigenic response (Fischer et al. 2007). Aptamers specific to biomolecules (Li et al. 2009), small organic molecules (Liu and Lu 2006b, Liu et al. 2006), and even heavy metals (Ye et al. 2012) have been isolated by SELEX and other related selection processes.

Because aptamers are oligonucleotides they are susceptible to nuclease (DNAse or RNAse) attack (Famulok et al. 2000, Tombelli et al. 2007, Mairal et al. 2008). To produce higher stability aptamers it is possible to modify functional groups on the aptamer or to replace their natural component D-ribose with its unnatural enantiomer L-ribose (Klussmann et al. 1996, Nolte et al. 1996, Famulok et al. 2000, Mairal et al. 2008). Nucleases cannot recognize these unnatural oligonucleotides and thus aptamer stability is improved. While aptamers are relatively new recognition elements, their application for analyte detection in environmental samples is generating a great deal of excitement (Sharma et al. 2015). The translation of the current excitement surrounding aptamers to commercially viable nanosensor products is a major goal of the nanosensor community.

Carbohydrates

Carbohydrates (oligosaccharides or polysaccharides), are a large and diverse class of biomolecules that often dictate pathogen (and toxin) recognition and attachment to human cells (Haseley 2002). Carbohydrates consist of variable sequences of monosaccharide units held together through glycosidic bonds (Haseley 2002). Within the human gut, carbohydrates are conjugated to proteins (glycoproteins) as well as lipids (glycolipids/lipopolysaccharides) in a dense glycocalyx layer. Many pathogens, as well as their toxins, recognize and bind to specific carbohydrate sequences in the glycocalyx. In contrast to more specific antibodies and aptamers, carbohydrates are broad targets with affinity for a variety of organisms and exogenous proteins (Shen et al. 2007). This broad specificity can be an advantage since it may make them useful for the identification of unanticipated targets (Haseley 2002). In the case of pathogen sensing, antibodies and aptamers can be too specific for a particular pathogen and may not have affinity for mutants that differ slightly from the original target. Recognizing the importance of these interactions, many recent efforts have been made to utilize carbohydrates as receptor elements for pathogen and toxin detection (Haseley 2002, De la Fuente and Penades 2006, Trungkathan et al. 2014).

Nanomaterial and Signal Transduction

In addition to the recognition element, the nanomaterial itself and the signal transduction method are the other two components of a nanosensor. Commonly used nanomaterials in biosensors include noble metal nanoparticles (Dasary et al. 2009), quantum dots (Hansen et al. 2006), carbonaceous nanomaterials

(Wang 2005), and magnetic nanoparticles (Haun et al. 2010). Based on the properties of a nanomaterial, the signal transduction method can generally be considered either as optical (Dasary et al. 2009), magnetic (Haun et al. 2010), or electrochemical (Wang 2005, Sadik et al. 2009). Because signal transduction is inextricably linked to the properties of a nanomaterial we discuss these two components simultaneously.

Optical Signal Transduction

Among the three major signal transduction methods, optical signal transduction is generally the most popular due to its general simplicity of operation, rapid readout, and high level of sensitivity. In most optical sensors, the optical signal is produced when the target analyte interacts with a recognition element. Fluorescence and surface plasmon–enabled spectroscopies are the dominant optical signal transduction approaches.

Fluorescence

Fluorescence is generated by fluorophores or chemical compounds that emit light of a different wavelength following light excitation (Comparelli et al. 2007). Quantification of fluorescence intensity following fluorophore excitation is widely used in many biosensors. For example, pathogens stained with fluorescent labels are readily enumerated using a fluorescence microscope or by relating the measured intensity of the emitted light to the pathogen concentration. For example, U.S. EPA method 1623 uses antibodies labelled with fluorescent dyes to detect *Cryptosporidium parvum* oocysts and *Giardia lamblia* cysts. Unfortunately, the fluorescent dyes often employed in these biosensors are subject to thermal fluctuations, self-absorption, self-fluorescence, and photobleaching. Fluorescent nanomaterials offer a number of advantages over the organic dyes traditionally used in these detection schemes. For instance, functionalization of silica nanoparticles with dye molecules stabilizes the fluorescent signal of the dye (Santra et al. 2001). Nanosensors functionalized in this manner have been successfully applied for DNA (Zhao et al. 2003), *E. coli* (Zhao et al. 2004), and *Mycobacterium* spp. (Qin et al. 2008) detection, among many others. However, functionalizing silica with dyes requires complicated reaction chemistry and numerous reagents (Hilliard et al. 2002, Bagwe et al. 2006, Wang et al. 2007, Yan et al. 2007), thus making the preparation process complex and potentially expensive.

Quantum dots (QDs) are inorganic semiconductor nanocrystals that, unlike most organic dyes, are not subject to photobleaching. QDs employed for sensing have a typical composition MX where M is Cd or Zn and X is Se, S, or Te. In many cases, these QDs are coated by a second MX alloy, or shell, to create core/shell QDs with highly tunable properties. QDs commonly employed in nanosensor applications include: CdSe (Algarra et al. 2012), CdSe/ZnS (Luan et al. 2012, Sung and Lo 2012, Li et al. 2013b), CdTe (Wu et al. 2007, Gan et al. 2012, Gui et al. 2012, Liu et al. 2012, Rane et al. 2012, Chao et al. 2013), CdTe/CdS (Gui et al. 2012), ZnS (Koneswaran and Narayanaswamy 2009), and ZnSe/ZnS (Ke et al. 2012). QDs exhibit narrow emission bands that are size, shape, and core/shell composition tunable (Alivisatos 2004). These features enable multiplex detection because a single excitation source can be used to produce multiple distinguishable signals (Han et al. 2001, Resch-Genger et al. 2008). Multiplex analyte detection using QDs has been accomplished for many pathogens, including *Cryptosporidium parvum* and *Giardialamblia*, *Bacillus anthracis* and *Yersinia pestus*, and *E. coli* and *Salmonella typhimurium* (Zhu et al. 2004, Yang and Li 2006, Zahavy et al. 2010).

Recently a novel application of QD fluorescence was developed by Yang and colleagues (Yang et al. 2011). Utilizing commercially available QDs the authors were able to probe temperature distributions within living cells. This achievement was made possible by monitoring the QD emission peak, that red shifts in response to an increase in temperature. The red shift is a result of the expansion of the QD crystalline lattice at higher temperature.

Surface Plasmon–Enabled Spectroscopies

Noble metal nanoparticles such as gold and silver can support surface plasmons (e.g., coherently oscillating surface electrons). When excited by light, surface conduction electrons oscillate collectively and generate strong electromagnetic fields within the vicinity (i.e., several nanometers) of the nanoparticle surface (Link and El-Sayed 1999, Haes and Van Duyne 2004, Haes et al. 2005). The frequency at which the surface electrons resonate results in a band in the absorption spectra of the nanoparticles that is referred to as the localized surface plasmon resonance (LSPR) band (Kelly et al. 2003, Jain et al. 2008, Murphy et al. 2008). Two categories of signal transduction rely on LSPR—colorimetric and surface-enhanced Raman spectroscopy (Wei et al. 2015a).

Colorimetric Assay   The LSPR of spherical gold/silver nanoparticles generally lies within the visible light region of the spectrum and thus visible light is scattered/absorbed by these nanoparticles. This process results in the red/green color of gold/silver nanoparticle suspensions, respectively (Mock, 2002, Zielińska, 2009). The location of the LSPR is determined by the suspension chemistry as well as the nanoparticle shape, size, and aggregation state (Mock, 2002, Sun, 2003). A change in the nanoparticle suspension color is expected when the location of the nanoparticle LSPR shifts due to recognition of a target analyte. This color change can be identified both by the naked eye as well as by UV-vis spectroscopy and can be quantitatively correlated with analyte concentration. Such a signal transduction approach is termed a *colorimetric assay* (Daniel et al. 2009, Xia et al. 2010, Du et al. 2013).

For a given nanoparticle suspension, a change in aggregation state results in a quantifiable LSPR band shift or the development of a new LSPR band (Figure 10.4a). Most LSPR-based colorimetric nanosensors are based around LSPR shifts that are induced by nanoparticle aggregation (Storhoff et al. 1998, Liu and Lu 2006a, Huang and Chang 2007, Xia et al. 2010). For example, following the addition of *E. coli* to a suspension of anti–*E. coli* antibody-functionalized silver nanoparticles, the bacteria induce a change in suspension color because they simultaneously bind many different nanoparticles and cause their aggregation (Kalele et al. 2006). Similar protocols have been developed for influenza virus (H_3N_2, Figure 10.4a) (Liu et al. 2015, Poonthiyil et al. 2015) and T7 bacteriophage (Lesniewski et al. 2014) using antibody- or carbohydrate-functionalized gold nanoparticles.

In addition to pathogens, another popular application of LSPR-based colorimetric sensors is for heavy metal detection (Li et al. 2013a). Heavy metals form complexes with surface associated recognition elements and induce nanoparticle aggregation. For example, Hg^{2+} forms complexes with mercaptopropionic acid–functionalized gold nanoparticles and induces their aggregation. The quantifiable change in the LSPR bands results in a LOD for Hg^{2+} of 100 nm (Huang and Chang 2007). More recently, oligonucleotide-functionalized colorimetric Hg^{2+} sensors have been developed that rely on the selective binding of Hg^{2+} to T-T mismatches (Xue et al. 2008). This binding in turn causes nanoparticle aggregation. Via this approach a LOD of 3 µm was reported.

A developing area of research is the use of aptamer-functionalized nanoparticles for colorimetric detection of pesticides. While numerous aptamer-based nanosensors have been used for heavy metals, biological toxins, and others pathogens there have been to date few applications of this technology toward pesticide detection. Recently, a novel colorimetric approach was developed (Bai et al. 2015). In this effort (Figure 10.4b), gold nanoparticles nonspecifically functionalized with single-strand aptamers were stable against salt-mediated aggregation. In the presence of the organophosphorous pesticide targets, however, the aptamer detaches from the nanoparticle surface and the suspension color changes from red to blue. Via this approach mixtures of isocarbophos, methamidophos, acephate, dursban, trichlorfon, and phosalone could be detected in both lab and river water.

One drawback of suspension-based colorimetric assays is their poor reproducibility, stability, and the inconvenience associated with storage and transport. Recently, paper-based colorimetric assays have been proposed with the capacity to overcome these limitations. For example, a sensor for Cu^{2+} was produced by functionalizing silver nanoparticles with homocysteine and dithiothreitoland integrating them into a paper matrix, to which a droplet containing Cu^{2+} is subsequently applied. The color of the paper changed immediately after the addition of Cu^{2+} and was quantified with UV-vis spectroscopy over the copper concentration range of 7.8 to 62.8 µm (Ratnarathorn et al. 2012).

Surface-Enhanced Raman Spectroscopy   SERS is another optical phenomenon that originates from the LSPR of noble metal nanoparticles. As stated previously, LSPR generates an enhanced electromagnetic field near the noble metal surface. When target analytes locate near the nanoparticle surface, their Raman cross sections are enhanced by several orders of magnitude due to the enhanced electromagnetic field (Figure 10.5a) (Mrozek and Otto 1990, Compagnini et al. 1999). This electromagnetic enhancement is the main mechanism for SERS, while chemical enhancement is believed to be the other, generally less important, mechanism for SERS (Moskovits 2005). Chemical enhancement is the result of charge transfer between the analyte and the nanoparticle and thus only occurs when the analyte is in direct contact with the nanoparticle surface (Mrozek and Otto 1990). In addition to these two mechanisms, some analytes themselves also contribute to the overall Raman enhancement if they strongly absorb incident light (Stacy and Van Duyne 1983). Under this situation, SERS is called surface-enhanced resonance Raman spectroscopy (SERRS). Single molecule detection can be achieved with SERRS (Kneipp et al. 1995, Nie and Emery 1997).

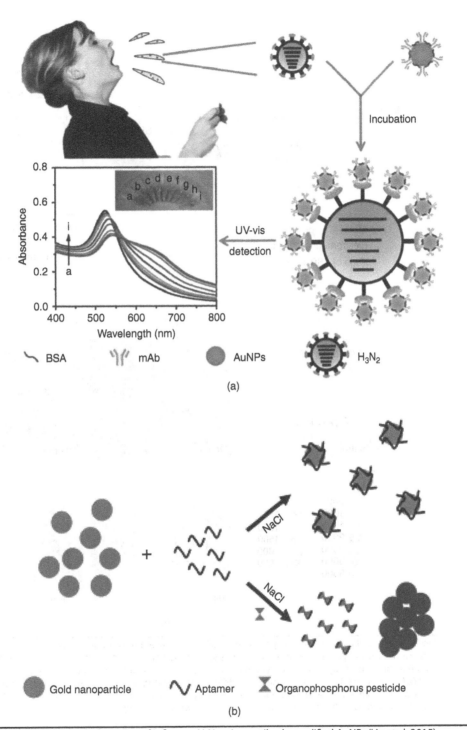

FIGURE 10.4   (a) Colorimetric detection of influenza H₃N₂ using antibody-modified AuNPs (Liu et al. 2015). (b) Colorimetric assay for the detection of organophosphorous pesticides (Bai et al. 2015). *Image (a) reprinted with permission from the Royal Society of Chemistry; image (b) reprinted with permission from Wiley-Blackwell.*

A potential advantage of SERS is that it does not require the use of recognition elements to achieve specificity. Because it is a vibrational spectroscopic technique, the characteristic covalent bonds of an analyte are displayed in its Raman spectra. While specificity is an advantage of SERS, sensitivity is the opposite. Although single molecule detection has been achieved with SERRS, many environmental organic pollutants do not strongly absorb visible light and are thus not resonant. This fact means that their Raman cross sections are three orders of magnitude lower than resonant ones. It has been found that the maximum

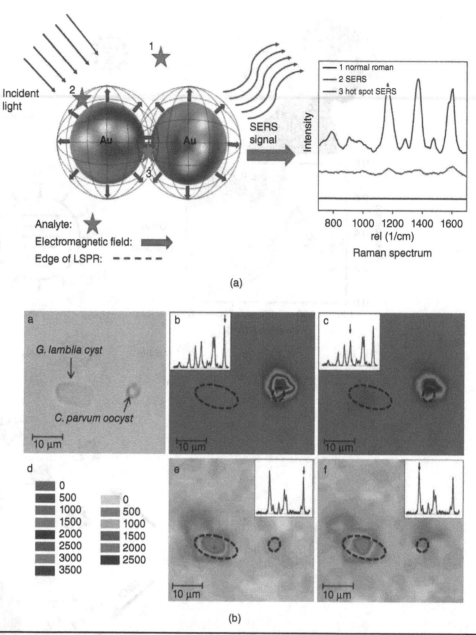

FIGURE 10.5 (a) Illustration of hot spot–mediated SERS signal enhancement. An analyte located at position 1 has no SERS signal. An analyte located at position 2 has some signal enhancement. An analyte located at position 3 in the hot spot has the highest signal intensity. (*Source:* Wei et al. 2015a.) (b) Multiplex detection of *C. parvum* oocysts (top row) and *G. lamblia* cysts (bottom row) using two different dyes (RBITC and MGITC, respectively) with gold nanoparticle SERS labels. (*Source:* Rule and Vikesland 2009.) *Image (a) reprinted with permission from the Royal Society of Chemistry; image (b) reprinted with permission from the American Chemical Society.*

electromagnetic field is located in the small gap between two nanoparticles, the so-called SERS hot spot (Figure 10.5a) (Basu et al. 2008, Lim et al. 2009, Lim et al. 2011, Taylor et al. 2011, Taylor et al. 2013, Yang et al. 2013). Since the gap is very small (<5 nm), many large molecules cannot readily access it, let alone pathogens. Therefore, the direct application of SERS is generally limited to small inorganic and organic chemicals (Allain and Vo-Dinh 2002, Schwartzberg et al. 2004, Chaney et al. 2005, Shanmukh et al. 2006, Banholzer et al. 2008).

Another big limitation for SERS is the low affinity of many environmental pollutants for the noble nanoparticle surface. Electromagnetic intensity decays exponentially with distance from the nanoparticle surface (Compagnini et al. 1999, Su et al. 2003, Qian et al. 2008). Because of charge-charge repulsion and steric effects many environmental pollutants do not accumulate on noble metal surfaces and thus their SERS intensity is far from maximized and may even be zero. One solution to this problem is to modify the nanoparticle surface with chemicals that have high affinity to the target analyte (Leyton et al. 2005, Guerrini et al. 2006, Álvarez-Puebla et al. 2009, Dasary et al. 2009, Guerrini et al. 2009a, Guerrini et al. 2009b, Alvarez-Puebla and Liz-Marzan 2012). Chemicals grafted to the nanoparticle surface are a special category of recognition element that generally show much lower specificity than either antibodies or aptamers. For example, macromolecules such as dithiolcarbamate calix[4]arene (Guerrini et al. 2009b), viologen (Guerrini et al. 2009a), and cyclodextrin (Xie et al. 2010) have been grafted on gold/silver nanoparticle surfaces. These molecules have high affinity to the gold/silver surface while also containing hydrophobic portions that can capture hydrophobic polycyclic aromatic hydrocarbons (PAHs). A nanomolar level of limit of detection for PAHs has been achieved via this approach. However, functionalization of the nanoparticle surface adds complexity to both the preparation procedure and spectal analysis (some linker molecules are SERS active). Therefore, it is still a challenge to find a better way to enhance the affinity of organic pollutants to the nanoparticle surface.

For larger size analytes (e.g., biomolecules and pathogens), their intrinsic SERS spectra have been reported (Allain and Vo-Dinh 2002, Zeiri et al. 2002, Zeiri et al. 2004, Sengupta et al. 2005, Zeiri and Efrima 2005, Escoriza et al. 2006, Shanmukh et al. 2006, Shanmukh et al. 2008, Tripp et al. 2008, Abell et al. 2009). However, these analytes cannot enter SERS hot spots, making the SERS intensity very low and their reproducible and quantitative analysis difficult. The most successful approach for pathogen detection by SERS is to utilize a SERS label (Figure 10.5b). SERS labels are usually resonant dye molecules that exhibit high signal intensities. By incorporation of a SERS label and recognition elements (antibody or aptamer) on the noble metal nanoparticle surface, the nanosensor is able to capture the pathogen as well as generate a strong SERS signal. The detection of the label molecule via SERS indicates the presence of the target analyte. This protocol has been applied for the detection of a diverse array of biomolecules (Fang et al. 2008, Kang et al. 2010), bacteria (Huang et al. 2009, Khan et al. 2011), and protozoa (Figure 10.5b) (Rule and Vikesland 2009). Multiplex detection have been readily achieved by the use of two or more SERS labels (Ravindranath et al. 2011). The disadvantage of SERS labels for pathogen detection is the potentially high background interference. Because it is the SERS label, and not the target that is detected, this nanosensor requires the removal of all unbound SERS labels or else a false positive will occur.

Electrochemical Methods and Novel Sensing Platforms

Electrochemical methods are another signal transduction approach. These methods take advantage of the high conductivity and electrochemical stability of noble metal nanoparticles (Pt, Ag, Au) or carbonaceous nanomaterials (graphene, carbon nanotube) and rely on the measurement of changes in an electrochemical signal following exposure to an analyte (Park et al. 2002, Merkoci 2007, Wang 2007, Sadik et al. 2009). These changes can be displayed as current change (amperometric), potential change (potentiometric), or conductivity change (conductometric) (Grieshaber et al. 2008). By modifying electrodes with recognition elements (antibody or aptamer), *E. coli* (So et al. 2008), hepatitis C (Peterson et al. 2000) and a variety of cancer biomarkers such as prostate specific antigen (Sarkar et al. 2002) and carcinoembryonic antigen (Dai et al. 2004) have been detected.

Novel nanostructures include anisotropic nanoparticles such as nanorods (Norman et al. 2008), nanowires (Patolsky et al. 2004, Patolsky et al. 2006), nanostars (Hao et al. 2007), and nanoflowers (Jiang et al. 2007) as well as nanoarrays that can be engineered by a variety of lithography methods (Figure 10.6a and 10.6b). Due to the extraordinary optical and electrochemical properties of these nanostructures, new opportunities of producing novel biosensors are created. For example, the tips of gold nanorods show much stronger electromagnetic fields than traditional gold nanospheres and can be considered a new category of SERS hot spots (McLintock et al. 2014).

Nanocomposites consisting of two or more components show extraordinary optical, electrical, and magnetical properties and have been recognized as a new category of biosensor (Wei et al. 2014, Wei et al. 2015b). For example, by integrating gold nanoparticles and nanocellulose into an electrode, a conductive (gold nanoparticle) and biocompatible (nanocellulose) electrochemical biosensor has been developed for the detection of glucose and hydrogen peroxide (Wang et al. 2010, Zhang et al. 2010, Wang et al. 2011). By combining the catalytic properties of palladium nanoparticles and the plasmonic properties of silver nanoparticles, the reduction reaction of 4-nitrothiolphenol to 4-aminothiolphenol can be monitored (Li et al. 2015). Recently, bacterial cellulose has been shown to be an ideal scaffold for housing gold nanoparticles and the as-prepared nanocomposites demonstrated ultrasensitive SERS detection capability (Figure 10.6c).

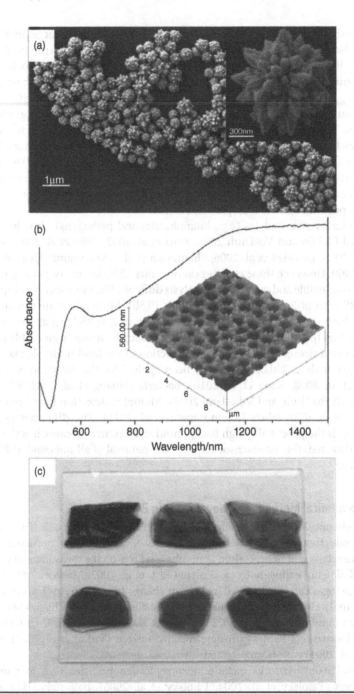

Figure 10.6 (a) Scanning electron microscopy image of gold nanoflowers (Yi et al. 2013). (b) Atomic force microscopy image of gold honeycomb and its UV-vis extinction spectrum (Leng and Vikesland 2013). (c) Photo of gold nanoparticle/bacterial cellulose nanocomposites prepared at different conditions (Wei et al. 2015b).

Conclusions and Future Outlook

The extraordinary optical, electrochemical, and magnetic properties of nanomaterials opens an avenue for the design of novel biosensors. The literature illustrates their capability for ultrasensitive detection of heavy metal, organic pollutant, and waterborne pathogens. Compared with traditional water monitoring techniques, nanomaterial-enabled biosensors do not require complex instrumentation, laborious sample pretreatment, or long detection times. Despite these advantages, the majority of the nanosensors developed to date are viable only in the laboratory. Whether these nanosensors can maintain their stability, sensitivity, and specificity when applied to real environmental samples is a remaining question. To address these challenges, additional effort should be paid on the interactions of nanosensors within complex environmental samples. The influence of important factors in water chemistry, such as natural organic matter (NOM), pH, ionic strength on the performance of biosensors must be elucidated.

References

1. Abdel-Hamid, I., D. Ivnitski, P. Atanasov, and E. Wilkins (1999). "Highly sensitive flow-injection immunoassay system for rapid detection of bacteria." *Analytica Chimica Acta* 399(1–2): 99–108.

2. Abell, J., J. Driskell, R. Dluhy, R. Tripp, and Y.-P. Zhao (2009). "Fabrication and characterization of a multiwell array SERS chip with biological applications." *Biosensors and Bioelectronics* 24(12): 3663–3670.

3. Algarra, M., B. B. Campos, B. Alonso, M. S. Miranda, Á. M. Martínez, C. M. Casado, and J. C. G. Esteves da Silva (2012). "Thiolated DAB dendrimers and CdSe quantum dots nanocomposites for Cd(II) or Pb(II) sensing." *Talanta* 88(0): 403–407.

4. Alivisatos, P. (2004). "The use of nanocrystals in biological detection." *Nature Biotechnology* 22(1): 47–52.

5. Allain, L. R. and T. Vo-Dinh (2002). "Surface-enhanced Raman scattering detection of the breast cancer susceptibility gene BRCA1 using a silver-coated microarray platform." *Analytica Chimica Acta* 469(1): 149–154.

6. Alvarez-Puebla, R. A. and L. M. Liz-Marzan (2012). "Traps and cages for universal SERS detection." *Chemical Society Reviews* 41(1): 43–51.

7. Álvarez-Puebla, R. A., R. Contreras-Cáceres, I. Pastoriza-Santos, J. Pérez-Juste, and L. M. Liz-Marzán (2009). "Au@pNIPAM colloids as molecular traps for surface-enhanced, spectroscopic, ultra-sensitive analysis." *Angewandte Chemie International Edition* 48(1): 138–143.

8. Bagwe, R. P., L. R. Hilliard, and W. H. Tan (2006). "Surface modification of silica nanoparticles to reduce aggregation and nonspecific binding." *Langmuir* 22(9): 4357–4362.

9. Bai, W., C. Zhu, J. Liu, M. Yan, S. Yang, and A. Chen (2015). "Gold nanoparticle-based colorimetric aptasensor for rapid detection of six organophosphorous pesticides." *Environmental Toxicology and Chemistry* 34(10): 2244–2249.

10. Banholzer, M. J., J. E. Millstone, L. D. Qin, and C. A. Mirkin (2008). "Rationally designed nanostructures for surface-enhanced Raman spectroscopy." *Chemical Society Reviews* 37(5): 885–897.

11. Basu, S., S. Pande, S. Jana, S. Bolisetty, and T. Pal (2008). "Controlled interparticle spacing for surface-modified gold nanoparticle aggregates." *Langmuir* 24(10): 5562–5568.

12. Bordeaux, J., A. W. Welsh, S. Agarwal, E. Killiam, M. T. Baquero, J. A. Hanna, V. K. Anagnostou, and D. L. Rimm (2010). "Antibody validation." *BioTechniques* 48(3): 197–209.

13. Campbell, G. A. and R. Mutharasan (2008). "Near real-time detection of Cryptosporidium parvum oocyst by IgM-functionalized piezoelectric-excited millimeter-sized cantilever biosensor." *Biosensors and Bioelectronics* 23(7): 1039–1045.

14. Chaney, S. B., S. Shanmukh, R. A. Dluhy, and Y. P. Zhao (2005). "Aligned silver nanorod arrays produce high sensitivity surface-enhanced Raman spectroscopy substrates." *Applied Physics Letters* 87(3): 1–3.

15. Chao, M. R., Y. Z. Chang, and J. L. Chen (2013). "Hydrophilic ionic liquid-passivated CdTe quantum dots for mercury ion detection." *Biosensors and Bioelectronics* 42: 397–402.

16. Compagnini, G., C. Galati, and S. Pignataro (1999). "Distance dependence of surface enhanced Raman scattering probed by alkanethiol self-assembled monolayers." *Physical Chemistry Chemical Physics* 1(9): 2351–2353.

17. Comparelli, R., M. L. Curri, P. D. Cozzoli, and M. Striccoli (2007). "Optical biosensing based on metal and semiconductor colloidal nanocrystals." *Nanotechnologies for the Life Sciences: Nanomaterials for Biosensors*. C. S. S. R. Kumar, Editor. Wiley-VCH, Weinheim. Vol. 8: 123–174.

18. Cui, Y., B. Ren, J. L. Yao, R. A. Gu, and Z. Q. Tian (2006). "Synthesis of AgcoreAushell bimetallic nanoparticles for immunoassay based on surface-enhanced Raman spectroscopy." *Journal of Physical Chemistry B* 110(9): 4002–4006.

19. Dai, Z., F. Yan, H. Yu, X. Hu, and H. Ju (2004). "Novel amperometric immunosensor for rapid separation-free immunoassay of carcinoembryonic antigen." *Journal of Immunological Methods* 287(1): 13–20.

20. Daniel, W. L., M. S. Han, J.-S. Lee, and C. A. Mirkin (2009). "Colorimetric nitrite and nitrate detection with gold nanoparticle probes and kinetic end points." *Journal of the American Chemical Society* 131(18): 6362–6363.

21. Dasary, S. S., A. K. Singh, D. Senapati, H. Yu, and P. C. Ray (2009). "Gold nanoparticle based label-free SERS probe for ultrasensitive and selective detection of trinitrotoluene." *Journal of the American Chemical Society* 131(38): 13806–13812.

22. De la Fuente, J. M. and S. Penades (2006). "Glyconanoparticles: types, synthesis and applications in glycoscience, biomedicine and material science." *Biochimica Et Biophysica Acta-General Subjects* 1760(4): 636–651.

23. Desphande, S. S. (1996). *Enzyme Immunoassays: From Concept to Product Development*. Chapman & Hall, New York.

24. Dou, X., T. Takama, Y. Yamaguchi, H. Yamamoto, and Y. Ozaki (1997). "Enzyme immunoassay utilizing surface-enhanced Raman scattering of the enzyme reaction product." *Analytical Chemistry* 69(8): 1492–1495.

25. Dowd, A., D. Pissuwan, and M. B. Cortie (2014). "Optical readout of the intracellular environment using nanoparticle transducers." *Trends in Biotechnology* 32(11): 571–577.

26. Du, J., B. Zhu, and X. Chen (2013). "Urine for plasmonic nanoparticle-based colorimetric detection of mercury ion." *Small* 9(24): 4104–4111.

27. El-Boubbou, K., C. Gruden, and X. Huang (2007). "Magnetic glyco-nanoparticles: A unique tool for rapid pathogen detection, decontamination, and strain differentiation." *Journal of the American Chemical Society* 129(44): 13392–13393.

28. Ellington, A. D. and J. W. Szostak (1990). "In vitro selection of RNA molecules that bind specific ligands." *Nature* 346(6287): 818–822.

29. Escoriza, M. F., J. M. VanBriesen, S. Stewart, J. Maier, and P. J. Treado (2006). "Raman spectroscopy and chemical imaging for quantification of filtered waterborne bacteria." *Journal of Microbiological Methods* 66(1): 63–72.

30. Famulok, M., G. Mayer, and M. Blind (2000). "Nucleic acid aptamers—from selection in vitro to applications in vivo." *Accounts of Chemical Research* 33(9): 591–599.

31. Fang, C., A. Agarwal, K. D. Buddharaju, N. M. Khalid, S. M. Salim, E. Widjaja, M. V. Garland, N. Balasubramanian, and D.-L. Kwong (2008). "DNA detection using nanostructured SERS substrates with Rhodamine B as Raman label." *Biosensors and Bioelectronics* 24(2): 216–221.

32. Fischer, N., T. M. Tarasow, and J. B. H. Tok (2007). "Aptasensors for biosecurity applications." *Current Opinion in Chemical Biology* 11(3): 316–328.

33. Gan, T. T., Y. J. Zhang, N. J. Zhao, X. Xiao, G. F. Yin, S. H. Yu, H. B. Wang, J. B. Duan, C. Y. Shi, and W. Q. Liu (2012). "Hydrothermal synthetic mercaptopropionic acid stabled CdTe quantum dots as fluorescent probes for detection of Ag(+)." *Spectrochimica Acta Part A Molecular Biomolecular Spectroscopy* 99: 62–68.

34. Grieshaber, D., R. MacKenzie, J. Vörös, and E. Reimhult (2008). "Electrochemical biosensors—sensor principles and architectures." *Sensors* 8(3): 1400–1458.

35. Guerrini, L., J. V. Garcia-Ramos, C. Domingo, and S. Sanchez-Cortes (2006). "Functionalization of Ag nanoparticles with dithiocarbamate calix [4] arene as an effective supramolecular host for the surface-enhanced Raman scattering detection of polycyclic aromatic hydrocarbons." *Langmuir* 22(26): 10924–10926.

36. Guerrini, L., J. V. Garcia-Ramos, C. Domingo, and S. Sanchez-Cortes (2009a). "Nanosensors based on viologen functionalized silver nanoparticles: few molecules surface-enhanced Raman spectroscopy detection of polycyclic aromatic hydrocarbons in interparticle hot spots." *Analytical Chemistry* 81(4): 1418–1425.

37. Guerrini, L., J. V. Garcia-Ramos, C. Domingo, and S. Sanchez-Cortes (2009b). "Sensing polycyclic aromatic hydrocarbons with dithiocarbamate-functionalized Ag nanoparticles by surface-enhanced Raman scattering." *Analytical Chemistry* 81(3): 953–960.

38. Gui, R., X. An, H. Su, W. Shen, Z. Chen, and X. Wang (2012). "A near-infrared-emitting CdTe/CdS core/shell quantum dots-based OFF-ON fluorescence sensor for highly selective and sensitive detection of Cd2+." *Talanta* 94: 257–262.

39. Haes, A. J., C. L. Haynes, A. D. McFarland, G. C. Schatz, R. P. Van Duyne, and S. Zou (2005). "Plasmonic materials for surface-enhanced sensing and spectroscopy." *MRS Bulletin* 30(5): 368–375.

40. Haes, A. J. and R. P. Van Duyne (2004). "A unified view of propagating and localized surface plasmon resonance biosensors." *Analytical and Bioanalytical Chemistry* 379(7–8): 920–930.

41. Han, M. Y., X. H. Gao, J. Z. Su, and S. Nie (2001). "Quantum-dot-tagged microbeads for multiplexed optical coding of biomolecules." *Nature Biotechnology* 19(7): 631–635.

42. Hansen, J. A., J. Wang, A.-N. Kawde, Y. Xiang, K. V. Gothelf, and G. Collins (2006). "Quantum-dot/aptamer-based ultrasensitive multi-analyte electrochemical biosensor." *Journal of the American Chemical Society* 128(7): 2228–2229.

43. Hao, F., C. L. Nehl, J. H. Hafner, and P. Nordlander (2007). "Plasmon resonances of a gold nanostar." *Nano letters* 7(3): 729–732.

44. Haseley, S. R. (2002). "Carbohydrate recognition: a nascent technology for the detection of bioanalytes." *Analytica Chimica Acta* 457(1): 39–45.

45. Hassan, A. R., A. de la Escosura-Muniz, and A. Merkoci (2015). "Highly sensitive and rapid determination of *Escherichia coli* O157:H7 in minced beef and water using electrocatalytic gold nanoparticle tags." *Biosensors and Bioelectronics* 67: 511–515.

46. Haun, J. B., T. J. Yoon, H. Lee, and R. Weissleder (2010). "Magnetic nanoparticle biosensors." *Wiley Interdisciplinary Reviews: Nanomedicine and Nanobiotechnology* 2(3): 291–304.

47. Hilliard, L. R., X. J. Zhao, and W. H. Tan (2002). "Immobilization of oligonucleotides onto silica nanoparticles for DNA hybridization studies." *Analytica Chimica Acta* 470(1): 51–56.

48. Huang, C.-C. and H.-T. Chang (2007). "Parameters for selective colorimetric sensing of mercury (II) in aqueous solutions using mercaptopropionic acid-modified gold nanoparticles." *Chemical Communications* 12: 1215–1217.

49. Huang, P. J., L. L. Tay, J. Tanha, S. Ryan, and L. K. Chau (2009). "Single-domain antibody-conjugated nanoaggregate-embedded beads for targeted detection of pathogenic bacteria." *Chemistry-A European Journal* 15(37): 9330–9334.

50. Iqbal, S. S., M. W. Mayo, J. G. Bruno, B. V. Bronk, C. A. Batt, and J. P. Chambers (2000). "A review of molecular recognition technologies for detection of biological threat agents." *Biosensors and Bioelectronics* 15(11–12): 549–578.

51. Jain, P. K., X. H. Huang, I. H. El-Sayed, and M. A. El-Sayed (2008). "Noble metals on the nanoscale: optical and photothermal properties and some applications in imaging, sensing, biology, and medicine." *Accounts of Chemical Research* 41(12): 1578–1586.

52. Jayasena, S. D. (1999). "Aptamers: an emerging class of molecules that rival antibodies in diagnostics." *Clinical Chemistry* 45(9): 1628–1650.

53. Jenison, R. D., S. C. Gill, A. Pardi, and B. Polisky (1994). "High-resolution molecular discrimination by RNA." *Science* 263(5152): 1425–1429.

54. Jiang, C., X. Sun, G. Lo, D. Kwong, and J. Wang (2007). "Improved dye-sensitized solar cells with a ZnO-nanoflower photoanode." *Applied Physics Letters* 90(26): 263501.

55. Kalele, S. A., A. A. Kundu, S. W. Gosavi, D. N. Deobagkar, D. D. Deobagkar, and S. K. Kulkarni (2006). "Rapid detection of *Escherichia coli* by using anti body-conjugated silver nanoshells." *Small* 2(3): 335–338.

56. Kang, T., S. M. Yoo, I. Yoon, S. Y. Lee, and B. Kim (2010). "Patterned multiplex pathogen DNA detection by Au particle-on-wire SERS sensor." *Nano Letters* 10(4): 1189–1193.

57. Ke, J., X. Li, Y. Shi, Q. Zhao, and X. Jiang (2012). "A facile and highly sensitive probe for Hg(II) based on metal-induced aggregation of ZnSe/ZnS quantum dots." *Nanoscale* 4(16): 4996–5001.

58. Kelly, K. L., E. Coronado, L. L. Zhao, and G. C. Schatz (2003). "The optical properties of metal nanoparticles: the influence of size, shape, and dielectric environment." *Journal of Physical Chemistry B* 107(3): 668–677.

59. Khan, S. A., A. K. Singh, D. Senapati, Z. Fan, and P. C. Ray (2011). "Targeted highly sensitive detection of multi-drug resistant Salmonella DT104 using gold nanoparticles." *Chemical Communications* 47(33): 9444–9446.

60. Klussmann, S., A. Nolte, R. Bald, V. A. Erdmann, and J. P. Furste (1996). "Mirror-image RNA that binds D-adenosine." *Nature Biotechnology* 14(9): 1112–1115.

61. Kneipp, K., Y. Wang, R. R. Dasari, and M. S. Feld (1995). "Approach to single-molecule detection using surface-enhanced resonance Raman scattering (SERRS)—A study using rhodamine 6G on colloidal silver." *Applied Spectroscopy* 49(6): 780–784.

62. Koneswaran, M. and R. Narayanaswamy (2009). "l-Cysteine-capped ZnS quantum dots based fluorescence sensor for Cu2+ ion." *Sensors and Actuators B: Chemical* 139(1): 104–109.

63. Leng, W. and P. J. Vikesland (2013). "Nanoclustered gold honeycombs for surface-enhanced Raman scattering." *Analytical Chemistry* 85(3): 1342–1349.

64. Leonard, P., S. Hearty, J. Brennan, L. Dunne, J. Quinn, T. Chakraborty, and R. O'Kennedy (2003). "Advances in biosensors for detection of pathogens in food and water." *Enzyme and Microbial Technology* 32(1): 3–13.

65. Lesniewski, A., M. Los, M. Jonsson-Niedziółka, A. Krajewska, K. Szot, J. M. Los, and J. Niedziolka-Jonsson (2014). "Antibody modified gold nanoparticles for fast and selective, colorimetric T7 bacteriophage detection." *Bioconjugate Chemistry* 25(4): 644–648.

66. Leyton, P., C. Domingo, S. Sanchez-Cortes, M. Campos-Vallette, and J. Garcia-Ramos (2005). "Surface enhanced vibrational (IR and Raman) spectroscopy in the design of chemosensors based on ester functionalized p-tert-butylcalix [4] arene hosts." *Langmuir* 21(25): 11814–11820.

67. Li, J., J. Liu, Y. Yang, and D. Qin (2015). "Bifunctional Ag@ Pd-Ag nanocubes for highly sensitive monitoring of catalytic reactions by surface-enhanced Raman spectroscopy." *Journal of the American Chemical Society* 137(22): 7039–7042.

68. Li, M., H. Gou, I. Al-Ogaidi, and N. Wu (2013a). "Nanostructured sensors for detection of heavy metals: a review." *ACS Sustainable Chemistry and Engineering* 1(7): 713–723.

69. Li, M., X. Zhou, S. Guo, and N. Wu (2013b). "Detection of lead (II) with a 'turn-o' fluorescent biosensor based on energy transfer from CdSe/ZnS quantum dots to graphene oxide." *Biosensors and Bioelectronics* 43: 69–74.

70. Li, T., L. L. Shi, E. K. Wang, and S. J. Dong (2009). "Multifunctional G-quadruplex aptamers and their application to protein detection." *Chemistry—A European Journal* 15(4): 1036–1042.

71. Lim, D.-K., K.-S. Jeon, J.-H. Hwang, H. Kim, S. Kwon, Y. D. Suh, and J.-M. Nam (2011). "Highly uniform and reproducible surface-enhanced Raman scattering from DNA-tailorable nanoparticles with 1-nm interior gap." *Nature Nanotechnology* 6(7): 452–460.

72. Lim, D.-K., K.-S. Jeon, H. M. Kim, J.-M. Nam, and Y. D. Suh (2009). "Nanogap-engineerable Raman-active nanodumbbells for single-molecule detection." *Nature Materials* 9(1): 60–67.

73. Link, S. and M. A. El-Sayed (1999). "Spectral properties and relaxation dynamics of surface plasmon electronic oscillations in gold and silver nanodots and nanorods." *The Journal of Physical Chemistry B* 103(40): 8410–8426.

74. Liu, J. and Y. Lu (2006a). "Fast colorimetric sensing of adenosine and cocaine based on a general sensor design involving aptamers and nanoparticles." *Angewandte Chemie* 118(1): 96–100.

75. Liu, J. and Y. Lu (2006b). "Preparation of aptamer-linked gold nanoparticle purple aggregates for colorimetric sensing of analytes." *Nature Protocols* 1(1): 246–252.

76. Liu, J. W., D. Mazumdar, and Y. Lu (2006). "A simple and sensitive 'dipstick' test in serum based on lateral flow separation of aptamer-linked nanostructures." *Angewandte Chemie—International Edition* 45(47): 7955–7959.

77. Liu, X.-Q., Q. Liu, S.-H. Cao, W.-P. Cai, Y.-H. Weng, K.-X. Xie, and Y.-Q. Li (2012). "Directional surface plasmon-coupled emission of CdTe quantum dots and its application in Hg(ii) sensing." *Analytical Methods* 4(12): 3956.

78. Liu, Y., L. Zhang, W. Wei, H. Zhao, Z. Zhou, Y. Zhang, and S. Liu (2015). "Colorimetric detection of influenza A virus using antibody-functionalized gold nanoparticles." *Analyst* 140(12): 3989–3995.

79. Low, S. Y., J. E. Hill, and J. Peccia (2009). "DNA aptamers bind specifically and selectively to (1→3)-a0D-glucans." *Biochemical and Biophysical Research Communications* 378(4): 701–705.

80. Luan, W., H. Yang, Z. Wan, B. Yuan, X. Yu, and S.-t. Tu (2012). "Mercaptopropionic acid capped CdSe/ZnS quantum dots as fluorescence probe for lead(II)." *Journal of Nanoparticle Research* 14(3): 1–8.

81. Mairal, T., V. C. Ozalp, P. L. Sanchez, M. Mir, I. Katakis, and C. K. O'Sullivan (2008). "Aptamers: molecular tools for analytical applications." *Analytical and Bioanalytical Chemistry* 390(4): 989–1007.

82. McLintock, A., C. A. Cunha-Matos, M. Zagnoni, O. R. Millington, and A. W. Wark (2014). "Universal surface-enhanced Raman tags: individual nanorods for measurements from the visible to the infrared (514–1064 nm)." *ACS Nano* 8(8): 8600–8609.

83. Merkoci, A. (2007). "Electrochemical biosensing with nanoparticles." *FEBS Journal* 274(2): 310–316.

84. Moskovits, M. (2005). "Surface-enhanced Raman spectroscopy: a brief retrospective." *Journal of Raman Spectroscopy* 36(6–7): 485–496.

85. Mrozek, I. and A. Otto (1990). "Quantitative separation of the 'classical' electromagnetic and the 'chemical' contribution to surface enhanced Raman scattering." *Journal of Electron Spectroscopy and Related Phenomena* 54: 895–911.

86. Murphy, C. J., A. M. Gole, J. W. Stone, P. N. Sisco, A. M. Alkilany, E. C. Goldsmith, and S. C. Baxter (2008). "Gold nanoparticles in biology: beyond toxicity to cellular imaging." *Accounts of Chemical Research* 41(12): 1721–1730.

87. Nie, S. M. and S. R. Emery (1997). "Probing single molecules and single nanoparticles by surface-enhanced Raman scattering." *Science* 275(5303): 1102–1106.

88. Nolte, A., S. Klussmann, R. Bald, V. A. Erdmann, and J. P. Furste (1996). "Mirror-design of L-oligonucleotide ligands binding to L-arginine." *Nature Biotechnology* 14(9): 1116–1119.

89. Norman, R. S., J. W. Stone, A. Gole, C. J. Murphy, and T. L. Sabo-Attwood (2008). "Targeted photothermal lysis of the pathogenic bacteria, *Pseudomonas aeruginosa*, with gold nanorods." *Nano Letters* 8: 302–306.

90. Pandey, S. and S. B. Mishra (2014). "Chemical nanosensors for monitoring environmental pollution." *Application of Nanotechnology in Water Research*: 309–332.

91. Park, S. J., T. A. Taton, and C. A. Mirkin (2002). "Array-based electrical detection of DNA with nanoparticle probes." *Science* 295(5559): 1503–1506.

92. Patolsky, F., G. F. Zheng, O. Hayden, M. Lakadamyali, X. W. Zhuang, and C. M. Lieber (2004). "Electrical detection of single viruses." *Proceedings of the National Academy of Sciences of the United States of America* 101(39): 14017–14022.

93. Patolsky, F., G. F. Zheng, and C. M. Lieber (2006). "Fabrication of silicon nanowire devices for ultrasensitive, label-free, real-time detection of biological and chemical species." *Nature Protocols* 1(4): 1711–1724.

94. Peterson, J., G. Green, K. Iida, B. Caldwell, P. Kerrison, S. Bernich, K. Aoyagi, and S. Lee (2000). "Detection of hepatitis C core antigen in the antibody negative 'window'phase of hepatitis C infection." *Vox Sanguinis* 78(2): 80–85.

95. Poonthiyil, V., P. T. Nagesh, M. Husain, V. B. Golovko, and A. J. Fairbanks (2015). "Gold nanoparticles decorated with sialic acid terminated bi-antennary N-glycans for the detection of influenza virus at nanomolar concentrations." *ChemistryOpen* 4(6): 708–716.

96. Qian, X., X. Zhou, and S. Nie (2008). "Surface-enhanced Raman nanoparticle beacons based on bioconjugated gold nanocrystals and long range plasmonic coupling." *Journal of the American Chemical Society* 130(45): 14934–14935.

97. Qin, D. L., X. X. He, K. M. Wang, and W. H. Tan (2008). "Using fluorescent nanoparticles and SYBR Green I based two-color flow cytometry to determine *Mycobacterium tuberculosis* avoiding false positives." *Biosensors and Bioelectronics* 24(4): 626–631.

98. Rane, T. D., H. C. Zec, C. Puleo, A. P. Lee, and T. H. Wang (2012). "Droplet microfluidics for amplification-free genetic detection of single cells." *Lab on a Chip* 12(18): 3341–3347.

99. Ratnarathorn, N., O. Chailapakul, C. S. Henry, and W. Dungchai (2012). "Simple silver nanoparticle colorimetric sensing for copper by paper-based devices." *Talanta* 99: 552–557.

100. Ravindranath, S. P., Y. Wang, and J. Irudayaraj (2011). "SERS driven cross-platform based multiplex pathogen detection." *Sensors and Actuators B: Chemical* 152(2): 183–190.

101. Resch-Genger, U., M. Grabolle, S. Cavaliere-Jaricot, R. Nitschke, and T. Nann (2008). "Quantum dots versus organic dyes as fluorescent labels." *Nature Methods* 5(9): 763–775.

102. Robertson, D. L. and G. F. Joyce (1990). "Selection in vitro of an RNA enzyme that specifically cleaves single-stranded-DNA." *Nature* 344(6265): 467–468.

103. Rule, K. L. and P. J. Vikesland (2009). "Surface-enhanced resonance Raman spectroscopy for the rapid detection of *Cryptosporidium parvum* and *Giardia lamblia*." *Environmental Science and Technology* 43(4): 1147–1152.

104. Sadik, O. A., A. O. Aluoch, and A. Zhou (2009). "Status of biomolecular recognition using electrochemical techniques." *Biosensors and Bioelectronics* 24(9): 2749–2765.

105. Santra, S., P. Zhang, K. Wang, R. Tapec, and W. Tan (2001). "Conjugation of biomolecules with luminophore-doped silica nanoparticles for photostable biomarkers." *Analytical Chemistry* 73(20): 4988–4993.

106. Sarkar, P., P. S. Pal, D. Ghosh, S. J. Setford, and I. E. Tothill (2002). "Amperometric biosensors for detection of the prostate cancer marker (PSA)." *International Journal of Pharmaceutics* 238(1): 1–9.

107. Schwartzberg, A. M., C. D. Grant, A. Wolcott, C. E. Talley, T. R. Huser, R. Bogomolni, and J. Z. Zhang (2004). "Unique gold nanoparticle aggregates as a highly active surface-enhanced Raman scattering substrate." *Journal of Physical Chemistry B* 108(50): 19191–19197.

108. Sengupta, A., M. L. Laucks, N. Dildine, E. Drapala, and E. J. Davis (2005). "Bioaerosol characterization by surface-enhanced Raman spectroscopy (SERS)." *Journal of Aerosol Science* 36(5–6): 651–664.

109. Shanmukh, S., L. Jones, J. Driskell, Y. Zhao, R. Dluhy, and R. A. Tripp (2006). "Rapid and sensitive detection of respiratory virus molecular signatures using a silver nanorod array SERS substrate." *Nano Letters* 6(11): 2630–2636.

110. Shanmukh, S., L. Jones, Y.-P. Zhao, J. Driskell, R. Tripp, and R. Dluhy (2008). "Identification and classification of respiratory syncytial virus (RSV) strains by surface-enhanced Raman spectroscopy and multivariate statistical techniques." *Analytical and Bioanalytical Chemistry* 390(6): 1551–1555.

111. Sharma, R., K. V. Ragavan, M. S. Thakur, and K. S. M. S. Raghavarao (2015). "Recent advances in nanoparticle based aptasensors for food contaminants." *Biosensors and Bioelectronics* 74: 612–627.

112. Shen, Z. H., M. C. Huang, C. D. Xiao, Y. Zhang, X. Q. Zeng, and P. G. Wang (2007). "Nonlabeled quartz crystal microbalance biosensor for bacterial detection using carbohydrate and lectin recognitions." *Analytical Chemistry* 79(6): 2312–2319.

113. Singh, I. (2015). "A review on outlook on future prospects of nanosensors." *International Journals of Nanosensors* 1(1): 1–7.

114. So, H. M., D. W. Park, E. K. Jeon, Y. H. Kim, B. S. Kim, C. K. Lee, S. Y. Choi, S. C. Kim, H. Chang, and J. O. Lee (2008). "Detection and titer estimation of *Escherichia coli* using aptamer-functionalized single-walled carbon-nanotube field-effect transistors." *Small* 4(2): 197–201.

115. Stacy, A. M. and R. P. Vanduyne (1983). "Surface enhanced Raman and resonance Raman-spectroscopy in a non-aqueous electrochemical environment- Tris(2,2'-Bipyridine)Ruthenium(II) adsorbed on silver from acetonitrile." *Chemical Physics Letters* 102(4): 365–370.

116. Storhoff, J. J., R. Elghanian, R. C. Mucic, C. A. Mirkin, and R. L. Letsinger (1998). "One-pot colorimetric differentiation of polynucleotides with single base imperfections using gold nanoparticle probes." *Journal of the American Chemical Society* 120(9): 1959–1964.

117. Su, K.-H., Q.-H. Wei, X. Zhang, J. Mock, D. R. Smith, and S. Schultz (2003). "Interparticle coupling effects on plasmon resonances of nanogold particles." *Nano Letters* 3(8): 1087–1090.

118. Sung, T.-W. and Y.-L. Lo (2012). "Highly sensitive and selective sensor based on silica-coated CdSe/ZnS nanoparticles for Cu2+ ion detection." *Sensors and Actuators B: Chemical* 165(1): 119–125.

119. Swierczewska, M., G. Liu, S. Lee, and X. Chen (2012). "High-sensitivity nanosensors for biomarker detection." *Chemical Society Reviews* 41(7): 2641–2655.

120. Taylor, R. W., R. J. Coulston, F. Biedermann, S. Mahajan, J. J. Baumberg, and O. A. Scherman (2013). "In situ SERS monitoring of photochemistry within a nanojunction reactor." *Nano Letters* 13(12): 5985–5990.

121. Taylor, R. W., T.-C. Lee, O. A. Scherman, R. Esteban, J. Aizpurua, F. M. Huang, J. J. Baumberg, and S. Mahajan (2011). "Precise subnanometer plasmonic junctions for SERS within gold nanoparticle assemblies using cucurbit [n] uril 'glue'." *ACS Nano* 5(5): 3878–3887.

122. Tombelli, S., M. Minunni, and M. Mascini (2007). "Aptamers-based assays for diagnostics, environmental and food analysis." *Biomolecular Engineering* 24(2): 191–200.

123. Trilling, A. K., J. Beekwilder, and H. Zuilhof (2013). "Antibody orientation on biosensor surfaces: a minireview." *Analyst* 138(6): 1619–1627.

124. Tripp, R. A., R. A. Dluhy, and Y. Zhao (2008). "Novel nanostructures for SERS biosensing." *Nano Today* 3(3): 31–37.

125. Trungkathan, S., D. Polpanich, S. Smanmoo, and P. Tangboriboonrat (2014). "Magnetic polymeric nanoparticles functionalized by mannose-rhodamine conjugate for detection of *E. coli*." *Journal of Applied Polymer Science* 131(6): 1–9.

126. Tuerk, C. and L. Gold (1990). "Systematic evolution of ligands by exponential enrichment—RNA ligands to bacteriophage-T4 DNA-polymerase." *Science* 249(4968): 505–510.

127. Vikesland, P. J. and K. R. Wigginton (2010). "Nanomaterial enabled biosensors for pathogen monitoring-a review." *Environmental Science and Technology* 44(10): 3656–3669.

128. Wang, J. (2005). "Carbon-nanotube based electrochemical biosensors: a review." *Electroanalysis* 17(1): 7–14.

129. Wang, J. (2007). "Nanoparticle-based electrochemical bioassays of proteins." *Electroanalysis* 19(7–8): 769–776.

130. Wang, L., W. J. Zhao, M. B. O'Donoghue, and W. H. Tan (2007). "Fluorescent nanoparticles for multiplexed bacteria monitoring." *Bioconjugate Chemistry* 18(2): 297–301.

131. Wang, W., H. Y. Li, D. W. Zhang, J. Jiang, Y. R. Cui, S. Qiu, Y. L. Zhou, and X. X. Zhang (2010). "Fabrication of bienzymatic glucose biosensor based on novel gold nanoparticles-bacteria cellulose nanofibers nanocomposite." *Electroanalysis* 22(21): 2543–2550.

132. Wang, W., T.-J. Zhang, D.-W. Zhang, H.-Y. Li, Y.-R. Ma, L.-M. Qi, Y.-L. Zhou, and X.-X. Zhang (2011). "Amperometric hydrogen peroxide biosensor based on the immobilization of heme proteins on gold nanoparticles–bacteria cellulose nanofibers nanocomposite." *Talanta* 84(1): 71–77.

133. Wei, H., S. M. H. Abtahi, and P. J. Vikesland (2015a). "Plasmonic colorimetric and SERS sensors for environmental analysis." *Environmental Science: Nano* 2(2): 120–135.

134. Wei, H., K. Rodriguez, S. Renneckar, W. Leng, and P. J. Vikesland (2015b). "Preparation and evaluation of nanocellulose-gold nanoparticle nanocomposites for SERS applications." *Analyst* 140(16): 5640–5649.

135. Wei, H., K. Rodriguez, S. Renneckar, and P. J. Vikesland (2014). "Environmental science and engineering applications of nanocellulose-based nanocomposites." *Environmental Science: Nano* 1(4): 302–316.

136. Win, M. N., J. S. Klein, and C. D. Smolke (2006). "Codeine-binding RNA aptamers and rapid determination of their binding constants using a direct coupling surface plasmon resonance assay." *Nucleic Acids Research* 34(19): 5670–5682.

137. Wu, H., J. Liang, and H. Han (2007). "A novel method for the determination of Pb2+ based on the quenching of the fluorescence of CdTe quantum dots." *Microchimica Acta* 161(1–2): 81–86.

138. Wujcik, E., H. Wei, X. Zhang, J. Guo, X. Yan, N. Sutrave, S. Wei, and Z. Guo (2014). "Antibody nanosensors: a detailed review." *RSC Advances* 4(82): 43725–43745.

139. Xia, F., X. Zuo, R. Yang, Y. Xiao, D. Kang, A. Vallée-Bélisle, X. Gong, J. D. Yuen, B. B. Hsu, and A. J. Heeger (2010). "Colorimetric detection of DNA, small molecules, proteins, and ions using unmodified gold nanoparticles and conjugated polyelectrolytes." *Proceedings of the National Academy of Sciences* 107(24): 10837–10841.

140. Xie, Y., X. Wang, X. Han, X. Xue, W. Ji, Z. Qi, J. Liu, B. Zhao, and Y. Ozaki (2010). "Sensing of polycyclic aromatic hydrocarbons with cyclodextrin inclusion complexes on silver nanoparticles by surface-enhanced Raman scattering." *Analyst* 135(6): 1389–1394.

141. Xue, X., F. Wang, and X. Liu (2008). "One-step, room temperature, colorimetric detection of mercury (Hg2+) using DNA/nanoparticle conjugates." *Journal of the American Chemical Society* 130: 3244–3245.

142. Yan, J. L., M. C. Estevez, J. E. Smith, K. M. Wang, X. X. He, L. Wang, and W. H. Tan (2007). "Dye-doped nanoparticles for bioanalysis." *Nano Today* 2(3): 44–50.

143. Yang, J., M. Palla, F. G. Bosco, T. Rindzevicius, T. S. Alstrøm, M. S. Schmidt, A. Boisen, J. Ju, and Q. Lin (2013). "Surface-enhanced Raman spectroscopy based quantitative bioassay on aptamer-functionalized nanopillars using large-area Raman mapping." *ACS Nano* 7(6): 5350–5359.

144. Yang, J.-M., H. Yang, and L. Lin (2011). "Quantum dot nano thermometers reveal heterogeneous local thermogenesis in living cells." *ACS Nano* 5(6): 5067–5071.

145. Yang, L. J. and Y. B. Li (2006). "Simultaneous detection of *Escherichia coli* O157: H7 and *Salmonella typhimurium* using quantum dots as fluorescence labels." *Analyst* 131(3): 394–401.

146. Ye, B.-F., Y.-J. Zhao, Y. Cheng, T.-T. Li, Z.-Y. Xie, X.-W. Zhao, and Z.-Z. Gu (2012). "Colorimetric photonic hydrogel aptasensor for the screening of heavy metal ions." *Nanoscale* 4(19): 5998–6003.

147. Yi, S., L. Sun, S. C. Lenaghan, Y. Wang, X. Chong, Z. Zhang, and M. Zhang (2013). "One-step synthesis of dendritic gold nanoflowers with high surface-enhanced Raman scattering (SERS) properties." *RSC Advances* 3(26): 10139–10144.

148. Zahavy, E., V. Heleg-Shabtai, Y. Zafrani, D. Marciano, and S. Yitzhaki (2010). "Application of fluorescent nanocrystals (q-dots) for the detection of pathogenic bacteria by flow-cytometry." *Journal of Fluorescence* 20(1): 389–399.

149. Zeiri, L., B. V. Bronk, Y. Shabtai, J. Czege, and S. Efrima (2002). "Silver metal induced surface enhanced Raman of bacteria." *Colloids and Surfaces A: Physicochemical and Engineering Aspects* 208(1–3): 357–362.

150. Zeiri, L., B. V. Bronk, Y. Shabtai, J. Eichler, and S. Efrima (2004). "Surface-enhanced Raman spectroscopy as a tool for probing specific biochemical components in bacteria." *Applied Spectroscopy* 58(1): 33–40.

151. Zeiri, L. and S. Efrima (2005). "Surface-enhanced Raman spectroscopy of bacteria: the effect of excitation wavelength and chemical modification of the colloidal milieu." *Journal of Raman Spectroscopy* 36(6–7): 667–675.

152. Zhang, T., W. Wang, D. Zhang, X. Zhang, Y. Ma, Y. Zhou, and L. Qi (2010). "Biotemplated synthesis of gold nanoparticle-bacteria cellulose nanofiber nanocomposites and their application in biosensing." *Advanced Functional Materials* 20(7): 1152–1160.

153. Zhao, X., L. R. Hilliard, S. J. Mechery, Y. Wang, R. P. Bagwe, S. Jin, and W. Tan (2004). "A rapid bioassay for single bacterial cell quantitation using bioconjugated nanoparticles." *Proceedings of the National Academy of Sciences of the United States of America* 101(42): 15027–15032.

154. Zhao, X., R. Tapec-Dytioco, and W. Tan (2003). "Ultrasensitive DNA detection using highly fluorescent bioconjugated nanoparticles." *Journal of the American Chemical Society* 125(38): 11474–11475.

155. Zhu, L., S. Ang, and W. T. Liu (2004). "Quantum dots as a novel immunofluorescent detection system for *Cryptosporidium parvum* and *Giardia lamblia*." *Applied and Environmental Microbiology* 70(1): 597–598.

PART 4

Environmental Implications

CHAPTER 11

Toxicological Impacts of Nanomaterials

Sophie Lanone

Inserm, U955, Equipe 04, Créteil, France

Nancy Ann Monteiro-Riviere

Nanotechnology Innovation Center, Kansas State University, Manhattan, Kansas, USA

Overview

Many nano-enabled products entail direct human contact, such as toothpaste, food additives, shampoos, and sunscreens. The expansion of nanomaterial applications, including in the field of nanomedicine, must be accompanied by a continuous evaluation of the possible impacts of these products on human health. In the last decade, almost 10,000 publications have been referenced in PubMed with the search terms "nanomaterial" and "toxicity", with an increasing publishing rate over the years: from 36 publications in 2004 to more than 2000 in 2013. To date, there are no known documented cases of adverse effects on human health following exposure to engineered nanomaterials (ENMs), with perhaps the caveat that the ENMs are not composed of previously known toxicants such as Cd or Hg. The LD50 for some of the highest volume produced ENMs, (CNTs, TiO_2 NPs, and Ag NPs) are generally reported to be greater than 2000 mg/kg based on oral exposure to rats or mice; comparable to that for vinegar, and NaCl. Judgments on the potential toxicity of nanomaterials are drawn primarily from tissue models and *in vitro* cell culture studies. From these studies, it appears that many ENMs can exert toxic effects, and the research community is now beginning to discern how nanomaterials interact with cells, biological tissues, and their components. Thus, several biological mechanisms such as inflammation, oxidative stress, and more recently, autophagy, have been proposed as underlying processes that could explain ENM toxicity. However, not all ENMs induce toxic effects, and although the same properties that make them so special to manufacturers are thought to be important determinants of their toxic effects, the exact ins and outs of ENM toxicity are not entirely understood yet. Moreover, there are no clear indications of the relevance of *in vitro* studies using ENMs to whole animal studies. As is problematic in all toxicology studies, extrapolation from animal studies to human effects, extrapolation from high to low concentration, and differences between acute and chronic exposure further cloud one's ability to make definitive statements regarding the toxicity of new materials for which there is limited to no real exposure at the current time.

In this chapter, we review the current knowledge on the toxicological impacts of ENMs, the essential molecular pathways modulated by ENMs, potential mechanisms of action, and the physico-chemical determinants that may govern their toxicological effects. The first portion of this chapter presents information based on selected mechanisms. The second portion considers the toxicity studies for several specific classes of nanomaterials.

Airway Remodeling

Many early concerns regarding potential ENM impacts on human health were grounded in experience with particle and fiber inhalation and the potential for damage such as silicosis and asbestos. The existing literature regarding the pulmonary effects of NM primarily relates experiments performed with mice or rats mainly exposed by intratracheal or oropharyngeal administration, but also by aerosol inhalation [1]. Among the various classes of ENMs, two classes stand out based on their present and predicted production amounts and use in industry: (1) carbonaceous ENMs, which essentially includes carbon black (CB), carbon nanotubes (CNT) and, fullerenes and (2) metal and metal oxide nanoparticles (NP), such as those obtained from iron (Fe), titanium (Ti), manganese (Mn), cerium (Ce), chromium (Cr), and copper (Cu) and metal nanoparticles, primarily gold (Au) and silver (Ag). Two major pulmonary effects of such exposures have been described: the formation of granuloma and the development of pulmonary fibrosis [2–9]. Granuloma formation, consistent with foreign body response [10–12], has been described within hours after the initial ENM administration [8, 9] and can persist for weeks to months [6, 7, 9]. Fibrosis on the other hand generally takes a few weeks to develop. Granulomas formed may persist even after several months [9, 12, 13]. This histopathological modification can occur either within granulomas or as diffuse interstitial and septal fibrosis, distal to granulomas. These pulmonary manifestations do not always occur, and some ENMs, in some experimental conditions have not been observed to induce measurable lung remodeling effects [6, 14, 15].

Besides the possibility of these well-described lung remodeling effects, ENM administration has been in some cases associated with the development of emphysema-like alterations, for example, as observed after administration of TiO_2 or Fe_2O_3 NP in mice [16, 17]. Moreover, the development of mesothelioma, a malignant cancer of the mesothelium most commonly caused in response to asbestos exposure [18], has been reported after exposure to some CNTs or carbon nanofibers (CNF) [19–21]. Mesothelioma formation was observed in the visceral mesothelium, accompanied by hyperplastic proliferative lesions, inflammatory cell infiltration and inflammation-induced fibrotic lesions of pleural tissue in the lungs of rats exposed to multi-walled CNTs (MWCNTs) [22].

More recently, studies have been performed to evaluate pulmonary effects of ENMs in the context of a preexistent respiratory disease such as asthma or bacterial infection. As an example, in a murine model of asthma, concomitant exposure to CNTs can enhance the susceptibility of mice to develop airway fibrosis [23, 24]. The same exacerbated fibrotic response has also been described when CNTs were administrated concomitantly to Gram positive (*Listeria monocytogenes*) or Gram negative (*Escherichia coli* lipopolysaccharide) bacteria; in both cases, CNT exposure in combination with bacterial infection was able to induce an exaggerated airway fibrosis as compared to what was observed after the bacterial infection alone [25, 26]. Using transgenic mice, Glista-Baker and colleagues recently demonstrated that the deletion of the T-box transcription factor Tbx21 (T-bet), was associated with an increased pulmonary response to nickel (Ni) NP oropharyngeal aspiration in terms of mucus metaplasia, total lung Muc5AC and Muc5B mRNA expression, chronic alveolitis, IL-13 and CCL-2 secretion, and number of eosinophils in bronchoalveolar lavage (BAL) fluid [27]. Similar results, although to a lesser extent, were obtained after CNT exposure. Such results are important in the context of asthma since loss of T-bet is associated with a shift toward Th2 type allergic airway inflammation that characterizes asthma [28]. It must be noted that, in some cases, NM administration can protect from airway inflammation, mucus secretion, and hyperresponsiveness in murin model of asthma. This has been described for TiO_2 NP, depending on the dose and timing of exposure [29, 30], and for Ag NP via the suppression of PI3K/HIF-1alpha/VEGF signaling pathway [31, 32].

Overall, the pulmonary effects of NMs in susceptible individuals are not entirely understood and, even in healthy individuals, deserve further studies to understand the overall (respiratory) toxicity of nanoscale particles.

General Biological Effects

Several biological effects have been described as part of potential pathways of ENM toxicity. The generation of oxidative stress and/or the induction of an inflammatory response are the most frequently evoked, as well as the occurrence of genotoxicity.

Oxidative Stress

Oxidative stress is defined as the imbalance between the production of reactive oxygen species (ROS) and antioxidant defenses, where the pro-oxidant forces exceed the antioxidant one. Oxidative stress has long been proposed to be a common mechanism in ENM pathogenicity [33, 34]. Several footprints of oxidative stress have been detected in the BAL fluid and/or in the lung of mice or rats exposed to various ENMs, such as CNT [12, 35], TiO_2 NP [36], or Fe oxides [37] (see [38] for review). As examples, this includes the pulmonary expression of heme oxygenase-1 (HO-1), a potent antioxidant used as a marker of oxidative insult, the presence of lipid peroxidation products such as 4-hydroxy-nonenal (4 HNE) or 8-isoprostane, and the depletion of glutathione in BAL fluid. These *in vivo* observations were also confirmed *in vitro* in various cell types (either primary cells or cell lines) exposed to ENMs [39, 40].

The contribution of oxidative stress to ENM toxicity is further exemplified by the fact that oxidant generation is important to the development of a toxic response after ENM exposure. Shvedova and coworkers demonstrated that mice maintained on a vitamin E–deficient diet showed an increased oxidative stress following single-wall CNT (SWCNT) exposure [41]. This was associated with a higher sensitivity to SWCNT-induced acute inflammation: increased number of inflammatory cells and pro-inflammatory cytokine production [tumor necrosis factor-α (TNF-α), interleukin-6 (IL-6)] and pro-fibrotic response. Similar results were obtained with positively charged Si-core NP [42]. Interestingly, a vitamin E–rich diet was observed to protect asthmatic rats from the exacerbation observed in response to CNTs [43]. Finally, it has been shown that mice lacking Nicotinamide Adenine Dinucleotide PHosphate-oxidase (NADPH) oxidase, a pro-oxidant enzyme that generates superoxide radicals, exhibited a decreased superoxide anion production by neutrophils, and an enhanced acute inflammatory response after SWCNT exposure by pharyngeal aspiration, together with a decreased profibrotic response [44]. Together, these results strongly suggest the role of oxidative stress as one important mechanism underlying NM toxicity.

Inflammation

The induction of an inflammatory response has been described in numerous *in vivo* studies, after exposure of rats or mice to various ENMs [45–47] (see [48] for review). This inflammation is characterized by an early onset, with the recruitment of neutrophils and macrophages essentially, in the BAL fluid only a few hours after NM exposure [47]. This tissue infiltration is usually diminished a few weeks after the initial exposure, although it can persist up to 1 month [6, 49]. The recruitment of inflammatory cells is accompanied by the release of pro-inflammatory cytokines, such as IL1-β, IL-6, monocyte chemoattractant protein-1 (MCP-1), macrophage inflammatory protein-2 (MIP-2), or TNF-α in the BAL fluid as well as in the lung tissue [46, 49, 50]. *In vitro* studies further identified at least macrophages [51], fibroblasts [52], epithelial [53], and mesothelial cells [54], as potent inflammatory cytokine producers in response to NM.

Genotoxicity

Due to the size of NMs, the probability of their internalization into the cells and their interaction with the intracellular environment such as the nucleus is very high [55]. These interactions can damage the genetic material and lead to genotoxic responses characterized by DNA damage and mutations that compromise the efficiency of the cells and therefore their survival. Genotoxicity can happen because of a direct interaction of NM with the genetic material of the cells or it can be indirect genotoxicity due to the generation of oxidative stress that in turn will induce oxidative damage to the genetic material [56, 57].

Several studies [55] have demonstrated the existence of DNA damage, chromosomal aberrations, or micronucleus induction after exposure to ENMs, including metal-based NPs and CNTs, *in vitro* as well as *in vivo* [58–61]. For example, An and coworkers demonstrated that the interaction of DNA with carbonaceous NPs resulted in DNA binding and aggregation both *in vitro* and *in vivo* in a dose-dependent manner [62]. Another study showed that metal-based NPs could tightly bind to DNA nucleobases, but also to Watson-Crick base pairs AT and GC [63]. Besides a direct interaction with DNA, ENMs are also able to bind to the active site of proteins implied in DNA "care" leading to their conformational or structural changes, or resulting in a competitive inhibition of the enzyme [64]. Such event has been described for C_{60} fullerene that interacts with the human DNA topoisomerase II α, leading to the inhibition of the enzyme activity [65]. Moreover, this NM may also interact with several proteins involved in the DNA mismatch repair pathway [66].

Oxidative stress may also be a possible mechanism behind ENM genotoxicity. It is known that ROS can directly attack DNA and generate modified DNA bases. ENMs such as TiO_2 NP can induce genotoxicity and impair DNA repair activity in cells, *via* their production of ROS [67], although it is not always true [60, 68]. Moreover, pretreatment with the free radical scavenger N-acetyl-L-cystein (NAC) lead to the inhibition of CNT or zinc oxide NP–induced genotoxicity [69, 70], demonstrating a role for oxidative stress in NM-induced genotoxicity.

Autophagy

Autophagy, derived from the Greek roots for "self-eating", is a general term standing for a cellular catabolic process in which cellular components, including organelles and macromolecules, are delivered to the lysosomes for degradation [71]. The resulting degradation products are recycled to maintain nutrient and energy homeostasis. Autophagy is characterized by the formation of a double membrane vesicle called autophagosome, which sequesters the cytoplasmic material to be degraded. At that point, the autophagosome fuses with a lysosome to form an autolysosome in which the lysosomal enzymes will degrade the cargo. In most cell types, autophagy occurs at a basal rate to maintain normal cellular homeostasis by eliminating misfolded proteins and damaged organelles. This process can also be induced under stress conditions, such as metabolic stress (amino acid or growth factors deficiency), hypoxia, or reticulum stress, to allow cell survival. As such, autophagy has been shown to play a key role in diverse pathologies, such as cancer, neurodegenerative, pulmonary, or inflammatory diseases [72, 73]. The involvement of autophagy in these pathologies can be associated to its role in the negative regulation of oxidative stress and inflammatory responses.

Autophagy, Oxidative Stress, and Inflammation

Autophagy can suppress ROS production. Indeed, impairment in the autophagy process leads to increased oxidative stress [74–78]. Moreover, autophagy plays a crucial role for the degradation of damaged mitochondria, which are the main sources of ROS generation. Additionally, several studies suggest that autophagy may have a key role in the degradation of oxidized proteins [79, 80]. In response to oxidative stress, the nuclear factor (erythroid-derived 2)-like 2 transcription factor (Nrf2), involved in the transcription of antioxidants genes such as hemeoxygenase, can induce p62 expression which in turn, activates Nrf2, subsequently forming a positive feedback loop [81, 82] to reduce the oxidative response.

Basal autophagy, by degrading cells debris or defective organelles which can activate the inflammasome, can negatively regulate inflammasome activation [83, 84]. For example, an autophagy blockade leads to the accumulation of damaged mitochondria, which in turn, via their ROS production, can activate the NLRP3 inflammasome [84]. In addition to control the production of cytokines by regulating the activation of inflammasomes, autophagy can also directly target pro-IL-1β for lysosomal degradation [85]. Importantly, Shi and colleagues, showed that the activation of inflammasomes in macrophages leads to the formation of autophagosomes and that a blockade of autophagy exacerbated the inflammasome activation [86]. Similarly, mice deficient for LC3b were found to be more susceptible to lipopolysaccharide than wild-type mice, with higher serum concentrations of IL-1β and IL-18 [83]. Importantly, autophagy is also involved in the transcriptional regulation of genes involved in the inflammatory response. Indeed, when autophagy is deficient, p62 protein, a substrate of autophagy, accumulates and leads to TRAF6 (tumor necrosis factor receptor-associated factor 6) oligomerization and to the further activation of NF-kB, a transcription factor involved in inflammation [87–89]. Furthermore, autophagy genes are associated with inflammatory disorders such as Crohn's disease, a chronic inflammatory disease of the intestine. Indeed, polymorphisms in autophagy-associated genes, such as Atg16L1, Irgm1 (immune-related GTPase M-1) but also Ulk1, are associated with Crohn's disease [90, 91].

As inflammation and oxidative stress are the most widely described mechanisms underlying NM-mediated toxicity, a growing amount of studies suggest that autophagy could be a potential new mechanism explaining, at least in part, the toxicity of NM.

Evidence of Autophagy Perturbations by NM

The literature on autophagy perturbations by NM is relatively new and remains patchy. The large majority of the studies reported the use of cell lines, and only rare studies showed data on animal models [92], probably because of the difficulty to use relevant methods to assess autophagy *in vivo*. A large part of these experiments reported increased numbers of autophagosomes in response to various NM, such as metal oxides NP [93, 94], graphene nanosheets [95], or Ag nanowires [96]. Some studies demonstrated an implication of the Akt/mTOR pathway in these effects. For example, Roy and coworkers described an enhancement of autophagosome formation in mouse peritoneal macrophages exposed to zinc oxide NP, through the inhibition of the Akt/mTOR pathway, ultimately leading to apoptosis [97]. Beside the modulation of early steps of autophagy, some studies also demonstrated that the accumulation of autophagosomes could be due to a blockage of autophagy flux. This has been described for various NM such as metal oxides [98, 99], dendrimers [92], or CNT [100, 101]. Interestingly, Sun and colleagues reported the only study so far demonstrating an increased autophagosome formation along with an increase of autophagy flux in A549 lung epithelial cells exposed to copper oxide NP [102].

Underlying Mechanisms of ENM Toxicity

Oxidative stress, inflammation, autophagy, and genotoxicity are examples of phenomena that may be observed in response to ENM exposure. The underlying mechanisms responsible for producing these responses are no doubt linked to the physical-chemistry of the nanomaterials and their biomolecular interactions with specific proteins, lipids, and other entities making up cells, membranes, and organelles. The following section details examples of such interactions.

Interactions with Protein Corona

When in contact with a biological environment, nanomaterials are rapidly coated with biomolecules that may modify their properties and the way in which they interact with cells [103]. This surface-bound coating is a dynamic mixture of proteins and lipids, called the protein corona. It has been argued that the interaction unit with the cell is not the nanomaterial by itself but the nanomaterial together with its corona of proteins issued from serum and other body fluids [104, 105]. Several characteristics of the corona are extremely important and are routinely measured; its thickness, density, identity, quantity, conformation, and affinity [106]. These measurements are achievable via numerous techniques including dynamic light scattering (DLS), transmission electron microscopy (TEM), polyacrylamide gel electrophoresis, circular dichroism, computational simulation, proteomics, or surface plasmon resonance usually performed *in situ* or *ex situ*. Generally, *in situ* techniques measure the corona with the ENMs still dispersed in the physiological environment. These techniques are considered the most relevant in the field of nanotoxicology. However, these *in situ* studies are rare and provide the least amount of information. On the other hand, data from *ex situ* techniques are numerous, but remain less relevant because of the need to isolate the nanomaterials from the physiological environment. Careful studies have revealed that the composition of this corona is dynamic [107], reflects the size, shape, and surface properties of the nanomaterials, and finally, that it is a major determinant of the localization and subsequent effects of nanomaterials *in vivo* [108]. The two main consequences of the formation of this protein corona can be (1) the modification of the nanomaterials (surface) characteristics and further reactivity [109]; and (2) the modification of the proteins that interact with the ENMs, possibly leading to their altered structural conformation and functionality. Both of these events can be important to NM toxicity [110, 111].

Until recently, studies on the interactions between biomolecules and ENMs have been limited to the use of purified proteins (e.g., chymotrypsin, fibrinogen, albumin) or of a so-called representative media [104, 112–115]. Interestingly, as ENM applications in nanomedicine are very promising, the blood takes on extra relevance as it is the first physiological environment that these ENMs will come in contact with after intravenous administration. It is not then surprising to find that a large proportion of studies have focused on isolated blood proteins or isolated plasma or serum as models for blood protein adsorption (fibrinogen, gamma globulin-transferrin, serum albumin, complement proteins, etc.) [104, 105, 111, 114]. From these studies, it is clear that ENM can adsorb proteins to their surface, although this binding depends on protein-specific as well as NM-specific characteristics. This is best exemplified in CNT interactions with proteins. With this in mind, Shannahan et al. assessed the protein corona associated to SWCNTs and MWCNTs presenting various functionalized groups [polyvinylpyrrolidone (PVP) or COOH], after a 1 hour incubation in culture medium supplemented with 10% fetal bovine serum [104]. The authors identified and quantified 366 different protein components in the various CNT coronas, (total of 2507 present in the culture medium). The highest number of proteins were bound to COOH-SWCNT > COOH-MWCNT > MWCNT-PVP > raw SWCNT and raw MWCNT. The large number of proteins bound to COOH-CNT is likely due to the abundance of protein amines in the medium, which could associate with the carboxyls through electrostatic interactions. Interestingly, only 14 proteins bound to all CNT [104]. In a similar experiment utilizing a proteomic approach with HeLa cells homogenates, as archetypal of human proteome, it was demonstrated that cytoskeletal proteins were dramatically overrepresented on the MWCNT biocorona [112]. By contrast, nuclear proteins (chromatin, ribosomal and sliceosomal proteins) were heavily represented in the fraction of proteins that did not bind effectively to CNT. The authors suggest that the preferential binding of CNT to cytoskeletal proteins might represent a biochemical basis for CNT toxicity. Interestingly, CNT diameter was also an important factor of efficiency for proteins: CNT with a diameter of 20 to 40 nm or above bound a greater amount of proteins than materials with a diameter less than 10 nm. The size and curvature of NP also represent major characteristics affecting the composition of bound protein in the corona. This has been recently described by Sanfins and colleagues [116]. Altogether, these authors demonstrate that NPs induce significant alteration of the enzymatic activity in a size-dependent manner, whereas enzyme kinetics studies show a critical role for NP surface area and curvature.

Surprisingly, despite of being the major route of entry for nanomaterials, only a few corona studies have been conducted with proteins relevant to pulmonary exposure [113, 115, 117]. For example, Kapralov and colleagues exposed mice to SWCNT by pharyngeal aspiration, and recovered the BAL fluid of the animals 24 hours postexposure [118]. The CNT were then extracted from the BAL. The team then utilized liquid chromatography mass spectrometry to investigate the adsorption of phospholipids by SWCNT. The data demonstrated that there was a selective adsorption of phosphatidylglycerols and phosphatidylcholines onto the surface of the CNT. Importantly, the presence of surfactant protein (SP) A, B, and D was identified. These proteins are secreted by airway epithelial cells and play an important role in the first line of defense against infection within the lungs.

An important issue of protein corona formation is the resulting modification, conformation and/or activity of the bound proteins. To address this issue, Banerjee and coworkers investigated the conformational and functional properties of a large multimeric protein, α-crystallin, absorbed on silver NP surface [119]. The authors demonstrated that the chaperone function and the refolding capacity of the protein, which is primarily governed by the α-crystallin domain, were lost to a significant extent when adsorbed onto silver NP surface, because of the selective alkylation of two cystein residues at the α-crystallin domain. Nonetheless, the secondary structure of α-crystallin was mostly retained. Another evidence of such protein modification by NM interaction has been given by **Zhang** and colleagues who showed that when CNT were bound to α-chymotrypsin, the complex could inhibit the enzymatic activity [120]. Finally, a study measured the extent and kinetics of internalization of acid-coated quantum dots, with and without adsorbed native and modified human serum albumin by HeLa cells [121]. Pronounced variations were observed, indicating that even small physico-chemical changes of the protein corona may affect biological responses. Despite uncertainties in the underlying mechanisms of protein interactions with NM, there is no doubt that this will result in the modification of toxicity *in vitro* [111], as well as NM uptake by macrophages [118], and targeting to professional phagocytes *in vitro* and *in vivo* [122]. It is indeed important to understand that a NM entering the pulmonary system may pass through the mucosal layer and enter into the blood stream. At the cellular level, the NM may moreover be phagocytized and taken to the endosomes that ultimately fuse with lysosomes. Each of these steps represents unique environments, with specific characteristics, that could cause NM modifications, and therefore lead to the modification of the protein corona formation [123]. NM entering the body have thus to be considered as evolving systems, that are far from being understood yet.

Biopersistence

Another critical issue in the understanding of ENM toxicity is their biopersistence in the organism. It had been observed that some ENM, and particularly CNTs, are relatively stable materials and can be detected up to 24 months after the initial exposure, at the site of exposure or distributed throughout the body [13, 124–126]. Clusters of CNTs often surrounded by macrophages were still detectable by optical microscopy up to 2 years after an initial pulmonary exposure—demonstrating the persistence of these nanomaterials [126]. Despite this clear indication of NM biopersistence and widespread evidence of their cellular internalization [127, 128], little information is available on the biopersistence or modifications of ENMs after cellular uptake. Experiments by Allen et al. were the first to demonstrate the degradation of SWCNT through enzymatic catalysis in abiotic conditions [129]. The authors incubated CNTs with horseradish peroxidase and low concentrations of hydrogen peroxide (40 μM), for 12 weeks, at 4°C. TEM images illustrated that CNT length decreased by 45% at week 8, with some globular materials visible, while at 12 weeks, mostly globular material were present. UV-vis-NIR spectroscopy measurements further demonstrated a diminution of CNT diameter over time. In contrast to these findings, the same group observed that pristine SWCNTs failed to degrade when incubated in the same conditions (hydrogen peroxide, 4°C), which lead the authors to hypothesize that defects or functionalization sites are important facilitators of enzymatic action [130]. These findings were further corroborated by a study showing the faster degradation of carboxylated MWCNTs when they contained more defects [131]. It is proposed that the presence of carboxylic functions as well as defects on the graphitic surface of the CNT is likely to offer sites for interactions with the oxidative agents which are responsible for the degradation of the CNT, therefore acting as facilitators of enzymatic action. Similar indications of NM degradation were reported by Kagan et al. after *ex vivo* incubation of SWCNT with myeloperoxidase (MPO) [132]. The authors observed a decrease in the quantity of graphitic material as early as 24 hours after the initial incubation. With the exception of these reports of enzymatic degradation *ex vivo*, the literature is very sparse regarding the eventual fate of NM *in vitro* or *in vivo*. However, two recent studies buck this trend by reporting the partial degradation of MWCNT functionalized with amine groups, as early as 2 days post cortical administration [133] and the degradation of SWCNT in the lung of mice after pharyngeal administration [33].

These studies pinpoint the importance of NM functionalization for the effective degradation of the carbon lattice. However, the exact mechanisms underlying these effects are currently unclear. A recent study attempting to answer this question demonstrates that the addition of antioxidants (namely L-ascorbic acid and glutathione) significantly decrease the biodegradation of SWCNT by MPO. The authors therefore suggest that CNT biodegradation is not only enzyme-dependent but also needs strong oxidants to take place [134], the antioxidants being therefore secondarily important, as their amount will determine the level of degradation possible. Finally it seems crucial to note the importance of the fate and biological impact of the by-products of CNT biodegradation as they might very well contribute to the overall adverse properties accredited to CNT. A recent study by Bussy and colleagues addressed this issue [135]. Indeed, the authors showed the rapid detachment of part of the iron NP initially attached to SWCNT which appeared as free iron NP in the cytoplasm and nucleus of CNT-exposed murine macrophages, and also that the blockade of intracellular lysosomal acidification prevented iron NP detachment from CNT bundles and protected cells from CNT downstream toxicity [135].

Mechanisms of Autophagy Perturbations by NM

The exact mechanisms underlying NM-induced autophagy perturbation are not completely understood yet, but an impairment of the autophagosome-lysosome fusion and/or a defect in lysosome function appear as potential targets.

The integrity of the cytoskeleton, a highly dynamic cellular scaffold that supports cell shape and regulates the intracellular trafficking has an important role in autophagy. Indeed, several studies revealed the importance of the microtubular network, and to a lesser extent, of the actin cytoskeleton in the formation and the fusion of autophagosomes with lysosomes [136–138]. Some studies also showed that, once formed, autophagosomes move along microtubules to concentrate at the perinuclear region around the microtubule organizing center (MTOC), where the majority of the lysosomes are found, to fuse with them [139, 140].

After passing the cell membrane, NM could interact with the proteins of the cytoskeleton, affect their functions, and then, as described, potentially lead to an impairment of the autophagy process. Indeed, the inhibition of tubulin polymerization by gold NP has been shown in a cell-free system [141]. Moreover, fullerene derivative carbon NP and TiO_2 NP were found to inhibit microtubule polymerization, potentially by a hydrogen bond between NP and the tubulin heterodimer [142, 143]. More recently, a study described that SWCNT can directly bind to actin via hydrophobic interactions which leads to changes in actin structure [144]. Gold NM have been shown to have a dose-dependent effect on actin stress fibers in human dermal fibroblasts, thereby inducing cytotoxicity [145]. Furthermore, in the same cell type, gold NPs have been described to induce a disruption of the cytoskeleton, despite no change in actin and β-tubulin protein expression [146]. However, the disruption of the cytoskeleton was reversible given that the cytoskeleton could reconstitute following NP removal. Exposure to magnetic NP, such as iron oxide (Fe_2O_3) ones, could also alter cell function in pheocromocytoma neuronal cells by decreasing the number of actin filaments [147]. Consequently, these cells were not able to extend neurites in response to nerve growth factor. In PC12 cells, ferromagnetic mineral magnetite (Fe_3O_4) NP lead to alterations in microtubule polymerization, potentially induced by a direct bind to tubulin dimer [148]. Moreover, exposure to iron oxide nanomaterials on human umbilical vein endothelial cells leads to a significant disruption of cytoskeletal structures, with diminished vinculin spots, and disorganized actin and tubulin networks. Interestingly, in addition to the observed cytoskeleton disruption, this study also suggests an autophagy dysfunction that could explain the toxic effects of particles [149, 150].

Because of the involvement of lysosomes in the final steps of the autophagy process, a lysosome dysfunction could also be a mechanism explaining a defect of the autophagy pathway leading to NM-induced toxicity. Indeed, several types of nanomaterials have been recognized as being associated to lysosomal dysfunction [151, 152]. For example, MWCNT, with a diameter < 8 nm, induced lysosomal membrane destabilization (LMD) in 3T3 fibroblasts, leading to the release of lysosomal contents inside the cytoplasm. This was associated with an increased ROS production [153]. However, no or minor lysosomal damage were observed with larger MWCNT, or with NM of different composition (TiO_2, SiO_2) and in other cell types (telomerase-immortalized human bronchiolar epithelial cells and RAW 264.7 macrophages). In another study, G5-polyamidoamine dendrimers have been shown to be taken up into the lysosomal compartment and to modify the lysosomal pH, increasing the cytotoxicity [154]. Likewise, gold NP can accumulate in lysosomes and cause lysosomal dysfunction by increasing the lysosomal pH in rat kidney cells [99]. Interestingly, a blockade of the autophagy flux was also observed in these cells, suggesting lysosomal dysfunction as a likely mechanism of autophagy blockade. Moreover, it has been shown that exposure to zinc oxide NP in human monocytic THP-1 cells induced a decrease of lysosomal stability together with a loss of viability [155]. The authors suggest that the lysosomes were destabilized by the production of Zn^{2+} ions formed by the dissolution of zinc oxide NP into the acidic lysosomes. The release of the lysosomal content and Zn^{2+} ions into the cytoplasm may damage other organelles and lead to cell death. Likewise, TiO_2 nanobelts and amino-functionalized polystyrene NP have been shown to induce toxicity by a loss of lysosomal integrity and a subsequent release of cathepsins which could lead to cell death, oxidative stress, and inflammation [156, 157]. Recently, in mouse peritoneal macrophages, CNTs have been shown to induce lysosome impairment, characterized by an overload of lysosomes by nanomaterials and a decreased lysosomal stability and biogenesis, associated with a dysfunction of autophagy [101].

Taken together, the disruption of cytoskeleton together with a defect in the lysosome function could represent essential mechanisms explaining how nanomaterials could perturbate the autophagy process. However, these mechanisms deserve further attention, as they are far from being completely understood yet.

Determinants of NM Toxicity

Nanoparticle composition is perhaps the single most important factor determining acute toxicity; nanoparticles made of known toxic materials such as heavy metals or persistent inhalable fibers are likely to produce adverse effects *in vivo* or *in vitro*. ENMs do not systematically induce identical effects, whether they are measured in terms of toxicity, lung remodeling, or oxidative stress. The same is true for the biological mechanisms underlying these effects. Having a size in the nanometer range is not the **unic** physico-chemical property determining the ENM toxicity [48]. Shape, crystalline structure, surface chemistry, and charge are at least as important. The latter cases of surface chemistry and charge are examples of characteristics that are determined by the system into which ENMs are introduced. The formation of a protein corona is but one example that also raises the question of methods used to stabilize nanomaterials in toxicity studies and the relevance of the data produced.

A study from Tabet et al. demonstrated that MWCNT-induced cytotoxicity, oxidative stress, and inflammation were increased by acid-based and decreased by polystyrene-based polymer coating both *in vitro* in murine macrophages and *in vivo* in lung of mice monitored for 6 months [6]. Similarly, another study observed that polyacrylate-coated TiO_2 NP exhibited less cytotoxicity and induced no DNA damage on lung fibroblasts compared to their noncoated counterparts [158]. These findings suggest that surface chemistry of NM has the ability to modify their behavior and subsequent toxicity.

Alterations of ENM surfaces for the convenience of laboratory manipulations, such as stabilizing nanoparticles in suspension using bovine serum protein, may introduce biological incompatibilities that may produce their own toxic responses. Importantly, ENM characteristics, alone or because of interactions among them, will condition the overall behavior of NM, including their ability to form aggregates/agglomerates, dissolution rates, antibody responses and other factors that may determine ENM toxicity (Figure 11.1).

ENMs, whether they are carbon-based, TiO_2, iron NP, or other types of NMs, can induce different biological effects depending on their chemical composition [159]. Moreover, NMs of the same chemical composition, can also exhibit different behaviors [160, 161], depending for example on the crystalline structure [52, 162], the shape [52], or the number of walls for CNT [161]. For these latter NMs, the length also seems to be important. Indeed, over the past decade, several studies demonstrated that long CNTs were more pathogenic than the short ones. However, a more recent investigation observed a higher inflammogenic reactivity for short MWCNTs as compared to long MWCNTs, most likely due not only to the length reduction but also because of the accompanying surface modifications induced by the length-reduction process [163].

The apparent toxicity of an ENM may be linked to impurities remaining from the fabrication process rather than the underlying ENM. For example, CNTs often include residual metal catalysts (such as iron) that are used during their manufacturing process. These transition metals can induce toxicity by the production of ROS, and Kagan and co-workers demonstrated that, in conditions without cells, SWCNT with 26% iron had a greater potential to produce free radicals than their iron-depleted counterparts (0.23%) [164]. In cellular conditions, the iron-rich SWCNTs were able to induce more oxidative stress than the iron-depleted SWCNTs on murine macrophages. Moreover, metal chelators were able to reduce the

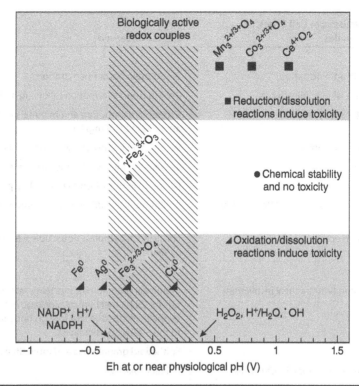

Figure 11.1 Relationship between the chemical behavior of metallic ENMs and reduction-oxidation potentials of biomolecules (from [168]).

toxicity observed on keratinocytes exposed to iron-rich SWCNTs (30%) [165], further demonstrating the importance of residual metal contaminants in NM toxicity. Interestingly, it was recently shown that iron NP can get detached from SWCNT (initially containing 25% iron) in murine macrophages, possibly via a pH-dependent mechanism [135]; the blockage of lysosomal acidification prevented this detachment and protected the cells against SWCNT toxicity. All these data suggest that the remaining contaminants play an important role in NM toxicity, and that they should be taken into account while assessing nanomaterial toxicity.

The aggregation or agglomeration state of NM also has to be taken into account when looking at NM effects [166]. Kreyling and coworkers showed that the translocation, from the lung of rats to their blood and other organs, is higher for clusters composed of 20 nm primary diameter iridium NP as compared to NP of 80 nm ones [167]. On the contrary, Noel and colleagues demonstrated a similar pattern between ENM clusters of different sizes; after their inhalation, TiO_2 agglomerates of 30 and 185 nm (primary diameter of NP) showed a similar lung deposition and both agglomerates resulted in comparable adverse effects in rats [36].

The redox couples that may be presented in cells representing in the case of metal-containing nanoparticles and the biomolecules present does appear to provide a basis for screening possible *in vitro* toxicity induced by these metallic nanoparticles *in vitro* [168]. Since redox calculations signal the potential for nanoparticle oxidation, reduction, or dissolution in biological media, they are likely basis for predicting when reactivity may lead to a toxic outcome. Auffan and coworkers first suggested that chemically stable metal-containing nanoparticles such as Fe_2O_3 in physiological redox conditions do not appear to exhibit cytotoxicity *in vitro*, while metal-containing nanoparticles with oxidant (e.g., CeO_2, Mn_3O_4, and Co_3O_4 nanoparticles) or reductive [e.g., Fe(0), Fe_3O_4, Ag(0), and Cu(0) nanoparticles] power can be cytotoxic and genotoxic toward biological targets *in vitro* [168].

Studies on Specific Nanomaterials

While the number of ENMs is large, research on potential health effects of ENMs has focused on a relatively small number of materials. This can be justified based on the production quantities of these materials and their potential for exposure as well as the availability of the materials in question. Tables 11.1, 11.2, and 11.3 summarize examples of research on health effects for several key, highly studied nanoparticles.

Type of Nanomaterials	Organisms, or Cell Types, or Organelles	Effects Observed	References
Fullerene			
C_{60} water suspension	Rat (IV administration)	Blood-brain barrier penetration	[172]
	Mouse (skin applications)	Benign tumors formation following initiation with TPA	[176]
MSAD-C_{60}	Rat (IV administration)	Extensive distribution and minimal clearance, death after 5 min at 25 mg/kg	[177]
$C_{60}(OH)_x$	Mice and rabbits	Distribution in the kidneys, bone, spleen, and liver by 48 h	[178]
Fullerenol-1	Mice	Decrease in the cytochrome P450 monooxygenase levels	[179]
	Liver microsomes	Decrease in mitochondrial oxidative phosphorylation	[179]
C_{60}-PVP	Mouse midbrain cell differentiation system	Inhibition of cell differentiation and proliferation	[180–182]
Functionalized fullerenes	Human epidermal keratinocytes	Localized in cytoplasmic vacuoles; expression of cytokines	[183]
SWCNT			
Underivatized SWCNT	Rat lung (intratracheal instillation)	Multifocal granulomas, mortality rate of 15% (5 mg/kg in rats for 24 h): mechanical blockage of the upper airways by the instillate due to a foreign body reaction and not to the SWCNT particulate	[184]
	Mice	Epithelioid granulomas and interstitial inflammation	[1]
	Human epidermal (HaCaT) cells	Cellular toxicity	[165]
	Human epidermal keratinocytes	Genes induction (structural proteins and cytokines)	[185, 186]
	Human embryo kidney HEK293 cells	Decrease in cell proliferation and cell adhesion	[187]
SWCNT–biotin-streptavidin complex	Human promyelocytic leukemia (HL60) cells and human T (Jurkat) cells	Endocytosis, intracytoplasmic localization, extensive cell death	[188]
Functionalized carbon nanotubes	Fibroblasts	Increase in the side wall functionalization decrease cytotoxicity	[189]
Iron-rich SWCNT	Macrophages	Loss of intracellular thiols and accumulation of lipid peroxides	[164]
	Immortalized keratinocytes	Activation of oxidative stress and nuclear transcription factor-kB	[190]
MWCNT	Human keratinocytes	Internalization in vacuoles; concentration dependant decrease in viability; concentration- and time-dependant increase in IL-8 production	[191]
		Proteomic analysis: dysregulation of intermediate filament expression, cell cycle inhibition, altered vesicular trafficking/ exocytosis, and membrane scaffold protein down-regulation	[192] [193]
	Intratracheal administration in rats	Agglomeration in the airways along with granulomas in the bronchial lumen and inflammation of the alveoli	[13, 194, 195]
	Alveolar macrophages	Necrosis, degeneration, and rarefaction of the nuclear matrix	[196]
Hat-stacked carbon nanofibers	Implantation in rat subcutaneous	Granulation and an inflammatory response; translocation to macrophages without severe inflammation, necrosis, or degeneration of tissue	[197]
Oxidized MWCNT	Human T cells	Toxicity and decrease in cell viability	[198]
Ground MWCNT	Intratracheal administration in rats	Dispersion in the lung and induction of an inflammatory and fibrotic response	[13, 194, 195]

*Negative impacts of nanomaterials on living microorganisms are discussed in Chapter 5.

IV: intravenous; MSAD-C_{60}: bis (monosuccinimide) derivative of p,p'-bis(2-amino-ethyl)-diphenyl-C_{60}; C_{60}-PVP: C_{60} solubilized by PVP coating in water; TPA: 12-0-tetradecanoylphorbol-13-acetate.

TABLE 11.1 Negative Impacts of Several Carbon-Based Nanomaterials on Living Mammalian Cells and Organisms*

Type of Nanomaterials	Organisms, or Cell Types, or Organelles	Effects Observed	References
Titanium dioxide			
	Human endothelial cells	Proinflammatory effects	[208]
	Human cells	DNA damage induction by sunlight-illuminated TiO_2	[207]
	Syrian hamster embryo fibroblast	Apoptosis induction and micronuclei formation	[211]
	Laboratory animals	Pulmonary inflammation	[214–217, 232]
	Rat and human alveolar macrophages	Production of ROS	[233]
Anastase TiO_2	Human bronchial epithelial cell line	Lipid peroxidation and oxidative DNA damage; increase in cellular nitric oxide and hydrogen peroxide levels; increase in cellular MDA levels	[213]
Iron oxide			
	Mouse macrophages or human ovarian tumor HeLa cells	Adhesion to plasma membrane and internalization efficiency of dextran-, albumin-, and DMSA-coated iron oxide nanoparticles	[234]
	Human CNS	Detection in CNS and staining of CNS parenchymal cells	[225]
	Human and rodent cells	Decreased MTT activity and DNA content	[227]
Magnetite	Human fibroblasts	Comparison between dextran-, albumin-derivatized and underivatized magnetite cytotoxicity (inhibition of the BrdU incorporation, disruption of the F-actin and vinculin filaments)	[235]
Transferrin-derivatized magnetite	Human fibroblasts	Increase in cell proliferation, alteration of cytoskeletal organization, and upregulation of proteins involved in cell signalization and extracellular matrix	[236, 237]
Ferumoxides	Intravenous injection	Accumulation in the Kupffer cells in the liver and in the reticuloendothelial system in the spleen	[238, 239]
Silica-overcoated magnetic nanoparticles containing RITC within the silica shell	Intraperitoneal injection in mice	Penetration through the blood-brain barrier without disturbing its function or producing apparent toxicity	[226]
Cerium oxide			
	Human lung fibroblasts	Size-dependent internalization of the particles	[226]
	Human and rodent cells	Decreased MTT activity and DNA content	[227]
Copper			
	Epithelial cell line	Inducer in production of IL-8	[219]
	Mice (gastrointestinal exposure)	Target organs of copper nanoparticles: kidney, spleen, and liver	[240]
Gold			
	Red blood cells, Cos-1 cells and *E. coli*	*In vitro* comparison of cationic (quaternary ammonium functionalized) and anionic (carboxylate-substituted) gold nanoparticles	[241]
	Human fibroblast cell line	Slight decrease in cell metabolic activity and/ or proliferation induced by a water-soluble gold nanoparticles functionalized with a Tat protein–derived peptide sequence	[242]

*Negative impacts of nanomaterials on living microorganisms are discussed in Chapter 5.

ROS: reactive oxygen species; MDA: melanodialdehyde; DMSA: meso-2,3-dimercaptosuccinic acid; CNS: central nervous system; MTT: 3-[4,5-dimethylthiazol-2-yl]-2,5-diphenyltetrazolium bromide.

TABLE 11.2 Negative Impacts of Several Metal-Containing Nanomaterials on Living Mammalian Cells and Organisms*

Type of Nanomaterials	Organisms, or Cell Types, or Organelles	Effects Observed	References
Quantum dots	Porcine skin	Penetration of the stratum corneum barrier with localization of QD in the underlying epidermal and dermal layers	[243]
	Mouse fibroblast (3T3) cells	Nuclear localization of cationic QD	[244]
QD 565 and QD 655 coated with PEG, PEG-amines, or carboxylic acids	Human epidermal keratinocytes	Nondependant size and surface coating internalization; surface coating–dependent cytotoxicity; size- and surface coating–dependant cytokine release; anionic QDs accumulated the most in cells and were most cytotoxic	[243]
PEG-silane-QD	Human fibroblasts	Downregulation of genes involved in controlling the mitotic M-phase spindle formation and cytokinesis; no activation of genes associated with immune and inflammatory responses	[245]
CdSe QD	Human hepatocytes	Decrease in MTT viability after treatment with uncoated UV irradiated or chemically oxidized CdSe QD	[246]
Mercaptoundecanoic acid–coated CdSe QD	Vero cells, HeLa cells, human hepatocytes	Increase in MTT cytotoxicity	[247]
Cadmium tellerium(CdTe) QD coated with cysteineamine	N9 murine microglial and rat pheochromocytoma (PC12) cells	Cytotoxicity induced by positively charged QDs; nuclear localization of the 2.3 nm but not the 5.2nm QD	[248]

*Negative impacts of nanomaterials on living microorganisms are discussed in Chapter 5.
MTT: 3-[4,5-dimethylthiazol-2-yl]-2,5-diphenyltetrazolium bromide; PEG: polyethylene glycol.

TABLE 11.3 Negative Impacts of Quantum Dots on Living Mammalian Cells and Organisms*

Fullerenes

Fullerenes, or buckyballs (short for buckminsterfullerene), are molecular structures containing 60 carbon atoms (C_{60}) in a spherical configuration, about 1 nm in diameter. Carbon atoms may be functionalized to yield hydroxylated, carboxylated, or other derivatized variations on C_{60} (see Chapter 2). There are conflicting reports as to the potential toxicity of fullerenes such as C_{60}. While C_{60} itself has essentially no solubility in water, it does have the ability to form aggregates, referred to as nC_{60}. Early work using organic solvent inclusion to produce nC_{60} yielded the conclusion that this material was more toxic than the herbicide paraquat [169]. However, this work was entirely discredited when it was shown that the toxicity was due to byproducts of the solvent used to produce the nC_{60} [170]. Carboxy-fullerenes are potent free radical scavengers and have been both effective in reducing neuronal cell death and suggested to act as neuroprotective agents in mice and in mouse neocortical cultures. A water-soluble carboxylic acid derivative fullerene (carboxyfullerene) has been shown to possess antioxidative properties that can suppress iron-induced lipid peroxidation in rat brains [171]. Study of the biodistribution and metabolism of C_{60} is complicated by the low solubility of C_{60} in water. These studies tend to show extensive tissue distribution and minimal clearance from the body [172].

Single-Walled Carbon Nanotubes

Carbon nanotubes, also known as buckytubes, are made up of a seamless single sheet of graphite rolled into cylindrical shells that range from one to tens of nanometers in diameter and up to several micrometers in length. They have numerous novel electrical and mechanical properties. They can be synthesized by electric arc discharge, laser ablation, or chemical vapor deposition [173–175] (see Chapter 2). Due to the fibrous structure of CNTs, they have been proposed for use in many tissue engineering applications. Variations on CNTs include the single-, double-, and multi-wall CNTs (SWCNT, DWCNT, and MWCNT). Like C_{60}, CNTs may be derivatized, resulting in very different affinities for suspending media and potentially different exposure or toxicity scenarios.

Titanium Dioxides

Titanium dioxide (TiO_2) particles larger than 100 nm are generally considered to be biologically inert in both humans and animals [199–203]. Thus, they have been widely used as a food colorant [204] and as a white pigment [205] in sunscreens and in cosmetic creams [206]. On the other hand, TiO_2 is also a well-known photocatalyst (see Chapter 4). Pure TiO_2 nanoparticles are typically either anatase, rutile, or a combination of the crystalline forms. TiO_2 absorbs UVA light, catalyzing the generation of ROS, such as superoxide anion radicals, hydrogen peroxide, free hydroxyl radicals, and singlet oxygen in aqueous media. These hydroxyl radicals are known to initiate oxidation. As TiO_2 reflects and scatters UVB and UVA in sunlight, nano-sized TiO_2 is used in numerous sunscreens [207]. Several studies have shown that the cytotoxicity of nano-sized TiO_2 was very low or negligible as compared with other nanoparticles [208–210], and size alone was not the effective predictor of cytotoxicity [208, 211]. In the absence of photoactivation, nanoscale TiO_2 (10 and 20 nm) in the anatase form can induce lipid peroxidation and oxidative DNA damage, and increase cellular nitric oxide and hydrogen peroxide levels in BEAS-2B, a human bronchial epithelial cell line [212, 213]. The pulmonary effects of TiO_2 particles have been well documented *in vivo*. Several investigators have shown that 20 to 30 nm TiO_2 particles cause pulmonary inflammation in laboratory animals [214–217]. Rat cells did not respond to 1 μm TiO_2 particles in suspension [218], which strongly suggested that an enhancement of biological reactivity and/or cytotoxicity occurs as the particle size decreased from the micrometer to the nanometer range [210, 219–222].

Iron Oxides

Iron in its cationic states (Fe^{2+} and Fe^{3+}) is essential for normal cell function and growth. However, chelation of intracellular iron can result in increased apoptosis and DNA fragmentation. Increases in intracellular unbound iron result in oxidative stress and injury to the cells by causing the formation of ROS, which may lead to cell death. Though iron plays an important role in virtually all living organisms—primarily through electron transport due to its ability to change valence—it has a rather limited bioavailability and, in some situations, it can be toxic to cells. For this reason, it is necessary for organisms to sequester iron in a nontoxic form. In the human body (including the brain), as well as in most organisms, iron is stored primarily in the core of the iron storage protein ferritin. The ferritin protein is a hollow spheroid shell 12 nm in diameter made up of 24 subunits, and is normally occupied by the iron biomineral ferrihydrite—a hydrated iron oxide ($5Fe_2O_3$ $9H_2O$) that generally contains only Fe (III). It is in this form that most of the iron in the body is stored [223]. It has been suggested that biogenic magnetite may be present in Alzheimer's disease plaques, senile plaques, and aberrant tau filaments extracted from progressive supranuclear palsy tissue [224]. Chronic exposure to iron nanoparticles over time is of concern. It has been thought that the particles would dissolve and superparamagnetic iron oxide signal would decrease through natural metabolic pathways [225]. It was shown that different iron oxide particles may be detected in the central nervous system (CNS) over extended periods, including strong staining in CNS parenchymal cells [225, 226].

Cerium Oxide

Cerium oxide has been widely investigated because of its multiple applications—as a catalyst, an electrolyte material of solid oxide fuel cells, a material of high refractive index, and an insulating layer on silicon substrates. The transport and uptake of industrially important cerium oxide nanoparticles into human lung fibroblasts has been measured *in vitro* after exposing thoroughly characterized particle suspensions to fibroblasts with four separate size fractions and concentrations ranging from 100 ng/g to 100 μg/g of fluid (100 ppb–100 ppm). At physiologically relevant concentrations, a strong dependence of the amount of incorporated ceria on particle size was reported, while nanoparticle number density or total particle surface area was of minor importance [227, 228].

While most of the *in vitro* studies describe the toxicity of nanoparticles, some attention has been given to some types of particles that can protect cells from various forms of lethal stress. It has been shown that nanoparticles composed of cerium oxide or yttrium oxide can protect nerve cells from oxidative stress and that the neuroprotection is independent of particle size [229]. Thus, ceria and yttria nanoparticles act as direct antioxidants to limit the amount of ROS required to kill the cells.

Copper Nanoparticles

Manufactured copper nanoparticles are industrially produced and are mainly used as inks, metallic coating, polymers/plastics, and additives in lubricants. They have been added to lubricant oil to reduce friction and wear or deposited on the surface of graphite to improve the charge-discharge property.

Fluoropolymer-conjugated copper nanoparticles have also been employed as bioactive coatings that can cause inhibition of microbial growth (*Saccharomyces cerevisiae*, *E. coli*, *Staphylococcus aureus*, and *Listeria*).

Excessive intake of copper would lead to hemolysis, jaundice, hepatocirrhosis, change in lipid profile, oxidative stress, renal dysfunction, and even death [230]. Consequently, in the human body, copper is maintained in homeostasis. Nevertheless, there is very little published data related to the toxicity of the copper nanoparticles on mammalian systems [219]. When copper nanoparticles reach the stomach, they react drastically with hydrogen ions (H^+) of gastric juices and can be quickly transformed into ionic states. This chemical process undoubtedly results in an overload of ionic copper *in vivo*. Finally, it has been shown that kidney, liver, and spleen are target organs for copper nanoparticles.

Gold Nanoparticles

Many biological applications using gold nanoparticles have emerged during the last decades. Indeed, these nanoparticles could be used as transfection vectors, DNA-binding agents, protein inhibitors, and spectroscopic markers. These nanoparticle systems, however, have not been well evaluated to determine their interactions with cells beyond the designated functions.

Cationically functionalized mixed monolayer protected gold clusters can be used to mediate DNA translocation across the cell membrane in mammalian cells at levels much higher than polyethyleneimine, a widely used transfection vector [231]. But globally, toxicity associated with the gold nanoparticles is expected to be minimal given the long history of gold as well as silver colloids used in medicines and natural therapeutics.

Quantum Dots

Quantum dots (QD) have widespread potential applications in biology and medicine, including semiconductor nanocrystals used as fluorescent probes in ultrasensitive bioassays, biological staining, and diagnostics. QD can be synthesized with different core semiconductor materials, including cadmium selenide (CdSe), cadmium telluride, indium phosphide, or indium arsenide, however, the latter three have not been shown to be useful conjugates for biological purposes. Because many QDs are made of known toxic metals such as Cd, it is not surprising that the literature contains many reports of toxicity from QDs.

Many of the studies discussed in this chapter clearly indicate that nanomaterials interact with biological systems. Composition, size, and surface property are important attributes that are needed to predict biological effects. It is intriguing to realize that although materials of vastly different chemical compositions have similar interactions with biological systems—the clearest example being preferential cellular uptake of anionic iron oxide nanoparticles and QD. Size also seems to be an important factor. The precise nature of toxicity seen is a function of the chemistry of the particle. The physical-chemical properties of many nanoparticle surface modifications may be the factor that determines their ultimate safety. However, much of the work cited here has been done using pristine nanomaterials. As described in Chapter 7, nanomaterials will be altered by the environmental and physiological systems they pass through. In subsequent chapters we explore some of the complexities that, for the time being, prevent us from fully and generally predicting nanomaterial effects from physical-chemical properties alone.

References

1. Lam, C.W., et al., "Pulmonary toxicity of single-wall carbon nanotubes in mice 7 and 90 days after intratracheal instillation." *Toxicol Sci*, 2004. **77**: p. 126–34.
2. Ho, C.-C., et al., "Quantum dot 705, a cadmium-based nanoparticle, induces persistent inflammation and granuloma formation in the mouse lung." *Nanotoxicology*, 2013. **7**(1): p. 105–115.
3. Mercer, R.R., et al., "Distribution and persistence of pleural penetrations by multi-walled carbon nanotubes." *Part Fibre Toxicol*, 2011. **7**: p. 28.
4. Murphy, F.A., et al., "Length-dependent retention of carbon nanotubes in the pleural space of mice initiates sustained inflammation and progressive fibrosis on the parietal pleura." *Am J Pathol*, 2011. **178**(6): p. 2587–600.
5. Park, E.J., et al., "Induction of chronic inflammation in mice treated with titanium dioxide nanoparticles by intratracheal instillation." *Toxicology*, 2009. **260**(1–3): p. 37–46.
6. Tabet, L., et al., "Coating carbon nanotubes with a polystyrene-based polymer protects against pulmonary toxicity." *Part Fibre Toxicol*, 2011. **8**: p. 3.
7. Tada, Y., et al., "Acute phase pulmonary responses to a single intratracheal spray instillation of magnetite [fe(3)o(4)] nanoparticles in Fischer 344 rats." *J Toxicol Pathol*, 2012. **25**(4): p. 233–9.
8. Coccini, T., et al., "Pulmonary toxicity of instilled cadmium-doped silica nanoparticles during acute and subacute stages in rats." *Histol Histopathol*, 2013. **28**(2): p. 195–209.

9. Tada, Y., et al., "Long-term pulmonary responses to quadweekly intermittent intratracheal spray instillations of magnetite (Fe3O4) nanoparticles for 52 weeks in fischer 344 rats." *J Toxicol Pathol*, 2013. **26**(4): p. 393–403.

10. Crouzier, D., et al., "Carbon nanotubes induce inflammation but decrease the production of reactive oxygen species in lung." *Toxicology*, 2010. **272**(1–3): p. 39–45.

11. Lam, C.W., et al., "A review of carbon nanotube toxicity and assessment of potential occupational and environmental health risks." *Crit Rev Toxicol*, 2006. **36**(3): p. 189–217.

12. Shvedova, A.A., et al., "Unusual inflammatory and fibrogenic pulmonary responses to single-walled carbon nanotubes in mice." *Am J Physiol Lung Cell Mol Physiol*, 2005. **289**(5): p. L698–708.

13. Muller, J., et al., "Respiratory toxicity of multi-wall carbon nanotubes." *Toxicol Appl Pharmacol*, 2005. **207**(3): p. 221–31.

14. Elgrabli, D., et al., "Induction of apoptosis and absence of inflammation in rat lung after intratracheal instillation of multiwalled carbon nanotubes." *Toxicology*, 2008. **253**: p. 131–6.

15. Mercer, R.R., et al., "Alteration of deposition pattern and pulmonary response as a result of improved dispersion of aspirated single-walled carbon nanotubes in a mouse model." *Am J Physiol Lung Cell Mol Physiol*, 2008. **294**(1): p. L87–97.

16. Chen, H.W., et al., "Titanium dioxide nanoparticles induce emphysema-like lung injury in mice." *Faseb J*, 2006. **20**(13): p. 2393–5.

17. Zhu, M.T., et al., "Particokinetics and extrapulmonary translocation of intratracheally instilled ferric oxide nanoparticles in rats and the potential health risk assessment." *Toxicol Sci*, 2009. **107**(2): p. 342–51.

18. Mossman, B.T., et al., "New insights into understanding the mechanisms, pathogenesis, and management of malignant mesotheliomas." *Am J Pathol*, 2013. **182**(4): p. 1065–77.

19. Donaldson, K., et al., "Asbestos, carbon nanotubes and the pleural mesothelium: a review of the hypothesis regarding the role of long fibre retention in the parietal pleura, inflammation and mesothelioma." *Part Fibre Toxicol*, 2010. **7**: p. 5.

20. Castranova, V., P.A. Schulte, and R.D. Zumwalde, "Occupational nanosafety considerations for carbon nanotubes and carbon nanofibers." *Acc Chem Res*, 2013. **46**(3): p. 642–9.

21. Donaldson, K., et al., "Pulmonary toxicity of carbon nanotubes and asbestos—similarities and differences." *Adv Drug Deliv Rev*, 2013. **65**(15): p. 2078–86.

22. Xu, J., et al., "Multi-walled carbon nanotubes translocate into the pleural cavity and induce visceral mesothelial proliferation in rats." *Cancer Sci*, 2012. **103**(12): p. 2045–50.

23. Ryman-Rasmussen, J.P., et al., "Inhaled carbon nanotubes reach the subpleural tissue in mice." *Nature Nano*, 2009. **4**: p. 747–51.

24. Inoue, K., et al., "Effects of multi-walled carbon nanotubes on a murine allergic airway inflammation model." *Toxicol Appl Pharmacol*, 2009. **237**(3): p. 306–16.

25. Cesta, M.F., et al., "Bacterial lipopolysaccharide enhances PDGF signaling and pulmonary fibrosis in rats exposed to carbon nanotubes." *Am J Respir Cell Mol Biol*, 2010. **43**(2): p. 142–51.

26. Shvedova, A.A., et al., "Sequential exposure to carbon nanotubes and bacteria enhances pulmonary inflammation and infectivity." *Am J Respir Cell Mol Biol*, 2008. **38**(5): p. 579–90.

27. Glista-Baker, E.E., et al., "Nickel nanoparticles cause exaggerated lung and airway remodeling in mice lacking the T-box transcription factor, TBX21 (T-bet)." *Part Fibre Toxicol*, 2014. **11**: p. 7.

28. Finotto, S., et al., "Development of spontaneous airway changes consistent with human asthma in mice lacking T-bet." *Science*, 2002. **295**(5553): p. 336–8.

29. Jonasson, S., et al., "Inhalation exposure of nano-scaled titanium dioxide (TiO₂) particles alters the inflammatory responses in asthmatic mice." *Inhal Toxicol*, 2013. **25**(4): p. 179–91.

30. Scarino, A., et al., "Impact of emerging pollutants on pulmonary inflammation in asthmatic rats: ethanol vapors and agglomerated TiO₂ nanoparticles." *Inhal Toxicol*, 2012. **24**(8): p. 528–38.

31. Jang, S., et al., "Silver nanoparticles modify VEGF signaling pathway and mucus hypersecretion in allergic airway inflammation." *Int J Nanomedicine*, 2012. **7**: p. 1329–43.

32. Park, H.S., et al., "Attenuation of allergic airway inflammation and hyperresponsiveness in a murine model of asthma by silver nanoparticles." *Int J Nanomedicine*, 2010. **5**: p. 505–15.

33. Shvedova, A.A., et al., "Mechanisms of carbon nanotube-induced toxicity: focus on oxidative stress." *Toxicol Appl Pharmacol*, 2012. **261**(2): p. 121–33.

34. Ayres, J.G., et al., "Evaluating the toxicity of airborne particulate matter and nanoparticles by measuring oxidative stress potential—a workshop report and consensus statement." *Inhal Toxicol*, 2008. **20**(1): p. 75–99.

35. Li, J.G., et al., "Comparative study of pathological lesions induced by multiwalled carbon nanotubes in lungs of mice by intratracheal instillation and inhalation." *Environ Toxicol*, 2007. **22**(4): p. 415–21.

36. Noel, A., et al., "Effects of inhaled nano-TiO₂ aerosols showing two distinct agglomeration states on rat lungs." *Toxicol Lett*, 2012. **214**(2): p. 109–19.

37. Park, E.J., et al., "Inflammatory responses may be induced by a single intratracheal instillation of iron nanoparticles in mice." *Toxicology*, 2010. **275**(1–3): p. 65–71.

38. Sarkar, A., M. Ghosh, and P.C. Sil, "Nanotoxicity: oxidative stress mediated toxicity of metal and metal oxide nanoparticles." *J Nanosci Nanotechnol*, 2014. **14**(1): p. 730–43.

39. De Angelis, I., et al., "Comparative study of ZnO and TiO₂ nanoparticles: physicochemical characterisation and toxicological effects on human colon carcinoma cells." *Nanotoxicology*, 2013. **7**(8): p. 1361–72.

40. Foldbjerg, R., D.A. Dang, and H. Autrup, "Cytotoxicity and genotoxicity of silver nanoparticles in the human lung cancer cell line, A549." *Arch Toxicol*, 2011. **85**(7): p. 743–50.

41. Shvedova, A.A., et al., "Vitamin E deficiency enhances pulmonary inflammatory response and oxidative stress induced by single-walled carbon nanotubes in C57BL/6 mice." *Toxicol Appl Pharmacol*, 2007. **221**(3): p. 339–48.

42. Bhattacharjee, S., et al., "Role of surface charge and oxidative stress in cytotoxicity of organic monolayer-coated silicon nanoparticles towards macrophage NR8383 cells." *Part Fibre Toxicol*, 2010. **7**: p. 25.

43. Li, J., et al., "Application of vitamin E to antagonize SWCNTs-induced exacerbation of allergic asthma." *Sci Rep*, 2014. **4**: p. 4275.

44. Shvedova, A.A., et al., "Increased accumulation of neutrophils and decreased fibrosis in the lung of NADPH oxidase-deficient C57BL/6 mice exposed to carbon nanotubes." *Toxicol Appl Pharmacol*, 2008. **231**(2): p. 235–40.

45. Aalapati, S., et al., "Toxicity and bio-accumulation of inhaled cerium oxide nanoparticles in CD1 mice." *Nanotoxicology*, 2014. **8**(7): p. 786–98.

46. Baisch, B.L., et al., "Equivalent titanium dioxide nanoparticle deposition by intratracheal instillation and whole body inhalation: the effect of dose rate on acute respiratory tract inflammation." *Part Fibre Toxicol*, 2014. **11**: p. 5.

47. Brown, D.M., et al., "Inflammation and gene expression in the rat lung after instillation of silica nanoparticles: effect of size, dispersion medium and particle surface charge." *Toxicol Lett*, 2014. **224**(1): p. 147–56.

48. Braakhuis, H., et al., "Physicochemical characteristics of nanomaterials that affect pulmonary inflammation." *Part Fibre Toxicol*, 2014. **11**(1): p. 18.

49. Di, Y.P., et al., "Dual acute proinflammatory and antifibrotic pulmonary effects of short palate, lung, and nasal epithelium clone-1 after exposure to carbon nanotubes." *Am J Respir Cell Mol Biol*, 2013. **49**(5): p. 759–67.

50. Blum, J.L., et al., "Short-term inhalation of cadmium oxide nanoparticles alters pulmonary dynamics associated with lung injury, inflammation, and repair in a mouse model." *Inhal Toxicol*, 2014. **26**(1): p. 48–58.

51. Lee, J.K., et al., "Multi-walled carbon nanotubes induce COX-2 and iNOS expression via MAP kinase-dependent and -independent mechanisms in mouse RAW264.7 macrophages." *Part Fibre Toxicol*, 2012. **9**: p. 14.

52. Armand, L., et al., "Titanium dioxide nanoparticles induce matrix metalloprotease 1 in human pulmonary fibroblasts partly via an interleukin-1beta-dependent mechanism." *Am J Respir Cell Mol Biol*, 2013. **48**(3): p. 354–63.

53. Abbott Chalew, T.E. and K.J. Schwab, "Toxicity of commercially available engineered nanoparticles to Caco-2 and SW480 human intestinal epithelial cells." *Cell Biol Toxicol*, 2013. **29**(2): p. 101–16.

54. Murphy, F.A., et al., "The mechanism of pleural inflammation by long carbon nanotubes: interaction of long fibres with macrophages stimulates them to amplify pro-inflammatory responses in mesothelial cells." *Part Fibre Toxicol*, 2012. **9**: p. 8.

55. Kumar, A. and A. Dhawan, "Genotoxic and carcinogenic potential of engineered nanoparticles: an update." *Arch Toxicol*, 2013. **87**(11): p. 1883–900.

56. Donaldson, K., C.A. Poland, and R.P. Schins, "Possible genotoxic mechanisms of nanoparticles: criteria for improved test strategies." *Nanotoxicology*, 2010. **4**: p. 414–20.

57. Hubbs, A.F., et al., "Nanotoxicology—a pathologist's perspective." *Toxicol Pathol*, 2011. **39**(2): p. 301–24.

58. Muller, J., et al., "Clastogenic and aneugenic effects of multi-wall carbon nanotubes in epithelial cells." *Carcinogenesis*, 2008. **29**(2): p. 427–33.

59. Trouiller, B., et al., "Titanium dioxide nanoparticles induce DNA damage and genetic instability in vivo in mice." *Cancer Res*, 2009.

60. Hackenberg, S., et al., "Silver nanoparticles: evaluation of DNA damage, toxicity and functional impairment in human mesenchymal stem cells." *Toxicol Lett*, 2011. **201**(1): p. 27–33.

61. Di Virgilio, A.L., et al., "Comparative study of the cytotoxic and genotoxic effects of titanium oxide and aluminium oxide nanoparticles in Chinese hamster ovary (CHO-K1) cells." *J Hazard Mater*, 2010. **177**(1–3): p. 711–8.

62. An, H., et al., "DNA binding and aggregation by carbon nanoparticles." *Biochem Biophys Res Commun*, **In Press, Uncorrected Proof**.

63. Jin, P., et al., "Interactions between Al(1)(2)X (X = Al, C, N and P) nanoparticles and DNA nucleobases/base pairs: implications for nanotoxicity." *J Mol Model*, 2012. **18**(2): p. 559–68.

64. Kain, J., H.L. Karlsson, and L. Moller, "DNA damage induced by micro- and nanoparticles—interaction with FPG influences the detection of DNA oxidation in the comet assay." *Mutagenesis*, 2012. **27**(4): p. 491–500.

65. Baweja, L., et al., "C60-fullerene binds with the ATP binding domain of human DNA topoiosmerase II alpha." *J Biomed Nanotechnol*, 2011. **7**(1): p. 177–8.

66. Gupta, S.K., et al., "Interaction of C60 fullerene with the proteins involved in DNA mismatch repair pathway." *J Biomed Nanotechnol*, 2011. **7**(1): p. 179–80.

67. Jugan, M.L., et al., "Titanium dioxide nanoparticles exhibit genotoxicity and impair DNA repair activity in A549 cells." *Nanotoxicology*, 2012. **6**(5): p. 501–13.

68. Wirnitzer, U., et al., "Studies on the in vitro genotoxicity of baytubes, agglomerates of engineered multi-walled carbon-nanotubes (MWCNT)." *Toxicol Lett*, 2009. **186**(3): p. 160–5.

69. Guo, Y.Y., et al., "Cytotoxic and genotoxic effects of multi-wall carbon nanotubes on human umbilical vein endothelial cells in vitro." *Mutat Res*, 2011. **721**(2): p. 184–91.

70. Sharma, V., D. Anderson, and A. Dhawan, "Zinc oxide nanoparticles induce oxidative DNA damage and ROS-triggered mitochondria mediated apoptosis in human liver cells (HepG2)." *Apoptosis*, 2012. **17**(8): p. 852–70.

71. Nakahira, K. and A.M. Choi, "Autophagy: a potential therapeutic target in lung diseases." *Am J Physiol Lung Cell Mol Physiol*, 2013. **305**(2): p. L93–107.

72. Levine, B. and G. Kroemer, "Autophagy in the pathogenesis of disease." *Cell*, 2008. **132**: p. 27–42.

73. Ryter, S.W., et al., "Autophagy in pulmonary diseases." *Annu Rev Physiol*, 2012.

74. Farombi, E.O., "Genotoxicity of chloroquine in rat liver cells: protective role of free radical scavengers." *Cell Biol Toxicol*, 2006. **22**(3): p. 159–67.

75. Hara, T., et al., "Suppression of basal autophagy in neural cells causes neurodegenerative disease in mice." *Nature*, 2006. **441**(7095): p. 885–9.

76. Park, E.J., et al., "Oxidative stress and apoptosis induced by titanium dioxide nanoparticles in cultured BEAS-2B cells." *Toxicol Lett*, 2008. **180**(3): p. 222–9.

77. Park, J., et al., "Reactive oxygen species mediate chloroquine-induced expression of chemokines by human astroglial cells." *Glia*, 2004. **47**(1): p. 9–20.

78. Yamasaki, R., et al., "Involvement of lysosomal storage-induced p38 MAP kinase activation in the overproduction of nitric oxide by microglia in cathepsin D-deficient mice." *Mol Cell Neurosci*, 2007. **35**(4): p. 573–84.

79. Kaushik, S. and A.M. Cuervo, "Autophagy as a cell-repair mechanism: activation of chaperone-mediated autophagy during oxidative stress." *Mol Aspects Med*, 2006. **27**(5–6): p. 444–54.

80. Kiffin, R., et al., "Activation of chaperone-mediated autophagy during oxidative stress." *Mol Biol Cell*, 2004. **15**(11): p. 4829–40.

81. Jain, A., et al., "p62/SQSTM1 is a target gene for transcription factor NRF2 and creates a positive feedback loop by inducing antioxidant response element-driven gene transcription." *J Biol Chem*, 2010. **285**(29): p. 22576–91.

82. Komatsu, M., et al., "The selective autophagy substrate p62 activates the stress responsive transcription factor Nrf2 through inactivation of Keap1." *Nat Cell Biol*, 2010. **12**(3): p. 213–23.

83. Nakahira, K., et al., "Autophagy proteins regulate innate immune responses by inhibiting the release of mitochondrial DNA mediated by the NALP3 inflammasome." *Nat Immunol*, 2011. **12**(3): p. 222–30.

84. Zhou, R., et al., "A role for mitochondria in NLRP3 inflammasome activation." *Nature*, 2011. **469**(7329): p. 221–5.

85. Harris, J., et al., "Autophagy controls IL-1beta secretion by targeting pro-IL-1beta for degradation." *J Biol Chem*, 2011. **286**(11): p. 9587–97.

86. Shi, C.S., et al., "Activation of autophagy by inflammatory signals limits IL-1beta production by targeting ubiquitinated inflammasomes for destruction." *Nat Immunol*, 2012. **13**(3): p. 255–63.

87. Ling, J., et al., "KrasG12D-induced IKK2/beta/NF-kappaB activation by IL-1alpha and p62 feedforward loops is required for development of pancreatic ductal adenocarcinoma." *Cancer Cell*, 2012. **21**(1): p. 105–20.

88. Mathew, R., et al., "Autophagy suppresses tumorigenesis through elimination of p62." *Cell*, 2009. **137**(6): p. 1062–75.

89. Moscat, J. and M.T. Diaz-Meco, "p62: a versatile multitasker takes on cancer." *Trends Biochem Sci*, 2012. **37**(6): p. 230–6.

90. Consortium., W.T.C.C "Genome-wide association study of 14,000 cases of seven common diseases and 3,000 shared controls." *Nature*, 2007. **447**(7145): p. 661–78.

91. Henckaerts, L., et al., "Genetic variation in the autophagy gene ULK1 and risk of Crohn's disease." *Inflamm Bowel Dis*, 2011. **17**(6): p. 1392–7.

92. Li, C., et al., "PAMAM nanoparticles promote acute lung injury by inducing autophagic cell death through the Akt-TSC2-mTOR signaling pathway." *J Mol Cell Biol*, 2009. **1**(1): p. 37–45.

93. Lee, Y.H., et al., "Cytotoxicity, oxidative stress, apoptosis and the autophagic effects of silver nanoparticles in mouse embryonic fibroblasts." *Biomaterials*, 2014. **35**(16): p. 4706–15.

94. Parker, M.W., et al., "Fibrotic extracellular matrix activates a profibrotic positive feedback loop." *J Clin Invest*, 2014. **124**(4): p. 1622–35.

95. Chen, R., et al., "Formation and cell translocation of carbon nanotube-fibrinogen protein corona." *Appl Phys Lett*, 2012. **101**(13): p. 133702.

96. Verma, N.K., et al., "Autophagy induction by silver nanowires: a new aspect in the biocompatibility assessment of nano-composite thin films." *Toxicol Appl Pharmacol*, 2012. **264**(3): p. 451–61.

97. Roy, R., et al., "ZnO nanoparticles induced adjuvant effect via toll-like receptors and Src signaling in Balb/c mice." *Toxicol Lett*, 2014. **230**(3): p. 421–33.

98. Khan, M.I., et al., "Induction of ROS, mitochondrial damage and autophagy in lung epithelial cancer cells by iron oxide nanoparticles." *Biomaterials*, 2012. **33**(5): p. 1477–88.

99. Ma, X., et al., "Gold nanoparticles induce autophagosome accumulation through size-dependent nanoparticle uptake and lysosome impairment." *ACS Nano*, 2011. **5**(11): p. 8629–39.

100. Orecna, M., et al., "Toxicity of carboxylated carbon nanotubes in endothelial cells is attenuated by stimulation of the autophagic flux with the release of nanomaterial in autophagic vesicles." *Nanomedicine*, 2014. **10**(5): p. 939–48.

101. Wan, B., et al., "Single-walled carbon nanotubes and graphene oxides induce autophagosome accumulation and lysosome impairment in primarily cultured murine peritoneal macrophages." *Toxicol Lett*, 2013. **221**(2): p. 118–27.

102. Sun, T., et al., "Copper oxide nanoparticles induce autophagic cell death in A549 cells." *PLoS One*, 7(8): p. e43442.

103. Lynch, I., et al., "The nanoparticle-protein complex as a biological entity; a complex fluids and surface science challenge for the 21st century." *Adv Colloid Interface Sci*, 2007. **134–135**: p. 167–74.

104. Shannahan, J.H., et al., "Comparison of Nanotube-Protein Corona Composition in Cell Culture Media." *Small*, 2013. **9**: p. 2171–81.

105. Podila, R., et al., "Evidences for charge transfer-induced conformational changes in carbon nanostructure-protein corona." *J Phys Chem C Nanomater Interfaces*, 2012. **116**(41): p. 22098–103.

106. Walkey, C.D. and W.C. Chan, "Understanding and controlling the interaction of nanomaterials with proteins in a physiological environment." *Chem Soc Rev*, 2012. **41**(7): p. 2780–99.

107. Cedervall, T., et al., "Detailed identification of plasma proteins adsorbed on copolymer nanoparticles." *Angew Chem Int Ed Engl*, 2007. **46**(30): p. 5754–6.

108. Lundqvist, M., et al., "Nanoparticle size and surface properties determine the protein corona with possible implications for biological impacts." *Proc Natl Acad Sci USA*, 2008. **105**: p. 14265–70.

109. Nel, A., et al., "Understanding biophysicochemical interactions at the nano-bio interface." *Nat Mater*, 2009. **8**: p. 543–57.

110. Johnston, H., et al., "Investigating the relationship between nanomaterial hazard and physicochemical properties: informing the exploitation of nanomaterials within therapeutic and diagnostic applications." *J Control Release*, 2012. **164**(3): p. 307–13.

111. Ge, C., et al., "Binding of blood proteins to carbon nanotubes reduces cytotoxicity." *Proc Natl Acad Sci USA*, 2011. **108**(41): p. 16968–73.

112. Cai, X., et al., "Characterization of carbon nanotube protein corona by using quantitative proteomics." *Nanomedicine*, 2012. **9**: p. 583–93.

113. Karajanagi, S.S., et al., "Structure and function of enzymes adsorbed onto single-walled carbon nanotubes." *Langmuir*, 2004. **20**(26): p. 11594–9.

114. Salvador-Morales, C., et al., "Complement activation and protein adsorption by carbon nanotubes." *Mol Immunol*, 2006. **43**(3): p. 193–201.

115. Salvador-Morales, C., et al., "Binding of pulmonary surfactant proteins to carbon nanotubes; potential for damage to lung immune defense mechanisms." *Carbon*, 2007. **45**(3): p. 607–17.

116. Sanfins, E., et al., "Size-dependent effects of nanoparticles on enzymes in the blood coagulation cascade." *Nano Lett*, 2014. **14**(8): p. 4736–44.

117. Zhang, B., et al., "Functionalized carbon nanotubes specifically bind to alpha-chymotrypsin's catalytic site and regulate its enzymatic function." *Nano Lett*, 2009. **9**(6): p. 2280–4.

118. Kapralov, A.A., et al., "Adsorption of surfactant lipids by single-walled carbon nanotubes in mouse lung upon pharyngeal aspiration." *ACS Nano*, 2012. **6**(5): p. 4147–56.

119. Banerjee, R. and D. Dhara, "Functional group-dependent self-assembled nanostructures from thermo-responsive triblock copolymers." *Langmuir*, 2014. **30**(14): p. 4137–46.

120. Zhang, Q., et al., "Autophagy-mediated chemosensitization in cancer cells by fullerene C60 nanocrystal." *Autophagy*, 2009. **5**(8): p. 1107–17.

121. Treuel, L., et al., "Impact of protein modification on the protein corona on nanoparticles and nanoparticle-cell interactions." *ACS Nano*, 2014. **8**(1): p. 503–13.

122. Konduru, N.V., et al., "Phosphatidylserine targets single-walled carbon nanotubes to professional phagocytes in vitro and in vivo." *PLoS One*, 2009. **4**(2): p. e4398.

123. Saptarshi, S., A. Duschl, and A. Lopata, "Interaction of nanoparticles with proteins: relation to bio-reactivity of the nanoparticle." *J Nanobiotech*, 2013. **11**(1): p. 26.

124. Deng, X., et al., "Translocation and fate of multi-walled carbon nanotubes in vivo." *Carbon*, 2007. **45**: p. 1419–24.

125. Elgrabli, D., et al., "Biodistribution and clearance of instilled carbon nanotubes in rat lung." *Part Fibre Toxicol*, 2008. **5**: p. 20.

126. Muller, J., et al., "Absence of carcinogenic response to multiwall carbon nanotubes in a 2-year bioassay in the peritoneal cavity of the rat." *Toxicol Sci*, 2009. **110**(2): p. 442–8.

127. Raffa, V., et al., "Physicochemical properties affecting cellular uptake of carbon nanotubes." *Nanomedicine (Lond)*, 2010. **5**(1): p. 89–97.

128. Bussy, C., et al., "Carbon nanotubes in macrophages: imaging and chemical analysis by X-ray fluorescence microscopy." *Nano Lett*, 2008. **8**: p. 2659–63.

129. Allen, B.L., et al., "Biodegradation of single-walled carbon nanotubes through enzymatic catalysis." *Nano Lett*, 2008. **8**(11): p. 3899–903.

130. Allen, B.L., et al., "Mechanistic investigations of horseradish peroxidase-catalyzed degradation of single-walled carbon nanotubes." *J Am Chem Soc*, 2009. **131**(47): p. 17194–205.

131. Russier, J., et al., "Oxidative biodegradation of single- and multi-walled carbon nanotubes." *Nanoscale*, 2011. **3**: p. 893–6.

132. Kagan, V.E., et al., "Carbon nanotubes degraded by neutrophil myeloperoxidase induce less pulmonary inflammation." *Nat Nano*, 2010. **5**(5): p. 354–9.

133. Nunes, A., et al., "In vivo degradation of functionalized carbon nanotubes after stereotactic administration in the brain cortex." *Nanomedicine (Lond)*, 2012. **7**(10): p. 1485–94.

134. Kotchey, G.P., et al., "Effect of antioxidants on enzyme-catalysed biodegradation of carbon nanotubes." *J Mater Chem B Mater Biol Med*, 2013. **1**(3): p. 302–9.

135. Bussy, C., et al., "Intracellular fate of carbon nanotubes inside murine macrophages: pH-dependent detachment of iron catalyst nanoparticles." *Part Fibre Toxicol*, 2013. **10**(1): p. 24.

136. Monastyrska, I., et al., "Multiple roles of the cytoskeleton in autophagy." *Biol Rev Camb Philos Soc*, 2009. **84**(3): p. 431–48.

137. Aplin, A., et al., "Cytoskeletal elements are required for the formation and maturation of autophagic vacuoles." *J Cell Physiol*, 1992. **152**(3): p. 458–66.

138. Seglen, P.O., et al., "Structural aspects of autophagy." *Adv Exp Med Biol*, 1996. **389**: p. 103–11.

139. Jahreiss, L., F.M. Menzies, and D.C. Rubinsztein, "The itinerary of autophagosomes: from peripheral formation to kiss-and-run fusion with lysosomes." *Traffic*, 2008. **9**(4): p. 574–87.

140. Young, A.R., et al., "Starvation and ULK1-dependent cycling of mammalian Atg9 between the TGN and endosomes." *J Cell Sci*, 2006. **119**(Pt 18): p. 3888–900.

141. Choudhury, D., et al., "Unprecedented inhibition of tubulin polymerization directed by gold nanoparticles inducing cell cycle arrest and apoptosis." *Nanoscale*, 2013. **5**(10): p. 4476–89.

142. Gheshlaghi, Z.N., et al., "Toxicity and interaction of titanium dioxide nanoparticles with microtubule protein." *Acta Biochim Biophys Sin (Shanghai)*, 2008. **40**(9): p. 777–82.

143. Ratnikova, T.A., et al., "In vitro polymerization of microtubules with a fullerene derivative." *ACS Nano*, 2011. **5**(8): p. 6306–14.

144. Shams, H., et al., "Actin reorganization through dynamic interactions with single-wall carbon nanotubes." *ACS Nano*, 2014. **8**(1): p. 188–97.

145. Pernodet, N., et al., "Adverse effects of citrate/gold nanoparticles on human dermal fibroblasts." *Small*, 2006. **2**(6): p. 766–73.

146. Mironava, T., et al., "Gold nanoparticles cellular toxicity and recovery: effect of size, concentration and exposure time." *Nanotoxicology*, 2010. **4**(1): p. 120–37.

147. Pisanic, T.R., 2nd, et al., "Nanotoxicity of iron oxide nanoparticle internalization in growing neurons." *Biomaterials*, 2007. **28**(16): p. 2572–81.

148. Dadras, A., et al., "In vitro study on the alterations of brain tubulin structure and assembly affected by magnetite nanoparticles." *J Biol Inorg Chem*, 2013. **18**(3): p. 357–69.

149. Wu, Y.T., et al., "Dual role of 3-methyladenine in modulation of autophagy via different temporal patterns of inhibition on class I and III phosphoinositide 3-kinase." *J Biol Chem*, 2010. **285**(14): p. 10850–61.

150. Johnson-Lyles, D.N., et al., "Fullerenol cytotoxicity in kidney cells is associated with cytoskeleton disruption, autophagic vacuole accumulation, and mitochondrial dysfunction." *Toxicol Appl Pharmacol*, 2011. **248**(3): p. 249–58.

151. Hussain, S., et al., "Carbon black and titanium dioxide nanoparticles elicit distinct apoptotic pathways in bronchial epithelial cells." *Part Fibre Toxicol*, 2010. **7**(1): p. 10.

152. Jin, C.Y., et al., "Cytotoxicity of titanium dioxide nanoparticles in mouse fibroblast cells." *Chem Res Toxicol*, 2008. **21**: p. 1871.

153. Sohaebuddin, S.K., et al., "Nanomaterial cytotoxicity is composition, size, and cell type dependent." *Part Fibre Toxicol*, 2010. **7**: p. 22.

154. Thomas, T.P., et al., "Cationic poly(amidoamine) dendrimer induces lysosomal apoptotic pathway at therapeutically relevant concentrations." *Biomacromolecules*, 2009. **10**(12): p. 3207–14.

155. Cho, W., et al., "Progressive severe lung injury by zinc oxide nanoparticles; the role of Zn2+ dissolution inside lysosomes." *Part Fibre Toxicol*, 2011. **8**: p. 27.

156. Hamilton, R.F., Jr., et al., "Particle length-dependent titanium dioxide nanomaterials' toxicity and bioactivity." *Part Fibre Toxicol*, 2009. **6**(1): p. 35.

157. Lunov, O., et al., "Differential uptake of functionalized polystyrene nanoparticles by human macrophages and a monocytic cell line." *ACS Nano*, 2011. **5**(3): p. 1657–69.

158. Hamzeh, M. and G.I. Sunahara, "In vitro cytotoxicity and genotoxicity studies of titanium dioxide (TiO$_2$) nanoparticles in Chinese hamster lung fibroblast cells." *Toxicol In Vitro*, 2013. **27**(2): p. 864–73.

159. Donaldson, K. and P. Borm, "Interactions between macrophages and epithelial cells." *Particle Toxicology*, 2006: p. 190.

160. Lanone, S., et al., "Comparative toxicity of 24 manufactured nanoparticles in human alveolar epithelial and macrophage cell lines." *Part Fibre Toxicol*, 2009. **6**(14): p. 14.

161. Tian, F., et al., "Cytotoxicity of single-wall carbon nanotubes on human fibroblasts." *Toxicol In Vitro*, 2006. **20**(7): p. 1202–12.

162. Petkovic, J., et al., "DNA damage and alterations in expression of DNA damage responsive genes induced by TiO$_2$ nanoparticles in human hepatoma HepG2 cells." *Nanotoxicology*, 2011. **5**(3): p. 341–53.

163. Bussy, C., et al., "Critical role of surface chemical modifications induced by length shortening on multi-walled carbon nanotubes-induced toxicity." *Part Fibre Toxicol*, 2013. **9**: p. 46.

164. Kagan, V.E., et al., "Direct and indirect effects of single walled carbon nanotubes on RAW 264.7 macrophages: role of iron." *Toxicol Lett*, 2006. **165**: p. 88–100.

165. Shvedova, A.A., et al., "Exposure to carbon nanotube material: assessment of nanotube cytotoxicity using human keratinocyte cells." *J Toxicol Environ Health A*, 2003. **66**(20): p. 1909–26.

166. Patnaik, A., "Structure and dynamics in self-organized C60 fullerenes." *J Nanosci Nanotechnol*, 2007. **7**(4–5): p. 1111–50.

167. Kreyling, W., et al., "Size dependence of the translocation of inhaled iridium and carbon nanoparticle aggregates from the lung of rats to the blood and secondary target organs." *Inhal Toxicol*, 2009. **21**(Suppl 1): p. 55–60.

168. Auffan, M., et al., "Chemical stability of metallic nanoparticles: a parameter controlling their potential cellular toxicity in vitro." *Environ Pollut*, 2009. **157**(4): p. 1127–33.

169. Sayes, C.M., et al., "The differential cytotoxicity of water-soluble fullerenes." *Nano Lett*, 2004. **4**(10): p. 1881–7.

170. Kovochich, M., et al., "Comparative toxicity of C60 aggregates towards mammalian cells: role of the tetrahydrofuran (THF) decomposition." *Environ Sci Technol*, 2009. **43**(16): p. 6378–84.

171. Lin, A.M., et al., "Carboxyfullerene prevents iron-induced oxidative stress in rat brain." *J Neurochem*, 1999. **72**(4): p. 1634–40.

172. Yamago, S., et al., "In-vivo biological behavior of a water-miscible fullerene—C-14 labeling, absorption, distribution, excretion and acute toxicity." *Chem Biol*, 1995. **2**(6): p. 385–9.

173. Ebbesen, T. and P. Ajayan, "Large-scale synthesis of carbon nanotubes." *Nature*, 1992. **358**(6383): p. 220–2.

174. Thess, A., et al., "Crystalline ropes of metallic carbon nanotubes." *Science-AAAS-Weekly Paper Edition*, 1996. **273**(5274): p. 483–7.

175. Willems, I., et al., "Control of the outer diameter of thin carbon nanotubes synthesized by catalytic decomposition of hydrocarbons." *Chem Phys Lett*, 2000. **317**(1): p. 71–6.

176. Nelson, M., et al., "Effects of acute and subchronic exposure of topically applied fullerene extracts on the mouse skin." *Toxicol Ind Health*, 1992. **9**(4): p. 623–30.

177. Rajagopalan, P., et al., "Pharmacokinetics of a water-soluble fullerene in rats." *Antimicrob Agents Chemother*, 1996. **40**(10): p. 2262–5.

178. Qingnuan, L., et al., "Preparation of (99m)Tc-C(60)(OH)(x) and its biodistribution studies." *Nucl Med Biol*, 2002. **29**(6): p. 707–10.

179. Ueng, T.-H., et al., "Suppression of microsomal cytochrome P450-dependent monooxygenases and mitochondrial oxidative phosphorylation by fullerenol, a polyhydroxylated fullerene C60." *Toxicol Lett*, 1997. **93**(1): p. 29–37.

180. Friedman, S.H., et al., "Inhibition of the HIV-1 protease by fullerene derivatives: model building studies and experimental verification." *J Am Chem Soc*, 1993. **115**(15): p. 6506–9.

181. Tsuchiya, T., et al., "Novel harmful effects of [60] fullerene on mouse embryos in vitro and in vivo." *FEBS Lett*, 1996. **393**(1): p. 139–45.

182. Tsuchiya, T., Y.N. Yamakoshi, and N. Miyata, "A novel promoting action of fullerene C60 on the chondrogenesis in rat embryonic limb bud cell culture system." *Biochem Biophys Res Commun*, 1995. **206**(3): p. 885–94.

183. Rouse, J.G., et al., "Fullerene-based amino acid nanoparticle interactions with human epidermal keratinocytes." *Toxicol In Vitro*, 2006. **20**(8): p. 1313–20.

184. Warheit, D.B., et al., "Comparative pulmonary toxicity assessment of single-wall carbon nanotubes in rats." *Toxicol Sci*, 2004. **77**: p. 117–25.

185. Allen, D., J. Riviere, and N. Monteiro-Riviere, "Cytokine induction as a measure of cutaneous toxicity in primary and immortalized porcine keratinocytes exposed to jet fuels, and their relationship to normal human epidermal keratinocytes." *Toxicol Lett*, 2001. **119**(3): p. 209–17.

186. Cunningham, M.J., S.R. Magnuson, and M.T. Falduto, "Gene expression profiling of nanoscale materials using a systems biology approach." *Toxicol Sci*, 2005. **84**(S-1): p. 9.

187. Cui, D., et al., "Effect of single wall carbon nanotubes on human HEK293 cells." *Toxicol Lett*, 2005. **155**: p. 73–85.

188. Kam, N.W.S., et al., "Nanotube molecular transporters: internalization of carbon nanotube-protein conjugates into mammalian cells." *J Am Chem Soc*, 2004. **126**: p. 6850–1.

189. Sayes, C.M., et al., "Functionalized density dependence of single-walled carbon nanotubes cytotoxicity in vitro." *Toxicol Lett*, 2006. **161**(2): p. 135–42.

190. Manna, S.K., et al., "Single-walled carbon nanotube induces oxidative stress and activates nuclear transcription factor-κB in human keratinocytes." *Nano Lett*, 2005. **5**(9): p. 1676–84.

191. Monteiro-Riviere, N.A., et al., "Multi-walled carbon nanotube interactions with human epidermal keratinocytes." *Toxicol Lett*, 2005. **155**(3): p. 377–84.

192. Monteiro-Riviere, N.A., et al., "Proteomic analysis of nanoparticle exposure in human keratinocyte cell culture." *Toxicol Sci*, 2005. **84**(S-1): p. 447.

193. Monteiro-Riviere, N.A. and A.O. Inman, "Challenges for assessing carbon nanomaterial toxicity to the skin." *Carbon*, 2006. **44**(6): p. 1070–8.

194. Muller, J., F. Huaux, and D. Lison, "Respiratory toxicity of carbon nanotubes: how worried should we be?" *Carbon*, 2006. **44**(6): p. 1048–56.

195. Muller, J., et al., "Respiratory toxicity of multi-wall carbon nanotubes." *Toxicol Appl Pharmacol*, 2005. **207**(3): p. 221–31.

196. Jia, G., et al., "Cytotoxicity of carbon nanomaterials: single-wall nanotube, multi-wall nanotube, and fullerene." *Environ Sci Technol*, 2005. **39**: p. 1378–83.

197. Yokoyama, A., et al., "Biological behavior of hat-stacked carbon nanofibers in the subcutaneous tissue in rats." *Nano Lett*, 2005. **5**(1): p. 157–61.

198. Bottini, M., et al., "Multi-walled carbon nanotubes induce T lymphocyte apoptosis." *Toxicol Lett*, 2006. **160**(2): p. 121–6.

199. Bernard, B.K., et al., "Toxicology and carcinogenesis studies of dietary titanium dioxide-coated mica in male and female Fischer 344 rats." *J Toxicol Environ Health*, 1990. **29**(4): p. 417–29.

200. Chen, J.L. and W.E. Fayerweather, "Epidemiologic study of workers exposed to titanium dioxide," *J Occup Med*, 1988. **30**(12): p. 937–42.

201. Hart, G.A. and T.W. Hesterberg, "In vitro toxicity of respirable-size particles of diatomaceous earth and crystalline silica compared with asbestos and titanium dioxide." *J Occup Environ Med*, 1998. **40**(1): p. 29–42.

202. Lindenschmidt, R.C., et al., "The comparison of a fibrogenic and two nonfibrogenic dusts by bronchoalveolar lavage." *Toxicol Appl Pharmacol*, 1990. **102**(2): p. 268–81.

203. Ophus, E.M., et al., "Analysis of titanium pigments in human lung tissue." *Scand J Work Environ Health*, 1979. **5**: p. 290–6.

204. Lomer, M.C., R.P. Thompson, and J.J. Powell, "Fine and ultrafine particles of the diet: influence on the mucosal immune response and association with Crohn's disease." *Proc Nutr Soc*, 2002. **61**(1): p. 123–30.

205. Nordman, H. and M. Berlin, "Titanium," in *Handbook on the Toxicology of Metals, vol. II.*, L. Friberg, G.F. Nordberg, and V.B. Vouk, Editors. 1986, Elsevier: Amsterdam. p. 595–609.

206. Gelis, C., et al., "Assessment of the skin photoprotective capacities of an organo-mineral broad-spectrum sunblock on two ex vivo skin models." *Photodermatol Photoimmunol Photomed*, 2003. **19**(5): p. 242–53.

207. Dunford, R., et al., "Chemical oxidation and DNA damage catalysed by inorganic sunscreen ingredients." *FEBS Lett*, 1997. **418**(1–2): p. 87–90.

208. Peters, K., et al., "Effects of nanoscaled particles on endothelial cell function in vitro: studies on viability, proliferation and inflammation." *J Mater Sci Mater Med*, 2004. **15**: p. 321–5.

209. Yamamoto, A., et al., "Cytotoxicity evaluation of ceramic particles of different sizes and shapes." *J Biomed Mater Res A*, 2004. **68A**(2): p. 244–56.

210. Zhang, Q., et al., "Differences in the extent of inflammation caused by intratracheal exposure to three ultrafine metals: role of free radicals." *J Toxicol Environ Health*, 1998. **53**(6): p. 423–38.

211. Rahman, Q., et al., "Evidence that ultrafine titanium dioxide induces micronuclei and apoptosis in Syrian hamster embryo fibroblasts." *Environ Health Perspect*, 2002. **110**(8): p. 797–800.

212. Long, T., et al. "Metal oxide nanoparticles produce oxidative stress in CNS microglia and neurons." in *Society of Toxicology Annual Meeting*. 2006. San Diego, CA.

213. Gurr, J.-R., et al., "Ultrafine titanium dioxide particles in the absence of photoactivation can induce oxidative damage to human bronchial epithelial cells." *Toxicology*, 2005. 213(1-2): p. 66–73.

214. Ferin, J., et al., "Increased pulmonary toxicity of ultrafine particles? 1. Particles clearance, translocation, morphology." *J Aerosol Sci*, 1990. **21**: p. 384–7.

215. Ferin, J., G. Oberdoerster, and D.P. Penney, "Pulmonary retention of ultrafine and fine particles in rats." *Am J Resp Cell Mol Biol*, 1992. **6**: p. 535–42.

216. Oberdörster, G., J. Ferin, and B.E. Lehnert, "Correlation between particle size, in vivo particle persistence, and lung injury." *Environ Health Perspect*, 1994. **102**(suppl 5): p. 173–9.

217. Oberdoerster, G., et al., "Association of particulate air-pollution and acute mortality-involvement of ultrafine particles." *Inhahalational Toxicol*, 1995. **7**(1): p. 111–24.

218. Stringer, B.K., A. Imrich, and L. Kobzik, "Lung epithelial cells (A549) interaction with unopsonized environmental particulates: quantitation of particle-specific binding and IL-8 production." *Exp Lung Res*, 1996. **22**: p. 495–508.

219. Cheng, M.D., "Effects of nanophase materials (≤20 nm) on biological responses." *J Environ Sci Health*, 2004. **A39**: p. 2691–705.

220. Rehn, B., et al., "Investigations on the inflammatory and genotoxic lung effects of two types of titanium dioxide: untreated and surface treated." *Toxicol Appl Pharmacol*, 2003. **189**(2): p. 84–95.
221. Tsuji, J.S., et al., "Research strategies for safety evaluation of nanomaterials, Part IV: risk assessment of nanoparticles." *Toxicol Sci*, 2006. **89**(1): p. 42–50.
222. Warheit, D.B., et al., "Pulmonary instillation studies with nanoscale TiO_2 rods and dots in rats: toxicity is not dependent upon particle size and surface area." *Toxicol Sci*, 2006. **91**(1): 227–236.
223. Dobson, J., "Magnetic iron compounds in neurological disorders." *Ann NY Acad Sci*, 2004. **1012**: p. 183–92.
224. Quintana, C., et al., "Initial studies with high resolution TEM and electron energy loss spectroscopy studies of ferritin cores extracted from brains of patients with progressive supranuclear palsy and Alzheimer disease." *Cell Mol Biol*, 2000. **46**: p. 807–20.
225. Muldoon, L.L., et al., "Imaging, distribution, and toxicity of superparamagnetic iron oxide magnetic nanoparticles in the rat brain and cerebral tumor." *Neurosurgery*, 2005. **57**: p. 785–96.
226. Kim, J.S., et al., "Toxicity and tissue distribution of magnetic nanoparticles in mice." *Toxicol Sci*, 2006. **89**: p. 338–47.
227. Brunner, T., et al., "In vitro cytotoxicity of oxide nanoparticles: comparison to asbestos, silica, and the effect of particle solubility." *Environ Sci Technol*, 2006. **40**(14), p. 4374–81.
228. Limbach, L.K., et al., "Oxide nanoparticle uptake in human lung fibroblasts: effects of particle size, agglomeration, and diffusion at low concentrations." *Environ Sci Technol*, 2005. **39**(23): p. 9370–6.
229. Schubert, D., et al., "Cerium and yttrium oxide nanoparticles are neuroprotective." *Biochem Biophys Res Commun*, 2006. **342**: p. 86–91.
230. Björn, P.Z., et al., "Epidemiological investigation on chronic copper toxicity to children exposed via the public drinking water supply." *Sci Total Environ*, 2003. **302**: p. 127–44.
231. Sandhu, K.K., et al., "Gold nanoparticle-mediated transfection of mammalian cells." *Bioconjugate Chem*, 2002. **13**: p. 3–6.
232. Oberdörster, G., et al., "Role of the alveolar macrophage in lung injury: studies with ultrafine particles." *Environ Health Perspect*, 1992. **97**: p. 193–9.
233. Rahman, Q., J. Norwood, and G. Hatch, "Evidence that exposure of particulate air pollutants to human and rat alveolar macrophages lead to different oxidative stress." *Biochem Biophys Res Commun*, 1997. **240**: p. 669–72.
234. Wilhelm, C., et al., "Intracellular uptake of anionic superparamagnetic nanoparticles as a function of their surface coating." *Biomaterials*, 2003. **24**: p. 1001–11.
235. Berry, C., et al., "Dextran and albumin derivatized iron oxide nanoparticles: influence on fibroblasts in vitro." *Biomaterials*, 2003. **24**: p. 4551–7.
236. Berry, C., et al., "The influence of transferrin stabilised magnetic nanoparticles on human dermal fibroblasts in culture." *Int J Pharm*, 2004. **269**: p. 211–25.
237. Berry, C., et al., "Cell response to dextran-derivatized iron oxide nanoparticles post internalisation." *Biomaterials*, 2004. **25**(23): p. 5405–13.
238. Ferrucci, J.T. and D.D. Stark, "Iron oxide-enhanced MR imaging of the liver and spleen: review of the first 5 years." *Am J Roentgenol*, 1990. **155**: p. 943–50.
239. Pouliquen, D., et al., "Iron oxide nanoparticles for use as an MRI contrast agent: pharmacokinetics and metabolism." *Magn Reson Imaging*, 1991. **9**: p. 275–83.
240. Cheng, M.D., "Effects of nanophase materials (≤20 nm) on biological responses." *J Environ Sci Health*, 2004. **A39**: p. 2691–705.
241. Goodman, C., et al., "Toxicity of gold nanoparticles functionalized with cationic and anionic side chains." *Bioconjugate Chem*, 2004. **15**: p. 897–900.
242. de la Feunte, J.M., et al., "Nanoprticle targeting at cells." *Langmuir*, 2006. **22**: p. 3286–93.
243. Ryman-Rasmussen, J.P., J.E. Riviere, and N.A. Monteiro-Riviere, "Surfacecoatings determine cytotoxicity and irritation potential of quantum dot nanopaticles in epidermal keratinocytes," *J Invest Dermatol*, 2007. **127**: p. 143–53.
244. Bruchez, M., et al., "Semiconductor nanocrystals as fluorescent biological labels." *Science*, 1998. **281**: p. 2013–6.
245. Zhang, T., et al., "Cellular effect of high doses of silica-coated quantum dot profiled with high throughput gene expression analysis and high content cellomics measurements." *Nano Letters*, 2006. **6**(4), p. 800–808.
246. Derfus, A.M., W.C.W. Chan, and S. Bhatia, "Probing the cytotoxicity of semiconductor quantum dots." *Nano Letters*, 2004. **4**(1): p. 11–8.
247. Shiosahara, A., et al., "On the cytotoxicity caused by quantum dots." *Microbiology. Immunology*, 2004. **48**(9): p. 669–75.
248. Lovrić, J., et al., "Differences in subcellular distribution and toxicity of green and red emitting CdTe quantum dots." *J Molecular Med*, 2005. **83**(5): 377-85.

CHAPTER 12
Nanomaterials in Ecosystems

Mélanie Auffan

CNRS–Aix-Marseille Université, Aix-en-Provence, France, and
iCEINT, Aix-en-Provence, France, and
CEINT, Duke University, Durham, North Carolina, USA

Fabienne Schwab

CNRS–Aix-Marseille Université, Aix-en-Provence, France, and
iCEINT, Aix-en-Provence, France, and
CEINT, Duke University, Durham, North Carolina, USA

Alain Thiéry

iCEINT, Aix-en-Provence, France, and
Aix-Marseille Université, Aix-en-Provence, France, and
IMBE, CNRS-IRD, Avignon Université, Marseille, France

Benjamin P. Colman

CEINT, Duke University, Durham, North Carolina, USA, and
Department of Ecosystem and Conservation Sciences, University of Montana, Missoula, Montana, USA

Jean-Yves Bottero

CNRS–Aix-Marseille Université, Aix-en-Provence, France, and
iCEINT, Aix-en-Provence, France, and
CEINT, Duke University, Durham, North Carolina, USA

Mark R. Wiesner

CNRS–Aix-Marseille Université, Aix-en-Provence, France, and
iCEINT, Aix-en-Provence, France, and
CEINT, Duke University, Durham, North Carolina, USA

Current strategies for assessing the environmental safety of engineered nanomaterials (ENMs) are largely based on nanoecotoxicology approaches [1] focusing on a single species or strain of model organisms. Such experiments are needed to gain knowledge on the mechanisms and thresholds of toxicity (hazard), but they separate the organism from the environment and community of organisms (i.e., the ecosystem), in which the environmental risks of ENMs ultimately need to be understood. Moreover, while the hazard to cells and organisms is extensively investigated, relatively little attention is paid to the exposure of organisms to ENMs despite the pivotal role of both exposure and hazard in together driving risk. The exposure of ENMs to organisms in the environment is controlled by the interplay of parameters including aggregation state and sorption of (in)organic substances [2], oxidation-reduction potential, as well as ecological factors such as interacting organisms [e.g., microorganisms, (in)vertebrates, plants, fungi], trophic levels present (e.g., primary producer, primary consumer, secondary consumer) and trophic transfer potential [3, 4].

Abundant literature exists on the effects of environmental parameters taken separately, but for a robust characterization of risk, the complex ecosystem-level interplay between the organisms, their environment, and the ENMs needs to be taken into account. Such experimental studies on the risk of ENMs under environmentally realistic conditions require a diverse collection of expertise—including but not limited to physical chemistry, (micro)biology, and ecology. Particularly well-suited experimental units with which to engage such multidisciplinary teams are engineered ecosystems that function as controlled release facilities, where replicated studies of some of the complex interactions that occur in full ecosystems can be performed [5–7]. These facilities may take the form of true mesocosms, which are exposed to the variability of weather and may be partially open to the surrounding environment, or more rigorously controlled indoor facilities. Both indoor and outdoor experimental systems contain a portion of the natural environment that is (1) self-sustaining once set up and acclimatized, without any additional input of nutrients or resources, and (2) monitored in detail for all pertinent variables, but in particular, allowing for quantification of inputs and outputs such that mass balances can be determined [8]. The experimental design of engineered ecosystems is an invaluable tool for addressing the complex issue of exposure during nano-ecotoxicological testing [9, 10]. The main goal of ENM studies with engineered ecosystems is to measure changes in the environment and in the concentration and speciation of ENMs over time, space, and ecosystem compartment. This information is then used to discern the kinetics of transformation, distribution, and impacts of the ENMs under investigation to better inform modeling of their environmental risks (Chapter 14).

Engineered ecosystems have already been used to study the behavior or impacts of ENMs, for example, in refs. [8, 16–22] (Table 12.1). These studies include an examination of the transfer of gold ENMs from the water column to other compartments in an estuarine food web [11], the long-term (18 months) distribution and transformation of silver NPs in mesocosms mimicking freshwater emergent wetlands [14], the Ag ENM aggregation in the water column [23, 24], as well as the leaching of Ag from ENM-containing consumer products over 60 days [13]. Such experiments require a strong integration across disciplinary boundaries. They offer physical chemists, (micro)biologists, and ecologists the possibility of conceiving robust experiments for the study of exposure and impacts of ENMs at low doses and over month to year time scales while testing and generating hypotheses about factors driving observed patterns of exposure and hazard across space and time. In this chapter, the basic setup of experiments involving the controlled release of ENMs to ecosystems is explained, and illustrated by two studies assessing the distribution and impact of CeO_2 and Ag-based ENMs.

Experimental Setup, Monitoring, Dosing

A broad diversity of ecosystems may be engineered in term of dimension, location (indoor, outdoor) and the type of ecosystem simulated (estuarine, freshwater, and terrestrial) [6, 25–27]. A common factor of all these studies is that a small portion of natural ecosystems is brought under varying degrees of control. The main difference highlighted here are indoor versus outdoor facilities, which present important tradeoffs between control and realism. While true mesocosms are outdoor facilities, we will use the term *mesocosm* here to refer to both indoor and outdoor facilities.

Indoor Aquatic Mesocosms

The modular nature, small size, and high degree of control over light, temperature, and organisms used in indoor aquatic facilities allows for simultaneous monitoring of a number of parameters (e.g., aggregation, settling, mass balance, trophic transfer, biotransformation, oxidative stress, microbial diversity) under environmentally relevant conditions. These systems can be reconfigured to represent several types of ecosystems including stream, lake, estuary, or lagoon, without requiring expensive and/or cumbersome infrastructures. These versatile tools serve interdisciplinary teams to study the exposure and impacts of ENMs (low doses, chronic contamination) as well as more mechanistic concepts at various temporal and spatial scales.

Experimental Setups

Ferry et al. [11] used multiple estuarine mesocosms (volume, 366 L each) simulating the edge of a tidal marsh creek, and maintained in a greenhouse [11]. Each tank contained natural seawater, sediment, biofilms, flora, and fauna. Sediments were periodically submerged in the primary tank using an attached water reservoir to simulate tides using submersible pumps set to timers. Natural coastal sediments were sieved with a 3-mm sieve, homogenized and dispensed into sediment trays. Four sediment trays were placed into each tank and elevated to allow for drainage. Tanks were monitored continuously for several water quality parameters, including temperature, pH, dissolved oxygen, and salinity, which varied in accordance with the tides, daytime heating, and photosynthetic activity [11].

Ref.	Location	Environment	Size	Nanoparticles	Dosing		Duration (days)
8	Indoor	Pond	60 L	Citrate-coated CeO_2	1.5 mg·L^{-1} 11×0.127 mg·L^{-1}	One pulse or Multiple pulses	28
19		Pond	60 L	Citrate-coated CeO_2 Uncoated CeO_2	1.5 mg·L^{-1}	One pulse	28
10		Pond	60 L	noncoated CeO_2	1 mg·L^{-1}	Multiple pulses	28
11		Estuary	366 L	Cetyltrimethylammonium bromide-coated Au	7.08 × 10⁸ NP·mL^{-1}	One pulse	12
13		Estuary	366 L	Ag, and nano-enhanced products†	~273 mg·L^{-1}	One pulse	60
20		Stream	5 L	TiO_2†	0.1 and 1.0 mg·L^{-1}	One pulse	32
18	Outdoor	Terrestrial	81 kg	PVP-coated Ag	0.14 mg·kg^{-1} soil	One pulse (sewage biosolid application)	55
14		Freshwater emergent wetland	1200 L*	PVP-coated Ag	25 mg·L^{-1} ~1 880 mg·m^{-2}	One pulse (water column) or One pulse (soil)	540
15		Freshwater emergent wetland	1200 L*	PVP-coated Ag Gum arabic–coated Ag	2.5 mg·L^{-1}	One pulse (water column)	30
16		Freshwater emergent wetland	1200 L*	Gum arabic–coated single-walled carbon nanotubes	2.5 mg·L^{-1}	One pulse	365
12		Estuary	125 L	CuO†	10 µg·L^{-1}	One pulse	21
21		Estuary	125 L	Maltose-coated Ag	10 µg·L^{-1}	One pulse	21
17		Littoral lake	4000 L	PVP-coated Ag	60 µg·L^{-1}	One pulse (water column)	33
22		Littoral lake	2392–3475 L	PVP-coated Ag Citrate-coated Ag	60 µg·L^{-1} 6, 23, and 93 µg·L^{-1}	One pulse or Multiple pulses Multiple pulses	52

*Maximum volume; water level variable depending on rainfall, evaporation, and transpiration.

†Unspecified coating.

PVP: Polyvinylpyrrolidone.

TABLE 12.1 List of Key Studies on the Exposure and Impact of ENMs in Engineered Ecosystems, as of August 2015

Simulating a pond ecosystem, a mesocosm platform was designed consisting of nine tanks of 750 × 200 × 600 mm (60 L) [8, 19]. This allows three replicates per experimental condition (i.e., two different contamination scenarios and control). Each mesocosm is made of monolithic glass panels with five holes (ø 15 mm) drilled at mid height on both of the long sides, which were connected to a pump using silicone tubing (Figure 12.2). The mesocosms were filled with a few centimeters of artificial sediment made of quartz, clays, and calcium carbonate. A natural sediment inoculum collected in the pond was sieved at 0.2 mm and laid at the surface of the artificial sediment (few mm-thick). This inoculum brings primary producers and consumers into the mesocosms (diatoms, green algae, bacteria, fungi, etc.). Each mesocosm contained 55 L of water. The initial setup consists in introducing the sediment and filling the tank with water. The water is chosen to be close to the solution chemistry of the natural ecosystem of interest where macro- and microorganisms are collected. From a variety of commercially available waters, Volvic water was selected as representative of a temporary pond on calcareous ground, having calcium (12 mg·L^{-1}), magnesium (8 mg·L^{-1}), sodium (12 mg·L^{-1}), and potassium (6 mg·L^{-1}) (Figure 12.1). This water is compatible with the osmoregulation of freshwater invertebrates, and is representative of a temporary pond on calcareous ground.

Temperature (°C), pH, conductivity (µS·cm^{-1}), redox potential (mV), and dissolved O$_2$ (mg·L^{-1}) were measured every 5 min at mid height of the water column using multiparameter probes and at the water/sediment interface (up to 10 mm below surficial sediment) and mid height of the sediment using platinum-tipped redox probes [14, 28]. A day/night cycle was applied using full spectrum light, and room temperature was kept constant (Figure 12.2).

Establishment and Contamination Phases

An experiment performed in the platform used in [8, 19] involves two phases: the equilibration of the mesocosms (phase I), and the contamination experiment itself (phase II). During phase I, the particles suspended by the addition of water to the sediment are given time to settle, the pH, conductivity, O$_2$, turbidity, and redox potential stabilize, and the biological community becomes established. The duration of this phase depends on the target values (and acceptable variation around them) defined for each key parameter (e.g., ΔpH, ΔT, turbidity) (Figure 12.3). Then the organisms are allowed to acclimatize (e.g., primary, secondary consumers) and the water pumps are turned on. The selected organisms are

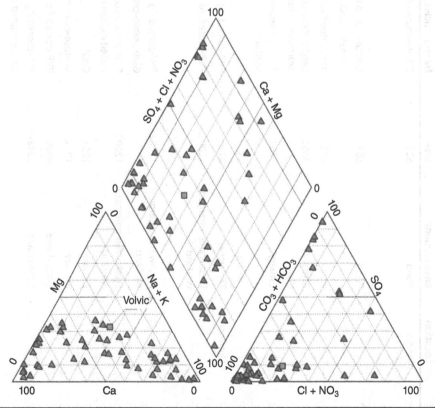

FIGURE 12.1 Piper diagram with 50 different French waters plotted. The Volvic water is one of the most well-balanced waters.

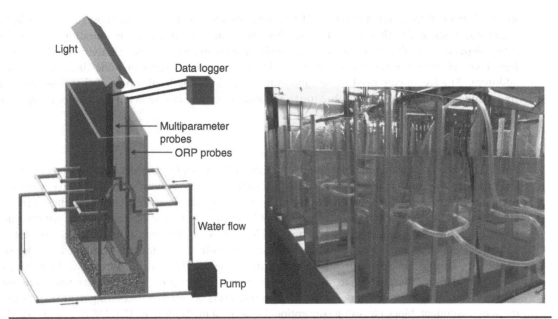

Figure 12.2   Small size (60 L) indoor aquatic mesocosms mimicking a pond ecosystem. (Adapted from [8].)

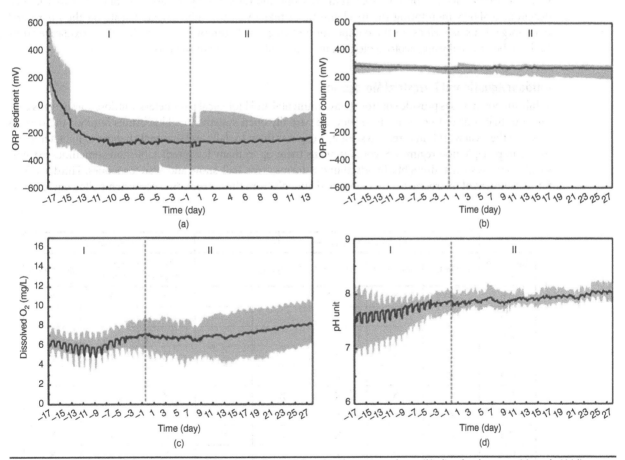

Figure 12.3   Evolution of the redox potential in the sediment (*a*) and in the water column (*b*), dissolved oxygen (*c*), and pH (*d*) measured during phases I and II. The grey surface is defined by the maximum and minimum values of each parameter, and the dark line corresponds to the average values of the nine mesocosms. One measurement was performed every 5 min. (Adapted from [8].)

involved in a real food web and have different habitats and functions in the ecosystems. The density of organisms is adjusted to their respective densities in natural environment. The duration of the acclimatization depends on biological features of the species as growth rate, metabolic activity, and life cycle duration. Phase II corresponds to the ENM contamination period consisting of a one-time pulse or multiple additions. The treatments need to be randomized based on the localization of the mesocosms in the room, and on the picoplankton (0.2–2 µm) and algae concentrations to avoid any confounding variation among replicates.

Sampling and Analysis

Several physico-chemical, microbial, and biological analyses can be performed to adequately assess both the exposure and impacts of ENMs on a designated trophic link. A number of parameters can be monitored continuously with the appropriate probe (e.g., pH, temperature, redox). Other parameters [e.g., metal concentration, number of colloids, picoplankton (0.2–2 µm) and algae concentrations] require sampling (an example for a sampling schedule is shown in Figure 12.4). Using such a small mesocosm setup, water, superficial sediments, cores, picoplankton, algae, and macro-invertebrates can be sampled with any desired periodicity. During sampling, special attention must be given (1) to avoid disturbing the sediment and water column properties, and (2) to keep microorganism densities and ENM concentrations constant. The distribution of the ENMs or their degradation by-products is assessed by measuring their concentration in the water, sediment, biota, etc. using conventional analytical methods (e.g., ICP-MS or ICP-AES). The dissolution of ENMs can be measured separately by placing sealed dialysis bags in the mesocosms. In-depth characterization of the speciation, (bio)transformation, and (bio)distribution of the ENMs in the water, sediment, or biota can be performed using X-ray, IR, and Raman spectroscopy, nuclear magnetic resonance, as well as electron or X-ray-based microscopy and tomography (Chapter 3). Such an experimental design also allows monitoring the mechanisms of toxicity at the subindividual scale on the micro and macroorganisms as well as on the composition of microbial communities. For instance, oxidative stress [29] can be assessed using ecotoxicological markers and ecophysiological processes [30–33].

Outdoor Aquatic and Terrestrial Mesocosms

While indoor facilities provide control of environmental and biological parameters, outdoor mesocosms cede much of that control to nature where they are partially open systems [6]. This brings outdoor mesocosms closer to the "real world" in three ways. First, it exposes them to daily and seasonal changes in light, temperature, and precipitation regimes. Second, it opens them up to many less predictable and sometimes extreme weather events such as droughts, floods, thunderstorms, frost, hail, snow, and even tornadoes. Third, it allows for recruitment of microorganisms, plants, and animals from the surrounding ecosystems allowing for more

Sampling time	Environmental compartments					Organisms		
	Seawater	Sediment	Biofilms	Cordgrass	Sand	Hard clams	Grass shrimp	Mud snails
0 h	X	X	X	X		X	X	X
2 h	X							
6 h	X							
12 h	X	X	X			X	X	X
24 h	X	X	X			X	X	X
48 h	X	X	X	X		X	X	X
96 h	X	X	X	X		X	X	X
7 d	X	X	X	X	X	X	X	X
13 d	X	X	X	X		X	X	X
30 d	X	X	X	X	X	X	X	X
60 d	X	X	X	X	X	X	X	X

Figure 12.4   Example for a sampling schedule for a mesocosm study. "X" denotes that a sample was collected; grey cells denote that the compartment was not collected at the corresponding time point. (Adapted from [13].)

complex food webs than could easily be assembled in a laboratory [6, 7]. By using these open systems, outdoor mesocosm experiments are a useful tool to test how natural variability effects ENM fate and transport, and whether ENM effects on the ecosystem surpass the effects caused by natural variability.

Experimental Setup

Briefly, ultraviolet resistant enclosures in the range of several tens of centimeters to several meters in length, width, and height, are filled with several tens to hundred kilograms or liters of soil, sediment, and water to simulate the respective environmental compartments [6, 7, 14] (Table 12.1). As with indoor mesocosms, using replicated mesocosms for each treatment (in practice often 2–3) is important. The number of replicates needs to be adjusted to the natural variability of the organism and ecosystem-level impacts under investigation [6]. Soil, sediment, and water compartments can be arranged in a variety of configurations to create strictly aquatic or terrestrial ecosystems, or incorporate a range of ecosystems by including a permanently flooded portion, a periodically flooded portion, and/or a rarely flooded portion [15], (Figure 12.5).

Cylinder-shaped mesocosm enclosures several meters in diameter have been used in lake setups [17]. For the study of ENMs on land, more compact rectangular mesocosms are used more frequently [12, 15, 16, 21]. Although small mesocosm enclosures containing just 81 kg of soil have been successfully used for terrestrial experiments [18], larger mesocosms several meters long and wide allow for longer term experiments with a greater extent of sampling. Larger mesocosm enclosures also have greater degree of environmental realism [6]. Smaller enclosures can compromise predator-prey interactions [34], and may yield unrealistically high dissipation rates for chemicals tending to bind to organic matter [6]. Moreover, the root/soil ratio, shading by the enclosure walls (both commonly referred to as edge effects), and the mixing rate of the water column can be substantially different from that in larger bodies of water [6, 35]. Consequently, the transport rates to the sediment (aggregation, agglomeration, sedimentation), and the biodegradation rates can be different in small enclosures. The size of outdoor mesocosms is mainly limited by the number of desired replicates and the material costs [6], that includes the ENMs used for the exposure experiment as well as disposal of materials at the conclusion of the experiment.

It is noteworthy that the natural solar spectrum, especially in the UV range, provides a spatially homogeneous, temporally dynamic irradiation that is still very difficult to imitate by artificial light [35]. Phototransformation processes of ENMs include many redox reactions involving light sensitive macromolecules such as dissolved organic matter (DOM); dissolution and (re)precipitation reactions; ENM photooxidation and photoreduction; and reactive (oxygen) species generation (e.g., TiO_2 NPs and Ag NPs) [36–38]. Shading and illumination can thus affect ENM heteroaggregation processes with background particles, for example, DOM, green algae, and clays [24, 39]. Consequently, outdoor mesocosms are particularly suitable for the study of light-dependent ENM aging and degradation processes, sedimentation rates, and ENM-plant interactions.

Figure 12.5 Large size (1200 L) outdoor mesocosm enclosures mimicking an emergent wetland. (*a*) Mesocosm enclosure in initial stage filled with sediment only. (*b*) The soil and water compartments form an aquatic, a transition, and an upland zone. (*a* and *b* adapted from [14].) (*c*) Photos of the mesocosms several months after planting; adapted from [15].)

Equilibration and Contamination Phases

Like in indoor mesocosms, experiments in outdoor mesocosms are performed in two phases: equilibration and contamination. As soon as the planted vegetation thrives, the artificial colonization by selected local animal species follows like in indoor mesocosms [8, 14, 16]. In addition to the artificial colonization, plants, microorganisms, fungi, and animals living in proximity to the mesocosm enclosures are allowed to naturally colonize the outdoor mesocosms to enrich the biodiversity. The only exceptions to this are certain vertebrates with high feeding and burrow activity [15], which need to be excluded through fencing and/or removed if colonization occurs.

Equilibration Phase During the equilibration phase in aquatic and wetland mesocosms, the water of different mesocosm enclosures can be homogenized using pumps to force convergence of microorganisms [15]. These pumps can later serve to prevent overflows and spillages of ENM contaminated water to the environment. In case of extreme rainfall, excess water can be pumped into a reserve pool. The mesocosms shown in Figure 12.5 were equilibrated for 3 to 5 months to reach a biodiversity that was reasonably comparable to the surrounding environment [15, 16]. As in indoor mesocosms, a set of parameters should be monitored to assess the establishment of parameters such as pH, oxidation-reduction potential, turbidity, and dissolved oxygen. Due to the seasons and weather changes, outdoor mesocosms are, like the natural environment, never truly equilibrated. Outdoor mesocosms have been considered to be under sufficiently stable conditions and ready for dosing after a thriving plant cover had grown, and the pH and turbidity measurements had been stable with time and rain events [16].

Contamination Phase Outdoor mesocosms have often been dosed with ENMs into the water column in a single pulse dose [8, 15, 17, 21]. However, as the dosing should mimic a possible exposure pathway, chronic doses [8], and other application methods can be used depending on the research question, for example, spraying the EMN suspension on the upland zone of mesocosms [14], or soaking EMN-containing consumer products in the water column, such as socks, wound dressings, and toy bears [13].

Sampling and Analysis

The same parameters and endpoints analyzed for indoor mesocosms can be measured during the equilibration and contamination phase in outdoor mesocosms (refer to section Indoor Aquatic Mesocosms), where the sampling and the parameters to measure are already described, and Figure 12.4). In practice, this set of monitoring parameters and the sampling frequency is often limited by the higher costs associated with weatherproof equipment, site access, sample transport, and storage. The most crucial monitoring parameters to quantify depend on the ecosystem to be studied (aquatic or terrestrial), but can include basic ecosystem-level characteristics like: plant cover and composition; water quality parameters such as pH, conductivity, turbidity, and temperature; and trace gas concentrations and fluxes into and out of the system [11, 16].

Selected Mesocosm Studies of ENM Exposure and Impacts

The first short-term mesocosm study on ENMs was performed in 2009 [11]. Since then, slightly more than a dozen mesocosm studies on ENMs have been performed (Table 12.1), and a greater number of ENM mesocosm studies are underway (e.g., [40], where a whole lake was dosed with silver NPs). In the following two sections, the results of three representative indoor and outdoor mesocosm studies will be discussed.

Indoor Mesocosms to Study CeO$_2$ ENMs in Pond Ecosystems

Two experiments were performed in the indoor facilities described above and shown in Figure 12.2 [8, 19] to assess the environmental risks of a CeO$_2$-based ENM that is included on the OECD list for ENMs requiring (eco)toxicological testing [41]. Citrate-coated CeO$_2$ nanoparticles (8 nm in hydrodynamic diameter) sold (Nanobyk 3810, Byk [42]) as long-term UV stabilizers, were used in this work [19, 43]. The mesocosm platform was first configured to simulate a pond ecosystem using the common ramshorn snail (*Planorbarius corneus* L., 1758) and a natural sediment inoculum coming from a noncontaminated pond (43.34361 N, 6.259663 E, and 107 m above sea level). The study proceeded through establishment and contamination phases with respective periods of 14 and 28 days. Two weeks were necessary to reach a state of equilibrium with anoxic conditions in the sediments, the sedimentation of the particles, and the homogenization of the microbial community composition. Such a 14-day stabilization time is consistent with earlier mesocosm studies for estuarine ecosystems [44, 45]. The second experiment (detailed in the Trophic Interactions and Trophic Transfer section below) used as primary consumers/detritivores by the midge, *Chironomus riparius* Meigen, 1804, and as secondary consumer the Spanish ribbed newt (*Pleurodeles waltl* Michahelles, 1830).

In the first experiment, two contamination scenarios were simulated: the response of the mesocosms to a single mass addition (pulse dose) of 69 mg to achieve an initial concentration of 1.1 mg L^{-1} of CeO$_2$ ENMs at day 0; and the addition of CeO$_2$ added as 11 smaller doses (chronic doses) of 5.2 mg each administered three times per week during 4 week experiment.

Ce Distribution in Environmental Compartments

At the end of the 28-day contamination phase, total mass CeO$_2$ recovered were about 115 ± 18 percent of the Ce (following a pulse dosing) and 60 ± 30 percent of the Ce (following a multiple dosing) which is in agreement with 84 percent of recovery observed in [11] or 68 to 76 percent obtained in [14]. Of the total Ce recovered, 89.2 ± 5 to 99.2 ± 0.2 percent was found in the surficial sediments (for multiple and single dosing respectively), 10.8 ± 5 to 0.8 ± 0.2 percent in the water column, and less than 0.1 percent in *P. corneus*. Moreover, transfer of Ce from the surficial sediment to deeper layers of sediment was not observed. Mobility of nanoparticles in saturated sandy porous media is expected to be strongly limited given surface chemistry favored nanoparticle—nanoparticle or nanoparticle-sediment interactions (especially with sand [46] and clay [47, 48]). Thus, the presence of both kaolinite and sand in our artificial sediment in the absence of bioturbating organisms maintained the CeO$_2$ ENMs at the surface of the sediment.

Different distributions of Ce were observed over time depending on contamination scenario (pulse versus chronic). After the pulse dosing, the ENMs aggregated in 1 week as apparent from the decrease of the total Ce concentration in the water column (to 15 µg·L^{-1}) (Figure 12.6) and concomitantly increased in the surficial sediments (to 540 mg·kg^{-1}). In contrast, in the chronic treatment total Ce concentration in the water column remained almost constant (50 ± 10 µg·L^{-1}) while increasing slightly in the sediment (100 mg·kg^{-1}). Sedimentation of Ce was due in both cases was due to homoaggregation of the ENMs and their heteroaggregation with natural colloidal particles. In this study, the initial number of colloidal particles (clay, bacteria, etc.) was low (~10^5 particles mL^{-1}). Hence, the final addition of 1.4 mg·L^{-1} of ENMs corresponding to 6×10^5 particles mL^{-1} could lead to both homoaggregation of CeO$_2$ ENMs and to their heteroaggregation with other particles [19].

Ce Interactions with Organisms

Sedimentation appeared to have favored the ingestion of Ce by adult snails since at day 28, 104 ± 75 mg·kg^{-1} and 60 ± 40 mg·kg^{-1} (dry weight of digestive gland) were assimilated following a pulse versus multiple dosing. Cross-sections of digestive gland of snail exposed to CeO$_2$ ENMs for 4 weeks were analyzed using

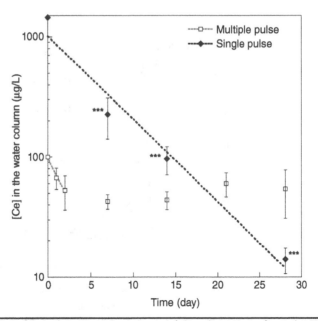

FIGURE 12.6 Concentration of Ce in the water column in µg·L^{-1}. Sampling was performed at 10 cm from the water surface. Values are means ± standard deviation. Dotted lines are the exponential fits of the experimental data. (∗∗∗) is statistical difference between multiple and pulse dosing at a given time ($p < 0.001$). (Adapted from [8].)

FIGURE 12.7 Laser ablation–inductively coupled plasma mass spectrometry on a crossed section of digestive gland of snails exposed to CeO$_2$ ENMs for 4 weeks in mesocosms. Intensity in legend corresponds to the abundance of Ce isotopes analyzed by ICP-MS. (Adapted from [19].)

laser ablation ICP-MS and X-ray tomography to assess the Ce distribution. Several spots rich in Ce were found in the digestive gland with an average spot size of 80 ± 20 µm (Figure 12.7). The elevated concentrations of Ce in the digestive gland of adults *P. corneus* highlights that feeding was one of the major route of transfer of Ce into adult snails. But cerium was also taken up or adsorbed externally on *P. corneus* at all three life stages (i.e., egg-mass, juvenile, and adult). In eggs, CeO$_2$-NPs were likely adsorbed on the mucopolysaccharides or chitin-like membranes composing the organic matrix [49]. Such an adsorption of NPs has previously been observed in other taxa, such as the exopolysaccharide layer surrounding bacteria [50], algae [40], or the chorion surrounding fish embryos [51, 52].

The digestive gland of mollusks is the center for nutrient metabolism and storage, and produces diverse enzymes and metalloproteins containing high amounts of thiol groups (e.g., digestive cysteine proteinase and metallothioneins [53, 54]) that might be responsible for Ce reduction from Ce^{IV} to Ce^{III} observed in the snails (78 ± 15% of Ce was under the form of Ce^{III} in the digestive gland) [19] (Figure 12.8). Such a reduction to Ce^{III} has been previously observed with bacteria [56], human dermal fibroblasts [50], nematodes (roundworms) [56] and plants [57]. This reduction is important as it has been linked in the literature to cytotoxic, genotoxic, phytotoxic effects [50, 55, 58]. In this study, a transitory oxidative stress response was observed for *P. corneus* digestive glands after 3 weeks exposure to CeO$_2$ ENMs. This was then followed by production of antioxidant compounds, a possibly adaptive response of snails that may have limited the potential for oxidative damage from the production of Ce^{III}. Surprisingly, though organisms showed an increase in oxidative damage, reproduction was observed during the experiment. This

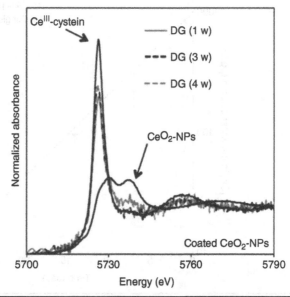

FIGURE 12.8 Ce L$_3$-edge XANES spectra of the Ce in the digestive gland (DG) of snail exposed to CeO$_2$ ENMs for 4 weeks in mesocosms. XANES spectra of the initial ENMs and of the Ce(III)-cysteine reference compound are displayed. 1 w = 1 week. 3 w = 3 weeks. 4 w = 4 weeks. (Adapted from [19].)

result may highlight an allocation to reproduction under stressful conditions, which is a typical ecological trait of pond-living organisms [19].

Trophic Interactions and Trophic Transfer

In a second indoor mesocosm experiment, the same mesocosm setup (Figure 12.2) [10] was used to examine the trophic transfer and toxicity of CeO_2 ENMs in an experimental freshwater ecosystem with three trophic levels: primary producers (algae, diatoms, and cyanobacteria); primary consumers (fungi; bacteria; and midge larvae, *C. riparius*); and secondary consumers (newt larvae, *P. waltl*). The ENM contamination consisted of repeated additions of CeO_2 ENMs over 4 weeks to obtain a final concentration of approximately $1 \ mg \cdot L^{-1}$. No effects were observed on litter decomposition or associated fungal biomass. Changes in bacterial communities were observed from the third week of ENM contamination. No toxicity was recorded in midge larvae, despite substantial Ce accumulation ($265.8 \pm 14.1 \ mg \ Ce \cdot kg^{-1}$). Newt larvae accumulated far less Ce ($13.5 \pm 3.9 \ mg \cdot kg^{-1}$) but had elevated mortality ($35.3 \pm 6.8\%$). The authors hypothesized that the toxicity observed in newts was indirect, due to either microorganism's interaction with CeO_2 ENMs, or to the dissolution of CeO_2 [10].

Outdoor Mesocosms to Study Ag ENMs in Emergent Wetlands

In the following example, the results of a large-scale, 30-day mesocosm study on Ag ENMs in emergent freshwater wetlands will be discussed [15]. It is one of the most complete and realistic fully replicated outdoor ENM exposure experiments to date. The aim of this study was to compare the fate and impacts of medium concentrations ($2.5 \ mg \cdot L^{-1}$) of two fundamentally different silver nanoparticles: gum arabic coated Ag ENMs (GA-AgNPs) and polyvinylpyrrolidone coated Ag ENMs (PVP-AgNPs). Several controls were used consisting of mesocosms that were dosed with water only, dosed with the coatings alone, and yet others that were dosed with silver salt (Ag^+; added as $AgNO_3$). The two silver nanoparticles differed in their size, coating, and behavior in previous laboratory experiments: The 12 nm GA-AgNPs were better dispersed and more toxic than the larger 49 nm PVP-AgNPs [15].

The mesocosms were located outdoors in a clearing in the Duke forest, a temperate forest located near Duke University in Durham NC, USA (Figure 12.5c). The $3.66 \times 1.22 \times 0.81$-m mesocosms were filled using a local soil blend with a texture of approximately 60 percent sand, approximately 30 percent silt, and approximately 10 percent clay, with 5.1 percent organic matter. The soil and water were added as shown in (Figure 12.5a, b), to create an emergent wetland mesocosm with a permanently submerged aquatic zone, a periodically flooded transition zone, and a rarely flooded upland zone.

The mesocosms were planted in March of 2010. Three lead organisms were planted in the water compartment: the submerged macrophyte *Egeria densa* (Brazilian waterweed), the half-submerged *Potamogeton diversifolius* (diverse-leaved pondweed), and the floating *Lemna punctata* (dotted duck meat). The transition and upland zones were planted predominantly with grasses (order *Poales*). A wide range of benthic macroinvertebrates was allowed to recruit independently. With the exception of mosquitofishes *Gambusia holbrooki*, vertebrates including raccoons, deer, frogs, and toads were excluded and/or removed from the mesocosms. After 5 months of equilibration, the mesocosms were ready for dosing in mid August 2010 [15].

Each of the six treatments (GA-AgNPs, PVP-AgNPs, GA, PVP, $AgNO_3$, and control) was run in triplicate mesocosm enclosures, except of the control, which was run in quadruplicate. This resulted in a total of 19 mesocosm enclosures (Figure 12.5c). The mesocosms were contaminated via the water column in one pulse. Over the next 30 days after contamination, changes in the dissolved gases and solutes were observed [Ag, Cl^-, SO_4^{2-}, Br^-, NO_3^-, NH_4^+, o-PO_4; and dissolved organic carbon (DOC), total nitrogen, O_2, CH_4, and CO_2], and the Ag concentration in the water column and the plants was measured.

Water Column Ag Concentrations

Within 24 h of dosing, total water column Ag concentrations (Figure 12.9) dropped from the initial concentration of $2.5 \ mg \ L^{-1}$ down to 2.02 ± 0.12, 1.99 ± 0.07, and $0.85 \pm 0.24 \ mg \cdot L^{-1}$ of Ag for the GA-AgNPs, PVP-AgNPs, and $AgNO_3$, respectively (mean $\pm$ standard error of the mean). This revealed an unexpected behavior of both the Ag^+ and the AgNPs. Instead of remaining in the water column, the Ag in the $AgNO_3$ treatment disappeared more rapidly than the AgNPs. The Ag concentrations of the $AgNO_3$ treatment remained significantly lower than in the AgNP treatments in the first 4 days following dosing ($p < 0.05$, days 1–4). Despite different sizes and coatings and behavior in the lab, the water column Ag concentration was not significantly different for the two AgNP treatments on any individual date. By day 6,

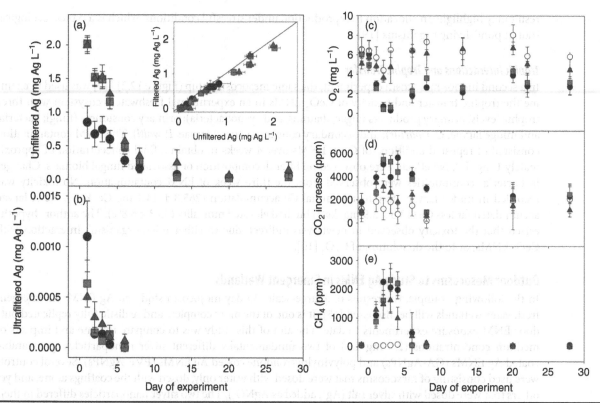

FIGURE 12.9 Water chemistry of the outdoor mesocosms dosed with silver nanoparticles (Ag NPs) dosed on day 1. Empty circles: controls; black circles: AgNO₃; red squares: gum arabic coated AgNPs; and blue triangles: polyvinylpyrrolidone AgNPs. (a) and (b): Ag concentration of unfiltered (a) and filtered (b) mesocosm water. Note the exponential decay of the water concentration in all treatments between day 1 to 8 following first order kinetics [Equation (12.1), average $k = 0.21 \pm 0.066$ d⁻¹]. Insert in (a): Linear regression between Ag concentrations in filtered and unfiltered water (R₂ = 0.993). (c): Dissolved oxygen. (d) and (e): Dissolved gases CH₄ and CO₂. Values are mean ± standard error of the mean.

Ag concentrations in all Ag treatments had converged to 0.44 ± 0.06 mg·Ag·L⁻¹. The removal kinetics of Ag from the water column followed first order kinetics

$$[Ag]_t = [Ag]_0 \cdot e^{-kt} \tag{12.1}$$

where $[Ag]_t$ is the Ag concentration at time t, and $[Ag]_0$ is the initial Ag concentration. The average removal rate constant k was 0.21 ± 0.066 d⁻¹ (day 1–8). After day 8, 2 days after several intense rain events starting from day 6, and until the end of the measurements, the Ag concentration in all Ag treatments remained fairly stable and averaged 0.13 ± 0.02 mg·Ag·L⁻¹. Most of the Ag still present in the water column remained dissolved or well dispersed particles, as a linear regression between the Ag concentrations of filtered and unfiltered mesocosm water showed (slope indistinct from 1, $R^2 = 0.993$, Figure 12.9a, insert).

This unexpectedly similar behavior of the three different Ag treatments can be explained in several ways. After the dosing, large quantities of DOC and Cl⁻ were measured, likely released by the stressed floating and submersed aquatic plants [15]. Ag⁺ can be transformed in the presence of this Cl⁻, or DOC and/or sunlight into AgCl, or Ag(0) NPs, respectively [37]. Solution chemistry calculations predicted that nearly 82 percent of the Ag could form AgCl₍ₛ₎ precipitates or colloids at post-treatment Cl⁻ concentrations [15]. Another possible explanation is that mechanistically distinct, but kinetically similar factors drove the Ag⁺ removal. Potential removal mechanisms for both Ag and AgNPs include aggregation and sedimentation [24]; and sorption or uptake by plants [4] or biofilms [20].

Accumulation of Ag in Aquatic Plants
Aquatic plants, especially when submerged, rapidly accumulated high concentrations of Ag (Figure 12.10). Within 7 days, *E. densa* accumulated up to 4180 ± 250 mg·Ag·kg⁻¹ in the AgNO₃ treatment. The highest Ag concentration for AgNPs was for the GA-AgNPs treatments with *E. densa* at 3060 ± 250 mg·Ag·L⁻¹. Concentrations of Ag in plants in PVP-AgNP were generally half those found for the GA-AgNP, which may

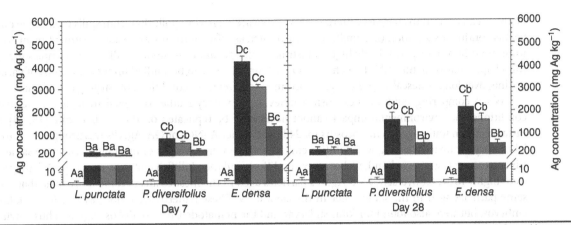

Figure 12.10 Silver concentrations in plants by species, treatment, and date. White bars: Controls; black bars: AgNO₃; red bars: GA-AgNPs; blue bars: PVP-AgNPs. The bars with the same capital letter (A, B, and C) are not significantly different (″ = 0.05) among treatments, and means with the same lowercase letter (a, b, and c) are not significantly different (″ = 0.05) among species. Values are mean ± standard error of the mean.

be due to higher reactivity and bioavailability of the smaller GA-AgNPs. A simple explanation for the rapid accumulation of Ag and AgNPs by the completely submerged *E. densa* is these plants' anatomy featuring a large surface of hydrophilic, highly metabolically active, gas and water permeable leaves [4]. Plant uptake could at least partially explain the exponential decay of Ag in the water column [Equation (12.1)], which is similar to the removal kinetics of plants for other ENMs, metal complexes, and organic pollutants [4, 59, 60]. Regardless of whether it was due to uptake, or other processes involving epiphytic biofilms or sorption, plants were an important early sink of Ag.

Cascade of Organismal and Ecosystem-Level Impacts

A series of complex and unexpected interactions between organisms was observed in the dosed mesocosms. First, within 2 days of adding Ag, the aquatic plants of all Ag dosed mesocosms showed signs of toxicity, which is in line with the high Ag tissue concentrations. The plants suffered widespread leaf loss and a loss of photosynthetic pigments from the foliage (browning), coinciding with an increase of DOC in the water column. Similar trends were observed for phytoplankton, where cell densities declined in the three Ag treatments over the first 4 days. This decline in the viability of plants and phytoplankton, and the increase of DOC, fueled a short-lived bloom of heterotrophic prokaryotes (bacteria and archaea). Water column cell densities increased by three to fourfold within a day of dosing in all Ag treatments. The prokaryote bloom was partially attributable to the rain events after day 6, due to microbe-containing soil particles from the upland zone rinsed into the submerged zone. Increased respiration of the blooming heterotrophic prokaryotes consumed dissolved oxygen (Figure 12.9c). Combined with decreased photosynthetic capacity due to declines in algal and macrophyte biomass and viability respectively, this drove the water column to hypoxia (rapid decline of O_2) and significantly increased the water column concentrations of CO_2 (Figure 12.9c, d). In both the AgNO₃ and GA-AgNP treatments, this cascade of ecosystem-level impacts occurred with roughly equivalent magnitude. The PVP-AgNPs had a much more modest or nonsignificant impact on DOC, CO_2, and O_2, but trends were in the same direction as for the GA-AgNPs and AgNO₃ (Figure 12.9c–e). The most striking ecosystem-level impact was nearly identical in all three Ag treatments: Dissolved methane increased 40-fold within 2 to 3 days after Ag addition (Figure 12.9c–e). This may have been fueled by the release of DOC and CO_2 (both CH_4 precursors), and the decline in O_2 which led to a combination of decreased CH_4 consumption (largely aerobic) and increased CH_4 production (strictly anaerobic). Other hypotheses for this CH_4 overproduction include effects of Ag on CH_4-metabolizing microbes [61], or the release of dissolved CH_4 from plant biomass [62]. In summary, there were dramatic impacts of the outdoor mesocosm environment on Ag ENM fate and impacts, few of them having been predictable by the previous laboratory experiments.

Summary

Experimental setups representing complex ecological interactions are important tools for understanding how nanomaterials are transferred between environmental compartment, how ENMS may be transformed, and the nature of impacts on individual organisms, populations, and ecosystem functions. Such pilot facilities can be used to compare different ENM exposure scenarios: pollution from spills (i.e., single pulse) versus continuous addition (i.e., multiple pulses), such as effluent from wastewater treatment plants (continuous addition), or periodically high industrial discharge and rain runoff (spill). The examples given in this chapter illustrate how ENMs such as CeO_2 NPs or Ag NPs can be studied under conditions of variable dosing, over long timescales, and under environmentally relevant conditions of varying complexity.

When comparing outdoor mesocosm studies to laboratory studies designed to replicate mesocosm conditions, the environmental impact cannot necessarily be replicated on the smaller scale [15, 63]. For instance, when water aged in mesocosms for 24 h with either $AgNO_3$ or GA-AgNPs was brought back to the laboratory for toxicity testing with killifish embryos, there were no signs of mortality. When the same two silver treatments were aged under laboratory conditions and then killifish embryos were exposed, they both were highly toxic. Conversely, PVP-AgNPs particles aged in the mesocosms became more toxic than those same particles aged in the same water in the laboratory. These patterns were not just isolated to killifish embryos, but were also found for killifish larvae and the nematode *Caenorhabditis elegans*. Thus, environmental factors unique to ecosystems can strongly affect the behavior of ENMs, which then influences the effects of those ENMs on ecosystems [63].

Clearly, indoor and outdoor mesocosms provide more realistic conditions for the study of ENM exposure that go beyond laboratory studies of ENM impacts on individual organisms or cell cultures. This environmental realism is critical for the assessment of the potential hazards presented by ENMs and "ground-truthing" laboratory and high throughput data. Indoor mesocosm experiments allow for simultaneous and temporally highly resolved evaluation of physico-chemical properties, and their relationship to the biological systems in situ as the ecosystem and ENM properties develop and change. On the other hand, outdoor mesocosm experiments take into consideration a higher variability of the natural environment induced by wind, direct sunlight, weather events such as drought and rain, spatial migration. Outdoor mesocosm may also include a greater variation organismal diversity and number. Thus, outdoor mesocosm experiments provide a highly representative simulation of natural ecosystems, but are also more difficult to analyze given the natural variability.

Depending on the ENMs and the contamination scenarios, mesocosm studies can be designed to contain a range of environmental compartments with different physico-chemical features (e.g., water quality and depth, sediment mineralogy and depth, current velocity, tidal reservoirs, etc.) and biota. "Indoor" and "outdoor" approaches are complementary and can be adapted to the ecotoxicological benchmarks under investigation. This allows for testing whether an ENM is a Contaminant of Potential Ecological Concern (COPEC) that may or may not cause risk or adverse effects to biota at a site. By simultaneously creating representative conditions for environmental transformation and ecosystem exposure, these engineered platforms facilitate the integration of reliable exposure and hazard data into an environmental risk assessment framework.

References

1. Kahru, A.; Dubourguier, H.-C., "From ecotoxicology to nanoecotoxicology." *Toxicology* **2010**, *269* (2–3), 105–119.
2. Schwab, F.; Camenzuli, L.; Knauer, K.; Nowack, B.; Magrez, A.; Sigg, L.; Bucheli, T. D., "Sorption kinetics and equilibrium of the herbicide diuron to carbon nanotubes or soot in absence and presence of algae." *Environ. Pollut. (Oxford, U.K.)* **2014**, *192*, 147–153.
3. Auffan, M.; Rose, J.; Bottero, J. Y.; Lowry, G.; Jolivet, J. P.; Wiesner, M. R., "Towards a definition of inorganic nanoparticles from an environmental, health, and safety perspective." *Nature Nanotechnology* **2009**, *4*, 634–641.
4. Schwab, F.; Zhai, G.; Kern, M.; Turner, A.; Schnoor, J. L.; Wiesner, M. R., Barriers, "Pathways and processes for uptake, translocation, and accumulation of nanomaterials in plants—critical review." *Nanotoxicology* **2015**, *Early Online*: 1–22.
5. FAO, Food and agriculture organization of the united nations. In *Biosafety of genetically modified organisms: basic concepts, methods and issues*, FAO: Rome, 2009.
6. Crossland, N. O.; La Point, T. W., The design of mesocosm experiments. *Environ. Toxicol. Chem.* **1992**, *11* (1), 1–4.
7. Shaw, J. L.; Kennedy, J. H., "The use of aquatic field mesocosm studies in risk assessment." *Environ. Toxicol. Chem.* **1996**, *15* (5), 605–607.
8. Auffan, M.; Tella, M.; Santaella, C.; Brousset, L.; Pailles, C.; Barakat, M.; Espinasse, B.; et al., "An adaptable mesocosm platform for performing integrated assessments of nanomaterial risk in complex environmental systems." *Scientific Reports* **2014**, *4*, 5608.
9. Bour, A.; Mouchet, F.; Silvestre, J.; Gauthier, L.; Pinelli, E., "Environmentally relevant approaches to assess nanoparticles ecotoxicity: A review." *J. Hazard. Mater.* **2015**, *283*, 764–777.
10. Bour, A.; Mouchet, F.; Cadarsi, S.; Silvestre, J.; Verneuil, L.; Baqué, D.; Chauvet, E.; et al., "Toxicity of CeO_2 nanoparticles on a freshwater experimental trophic chain: A study in environmentally relevant conditions through the use of mesocosms." *Nanotoxicology* **2016**, *10* (2), 245–255.
11. Ferry, J. L.; Craig, P.; Hexel, C.; Sisco, P.; Frey, R.; Pennington, P. L.; Fulton, M. H.; et al., "Transfer of gold nanoparticles from the water column to the estuarine food web." *Nat. Nano.* **2009**, *4* (7), 441–444.

12. Buffet, P.-E.; Richard, M.; Caupos, F.; Vergnoux, A.; Perrein-Ettajani, H.; Luna-Acosta, A.; Akcha, F.; et al., "A mesocosm study of fate and effects of CuO nanoparticles on endobenthic species (*Scrobicularia plana, Hediste diversicolor*)." *Environ. Sci. Technol.* **2013,** *47* (3), 1620–1628.

13. Cleveland, D.; Long, S. E.; Pennington, P. L.; Cooper, E.; Fulton, M. H.; Scott, G. I.; Brewer, T.; et al., "Pilot estuarine mesocosm study on the environmental fate of silver nanomaterials leached from consumer products." *Sci. Total Environ.* **2012,** *421–422*, 267–272.

14. Lowry, G. V.; Espinasse, B. P.; Badireddy, A. R.; Richardson, C. J.; Reinsch, B. C.; Bryant, L. D.; Bone, A. J.; et al., "Long-term transformation and fate of manufactured Ag nanoparticles in a simulated large scale freshwater emergent wetland." *Environ. Sci Technol.* **2012,** *46* (13), 7027–7036.

15. Colman, B. P.; Espinasse, B.; Richardson, C. J.; Matson, C. W.; Lowry, G. V.; Hunt, D. E.; Wiesner, M. R.; et al., "Emerging contaminant or an old toxin in disguise? Silver nanoparticle impacts on ecosystems." *Environ. Sci. Technol.* **2014,** *48* (9), 5229–5236.

16. Schierz, A.; Espinasse, B.; Wiesner, M. R.; Bisesi, J. H.; Sabo-Attwood, T.; Ferguson, P. L., "Fate of single walled carbon nanotubes in wetland ecosystems." *Environ. Sci. Nano* **2014,** *1* (6), 574–583.

17. Furtado, L. M.; Hoque, M. E.; Mitrano, D. M.; Ranville, J. F.; Cheever, B.; Frost, P. C.; Xenopoulos, M. A.; et al., "The persistence and transformation of silver nanoparticles in littoral lake mesocosms monitored using various analytical techniques." *Environ. Chem.* **2014,** *11* (4), 419–430.

18. Colman, B. P.; Arnaout, C. L.; Anciaux, S.; Gunsch, C. K.; Hochella, M. F., Jr.; Kim, B.; Lowry, G. V.; et al., "Low concentrations of silver nanoparticles in biosolids cause adverse ecosystem responses under realistic field scenario." *PLoS ONE* **2013,** *8* (2), e57189.

19. Tella, M.; Auffan, M.; Brousset, L.; Issartel, J.; Kieffer, I.; Pailles, C.; Morel, E.; et al., "Transfer, transformation and impacts of ceria nanomaterials in aquatic mesocosms simulating a pond ecosystem." *Environ Sci Technol.* **2014,** *48* (16), 9004–9013.

20. Kulacki, K. J.; Cardinale, B. J.; Keller, A. A.; Bier, R.; Dickson, H., "How do stream organisms respond to, and influence, the concentration of titanium dioxide nanoparticles? A mesocosm study with algae and herbivores." *Environ. Toxicol. Chem.* **2012,** *31* (10), 2414–2422.

21. Buffet, P.-E.; Zalouk-Vergnoux, A.; Châtel, A.; Berthet, B.; Métais, I.; Perrein-Ettajani, H.; Poirier, L.; et al., "A marine mesocosm study on the environmental fate of silver nanoparticles and toxicity effects on two endobenthic species: The ragworm *Hediste diversicolor* and the bivalve mollusc *Scrobicularia plana*." *Sci. Total Environ.* **2014,** *470–471*, 1151–1159.

22. Furtado, L. M.; Norman, B. C.; Xenopoulos, M. A.; Frost, P. C.; Metcalfe, C. D.; Hintelmann, H., "Environmental fate of silver nanoparticles in boreal lake ecosystems." *Environ. Sci. Technol.* **2015,** *49* (14), 8441–8450.

23. Badireddy, A. R.; Wiesner, M. R.; Liu, J., "Detection, characterization, and abundance of engineered nanoparticles in complex waters by hyperspectral imagery with enhanced darkfield microscopy." *Environ. Sci. Technol.* **2012,** *46* (18), 10081–10088.

24. Therezien, M.; Thill, A.; Wiesner, M. R., "Importance of heterogeneous aggregation for NP fate in natural and engineered systems." *Sci. Total Environ.* **2014,** *485–486*, 309–318.

25. Shriner, C.; Gregory, T.; Brocksen, R. W., "Use of artificial streams for toxicological research." *Critical Reviews in Toxicology* **1984,** *13* (3), 253–281.

26. Sommer, U.; Aberle, N.; Engel, A.; Hansen, T.; Lengfellner, K.; Sandow, M.; Wohlers, J.; et al., "An indoor mesocosm system to study the effect of climate change on the late winter and spring succession of Baltic Sea phyto- and zooplankton." *Oecologia* **2007,** *150* (4), 655–667.

27. Mohr, S.; Feibicke, M.; Ottenströer, T.; Meinecke, S.; Berghahn, R.; Schmidt, R., "Enhanced experimental flexibility and control in ecotoxicological mesocosm experiments: A new outdoor and indoor pond and stream system (3 pp)." *Environ. Sci. Pol. Res.* **2005,** *12* (1), 5–7.

28. Wafer, C. C.; Richards, J. B.; Osmond, D. L., "Construction of platinum-tipped redox probes for determining soil redox potential." *J. Environ. Qual.* **2004,** *33*, 2375–2379.

29. Xia, T.; Kovochich, M.; Brant, J.; Hotze, M.; Sempf, J.; Oberley, T.; Sioutas, C.; et al., "Comparison of the abilities of ambient and manufactured nanoparticles to induce cellular toxicity according to an oxidative stress paradigm." *Nano Lett.* **2006,** *6* (8), 1794–1807.

30. Fan, W.; Shi, Z.; Yang, X.; Cui, M.; Wang, X.; Zhang, D.; Liu, H.; et al., "Bioaccumulation and biomarker responses of cubic and octahedral Cu_2O micro/nanocrystals in *Daphnia magna*." *Water Res.* **2012,** *46* (18), 5981–5988.

31. Buffet, P.-E.; Tankoua, O. F.; Pan, J.-F.; Berhanu, D.; Herrenknecht, C.; Poirier, L.; Amiard-Triquet, C.; et al., "Behavioural and biochemical responses of two marine invertebrates *Scrobicularia plana* and *Hediste diversicolor* to copper oxide nanoparticles." *Chemosphere* **2011,** *84* (1), 166–174.

32. Chae, Y. J.; Pham, C. H.; Lee, J.; Bae, E.; Yi, J.; Gu, M. B., "Evaluation of the toxic impact of silver nanoparticles on Japanese medaka (*Oryzias latipes*)." *Aquat. Toxicol.* **2009,** *94* (4), 320–327.

33. Zhu, S.; Oberdörster, E.; Haasch, M. L., "Toxicity of an engineered nanoparticle (fullerene, C_{60}) in two aquatic species, *Daphnia* and fathead minnow." *Marine Environ. Res.* **2006,** *62*, Supplement 1, S5–S9.

34. Martin, L. E., "Limitations on the use of impermeable mesocosms for ecological experiments involving *Aurelia* sp. (Scyphozoa: Semaeostomeae)." *Journal of Plankton Research* **2001,** *23* (1), 1–10.

35. Poorter, H.; Fiorani, F.; Stitt, M.; Schurr, U.; Finck, A.; Gibon, Y.; Usadel, B.; et al., "The art of growing plants for experimental purposes: A practical guide for the plant biologist." *Funct. Plant Biol.* **2012,** *39* (11), 821–838.

36. Badireddy, A. R.; Farner Budarz, J.; Marinakos, S. M.; Chellam, S.; Wiesner, M. R., "Formation of silver nanoparticles in visible light–illuminated waters: Mechanism and possible impacts on the persistence of AgNPs and bacterial lysis." *Environ. Eng. Sci.* **2014,** *31* (7), 338–349.

37. Yin, Y.; Yu, S.; Liu, J.; Jiang, G., "Thermal and photoinduced reduction of ionic Au(III) to elemental Au nanoparticles by dissolved organic matter in water: Possible source of naturally occurring Au nanoparticles." *Environ. Sci. Technol.* **2014,** *48* (5), 2671–2679.

38. Hong, F. H.; Yang, F.; Liu, C.; Gao, Q.; Wan, Z. G.; Gu, F. G.; Wu, C.; et al., "Influences of nano-TiO_2 on the chloroplast aging of spinach under light." *Biol. Trace Elem. Res.* **2005,** *104* (3), 249–260.

39. Schwab, F.; Bucheli, T. D.; Lukhele, L. P.; Magrez, A.; Nowack, B.; Sigg, L.; Knauer, K., "Are carbon nanotube effects on green algae caused by shading and agglomeration?" *Environ. Sci. Technol.* **2011**, *45* (14), 6136–6144.

40. Xenopoulos, M. A.; Frost, P. C., "The dawn of a new era for the experimental lakes area." *Limnol. Oceanogr. Bull.* **2015**, *24* (3), 85–87.

41. OECD, List of Manufactured Nanomaterials and List of Endpoints for Phase One of the OECD Testing Programme. In *Series on the Safety of Manufactured Nanomaterials No. 6*, Organization for Economic Cooperation and Development: Paris, France, 2008.

42. BYK, Nanobyk-3810 Nanoparticle dispersion for long-term UV-stability of water-borne systems. *Data Sheet E207 03/10.*

43. Auffan, M.; Masion, A.; Labille, J.; Diot, M.-A.; Liu, W.; Olivi, L.; Proux, O.; et al., "Long-term aging of a CeO_2 based nanocomposite used for wood protection." *Environ. Pollut. (Oxford, U.K.)* **2014**, *188*, 1–7.

44. Lauth, J. R.; Cherry, D. S.; Buikema, A. L.; Scott, G. I., "A modular estuarine mesocosm." *Environ. Toxicol. Chem.* **1996**, *15* (5), 630–637.

45. Graney, R. L.; Kennedy, J. H.; Rodgers, J. H., *Aquatic Mesocosm Studies in Ecological Risk Assessment*. CRC Press Inc: Boca Raton, Florida, US, 1993.

46. Solovitch, N.; Labille, J.; Rose, J.; Chaurand, P.; Borschneck, D.; Wiesner, M. R.; Bottero, J. Y., "Concurrent aggregation and deposition of TiO_2 nanoparticles in a sandy porous media." *Environ. Sci. Technol.* **2010**, *44* (13), 4897–4902.

47. El Badawy, A. M.; Aly Hassan, A.; Scheckel, K. G.; Suidan, M. T.; Tolaymat, T. M., "Key factors controlling the transport of silver nanoparticles in porous media." *Environ. Sci. Technol.* **2013**, *47* (9), 4039–4045.

48. Cornelis, G.; Ryan, B.; McLaughlin, M. J.; Kirby, J. K.; Beak, D.; Chittleborough, D., "Solubility and batch retention of CeO_2 nanoparticles in soils." *Environ. Sci. Technol.* **2011**, *45* (7), 2777–2782.

49. Ross, L. F.; Harrison, A. D., "Effects of environmental calcium deprivation on the egg masses of *Physa marmorata* Guilding (Gastropoda: Physidae) and *Biomphalaria glabrata* Say (Gastropoda: Planorbidae)." *Hydrobiologia* **1977**, *55*, 45–48.

50. Zeyons, O.; Thill, A.; Chauvat, F.; Menguy, N.; Cassier-Chauvat, C.; Orear, C.; Daraspe, J.; et al., "Direct and indirect CeO_2 nanoparticles toxicity for *Escherichia coli* and *Synechocystis.*" *Nanotoxicology* **2009**, *3* (4), 284–295.

51. Auffan, M.; Matson, C. W.; Rose, J.; Arnold, M.; Proux, O.; Fayard, B.; Liu, W.; et al., "Salinity-dependent silver nanoparticle uptake and transformation by Atlantic killifish (*Fundulus heteroclitus*) embryos." *Nanotoxicology* **2014**, *8* (suppl. 1), 167–176.

52. Auffan, M.; Bertin, D.; Chaurand, P.; Pailles, C.; Dominici, C.; Rose, J.; Bottero, J. Y.; et al., "Role of molting on the biodistribution of CeO_2 nanoparticles within *Daphnia pulex.*" *Water Res.* **2013**, *47* (12), 3921–3930.

53. Walker, A. J.; Glen, D. M.; Shewry, P. R., "Purification and characterization of a digestive cysteine proteinase from the field slug (*Deroceras reticulatum*): A potential target for slug control." *J. Agr. Food Chem.* **1998**, *46*, 2873–2881.

54. Dallinger, R.; Berger, B.; Hunziger, P.; Kägi, J. H. R., "Metallothionein in snail Cd and Cu metabolism." *Nature* **1997**, *38*, 237–238.

55. Thill, A.; Zeyons, O.; Spalla, O.; Chauvat, F.; Rose, J.; Auffan, M.; Flank, A. M., "Cytotoxicity of CeO_2 nanoparticles for *Escherichia coli*. Physico-chemical insight of the cytotoxicity mechanism." *Environ. Sci. Technol.* **2006**, *40* (19), 6151–6156.

56. Collin, B.; Oostveen, E.; Tsyusko, O. V.; Unrine, J. M., "Influence of natural organic matter and surface charge on the toxicity and bioaccumulation of functionalized ceria nanoparticles in *Caenorhabditis elegans.*" *Environ. Sci. Technol.* **2014**, *48* (2), 1280–1289.

57. Zhang, P.; Ma, Y.; Zhang, Z.; He, X.; Zhang, J.; Guo, Z.; Tai, R.; et al., "Biotransformation of ceria nanoparticles in cucumber plants." *ACS Nano* **2012**, *6* (11), 9943–9950.

58. Yokel, R. A.; Florence, R. L.; Unrine, J. M.; Tseng, M. T.; Graham, U. M.; Wu, P.; Grulke, E. A.; et al., "Biodistribution and oxidative stress effects of a systemically-introduced commercial ceria engineered nanomaterial." *Nanotoxicology* **2009**, *3* (3), 234–248.

59. Nowack, B.; Schulin, R.; Robinson, B. H., "Critical assessment of chelant–enhanced metal phytoextraction." *Environ. Sci. Technol.* **2006**, *40* (17), 5225–5232.

60. Burken, J. G.; Schnoor, J. L., "Uptake and metabolism of atrazine by poplar trees." *Environ. Sci. Technol.* **1997**, *31* (5), 1399–1406.

61. Arnaout, C. L.; Gunsch, C. K., "Impacts of silver nanoparticle coating on the nitrification potential of *Nitrosomonas europaea.*" *Environ. Sci. Technol.* **2012**, *46* (10), 5387–5395.

62. Sorrell, B. K.; Downes, M. T., "Water velocity and irradiance effects on internal transport and metabolism of methane in submerged *Isoetes alpinus* and *Potamogeton crispus.*" *Aquatic Botany* **2004**, *79* (2), 189–202.

63. Bone, A. J.; Matson, C. W.; Colman, B. P.; Yang, X.; Meyer, J. N.; Di Giulio, R. T., "Silver nanoparticle toxicity to Atlantic killifish (*Fundulus heteroclitus*) and *Caenorhabditis elegans*: A comparison of mesocosm, microcosm, and conventional laboratory studies." *Environ. Toxicol. Chem.* **2015**, *34* (2), 275–282.

CHAPTER 13

Risk Forecasting and Life-Cycle Considerations

Christine Ogilvie Hendren

Department of Civil & Environmental Engineering, Duke University, Durham, North Carolina, USA

Mark R. Wiesner

Department of Civil & Environmental Engineering, Duke University, Durham, North Carolina, USA

Life-Cycle Impacts and Sustainability

The applications of nanomaterials discussed in previous chapters illustrate some of nanotechnology's many positive contributions to sustainability—an integrated approach to actions that takes into account the social, economic, and environmental consequences of these actions for current and future generations. The design of nanomaterials and the infrastructure for fabricating and using these materials require a systems approach that extends environmental responsibility upstream from the end-use of the material to the manufacturing of the nanomaterials. In its first nearly two decades of existence, work on the environmental, health, and safety (nanoEHS) dimensions of nanotechnology, has made great advances toward identifying key questions to be asked in evaluating the risks of nanomaterials, and how to go about addressing these questions to support risk-based decisions. However, nanotechnology is merely a dress rehearsal for the next emerging technology. In the process of making progress in proactively addressing nanomaterial risk, much has been learned about performing interdisciplinary research, pulling together data from disparate laboratories and methodologies, developing integrated risk forecasting models, and bridging laboratory and field experiments. The ceaseless development of new technologies and the desire to ensure that these new technologies emerge as tools for sustainability rather than environmental liabilities, underscores the need to generalize what we are learning about the process of evaluating nanotechnology risk and uncertainty to inform decision-making surrounding future technologies.

Decision making on emerging technologies must adapt to an evolving profile of known and perceived the risks and benefits associated with the technology [1]. These risks and benefits are associated with evolving levels of uncertainty (Figure 13.1). For example, the need for a given technology may be well known, entailing a relatively small margin of uncertainty associated with estimating the benefits associated with the life cycle of that technology. In contrast, the level of uncertainty surrounding risk may be quite high. A precautionary approach to decision making may imply high levels of uncertainty, and a well-identified source of risk, but may also come into play when the potential sources of risk are unknown. In the case shown in Figure 13.1, additional information yields improved estimates of both benefit and risk such that as the system moves toward more perfect information, risk and benefit are indistinguishable within the bounds of the uncertainty. In these cases, risks are initially estimated to be lower than benefits, but are indistinguishable within the margins of uncertainty. Similar scenarios might be constructed where the risks clearly outweigh benefits and vice versa. In the case where benefits are comparable to or greater than risks, the possibility arises that early overestimates of risk may stifle the development of what would ultimately be determined to be a beneficial technology. In these cases, technology developers must climb this "mountain of uncertainty" to bring ultimately sustainable products to market. Uncertainty at the early stages of a technology's trajectory may be reflected in a reluctance of venture capital to invest in the technology, inappropriately high insurance premiums, and hesitation to adopt the technology. Timely production and updates of risk information are therefore critical to guiding nanotechnology development at early, sensitive stages in the trajectory of their development.

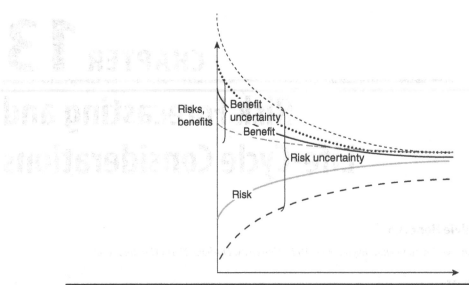

FIGURE 13.1 One example of a changing risk-benefit profile for a new technology with time and associated levels of uncertainty. (Adapted from [1].)

Applying Life-Cycle Thinking and Risk Assessment to Nanomaterials

In this chapter, we consider ways to ascertain the life-cycle impacts and risks of nanomaterials. In addition to consideration of a changing risk-benefit profile over time, the evaluation of a new technology requires methods to forecast benefit and risk for systems of production, distribution, use, and disposal that may not yet exist. To this end, we begin with an introduction to the concepts and goals of life-cycle assessment (LCA) and risk assessment developed largely for existing systems. We review varying approaches to framing a risk forecasting, briefly address the important ways approaches have advanced and changed over the first decade of the nanoEHS field, and present an example of how an existing risk assessment tool can be applied to look at impacts based on the pieces of information that are already known.

This example application will illustrate how employing risk assessment methods used in industry can be directionally helpful; however, it is important to recognize that full impacts will not be adequately captured solely via application of existing assessment methods to these novel materials. The composition of nanomaterials may closely resemble known bulk materials, but their properties at the nanoscale may differ from those of bulk materials. It is precisely because of these useful properties that nanomaterials are used in the first place. In some cases, it may be that the particular properties and behaviors used until now to characterize the impacts of a given material are inadequate in describing impacts of nanomaterials.

Indeed, the elemental atomic composition and structure of a substance appear to be insufficient identifiers for many nanomaterials. As a result, the risk assessment and life-cycle impact communities must grapple with assessment criteria that will more adequately account for factors unique to nanomaterials and any risk these materials may pose. The very notion of grouping materials under the heading "nanomaterials" may be less useful from a decision-making standpoint than a grouping based on the useful properties that inspire nanomaterial applications as discussed later in this chapter.

Life-cycle and risk assessments are instruments that tie together the insights of several disciplines in order to provide a framework for decision making. Now, just as the nano-industry is burgeoning, is the time to employ such decision-making tools. Information is now emerging that allows for meaningful assessments of the risks and benefits posed throughout the life cycle of these materials, but there are still many choices to be made. At this early stage, the industry still has the flexibility to intelligently and responsibly apply these technologies while minimizing negative environmental impact.

Concepts from LCA and Risk Assessment

An examination of the repercussions of activities and products throughout a life cycle of production, use, disposal, and reuse is at the heart of approaches for sustainability planning and decision making. For any given industrial product, the life-cycle stages of resource extraction, raw material production, product manufacturing, transportation, use, and end-of-life can all be associated with significant costs and benefits to the manufacturers, customers, environment, and other stakeholders. We distinguish the "value chain"

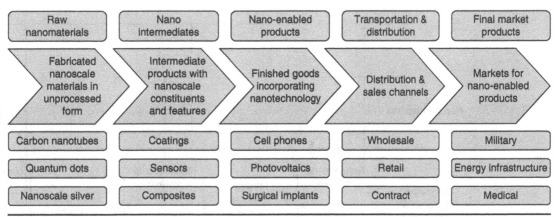

Figure 13.2. A generic value chain for nanomaterials. (Adapted from [2].)

here as an important element of the life cycle of engineered nanomaterials that affects the potential for exposure to human and ecological receptors.

The value chain for a given nanomaterial considers the initial production of the nanomaterial, its incorporation into subsequent products, and distribution, use and disposal of these products (Figure 13.2). The nanomaterial value chain involves the production of basic building blocks of nanomaterials (often nanoparticles) and their incorporation in subsequent stages into products of increasing complexity. For example, engineered nanomaterials such as carbon nanotubes might be modified to yield an intermediate product such as a nanotube suspension with a tailored surface chemistry. These suspensions might then be used to create another intermediate product such as a display screen that is then incorporated into a cell phone, that is packaged, distributed sold, and used. While this example describes a sequential flow of nanomaterials, the value "chain" may be a network of flows of materials and products.

At each stage in the value chain, there exists the possibility of nanomaterial release and subsequent exposure to humans or ecosystems through the production, transport, use and disposal of nanomaterials, and nanomaterial-containing products. Important factors to be identified in evaluating potential nanomaterial exposure at each stage in the value chain are the format that nanomaterials will be present in as commercial products, the potential for these materials to be released to the environment, and the transformations that those materials may undergo that may affect their subsequent potential for exposure. Indeed, due to modifications along the value chain or environmental transformations, the potential contact between humans and ecosystems outside of the work place will most likely involved nanomaterials there bear little resemblance to the initial material [1].

LCA involves an analysis of the materials and energy inputs, the wastes produced, and nanomaterials released at one or more stages of the value chain (Figure 13.3). The repercussions of energy and materials use, wastes produced and nanomaterials released can be measured in many different manners. Among the metrics of potential interest are the positive benefits (e.g., lives saved, improved economic conditions or societal improvements) and the negative benefits (e.g., cancer risk, increased CO_2 emissions and global warming, or money spent on health care). The potential for nanomaterial release associated with the production use and disposal of products, generally tends to decrease as one moves along the value chain. However, the environmental media to which nanomaterials are released may also be considered as "products" in the sense that air, water and land are public goods that may be degraded or improved by the presence of nanomaterials. These latter environmental stages of the value chain in particular may involve an increased potential in exposure. Examples of increased exposure to humans or ecosystems along the value chain include some medical uses of nanomaterials where patients may have direct contact or the injection of nanomaterials to ground water for contaminant remediation (see Chapter 8).

A quantitative accounting of nanomaterials across the value chain and all receiving environmental compartments can be mathematically represented as a series of linear equations. In the generalized form where products along the value chain and environmental compartments alike are considered as compartments that may be sources or recipients of nanomaterials flowing through the network, each product i is a possible source of nanomaterials to another product j, where $i = 1$ refers to the inventory, S, of nanomaterial at the source of fabrication, and where the flow occurs at a rate F, having units for example of mass per time (see section on Source Estimates later in this chapter). At S there is a singular "product," which is the raw

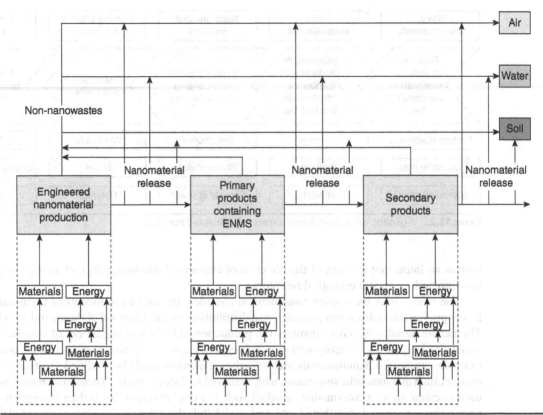

FIGURE 13.3 Energy and materials inputs, wastes produced, and nanomaterials released along the value chain.

nanomaterial; this source material may next be incorporated into any number of nano-enabled intermediary and final products, where the individual nano-enabled products are indexed as the jth product in network. For example, S may be incorporated into products 2 and 3 that may then be used in fabrication of products 4 and 5, which enter into nano-enabled product 6, and so forth. Any of these products may be sources of nanomaterials to the "products" of water, air, etc. For any product i, the quantity of nanomaterial in the product inventory j at any time is designated as P_i, (under this formulation the quantity of source material is designated as $S = P_1$). The fraction of nanomaterials in the inventory of product P_i, incorporated per time into downstream product P_j is designated as $f_{i,j}$.

For a given production rate, F, the source term S is given as

$$S = \frac{F}{\sum_i f_{1i}} \tag{13.1}$$

The flows out of a given node must sum to 100 percent per unit of time, and this S and F will have the same numerical value with the denominator in Equation (13.1) introducing the appropriate unit of time.

Using this notation, the accumulation nanomaterials present in any product inventory j of the value chain, P_j can be represented as a differential equation describing materials flowing in from all other nodes of the value chain minus those flowing out:

$$\frac{dP_j}{dt} = \sum_i P_i f_{i,j} - P_i \sum_k f_{j,k} \tag{13.2}$$

At steady state, the inventory for node j is given as

$$P_j = \frac{\sum_i P_i f_{i,j}}{\sum_k f_{j,k}} \tag{13.3}$$

If we assume that only lower-order nodes contribute to any given node in the value chain, Equation (13.3) for $j = 2$ reduces to

$$P_2 = \frac{Sf_{1,2}}{\sum_{k} f_{2,k}} \qquad (13.4)$$

Generalizing Equation (13.4) to any given stage, we conclude that the release to any given compartment can be expressed as the product of the source term S and a function of the $f_{i,j}$. In other words, the rate of release of nanomaterials to any given compartment and, of particular interest in exposure assessment, any environmental compartment, is proportional to the production rate S. This result provides some basis for rationalizing regulatory exclusions based on small production quantities.

Conceptually, the description of all flows within the value chain network, that is, the specification of all the $f_{i,j}$ coefficients, represents a very high demand for information. The $f_{i,j}$ contain information on trends in commercialization, product use, human behavior, product degradability, the nature of receiving environments at associated stages of the value chain, nanomaterial transformations, and transport. While the many $f_{i,j}$ that contribute to the constant of proportionality may not be known or even knowable, summation across the value chain, or across receiving compartments such as wastewater, water or air, reduces the number of unknowns at the cost of a loss of detail. For example, the amount of nanomaterial entering the wastewater compartment can be expressed as a product of the source of nanomaterials produced S and a single constant $f_{\text{wastewater}}$, which captures nanomaterial production and use profiles.

The value of $f_{\text{wastewater}}$ may be estimated from a simplified value chain analysis, or from direct measurements of the quantities of nanomaterials in wastewater, may be extrapolated from previous experience with persistent materials released to the environment, or simply treated as a constant that is adjusted to explore various release scenarios.

LCA and Life-Cycle Thinking

LCA provides a framework for identifying and evaluating the life-cycle impacts of a product, process, or activity. Typically, impacts on human health, ecosystem health, and effects of pollutant depositions in all environmental media are evaluated for each stage of the life cycle. For a new technology application in an area such as nanotechnology, LCA offers a context for looking at potential life-cycle implications, optimizing its economic, environmental, and societal benefits, and minimizing risks.

Formal LCA

While there are many variations in LCA methodologies, arguably the most broadly accepted is one formalized in the ISO-14040 series of standards [3, 4]. Often referred to as the "formal LCA" or "full LCA," the ISO methodology guides the quantitative assessment of environmental impacts throughout a product's life cycle. Typically used in comparing product or technology alternatives, the formal LCA involves the following phases:

1. *Goal and scope definition:* Defining purpose, comparison basis, and boundaries, including the life-cycle stages and impacts that require detailed assessment.

2. *Inventory:* Developing "life-cycle inventory" (LCI) where information on the inflows and outflows of materials and energy (including resource use and emissions) for each relevant life-cycle stage are collected and tabulated.

3. *Impact assessment:* Analyzing LCI data and grouping the impacts into several impact categories, such as resource depletion, human and ecosystem toxicity, global warming potential, smog formation potential, eutrophication potential, and so on—this phase is often termed "life-cycle impact assessment" (LCIA).

4. *Interpretation:* Evaluating, identifying, and reporting the needs and opportunities to reduce negative impacts and optimize the environmental benefits of the products.

A major challenge in conducting the formal LCA is to obtain reliable and available data for LCI. Data collection can easily become an enormous and expensive task, especially if data specific for the exact materials and processes involved in the product life cycle are sought. Generic data for various raw material classes and life-cycle activities, however, are available in much of the commercially available LCA software.

Once the inventory is developed, the life-cycle environmental impacts can be determined using various existing LCIA methodologies [5–7]. The life-cycle impacts can then be weighted based on criteria such as geographical relevance, relevance to the decision, and stakeholder value.

Although not explicitly part of the formal LCA, life-cycle economic implications are often considered. The "eco-efficiency assessment" developed by BASF [8], for example, balances weighted environmental impacts with the "total cost of ownership," which covers the costs of production, use, and disposal of the product. Results from LCA can also be expressed in eco-efficiency metrics that normalize impacts by revenue or monetary value-added [9, 10].

Efforts have also been made to integrate social impact considerations into LCA. The eco-efficiency assessment of BASF has recently been extended to include social impacts [11]. Individual indicators on a product's impacts on human health and safety, nutrition, living conditions, education, workplace conditions, and other social factors are assessed and scored relative to a reference (usually the product being replaced).

Streamlined LCA

The life-cycle thinking formalized in LCA also lies behind strategies, guidelines, and screening methodologies that may be used in the early stages of product and process design. Often called "streamlined LCA" [12], they typically rely on the use of a two-dimensional matrix, with life-cycle stages on one dimension and impacts on the other (Table 13.1). The streamlined LCA is typically qualitative, with a relative score or color-coded rating assigned for each impact at each stage of the life cycle. The qualitative assessment allows the consideration of impacts not typically considered in the formal LCA, such as nuisance, public perception, market position, and so on. Supported by a question-driven approach, such assessment can help identify problems and point out where improvements are most needed [13].

Application to Emerging Technologies

LCA in its various forms has been used to identify and communicate the potential costs and benefits of new technology applications. For example, formal LCA was applied to compare the environmental and human health impacts of genetically modified herbicide-tolerant sugar-beet crop with those of the conventional crop [14]. The study demonstrated the benefits of the herbicide-tolerant crop in lowering emissions and their associated environmental and human health impacts in various stages of the life cycle (primarily herbicide manufacturing, transport, and field operations), while recognizing that the life-cycle costs and benefits of genetically modified crops in general need to be assessed on a case-by-case basis.

LCA has also been used to examine the impacts of replacing traditional materials with engineered nanoparticles. Lloyd et al. [15, 16] first applied LCA to examine the potential economic cost and environmental implications of nanotechnology applications in automobiles, specifically the use of nanocomposites [15] and the use of nanotechnology to stabilize platinum-group metal particles in automotive catalysts [16]. The authors utilized a variant of LCA called "economic input-output LCA" (EIO-LCA) developed by their group at Carnegie Mellon University, where environmental impacts are estimated using publicly available U.S. data on various industrial sectors involved in the product's life cycle (see www.eiolca.net). Both studies demonstrated the benefits of the nanotechnology applications in reducing resource use and emissions along the life cycle compared to conventional practices. Nevertheless, these applications of LCA toward nanotechnology applications focus primarily on their potential benefits. Risks posed by the nanomaterials themselves were not addressed.

Life-Cycle Stages	Materials Choice	Energy Use	Solid Emissions	Liquid Emissions	Gaseous Emissions
Resource extraction and raw material production	3	2	1	4	2
Manufacturing	3	4	3	4	4
Product delivery	3	4	2	4	4
Product use	4	3	4	4	4
Refurbishment, recycling, disposal	2	1	1	4	4

Adapted from Graedel (1998) [12].

TABLE 13.1   Example of a Streamlined LCA Matrix for Automobiles along Life-Cycle Stages Comparing Today's Cars with Those in the 1950s (Higher scores indicate larger improvements, while low scores indicate where attention is needed.)

Since that time, LCA approaches have been applied by a variety of researchers to assess specific nanomaterial risks as more data become available, and as more materials are adopted into specific production methods that can be practically assessed [17–19]. One 2012 review of LCA applications to nanoscale materials production indicated that multiple studies have applied LCA methods to evaluate nanomaterial production methods, but that these applications remain exploratory and often limited in scope [20]. Of the 18 applications of LCA studies to ENMs, all but one evaluated the manufacturing phase and 14 evaluated the extraction phase, but only three addressed the use phase of the life cycle and only two addressed end-of-life, ostensibly due to insurmountable uncertainties associated with these phases. Many of the evaluations were qualitative rather than quantitative. Another review published the same year similarly concluded that most studies were "cradle-to-gate" meaning that the assessment included activities up to the products leaving the factory gate, as opposed to "cradle-to-grave" [21]. To advance this field, these reviews recommend focusing on generating ENM-specific LCI data as far as scientific advances will allow, including protocols and models for how ENM properties are determined. They stress the importance of collecting information comprehensively and transparently, and advise employing qualitative or screening-level assessments where quantitative data remain unavailable. Finally, they advise the community to leverage existing modeling tools modified with nano-specific parameters, and to utilize additional decision-support tools along with LCA such as risk assessment in order to incorporate other important factors like location-specificity, etc. in an ultimate decision [20, 21].

The results of a LCA are helpful in understanding the impacts of a particular process, but the ability of these assessments to directly inform decisions often depends on contextual information and the underlying values of the decision makers. Uncertainty in model parameters, their values, and decision-maker preferences affect the manner in which LCA data are used to support decisions ranging from product development to regulatory responses. In an effort to better connect the results of such assessments to the decision-making contexts they are intended to inform, LCA has been coupled with decision analytic methods such as Multi-Criteria Decision Analysis (MCDA). By clearly defining the decision being supported by the life-cycle analysis data and articulating a clear scoring system that weights the various attributes of the decision according to decision-maker preferences, the LCA-MCDA approach provides a framework for interpretation of the data in guiding decisions on nanomaterials [22, 23].

Risk Assessment

Oxford Dictionary defines risk assessment as a systematic process of evaluating the potential risks of a projected activity, in this case, the introduction of a new platform technology. Risk assessments characterize levels of risk, usually in terms of a relative score or ranking, and in some established cases in terms of probabilistic calculations of specific outcomes (e.g., the increased risk of cancer due to exposure to a substance). As with LCA, the goal is to provide information that will help evaluate alternatives and inform decisions; indeed, categories of risk may be included as elements of an LCA. Generally speaking, hazards are identified and characterized, exposure likelihood is evaluated, and both are quantified. These risk-determining factors are then combined to characterize relative risk [24].

Defining Risk

Risk has many context-specific definitions, but can be generally described as the probability of exposure to a hazard. One definition from a business perspective is that a probability or threat of damage, injury, liability, loss or any other negative occurrence (the hazard) that is caused by external or internal vulnerabilities (exposure to the hazard). From an expert's point of view, risk has a statistical meaning: the product of the probability of an undesirable event and some measure of the magnitude of its negative consequences. Hazard (the potential to cause harm) is combined with probability of exposure to arrive at risk, a quantification of the likelihood that such harm will occur [25]. A communal perception of risk, on the other hand, is more subjective. People combine their knowledge, experiences, and anecdotal evidence to make decisions that can be related afterward to a perception of the likelihood and unpleasantness of a particular outcome [26]. Both objective and subjective views affect the public's acceptance of a new technology and should be considered in decision making.

This chapter is primarily concerned with the quantitative dimensions of risk. However, as the subjects of hazard identification and risk assessment are explored, it is important to maintain an appreciation for the role of risk perception and public opinion on the decision-making process.

Systemic Bias

The kernel of the risk assessment is the definition and mathematical combination of exposure and hazard, weighted so that risk factors considered as most important are valued higher and a relevant ranking system can be created. It is easy to see how such a process of expressing some objective reality in terms of a score necessarily imparts a subjective filter, or value system, in order to weight and combine the factors that determine the risk.

Biases are a natural part of risk assessment and need to be understood and made transparent for the results of the assessment to adequately inform a decision-making process. The parameters chosen to represent hazard and exposure, as well as the way in which the collected values for these parameters are combined, may vary between systems and significantly affect the conclusions of the entire assessment. These inherent biases are depicted in Figure 13.4. Out of the universe of hazards potentially posed by nanomaterials and factors that may affect exposure, one inevitably must choose those to be included in the risk assessment model. While they may be selected because they are believed to be most critical to accurately capturing the hazard or exposure, they are often more practically selected on the basis of data availability or importance to specific group of stakeholders. Once selected, all the risk factors are defined in terms of measurable parameters, entered into the risk assessment methodology, and combined mathematically to produce a relative risk rating. The process of combining the parameters introduces another layer of bias because the relative weight of the different values decides which parameters have the greatest influence on final risk.

For example, an important factor that defines the hazard of a material is toxicity, a complex concept that can range in interpretation and can refer to multiple specific effects such as primary irritation, cytotoxicity, or genotoxicity. In many cases the toxicity is measured by recording the LC_{50}, which is the lethal concentration of a material administered to a population of, for example, rats or fish at which half of them die. However, this response represents a culmination of responses at the level of the cell and organisms, any one of which may have relevance in characterizing toxicity (discussed further later in this chapter, and in detail in Chapter 5). Selecting, for example, LC_{50} as the parameter to represent toxicity via mortality may often be the most straightforward, but fails to address sublethal endpoints. In fact, utilizing toxicity to represent hazard is also limited and fails to address more subtle but potentially important hazards such as, for example, nutrient perturbation in an ecosystem. Whether the parameter is appropriate and adequate depends on the nature of the hazard and its utility in decision making.

We will further consider the manner in which we parameterize and score risk. However, as we explore the idea of risk assessment pertaining to nanomaterials it is, at the outset, easy to imagine a variety of risk assessment methods that deliver equally valid results. Depending on the decision-making process being affected, a risk assessment protocol might be designed to address an economic, legal, political, engineering, or social perspective [24]. Whatever the perspective, the goals of a quantitative, statistical risk assessment are to inform decision making by allowing comparison between different options, and to serve as a critical input for a plan to communicate with the appropriate stakeholders [27].

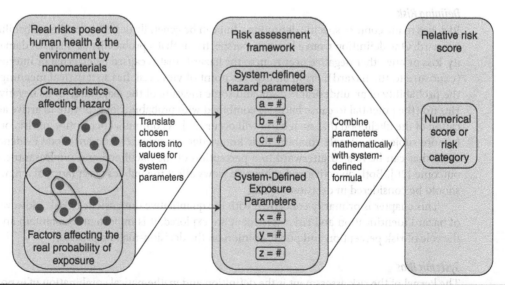

Figure 13.4   Pictorial representation of the steps and biases involved with mapping real-world risks to a relative risk score via a risk assessment framework.

Stakeholder Bias

Numerous stakeholders may be affected by and/or influence the application and commercialization of engineered nanomaterials. Each of these groups may characterize the importance of various parameters based on their respective value systems, none of which are without bias [28]. Startup nanotechnology companies see the great benefits of the growing markets for their products. Users of products containing nanomaterials will enjoy the benefit of the new products but will be sure to care deeply about the safety of being exposed to them. Insurance companies will need to know how much to charge for premiums based on the risk they are taking on when underwriting the nanotechnology industry. The general public will both experience positive changes enabled by nanomaterials and be impacted in some way by nanomaterials entering the air, water, soil, and organisms around them. Different stakeholder groups with their varying value systems and priorities see the costs and benefits through different lenses.

The public at large, including even a few influential groups, can be particularly effective in determining the trajectory of nanotechnologies. In the case of another emerging technology full of unknowns—genetically modified foods (GMOs)—market leaders failed to anticipate and react to societal perceptions of the risks (largely in Europe and Canada) versus benefits from these products. The resulting public backlash led to regulatory limitations that continue to inhibit the development of this technology and have brought genetic modification of food to a virtual standstill in Europe. This example is not intended to be illustrative of an extended analogy between the technology of nanomaterials and GMOs, but to highlight the crucial role of public perception and communication, especially in the case of uncertainty. Financial industry forecasters, yet another stakeholder group, recognize this importance; even as one strategic advising firm predicted the immense economic impact of nanotechnology in a 2005 report, they cautioned that "poor handling of risk by any player could result in perception problems that would affect entire markets" [29].

Regardless of which lens is chosen to view the risk, there is a need for information on the likelihoods of hazard and exposure associated with nanomaterials. Once quantified, additional challenges lie in interpreting and managing risk. Acceptance and risk avoidance behaviors often depend on the nature of the risk itself and may vary over time. For example, "publicity bias" refers to a more recently realized risk, such as a tornado, that may be overestimated when compared to a greater risk whose last manifestation has faded from memory, such as a pandemic plague [26]. In this case, the perceived probability of exposure outweighs the influence of hazard magnitude on the perceived risk. Alternatively, the opposite can happen, with people focusing heavily on hazard while downplaying the influence of exposure. They overestimate their actual danger by highly emphasizing the severity of the potential effects without considering the likelihood that they will actually experience the effects. This propensity is termed "compression bias," wherein rare or dreaded risks are overestimated when compared with the impact of common risks [30]. A familiar example of such miscalculation would be the widespread fear of dying in a plane crash versus the relative acceptance of dangers inherent to the more mundane act of driving a car. Risk perception research has shown many other predictable tendencies, such as the preference for controllable risks over ambiguous, uncontrollable ones. An example of this would be the acceptance of comparatively large risk of food-borne illness from eating raw fish such as sushi or raw oysters, which many people purposely choose, as compared to the recent public outcry and aversion to eating beef due to the arguably less likely risk of contracting mad cow disease. Our very behavior reveals inherent calculations of the trade-offs between perceived risks and benefits that may or may not be consistent over time. We perform a rather intricate balancing act between factors such as temporality, ambiguity, controllability, and dread level to arrive at an instantaneous decision regarding preferences of risk. Better information on risk and reduced uncertainty are needed to improve the quality of decisions. However, generalizing individual preferences of risk to predict behavior may not be meaningful. Moreover, preferences for risk expressed in a political or social context may not coincide with decisions that would otherwise appear to be "optimal" based on available information. The public is becoming increasingly influential in shaping policy, and decreasingly reliant on a strictly science-based decision-making regime regarding the risks of new technologies [31]. Nonetheless, information on risk is needed whether the public makes informed choices between risks and benefits or chooses to put in place policies to avoid risk where information is thought to be lacking.

We will not attempt to condense the vast body of work addressing the complex behaviors and choices related to risk perception, but rather will simply summarize with the following two concepts:

1. "Riskiness means more to people than 'expected number of fatalities'" [30].

2. Because no stakeholder group is without its bias, an environmental risk assessment process needs to include iterative feedback loops to allow for the influence of a variety of valid viewpoints [28].

While there are major gaps in our ability to characterize the risks of nanomaterials, a lack of information will not preclude stakeholders from taking actions that will determine the future landscape of risk for better or worse. One editorial summarized the importance of active participation by researchers in helping to inform the decision-making process: "The time has come for the environmental research community to come together with the environmental regulators to address [impacts from the manufacture and use of nanomaterials]. Failure to undertake this will almost certainly ensure that the media steers public understanding of and confidence in nanotechnology, leading to unsubstantiated anecdotes and wild conjecture potentially forming the basis for an ill-informed debate with outcomes that may be wholly disproportionate to the risks" [32].

Assessing the Risks of Nanomaterials

A predicament is created by an urgent need for guidance in developing risk management strategies for a developing nanomaterials industry at a time when data to support decision making remain disparate and fraught with uncertainty. Our approach must therefore be incremental, allowing us to build toward the "ideal" risk assessment methodology as data become available.

Working with Existing Risk Models

Current tools for risk assessment and life-cycle analysis can provide useful guidance for short-term risk assessments for nanomaterials as we await the results of studies of exposure and hazards. Physical and chemical properties, transport and mobility behaviors, projected physical quantities, and manufacturing practices must be understood for chemicals and the materials used to produce them in order to evaluate their impacts. To date, a number of environmental exposure and hazard models developed for chemical substances have been applied to ENMs; indeed this is a sensible starting point, but a 2013 review of more than 30 exposure models revealed that only a small fraction accounted for the unique properties of ENMs or explicitly dealt with uncertainty [33]. As discussed later, risk evaluation for emerging technologies is a forecasting rather than a forensics effort. A significant challenge in developing robust exposure models for ENMs remains the inability to validate risk models with field level data on exposures and effects. As relevant data are generated from complex systems such as mesocosm studies, a balance must be sought between the use of quantitative mechanistic models and the inclusion of qualitative, expert-driven models that are able to acknowledge new information unique to ENMs and their associated uncertainties prior to the more ideal position of being able to support accurate mechanistic predictions [33].

In the spirit of placing nanomaterials into established impact evaluation systems, later in this chapter we will present an example of a risk assessment as seen through the lens of an insurance company. The relative risk assessment of five nanomaterial manufacturing processes will demonstrate how an existing insurance risk analysis tool can be applied to nanotechnology manufacturing [34].

Models for Risk

Some background on the underlying mathematical approaches to modeling physical phenomena is useful to understanding the various approaches to modeling risks of nanomaterials.

A deterministic system is one entirely without randomness, so that predictions of the outputs will be identical each time given the same initial parameter conditions. A stochastic system, in contrast, refers to a physical system with uncertainty about the values of input parameters that may depend on many random processes; predictions of the outputs of stochastic systems are determined probabilistically rather than on unique input values and therefore may not be predicted precisely. The interactions of chemicals and particles with environmental and biological systems may include some deterministic processes, but clearly include stochastic processes as well given the complex interactions of real systems. Selection of a modeling approach may depend on a mixture of what will provide the most objectively accurate representation of the system and on pragmatic constraints such as data availability and limited knowledge of initial parameter states. In environmental risk assessment, collecting detailed data specific to each individual scenario is typically impossible or prohibitively expensive, so data may be sparse. Add to this that in the specific case of emerging materials such as engineered nanomaterials, there are often great uncertainties associated with the parameters, their values, and their relationships [35]. For these reasons, a variety of modeling approaches are used to estimate the exposures, hazards, and resulting risks of engineered nanomaterials.

Forensics versus Forecasting   Risk assessment has historically been a forensics exercise. There are multiple unfortunate examples of potentially revolutionary technologies that have imparted regrettable collateral damage, a cycle which the nanoEHS community is working to break. The fuel additive MTBE (methyl tertiary butyl ether) replaced lead as an oxygenate in gasoline, but ended up ubiquitous in groundwaters due to leaking tanks and was found carcinogenic. The pesticide DDT (dichlorodiphenyltrichloroethane) was first used to control malaria and typhus, but eventually through widespread agricultural use caused ecosystem collapses, thinning the walls of birds' eggs and inspiring Rachel Carson's landmark book *Silent Spring*. Assessment of the risks of these contaminants happened after their production and release, through thorough detective work aimed at linking real environmental effects data to the upstream chemical causes. In the case of engineered nanomaterials, the existence and availability of such field level effects data is precisely what the community is working to avoid. Therefore, the methods used in these previous risk assessments must be mined for resulting wisdom (What endpoints were key ecological indicators? How do stressors proliferate through ecological systems and what are the mechanisms of their negative biological effects?). Models built to predict behavior of physical and chemical stressors must similarly be utilized where possible as resources to forecast ENM behavior. However, given that our mission is to forecast a future for which we hope there are newer data to fully test our estimation methods, particular attention must be paid to methods that allow for persistent uncertainties and for iterative updates with emerging knowledge on ENM-specific parameters and values.

Mechanistic Models   Mechanistic models are built on the assumption that complex systems can be represented by parameters representing the functioning of their individual components and the way these components are coupled.

Mechanistic effects models (MEMs) combine modeling approaches at multiple system levels to investigate the true drivers of observed effects. For example, combining individual-based population models (IBMs) with toxicokinetic/toxicodynamic (TK/TD) models link individual organismal level effects of individual toxicants to the effects of that toxicant at the population level. Various assumptions about organismal level effects of toxicants can be tested to explore which are the most predictive of effects at the population level [36].

Probabilistic Models   Probabilistic models predict the outcomes of stochastic processes, representing parameters with probability distributions rather than unique values. After a mathematical model is built to represent the physical system of interest as a function of the relevant parameters, Monte Carlo simulation methods solve the expression many times, each time with input values selected randomly from their probability distribution functions. By taking many repeated samples from defined distributions for model input parameters, Monte Carlo simulations generate an estimated probability distribution of different outcomes. The larger the number of simulation runs, the richer the PDF that is generated. This type of model, named after the famed Monoco casino, is useful in modeling systems characterized by uncertainty because by providing results in probability distributions rather than discrete values, they include evaluations of the certainty of calculated quantities [37]. This can be particularly helpful in modeling real environmental systems where it is necessary to grapple with both variability and potential inaccuracy, and has been applied in several approaches to modeling environmental concentrations of engineered nanomaterials [37–39].

Bayesian Models   Bayesian networks (also known as Bayesian belief networks or probability networks) are a modeling approach derived from Bayes' theorem, which describes the probability of an event based on conditions that may be related to that event. This approach allows for the use of new evidence to update prior beliefs about this conditional probability. In a Bayesian belief network, a system of potentially interconnected variables is linked to one another in directed, acyclic graphs to represent which variables influence the values or states of other variables. The strength of this conditionality can be explored in a quantitative approach to forecasting outcomes by drawing on existing information on quantitative and qualitative information, and be easily updated as new information becomes available, and allow examination of the origins of uncertainty within the models [1]. Bayesian networks offer a credible yet practical approach to modeling systems with pervasive uncertainties such as complex ecological systems [40, 41]. These features address many of the challenges presented both by complex stochastic systems and new potential stressors associated with uncertainty. This has resulted in increasing application of Bayesian network approaches to modeling ecological risks in complex environmental systems [42], and to several applications proposing and testing the outcomes of emerging technologies such as engineered nanomaterials [43–45].

QSAR Models   Quantitative structure activity relationship (QSAR) models are predictive statistical models developed to relate a set of physical-chemical and structural parameters to the values of select response variables in chemical and biological systems. The measurable biological activity of interest may vary, sometimes including undesirable toxic effects and sometimes including desirable effects as in the case of therapeutic pharmaceuticals, with the goal of identifying which properties of a molecule drive the outcomes of interest; once this relationship is mathematically established and validated, the intent is to utilize it to predict outcomes for molecules with similar properties without having to exhaustively test the activities for each new material [46]. Some researchers are developing QSARs for application to nanomaterials, also known as QNAR (quantitative nanostructure activity relationship) models, demonstrating their utility for predicting biological activity profiles and for screening level decisions in design and management of nanomaterials [47].

Actuarial Models   Actuarial risk assessments employ robust statistical models, developed when enough empirical evidence exists to validate specific parameters and their relationships, generating an analysis of the likelihood of a risk event. Based on this definition, it is clear that their information requirements do not align particularly well with the problem of forecasting risk outcomes of novel materials in complex real environments that we hope the materials never negatively affect. However, realistically an important mechanism of managing risk is via private insurance, which requires the calculation of premiums that rely on actuarial methods. Because insurance is an important and immediate lever for managing risk, approaches have been proposed to forecast and thus mitigate environmental risks via insurance calculations [48]. One study applying an existing insurance model to evaluate the relative risk of the manufacturing processes for various engineered nanomaterials is presented later in this chapter as an example [49].

Data Needs

With hundreds of nano-enabled products already on the market and ever more in development, near-term decisions must be made on how best to design nanomaterials to mitigate their exposure and hazard, and on how to prioritize focus for risk management and future research. Given the exhaustive number of materials, complex interactions and scenarios that will be encountered as nanomaterials enter the environment, it is clear that risk assessments cannot be taken on a case-by-case basis. In the end, we will need to parameterize predictive models for exposure, hazard, and ultimately risks in terms of a relatively small number of repeatedly measurable parameters. To support this, it is important to know what are the data needs to adequately forecast exposure and hazard; here we will discuss some of these data needs and considerations for which should be prioritized.

Searching for the True Risk-Determining Characteristics of Nanomaterials

Given the information intensive nature of robust risk assessments and the still nascent body of knowledge on nanomaterials, it is critical to identify which characteristics of nanomaterials and their interactions with their environments are most predictive of outcomes of interest. The required characteristics presented in Table 13.2 can initiate a discussion of the state of risk assessment knowledge, but this is by no means an exhaustive list of factors that may well be useful in describing the true risk and impacts of nanomaterials. It is important to note that often these parameter values depend greatly on modifications and speciation of the underlying material. Categorization schema and risk management approaches must account for these variations in modeling the risks of engineered nanomaterials. Should we characterize the solubility of a coated or uncoated nanoparticle? Should we adhere strictly to the pure nanomaterial or evaluate the more likely to occur aggregate as the base substance in a risk database? This underscores the importance, presented earlier in this chapter, of modeling risk in ways that accounts explicitly for uncertainties and for model sensitivity to particular parameters.

We must understand in which life-cycle phases the nanomaterials have the potential to be released to the environment and in which they may be safely bound; their mere existence does not ensure that they will pose a problem to human health or the environment. The Risk Assessment Unit of the European Commission and the U.S. EPA have both listed this distinction as key to enabling a practical assessment of nanomaterial risks [50, 51]. Such clarifications must be made for non-nano-materials already. For example, the LED screens in computer monitors and televisions are composed of hazardous heavy metals such as cadmium and selenium, and yet we live and work in very close proximity to the materials in their fixed form every day with no worry of their toxicity. There is a growing consensus concerning the priorities for efforts in evaluating the impacts on health and environment from nanomaterials that includes standardizing materials, developing or improving the ability to measure these materials, and agreement on what is

Intrinsic and Extrinsic Nanomaterial Characteristics	Exposure Indicators	Hazard Indicators
• Added molecular groups • Chemical composition • Number of particles • Free or bound in matrix • Size distribution • Shape • Solubility • Surface area • Surface charge • Surface coating • Surface reactivity • Thermal conductivity • Electrical conductivity • Tensile strength • Proportion of total number of atoms at the surface of a structure • Molecular weight • Boiling point • Optical behaviors • Magnetic behaviors	• Adsorption tendency • Ability to cross blood brain barrier • Ability to cross placenta • Degree of aggregation • Ability to travel to deep lung • Ability to travel to upper lung • Dispersibility • Interactions with naturally occurring chemical species in aqueous phase • Transformation to other compounds • Interactions with naturally occurring chemical species in soils • Bioaccumulation potential • Ability to travel to central nervous system	• ROS generation • Oxidative stress • Mitochondrial perturbation • Inflammation • Protein denaturation, degradation • Breakdown in immune tolerance • Allergenicity • Cellular/sub-cellular changes • Damage to eyes • Damage to lung • Damage to GI tract • Excreted/cleared from body • Irritation effect • Damage to central nervous system

TABLE 13.2 Nanomaterial Characteristics, Behaviors, and Effects That May Affect Their Relative Risk

important to measure [51–53]. An "ideal" list of parameters to be evaluated to enable an assessment of nanomaterial impacts will not necessarily coincide with the list of parameters that can actually be measured physically or feasibly. The following table summarizes suggestions from six different studies [25, 32, 45, 51, 54, 55].

As we know from some of the earlier chapters in this book, many of these descriptors, especially the physical characteristics, are already known for pristine nanomaterials, though in many cases they will need to be evaluated for relevant transformed nanomaterials in realistic systems. In Chapter 3, *Methods for Structural and Chemical Characterization of Nanomaterials*, methods to measure many physical chemical properties were discussed, including solubility, density, melting point, dielectric constants, size, and surface exchange capacity. The second and third categories of descriptors, which we present here as "exposure indicators" and "hazard indicators," are dependent on the physical-chemical characteristics of the nanoparticles as well as their interactions with the systems in which they are found. They are presented as two distinct groups to set apart the different contributions to risk. Exposure factors are primarily fate and transport variables, while the hazard factors are related to toxicity. While the toxicity factors here are presented in a single list, it is likely that for research and modeling purposes a more hierarchical approach will be taken to defining and testing for toxicity. For example, different enzymes may be present based on different levels of oxidative stress, so an assessment of the extent of damage may involve testing for the presence of a specific enzyme [55]. The overall question for the risk community is how to choose the most appropriate indicators, the most realistically measurable indicators, and how to combine these once chosen. This aspect of identifying data needs will be discussed in more detail below in the proposed functional assay based approach to assessing nanomaterial risk.

Nanomaterial Classification Schemata

Numerous schemata have been proposed for classifying nanomaterials, usually falling into one of three groups. These groups are categorization based on (1) physicochemical properties of nanomaterials, (2) factors affecting exposure and, (3) factors affecting hazard, corresponding in spirit, if not explicitly, to the three columns shown in Table 13.2. Perhaps the most prevalent scheme groups nanomaterials by composition (group 1) using categories such as metals, metal oxides, carbon nanotubes and fullerenes, metal sulfides, etc. Classification schemes in group 1 also include those based on redox properties and shape (spheres, clusters, nanowires, nanotubes, films, etc.), redox properties. Classifications in group 2 include mode of application (e.g., mixed-matrix nanocomposites, thin films, powders, dispersions) as well as bio-availability (respirable materials, bioaccumulation potential). Finally group 3 classifications differentiate nanomaterials by their interactions with biomolecules, cells and organisms (e.g., mutagens, measures of oxidative stress, genomic and proteomic measures).

While the objectives for grouping nanomaterials may vary, there is certainly strong interest in identifying of attributes of classes that might allow one to extrapolate to conclusions on the possible risks posed by materials within a given class. Thus, categorization schemata that combine attributes from groups 1, 2, and 3 incorporate notions of exposure and hazard with easily measured physical-chemical properties. By associating materials of known risk with similar attributes to those of unknown risk, the objective in this case is to reduce to reduce the number of measurements needed required to screen nanomaterials for risk in a process often referred to as "read-across." Such schemata are often presented as a tiered progression of evaluations that combines both the notion of classification with that of a decision tree for evaluating the potential risk of a new material.

A complementary approach to these schemata begins with a "zero-level" tier that differentiates nanomaterials based on the useful properties they exhibit. It is the useful properties seen at the nanoscale that are responsible for both the purposeful interactions driving the performance of these materials; it is also these properties that have motivated concern and investigation with regard to collateral nanomaterial environment, health and safety (nanoEHS). These properties will drive interactions and transformations within environmental compartments and biota, and they will be the basis of decisions to develop and utilize, or abandon, individual ENMs.

Categorization beginning with a consideration of the useful properties of nanomaterials immediately takes account of the manner in which nanomaterials are likely to be employed, allows for comparison with alternatives, and highlights the benefits that these materials provide. For example, nanomaterials that are of interest due to their photocatalytic properties will most likely be used as suspensions or surface films that may have greater potential for release to the environment or contact with humans than, for example, a material embedded within resin matrix. The photocatalytic property therefore suggests an exposure potential. The useful property (or properties) on nanomaterials dictate(s) the type of product that will incorporate these materials. Therefore, screening based on useful properties also suggests one or more exposure pathways. Continuing with the example of photocatalysts, the overall size of the market for photocatalysts implies a value of "F" for the maximum fabrication rate that also implies maximum exposures. On the hazard side of the risk equation, useful properties suggest potential pathways for toxicity (e.g., production of reactive oxygen species, size, charge, persistence upon inhalation, stiffness, etc.). Useful properties may be loosely correlated with factors known to affect exposure and hazard. For example, melting point often correlated with solubility, while interfacial reactivity may be related to ROS generation. Table 13.3 cross-lists several useful properties exploited in nanomaterial applications and factors known to affect nanomaterials transport and exposure.

Finally, classification by useful property facilitates a comparison across materials of similar functionality for their risk, efficacy and associated benefits. Because such a data categorization schema can be utilized to answer questions that inform decisions about performance (which material absorbs the widest spectrum

Fate Process	Dissolution	Deposition	Advection	Diffusion	Abrasion	Transformation
Useful Property						
Photocatalytic	☆			☆		
Interfacial Reactivity	☆	☆			☆	☆
Mechanical Strength	☆		☆		☆	
Melting Point	☆				☆	☆
Optical Properties	☆	☆		☆		
Magnetic Properties	☆	☆				☆

Table 13.3   Examples of Useful Properties of Materials That Might Be Used to Categorize Materials and Several Transport Processes That May Be Affected

of light?) as well as decisions about assessing and managing risks (which material generates the most net reactive oxygen species?), the useful property-based categorization provides a common basis for evaluating nano-enabled applications and their implications.

Functional Assays

While ENMs may present novel properties, and resulting new interactions with the environment and biota, the challenges associated with risk forecasting are similar to those confronted in other contexts, such as dealing with the impacts of municipal wastes on surface water quality. In contrast to the statistically validated, mechanistic QSAR approaches used in narrowly scoped data rich investigations, in these cases decision-making was aided by measurements of material behavior in prescribed conditions that represented useful simplifications of the environment(s) into which the materials were introduced. In the wastewater example, the degradation of complex mixtures of municipal wastes were causing oxygen depletion in surface waters that needed to be controlled. Predictions of oxygen depletion were indeed made, but not from a first principles analysis of the chemical composition of the waters or the microbial populations present in the waste and in surface water, even if these parameters were later investigated. Instead, oxygen consumption by the bacterial mixtures present in the actual waste entering actual surface waters was measured functionally, introducing the concept of biochemical oxygen demand (BOD). This example illustrates the utility of measuring key variables (such as BOD) in complex systems prior to seeking the full mechanistic understanding of the processes causing the behavior of interest.

As in the BOD case, the complex interactions of real environmental systems with nanomaterials, and the resulting transformations and modifications to their material properties and behaviors, can obfuscate direct linkages between intrinsic nanoparticle properties and the exposure and hazard endpoints we must forecast in order to characterize risk. A more practical and productive approach than directly linking ENM properties to their effects is to develop standardized measurements of nanomaterial behaviors in standardized, relevant systems, akin to the BOD measurement. *Functional assays* are procedures designed to produce these measurements, providing empirical parameter values that can parameterize predictive models to inform near-term decisions.

Functional assays are defined as procedures for quantifying parameters that describe a specific process or function within a given, often complex, system. They provide quantitative information on either rates of relevant processes (e.g., rates of aggregation, bioaccumulation or precipitation) or characterizing performance or final states [e.g., distribution between air/water, living/dead (LD_{50})]. These semi-empirical parameters inherently include the aggregate effects of relationships between properties and functions of nanomaterials, their surrounding system characteristics, and the exposure or hazard indicators of interest to predict [56]. The multitude of processes and interactions that occur affecting both the nanoparticles and their environments and lead to the state expressed by FA values may eventually be understood on a first principles basis, but prior to acquiring this fundamental knowledge these measurements provide a pragmatic approach to parameterizing models and forecasting transport, transformations, and effects of nanomaterials. Standardizing the protocols for FA measurements as well as the preparation and reporting of the ENMs and the systems in which they are measured will be the next step in a path toward amassing data that can both support near-term, realistic exposure and hazard evaluations while building an integrated dataset that can later be used to interrogate mechanistic relationships.

Information Management

This chapter, and indeed the book as a whole, has illustrated how the factors describing nanomaterials, their interactions, and their effects are innumerable. That they cannot be tested or modeled on an individual case-by-case basis is clear, as is the immediate need for near-term decisions on topics ranging from materials design to risk mitigation. The valuable resources necessary to support research must be prioritized and invested strategically to build a dataset that includes information on the most critical parameters for forecasting impacts (e.g., the proposal of functional assays), and that will allow comparisons between datasets, across multiple studies (hence, the focus on standardizing characterization of ENMs and their systems). Many of the previous sections in this chapter have stressed the need to minimize the numbers of measurements, of material categories, of standardized environmental systems, that must be taken in order to formulate a plausible approach to amassing enough data to reliably forecast the fate and effects of a multitude of materials in complex systems. One key component of leveraging each measurement and minimizing the need for case-by-case evaluations is by integrating datasets through information management practices that enable the analysis of large datasets. Only when data can be compared across multiple studies can the investments in individual experiments be leveraged to generate larger scale discoveries from existing data; to borrow a classic phrase from the knowledge management field, "If only we knew what we know" [57]. Nanoinformatics is a new and rapidly growing field focused on information management approaches and tools for integrating and analyzing nanomaterial data, with important implications for experimental and project design.

A concrete illustration of the value in developing ways to synthesize large, disparate datasets for collective analysis can be seen in the bioinformatics community, where integration and standardization of data collection has helped to sequence the human genome and to streamline drug development processes. If the nanoEHS and nanomaterial application R&D communities develop similar collaborative approaches to data synthesis, benefits could include leveraging investments across multiple efforts and multiple disciplines, trans-disciplinary sharing of information and methods, and more rapid enablement of risk assessment and technology commercialization.

The wide variety of nanomaterial physicochemical characteristics and potential applications has led to the rapid generation of disparate data in terms of pristine and modified materials, their interactions in laboratory-based and natural environments, and a wide array of potentially relevant biological interactions. The task of integrating nanomaterial datasets—pulling the data together in ways that allow us to know when the materials and experimental outcomes can be compared and analyzed together meaningfully and when they cannot—is daunting enough. This is further complicated by multiple confounding factors: the fact that protocols and standards for nanomaterial synthesis, characterization, and environmental and biological testing are still in development, and that these dynamic materials can transform dramatically when released into real environments, in ways that depend heavily on system characteristics. Achieving a level of comparability between datasets that will facilitate sharing and comparison for defined analytical purposes will require the commitment of the broader nanomaterial community, including nanoEHS researchers, product developers, and funding agencies, who will develop and adopt standards, cyberinfrastructures, and a culture of data sharing to support these goals.

Several groups have made progress in supporting the development of a nanoinformatics community and supporting tools. A first step was taken with the development of the NanoParticle Ontology (NPO), published in 2011 to introduce vocabulary standards. Since then, ISA-TAB-nano data exchange standards (ASTM International E2909-13 [58]) have been introduced along with tools to enable their public accessibility. ISA-TAB-nano terminology and formats have been adopted and consulted by a wide variety of stakeholders across the nanomaterials data community, including nanomedicine- and nanoEHS-related projects in the United States and the European Union.

Other groups have proposed controlled vocabularies and developed data sharing cyberinfrastructures at a variety of levels of specificity. The NIH-funded Nanomaterial Registry has been developed as a public data-sharing platform with minimal information requirements, with the mission to collect data from as many sources as possible [59]. The cancer Nanotechnology Laboratory describes itself as "a data sharing portal designed to facilitate information sharing across the international biomedical nanotechnology research community to expedite and validate the use of nanotechnology in biomedicine, [that] provides support for the annotation of nanomaterials with characterizations resulting from physicochemical, in vitro and in vivo assays and the sharing of these characterizations and associated nanotechnology protocols in a secure fashion" [60]. In the European Union, the eNanoMapper project addresses interoperability challenges of disparate datasets and models, developing application programming interfaces (APIs) that incorporate data sharing standards (including ISA-TAB-nano) wherever applicable [61]. International pre-standardization organization VAMAS has joined with international science and technology data organization CODATA to create an interdisciplinary and trans-sector working group to develop a uniform description system for nanomaterials [62]. Very specific databases have also been created, serving as case studies for the intricacies involved in integrating detailed datasets for multi-study analyses such as the Nanomaterial-Biological Interactions (NBI) Knowledgebase database, which houses hundreds of studies of zebrafish toxicity to nanomaterials and is open source [63]. Another key resource, nanoHUB, has been developed with a focus on enabling crowd-sourcing of nanomaterial simulation tools via hub technologies; in addition to providing a globally accessible platform and hosting a wide variety of tools, nanoHUB is collecting important social science data regarding the user communities and use patterns of constituent nanoinformatics tools [64].

A singular authoritative integrated nanoinformatics resource will never exist. Rather, the goal of nanoinformatics integration work must be to enable sharing of data between multiple resources across user communities. The design requirements guiding the development of an individual informatics resource, including aspects such as incorporated vocabularies, the level of granularity included in the cyberinfrastructure and the querying and visualization capabilities, are driven by the needs of its intended user community [65]. The dual missions of the Center for Environmental Implications of NanoTechnology (CEINT), are to (1) elucidate the mechanisms governing the behavior and effects of nanoscale materials, and (2) iteratively translate this understanding into guidance for risk-based decisions for the sustainable development of nanotechnology. In addition to guiding the interdisciplinary, integrated structure of the organization and research strategy, the mission has guided the development of the CEINT NanoInformatics Knowledge Commons (NIKC), a cyberinfrastructure with associated analytical tools built to capture and integrate datasets across the center and beyond. One requirement stemming from this mission is the ability to interrogate data for directional guidance in terms of which, for example, material characteristics, medium characteristics, and functional assay measurements are most predictive of endpoints of interest. Then once enough data are amassed, a subsequent requirement is that data must have been entered, or "curated" at a granular enough level to support mechanistic investigations capable of deriving theory to interrogate the data to test hypotheses and seek relationships between parameters.

The CEINT NIKC DB is designed with specific focus on inclusion of data and meta-data to as nearly as possible support reproducibility of studies, including fields such as: ENM physical-chemical characteristics, system parameters, system-dependent ENM characteristics, functional assay parameters, data types for each field, meta-data requirements including protocols, instrumentation used for characterization and experiments, and details behind capturing timing of measurements. The data fields selected for inclusion and the queries and graphs being built to analyze the combined dataset are designed to address the driving research questions, starting with: are functional assay measurements such as surface affinity and dissolution rate, as measured for ENMs in relevant standardized systems, predictive of ENM behavior in real, complex systems? With such a specific analytical mission, the granularity required to support these investigations is more detailed than the more generalized nanoinformatics approaches that have been built to address a multitude of potential user communities. In turn, curating a single study into the CEINT NIKC is relatively resource-intensive; future sustainability of nanoinformatics at a level that could support multi-study investigations into nanomaterial behavior in the environment will require the development and adoption of user-interfaces to allow researcher-based curation of data rather than retro-active curation from publications.

A sample visualization appears in a CEINT NIKC tool screenshot in Figure 13.5, showing dose-response curve formulated from multiple curated nano-Ag toxicity studies within the literature, with a forecasted value for a relevant nano-Ag exposure concentration superimposed onto the curve to project the associated response based on the data. The tool built to generate this plot incorporates user-entered estimates describing the exposure scenario and is customizable so that the user may adjust the base assumptions. As additional data are curated the tool can be made more robust, to include more points on the dose-response curve, more explicit consideration of uncertainty, and more data-driven estimates of the realistic exposures.

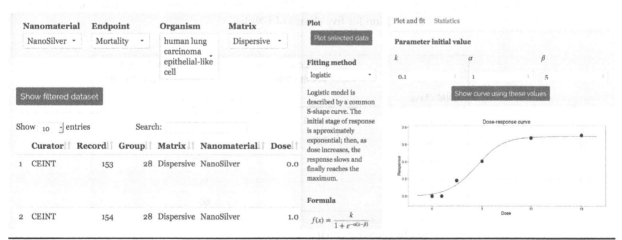

Figure 13.5 Screen shot of interactive literature-based CEINT NIKC effects forecasting tool.

Continued collaboration among the members of the nanoinformatics community, including diverse representation in terms of disciplines, sectors, funding sources and timelines, and driving research missions will be critical to achieving a level of functional interoperability that leverages progress across all groups and supports near-term discovery while investing in longer-term advancement of fundamental science [65].

Describing Exposure

The potential for exposure to engineered nanomaterials begins with the production of these materials. However, one consequence of the nascent stage of the nanomaterials industry is considerable uncertainty regarding the quantity of materials to be produced, the manner in which they will be handled in day-to day operations, and the size of potential markets. Compared to the work done to characterize hazards of nanomaterials, there has been very little work to date in characterizing exposures. It is widely agreed that development a risk management strategy based in understanding the possible pathways and for exposure to nanomaterials should be the primary criterion for prioritizing risk-based research on engineered nanomaterials [66]. However, the information requirements to forecast the releases and environmental exposures to nanomaterials are daunting, with only sparse data available to date. Generating quantitative estimates of nanomaterial sources and the relevant transformations and doses is difficult. Here again the use of flexible modeling approaches to incorporate emerging information and to manage uncertainties is important [1].

ENM Source Estimates

To project the physical amount of an ENM present, which will contribute to the magnitude of the exposure to a nanomaterial, some market data must be available to tell us the volume we can expect to be produced. The lion's share of this information remains proprietary, as companies rush to be among the early providers of reliable, industrial volumes of these much-anticipated materials. Companies such as Cientifica and Lux Research conduct detailed market research that can provide some idea of the volumes we can expect to see produced in the near term. Earlier market studies found that more than $30 billion in goods manufactured in 2005 incorporated emerging nanotechnology, and further projected that by 2014 a full 15 percent of total global manufacturing output would incorporate nanotechnology [67]. While these numbers do not directly translate into physical measurements describing the mass of nanomaterials in question, they offer an appreciation for the magnitude of exposure to be expected. As more companies develop methods for large-scale industrial production, further reviews of their patents will also offer quantitative insights. Information about the emissions of nanomaterials in the course of production will be necessary in evaluating this parameter as well, including amounts and states of the materials released.

The quantity of materials produced will also be related to the format that they are produced in. In some cases, "raw" nanomaterials such as C_{60} may be produced as feedstock to downstream processes that then incorporate these materials in composite materials, fluids, and devices. Incorporation of SWNTs into a plastic composite to improve the wear characteristics of the product will likely alter the availability of the nanomaterial (physically and biologically) while introducing a new step in processing with inherent risk of handling. Efforts to date have produced estimates for the production quantities of various classes of nanomaterials, to provide guidance on the relative physical relevance of different materials, which helps to prioritize future research, and with the understanding that the production quantity can serve as an absolute upper bound for exposure potential [68, 69]. Table 13.4 is adapted below from a Hendren et al. (2011) study estimating ranges of U.S. production for five classes of ENMs.

ENM Class	Lower Bound Production Estimate	Upper Bound Production Estimate
	Tons per Year	
nano-TiO$_2$	7,800	38,000
nano-Ag	2.8	20
nano-CeO$_2$	35	700
CNT	55	1101
Fullerenes	2	80

Adapted from Hendren et al. (2011).

TABLE 13.4 Annual Production Estimates (F) for Five Classes of ENMs

ENM Release Probability and Potency Factors

To understand the likelihood of a material being released from any life-cycle stage across the value chain, the possible pathways or events leading to that release must be identified and characterized. As discussed, exposure related data are sparse, and methods for generating these release estimates are emerging.

For the manufacturing stage, processing information must be available that reflects variations in emissions and sources of incident risk arising from variations in day-to-day operations such as extreme temperatures and pressures. These emissions will likely be the most predictable due to both the expense of nanomaterials and to tight manufacturing controls. Release probability due to an incident will also hinge on data regarding the fire and explosion dangers and catalytic reactivity levels of the nanomaterials [54]. In the case of fullerenes, the processes used to fabricate them do involve some extreme temperatures and high pressures, but thus far these process steps are carried out under extreme control with a great focus on collecting the nanomaterials yielded. It has also been suggested that the large specific surface area of some particles, especially easily oxidizable metal nanoparticles, could pose a risk of spontaneous flammability in air [70]. More work needs to be done to offer conclusive results on the degree and causes of ignition potential, but to date carbon-based fullerenes are not expected to be highly flammable, combustible, or likely to spontaneously catalyze chain reactions in air.

A number of efforts have been made to predict the releases of nanomaterials across the value chain with a focus on use and disposal, and to forecast which environmental compartments ENMs might move through and accumulate within. Keller et al. estimated that between 63 and 91 percent of all ENMs are eventually disposed of in landfills and that up to a quarter of global production might end up in soils, which highlights the potential implications for agriculture. Such studies, even given current uncertainties, can provide important directional guidance for where to focus future research on releases [71].

It is important to note, however, that ENM quantities alone may not be sufficient to forecast the relevant exposures based on releases. A combination of market research, materials research, and continued environmental science work is needed to fill all the data gaps necessary to estimate the quantities of ENMs incorporated into products, the quantities that may be released upon product transport, use and disposal, and the portion of nanomaterials in the released material that may be available for reaction. In the case of, for example, a nano-enabled thermoplastic, the abraded material that could be released during its use or disposal may have only a limited number of nanomaterials present at the surfaces of the abraded materials, and only some fraction of those may be able to react with their surroundings. Each of these categories of ENMs as modified by their inclusion in and release from matrices, as well as the resulting environmentally transformed versions of the materials, must be separately characterized to estimate the true availability of the ENMs for interaction [72]. Understanding such "potency factors" will require future research.

Use Phase ENM Releases

Releases of ENMs from the product use phase of the value chain are of particular interest to study both from an exposure scenario perspective and an information availability perspective. Releases from consumer products are directly relevant to consumer exposure estimates, and releases from many product categories present relevant exposure scenarios directly to ecosystems as well (e.g., run-off from TiO_2 enabled coatings). The information demands in understanding potential releases from the use phase of nanomaterials and nano-enabled products are particularly complex, involving not only a wide array of nanomaterials, variations including coatings and functionalizations, and products, but the complex social aspect of consumer behavior and use patterns that inevitably affect the release quantities, rates, and locations. A number of studies have investigated release potential from specific nano-enabled products including textiles, plastics, composites, personal care products and dispersive antimicrobial applications, confirming the release of nanoscale materials at various concentrations and suggesting the importance of allotting more resources to the topic [73, 74]. The NanoRelease project, an multi-stakeholder effort of the International Life Sciences Institute (ILSI), recognizing the "tremendous potential for nanotechnologies to create new materials and devices for use in medicine, electronics, energy production, cosmetics, packaging, food manufacturing, and many other fields," is specifically focused on creating standard methods for detecting and characterizing ENM releases from products with the recognition that reliable, consistent testing and reporting of ENM releases are critical for enabling protective risk management action. Initial products of these efforts include both investigations of potential release scenarios [72] and proposed characterization standards for relevant ENMs [74].

Persistence and Mobility in Air and Water

In traditional chemical risk models the persistence of a material in air is often measured based on photolytic half-life, or time until absorption of light will result in half of the mass of a substance being altered via cleavage of one or more covalent bonds within the material's molecular entity. Research on photoreactivity and photocatalysis, as well as studies of reactivity with other photoreaction by-products such as ozone, are being carried out that may deliver values for this parameter. To follow the thread of our fullerene example, most of the research into photoreactivity has focused on the effects of excited nanomaterials on other constituent chemicals in the air or solution rather than seeking the true photolytic half-life. To date, however, there does not seem to be evidence that these carbon-based nanomaterials would be rapidly altered by light absorption. Another factor likely to affect whether a material persists in the air will be aggregation; if, for example, fullerenes tend to form aggregates in the air, they will then potentially exhibit behaviors distinct from those of the individual particles, reacting differently with the environment and potentially even settling out of the media altogether.

With typical non-nanoscale substances the molecular weight is often of great importance for describing its mobility in air; for nanomaterials, this same parameter could potentially be represented instead by a group of descriptors such as particle size, density, and surface chemistry, which would allow determination of which forces dominate transport.

Similar parameters affecting persistence and mobility of nanoparticles in air, including photoreactivity, aggregation state, and size will apply to their partitioning from air to aqueous solutions, as well as to their behavior within aqueous solutions.

Aqueous and Soil Persistence and Mobility: Aggregation State and Rate

The molecular weight of many compounds has been related to their mobility, hydrophobicity, and propensity to bioaccumulate. Here again, particle characteristics like size, density and surface chemistry will play a key role. In addition to these important factors, the degree of aggregation (aggregation state) and the rate at which materials tend to aggregate is an important determinant of a material's persistence in the environment. Beyond determining whether the substance remains in a particular medium, the importance of aggregation with regard to reactivity and bioavailability was underscored in the earlier discussion of how aggregation state may affect ROS generation by nanomaterials. These parameters will be linked with the functionality of the nanomaterial. Aggregation and deposition on surfaces will affect not only the net exposure and persistence of these materials but perhaps their dose as well, as in the case of nanoparticle uptake by inhalation.

Chapter 6 talked in detail about processes governing nanoparticle stability and transport in the environment. We saw here that factors such as pH of the solution and nanomaterial functionalization drive the material's behavior in the water. Here again, studies on the effects of functionalized groups, coatings, and aggregation behavior promise to provide crucial information in predicting the behaviors of nanoparticles in solution. We know that buckyballs are virtually insoluble in water in their pristine form, with their true thermodynamic solubility estimated at −18 mol/L [75], but that they can be suspended with the additional of functionalized hydroxyl groups as they are in the case of fullerol. We know that the individual molecules have a log K_{ow} value near 12.6 [75], but the question remains as to whether we can really expect to find individual molecules free in the environment.

Recent research has revealed that the phenomenon dominating persistence behavior is often heteroaggregation of nanomaterials with larger background particles. These results are supporting the development of theoretical approaches to forecast the ENM concentration that might remain in a water column largely as a function of background particle concentration. It is these types of advances that reduce the number of necessary measurements and calculations necessary to forecast expected exposures and thus risks of engineered nanomaterials [76].

Aqueous and Soil Persistence and Mobility: Biological Interactions

Biodegradation rates determine the persistence of a material in soils within the sample risk assessment database. Most of the research to date has concentrated on establishing the potential for uptake, but whether the materials are metabolized or degraded has yet to be established. For example, one recent study shows the uptake of nanoiron into central nervous system microglia cells, proving that at least these particles are assimilated into the cell [77]. Some other biological interactions are being studied with a focus on reduction of metal ions, with one body of work establishing actual nanoparticle synthesis within alfalfa plants [78]. While this is not biodegradation, it has interesting implications on our understanding of metabolic pathways in relation to metal nanoparticles. Some of the work presented in Chapter 12 will lead to a future understanding of nanomaterials pathways in microbial environments. The same information that tells us about nanomaterial mobility in solutions, water solubility, and portioning behaviors will also help us understand behavior in soils and propensity for bioaccumulation.

Describing Hazard

Hazards such as toxicity and ecotoxicity are discussed in Chapters 5, 11 and 12. Despite the attention that early studies have given to this element of hazard, general principles relating the characteristics of a given nanomaterial and the hazards it may pose remain elusive. Toxicity can be measured with respect to multiple endpoints, including cytotoxicity, genotoxicity, toxicity to specific organs, or the onset of acute effects or chronic effects (Chapter 11). Interference with biological processes, such as endocrine system disruption, is be another potential category of hazard that should be studied and considered with respect to assessing risk. It may well be that research on an indicator parameter such as nanoparticle reactivity could assist in predicting the hazardous effects it may have on its various surroundings. Very little is known regarding fullerene metabolism or systemic elimination. Several studies have claimed various toxic effects may be induced by C_{60} fullerenes, while others have indicated there may even be therapeutic effects attributable to these materials [79–84]. In addition to the more empirical toxicity studies, priority is being given to understanding the toxicokinetics within the human body. The NanoSafe II initiative in Europe includes research on modeling transport within the body to various targets, including adsorption, distribution, metabolism, and excretion [52]. The intermediate step of modeling dose based on a given level of exposure requires similar information on the physical chemical behaviors of these materials as they are transported and transformed by organisms. In addition to data needs in modeling dose for individual organisms, interactions between organisms must be detailed to arrive a complete ecotoxicological assessment. By beginning with cases at the base of the food web, an initial assessment is greatly simplified.

Ultimately, an understanding of mechanisms that underpin any nanomaterial toxicity will be required if we aim beyond an empirical level toward predictive modeling. Work on fundamental characteristics and behavior of nanomaterials is beginning to tie into our understanding of toxicity to shed light on some of these mechanisms [85].

One challenge for risk assessment, particularly of ENMs, is the lack of data that would directly feed into risk assessments. For example, publications in toxicology are often skewed toward positive results, meaning it is easier to find reports of evidence of toxic effects based on exposure to a stressor than to find publications of low or no effects. However, to set protective guidelines for exposure to that stressor, the useful information is essentially negative results: the maximum exposure concentration or dose that can be expected to elicit NO effect (predicted no effects concentration in an exposure medium (PNEC), or no adverse effects level (NOAEL)), or in other cases the very minimum amount expected to result in negative bioactivity (lowest observed adverse effect level, or LOAEL). This disconnect between the results emerging from scientific disciplines whose expertise is needed to feed risk assessment, and the information requirements of the risk assessment paradigms, is a systemic problem for risk assessment that also affects abilities to forecast impacts of ENMs.

Organismal Effects

Organismal toxicity describes the adverse health effects that result from exposure to a stressor substance, as measured in terms of a variety of impacts that include death, changes in growth or behavior of individuals, or their propensity to accumulate or transfer the substance. These endpoints may be measured at different exposure levels and timescales, via different exposure pathways, and/or at different life stages of a given organism.

Mortality   Acute toxicity refers to consequences of a single exposure, whereas and chronic toxicity refers to the adverse health effects stemming from repeated, typically lower concentration, exposures over a longer period of time. Mortality measurements are often used to evaluate acute toxicity of substances, reporting the amount of a substance that causes the death of the test animals. The dominant practice is the utilization of the LD_{50} (as introduced earlier in the chapter) or LC_{50}, originating from the 1927 work of J. W. Trevan. The LC_{50} stands for "lethal concentration for 50 percent." and refers to the concentration in the surrounding medium (air or water) associated with the death of 50 percent of the individuals as measured under exposure protocols with standardized media characteristics and predefined time periods. The LD_{50} signifies the "lethal dose for 50 percent, and refers to the dose of the material, administered directly to the organism at one time via any route of entry, that causes the death of 50 percent of the individuals. To enable comparisons, mortality values are reported along with their routes of exposure, with the test animal, and where applicable with the relevant exposure time period. Because these methods require the use of large numbers of animals (50–200 per test), it is desirable to find alternative testing strategies to minimize the ethical and economic impacts of toxicity testing. In addition to developments of alternative measurement approaches within conventional toxicity [86], the NanoEHS community continues to pursue weight of evidence approaches that would allow for read-across from a smaller number of tested materials, reducing testing requirements [87].

Developmental Effects   Developmental effects test a variety of impacts a stressor may have on target organisms across spectra of exposures, across life-cycle stages of animals or plants, and across a wide range of organismal functions. Some examples include changes in growth (measured in a number of ways including changes in length, mass, or rate), reproductive impairment. Results of developmental effects tests can be utilized to inform estimates with flexible endpoint inputs such as lowest-observed–adverse-effect-level (LOAEL), which are used by Federal agencies to set protective limits.

Behavioral Effects   Other important measures of impacts can include behavioral changes induced by exposure to a stressor. This can involve a broad array of functional responses that may be complex and interlinked. Examples of these can include changes in feeding, mating, choices of where to lay eggs or give birth, changes in coordination, learning ability, sensory and motor integration. Interpreting these results in terms of relationships to toxicity can be challenging and requires multidisciplinary investigation [88].

Maternal Transfer   As the name suggests, this organismal affect describes the ability of a substance to be passed from mother to offspring. This can refer to transfer of the material during any stage in gestation, birth and feeding. The substance may transfer via yolk sack, the placental interface, or milk. Measuring maternal transfer is important to understand bioaccumulation and propagation of substances through an ecosystem, as well as to investigate teratogenic reproductive impacts (though teratogenic effects can be observed without the transfer of the substance itself). Early investigations reveal that maternal transfer of nanomaterials is possible in animal models as well as humans, and suggest that the governing mechanisms can differ from classical chemicals [89, 90].

Population and Ecosystem Effects

Population and ecosystem effects measure the impacts of stressors across multiple organisms or ecosystem functions, using endpoints such as changes in numbers of species or disruption of nutrient cycles. Direct attribution of these complex effects to a single stressor or event is difficult even in the assessment of conventional materials, and the uncertainty associated with forecasting emerging nanoscale materials is large. Nonetheless, it is crucial to develop methods to forecast these effects for a number of reasons. Risk management approaches typically orient around measurements at the population level, so estimates of population level effects are critical to connect ecological testing with associated risk assessment and risk management recommendations. Ecosystem effects are also important because this category of potential impacts includes complex ripple effects that would otherwise not be represented in risk forecasting; consider, for example, that no amount of thorough toxicity testing of chlorofluorocarbons (CFCs), the nontoxic, nonflammable alternative introduced to replace ammonia for use in used in refrigeration and compressed spray cans, would have predicted their atmospheric interactions that resulted in the dangerous depletion of the ozone layer. Ecosystem level effects, including nutrient cycling and environmental impacts beyond toxicity, are critical to consider in mitigating the collateral impacts of emerging technologies such as ENMs. A number of impacts that are commonly utilized in risk assessments are presented here.

Changes in Population Number and Composition   Populations may be measured in a number of ways including abundance, population growth, structure, and persistence; changes to any of these as a result of exposure to a stressor may be measured within a site-specific risk assessment and can indicate a population-level effect [91]. Understanding population number and compositional effects and the nanomaterial functions that may induce those effects will be important for informing risk assessments and their resulting recommended protective limits [66].

Trophic Transfer, Biomagnification, and Biominification   Trophic transfer refers to the transmission of substances between trophic levels, or the propagation through a food web. Starting with small single-cell organisms, as organisms are consumed by larger and larger organisms higher up the food chain, the substances they have internalized can be transferred upon their consumption. Trophic transfer can result in biomagnification (cumulatively higher concentrations at higher trophic levels), biominification (which is trophic dilution), or equal levels between predator and prey. Studies have shown that trophic transfer of nanomaterials is possible; the ability to forecast transfer and biomagnification in particular will be important to inform whether nanomaterial bioaccumulation results in concentrations that pose wither wildlife or human health risks [92, 93].

Nutrient Cycling   Ecosystems, as interconnected systems of organisms and the environments in which they live, cycle matter throughout via exchanges of materials and energy flows. These complex "nutrient cycling" (also referred to as "biogeochemical cycling") processes constantly redistribute organic and inorganic matter through ecosystems, as organisms and geological systems take in, transport, transform, and expel minerals and nutrients throughout the global system. Human activities have been shown to significantly alter some of Earth's naturally occurring nutrient cycles, with important consequences; the advances within industrial agriculture allowing fixation of nitrogen for use in fertilizers has increased the agricultural yields of the planet, with significant deleterious aquatic ecosystem and global climate change impacts [94]. Given the emerging evidence that chronic low dose releases of nano-Ag particles into aquatic ecosystems increases fluxes of N_2O to the atmosphere, an understanding of ENM effects on this and other key nutrient cycles is critical for forecasting their potential risks [95].

Applying the Concepts: Describing Life-Cycle Risks Associated with Nanomaterial Production through an Example Risk Assessment

To close out this chapter, the following section presents an example risk assessment that embodies many of the concepts introduced here. An existing risk assessment tool is employed in a preliminary relative risk assessment of the manufacturing processes for five classes of ENMs, compared with five non-ENM established processes. This illustrates the use of available models as well as the importance of considering life-cycle impacts beyond the direct effects from the ENMs themselves. The detailed presentation of the model algorithm throughout this use case also illustrates concrete examples of many of risk parameters discussed throughout this chapter including material characterization, exposure metrics, hazard metrics, and their weighted combination to arrive at the relative risk score.

Wastes and Impacts from ENM Production

As the impacts of engineered nanomaterials are studied it is increasingly clear that the environmental impacts resulting from the production phase of a nanomaterial are not limited to the material itself, but are also greatly determined by the inputs and by-products of the fabrication process. Indeed not only the material inputs and by-products, but the energy and water demands, contribute to the overall impact of producing an ENM (Fig. 13.3). In terms of potential ecotoxicity, one study concluded that the negative impacts from producing carbon nanotubes exceeded the negative ecotoxicological impacts from the direct exposure to the resulting nanotubes themselves [96]. Another study illustrated how changing the chemical vapor deposition process for synthesizing carbon nanotubes could significantly diminish the emissions of greenhouse gases, thus highlighting how research into the production impacts of ENMs has the potential to positively influence process design [97]. Such a green chemistry focus could be particularly impactful now at the early stages of nanomaterial adoption, when it could guide the investment and scale-up for nanomaterial production processes. The reader is encouraged to consider this while working through this example risk assessment, which includes five early fabrication processes for ENMs and illustrates the heavy influence of decisions on synthesis methods, with their associated chemical inputs and by-products, on resulting relative risk.

An Illustrative Relative Risk Assessment from an Insurance Industry Perspective

Insurance companies routinely assess the risks associated with industrial processing. In the chemical industry, the relative risks of substances and processes involved in fabricating a product are assessed by insurance companies as a basis for calculating the appropriate premiums to charge their customers. Nanotechnology companies, like all others, will have to carry insurance on manufacturing processes. They will be subjected to actuarial protocols that assess the risk posed by a given process and determine how much a company will have to pay to an insurance provider in exchange for liability coverage. The interface between developing nanoscience and existing risk assessment frameworks presents itself as a natural starting point for translating this emerging class of novel technologies into the existing language of the risk community. This is already an active area of interest and development among insurance companies worldwide.

XL Insurance is a Swiss insurance company that employs a numerical rating tool to combine risk factors and calculate premiums for the processes it underwrites. With a focus on environmental pollution and health risks, the aim of the algorithm, understandably, is to quantify the relative risk of manufacturing processes with respect to financial liability of the insurer. As we have discussed, a full LCA would build out from the impact of the manufacturing process itself to evaluate the effects of the product in question from cradle to grave. As a starting point, though, XL Insurance allowed the transparent use of their insurance protocol to rank the fabrication processes of nanomaterials. Because of the established dearth of conclusive data regarding nanomaterials, the fabrication processes are initially assessed independent of the nanomaterials themselves. In presenting the methodology and results, the specific information needs for model development will be explained. Thus, this exercise serves as a first pass at ranking the relative risk of nanomaterials based on the known and estimated physicochemical properties that are used in current assessment frameworks, but it will also highlight the unknowns in a specific sense and highlight why research in mobility, transport, aggregation, and toxicity (such as the work presented in this book) is so important in predicting the behavior and effects of nanomaterials

General Data Requirements of Risk Assessment Methodology

To evaluate environmental risk, a function of hazard and exposure, we have established that the effects of a substance must be combined with the likelihood that it will come in contact with the environment. The definition of risk makes clear that the level of hazard posed by a nanomaterial is a characteristic independent of the chance that it will ever realize its potential damage. Thus, in order to understand the impact a material will have on its surroundings, we have to know how dangerous it is as well as how it will interact when released to the environment. Where will it go, in which media will it tend to collect, with what organisms will it be likely to make contact, and to what extent will it have toxic effects on them when that contact occurs? We must also include information about the amount of the material and the conditions in which it is being handled or produced. We need to understand the probability of substance release and the quantity that is at stake in order to combine risk with exposure. An appropriate algorithm can then be used to combine the factors and determine the resulting relative impact of the material.

Just as in any risk assessment, choosing the appropriate parameters and the appropriate methodology for combining them is the crux of evaluating the environmental impact and insurance liability for this sample model. In the XL Insurance protocol, the hazard of each process is defined based on constituent substance characteristics such as carcinogenicity or lethal dose in rats, and on such relevant process characteristics as temperature, pressure, enthalpy, and fire hazard rating [98]. The exposure portion is quantified by incorporating persistence and mobility of constituent substances and by scoring expected emissions of the substances during the manufacturing process. These factors are combined with a focus on order of magnitude so that an idea of the risk range is what drives the premium. This focus on relative risk acknowledges the pragmatic reality of uncertainty inherent to the business of insuring a wide range of manufacturing industries, and a variety of specific processes within each industry. The fact that a level of uncertainty is built into the system by aiming at orders of magnitude may prove to narrow the knowledge gaps between traditional products and nano-products in practice. By applying an insurance database currently in commercial use, we are able to benchmark the risk of nanomaterials' fabrication processes against each other and against other non-nanomaterials from the perspective of an industrial insurer. This type of risk assessment, although not inclusive of all environmental impacts or life-cycle considerations, is representative of the type of risk assessment nanomaterials manufacturers will first encounter as insurers grapple with qualifying the relative risk of these new processes.

The sample relative risk assessment excerpted in this chapter was carried out to rank the fabrication processes for five different nanomaterials by their relative risk along with five established non-nanoscale manufacturing processes. Nanomaterials that exhibited foreseeable potential for production beyond laboratory scale were chosen for manufacturing evaluation: single-walled carbon nanotubes, buckyballs (C_{60}), quantum dots composed of zinc selenide, alumoxane nanoparticles, and nano-titanium dioxide. For each nanomaterial, a process flow chart was created to capture the published synthesis method that had the most potential for being scaled up for commercialization at the time of the study. An inventory of input materials, output materials less the final product, and waste streams for each step of fabrication was then created from the process flows. This method facilitated the collection of insurance protocol data requirements for all materials involved in the fabrication processes. Physicochemical properties and quantities of the inventoried materials were used to qualitatively assess relative risk based on factors such as toxicity, flammability, and persistence in the environment. The substance and process data were then quantitatively combined according to the actuarial protocol in the risk database to arrive at a final numerical score for the process. This relative risk score could then be compared with the risk scores of other commonplace production processes. It is relevant to mention that in this way the risk assessment was no different from that conducted in any other new chemical process. The exercise was carried out as a first step toward bridging the gap between novel materials and the established production industry into which they are already being incorporated. The process of listing all input and output streams was a thorough, yet straightforward exercise, so it is not detailed here. Instead we walk through the input parameters to the database.

System Data Requirements

To incorporate the risks of nanomaterials themselves using a model such as the XL Insurance database, information on material properties analogous to that entered for the reagents must be available for the nanomaterial product.

Appreciation of either the qualitative or quantitative conclusions of this model requires recognition of the two main sets of choices involved in setting up a risk assessment tool: which characteristics are chosen for inclusion in the tool, and how those parameters are then combined (Figure 13.4). In the case of environmental risk posed by an industrial process, the hazard component of risk is represented by characteristics of the materials involved as well as some characteristics of the process itself. The exposure risk is evaluated via three pathways: air, water, and soil. Information about the likelihood of the process to result in release of the substance is combined with its expected physical quantity and with characteristics that predict the fate and transport in various media to represent the risk of exposure to a given substance.

Figure 13.6 shows a conceptual diagram of the parameters required by the database, offering a first pass indication of how parameter values will affect the ultimate risk score. This simplified explanation of the selected parameters shows that they were included in the database because they capture pertinent material and process characteristics in a way that will allow resulting risk to be calculated.

It makes sense, for instance, that if a substance degrades slower in the air it will persist there longer and have more chance for exposure, resulting in higher risk in that medium. Readers are encouraged to review each parameter to convince themselves of the directional effects these values would have on overall risk.

We will then discuss the parameters in more detail, using fullerenes as a sample nanomaterial to show what is known thus far as well as to flesh out the details of what cannot yet be predicted. This table is useful in expressing the requirements of the insurance model, and in suggesting how properties of nanomaterials might be chosen to fit into these preset parameters; however, it may not be inclusive of all the relevant parameters needed to best capture nanoparticle-specific behaviors. Potential addition of more appropriate nanoparameters will be discussed later in the chapter.

Collecting, Weighting, Combining Data Parameters

With the chosen characteristics defined by measurable parameters in the system, some quantitative decisions must be made regarding how to mathematically combine the values to arrive at a final risk score. Finding real values for these parameters presents many challenges that will be expounded upon as the detailed data requirements are explained. For now, keep in mind that adapting our available information to the requirements of the system was an iterative process, resulting in the expression of each chemical and process according to the chemical and physical properties required by the insurance company's algorithm in determining its relative risk.

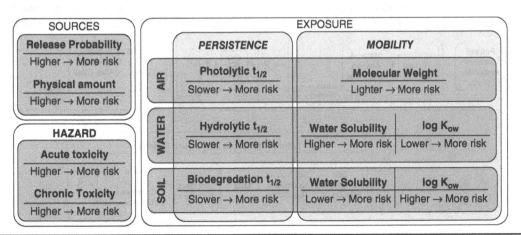

Figure 13.6 Pictorial representation of XL database required risk parameters.

Required Substance Characteristics

The substance characteristics utilized in the database, as well as the relationships used to combine the parameters, are detailed in Figure 13.7. The numerical values for each of the substance data fields are mapped to relative risk classes, which group levels of risk posed by orders of magnitude. For instance, the characteristic of toxicity is represented by the data parameter LC_{50} in the database, and the spectrum of possible values for the LC_{50} parameter of a given substance is represented in a scale of 1 through 4. For instances where a substance characteristic value was not known, which was often the case for persistence factors such as photolytic half-life, a relative risk class was directly assigned to at least approximate the parameter's order of magnitude. This assignment was often made via comparison with similar compounds. Risk scores corresponding to toxicity, persistence, and mobility data for each constituent substance were then combined to calculate the risk associated with that constituent's interaction with air, water, and soil; this cumulative risk score for each compound was denoted as its substance hazard risk class.

Required Process Characteristics

Each manufacturing process is defined within the database by entering its constituent substances and conditions such as temperature, pressure, and enthalpy. Figure 13.7 depicts the process characteristics utilized

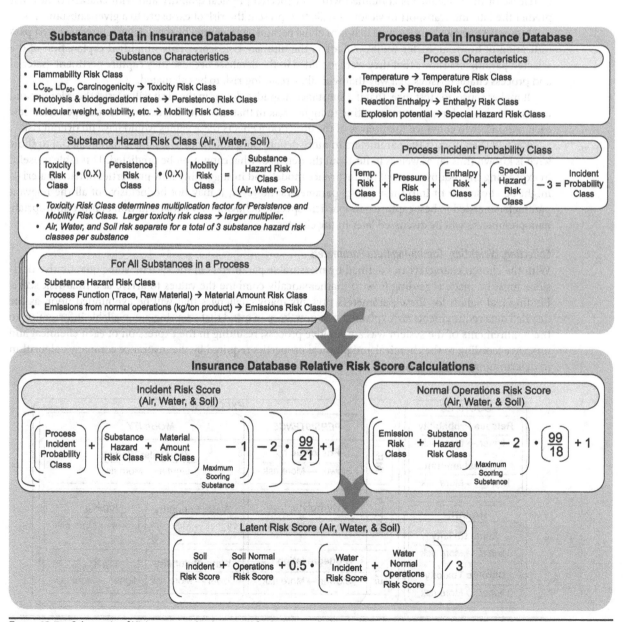

FIGURE 13.7 Schematic of XL Insurance database and formulation of risk scores. (Adapted from Robichaud et al., 2005.)

in the insurance algorithm. Maximum temperature, pressure, and enthalpy values for the process are entered to represent the highest risk step in each fabrication process, and the corresponding risk classes for those condition parameters were combined to calculate the probability of an accident occurring during the process. This value is referred to as the process incident probability class. Each substance involved in the fabrication process, with the exception of the final nanomaterial product, was defined individually in terms of its risk based on hazard and exposure variables. To quantify its contribution to risk within the particular fabrication process, the amount present, its role in the process, its physical phase at the temperature and pressure of the process, and any emissions were identified. Based on the relative amount of each material used in the process, an amount category was assigned so that a primary substance present in higher amounts represents greater risk than a sparsely used substance. By specifying substance role, the model determines an exposure probability for each substance based on whether such exposure would only be expected due to an in-process accident, whether normal operations would result in emissions, or whether both pathways would be possible. The physical phase of each substance if or when it is released must be included to apply the appropriate persistence and mobility parameters for the media of air, soil, and water. An emission factor is included for any substance emitted during normal production, indicating the order of magnitude of the substance released to the environment in kg per ton of product. Like the other physical values, these emission factors are mapped to a unitless scale representing orders of magnitude. Any normal releases of a material from a level of 0.00001 to 1000 kg/ton are represented by a set of emissions risk classes ranging from 1 to 10. The emissions were particularly challenging parameters to collect due to uncertainty of projecting industry scaled versions of published processes. Even on a smaller production scale insufficient data regarding detailed mass balances of waste streams made accurate estimation of emissions difficult. In cases where an emission risk class could not be sufficiently determined, one was approximated based on either a similar process that already existed within the commercial database or on a stoichiometric calculation of mass balance.

Comparison of Risk

At this point the assembled risk class information can be combined to arrive at some conclusions about relative risk, both qualitatively using some of the collected information, and quantitatively via the insurance protocol. Because we were starting with a populated insurance database that included many known fabrication processes, the nanomaterial manufacturing risk could be compared with each other and against other common processes previously defined in the system. For both our qualitative and quantitative results, nanomaterials risk scores are presented alongside the scores for six other familiar processes: silicon wafer (semiconductor) production, wine production, high density plastic (polyolefin) production, automotive lead-acid battery production, petroleum refining, and aspirin production. Each of these processes and their supporting substance data were taken from the insurance database; because they are used here as established in the XL database already for commercial use, their data and underlying assumptions were not addressed. The comparison processes were chosen to offer a benchmark of the risk posed by widely accepted or familiar operations. Silicon wafer production, an integral part of computer manufacturing, and automotive lead-acid battery production are both manufacturing processes found in or near many communities. Petrochemical complexes in industrial cities are characterized by polyolefin production and petroleum refining facilities. The well-established yet constantly evolving pharmaceutical industry is represented by aspirin production. Lastly, wine production was chosen as an interesting benchmark comparison since it would be considered by most a relatively benign, and even desirable, process to be in close proximity to a community. When including this comparison data, the only change made was to remove final products from the process as we did in the case of nanomaterial production, in order to consistently compare only the manufacturing processes' contribution to insurance risk.

Qualitative Risk Assessment

The nanomaterial fabrication data, once collected, were first organized qualitatively to characterize and compare the relative risks. This representation of the data focused on the relative risk posed by five key properties for each constituent material of each nanomaterial fabrication process. We collected data regarding each substance's toxicity (LC_{50} and LD_{50}), water solubility, log K_{ow}, flammability, and expected emissions. The risk classes in each of these five categories were represented visually in the form of squares, where filled-in squares represent higher risk and empty squares represent lower risk. Two sample resulting risk tables (Tables 13.5 and 13.6) provide a subjective sense of the risk posed by manufacturing a nanomaterial and a non-nanoscale material, based solely on the order of magnitude of selected parameters but without mathematically combining them. Tables 13.5 and 13.6 are adapted from Robichaud et al. (2005) [34].

C_{60}	Toxicity	Water Solubility	Log K_{ow} (Bio-accumulation)	Flammability	Emissions
Benzene	■■■	■■	□	■■	
Toluene	■■	■	■	■■	□ – ■
Argon	□	■	□	□	□
Nitrogen	□	■	□	□	□
Oxygen	□	■	□	■	□ – ■■
Soot	■■	□	■■	□	■■■
Activated carbon	□	□	■■	■	
Carbon dioxide	□	■■	□	□	■■
Water	□			□	□

TABLE 13.5 Qualitative Risk Rankings for C_{60} Production

Automotive Lead Battery	Toxicity	Water Solubility	Log K_{ow} (Bio-Accumulation)	Flammability	Emissions
Sulfuric acid	■■■	■■	□	■	■
Calcium sulfate	□	■■	□	□	■■■
Antimony	■■	□	■■	□	
Arsenic	■■■	□	■■	□	
Tin	■	□	■■	□	
Soot	■■	□	■■	□	
Lead	■■	□	■■	□	
Lead oxide	■■	□	■■	□	
Lead monoxide	■■	■	■■	□	■
Lead dioxide	■■	■	■■	■	
Sodium perchlorate	■	■■	□	■	
Barium sulfate	■■	□	■■	□	
Polyvinyl chloride	□	□	■■	□	
Hydrochloric acid	■■	■■	□	□	

TABLE 13.6 Qualitative Risk Rankings for Automotive Lead-Acid Battery Production

These qualitative graphs organize the information we know about manufacturing the chosen materials so as to allow some general observations. The nanomaterial fabrication processes appear overall to have fewer toxic ingredients and fewer total constituent materials, but they are also predicted to have relatively higher emissions compared to the non-nanoprocesses. A detailed discussion of the caveats behind these graphs and the takeaway insight on relative risk of all 10 processes is presented in Robichaud et al. (2005) [49].

Quantitative Risk Assessment

Three relative risk scores are calculated by the insurance protocol for each fabrication process: incident risk, normal operations risk, and latent contamination risk. The *incident risk score* represents the risk due to an in-process accident that leads to accidental exposure. Risk posed by substances that are expected to be emitted during the fabrication process are referred to as *normal operations risk*. To account for the potential long-term contamination of the operations site, a *latent risk score* is calculated by combining factors from the normal operations risk and incident risk scores.

The reader will recall that risk scores corresponding to conditions such as temperature and pressure were combined to calculate the probability of an accident occurring during the process, expressed as a process incident probability class. Risk scores corresponding to toxicity, persistence, and mobility data for each constituent substance were then combined to calculate the risk associated with that constituent's interaction with air, water, and soil; this cumulative risk score for each compound was denoted as its substance hazard risk class. An amount hazard risk class was assigned for each substance based on the relative amount used in the process, wherein more of a substance enhanced the level of risk posed. The process incident probability class, amount hazard risk class, and substance hazard risk class were used to compute final risk ratings for air, water, and soil pathways due to sudden release and due to normal emissions from a given process.

Incident risk was determined for each exposure pathway by combining the process hazard rating, the amount hazard ratings for each substance, the substance hazard risk classes, and an actuarial adjustment coefficient that forced the final risk score into a 1 to 100 distribution. Figure 13.7 provides a schematic of calculations, and the actuarial coefficient will be explained in the next section. Normal operations risk was calculated for each pathway by combining the constituent substances emission coefficient risk categories, their substance hazard risk classes, and another actuarial adjustment coefficient. The final incident and normal operations risk ratings calculated for soil and water were combined to arrive at a latent risk score. Air ratings are excluded from this calculation because by the time long-term latent contamination is of concern, airborne contaminants would either have already come in contact with an organism or settled into the water or soil.

Explaining the Actuarial Adjustment Coefficient

Mathematically combining the system parameters is the step that assimilates the collected information and imposes a weighting system to arrive at a relative score for each process. The actuarial coefficients chosen to combine the variables in the XL Insurance database are based on order of magnitude. Essentially, the rating is set up in such a way that the increase from one risk class to the next corresponds to an increase in risk by an order of magnitude. Theoretically the incident risk score is built from 21 risk classes. Fourteen classes are reserved for the characterization of the process risk exposure, which is comprised of temperature, pressure, enthalpy, and fire risk. Three classes (1–3) are for the amount categorization (main, secondary, trace). Ten classes (1–10) are for the substance property characterization, accounting for toxicity, persistence, and mobility. Adding up all these classes yields a range of a minimum of 6 to a maximum of 27 classes, so the risk range covers 21 "orders of magnitude." In essence, by adding up the classes we may say mathematically that we multiplied by adding up exponents. The company recognizes that 21 orders of magnitude in risk exposure is a wide range, and that this may not be a completely realistic view of the risk picture. However, since perfect data are almost never available with regard to real systems and the environment, organizing by order of magnitude provides a good way to arrive at useful conclusions given the available information. The choice was made to adjust the scale so that relative risk scores would range from 1 to 100. Since the risk classes are combined additively, the number of classes were intended to be set up so as to give equal weight to both the measure for severity of risk and the frequency or likelihood of its occurrence (risk = hazard × exposure). XL Insurance considers the 11 process related classes to represent exposure likelihood and the 12 substance related classes to represent the hazard posed by the processes. This choice in particular highlights how influential the decisions made in the "weighting filter" can be on the final relative risk. Other possibilities exist for combining the classes, of course. For example, some might argue that the substance characteristic of photolytic half-life really determines more about exposure than hazard because it describes how long the substance will remain in the air before chemically degrading.

The numerical factors added to the end of the combination expression are there to make sure that the lowest risk score in the scheme comes out as 1 and the highest as 100:

Case 1 (minimum risk score):

$$[((P[=1]+T[=1]+H[=1]+G[=1])+((S_{path}(=1)+M_{path}(=1))_{max}-1)-2)]*\left(\frac{99}{21}\right)+1=\left(0*\left(\frac{99}{21}\right)\right)+1=1$$

Case 2 (maximum risk score):

$$[(((P[=3]+T[=3]+H[=3]+G[=5])-3)+((S_i,L+(=3)+M_i(=10))_{max}-1)-2)]$$

$$*\left(\frac{99}{21}\right)+1=\left(21*\left(\frac{99}{21}\right)\right)+1=100$$

This calculation is done for every substance and for each of the three pathways considered: soil, water, and air. The maximum resulting value is taken as the overall risk score for the production process.

In the XL database, a final score in the incident risk and normal operations risk categories corresponds to the highest of the three pathway-specific scores out of air risk, water risk, and soil risk. For each of these three transport media-based risks, the score is based on the highest risk substance in that exposure pathway. This method of taking the highest risk medium based on its highest risk substance is another of the important choices involved with combining the system parameters. It is appropriate for insurance premiums, for as we have established, the overall goal in this system is to approximate the relative order of magnitude of the risks. The practice of choosing only the highest scoring constituent material instead of accounting for multiple substances admittedly precludes differentiating processes that involve multiple hazardous substances. The same score could be calculated for both a process that includes benzene and toluene as main ingredients as well as a process that uses only benzene, since benzene corresponds to the highest substance hazard risk class. Adding toluene to a process does add to the consequent risk, but it does not change it by a full order of magnitude from an insurance perspective.

Running the XL Insurance Database Protocol

The three types of relative risk scores were generated for each of the five chosen nanomaterial production processes. Due to the uncertain nature of predicting the final scaled-up manufacturing process volumes, expected emissions from normal operations were more difficult to estimate, and as such were only able to be approximated with the system. Three separate hypothetical cases are defined for each fabrication process, resulting in three sets of scores for normal operations and for latent risk, both of which incorporate the risk due to predicted emissions. The three cases are defined representing a range of manufacturing scenarios intended to estimate boundary conditions of "most materials emitted" and "least materials emitted." In processes for which the authors had mostly published academic work available, such as ZnSe quantum dots, we tried to vary the emission risk class was made to vary by a few orders of magnitude between the low risk and high risk cases. Much of the work for those nanomaterials has been carried out in the private sector, and as such less public information is available. In other processes for which more data were accessible, such as SWNT and alumoxane production, closer estimates could be made based on known practices or on reaction stoichiometry. For detailed assumptions related to all the nanomaterials fabrication processes' scores and the different cases considered, the reader is invited to refer to the 2005 *Environmental Science & Technology* article and its online supporting information.

Quantitative Insurance-Based Risk Results

Graphs were prepared comparing incident risk, normal operations risk, and latent risk as calculated by the insurance database across all of the nano and non-nanomanufacturing processes. Incident Risk scores showed that, by and large, the nanomaterials production processes score comparable to or lower than the traditional comparison processes. The Normal Operations scores incorporate risks posed by projected emissions of any materials during the fabrication process. The uncertainty in these emissions was accounted for by including a range of probable manufacturing scenarios, scoring three different cases defined for each nanomaterial. The assumptions upon which each of the three cases is built beget different risk scores for the nanomaterials' production processes. These scores are shown as a range of values, with the mid-risk case represented as the main data point. A single point denotes the score of each of the non-nanoproduction methods because they were already established in the database by XL Insurance as fully scaled-up manufacturing processes. The emissions data will not run the risk of varying by more than an order of magnitude as they would when projecting the scale-up of nanomaterial production processes, thus the variability already allowed by the protocol will suffice. For ZnSe quantum dots, changing the assumptions did not cause the final score to differ by orders of magnitude, so the range is very narrow. Alumoxane and SWNT production earned low normal operations risk scores, again proving to be on the order of the lowest scoring non-nanomaterial processes, wine and aspirin. Scores in this category for C_{60}, ZnSe quantum dots, and nano-TiO_2 placed these three nanomaterials in a similar range as silicon wafers and automotive lead-acid batteries. The highest normal operations risk scores were held by polyolefin production and petroleum refining, which were calculated as higher risk than all the nanomaterial manufacturing processes. When predicting industrial scale nanomaterial manufacturing practices, the emissions values were estimated conservatively by erring on the side of more emitted constituent materials and higher emissions levels for those substances. Therefore, it makes sense to predict that further clarification and streamlining of the processes would contribute to both reduced score variability and lower overall risk scores. For instance, with both SWNTs and alumoxane particles, more production information was accessible than for the other nanomaterials. The cases considered differed from each other by less, and the normal operations Risk scores for each of these prove to have both narrower margins of error and lower overall values. Mature processes with high Incident Risk scores such as polyethylene production and silicon wafer production offer another demonstration of the proposed trend toward lower emissions for more developed processes. These two fabrication processes employ many materials, including several very hazardous ones, yet they have very low expected emissions and thus lower normal operations risk scores.

Latent risk scores were also calculated for all 11 processes, ranking in an order similar to what we saw in the other two risk categories. This should be anticipated that latent risk is a function of the incident and normal operations risks. Except for C_{60}, all the nanomaterial fabrication processes' latent risk scores compare closely with silicon wafer, wine, and aspirin production, and are lower than polyolefin, automotive lead-acid battery, and refined petroleum production. While the production of C_{60} has a higher score than the other nanomaterials, it is still markedly lower than that of petroleum refining and could be considered comparable to the latent contamination score for polyolefin and automotive lead-acid battery production.

What Do We Learn from This Example?

This example illustrates how a slice of the nanomaterial risk landscape can be understood utilizing an existing methodology. From the point of view of insurance stakeholders, we conclude that there do not appear to be any abnormal pollution risks associated with the manufacturing portion of these five nanomaterials. Both the qualitative and quantitative insurance-based approaches suggest that differences in handling during operations affect the final risk scores as much as the identities of the constituent substances in a process. Such a conclusion offers some practical direction for manufacturers by pointing out that recycling and successful recapture of materials are viable ways to lower normal operations risk scores without altering the chemicals in the process.

This exercise highlights several challenges and limitations of developing a ranking of relative risk for nanomaterials with scarce information. One key challenge is the lack of real data regarding the mass balances and yield rates needed to estimate emissions values for the various substances in nanomaterial manufacturing processes. Especially in the case of new materials such as the five nanomaterials in our example, it becomes necessary to project from the lab scale to the industrial scale volumes handled in their manufacture. In the quantitative risk results, we dealt with this data deficiency by including a range of potential scores from a variety of manufacturing scenarios. As experience with larger scale fabrication yields better information on mass balances, the risk will be more accurately characterized.

The very manner in which the risk is calculated may be tailored for the end use. In the case of an insurance industry, it may be reasonable to evaluate risk from the standpoint of the largest single component, where from another perspective interactions between sources of risk must be accounted for. For example, the XL Insurance protocol applied here scores process risks by orders of magnitude, so it only accounts for the highest scoring material rather than accounting for multiple chemicals. This is a mathematical choice that results in acceptable accuracy for the company (e.g., if the plant blows up, do you need to worry about toxicity?). However, a model for toxicity that is focused on long-term health effects for workers may need to consider toxic interactions between chemicals.

Other Life-Cycle Stages

The manufacture of nanomaterials considered in the above analysis involves raw materials commonly used in the production of conventional materials, and the assessment concerned only the manufacturing portion of the life cycle without the materials themselves. The actuarial risk assessment protocol used for the manufacture of nanomaterials may also be extended to other stages of the life cycle. This is especially straightforward when the produced nanomaterials are in turn used in other manufacturing processes. Alumoxane nanoparticles, for example, are used in the production of ceramic membranes. Wiesner et al. pointed out that, compared to the more traditional sol-gel process, the use of alumoxane nanoparticles reduces the time and energy required and eliminates the use of organic solvents in the fabrication of ceramic membranes [99]. The risk assessment performed for alumoxane nanoparticles in the previous section can be extended to include their use in the manufacture of ceramic membranes. Such a comparison study was carried out to compare the risks associated with the production of ceramic membranes using alumoxane nanoparticles against the more traditional sol-gel process [100]. Again, the nanomaterials themselves could not be characterized, but the effects of their use on process parameters such as temperature and pressure, as well as the on the selection of other substances required for the manufacture of the membranes, could be captured. Although this work is not detailed here, Figure 13.8 is included to illustrate the concept of carrying a relative risk assessment through to additional life-cycle stages, showing the relative risk ranking on the XL scale of 1 to 100.

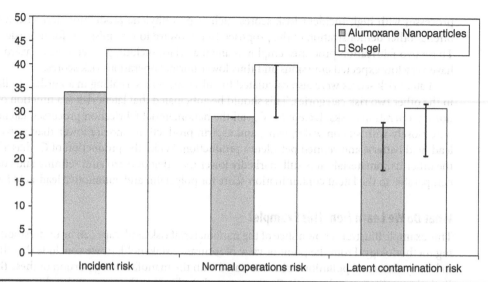

FIGURE 13.8 Relative risks in the manufacturing of ceramic membranes using alumoxane nanoparticles and sol-gel technologies, including risks in the production of the precursors. (Adapted from Beaver et al. 2005.)

Knowledge Gaps in the LCA of Nanomaterials Risks

Our ultimate goal is to incorporate information on the nanomaterials themselves into the risk assessment of the nanomaterials life cycle. The characteristics frequently used to estimate substance risks are often easily found in the literature for bulk substances, such as those used in producing the nanomaterials, but what about for the nanomaterials themselves? There is some temptation to use the information known about a bulk material that is considered closely related in composition. For instance, the MSDS for carbon black has been applied for C_{60} since they both consist of only carbon atoms. However, such descriptions are inappropriate because the size, shape, surface functionality, and structural uniformity of nanomaterials lend them unique attributes. If these engineered nanomaterials have truly unique properties that make them more desirable than their bulk counterparts for specific applications, then it is reasonable to assume that engineered nanomaterials may also have different physical chemical properties that may affect their risk profile. These distinctions may cause them to affect organisms in ways that differ from those of compositionally similar larger materials.

Although many of these engineered materials are new, exposure to nanoscale particles is not a new phenomenon for humans or the environment. For example, nanoscale particles are unintentional products of natural and anthropogenic combustion, and considerable research has been on these ultrafine particles and their effects [52]. The molecule C_{60} may be one of many combustion products making up atmospheric ultrafine particles in urban areas. It would be advantageous to group nanomaterials into classes that represent materials with similar properties, thereby avoiding a cases-by-case evaluation of every new nanomaterial. Rigorous nomenclature and pertinent criteria that might set one class of material apart from another do not yet exist. SWNTs exist in a variety of different lengths and chiralities and may be multi-walled or single-walled; fullerenes can be coated, uncoated, functionalized, aggregated, or free; quantum dots can be comprised of a wide array of different base metals. However, to assess the risks of classes of nanomaterials, these screening categories must first be established as a function of our improving knowledge of which parameters describing relevant materials, environments, interactions, and effects are most important to measure for forecasting outcomes of interest. The nanoEHS field continues to investigate these critical parameters and synthesize its findings for the advancement of both fundamental science and risk guidance.

References

1. Wiesner, M.R. and J.Y. Bottero, "A risk forecasting process for nanostructured materials, and nanomanufacturing." *Comptes Rendus Physique*, 2011; **12**(7): 659–668.
2. Dillemuth, J., et al., "Traveling technologies: Societal implications of nanotechnology through the global value chain." *Journal of Nano Education*, 2011; **3**(1–2): 36–44.

3. *International Standard 14040: Environmental Management—Life Cycle Assessment—Principles and Framework*. 1997, International Organisation for Standardisation (ISO).

4. Guinee, J.B., *Handbook on Life Cycle Assessment: Operational Guide to ISO Standards*. 2002: Dordrecht: Kluwer Academic Publishers.

5. Udo de Haes, H.A., et al., *Life-Cycle Impact Assessment: Striving Towards Best Practices*. 2002, Pensacola, FL: SETAC Press.

6. Goedkoop, M. and R. Spriensma, "The Eco-indicator 99: A damage oriented method for life cycle impact assessment." 1999, PRÉ Consultants, Amersfoort.

7. Bare, J.C., et al., "TRACI: The tool for the reduction and assessment of other environmental impacts." *Journal of Industrial Ecology*, 2003; **6**(3–4): 49–78.

8. Saling, P., et al., "Eco-efficiency analysis by BASF: The method." *International Journal of Life-Cycle Assessment*, 2002; **7**(4): 203–218.

9. Tanzil, D. and B. Beloff, "Assessing impacts: Indicators and metrics," In *Transforming Sustainability Strategy into Action: The Chemical Industry*, B. Beloff, M. Lines, and D. Tanzil, Editors. 2005, John Wiley & Sons, Inc.: Hoboken, NJ.

10. Tanzil, D. and B. Beloff, "Assessing impacts: Overview on sustainability indicators and metrics." *Environmental Quality Management*, 2006; **15**(4): 41–56.

11. Schmidt, I., et al., "Managing sustainability of products and processes with the socio-eco-efficiency analysis by BASF," In *Greener Management International*. 2004; 79–94.

12. Graedel, T.E., *Streamlined Life-Cycle Assessment*. 1998, Upper Saddle River, NJ: Prentice Hall.

13. Andersson, K., et al., "The feasibility of including sustainability in LCA for product development." *Journal of Cleaner Production*, 1998; **6**(3–4): 289–298.

14. Bennett, R., et al., "Environmental and human health impacts of growing genetically modified herbicide-tolerant sugar beet: A life-cycle assessment." *Plant Biotechnology Journal*, 2004; **2**(4): 273–278.

15. Lloyd, S.M. and L.B. Lave, "Life cycle economic and environmental implications of using nanocomposites in automobiles." *Environmental Science and Technology*, 2003; **37**: 3458–3466.

16. Lloyd, S.M., L.B. Lave, and H.S. Matthews, "Life cycle benefits of using nanotechnology to stabilize platinum-group metal particles in automotive catalysts." *Environmental Science and Technology*, 2005; **39**: 1384–1392.

17. Shatkin, J., "Chapter 3: Sustainable nanotechnology development using risk assessment and applying life cycle thinking," In *Nanotechnology: Health and Environmental Risks*, 2nd ed., J. Shatkin, Editor. 2013, CRC Press: Boca Raton, FL.

18. Wender, B. and T.P. Seager. "Towards prospective life cycle assessment: Single wall carbon nanotubes for lithium-ion batteries." In *Sustainable Systems and Technology (ISSST), 2011 IEEE International Symposium on*. 2011. IEEE.

19. Walser, T., et al., "Prospective environmental life cycle assessment of nanosilver T-shirts." *Environmental Science & Technology*, 2011; **45**(10): 4570–4578.

20. Gavankar, S., S. Suh, and A. Keller, "Life cycle assessment at nanoscale: Review and recommendations." *The International Journal of Life Cycle Assessment*, 2012; **17**(3): 295–303.

21. Hischier, R. and T. Walser, "Life cycle assessment of engineered nanomaterials: State of the art and strategies to overcome existing gaps." *Science of the Total Environment*, 2012; **425**(0): 271–282.

22. Linkov, I. and T.P. Seager, "Coupling multi-criteria decision analysis, life-cycle assessment, and risk assessment for emerging threats." *Environmental Science & Technology*, 2011; **45**(12): 5068–5074.

23. Seager, T.P. and I. Linkov, "Coupling multicriteria decision analysis and life cycle assessment for nanomaterials." *Journal of Industrial Ecology*, 2008; **12**(3): 282–285.

24. "Opinion on the appropriateness of existing methodologies to asses the potential risks associated with engineered and adventitious products of nanotechnologies." 2005, European Commission, Scientific Committee on Emerging and Newly Identified Health Risks. pp. 41–58.

25. Tran, C.L., et al., "A scoping study to identify hazard data needs for addressing the risks presented by nanparticles and nanotubes." 2005, Institute of Occupational Medicine. pp. 1–48.

26. Michalsen, A., "Risk assessment and perception." *Injury Control and Safety Promotion*, 2003; **10**(4): 201–204.

27. Biocca, M., "Risk communication and the precautionary principle." *Human and Ecological Risk Assessment*, 2005; **1**: 261–266.

28. Power, M. and L.S. McCarty, "Perspective: Environmental risk manageent decision-making in a societal context." *Human and Ecological Risk Assessment*, 2006; **12**: 18–27.

29. Langsner, H., *Nanotechnology: Non-Traditional Methods for Valuation of Nanotechnology Producers*. 2005, Innovest Strategic Value Advisors: New York City. pp. 13–95.

30. Slovic, P., "Perception of risk." *Science*, 1987; **236**(4799): 280–285.

31. Renn, O., "Towards an integrative approach." 2005, International Risk Governance Council: Geneva, Switzerland.

32. Owen, R. and M. Depledge, "Editorial: Nanotechnology and the environment: Risks and rewards," In *Marine Pollution Bulletin*. 2005. pp. 609–612.

33. Hendren, C.O., et al., "Modeling approaches for characterizing and evaluating environmental exposure to engineered nanomaterials in support of risk-based decision making." *Environmental Science & Technology*, 2013; **47**(3): 1190–1205.

34. Robichaud, C.O., et al., "Relative risk analysis of several manufactured nanomaterials: An insurance industry context." *Environmental Science & Technology*, 2005; **39**(22): 8985–8994.

35. Morgan, M.G., M. Henrion, and M. Small, "Uncertainty: A guide to dealing with uncertainty in quantitative risk and policy analysis," 1990, New York: Cambridge University Press. p. 332.

36. Gabsi, F., et al., "Coupling different mechanistic effect models for capturing individual- and population-level effects of chemicals: Lessons from a case where standard risk assessment failed." *Ecological Modelling*, 2014; **280**: 18–29.

37. Gilks, W.R., S. Richardson, and D.J. Spiegelhalter, "Markov chain monte carlo in practice: interdisciplinary statistics." 1996, Chapman & Hall/CRC. 481.

38. Gottschalk, F., et al., "Modeled environmental concentrations of engineered nanomaterials (TiO2, ZnO, Ag, CNT, Fullerenes) for different regions." *Environmental Science & Technology*, 2009; **43**(24): p. 9222.

39. Hendren, C.O., et al., "Modeling nanomaterial fate in wastewater treatment: Monte Carlo simulation of silver nanoparticles (nano-Ag)." *Science of the Total Environment*, 2013; **449**: 418–425.

40. Pollino, C.A. and B.T. Hart, "Bayesian approaches can help make better sense of ecotoxicological information in risk assessments." *Australasian Journal Ecotoxicology*, 2005; **11**: 57–58.

41. Hart, B.T. and C.A. Pollino, "Increased use of Bayesian network models will improve ecological risk assessments." *Human and Ecological Risk Assessment*, 2008; **14**(5): 851–853.

42. Uusitalo, L., "Advantages and challenges of Bayesian networks in environmental modelling." *Ecological Modelling*, 2007; **203**(3–4): 312–318.

43. Money, E.S., K.H. Reckhow, and M.R. Wiesner, "The use of Bayesian networks for nanoparticle risk forecasting: Model formulation and baseline evaluation." *Science of The Total Environment*, 2012; **426**: 436–445.

44. Money, E.S., et al., "Validation and sensitivity of the FINE Bayesian network for forecasting aquatic exposure to nano-silver." *Science of The Total Environment*, 2014; **473–474**: 685–691.

45. Morgan, K., "Development of a preliminary framework for informing the risk analysis and risk management of nanoparticles." *Risk Analysis*, 2005; **25**(6): 1621–1635.

46. Cronin, M.T.D., et al., "Use of QSARs in international decision-making frameworks to predict health effects of chemical substances." *Environmental Health Perspectives*, 2003; **111**(10): 1391–1401.

47. Fourches, D., et al., "Quantitative nanostructure-activity relationship (QNAR) modeling." *ACS Nano*, 2010; **4**(10): 5703–5712.

48. Freeman, P.K. and H. Kunreuther, *Managing Environmental Risk through Insurance*. Vol. 9. 1997: Springer Science & Business Media.

49. Robichaud, C.O., et al., "Relative risk analysis of several manufactured nanomaterials: An insurance industry context." *Environmental Science and Technology*, 2005; **39**(22): 8985–8994.

50. *Nanotechnologies: A Preliminary Risk Analysis on the Basis of a Workshop March 2004* in *Nanotechnologies: A Preliminary Risk Analysis*. 2004. Brussels: European Commission: Community Health and Consumer Protection Directorate General of the European Commission.

51. *Nanotechnology White Paper*, J. Morris and J. Willis, Editors. 2005, United States Environmental Protection Agency: Washington D.C.

52. *Characterising the Potential Risks Posed by Engineered Nanoparticles: A first UK Government Research Report*. 2005, Nanosafe II.

53. Roco, M.C. and B. Karn, "Environmentally responsible development of nanotechnology." *Environmental Science and Technology, A Pages*, 2005; **39**(5): 106A–112A.

54. Howard, J., *Approaches to Safe Nanotechnology*, N.I.f.O.S.a. Health, Editor. 2005, Centers for Disease Control and Prevention.

55. Nel, A., et al., "Toxic potential of materials at the nanolevel." *Science*, 2006; **311**: 622–627.

56. Hendren, C.O., et al., "A functional assay-based strategy for nanomaterial risk forecasting." *Science of the Total Environment*, In press.

57. O'Dell, C. and C.J. Grayson, "If only we knew what we know." *California Management Review*, 1998; **40**(3): 154–174.

58. Thomas, D.G., et al., "ISA-TAB-nano: a specification for sharing nanomaterial research data in spreadsheet-based format." *BMC Biotechnology*, 2013; **13**(1): 2.

59. Ostraat, M.L., et al., "The nanomaterial registry: Facilitating the sharing and analysis of data in the diverse nanomaterial community." *International Journal of Nanomedicine*, 2013; **8**(Suppl 1): 7.

60. Gaheen, S., et al., "caNanoLab: Data sharing to expedite the use of nanotechnology in biomedicine." *Computational Science & Discovery*, 2013; **6**(1): 014010.

61. Hastings, J., et al., "eNanoMapper: Harnessing ontologies to enable data integration for nanomaterial risk assessment." *Journal of Biomedical Semantics*, 2015; **6**(1): 10.

62. CODATA-VAMAS Working Group on the Description of Nanomaterials: A Uniform Description System for Materials. 2015.

63. Harper, S., "Nanomaterial-biological interactions knowledgebase," In *Nanoinformatics*. 2010, Arlington, VA.

64. Zentner, L., et al., "nanoHUB. org: Experiences and challenges in software sustainability for a large scientific community." *arXiv preprint arXiv:1309.1805*, 2013.

65. Hendren, C.O., et al., "The nanomaterial data curation initiative: A collaborative approach to assessing, evaluating, and advancing the state of the field." *Beilstein Journal of Nanotechnology*, In press.

66. NRC, *A Research Strategy for the Environmental, Health and Safety Aspects of Engineered Nanomaterials*, N.R. Council, Editor. 2012, The National Academies Press: Washington, D.C.

67. Holman, M.W., et al., *The Nanotech Report*, 4th ed. 2006, Lux Research.

68. Hendren, C.O., et al., "Estimating production data for five engineered nanomaterials as a basis for exposure assessment." *Environmental Science and Technology*, 2011; **45**(7): 2562–2569.

69. Piccinno, F., et al., "Industrial production quantities and uses of ten engineered nanomaterials in Europe and the world." *Journal of Nanoparticle Research*, 2012; **14**(9): 1–11.

70. *Explosion Hazards Associated with with Nanopowders*. 2005, United Kingdom Health and Safety Executive.

71. Keller, A., et al., "Global life cycle releases of engineered nanomaterials." *Journal of Nanoparticle Research*, 2013; **15**(6): 1–17.

72. Nowack, B., et al., "Potential scenarios for nanomaterial release and subsequent alteration in the environment." *Environmental Toxicology and Chemistry*, 2012; **31**(1): 50–59.

73. Benn, T., et al., "The release of nanosilver from consumer products used in the home." *Journal of Environmental Quality*, 2010; **39**(6): 1875–1882.

74. Nowack, B., et al., "Potential release scenarios for carbon nanotubes used in composites." *Environment International*, 2013; **59**: 1–11.

75. Abraham, M.H., C.E. Green, and J. Acree, "Correlation and prediction of the solubility of Buckminsterfullerene in organic solvents; estimation of some physicochemical properties." *The Royal Society of Chemistry, Journal of Chemistry Society, Perkin Transactions,* 2000, **41**(2): 281–286.

76. Therezien, M., A. Thill, and M.R. Wiesner, "Importance of heterogeneous aggregation for NP fate in natural and engineered systems." *Science of The Total Environment,* 2014; **485**: 309–318.

77. Wiesner, M.R., et al., "Assessing the risks of manufactured nanomaterials." *Environmental Science & Technology,* 2006; **40**(14): 4336–4345.

78. Gardea-Torresday, J.L., et al., "Formation and growth of Au nanoparticles inside live alfalfa plants." *Nano Letters,* 2002; **2**(4): 397–401.

79. Kasermann, F. and C. Kempf, "Buckminsterfullerene and photodynamic inactivation of viruses." *Reviews in Medical Virology,* 1998; **8**(3): 143–151.

80. Nakamura, E., et al., "Biological activity of water-soluble fullerenes. Structural dependence of DNA cleavage, cytotoxicity, and enzyme inhibitory activities including HIV-protease inhibition." *Bulletin of the Chemical Society of Japan,* 1996; **69**(8): 2143–2151.

81. Tokuyama, H., S. Yamago, and E. Nakamura, "Photoinduced biochemical activity of fullerene carboxylic acid." *Journal of the American Chemical Society,* 1993: **115**: 7918–7919.

82. Tabata, Y., Y. Murakami, and Y. Ikada, "Antitumor effect of poly(ethylene glycol)modified fullerene." *Fullerene Science and Technology,* 1997; **5**(5): 989–1007.

83. Tabata, Y., Y. Murakami, and Y. Ikada, "Photodynamic effect of polyethylene glycol-modified fullerene on tumor." *Japanese Journal of Cancer Research,* 1997; **88**: 1108–1116.

84. Tsao, N., et al., "Inhibition of *Escherichia coli*-induced meningitis by carboxyfullerence." *Antimicrob Agents Chemother,* 1999; **43**: 2273–2277.

85. Maynard, A.D., D.B. Warheit, and M.A. Philbert, "The new toxicology of sophisticated materials: Nanotoxicology and beyond." *Toxicological Sciences,* 2011; **120**(Suppl 1:S109–29).

86. Lipnick, R.L., et al., "Comparison of the up-and-down, conventional LD50, and fixed-dose acute toxicity procedures." *Food and Chemical Toxicology,* 1995; **33**(3): 223–231.

87. Nel, A., et al., "A multi-stakeholder perspective on the use of alternative test strategies for nanomaterial safety assessment." *ACS Nano,* 2013; **7**(8): 6422–6433.

88. NRC, *Effects of a Polluted Environment: Research and Development Needs: A Report of the Panel on Effects of Ambient Environmental Quality to the Environmental Research Assessment Committee.* 1977, Washington, D.C.: National Academy of Sciences.

89. Blickley, TM, et al., "Dietary CdSe/ZnS quantum dot exposure in estuarine fish: Bioavailability, oxidative stress responses, reproduction, and maternal transfer." *Aquatic Toxicology,* 2014; **148**: 27–39.

90. Juch, H., et al., "Nanomaterial interference with early human placenta: Sophisticated matter meets sophisticated tissues." *Reproductive Toxicology,* 2013; **41**(0): 73–79.

91. Wayne R. Munns, Jr. and M.G. Mitro, *Assessing risks to populations at superfund and RCRA sites: Characterizing effects on populations,* EPA, Editor. 2006: Narragansett, RI.

92. Judy, J.D., J.M. Unrine, and P.M. Bertsch, "Evidence for biomagnification of gold nanoparticles within a terrestrial food chain." *Environmental Science and Technology,* 2011; **45**(2): 776–781.

93. Unrine, J.M., et al., "Trophic transfer of Au nanoparticles from soil along a simulated terrestrial food chain." *Environmental Science and Technology,* 2012; **46** (17): 9753–9760.

94. Schlesinger, W.H., K.H. Reckhow, and E.S. Bernhardt, "Global change: The nitrogen cycle and rivers." *Water Resources Research,* 2006; **42**(3).

95. Bernhardt, E.S., et al., "An ecological perspective on nanomaterial impacts in the environment." *Journal of Environmental Quality,* 2010; **39**(6): 1954–1965.

96. Eckelman, M.J., et al., "New perspectives on nanomaterial aquatic ecotoxicity: Production impacts exceed direct exposure impacts for carbon nanotoubes." *Environmental Science & Technology,* 2012; **46**(5): 2902–2910.

97. Plata, D.L., et al., "Early evaluation of potential environmental impacts of carbon nanotube synthesis by chemical vapor deposition." *Environmental Science & Technology,* 2009; **43**(21): 8367–8373.

98. Robichaud, C.O., et al., "Relative risk analysis of several manufactured nanomaterials: An insurance industry context: Supporting information. *Environmental Science and Technology,* 2005; **39**(22): 8985–8994.

99. DeFriend, K.A., M.R. Wiesner, and A.R. Barron, "Alumina and aluminate ultrafiltration membranes derived from alumina nanoparticles." *Journal of Membrane Science,* 2003; **224**(1–2): 11–28.

100. Beaver, E., et al., *Implications of Nanomaterials Manufacture & Use.* 2005, BRIDGES to Sustainability.

CHAPTER 14

Nanotechnology Governance for Sustainable Science and Policy

Sally Tinkle

Science and Technology Policy Institute, Washington, D.C., USA

Claire Auplat

Novancia Business School, Paris, France

C ritical environmental policy issues are no longer easily defined and more often than not, defy straightforward solutions. Environmental issues have become increasingly complex due to the sophistication of technological solutions and the uncertainty surrounding the potential for unintended consequences. The impacts of climate change, globalization, economic competition, and social justice compound an environmental policy analysis by creating potential exceptions, modifications, and conflicting goals for the conditions of the problem and the solution. For these and other reasons, environmental policymaking is increasingly risk- and evidence-based, and increasingly participatory as a wide range of stakeholders seek a voice in the determination and management of environmental risks that affect their lives and communities (Mauelshagen et al., 2014). This evolution in the complexity of environmental policy- and rule-making has created the need for a problem-solving approach that expands the traditional framework of environmental risk assessment and management and adds the diversity of stakeholder needs, values, and perspectives to government regulatory and policy decisions.

This chapter serves as an introduction to the concept of risk governance, its elements and application to nanotechnology and environmental nanotechnology in particular, and it surveys the role of regulation and policy in nanotechnology risk governance.

What Is Governance? What Are the Characteristics of Governance?

In contrast to the concept of government as a hierarchical entity exerting control over geographical areas, the concept of governance has been explored by Rosenau and others to refer to collective actions of a horizontally organized system that influence, in a constraining or enabling way, the outcomes of other actors' actions across local, regional, and global scales. Rosenau (1990, 1995) suggested that the core of governance involved rule systems in which steering mechanisms are employed to frame and implement goals that move communities and nations in the directions they wish to go or that enable them to maintain the institutions and policies they wish to maintain.

The much-cited European Commission white paper (European Commission, 2001) expanded the concept of rule systems by outlining the essential principles of democratic governance as a response to a perceived growing divide between citizens and government on what constitutes appropriate regulations to protect public health and safety. The white paper identified five principles of governance—openness, participation, accountability, effectiveness, and policy coherence—to establish new ways of involving citizens in government policy decisions. Thus, governance implies the development of policies and regulations through a systematic process that reflects stakeholder perspectives about a collective problem. Given the potential breadth of policies and stakeholders, a governance process needs to be flexible and responsive to evolving events, opinions, and decisions. It needs to occur across the life cycle of a product or solution, from R&D to manufacture, consumer use, recycling, and reuse. It may occur through a formal or informal process, and although governments may participate in a governance process, governance does not inherently possess the authority to require compliance. At best, it is a self-regulating, consensus-based system.

What Is Risk? What Are the Characteristics of Risk?

Definitions of risk are varied and may be contextual, but all definitions encompass the notion of exposure to potential dangers that threaten to harm or destroy people or the environment. Risk is generally defined as being a function of both the expected losses which can be caused by an event and the probability of this event. The harsher the loss and the more likely the event, the greater the overall risk. These two factors, loss and the probability of the loss, are typically treated as sequential, independent events so that, in simplified terms, quantification of risk is expressed as:

$$Risk = (probability\ of\ an\ accident) \times (losses\ per\ accident)$$

Hood, Rothstein, and Baldwin (2001) studied the governance of risk and looked at a number of risk regulation regimes. They sought to explain why regulation varies importantly from one area to another, or why a lot of effort is spent on controlling some risks but not others, and they noted that in spite of globalization approaches to risk and regulation vary widely.

In order to regulate risk, one needs to be aware that there is some risk, and this is where the notion of risk assessment comes into play. One characteristic of risk assessment is that risk is always acknowledged and assessed *a posteriori*. Scientific proof can only be put forward when harm has already been done. This is the only way to obtain an estimation of an unobserved quantity on the basis of empirical data. It was only after the fire destroyed 13,200 houses, leaving some 70,000 Londoners homeless, that Nicolas Barbon's idea to insure buildings against fire could, in the first place, germinate and later become successful (Hanson, 2002; Wainwright, 1953). So there is a paradox that risk can only be envisaged in the light of past events and yet bears on the probability of the occurrence of future events. The difficulty with risk governance for nano-based materials is that there is no certainty that existing data that was used for other types of risks can be used to assess nanorisks. Assays and toxicity measurements, for one thing, cannot be simply transposed from the bulk size to the nano size for the same chemical elements and the results of available studies also diverge sometimes widely. This is one of the major reasons why the links between risk governance and regulation are so complex, and why the notion itself of nano-regulation is ill-defined. The development of risk assessment as used by the insurance sector may be a path to follow as actuaries have a long experience of risk management. In the case of nanoproducts, the notion of liability is taking increasing importance in the relations between regulation and risk governance, and the insurance sector's approach of risk management is one of the methodologies that are closely scrutinized.

Risk and Governance Translated to Risk Governance

Concepts of risk governance began to appear in the literature in the late 1990s as general principles of governance were integrated with concepts of science and technology risk, regulation, and policy. The Europeans first embraced risk governance as an overarching concept that integrated the traditional approaches of risk identification, assessment, management, and communication with its guiding principles for democratic governance (European Commission, 2001). In spite of these advances, several factors compelled an even more expanded risk governance paradigm. These factors include:

- Increasing globalization
- Rapidly accelerating development of new S&T materials and products
- The complexity of regulations and policies
- A larger role of civil society—NGOs and formal and informal civilian groups—in assessing the safety, suitability, and desirability of products for manufacture and use in their communities

The emerging risk governance paradigm was expanded to acknowledge the importance of stakeholder inclusion, the distribution of power, and shared decision-making.

In their 2006 and 2011 papers, Renn and Roco (2006), Roco (2011), and Roco, Harthorn, Guston, and Shapira (2011) applied risk governance to nanotechnology and expanded the risk governance concept to include conflict resolution in decision-making, the exercise of power over resources and interests, and mechanisms to insure stakeholder participation. They proposed adaptive management as a process through which risk governance could evaluate, modify, and apply existing risk frameworks to a technology across its life cycle. Adaptive management could also provide the context through which to consider the long-term effects of a technology on health and the environment; the different rates at which research data are generated for product development, risk assessment, and policy; and the applicability of regulatory legislation to risk management.

In addition to the flexibility implied by adaptation, a risk governance approach must be integrative, not only as a process that merges science and technology with regulation, policy, and ethics, but one that is also inclusive of government and nongovernment communities impacted by or responsible for the potential risks of a technology application. As noted in the section *The Role of Regulation in Risk Governance* later in this chapter, citizens and their nations bring perspectives and laws to the governance of risk that reflects national needs, values, and experiences that do not necessarily align. Thus the flexibility and inclusivity of risk governance also make possible considerations of social justice and provide a forum for meaningful involvement and treatment of all interested and impacted people regardless of origin or income. This critical yet often ignored component serves as the nexus for inclusive dialogue on societal values and experiences; ethics and fairness; economics; and local, national, and international interests that give as much weight to the community voice as to government or the industrial developer.

Renn and Roco (2006) also correctly noted many of the obstacles to the implementation of risk governance, such as the fragmented responsibility for environment, health, and safety research across government agencies; limited coordination of government and nongovernment communities; constrained resources for safety research, and the inability of governments to agree on the principles and elements of risk governance. The breadth and complexity of risk governance have also made it difficult to establish a single entity with the multidisciplinary expertise, neutrality, and objectivity needed to convene communities and governments within and across nations for consensus decisions.

Over the last decade, risk governance has continued to evolve from a theoretical construct to an organizational framework although many obstacles remain. The International Risk Governance Council (IRGC) was established in 2003 as a risk governance convening entity. IRGC considers risk governance a concept and a tool that promotes early adoption of technologies that solve critical local, regional, and global challenges, such as energy and water, and that allows for the economic advantages of innovation while protecting public health and the environment. Their white papers summarize and extend much of the early efforts by describing risk governance as a systematic approach to decision-making about natural and technological risks that incorporate principles of good governance, risk assessment, and risk management and that allow for the identification of gaps in risk policy and regulation. In addition to the key risk governance questions about the societal benefits and impacts of new technologies, the IRGC approach includes analysis of the societal, environmental, and economic values that affect the willingness to accept risk; the role of science and technology in managing risk-related policies and regulations; and the capacity and resilience of a society to manage those potential risks.

As a process that combines technology solutions with risk science, regulatory science and policy, and the social sciences, risk governance is, by need, a multidisciplinary series of dynamic and flexible networks. For environmental nanotechnology (Mauelshagen et al., 2014), risk governance attempts to balance problem-solving applications of technology with critical elements of life-cycle product stewardship and risk assessment, that is, hazard and exposure impacts for humans and the environment, with societal questions of cultural value, economic value, and long-term consequences.

Attributes of Risk Governance

To understand the theoretical underpinnings of risk governance, it is useful to examine its attributes and the underlying principles that distinguish it from the traditional risk assessment and risk management approaches.

The influence of the European Commission white paper (European Commission, 2001) on the early conceptualization of risk governance made consensus decision-making a central tenet, shifting the focus from government as the central, dominant actor establishing regulatory policy to consensus-building multiactor networks. Multiactor networks are inclusive of individuals and groups impacted by a technology-enabled solution or product at any point across the life cycle. Traditional risk assessment actors and groups—national and regional governments, academia, industry—are joined in risk governance approaches by municipal governments and the market sector, civil society such as NGOs, community groups, and concerned citizens acting locally and globally. This inclusive approach embraces the complexity of relationships between heterogeneous stakeholders, the dynamic nature of those relationships, and the diversity of perspectives on risk and options for managing benefits and unintended consequences.

As a consequence of this inclusivity, governance has both horizontal and vertical elements. Horizontal governance (Renn, Klinke, and van Asselt, 2011, p. 233) usually involves actors relevant to a functional, geographical, or political/administrative question or concern, whereas vertical governance involves institutional relationships at different levels of government. For example, horizontal governance would bring together local businesses and community leaders around a policy question, and vertical governance would engage local, regional, and national governments. Horizontal and vertical governance and multiactor networks are self-reinforcing as their interests and authorities will reflect their cultural and social perspectives, the point along the life cycle at which they engage the product or solution, and their tolerance for risk.

The changing nature of the risks requiring policy decisions also reinforces the risk governance approach. Historically, risk has been framed through a probabilistic model, that is, through a determination of the likelihood that a given dose of a substance will cause an adverse effect on human health and/or the environment. Risk was approached as cause and effect, and interventions and policies were developed to manage some combination of the hazard and exposure in order to disrupt the cause and effect relationship. As our understanding of technological solutions and environmental conditions has become more complex and multidimensional, simple cause and effect relationships no longer apply. Causality has become difficult to assign as exposure inputs and adverse outcomes have become intertwined. Simple risks evolved into systemic risks, or multidimensional risks with an inherent societal context that can be local or global (Renn, Klinke, and van Asselt, 2011). This characteristic is often described as complexity.

Ambiguity and uncertainty are also attributes of risk (Renn, Klinke, and van Asselt, 2011). Uncertainty is ascribed to imperfect scientific knowledge about complex technical solutions and products and the inability to know the full context of their consequences across their life cycle. Ambiguity derives from perceptions of potential risk that are described as context and experience based, reflecting a variety of cultural norms and value systems. These perceptions underlie a stakeholder's decisions about risk as low and acceptable, that is, the adverse impacts are negligible; tolerable, that is, the benefits of the technology outweigh the potential for adverse outcomes that can be managed; or high, that is, the technology should not to be implemented. As a process that brings stakeholders together in a horizontal network for shared decision-making, risk governance considers all viewpoints legitimate and worthy of consideration. Risk governance provides a context within which differing and contradictory perspectives on the tolerability of risk might be integrated.

These elements shaped perceptions about risk governance as it was translated from a theory to an organizational framework and highlight its multifactorial and complicated dynamics.

Modeling a Risk Governance Process

The very complexity of critical social issues, technologies, stakeholders, and governments makes the implementation of an orderly risk governance process near impossible to imagine. As an independent, international body established by the Swiss Parliament to advance the risk governance paradigm, the IRGC proposed its risk governance framework in 2005 (Renn and Graham, 2005) and it remains essentially unchanged in 2015 (http://www.irgc.org/risk-governance/irgc-risk-governance-framework/; http://www.irgc.org/risk-governance/irgc-risk-governance-framework/, accessed on 17 July 17, 2015).

This framework (Figure 14.1) has five components although the fifth component, risk communication is considered part of the other four elements as well as an independent element. The five elements are preassessment, appraisal, evaluation, management, and communication. These are described as follows:

- Preassessment serves as the framing stage of risk governance during which stakeholder perspectives on the potential risks are identified and the existing tools and research needs for risk assessment, regulation, and policy are determined. Preassessment creates the initial integrated context for risk associated with a given product or solution.

- Appraisal has two components: risk assessment and concern assessment. Risk assessment follows the traditional evaluation of measureable physical characteristics of hazard and exposure to estimate potential impacts. Concern assessment systematically analyzes stakeholder perspectives on the risks and consequences and estimates individual or community resilience in absorbing risks

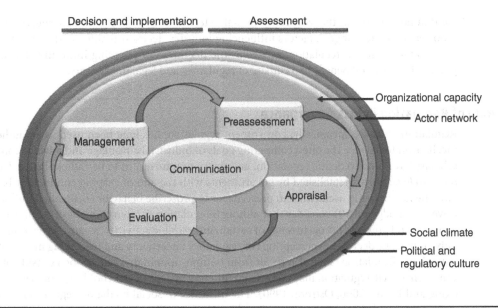

Figure 14.1 One possible risk governance model, adapted by the authors from IRGC (http://www.irgc.org/risk-governance/irgc-risk-governance-framework/, accessed July 17, 2015).

and benefits, including the possibility of political ramifications or unequal distribution of consequences. This is the stage at which the complexity, ambiguity, and uncertainty surrounding perceived risks are considered.

- Evaluation incorporates societal values, political and economic considerations, and ethical and moral principles into a determination of what is acceptable, tolerable, or intolerable risk and to whom is it acceptable, tolerable, or intolerable. The feasibility of alternative solutions and risk-reduction measures are part of this phase.

- Management of acceptable and tolerable risks is the phase in which knowledge is translated into an action plan to avoid, mitigate, transfer, or contain the risks. Management plans are frequently some combination of industry controls, voluntary and involuntary regulations, and government policies at local, regional, and global levels. Options are prioritized, where possible harmonized, and responsibilities for management and oversight assigned. Responsible risk governance dictates that a determination of intolerable risk stops the development of the technology or solution producing the intolerable risk.

- Communication is the critical linchpin that creates a multidirectional flow of information and knowledge between phases and stakeholders. It is the mechanism through which civil societies and concerned communities provide their perspectives and tolerance for risk to industries and governments and through which they understand the rationale for decisions, regulations, and policies.

The IRGC presents these elements as *phases* suggesting a linear progression, and they separate assessment, that is, the generation of data and knowledge, from management, or the development and implementation of decisions (Figure 14.1). This linear aspect of the model seems to lack the iterative flexibility implied by a complex, consensus-driven process, however, the IRGC places its framework within a risk governance context that acknowledges the critical importance of organizational capacity (assets, skills, and capabilities) needed to execute the process, the actor network, the political and regulatory culture, and the social climate. This suggests tacit understanding of the complexity and fluidity of the overall risk governance approach despite the linearity of the framework.

Others have added to, or modified elements of the IRGC framework, in order to clarify its elements or apply the process to a specific technology or aspect of risk management. Renn, Klinke, and van Asselt (2011) added a preestimation step prior to the IRGC preassessment step to allow for the screening of a large array of potential risks and actions and the down selection of a variety of "frames" within which to conceptualize the issues, risks, and possible solutions. Mauelshagen et al. (2014) examined effective risk governance for environmental policymaking. They emphasized the need for formal mechanisms for horizontal knowledge transfer in order to capture risk-relevant information from civil societies and local communities as well as the wide range of highly technical environmental data sources. Isaacs et al. (2015) reported a broader range of factors influencing industry decision-making and investment, public perception, and government regulation, as well as the dilemmas of pursuing innovative technologies with real potential for societal benefit in a climate of regulatory and legal uncertainty.

The Role of Regulation in Risk Governance

Regulation, or more specifically, government regulation, has long been used as an authoritative, vertical mechanism to balance the often competing goals of industrial advantage and adverse consequences. Many scholars have tried to define regulation, and the notion turns out to be rather elusive. Definitions usually relate to legal norms established by governments with the aim of shaping conduct for better societal welfare, the norms being used either to prescribe or to proscribe behavior (Brito and Dudley, 2012). Baldwin, Cave, and Lodge (2010) consider that although regulation is often thought of as an activity that restricts behavior and prevents the occurrence of certain undesirable activities, it can also have an enabling or facilitative impact by decreasing perceptions of regulatory uncertainty and increasing marketplace tolerance for risk taking. Regulations may also result from bottom-up initiatives developed outside of the government. This form of self-regulation may come from different groups of stakeholders, from civil society to industry (King and Lenox, 2000; Ostrom, 1990). The influence of social media on regulation, or as a form of self-regulation, is only beginning to be analyzed by scholars (Klischewski, 2014; Levine and Feigin, 2014; Lodge and Wegrich, 2015; McTaggart and Benina, 2014). This influence is likely to strongly modify existing methods of rule development and patterns of regulation.

Regulation occurs across the multiple vertical levels of risk governance. The terms "international," "regional," "national," and "local" usually refer to the bodies that enact the said regulations, not to the areas covered by them. For example, regional regulation refers to regulation passed by a regional institution, like the European Union, which includes regulatory bodies of different nations united by a specific agreement. On the whole, the geographic scope of the regulation corresponds to the level at which it is passed: national regulation will apply to a specific country, and local regulation to a local geographical area. However, the geographic scope of some regulations goes beyond that of the body that passed them. To continue with the example of E.U. regulation, when a specific law targets the products or substances manufactured or imported in the European Union, its scope may in effect be much larger since it may impact producers globally. Indeed, it is often too costly to run separate supply chains, and manufacturers who wish to do business with the European Union may have to adopt E.U. regulation for all their production processes.

Internationally, formal nongovernment entities such as the Strategic Approach to International Chemicals Management (SAICM) possess many of the elements of risk governance. SAICM is a global framework that aims to promote chemical safety around the world. It was introduced by the United Nations Environment Program (UNEP) at the World Summit on Sustainable Development in 2002, and its overall objective is to achieve the sound management of chemicals throughout their life cycle. In 2003, SAICM adopted its "2020 goal" as part of the Johannesburg Plan of Implementation to promote production and use of chemicals in ways that minimize significant adverse impacts on human health and the environment. UNEP and the World Health Organization (WHO) both have lead roles in the SAICM secretariat in their respective areas of expertise.

From the international level, one proceeds to the regional level, the location at which E.U. regulation applies to new chemicals and materials. The E.U. regulatory process is the primary vertical component of risk governance in the European Union as it is an authoritative and top-down mechanism.

The regulatory process is summarized briefly in Figure 14.2 and also in the section *The Role of Regulation in Nanotechnology Risk Governance* later in this chapter.

Influence of the Precautionary Principle on International Regulation and Risk Governance

Although, based on a common agreement, divergent perspectives on the precautionary principle have arisen and challenge consensus building in international risk governance and regulation. In 1992, the United Nations Conference on Environment and Development developed the Rio Declaration in which Principle 15 codifies as follows:

> for the first time at the global level the precautionary approach, which indicates that lack of scientific certainty is no reason to postpone action to avoid potentially serious or irreversible harm to the environment. Central to principle 15 is the element of anticipation, reflecting a requirement that effective environmental measures need

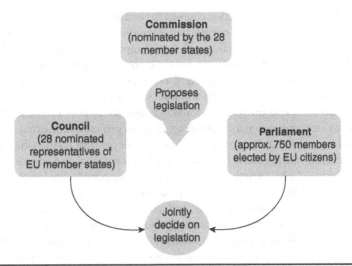

FIGURE 14.2   European Union law-making process.

to be based upon actions which take a long-term approach and which might anticipate changes on the basis of scientific knowledge." (The Global Development Research Center, http://www.gdrc.org/u-gov/precaution-7 .html, http://www.gdrc.org/u-gov/precaution-7.html, accessed July 17, 2015.)

In 2000, the European Union wrote a complete reference text that specifies the following:

> The absence of scientific proof of the existence of a cause-effect relationship, a quantifiable dose/response relationship or a quantitative evaluation of the probability of the emergence of adverse effects following exposure should not be used to justify inaction. (Commission of the European Communities, 2000).

The European Commission set out the specific cases for which this principle is applicable:

- The scientific data are insufficient, inconclusive, or uncertain

- A preliminary scientific evaluation shows that potentially dangerous effects for the environment and human, animal, or plant health can be reasonably inferred

These references now serve as a basis for all E.U. regulation on potentially hazardous chemicals including nanomaterials.

The United States accepts the Rio Declaration's language of a precautionary *approach* but does not support the E.U.'s translation of this concept to a precautionary *principle*. The United States considers a chemical or material safe until it is proved harmful. In addition to the U.S. government's position, an industry trade association, the U.S. Chamber of Commerce, specified in 2005 that it opposed the adoption of the precautionary principle as the basis for regulation, supporting instead "a science-based approach to risk management, where risk is assessed based on scientifically sound and technically rigorous standards" and stipulating its opposition to "domestic and international adoption of the precautionary principle as a basis for regulatory decision making" (U.S. Chamber of Commerce, 2005; U.S. Submission to the April 10–14 Codex Alimentarius Meeting in Paris Addressing U.S. Concerns Over EU Communication to the Codex on "Precautionary Principle", 2000). Wiesner and Bottero (2007) summarized these two approaches in Chapter 1: while the E.U. position is "no safety data, no market," the U.S. position is "no safety data, no regulation." Although these phrases generally stand true, they mask internal disagreements in the European Union and the underlying notion that the precautionary principle is accepted by the E.U. nations.

Nanotechnology Risk Governance

Nanotechnology sparked global excitement in the late twentieth century because of its potential to solve critical technological and social issues. The high surface reactivity and responsiveness of nanomaterials to their microenvironments, the novelty of their size-dependent properties, and the challenge of tracking and measuring these materials in the human body and the environment focused attention on the uncertainty, ambiguity, and complexity of nanomaterials and nanotechnology as an emerging field (van Asselt and

Renn, 2011). These characteristics, coupled to the ability to precisely engineer nanomaterials at the atomic level and the intriguing idea of *engineering in* nanoscale-dependent functionality and *engineering out* physical and chemical properties that cause adverse impacts (Subramanian et al., 2015) heightened sensitivities about the risk and benefit decisions across a nanotechnology-enabled product life cycle.

Nanotechnology has also challenged our global economy through rapid product development and commercialization while regulation and policy—critical elements of a stable marketplace—are dependent upon hazard and exposure research for which data are acquired more methodically and slowly. The rapid increase in the number of nanotechnology-enabled products in the first decade of the twenty-first century outpaced our understanding of the sensitive, size-dependent relationships of scale, composition, architecture, and functional attributes of nanomaterials to beneficial and adverse biological impacts.

Scientists and civil society quickly recognized that nanotechnology invokes many of the complex questions that are addressed in a risk governance framework. As a materials science, it has applications as a broad technology platform across the industrial sectors and includes nanotechnology-enabled products that interface directly with the human body and the environment. Examples of these products include food additives and food packaging, drug delivery platforms, personal care products such as toothpaste and creams, textile treatments, pesticides, and water and air filtration systems. As an interdisciplinary science, nanotechnology brings together researchers from disparate fields to launch new research directions and applications in robotics, neuroscience, and regenerative medicine—fields with many bioethical and dual use considerations. Nanomaterials also require the reengineering of manufacturing processes for which the occupational exposures are unknown. The potential for dispersal of nanomaterials into the environment and the behavior of the nanomaterials in water, soil, and air are also unknown. Although the paper was somewhat speculative, Renn and Roco applied the IRGC risk governance model to nanotechnology in 2006, creating separate frames for passive and active nanomaterials and outlining critical knowledge gaps for environment, health, and safety.

It is the diversity of applications and the global race to commercialize nanotechnology-enabled products; the number of stakeholder communities involved in or impacted by their development and use; the uncertainty surrounding hazard and exposure, including the potential for transformation and persistence in the body or the environment; and the potential need for regulation and policy that make risk governance economically and societally important for nanotechnology. The attention garnered by the rapid development of consumer products reinforced the need for an organized risk governance process as many in civil society and the public called for safety assessments on products still in development or early in their life cycle or a moratorium on use of nanomaterials until safety could be demonstrated. Few channels of communication existed between government and nongovernment groups. Concerns about the applicability of traditional chemical assessment and regulatory paradigms to nanomaterials were greatly debated, and regulators felt pressured to predict potential risks in the absence of sufficient data.

As described in Chapter 1, the nanomaterials industry has the potential to be an environmentally beneficial system throughout the life cycle of production, use, disposal, and reuse of nanomaterials and products. Environmental nanotechnology, as a socioindustrial design process, invokes the concept of sustainability: that is, the application of nanotechnology to create social benefits that do not compromise the future of the planet and its people. While innovation and technology ask what is possible for the technology, viable in the marketplace, and desirable to the user; sustainability invokes questions of social, environmental, and economic consequences—not just what is feasible, but what is societally equitable and bearable. The integration of environmental benefit and sustainability expands the concepts of interdisciplinarity and socioindustrial design (see Chapter 1) beyond academia and industry to include social scientists, regulators, policy makers, civil society, community leaders, and multiple levels of government. These factors underscore the need for an expanded environmental risk-policy paradigm that will sustain nanotechnology as an environmentally beneficial system. Risk governance contains many of the elements needed to address the global challenges that can be solved by—or created by—the use of emerging technologies.

The Role of Regulation in Nanotechnology Risk Governance

Enacting regulations for nanotechnology risk governance has been challenging for many years due, in part, to uncertainties that accompany the birth of a new technology. For example, the data and information needed for fundamental research, product development, risk assessment, regulation, and policy are acquired on differing timelines and timetables. This is evident, for example, in the mismatch between the rate at which products are developed and the rate at which data for policy and regulation are produced.

Products are developed and released to market when product testing is completed; however, regulations cannot be developed until there is a weight of evidence and a convergence of data support conclusions on safety. Regulatory communities have also been challenged by a confusion of definitions for nanomaterials and nanotechnology-enabled products and devices, the need to revise traditional toxicity assays and measurements for use with nanoscale materials, and limited standards nomenclature. Reference materials are needed to calibrate instruments and assays and few such materials existed for many years, further hampering the safety research needed for evidence-based rule-making. In addition, regulators grappled with the application of existing chemical regulations to nanoscale materials and devices with complex and frequently novel properties. As the complex relationship between risk governance and regulation continues to evolve, insurance liability practices are being examined as a possible methodological approach that could provide options to resolve the many divergent views.

The E.U. and U.S. Perspectives on Nanotechnology Regulation

To understand how E.U. regulation works at a regional level, a simplified account of the E.U. system of governance[1] is presented. The European Commission (EC) and the European Parliament (EP) are two institutions that are at the heart of the system of governance of the European Union. The EC is the executive body of the European Union. It is composed of 28 commissioners, one appointed to represent each member state. It can be said that it represents a collective voice, "the voice of the European Union" and that it acts as a guarantor of the E.U.'s general interest, including building up its capabilities and strengthening its international position. It is responsible for proposing legislation to the EP. The EP is directly elected every 5 years by universal suffrage. It is composed of about 750 members and represents over 375 million eligible voters. It can be said that it represents "the people" of the European Union. It does not have the power of legislative initiative but accepts, amends, or rejects legislation initiated by the EC.

The EC and the EP have opposite views on nanotechnology regulation and the shaping of formal nanotechnology governance. To simplify the issue, one may say that the EP insists much more on the precautionary principle as a backbone of E.U. regulation than the EC. The EP pushes for specific nanotechnology regulation while the EC considers existing legislation sufficiently robust for the regulation of nanomaterials and nanotechnology-enabled products.

This dynamic tension is illustrated in the EP's resolution of April 24, 2009 that addressed the EC's provisions for nanotechnology regulation (European Parliament, 2009). The resolution stated that the EP:

- Did not agree with the Commission's conclusions that current legislation covered in principle the relevant risks relating to nanomaterials.
- Considered that the concept of the "safe, responsible, and integrated approach" to nanotechnologies advocated by the European Union was jeopardized by the lack of information on the use and on the safety of nanomaterials that were already on the market.
- Called for the introduction of a comprehensive science-based definition of nanomaterials in community legislation.
- Called on the Commission to compile an inventory of the different types and uses of nanomaterials on the European market, while respecting justified commercial secrets such as recipes, and to make this inventory publicly available; furthermore called on the Commission to report on the safety of these nanomaterials at the same time.

The EC reacted to the resolution by introducing a definition of nanomaterials in 2010, but this definition was subject to criticism, and in 2015, the difficulties of finding the right regulatory compromise that would suit both parties had not been overcome.

The U.S. position has been established by the Emerging Technologies Interagency Policy Committee (ETIPC) Nanotechnology Working Group. The ETIPC issued two memos (March and June 2011) that stipulate broad principles to guide the development and implementation of policies for oversight across the U.S. government. These memos affirm that rule-making should be evidence-based and commensurate with risk and that nanomaterials as a class, should be presumed to be either benign or harmful.

---

[1]The reader has to keep in mind that the relationships between the various institutions of the European Union are much more intricate than described, and the value of the demonstration lies in a broad interpretation of the current interactions.

There is currently no comprehensive international regulation for nanotechnologies. In the early 2010s, SAICM meetings generated nonbinding resolutions on nanotechnology, including several recommendations like the requirement of a comprehensive hazard assessment prior to the introduction of nanomaterials into the market and the ban on shipments of waste containing nanomaterials unless the receiving country could adequately manage the waste. (See the report of the Fourth African Regional Meeting on SAICM and UNITAR/OECD Workshop on Nanotechnology and Manufactured Nanomaterials: http://www.saicm.org/index.php?content=meeting&mid=129&menuid=&def=1. [Accessed June 2, 2015]).

Alternative Approaches to Legislated Regulation

Apart from these inconsistent pieces of mandatory regulation and the vacillation of nations and regions between regulating and not regulating, other governance devices were examined to understand if they might successfully navigate these diverse perspectives and approaches. These alternatives could be divided into two broad categories, voluntary reporting schemes and codes of conduct, and standards and certifications.

Voluntary Reporting Schemes and Codes of Conducts

The Voluntary Reporting Scheme for Manufactured Nanomaterials (VRS) was set up in the United Kingdom in September 2006, as a temporary experiment with an expected life of 2 years. It was run by DEFRA, the Department for Environment, Food and Rural Affairs. The scheme aimed to provide an indication of the kinds of nanomaterials currently in development and production to help inform policymaking decisions and to focus efforts and funding on areas which were relevant to the U.K.'s current nanotechnology manufacturing and research base. The program asked for data that could be provided on manufactured nanomaterials from anyone involved in the manufacture or use of engineered nanomaterials, or anyone involved in nanoscience research or managing wastes consisting of engineered nanoscale materials. The program included regular updates to assess its implementation (DEFRA, 2008).

In 2008, the seventh such quarterly report indicated that after 22 months, a total of 11 submissions had been received since the scheme's launch in September 2006, 9 from industry and 2 from academia. These results appeared quite poor in view of the U.K. estimated nanotechnology production. For example, QinetiQ—said to be the U.K.'s first production facility dedicated to the volume production of specialist nanomaterials—stated on its website that it had two production rigs which were each capable of producing up to several kilos of material an hour, and that it was looking at over 25 key nanomaterial projects (retrieved from: http://www.qinetiq.com/home/newsroom/news_releases_homepage/2003/1st_quarter/nanomaterials0.html, accessed July 2007).

In January 2008, the U.S. Environmental Protection Agency (EPA) launched the Nanoscale Materials Stewardship Program (NMSP), a similar voluntary reporting scheme covering engineered nanoscale materials manufactured or imported for commercial purposes (EPA, 2007). The NMSP program invited interested parties to participate in a "basic" program by submitting existing data on the engineered nanoscale materials they manufactured, imported, processed, or used. The EPA also invited interested parties to participate in an "in-depth" program to test engineered nanoscale materials they manufactured, imported, processed, or used. This scheme seemed to meet the same fate as the U.K. VRS. When it launched the NMSP program, the EPA said it expected to receive 240 submissions from 180 companies under the basic program, and to attract 15 participants in the in-depth program. This projection was based on an estimate that in 2005 more than 600 companies were manufacturing and applying nanotechnology. However, in 2008, the EPA provided on its website a list of only 9 companies that had made submissions and 11 companies that intended to, and 2 more volunteering for the in-depth program component.

So it appeared that the voluntary reporting schemes were not enthusiastically received by industry. Interviews with industry representatives suggested that the major reason for this failure was that when a company attempted to give an earnest report of its nanotechnology activities, the very information it provided became a source for scrutiny—and sometimes criticism or concern. Companies felt that because of the current lack of data, the data they provided became the source of a benchmarking of nanotechnology risk evaluation for which they were accountable because they were the source of the data.

Approximately at the same time, a few initiatives were launched to design codes of conduct founded on voluntary implementation. One was the product of a public institution: the E.U. Code of Conduct for Responsible Nanoscience and Nanotechnology Research (European Commission, 2008). Based on the precautionary principle, this voluntary code covered seven general principles, including sustainability, precaution, inclusiveness, and accountability. Its main goal was to help research institutes, universities, and companies in the European Union ensure the safe development and use of nanotechnologies in the face of knowledge gaps and uncertainties about the future impact of these technologies on human health and the environment. The European Union launched a period of public consultation from October 2009 to January 2010 to evaluate the impact of this voluntary initiative and the revisions that might be introduced. The public consultation had been opened to all stakeholders directly or indirectly involved or interested in nanosciences and nanotechnologies research, but it received very little feedback. Only 49 valid answers were received from 17 countries including 4 which were outside the European Union. As of July 2015, the code of conduct had not been updated.

While the E.U. Code of Conduct for Responsible Nanoscience and Nanotechnology Research was an initiative coming from a public institution, the U.K. Responsible NanoCode, also released in 2008, was a joint production of several bodies including an asset management firm (Insight Investment) and a trade association (Nanotechnology Industries Association) as well as a research institution (The Royal Society) (Royal Society, Insight Investment, and Nanotechnology Industries Association, 2008). The project had started in 2006, and aimed to explore the societal and economic impacts of the technical, social and commercial uncertainties related to nanotechnologies. The idea had come from corporate codes of conduct that started being put in place by industry. The Responsible NanoCode had seven principles. It aimed to establish a consensus of what constituted good practice across the nanotechnology value chain (i.e., from R&D to manufacturing, distribution, and retailing) so that businesses could align their processes with emerging good practice and form the foundation for the development of indicators of compliance. It illustrated expected behaviors and processes, not standards of performance, and it did not aim to be an auditable standard. The scope of its application was limited and its 2008 update was not followed by any more recent document. [See the European Commission's Community of Practice (CoP) for better self- and co-regulation fourth plenary meeting, March 13, 2015, http://ec.europa.eu/digital-agenda/en/news/fourth-cop-meeting, accessed July 16, 2015.]

Another type of voluntary tool of governance was the Nano Risk Framework, a coproduction of industry and of an NGO. In June 2005, the chemical company DuPont and the environmental group, Environmental Defense Fund, began developing this initiative. Their objective was "to help answer questions an organization should consider in developing applications using nanomaterials, including providing a way to address areas of incomplete or uncertain information using 'reasonable assumptions and appropriate risk management practices'" (Environmental Defense and DuPont, 2007). The framework included guidance on how to communicate information and decisions to stakeholders. One year after its launch, the Nano Risk Framework was available in French, Mandarin, and Spanish, and DuPont made the framework mandatory for all its nanotechnology work. However, after two interactive workshops on nanotechnology risk management in 2008, the initiative also seemed to have gradually disappeared.

Finally, there was a purely corporate initiative. The chemical company BASF, which had developed, in 2004, the first corporate code of conduct outlining the company's duties to workers, investors and clients, introduced, in 2007, a specific code of conduct for nanotechnologies (http://www.nanotechnology.basf.com/group/corporate/nanotechnology/en/microsites/nanotechnology/safety/code-of-conduct, accessed 09 May 2016). Based on principles of responsible management, the code had four commitments detailing the company's approach, and included strong commitments to safety and transparency, among which one could quote:

- "Economic considerations do not take priority over safety and health issues and environmental protection."
- "We disclose new findings to the authorities and the public immediately."

The analysis of these voluntary codes showed that they had a limited impact. They were designed as voluntary, principles-based guidelines establishing a consensus of what constituted good practice in businesses across the nanotechnology value chain, so that these businesses could align their processes with emerging good practice. However, they dealt more with management principles than with the technology risks themselves. They illustrated expected behaviors and processes, not standards of performance; they did not aim to be auditable standards, nor did they include tools to evaluate their successful (or not) implementation. It was therefore not possible to measure their effectiveness in terms of regulation of the emergence of the nanotechnology industry.

Certifications and Standards

In 2008, Soil Association Certification Ltd., the United Kingdom's largest organic certification body, said it had become the first organization in the world to ban man-made nanomaterials from its certified organic products. The logo of Soil Association Certification Ltd. means that a product is free from nanoparticles, as detailed in the Soil Association organic standards food and drink Revision 17.4 of March 2015. Under the Soil Association standard, organic producers and processors must not use ingredients containing manufactured nanoparticles, where the mean particle size is 200 nm or smaller, and the minimum particle size is 125 nm or smaller (see https://www.soilassociation.org/media/4494/standards_food_drink.pdf accessed 09 May 2016). In 2015, the application of the Soil Association logo concerned mainly cosmetics, food, and clothing.

Apart from these voluntary labels and certifications, a very important trend concerns standardization organizations. Djelic and Sahlin-Andersson (2006) and Brunsson and Jacobsson (2002) already pointed that standardization was becoming a new transnational institution, and this was particularly true in the case of nanotechnologies.

The International Organization for Standardization (ISO) set up a special technical committee for nanotechnology, ISO TC-229, in May 2005. By 2008, each of the projects of ISO TC-229 had been categorized into one of four technical committees' working groups:

1. WG 1—Terminology and Nomenclature (convened by Canada)

2. WG 2—Measurement and Characterization (convened by Japan)

3. WG 3—Health, Safety, and Environment (convened by the USA)

4. WG 4—Material Specifications (convened by China)

As of June 2015, 43 international nanotechnology standards had been published by ISO TC-229, and 27 more were under development. Published standards related to terminology, definitions, health, and safety practices in occupational settings, methodology for the characterization and classification of nanomaterials and tests, as well as to risk evaluation and voluntary labeling.

While the European Union focuses on industrial sectors to delineate a case-by-case risk-centered nanotechnology regulation (cosmetics and novel food) ISO standardization remains transsectoral. The way standards shape regulation is to start from properties and to focus on exposure to determine risks. Standardization can be criticized for constraining innovation, and particularly for maintaining path dependency. However, their strong development seems to indicate that in spite of the additional cost of ISO certification for industry, they are becoming a popular solution to address the uncertainty of the development of nanotechnology markets. It is to be noted that the E.U. definition of nanomaterials was very close to that of ISO.

Who Should Managing Risk Governance?

As a first step to answer this tricky question, we outline below the main issues that are part of the framing of nanotechnology governance and that are currently unresolved.

First, should there be a general, formal, mandatory regulatory framework for nanomaterials and nanotechnology-enabled products? There is currently no such framework at the international or the regional level. On the national level, France introduced the first nation to introduce general mandatory regulations to govern nanotechnology materials and products in 2012 (French Parliament, 2012). All other general regulatory frameworks coming from official bodies are in the form of recommendations and guidelines.

Second, should there be a generally accepted definition of nanomaterials? The two opposing views are that nanomaterials have specific properties that differ from bulk materials and therefore should be defined, or that nanomaterials are, for the most part, smaller versions of existing materials, and therefore should not be defined. And if nanomaterials are defined, how should they be defined? The problem of defining nanomaterials is closely related to that of categorizing nanomaterials. Wiesner[2] summarizes commonly proposed schemes for categorizing nanomaterials as based on either (1) physical-chemical properties, (2) exposure potential, or (3) hazard potential, but suggests that instead of attempting to regulate a broad class

---

[2]Presentation to OECD workshop on categorizing nanomaterials, Washington D.C., September 17, 2014. Background document available on https://www.epa.gov/sites/production/files/2015-09/documents/oecd_expert_meeting_document_8.pdf accessed on 09 May 2016.

of materials, for which there is not likely to be a widely agreed upon, or enforceable definition, that materials should be categorized and regulated based on the functional properties that make them useful in commerce. Several approaches to nanomaterial definitions currently coexist:

- Physical properties–related definitions would be based on characteristics such as size, particle number size distribution, surface charge, or surface area. One proposal defined nanomaterials as having a specific surface area by volume that is greater than 60 m^2/cm^3.

- Size-related definitions, for example, at least one dimension of the nanomaterial is equal to or smaller than 100 nm.

- Particle number size distribution, that is, a definition stipulating that 50 percent or more of the particles are in a certain number size distribution, or 10 percent or more of the particles are in a certain number size distribution, depending on who is proposing the definition.

- Risk-related definition based on physical properties, such as solubility, agglomeration, or biopersistence. The difficulty with a physical property–based risk definition is that depending on the use of the nanomaterial and its microenvironment, a given physical property may be a source of increased risk under certain conditions and decreased risk under other conditions.

- Source-based definitions based on the three conditions under which nanoscale materials are produced, that is, intentionally designed and manufactured nanomaterials with specific properties, naturally occurring nanomaterials, and those produced unintentionally as by-products of other operations, such as combustion engines.

Alternatively, a nanomaterial case-by-case approach could be adopted, and two such distinct approaches co-exist:

1. Industrial sector case-by-case approach, for example, specific regulations have been introduced by the European Union for cosmetics and for novel food applications. In both of these cases, the E.U. regulation requires risk assessment data and registration of substances, as well as specific labeling of products containing nanomaterials to inform consumers of their presence.

2. Specific nanomaterials case-by-case approach, which is currently the most widespread type of mandatory regulation that applies in the European Union and in the United States for nanoscale carbon, silver, titanium dioxide, and silicium products.

We can conclude from the above that two main opposing views coexist to manage risk governance. The first, favored by the EC and United States, considers nanotechnology-enabled products on a case-by-case approach and bases regulation on the results of risk assessments. The other view, mainly that of the European Parliament, is to consider nanotechnology-enabled products as a new world of materials that need to be governed by an entirely new set of rules but within the general framework of the precautionary principle.

To summarize, while the IRGC framework provides one path forward, for some reason risk governance has not been widely embraced. There is also evidence for new forms of policymaking, where national decisions and regulatory frameworks may well be superseded by new forms of international regulation which are being put into place largely independent of individual nation states. This new system of governance has the following characteristics:

- It is global and transsectorial.

- It pushes back the notion of regulatory ownership. Traditionally, policymaking could have a national identification. With all the collective interactions mentioned above, the national identity of policymaking is becoming blurred and subsumed into global policy with a transnational and multilevel character.

- It has a form of autonomy that is conceptually similar to *autocatalysis*. In other words, policymaking and regulation are developing outside of the usual boundaries, creators, and contributors and evolving from its own input.

Nanotechnology governance is developing somewhat autonomously but sporadically with a self-organizing approach that frequently draws upon international standards and other voluntary initiatives. In parallel, large inter-government actors like the EU may be the first to act through market forces and effective policy implementation to shape future outcomes and responses from the industry, and thus provide the bases for nanotechnology policy and regulation. In the U.S., governance approaches to policy-making and regulation appear to have stabilized with existing laws and frameworks proving to be sufficiently robust to manage the challenges posed by nanotechnology.

Although the structure and role of risk governance is still very much evolving for emerging technologies, it is clear that nanotechnology, especially environmental nanotechnology, will be at the center of its development.

References

1. Baldwin, R., M. Cave, and M. Lodge, 2010. *The Oxford Handbook of Regulation,* Oxford University Press, Oxford.
2. Brito, J. and S.E. Dudley, 2012. *Regulation: A Primer,* Mercatus Center at George Mason University, Virginia (VA).
3. Brunsson, N., and B. Jacobsson, 2002. *A World of Standards,* Oxford University Press, Oxford, UK.
4. Commission of the European Communities, 2000. *Communication from the Commission on the Precautionary Principle,* COM (2000) 1, Commission of the European Communities, Brussels. Available at: http://www.gdrc.org/u-gov/precaution-7.html, accessed July 17, 2015.
5. Department for Environment Food and Rural Affairs. 2008. "UK voluntary reporting scheme for engineered nanoscale nanomaterials." Available at: http://webarchive.nationalarchives.gov.uk/20130123162956/, http://www.defra.gov.uk/environment/quality/nanotech/documents/vrs-nanoscale.pdf (accessed 09 May 2016).
6. Djelic, M.-L. and K. Sahlin-Andersson (Eds.), 2006. *Transnational Governance: Institutional Dynamics of Regulation,* Cambridge University Press, Cambridge.
7. Environmental Protection Agency. 2007. "Nanoscale Materials Stewardship Program and Inventory Status of Nanoscale Substances under the Toxic Substances Control Act." Available at: https://www.govinfo.gov/content/pkg/FR-2007-07-12/pdf/E7-13558.pdf (accessed 09 May 2016).
8. European Commission, 2001. *European Governance—A White Paper [COM(2001) 428 Final—Official Journal C 287 of 12.10.2001],* Communication from the Commission of July 25, 2001.
9. European Commission. 2008. "Code of conduct for responsible nanosciences and nanotechnologies research." C(2008) 424. Available at http://ec.europa.eu/research/participants/data/ref/fp7/89918/nanocode-recommendation_en.pdf (accessed 09 May 2016).
10. Environmental Defense and DuPont. 2007. "Nanorisk Framework." a draft version of the framework is available at: http://www.oecd.org/science/nanosafety/38173388.pdf (accessed 09 May 2016).
11. European Parliament, 2009. *Resolution of 24 April 2009 on Regulatory Aspects of Nanomaterials [2008/2208(INI)]. P6_TA(2009)0328.*
12. French Parliament. 2012. Décret n° 2012-232 du 17 février 2012 relatif à la déclaration annuelle des substances à ₽état nanoparticulaire pris en application de ₽article L. 523-4 du code de ₽environnement. *Journal Officiel de la République Française.*
13. Hanson, N., 2002. *The Dreadful Judgement: The True Story of the Great Fire of London,* Corgi Books, London.
14. Hood, C., H. Rothstein, and R. Baldwin, 2001. *The Government of Risk: Understanding Risk Regulation Regimes,* Oxford University Press, Oxford.
15. Isaacs, J.A., C.L. Alpert, M. Bates, C.J. Bosso, M.J. Eckelman, I. Linkov, and W.C. Walker, 2015. Engaging stakeholders in nano-EHS risk governance. *Environment Systems and Decisions,* 35(1): 24–28.
16. King, A.A. and M.J. Lenox, 2000. Industry self-regulation without sanctions: The chemical industry's responsible care program. *Academy of Management Journal,* 43(4): 698–716.
17. Klischewski, R., 2014. When virtual reality meets realpolitik: Social media shaping the Arab government-citizen relationship. *Government Information Quarterly,* 31(3): 358–364.
18. Levine, M.L. and P.A. Feigin, 2014. Crowdfunding provisions under the new rule 506(c). *CPA Journal,* 84(6): 46–51.
19. Lodge, M. and K. Wegrich, 2015. Crowdsourcing and regulatory reviews: A new way of challenging red tape in British government? *Regulation & Governance,* 9(1): 30–46.
20. Mauelshagen, C., M. Smith, F. Schiller, D. Denyer, S. Rocks, and S. Pollard, 2014. Effective risk governance for environmental policy making: A knowledge management perspective. *Environmental Science & Policy,* 41(0): 23–32.
21. McTaggart, T.R. and Y. Benina, 2014. Avoiding "epic fails": FFIEC supervisory guidance on financial institutions' use of social media. *Journal of Taxation & Regulation of Financial Institutions,* 28(2): 15–22.
22. Ostrom, E. 1990. *Governing the Commons: The Evolution of Institutions for Collective Action,* Cambridge University Press, Cambridge.
23. Renn, O. and P. Graham, 2005. White paper on *Risk Governance: Towards an Integrative Approach.* International Risk Governance Council (IRGC), Geneva. Available at: http://www.irgc.org/risk-governance/irgc-risk-governance-framework/ (accessed on July 17, 2015).
24. Renn, O., A. Klinke, and M. van Asselt, 2011. Coping with complexity, uncertainty and ambiguity in risk governance: a synthesis. *Ambio,* 40(2): 231–246.
25. Renn, O. and M.C. Roco, 2006. Nanotechnology and the need for risk governance. *Journal of Nanoparticle Research,* 8(2): 153–191.
26. Roco, M.C. 2011. The long view of nanotechnology development: the National Nanotechnology Initiative at 10 years. *Journal of Nanoparticle Research,* 13(2): 427–445.

27. Roco, M.C., B. Harthorn, D. Guston, and P. Shapira, 2011. Innovative and responsible governance of nanotechnology for societal development. In M.C. Roco, C.A. Mirkin, and M. C. Hersam (Eds.), *Nanotechnology Research Directions for Societal Needs in 2020*, Chapter 14, pp. 561–617, Springer, Netherlands.

28. Rosenau, James N. 1990. *Along the Domestic-Foreign Frontier: Exploring Governance in a Turbulent World*. New York: Cambridge University Press.

29. Rosenau, James N. 1995. "Governance in the 21st Century." *Global Governance* 1:13–43.

30. Royal Society, Insight Investment, and Nanotechnology Industries Association (NIA), 2008. *The Responsible NanoCode*.

31. Subramanian, V., E. Semenzin, D. Hristozov, E. Zondervan-van den Beuken, I. Linkov, and A. Marcomini, 2015. Review of decision analytic tools for sustainable nanotechnology. *Environment Systems and Decisions*, 35(1): 29–41.

32. U.S. Submission to the April 10–14 Codex Alimentarius Meeting in Paris Addressing U.S. Concerns over EU Communication to the Codex on "Precautionary Principle." 2000. *Joint FAO/WHO Food Standards Programme Codex Committee on General Principles*. Fifteenth Session, Paris, France.

33. U.S. Chamber of Commerce. 2005. *Regulatory Reform Issues, Precautionary Principle*.

34. van Asselt, M. and O. Renn, 2011. Risk governance. *Journal of Risk Research*, 14(4): 431–449.

35. Wainwright, N.B., 1953. Philadelphia's eighteenth-century fire insurance companies. *Transactions of the American Philosophical Society, New Ser.*, 43(1): 247–252.

36. Wiesner, M. and J.-Y. Bottero, 2007. Nanotechnology and the environment. In M. Wiesner and J.-Y. Bottero (Eds.), *Environmental Nanotechnology*, McGraw Hill, New York.

Index